FUNDAMENTALS
OF ENGINEERING
DRAWING

TENTH

FUNDAMENTALS
ENGINEERING

With an Introduction to Interactive Computer Graphics for Design and Production

Warren J. Luzadder, P.E.
Purdue University

Jon M. Duff, Ph.D.
Purdue University

OF
DRAWING

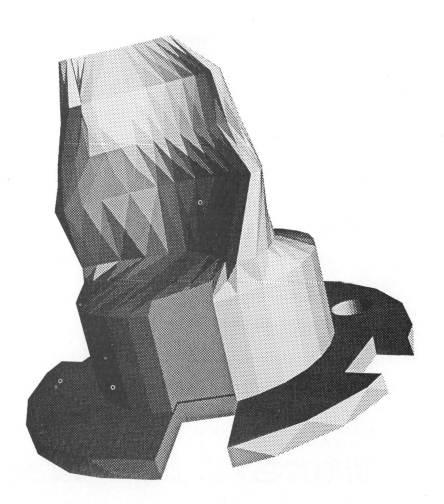

Prentice Hall, Englewood Cliffs, New Jersey 07632

Library of Congress Cataloging-in-Publication Data

LUZADDER, WARREN JACOB.
   Fundamentals of engineering drawing.

   Bibliography:
   Includes index.
   1. Mechanical drawing. 2. Computer graphics.
I. Duff, Jon M., (date)   II. Title.
T353.L88 1989   604.2'4   88–9826
ISBN 0–13–338443–8

Fundamentals of Engineering Drawing, Tenth Edition
Warren J. Luzadder, P.E., and Jon M. Duff, Ph.D.

Editorial/production supervision: Joan McCulley
Art direction and interior design: Anne T. Bonanno
Cover photo: Imtek Imagineering/Masterfile
Cover design: Anne T. Bonanno

 © 1989, 1986, 1981, 1977, 1971, 1965, 1959, 1952, 1946, 1943 by Prentice-Hall, Inc.
A Division of Simon & Schuster, Englewood Cliffs, New Jersey 07632

Portions of this text have been printed in previous editions.

other books by Duff

Industrial Technical Illustration, Van Nostrand Reinhold, 1982.
Introduction to Engineering Drawing, Prentice Hall, 1989.

other books by Luzadder

Innovative Design with an Introduction to Design Graphics, Prentice-Hall, Inc., 1975.
Basic Graphics for Design, Analysis, Communications and the Computer,
   2nd ed., Prentice-Hall, Inc., 1968.
Fundamentos De Dibujo Para Ingenieros, 6th ed., Compania Editoria Continental, 1977.
Graphics for Engineers, Prentice Hall of India Private Limited, 1964.
Technical Drafting Essentials, 2nd ed., Prentice-Hall, Inc., 1956.
Problems in Engineering Drawing, Vol. I, 9th ed., Prentice-Hall, Inc., 1986,
   with Botkin and Gross.
Problems in Engineering Drawing, Vol. II, 1st ed., Prentice-Hall, Inc., 1981,
   with Gross.
Engineering Graphics Problems for Design, Analysis and Communications, Prentice-Hall, Inc., 1968.
Problems in Drafting Fundamentals, Prentice-Hall, Inc., 1956.
Purdue University Engineering Drawing Films, with J. Rising et al.
Introduction to Engineering Drawing, Prentice Hall, 1989.

Printed in the United States of America
10 9 8 7 6 5 4 3 2

ISBN 0-13-338443-8

Prentice-Hall International (UK) Limited, *London*
Prentice-Hall of Australia Pty. Limited, *Sydney*
Prentice-Hall Canada Inc., *Toronto*
Prentice-Hall Hispanoamericana, S.A., *Mexico*
Prentice-Hall of India Private Limited, *New Delhi*
Prentice-Hall of Japan, Inc., *Tokyo*
Simon & Schuster Asia Pte. Ltd., *Singapore*
Editora Prentice-Hall do Brasil, Ltda., *Rio de Janeiro*

Contents

P A R T ■ 3
DESIGN 278

P A R T ■ 4
GRAPHICS FOR DESIGN AND COMMUNICATION 308

P A R T ■ 5

CAD/CAM: COMPUTER-AIDED DRAFTING AND COMPUTER-AIDED MANUFACTURING 454

P A R T ■ 6

GRAPHIC METHODS FOR ENGINEERING COMMUNICATION, DESIGN, AND COMPUTATION 490

P A R T ■ 7

DESIGN AND COMMUNICATION DRAWING IN SPECIALIZED FIELDS 532

Preface

This edition of *Fundamentals of Engineering Drawing* begins the most extensive process of evolution in its nine edition history. Computer-aided Drafting and Design (CADD) is a reality across American industry, signaling a new approach to the study and practice of engineering drawing. This, and subsequent editions, will be based on the authors' shared belief that the more powerful the tools available for making engineering drawings, the greater is the importance that the fundamentals of engineering drawing be stressed and understood. Powerful CADD equipment in the hands of engineers or technologists well trained in design and drafting can result in greater productivity, better response to changing needs, and, in the end, a more competitive workforce.

This new edition is not based on one particular CADD software product. There are daily developments in the areas of standardization of CADD drawings, CADD equipment, and the methods used for operating CADD programs. Since this is a textbook on engineering drawing and not computer graphics, *Fundamentals of Engineering Drawing* will, over the next editions, reflect developments in computer science and computer graphics as they apply to the making of engineering documents. The field of expert systems has a bright future in the selection, preparation, and use of CADD drawings, and we expect future editions to reflect this.

Engineering drawing and CADD are both moving toward being based on solid geometry and solid modeling. The time when a designer would lay out flat diagrams on paper and then try to visualize the design in three dimensions is rapidly coming to a close. This edition stresses developing an ability to manipulate three-dimensional geometry—whether on the surface of a drawing or as a solid computer model.

The promise of a Computer-Integrated Manufacturing Technology, where CADD and CAM and Computer-aided Engineering work as a team has been realized. In such an environment, CADD becomes an extension of the engineer's mind, allowing the design, testing, manufacture, and evaluation of products in the memory of the computer before the first part is ever made. It is now imperative that everyone involved in the engineering process understand the relationship of the computer to drawing and design.

This understanding makes each person a valuable member of the engineering team.

This edition has been extensively edited and revised, with outdated techniques eliminated and new material added in the crucial introductory chapters. The emphasis has been in integrating CADD into each chapter as it naturally occurs. Chapter 1 now includes an introduction to CADD as it logically fits into the field of modern engineering drawing. Chapter 2 introduces CADD equipment simply as additional tools available for making engineering drawings along with explaining traditional and CADD lettering techniques. Chapter 3 keeps many of the proven plane geometry constructions of previous editions and introduces new methods computers use for creating these same figures. Also in Chapter 3 is a presentation on solid modeling techniques, a subject critical to successful use of a CADD computer as a design tool. Chapter 4 begins a major expansion of traditional projection theory. This chapter now includes extensive sections on the coordinate axis systems used in CADD, including rotations. Chapters 5 and 9 integrate traditional and CADD techniques for achieving multiple principal and auxiliary views. It is for Chapter 17 then, after the theory has been presented in previous chapters, to apply the material presented on CADD and demonstrate how engineering drawings are actually made using computer devices. Chapter 17 leads the student through the steps necessary to complete an engineering drawing using CADD. Chapter 18 continues to present the latest developments in Computer-aided Manufacturing (CAM) and how CADD and CAM are integrated into a manufacturing system.

Elsewhere in this edition, several sections have been combined or deleted, most notably the inclusion of pipe fittings and welding in a chapter on joining and fastening, and the deletion of sections on reprographic techniques. The Glossary has been expanded into a full Appendix with shop and CADD terms, and the Bibliography has been updated.

To bring this text abreast of new technological developments, a number of leading industrial organizations have generously assisted the authors by supplying appropriate illustrations that were needed in developing specific topics. Every commercial illustration supplied by American industry has been identified using a courtesy line. The authors deeply appreciate the kindness and generosity of these many companies and the busy people in their employment who found the time to select these drawings and photographs that appear in almost every chapter.

The authors also would like to reaffirm their indebtedness to Professors George Shiers, W. L. Baldwin, R. H. Hammond, and Dennis R. Short. The chapter on electronics drawing has been used again in this edition. Professor Baldwin contributed the material on linkages in the chapter covering machine elements. Professor Hammond prepared the coverage on graphical algebra. Professor Short contributed computer-generated drawings which are displayed throughout the text as well as help on the preparation of Chapter 17. Commercial CADD drawings have been furnished by Computervision, IBM, Micro Control Systems, Hewlett-Packard, CalComp-Sanders, and the Ross Gear Division of TRW.

Professor Larry D. Goss of The University of Southern Indiana previously contributed material in Chapters 1 and 21, much of which has been carried over to this edition.

Special appreciation is extended for the assistance of Morris Buchanan and John Mitchell. Mr. Buchanan's con-

tribution of the IBM Metric Standards and drawings in the ninth edition appears again in this text. Mr. Mitchell of Cincinatti Milicron was instrumental in securing photographs of industrial robots and examples of numerical control programs for machine tools which continue to be used in this new edition.

The authors are grateful as well for the assistance and encouragement given by Professors Jerry V. Smith, Dennis R. Short, Mary A. Sadowski, and others of the Department of Technical Graphics at Purdue University. Many of the new illustrations and text passages were critiqued and evaluated by the faculty.

Last, we would like to acknowledge two people for their outstanding work on this text. The first, Joan McCulley, production editor, handled the task of coordinating new and previous sections and illustrations in this vanguard edition. The second, Anne Bonanno, book designer, executed a handsome book, easy to use by teachers and students alike. Our sincere thanks go to both of them.

W.J.L.
J.M.D.
Purdue University

FUNDAMENTALS OF ENGINEERING DRAWING

The IBM Interactive Graphics Display System unlocks man's imagination. With the IBM system product designs can be prepared and plots and maps displayed in their expected form. Under program control data can be entered, displayed, and modified as needed by using the light pen, program function keyboard, or alphanumeric keyboard.

Technical drawings may be created by techniques similar to those used at a regular drawing board. Drafting requirements for arcs, lines, dimensioning, notes, automatic scaling, parts lists, and three-dimensional pictorials are fully met in accordance with ANSI drafting standards by the applications program. (*Courtesy IBM Corporation*)

Introduction

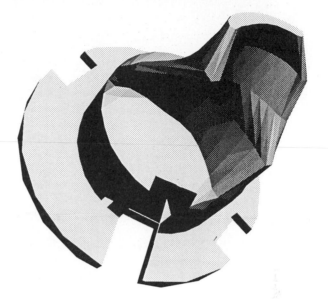

1.1 A Brief History of Drawing

For almost twenty thousand years a drawing has been the main way that ideas have been communicated. The first drawings were made even longer ago, when prehistoric man tried to communicate ideas by marking in the dirt floors of caves. It is natural for humans to graphically draw their ideas because *drawing is a universal language*. Even today, when some drawings are made by computers, this is true. The first permanent drawings, made on the rock walls of caves, depicted people, deer, buffalo, and other animals. These drawings were made to express emotion and record experiences, long before the development of writing. When writing developed, drawing came to be used primarily by artists and engineers as a means of showing design concepts for pyramids, war chariots, buildings, and simple mechanisms.

One of the earliest drawings, found in Mesopotamia, shows the use of the wheel about 3200 B.C. The drawing depicts a wheelbarrow-like mechanism being used by a man to transport his wife. This early drawing was a crude picture, without depth or perspective. Another example of early drawing is a floor plan of a fortress made on a clay tablet around 4000 B.C.

At the beginning of the Christian era, *Roman architects* had become skilled in making drawings of proposed buildings. They used straight edges and compasses to prepare plan (top), elevation (front and side), and perspective

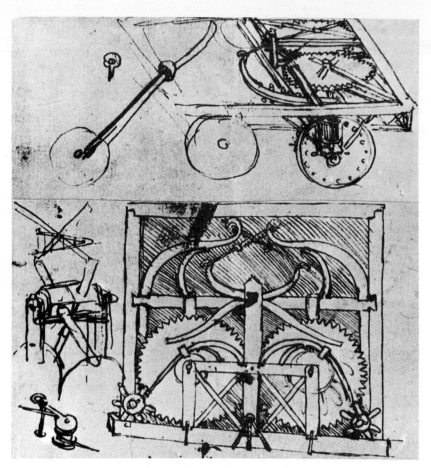

FIG. 1.1 Idea sketch prepared by Leonardo da Vinci (1452–1519). Leonardo's "automobile" was to have been powered by two giant springs and steered by the tiller, at the left in the picture, attached to the small wheel. (*From Collection of Fine Arts Department, IBM Corporation*)

views. However, the theory behind engineering drawing, the theory that enabled architects to depict views on various planes, was not developed until the Renaissance period. Even though *Leonardo da Vinci* was probably aware of these theories, his classical training as an artist influenced his engineering drawings like the one shown in Fig. 1.1. No multiview drawings (top, front, side views) made by Leonardo have been found. He knew the value of pictorial drawing in showing how parts of a mechanism fit together. It is interesting to note that even in this day of space travel, engineers continue to use pictorials to supplement multiview drawings (see the pictorial drawing of Skylab in Fig. 1.2).

Most of the very early drawings still existing today were made on parchment, a very durable type of paper. Although the paper that was developed in Europe in the twelfth century made the use of drawings more prevalent, the paper was too fragile. Of the thousands of engineering drawings of fortresses, buildings, and mechanisms made during the twelfth through fifteenth centuries, only a few exist today. Pictures made into pottery, carvings, and weavings were more permanent than paper drawings, and many of these artifacts have survived.

1.2 The Interrelationship of Engineering Drawing and Design

Engineering design uses engineering drawing as the way to communicate and document ideas. Engineers, drafters, and other members of the design team must work closely on the design project and speak the same language: *the language of engineering drawing*. A design engineer, though not responsible for the actual production of the drawings, must be able to read and understand all aspects of the drawing. At the very least, the design engineer must be able to make clear, concise sketches that can be given to technologists and drafting technicians for the actual preparation of the drawings.

Any person on the design team who makes engineering drawings must combine *classroom study* of engineering graphics with *practical experience* in company methods, standards, and practices. This knowledge may include such things as manufacturing methods, numerical control, and digital computer applications (Chapters 1–5 and 12–18).

Engineering drawings present technical information to tens or even hundreds of individuals who may be engineers, managers, suppliers, machinists, installers, or repairmen.

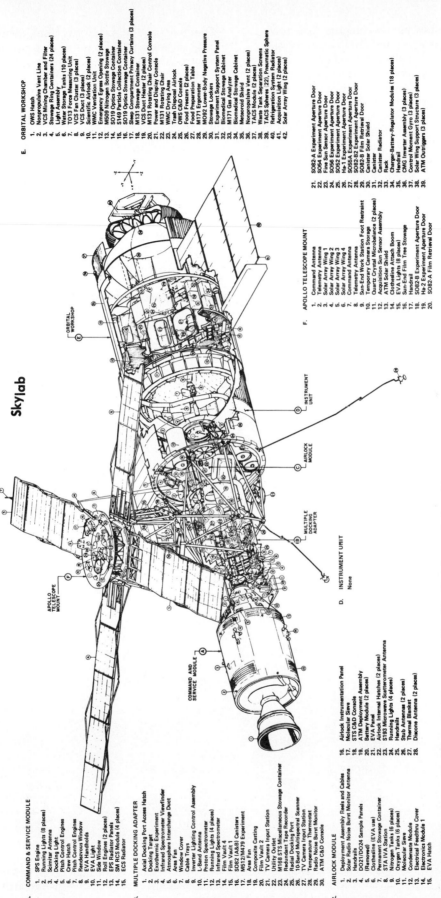

Skylab

FIG. 1.2 Skylab—manned orbital scientific space station. The Skylab was designed to expand our knowledge of manned earth-orbital operations and to accomplish carefully selected scientific, technological, and medical investigations. (*Courtesy National Aeronautics and Space Administration*)

All need engineering drawings to complete their tasks. For this reason drawings must conform to *exacting standards* for their preparation and understanding.

Engineering drawings are documents that change. Over the life of a drawing, there may be changes in the design, materials, suppliers, and uses of the product. These changes must be noted and recorded. Very often a single drawing may be used for a whole family of parts. Consequently, engineering drawings must be made with materials and equipment that allow for clean and efficient changes. (See Chapter 2, which covers equipment and materials.)

1.3 Engineering Drawing Today

The design team is often made up of individuals from many different disciplines. A project manager assembles a team to accomplish a goal such as building a bridge, designing a car, or placing scientific equipment on the moon (Fig. 1.3). In general, the design team is comprised of three groups of individuals, each having particular knowledge of engineering drawing: (1) engineering designers, (2) engineering technologists, and (3) engineering technicians.

Designers and *technologists* think about the way that a mechanism or product works and how it can be manufactured or built. These individuals must have a working knowledge of engineering drawing to be able to direct drafters in the preparation of final drawings. The final drawings follow preliminary design sketches that the designer may have made. Such design sketches must follow the basic rules of engineering drawing so that it will be easy for the drafter to turn rough sketches into final drawings.

To be effective, the engineering designer and technologist must have skill in three areas of communication: (1) English, both written and oral; (2) symbols, as used in science and mathematics; and (3) engineering drawing, both sketching and interpreting drawings.

A drafter must be able to assemble written, numeric, and drawing information and make final drawings. These drawings are called "working drawings" because they are the drawings from which the product is made. The drawing must convey *shape*, *size*, and *manufacturing information* needed to fabricate the parts and assemble the structure. This must be done with knowledge of company practices and national and international standards such as those of the *American National Standards Institute* (ANSI) or the *International Standards Organization* (ISO). Since

FIG. 1.3 The Lunar Rover. The photograph shows the Lunar Rover taken to the Moon by the Apollo 15 Mission. It is parked near Mount Hadley in the right center background. The Lunar Rover is one of man's most noted innovative designs. (*Courtesy National Aeronautics and Space Administration*)

ANSI standards govern the preparation of engineering drawings, drafters must keep up to date with the latest published standards.

Engineering technicians, assigned to aid the engineer or to work in production, must have considerable knowledge in preparing engineering drawings. Those working with the design engineer may be called upon to solve problems graphically or prepare working sketches within their own areas of expertise, such as electrical, structural, or mechanical systems.

1.4 Engineering Drawing and the Computer

Engineering drawing is a *graphic language* that allows humans and computers to work together. The computer has made the current period one of revolutionary change in how engineering drawings are made, stored, and printed. The emphasis in engineering drawing today is on conveying design information efficiently to a design team that may be in widely separated locations. The saying "a picture is worth a thousand words" has never had more meaning than today.

Computers understand numbers and are able to print information in the form of numbers—rows and rows of numbers, page after page of numbers. But humans, who find it easier to understand words and pictures, call upon graphic computers to take these numbers from the *data base* and show the drawing they represent (Fig. 1.4). Engineering drawings prepared with the aid of a computer are important as a means of visually checking the numbers

for correctness, since it is much easier to check a picture than 50 pages of numbers.

These computer-assisted drawings take on two general forms: (1) a chart (graph, plot, or diagram); or (2) a picture showing the shape or size of the object.

With either form, the designer is trying to determine whether or not the *computer model* satisfies the design criteria. To do this, the designer must be able to read the drawing and understand shape, size, and manufacturing information as it is represented. The principles of engineering drawing, the accepted ways of drawing standard features such as bolts and springs, and the methods of specifying size are discussed in Chapters 2 through 5 and in Chapters 13, 14, and 21.

With powerful computer-aided design (CAD) systems, drawings on paper may not exist. Instead, drawings may appear on computer monitors where they are corrected, updated, and evaluated.

1.5 Computer-Aided Design Drafting (CADD)

It might be helpful at this point to make a distinction between *computer-aided design* (CAD) and *computer-aided design drafting* (CADD). CAD is an analysis technique, a way of *modeling* the performance of a product before it is actually built. Paper drawings may not be necessary in the design phase. The ideal situation is one where the numbers controlling the machines produce the same geometric description for making the parts as was used to describe the design. To coordinate the workers manufacturing the product, engineering drawings are required.

Every point on an object has its own unique location, or "address," in space (see Chapter 4). Together, the addresses of all of the points on an object form the data base for that object. Two-dimensional objects have two-dimensional data bases. Three-dimensional objects have three-dimensional data bases. The size of the data base (the actual volume of numbers) and not the physical dimensions of the paper determines the size of the drawing. A CADD drawing's size is determined by

1. Dimensionality (3-D data bases are larger than 2-D data bases for the same object),

2. The size of the object (large objects contain more points),

3. The detail of the object (the more detail, the more points).

The following is a brief overview of important concepts in CADD. For more extensive information refer to Chapter 17.

With CADD, the computer can easily present individual components of a product separately by placing each component on a different layer, the computer version of clear plastic overlay sheets. Layers can be selected for viewing in any combination.

CADD programs allow the designer to select the units of measure (such as inches or millimeters) for the data

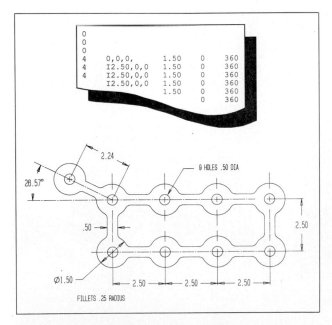

FIG. 1.4 Numeric data base and its graphic representation. The set of numbers which defines the geometry of an engineering design is "read" by the CADD program and displayed in the form of an engineering drawing.

base. The image can be displayed larger or smaller, depending on the level of detail desired. This does not affect the actual size of the object defined by the numbers in the data base; rather, it enables you to view the object close up or from farther away. If the data base itself is changed, the object is said to have been *scaled*.

CADD systems can be very accurate. Accuracy is determined by the total number of individual points that the computer can keep track of at one time. It is not uncommon for a CADD system to be capable of accuracy to 1/1000th of an inch and still allow a designer to work in a space 30 miles on each side. The greater the accuracy, the smaller the available work space. The smaller the accuracy, the larger the work space that can be defined.

This flexibility allows for the design of both large and small projects on the same system (Fig. 1.5). It is possible to work on the over-all design of a very large space station and a computer chip for the station's electrical system on the same computer terminal. Once it took hours to describe the geometry of even the simplest part. Today, with more sophisticated and "user-friendly" CADD programs, a complex design can be defined in a fraction of the time previously required.

The role that drawing plays in the design process is changing with the increased use of computer-aided machines and processes. Companies that continue to make engineering drawings without the aid of computers use drawings for the actual production of parts and for the assembly and checking of the final product. These manual drawings become the basis for all subsequent documents prepared for manufacture, distribution, and service of the company's product. There is no master set of numbers describing the design; so the drawing, in this case, is the data base.

Companies that build a computer data base of the part use that set of numbers to generate all engineering documents necessary during the lifetime of the design. This includes engineering drawings, instructions for numerically controlled machine tools, quality control, assembly, marketing, maintenance, inventory control, and accounting. There is a fundamental difference between manual and computer-aided design drafting. *In CADD, all documents go back to the same numeric data base for information*, assuring accuracy and completeness. In manual design, a drawing may be made from a previous drawing which was made from an earlier drawing, and so on. The chance for error with manual design and drawing is much greater.

1.6   How CADD Drawings Are Used

If the data base contains the representative of a part's geometry in the memory of the computer, *any and all views of that geometry are available* to the designer. To make an engineering drawing of the data base, the form must be displayed in a position that reveals the desired views (Fig. 1.6). When a CADD system is used, like lines will be the same thickness on every drawing, lettering will be the same regardless of the drafter, and other manual-skill operations will be made consistent. Prints are made only as a verification of the data base or for locations where a computer terminal is unavailable.

Although this difference appears to be a conflict of

FIG. 1.5 **Minicomputer interactive computer graphics system.** (*Courtesy Computervision Corporation*)

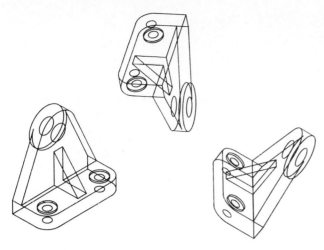

FIG. 1.6 Multiple views from the same data base. Once an object is described in a numeric data base, *any* view of the geometry is possible. (*Courtesy of Mike Gabel, Department of Technical Graphics, Purdue University*)

manual versus computer drawing, the use of engineering drawings is not an either-or situation. Industries will continue to use both paper and computer drawings side by side. Paying attention to drafting standards will assure that the two can be used interchangeably. The structural steel industry, for example, makes use of CADD data bases for design and analysis, but will continue to make extensive use of manual drawings or printed copies of computer drawings for the actual fabrication of structures. Why? Because a drawing is much easier to carry than a computer terminal as you climb the superstructure of a bridge.

1.7 Computer-Aided Design and Manufacturing (CAD/CAM)

Manufacturing companies are becoming increasingly interested in linking design and manufacturing in order to reduce production time, cut drawing room costs, simplify production planning, and increase accuracy. An integrated **CAD/CAM** system *shares the data base* created during the design phase (Fig. 1.7). **Computer Numerically Con-**

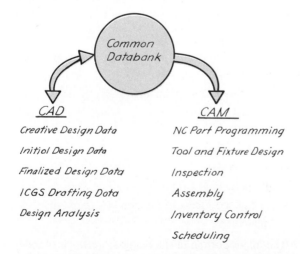

FIG. 1.7 Work flow in an integrated CAD/CAM system.

trolled (CNC) machines were developed before CAD and required a separate data base. If the machines could recognize this data base, why not create the original description of the part geometry in the same form, avoiding the costly process of translating manual drawings into a CNC data base? The coupling of these two activities became CAD/CAM—the marriage of computer-aided design and computer-aided manufacturing. Whenever computers are widely used for manufacture, the term CAM is used. This applies to the planning and production processes, inventory control, and for the programming of numerically controlled machine tools, to be discussed in Chapter 18.

CAD/CAM can be thought of as being divided into four general areas,

1. Engineering Design
2. Design Drawing or Design Drafting
3. Planning or Scheduling
4. Fabrication or Machining.

CAD/CAM is part of the complete integration of manufacturing, a process known as *Computer Integrated Manufacturing* (CIM). CIM integrates the manufacturing process from product planning through design, CADD, production, materials handling, inspection and quality control, and assembly.

In the future, small but powerful computers will control machines on the production line. These computers will in turn be controlled by larger computers which assign to them specific tasks, route materials to each work station, and keep track of inventory. These intermediate computers will be controlled by even larger computers which coordinate activities between departments.

The chapters in this text are limited to the computer-aided design drafting (CADD) aspects of CAD/CAM. Students who are interested in becoming drafting technicians must learn the fundamentals of engineering drawing before they enter the work force and become members of the total CAD/CAM team. As a member of that team, an understanding of the overall picture of CAD/CAM is essential.

1.8 International System of Measure

Although the *metric system* was legally approved for use in the United States by an act of Congress over one-hundred years ago (1866), it has not been widely adopted because its use was never made mandatory. Most industrial and trading nations use the metric system. To compete in a world market, many American companies had to convert to metric measurements. Companies such as General Motors, Ford, IBM, 3M, Navistar, John Deere, Caterpillar, Honeywell, McDonnel Douglas, Rockwell International, and TRW have or are currently converting to the SI (Système International d'Unités) or metric system.

Total conversion will not come about immediately for much of American industry. Acceptance of ISO standards is a slow process. The ANSI standards continue to be translated into metric; therefore this text will include problems in both the metric and English systems. Just as use

FIG. 1.8 Classroom laboratory showing microcomputer workstations for the use of beginning students. Computer, monitor, digitizing board, and digital plotter can be seen. (*Courtesy of Professors Watson and Sadowski and the Department of Technical Graphics, Purdue University*)

of computer graphics is not an either-or situation, many companies that have essentially "gone metric" continue to use the English system for some fasteners, pipes, drills, and bearings. The appendices of this text contain numerous conversion tables as well as with metric parts tables. In the revision of this text the authors have been guided by the current ANSI metric standards and by the metric design and drafting standards furnished by General Motors, John Deere, Navistar, and IBM.

1.9 The Educational Value of Engineering Drawing

By studying engineering drawing, a student becomes aware of how industry communicates technical information. Engineering drawing teaches the principles of accuracy and clarity in presenting the information necessary to produce products (Figs. 1.8 and 1.9). It develops the

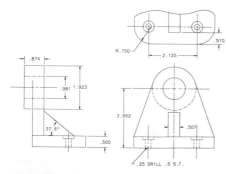

FIG. 1.9 An engineering detail drawing prepared using a CADD system. This drawing represents the mathematical description of a machine part in standard form, easily understood by engineers, technologists, and technicians.

engineering imagination that is so essential to a successful design. Finally, in learning the techniques of engineering drawing, you will find that something very important has happened: it has *changed the very way that you think about technical images* and that, more than facts and figures, will stay with you throughout your career. It is ideal to master the fundamentals of engineering drawing first and to later use these fundamentals for a particular application, such as CADD. The deeper your understanding of the fundamentals, the more command you will have over whatever engineering drawing tools you have available.

1.10 The Organization of this Text

The purpose of this text is to present the *grammar and composition* of engineering drawing, much as an English text presents the grammar and composition of our written language. With study, engineering and technology students will eventually be able to prepare satisfactory engineering drawings and, after some industrial experience, be capable of directing the work of others.

To organize study, this text has been separated into the following topics:

- Basic Graphical Techniques
- Spatial Graphics
- Design
- Graphics for Design
- CAD/CAM
- Computational Graphics
- Specialized Fields of Graphics
- Appendices

The major portion of the material in this text concerns the preparation of machine drawings. This may not seem to be of immediate interest to students studying construction or architectural or electrical technologies. However, knowing the methods used in machine drawings prepares the student for later specialization. A discussion of several of these specialized fields is presented in this text after the fundamentals of engineering drawing have been studied.

The same argument applies to computer-aided drafting and design. A student who is not thoroughly versed in the fundamentals of engineering drawing simply becomes an operator of the CADD system, unable to use the computer as a design tool. Space problems, such as the clearance between a wheel and a fender or the true angle between a turbine blade route and the axis of an engine, can be solved by computer solution only if the fundamentals of engineering drawing have been learned.

BASIC GRAPHICAL TECHNIQUES

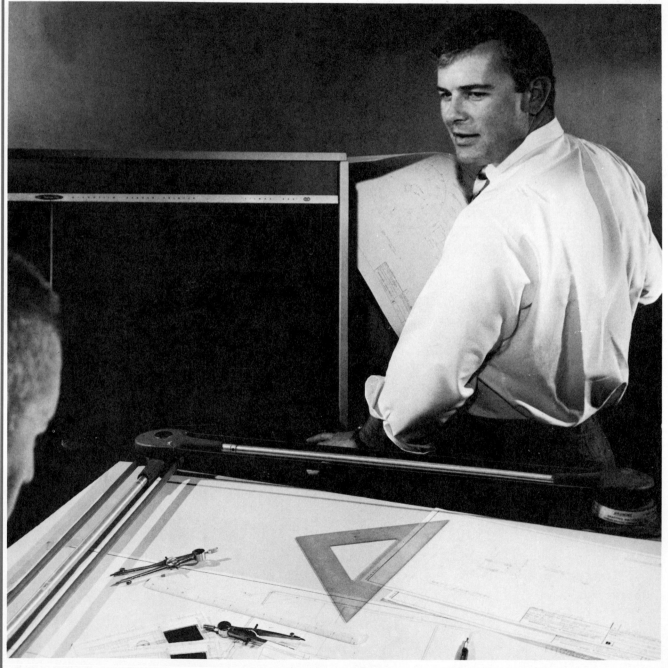

A draftsman discusses a problem with a designer. It should be noted that they are reviewing previous designs that have been filed on microfilm. The microfilm reader/printer can be seen in the background. (*Courtesy 3M Company*)

Drawing Instruments, Computer Drafting Equipment, and Drafting Techniques

A / MANUAL DRAWING EQUIPMENT AND ITS USE

2.1 Introduction

The instruments and materials needed for making engineering drawings are shown in Fig. 2.1. There will be individual differences among brands, but functionally the equipment will be the same. Drawing instruments should be well made. Inferior equipment often makes it difficult to produce drawings of professional quality and often costs more in the long run.

2.2 Basic List of Equipment and Materials

The following list contains the selection of equipment necessary for making instrument drawings.

1. Case of drawing instruments
2. Drawing surface (board or table)
3. Drafting edge (T-square, parallel edge, drafting machine)
4. Triangles (30, 45 degree or adjustable)
5. French curve
6. Scales (see 2.24 and 2.25)

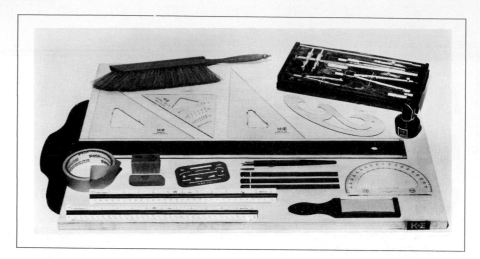

FIG. 2.1 **Manual drafting equipment.**

FIG. 2.2 **A drawing instrument set.**

7. Drawing pencils
8. Lead pointer
9. Drafting tape
10. Eraser
11. Dry cleaning pad
12. Erasing shield
13. Dusting brush
14. Drawing paper
15. Tracing paper or drafting film

2.3 *The Set of Drawing Instruments*

The minimum contents of a drawing instrument set should include a large bow compass, small bow compass, and set of frictional dividers (Fig. 2.2). Sets which contain the large bow are preferred by many persons, especially in the aircraft and automotive fields. The set often comes in a velvet-lined case, designed to protect the instruments. Some sets also contain a "beam compass" for constructing large arcs. The bow compass has a center wheel for adjustment and will hold its setting as the draftsman applies

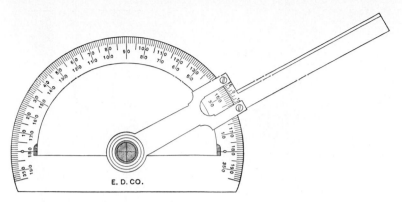

FIG. 2.3   Protractor.

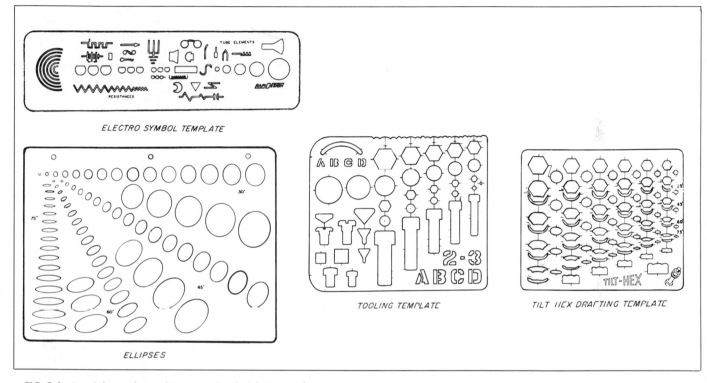

ELECTRO SYMBOL TEMPLATE

ELLIPSES

TOOLING TEMPLATE

TILT HEX DRAFTING TEMPLATE

FIG. 2.4   Special templates. (*Courtesy Frederick Post Co.*)

pressure to draw a line. Modern bow compasses often have a ''quick bow'' feature which allows rapid adjustment.

2.4   Protractor

The protractor (Fig. 2.3) is used for measuring and laying off angles. An alternative to this device is the protractor head of a drafting machine (Fig. 2.9).

2.5   Special Instruments and Templates

A few of the many special instruments available are shown in Figs. 2.4–2.8. *Drafting templates* are by far the most important special equipment. The use of templates (Fig. 2.4) can save valuable time in the drawing of standard figures and symbols. Electrical, hydraulic, and pneumatic

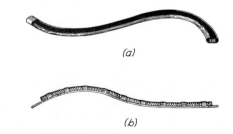

(a)

(b)

FIG. 2.5   Flexible curves.

equipment, springs, fasteners, gears, circles, ellipses, curves, boxes, and architectural symbols are all available on templates. Template manufacturers will make custom templates for specialized applications on request.

To supplement standard plastic curves, *flexible curves* (Fig. 2.5) provide limitless variations and are very con-

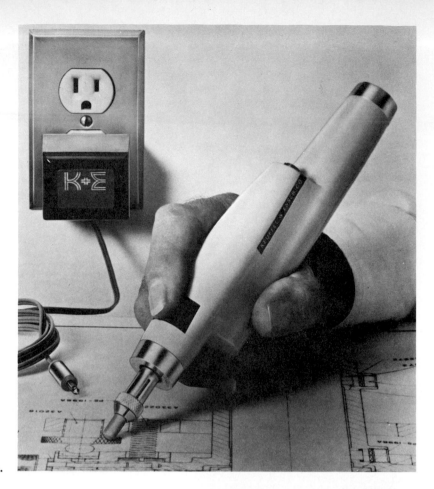

FIG. 2.6 Erasing machine.

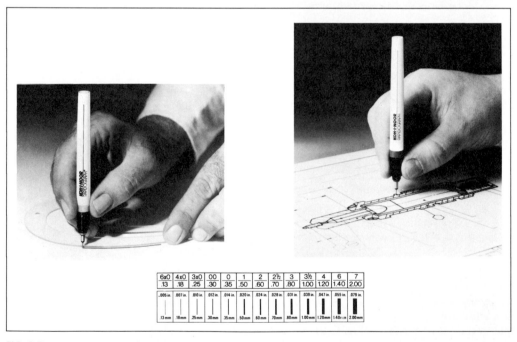

6x0	4x0	3x0	00	0	1	2	2½	3	3½	4	6	7
.13	.18	.25	.30	.35	.50	.60	.70	.80	1.00	1.20	1.40	2.00
.005 in	.007 in	.010 in	.012 in	.014 in	.020 in	.024 in	.028 in	.031 in	.039 in	.047 in	.055 in	.079 in
.13 mm	.18 mm	.25 mm	.30 mm	.35 mm	.50 mm	.60 mm	.70 mm	.80 mm	1.00 mm	1.20 mm	1.40 mm	2.00 mm

FIG. 2.7 **Technical fountain pen.** (*Courtesy Koh-I-Noor, Inc.*)

venient. The type shown in (a) is a lead bar enclosed in rubber. The more desirable one shown in (b) has a steel ruling edge attached to a spring with a lead core. Flexible curves are limited to gentle curves and are unsuitable for small radius arcs.

The *electric erasing machine* (Fig. 2.6) saves valuable drafting time for those persons who correct and update engineering drawings. An assortment of eraser stock is available for different papers and films.

When an ink drawing is required, the *technical pen* shown in Fig. 2.7 is available in the 13 metric line widths shown. These pens are suitable for instrument linework and for lettering using plastic lettering guides. For proper use, these pens should be held perpendicular to the drawing surface and moved with light downward pressure. Different inks and tips are available for different papers and films. These pens require occasional cleaning. These are precision drawing instruments and users are advised to follow the manufacturer's recommendations for filling, cleaning, and storing.

Proportional dividers (Fig. 2.8) supplement the dividers found in drawing instrument sets. They are used for reducing or enlarging drawings without converting numerical values.

2.6 The T-Square, Parallel Edge and Drafting Machine

The *T-Square* and *Parallel Edge* are used to draw horizontal lines and as a guide for triangles when drawing vertical and inclined lines (Figs. 2.25 and 2.26). Lines are drawn along the top of the straight edge.

The *drafting machine* (Fig. 2.9) combines straight edge, protractor, scale, and triangles. It is estimated that drafting machines lead to a 25–50% savings in time in commercial drafting rooms. Drafting machines may be of the parallel arm type or the more modern "track" design. Some include a digital electronic readout of protractor angles, much like a calculator. Many optional devices may be attached to the head of a drafting machine including electronic lettering machines and computer-interfaced digitizers (see Sec. 2.32).

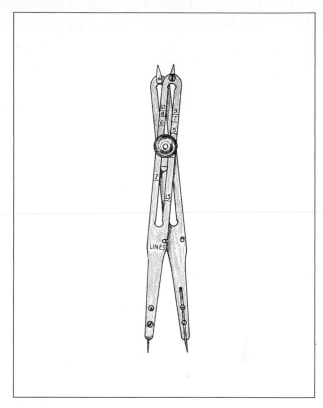

FIG. 2.8 **Proportional dividers.**

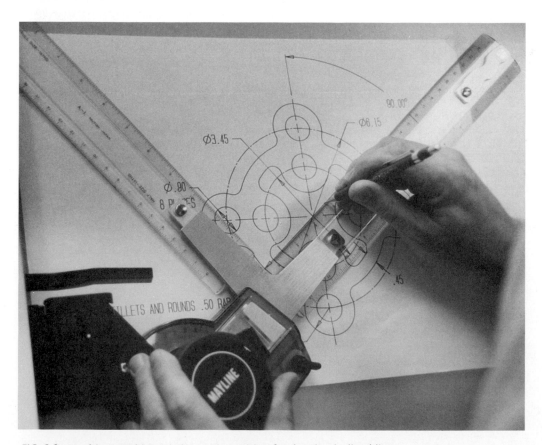

FIG. 2.9 **Drafting machine.** Blades are in position for drawing inclined lines.

2.7   *Tracing Paper and Drafting Film*

White lightweight tracing paper called *drafting vellum*, on which pencil drawings can be made and from which prints can be produced, is used in most commercial drafting rooms. Drafting vellum is not a permanent medium for drawings, however. Over time it will swell with humidity, crack with dryness, or yellow with age. It also does not keep its "dimensional stability," that is, exacting distances will change over time as vellum shrinks. Drawings done on vellum must be reduced and stored on microfilm for permanence.

A more expensive solution is polyester *drafting film*. This film is transluscent (frosted on one or both sides) and exhibits excellent dimensional stability and printing qualities. Drawings are made on drafting film either in ink or polymer (plastic) lead. Erasures can be made without leaving "ghost marks" that might show when making a print.

2.8   *Pencils*

Both the student and the professional should be equipped with a selection of pencils with leads of various degrees of hardness (Fig. 2.10). Examples of these leads are

FOR VELLUM		FOR FILM	
9H	Very hard		
6H	Hard	6P	Hard Plastic
2H	Medium	2P	Medium Plastic
H	Medium soft	P	Soft Plastic
F	Soft		

The grade of pencil chosen depends on the type of line desired, the kind of paper used, and humidity which affects the surface of the paper. *Hard leads* produce sharp but grey lines. *Soft leads* produce black but sometimes fuzzy lines. Standards for line quality will govern selection of lead (see Fig. 2.22). As a minimum, however, the student should have available a 6H pencil for light layout work

where accuracy is required, a 4H for reproducing light finished lines (dimension lines, center lines, and dashed lines), a 2H for visible object lines, and an F or H for all lettering and freehand work.

The most popular pencils use .7, .5, and .3 millimeter leads. These mechanical pencils are filled with lead stock of the desired hardness and should be held perpendicular to the drawing surface much like a technical ink pen. Graphite lead for vellum and polymer lead for drafting film are available. These leads do not need to be sharpened but may be "burnished" to a sharp point for particularly precise work.

When sharpening a wood-encased graphite pencil, the wood should be cut away (on the unlettered end) with a knife or a pencil sharpener equipped with draftsman's cutters. About 10 mm (.38 in.) of the lead should be exposed and should form a cut, including the wood, about 40 mm (1.5 in.) long. The lead then should be shaped to a conical point on a pointer or burnished on a piece of rough paper (see Fig. 2.11).

2.9   *Drawing Pencil Lines*

Pencil lines should be sharp and uniform along their entire length as well as uniform with similar lines elsewhere on the drawing. *Construction lines* (preliminary or layout lines) should be drawn *very* lightly so that they will not reproduce when copied. *Finished lines* should be made boldly and distinctly so that a distinction will be evident between solid and dashed object lines and *auxiliary lines* (dimension, center, and section lines). To give this contrast, necessary for clearness and ease in reading the drawing, object lines should be of medium width and very black; dashed lines, black and less wide; and auxiliary lines, dark and thin.

When drawing a line, a .3, .5, or .7 millimeter mechanical pencil should be held perpendicular to the drawing surface (Fig. 2.12). This type of pencil will produce a line consistent with the diameter of the lead. Traditional wooden drawing pencils or lead holders like those in Fig. 2.13 should be inclined slightly (about 60 degrees) in the direction that the line is being drawn. The pencil should be "pulled" (never pushed) at the same inclination for the full length of the line. If the pencil is rotated slowly between the fingers as the line is drawn, a symmetrical point will be maintained and a straight uniform line will be ensured.

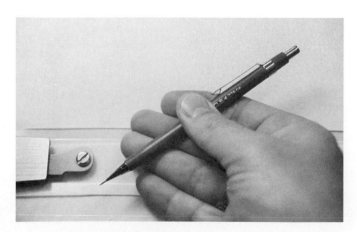

FIG. 2.10   Drafting pencil.

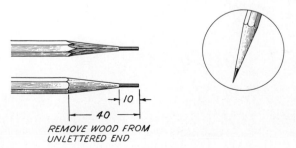

REMOVE WOOD FROM
UNLETTERED END

FIG. 2.11   Sharpening a wood-encased pencil.

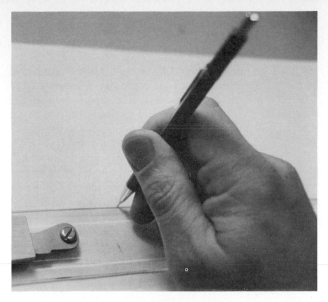

FIG. 2.12 Using the mechanical pencil.

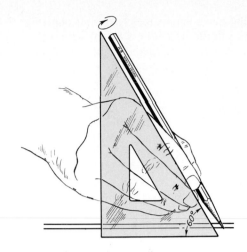

FIG. 2.13 Using the wooden pencil.

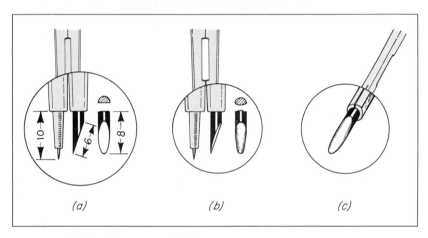

(a)   *(b)*   *(c)*

FIG. 2.14 Shaping the compass lead.

2.10 Placing and Fastening the Paper

The *drawing paper* should be placed on the board or table so that your drawing blade will be parallel to one edge of the paper and perpendicular to the other and so that areas at the top and bottom of the sheet can be used. Align the lower edge of the sheet (or the lower border line if the paper is preprinted) with a straight edge placed parallel to the edge of the board before securing the sheet at each corner with drafting tape. Tape opposite corners of the sheet, working the sheet flat from the center with the side of your hand.

2.11 The Large Bow Compass

The *compass* or *large bow* is used for drawing circles and circle arcs. For drawing pencil circles, the style of point illustrated in Fig. 2.14 should be used because it gives more accurate results and is easier to maintain than most other styles. This style of point is formed by first sharpening the outside of the lead, as shown in Fig. 2.15, to a

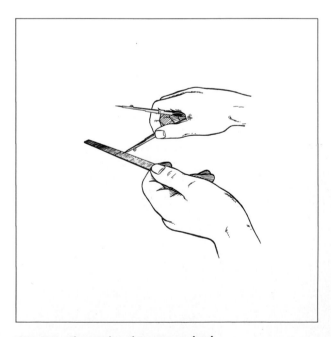

FIG. 2.15 Sharpening the compass lead.

long flat bevel approximately 6 mm long [Fig. 2.14(*a*)] and then finishing it [Fig. 2.14(*b*)] with a slight rocking motion to reduce the width of the point. Although a hard lead (4H–6H) will maintain a point longer without resharpening, it gives a finished object line that is too light in color. Soft lead (F or H) gives a darker line but quickly loses its edge and, on larger circles, gives a thicker line at the end than at the beginning. Some draftsmen have found that a medium-grade (2H–3H) lead is a satisfactory compromise for ordinary working drawings. For design drawings, layout work, and graphical solutions, however, a harder lead will give better results.

The needle point should have the shouldered end out and should be adjusted approximately 10 mm (.38 in.) beyond the end of the split sleeve [Fig. 2.14(*a*)].

2.12 Using the Compass

When possible, it is preferable to use a *circle template* to draw common diameter circles. Choose a lead hardness identical to that used for straight lines of similar intensity and sharpen the lead before drawing each arc (see Sec. 2.11). To draw a circle with a compass, it is first necessary to draw two lines intersecting at right angles; along one, mark off the radius. Place the compass point carefully on the point of intersection. Adjust the lead point to the radius mark and draw the circle in a clockwise direction if you are right-handed (Fig. 2.16). While drawing the circle, the instrument should be inclined slightly forward. If the line is not dark enough, draw over it again.

A beam compass is manipulated by steadying the instrument at the pivot leg with one hand while drawing the marking leg toward you with the other (Fig. 2.17).

2.13 Dividers

The *dividers* are used principally for dividing curved and straight lines into any number of equal parts and for transferring measurements. If the instrument is held with one leg between the forefinger and second finger, and the other leg between the thumb and third finger, as illustrated in Fig. 2.18, an adjustment may be made quickly and easily with one hand. The second and third fingers are used to ''open out'' the legs, and the thumb and forefinger to close them. This method of adjusting may seem awkward to the beginner at first, but with practice absolute control can be developed.

2.14 Using Dividers

The trial method is used to divide a line into a given number of equal parts (Fig. 2.19). To divide a line into a desired number of equal parts, open the dividers until the distance between the points is estimated to be equal to the length of a division, and step off the line *lightly*. If the last mark misses the end point, increase or decrease the setting by an amount estimated to be equal to the error divided by the number of divisions, before lifting the dividers from the paper. Step off the line again. Repeat this procedure until the dividers are correctly set, then space the line again and indent the division points. When stepping off a line, the dividers are rotated alternately in an opposite direction on either side of the line, each half-revolution, as shown in Fig. 2.19.

It is the common practice of many expert draftsmen to draw a small freehand circle around a very light indentation to establish location of the divider point.

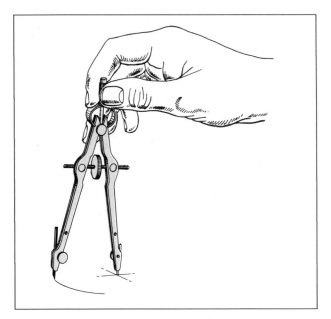

FIG. 2.16 Using the large bow (Vemco).

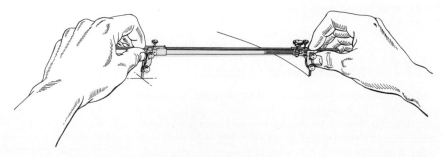

FIG. 2.17 Drawing large circles (Vemco beam compass).

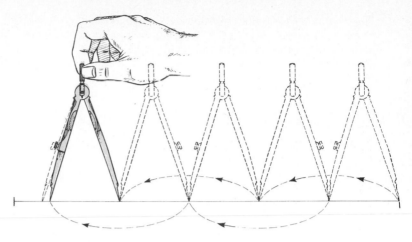

FIG. 2.18 To adjust the large dividers.

FIG. 2.19 Using the dividers.

2.15 Use of the French Curve, Adjustable Curve

A *French curve* is used for drawing irregular curves that are not circle arcs. After sufficient points have been located, the French curve is applied so that a portion of its ruling edge passes through at least *three points*, as shown in Fig. 2.20. It should be so placed that the increasing curvature of the section of the ruling edge being used follows the direction of that part of the curve which is changing most rapidly. To ensure that the finished curve will be free of humps and sharp breaks, the first line drawn should start and stop short of the first and last points to which the French curve has been fitted. Then the curve is adjusted in a new position with the ruling edge coinciding with a section of the line previously drawn. Each successive segment should stop short of the last point matched by the curve. In Fig. 2.20 the curve fits the three points, *A*, 1, and 2. A line is drawn from between point A and point 1 to between point 1 and point 2. Then the curve is shifted, as shown, to fit again points 1 and 2 with an additional point 3, and the line is extended to between point 2 and point 3.

Some people sketch a smooth continuous curve through the points in pencil before drawing the mechanical line. This procedure makes the task of drawing the curve less difficult, since it is easier to adjust the ruling edge to segments of the freehand curve than to the points.

The *adjustable curve* is laid flat on the drawing surface and gently bent until its edge follows as many points of the curve as possible (Fig. 2.5). If the curve becomes tight and the adjustable curve cannot be bent into position to follow it, finish the line with a small French curve.

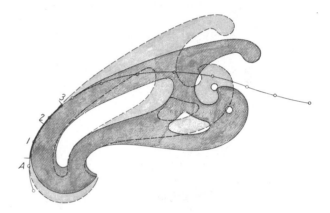

FIG. 2.20 Using the irregular curve.

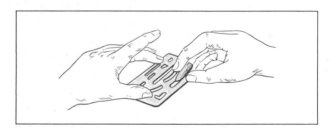

FIG. 2.21 Using the erasing shield.

used for the line (Fig. 2.21). Excessive pressure should not be applied to the eraser because the surface of the drawing is likely to be damaged. The fingers holding the erasing shield should rest partly on the drawing surface to prevent the shield from slipping.

2.16 Use of the Erasing Shield and Eraser

An erasure is made on an engineering drawing by placing an opening in the *erasing shield* over the work to be erased and rubbing with an eraser appropriate for the medium

2.17 Conventional Line Symbols

Symbolic lines of various weights are used in making technical drawings. The recommendations of the American National Standards Institute, as given in ANSI Y14.2M–1979, are the following:

Two widths of lines—thick and thin—are recommended for use on drawings (Fig. 2.22). Pencil lines in general should be in proportion to the ink lines except that the thicker pencil lines will be necessarily thinner than the corresponding ink lines but as thick as practicable for pencil work. Exact thicknesses may vary according to the size and type of drawing. For example, where lines are close together, the lines may be slightly thinner. The ratio of line thickness should be about two-to-one. The thin line width should be approximately 0.35 mm (.015 in.) and the thick line width approximately 0.7 mm (.03 in.).

Ink lines on drawings prepared for catalogs and books

FIG. 2.22 Alphabet of lines (finished weight).

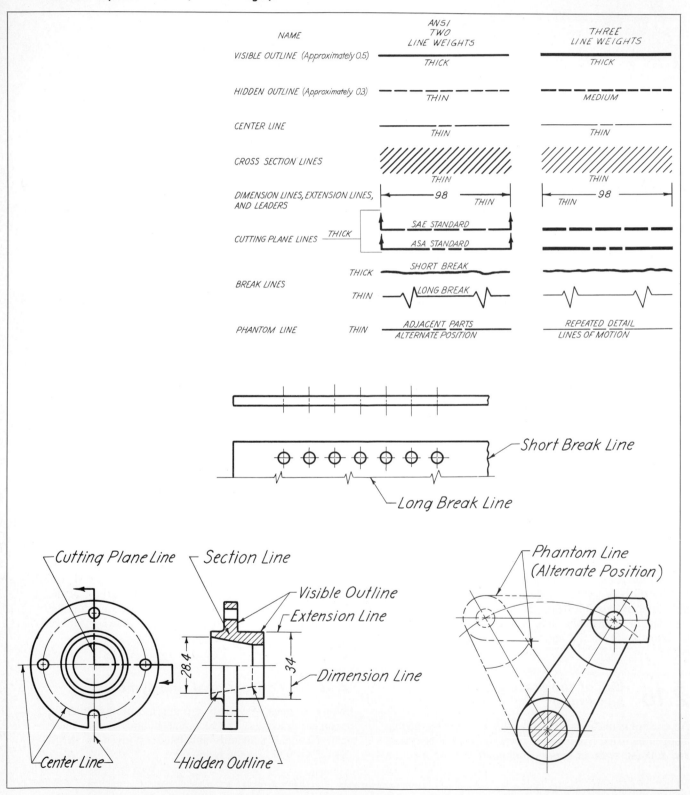

may be drawn using three widths—thick, medium, and thin. This provides greater contrast between types of lines and gives a better appearance.

The lines illustrated in Fig. 2.22 are shown full-size. When symbolic lines are used on a pencil drawing they should not vary in color. For example, center lines, extension lines, dimension lines, and section lines should differ from object lines only in width. The resulting contrast makes a drawing easier to read. All lines, except construction lines, should be very dark and bright to give the drawing the "snap" that is needed for good appearance. If the drawing is on tracing paper the lead must be "packed on" so that a satisfactory print can be obtained. Construction lines should be drawn *very* fine so as to be unnoticeable on the finished drawing. The lengths of the dashes and spaces shown in Figs. 5.31 and 7.21 are recommended for the hidden lines, center lines, and cutting-plane lines on average-size drawings.

2.18 Triangles

The 45 degree and 30 × 60 degree *right triangles* (Fig. 2.23) are commonly used to make engineering drawings. Even with a drafting machine, triangles are used to keep

drafting machine movement to a minimum. Triangles should receive special care to prevent nicks and chips. Never use a triangle with an art knife. A triangle may be checked for nicks by sliding your thumbnail along the ruling edges, as shown in Fig. 2.24. Some triangles have "inking edges" to keep ink from running under the plastic.

2.19 Horizontal and Vertical Lines

Horizontal lines are drawn along the top edge of the drafting straight edge, T-square, parallel bar, or drafting machine blade. A right-handed person draws horizontal lines left to right (Fig. 2.25). A left-handed person draws horizontal lines right to left.

Vertical lines are drawn upward along the vertical leg of a triangle whose adjacent leg is supported by the drafting straight edge. In the case of a right-handed person, the triangle should be to the right of the line being drawn

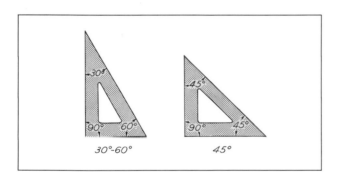

FIG. 2.23 Triangles.

FIG. 2.24 Testing a triangle for nicks.

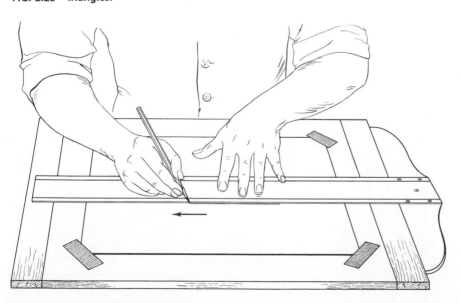

FIG. 2.25 Drawing horizontal lines.

(Fig. 2.26). Or, the vertical line may be drawn by setting the drafting-machine protractor head at zero degrees and bringing the vertical balde of the drafting machine into proper position (Fig. 2.9). Either the 30 × 60 or the 45 degree triangle may be used since both have a right angle. However the 30 × 60 is generally preferred because it usually has a longer perpendicular leg.

2.20 *Inclined Lines*

Triangles are also used for drawing *inclined lines* of 30, 60, and 45 degrees (Fig. 2.27). Angles in multiples of 15 degrees can be made with the two standard triangles and the drafting straight edge (Fig. 2.28). Triangles, used

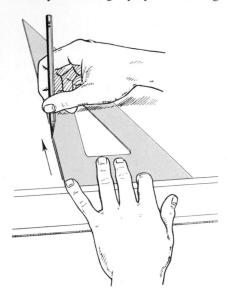

FIG. 2.26 Drawing vertical lines.

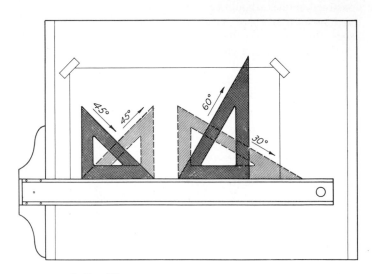

FIG. 2.27 Inclined lines.

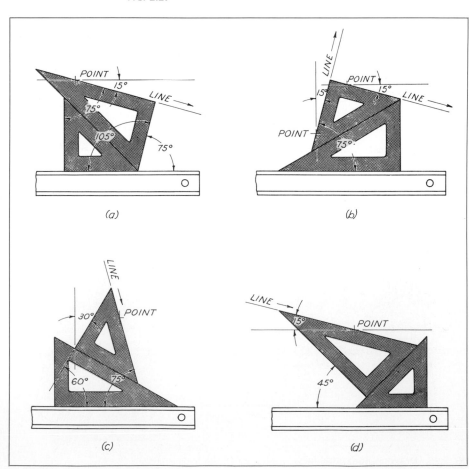

FIG. 2.28 Drawing inclined lines with triangles.

singly or in combination, provide a useful method for dividing a circle into 4, 6, 8, 12, or 24 equal parts (Fig. 2.29). These angles, as well as any others not divisible by 15, can also be drawn using the drafting machine. The protractor read-out on the drafting machine head registers deviation from zero degrees (horizontal). Incremental deviation from an inclined line can be laid out by resetting zero degrees equal to the inclined line.

2.21 *Parallel Lines*

Triangles are used in combination to draw a line parallel to a given line. To draw such a line, place an edge of a triangle, supported by a straight edge or another triangle, along the given line. Then, slide the triangle as shown in Fig. 2.30 to the new position and draw the parallel line along the triangle's same edge.

This process is greatly simplified by using a drafting machine. Unlock the protractor head and align the blade edge with the given line. Lock the protractor in this position. Move the drafting machine to the new position and draw the parallel line using the same edge (Fig. 2.9).

2.22 *Perpendicular Lines*

Either the *sliding triangle method* [Fig. 2.31(*a*)] or the *revolved triangle method* [Fig. 2.31(*b*)] may be used to draw a line perpendicular to a given line. When using the sliding triangle method, adjust to the given line a side of a triangle that is adjacent to the right angle. Guide the side opposite the right angle with a second triangle, as shown in Fig. 2.31(*a*); then slide the first triangle along the guiding triangle until it is in the required position for drawing the perpendicular along the other edge adjacent to the right angle.

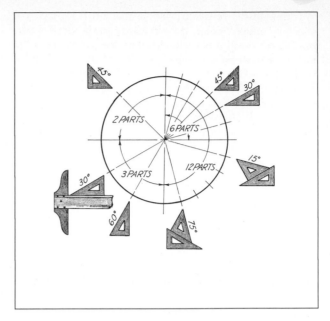

FIG. 2.29 **To divide a circle into 4, 6, 8, 12, or 24 equal parts.**

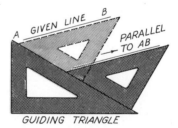

FIG. 2.30 **To draw a line parallel to a given line.**

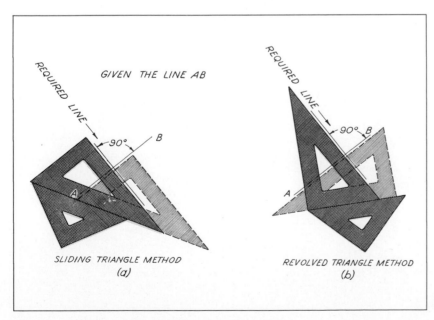

FIG. 2.31 **To draw a line perpendicular to another line.**

Although the revolved triangle method [Fig. 2.31(*b*)] is not so quickly done, it is widely used. To draw a perpendicular using this method, along the given line align the hypotenuse of a triangle, one leg of which is guided by a straightedge or another triangle; then hold the guiding member in position and revolve the triangle about the right angle until the other leg is against the guiding edge. The new position of the hypotenuse will be perpendicular to its previous location along the given line and, when moved to the required position, may be used as a ruling edge for the desired perpendicular.

2.23 Inclined Lines at 15°, 30°, 45°, 60°, or 75° with an Oblique Line

A line making an angle with an oblique line equal to any angle of a triangle may be drawn with the triangles. The two methods previously discussed for drawing perpendicular lines are applicable with slight modifications. To draw an oblique line using the revolved triangle method

[Fig. 2.32(*a*)], adjust the edge that is opposite the required angle along the given line; then revolve the triangle about the required angle, slide it into position, and draw the required line along the side opposite the required angle.

To use the sliding triangle method [Fig. 2.32(*b*)], adjust to the given line one of the edges adjacent to the required angle, and guide the side opposite the required angle with a straight edge; then slide the triangle into position and draw the required line along the other adjacent side.

To draw a line making a 75° angle with a given line, place the triangles together so that the sum of a pair of adjacent angles equals 75°, and adjust one side of the angle thus formed to the given line; then slide the triangle, whose leg forms the other side of the angle, across the given line into position, and draw the required line, as shown in Fig. 2.33(*a*).

To draw a line at 15° to a given line, select any two angles whose difference is 15°. Adjust to the given line a side adjacent to one of these angles, and guide the side adjacent with a straightedge. Remove the first triangle and substitute the other so that one adjacent side of the angle

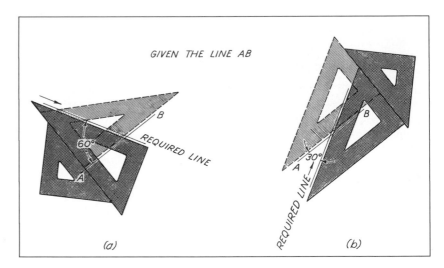

FIG. 2.32   To draw lines making 30°, 45°, or 60° with a given line.

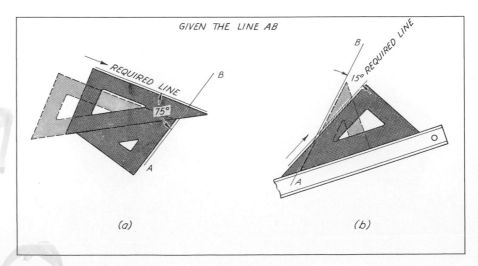

FIG. 2.33   To draw lines making 15° or 75° with a given line.

to be subtracted is along the guiding edge, as shown in Fig. 2.33(*b*); then slide the triangle into position and draw along the other adjacent side.

2.24 *Scales—Inches and Feet*

A number of *scales* are available for various types of engineering design. For convenience, however, all scales may be classified according to their use as mechanical engineers' scales (both fractional and decimal), civil engineers' scales, architects' scales, or metric scales.

The *mechanical engineers' scales* are generally of the full-divided type, graduated proportionately to give reductions based on inches. On one form (Fig. 2.34) the principal units are divided into the common fractions of an inch (4, 8, 16, and 32 parts). The scales are indicated as eighth size ($1\frac{1}{2}$ in. = 1 ft), quarter size (3 in. = 1 ft), half size (6 in. = 1 ft), and full size.

Decimal scales are more widely used in industrial draft-

ing rooms. The full size decimal scale shown in Fig. 2.35, which has the principal units (inches) divided into fiftieths, is particularly suited for use with the two-place decimal system. The half-size, three-eighths size, and quarter-size scales (Fig. 2.38) have the principal units divided into tenths.

The *civil engineers'* (chain) *scales* (Fig. 2.36) are full-divided and are graduated in decimal parts, usually 10, 20, 30, 40, 50, 60, 80, and 100 divisions to the inch.

Architects' scales (Fig. 2.37) differ from mechanical engineers' scales in that the divisions represent a foot, and the end units are divided into inches, half-inches, quarter-inches, and so forth (6, 12, 25, 48, or 96 parts). The usual scales are $\frac{1}{8}$ in. = 1 ft, $\frac{1}{4}$ in. = 1 ft, $\frac{3}{8}$ in. = 1 ft, $\frac{1}{2}$ in. = 1 ft, 1 in. = 1 ft, $1\frac{1}{2}$ in. = 1 ft, and 3 in. = 1 ft.

The sole purpose of the scale is to *reproduce the dimensions of an object full-size on a drawing or to reduce or enlarge them to some regular proportion*, such as eighth-size, quarter-size, half-size, or double-size. The scales of reduction most frequently used are as follows.

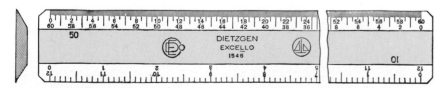

FIG. 2.34 **Mechanical engineers' scale, full-divided.**

FIG. 2.35 **Engineers' decimal scale.**

FIG. 2.36 **Civil engineers' scale.**

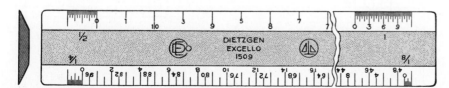

FIG. 2.37 **Architects' scale, open-divided.**

Fractional

Mechanical Engineers' Scales

Full-size	(1″ = 1″)
Half-size	($\frac{1}{2}$″ = 1″)
Quarter-size	($\frac{1}{4}$″ = 1″)
Eighth-size	($\frac{1}{8}$″ = 1″)

Architects' or Mechanical Engineers' Scales

Full-size	(12″ = 1′-0)
Half-size	(6″ = 1′-0)
Quarter-size	(3″ = 1′-0)
Eighth-size	(1$\frac{1}{2}$″ = 1′-0)
1″ = 1′-0	$\frac{1}{4}$″ = 1′-0
$\frac{3}{4}$″ = 1′-0	$\frac{3}{16}$″ = 1′-0
$\frac{1}{2}$″ = 1′-0	$\frac{1}{8}$″ = 1′-0
$\frac{3}{8}$″ = 1′-0	$\frac{3}{32}$″ = 1′-0

Decimal

Mechanical Engineers' Scales

Full-size	(1.00″ = 1.00″)
Half-size	(0.50″ = 1.00″)
Three-eighths-size	(0.375″ = 1.00″)
Quarter-size	(0.25″ = 1.00″)

Civil Engineers' Scales

10 scale:	1″ = 1′;	1″ = 10′;
	1″ = 100′;	1″ = 1000′
20 scale:	1″ = 2′;	1″ = 20′;
	1″ = 200′;	1″ = 2000′
30 scale:	1″ = 3′;	1″ = 30′;
	1″ = 300′;	1″ = 3000′
40 scale:	1″ = 4′;	1″ = 40′;
	1″ = 400′;	1″ = 4000′
50 scale:	1″ = 5′;	1″ = 50′;
	1″ = 500′;	1″ = 5000′
60 scale:		1″ = 60′; etc.
80 scale:		1″ = 80′; etc.

The first four scales, full-size, half-size, quarter-size, and eighth-size, are the ones most frequently selected for drawing machine parts, although other scales can be used. Since objects drawn by structural draftsmen and architects vary from small to very large, scales from full-size to $\frac{3}{32}$ in. = 1 ft ($\frac{1}{128}$-size) are commonly encountered. For maps, the civil engineers' decimal scales having 10, 20, 30, 40, 50, 60, and 80 divisions to the inch are used to represent 10, 20, 30 ft, and so forth, to the inch.

On a machine drawing, it is considered good practice to omit the inch marks (″) in a scale specification. For example, a scale may be specified as: FULL SIZE, 1.00 = 1.00, or 1 = 1; HALF SIZE, .50 = 1.00, or $\frac{1}{2}$ = 1; and so forth.

The decimal scales shown in Fig. 2.38 have been approved by the American National Standards Institute for making machine drawings when the decimal system is used.

It is essential that drafters always think and speak of each dimension as full-size when scaling measurements, because the dimension figures given on the finished drawing indicate full-size measurements of the finished piece, regardless of the scale used.

The reading of an open-divided scale is illustrated in Fig. 2.39 with the eighth-size (1$\frac{1}{2}$ in. = 1 ft) scale shown. The dimension can be read directly as 21 in., the 9 in. being read in the divided segment to the left of the cipher. Each long open division represents 12 in. (1 ft).

The reading of the full-size decimal scale is illustrated in Fig. 2.40. The largest division indicated in the illustration represents 1 in., which is subdivided into tenths and fiftieths (.02 in.). In Fig. 2.38 the largest divisions on the half-size, three-eighths size, and quarter-size decimal scales represent 1 in.

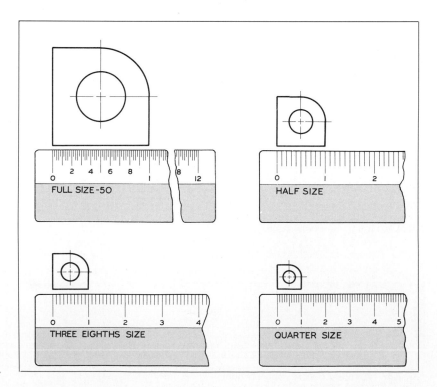

FIG. 2.38 Decimal scales.

2.25 *Metric Scales*

The *metric scale* (Figs. 2.41 and 2.42) is used in those countries where the meter is the standard of linear measurement.

Since the millimeter is the standard SI unit for linear dimensions, scales for drawing are marked in millimeters so that they can be read directly. The following scales for reduction and enlargement are recommended for the drawing of machine parts.

Reduced Scales	Enlarged Scales
1:2	2:1
1:3	3:1
1:5	5:1
1:10	10:1

1:2.5 and 2.5:1 scales may be used where available. The scales commonly used in the several fields are

Mechanical Engineers' Scales

1:2.5	1:25
1:5	1:33⅓
1:10	1:50
1:15	1:80
1:20	1:100

Civil Engineers' Scales

1:100	1:750
1:200	1:1000
1:300	1:1250
1:400	1:1500
1:500	1:2000
1:600	1:2500
1:625	1:3000

Architects' Scales

1:1	1:125
1:2	1:200
1:5	1:250
1:10	1:300
1:20	1:400
1:25	
1:50	1:500
1:100	

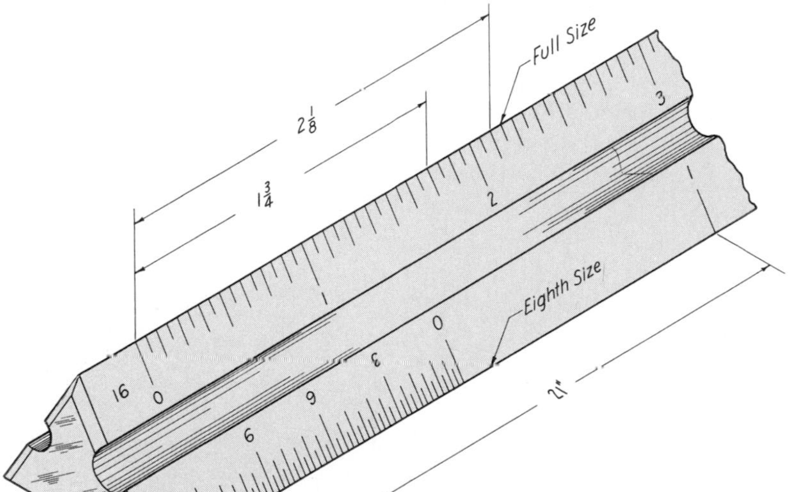

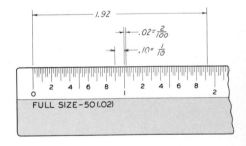

FIG. 2.39 Reading a scale.

FIG. 2.40 Reading a decimal scale.

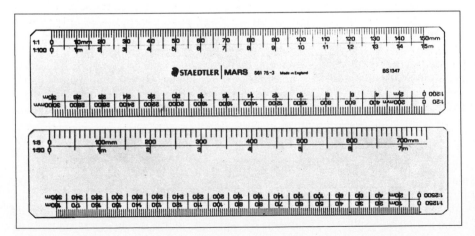

FIG. 2.41 Flat metric scale (front and reverse sides). *(Courtesy J. S. Staedtler, Inc.)*

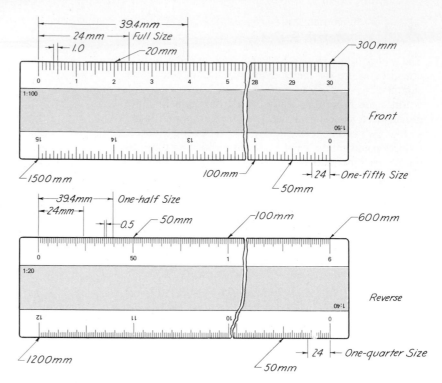

FIG. 2.42   Reading metric scales.

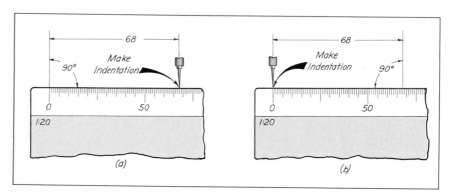

FIG. 2.43   To lay off a measurement—metric scale.

The reading of distance values along the four scales of a flat scale (front and reverse sides) are shown in Fig. 2.42.

To lay off a measurement, using a scale starting at the left of the rule, align the scale in the direction of the measurement with the zero of the scale being used toward the left. After it has been adjusted to the correct location, make short marks opposite the divisions on the scale that establish the desired distance (Fig. 2.43). For ordinary work most draftsmen use the same pencil used for the layout. When extreme accuracy is necessary, however, it is better practice to use a sharp point and make slight indentations (not holes) at the required points. If a regular point is not available, the dividers may be opened to approximately 60° and the point of one leg used as a substitute.

To ensure accuracy, place the eye directly over the division to be marked, hold the marking instrument perpendicular to the paper directly in front of the scale division, and mark the point. Always check the location of the point before removing the scale. If a slight indentation is made, it will be covered by the finished line; if a short mark is made and it is *very* light, it will be unnoticeable on the finished drawing.

To set off a measurement (say 68 mm) to half-size, the scale indicated either as 1:2 or 1:20 should be used. If the measurement is to be made from left to right, place the zero (0) division mark on the given line, and make an indentation (or mark) opposite the 68 division point [Fig. 2.43(a)]. The distance from the line to the point represents 68 mm, although it is actually 34 mm. To set off the same measurement from right to left, place the 68 mark on the given line, and make an indentation opposite the zero division mark [Fig. 2.43(b)].

B / COMPUTER-AIDED DRAWING EQUIPMENT

2.26 Introduction to Graphic Computers

Computers are being widely accepted by business, industry, science, and education as tools for problem solving and information processing. Some computers are capable of bringing major computing power to users who have not been able in the past to afford any type of computer. The computer that only a few years ago occupied an entire room now sits on a desk. In order to discuss computers, certain abbreviations are used for long and difficult terms. The first time such terms are presented in this text they are spelled out in full with their common abbreviations given. Subsequent references use only the abbreviations. (See the Glossary of computer graphic terms in the Appendix.)

CADD computers may be defined by the number of users each can support. *Mainframe* CADD computers may support hundreds of users. *Minicomputers* allow up to a dozen designers to share information and divide workload. *Microcomputers* are personal design tools that may or may not be connected to other microcomputers, minicomputers,

or mainframes. The system shown in Fig. 2.44 supports several CADD workstations.

Within the last few years microcomputers have become widely used for CADD, CAD, and CAM. These relatively low-cost machines have been made possible by the development of the *integrated circuit* (IC). It may be hard to believe, but it is possible for over 1,000,000 electronic components, like transistors, resistors, and capacitors, to be assembled on an integrated circuit less than half the size of a person's thumbnail.

Since the microcomputer will be the basis for most engineering workstations, it is important to understand the common components found in these units. Microcomputers are composed of

- the *microprocessor* or CPU,
- a number of *specialized integrated circuits* that provide memory, control, sound, and graphics,
- devices that allow *interaction* with the computer (input and output, called I/O).

The CPU functions include entering data, calling data from storage, executing the instructions of a program, and transferring information between storage and I/O devices. In addition, the CPU controls the logic section which performs math functions, makes decisions, and selects and converts data. The CPU is the "smart" section of the computer.

The microcomputer has two general types of memory:

- RAM or random-access memory
- ROM or read-only memory.

FIG. 2.44 Computer-aided drafting system. The unit appearing in the left foreground is the graphics processor. Behind the processor is a high-speed pen plotter.
(*Courtesy Computervision Corporation*)

Information in RAM is changeable. It is lost by storing new information in the same location. Information in ROM provides permanent storage of instructions. This information cannot be changed nor erased. Instructions stored in ROM are called *firmware*. Programs stored in RAM are called *software*. There are specialized types of ROM that can be programmed (PROM) or erased and reprogrammed (EPROM) by the user.

2.27 Input/Output (I/O) Devices

Using a graphics computer involves three processes:

1. *Input* of instructions and geometric data,

2. *Processing* of data,

3. *Output* in a visual form that can be understood by humans.

In the case of a CADD system, input data describes the shape of the part, that is the spatial relationship of planes, holes, and other features. The *instructions* are the commands the operator gives the computer to view the geometry from different vantage points (views) or to combine, delete, modify, or move the image.

Once the data has been entered and the instructions given, the computer acts on those instructions, altering the description of the part in computer memory. The more detailed and nonlinear the design, the longer it takes the computer to do its job. Remember, part geometry is held in the computer as a set of numbers. To change the design, the computer changes the numbers and their relationship to one another.

Once this processing is completed, the operator needs a visual indication of the changes that were made. It would be too difficult (even impossible) to review the numeric data base for anything other than the most simple geometry. For this reason, graphic computers display the image defined by the data base on a display device called a monitor. If a more transportable and permanent record of the data is needed, a print of the image or a *hard copy* can be made.

The process of getting information in and out of a computer is called *input/output* or I/O.

2.28 The Monitor

The *monitor* is the most crucial I/O device in a graphic computer system. By displaying text and the representation of graphic shapes and designs on the screen, the computer communicates with its human partner. In turn, the operator can communicate by giving instructions, verifying data, or adding new features to the design.

High quality monitors present the image just as it would appear on a finished drawing. The sharpness of a monitor is measured in *resolution*, the number of horizontal and vertical dots that make up the screen (for example 1,024 dots vertical by 780 dots horizontal). Of more importance is the number of dots in each inch (dpi). The greater the number of dots per inch, the finer the resolution.

There are two general types of monitors, each designed for specific applications.

1. Direct view storage tube (DVST) displays are very sharp but have long redisplay times,

2. Raster displays are like TVs, with constantly changing scan lines and patterns of dots.

2.29 The CADD Workstation

CADD activity is centered around the CADD workstation. The workstation may be part of several monitors arranged in a group, sharing the same graphics computer (Fig. 2.45). Or, the workstation may be a single unit connected to a *host* computer that is some distance away. More likely, the modern CADD workstation is a desktop microcomputer connected in a network to a larger CADD computer. This provides the engineer or technologist with both a personal productivity tool and an entry point (*node*) to a larger computing environment.

The key to this distributed graphics system is the sharing of *I/O peripherals*: The printers, plotters, digitizers, and scanners are not at the workstation but are on the CADD network, available when needed. The typical CADD workstation consists of

1. A powerful microcomputer,

2. Graphics and application software,

3. An input device like a mouse or digitizing pad with pen,

4. Communications hardware and software for networking.

2.30 Printers and Plotters

A *printer* gives a computer the capability to create paper or hard copy of textual material such as bills of material, parts lists, and schedules. Most computer printers are *dot matrix printers* that create characters with small dots arranged in a way that forms a letter or a numeral. Many dot matrix printers can also print drawings, but at a low, 75 to 150 dpi resolution. *Laser printers* use much more sophisticated technology to print text and drawings at an acceptable 300 to 600 dpi. Some graphic computers have hard copy units attached directly to each monitor so that quick copies can be made for review. In the engineering design room, a hard copy might be requested by a designer when he feels that he should check the dimensions and notations on a drawing executed from a sketch. A hard copy printer can be used by a design group during the early stages of a design project. Copies can continue to be called for at later stages of the design to document the development of the project (see Chapter 12). Some of these

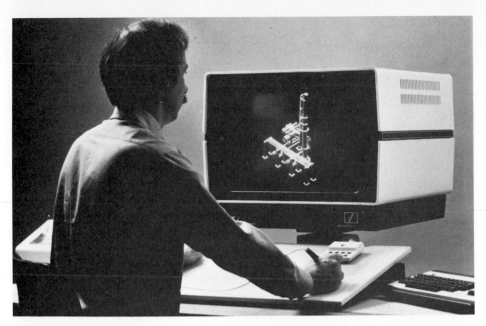

FIG. 2.45 CRT terminal. (*Courtesy Computervision Corporation*)

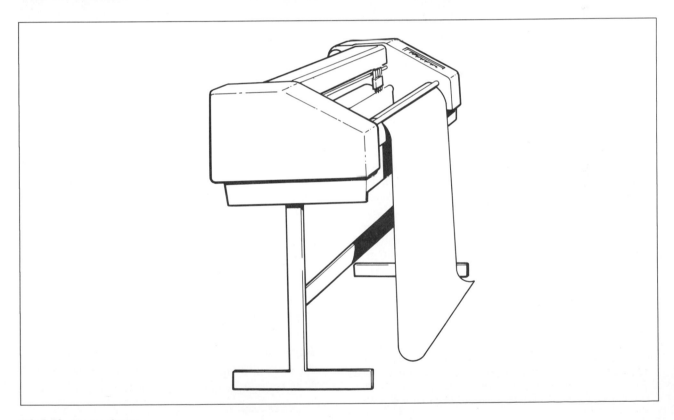

FIG. 2.46 Drum plotter.

prints may be included in both preliminary and final design reports.

When high-quality graphic prints are needed in sizes larger than that produced by a laser printer, plots are made by a *digital plotter*. Plotters use wet ink, felt tip, or ball point pens to draw on paper or polyester drafting film. Commercial plotters accommodate multiple pens, making possible varying line color. Plotters are available in several forms, commonly identified as

- Drum plotters (Fig. 2.46),
- Beltbed plotters (Fig. 2.47),
- Flatbed plotters (Fig. 2.48).

FIG. 2.47 Beltbed plotters. (*Courtesy CalComp*)

The least expensive is the *drum plotter*, which has a drum on which paper is placed (Fig. 2.46). The paper, with perforated holes along the edges, engages sprockets at both ends of the drum. A large drum plotter can take paper as wide as 36 inches.

The *beltbed plotter* makes possible the use of *cut sheet* paper or polyester film, while providing the compactness of the drum plotter. Four or more pens can be resident in the plotter at one time, and users can select from 16 pen widths. All that the operator has to do is select the color and width of the pens to be installed in the pen carriage. Smaller models can handle up to ANSI size D or ISO size A1. Larger models are for bigger drawings, up to ANSI size E or ISO size AO (Fig. 2.47).*

There are a number of small, inexpensive *flatbed plotters* available for CADD systems. More expensive flatbed plotters feature a carriage designed for motion in two directions (Fig. 2.48). Depending on the model, a flatbed plotter can have a horizontal, vertical, or tilting table. Although the horizontal table requires more floor space than a vertical table and cannot be viewed as easily, it has advantages that cause it to be favored when very accurate

* Standard paper sizes.

American Standard in inches	International Standard in millimeters
A 8.5 x 11.0	A4 210 x 297
B 11.0 x 17.0	A3 297 x 420
C 17.0 x 22.0	A2 420 x 594
D 22.0 x 34.0	A1 594 x 841
E 34.0 x 44.0	AO 841 x 1189

work is required. A tilting table can be horizontal or can be adjusted to almost vertical.

Commercial plotters often have a pen carriage that revolves, bringing the desired pen station into position. Each station can accommodate the types of pens mentioned earlier or can be loaded with pencils or scribes for cutting film.

The plotter shown in Fig. 2.49 combines automatic digitizing (a process of identifying X and Y points on an existing drawing) and high-speed plotting in one complete system. Its capabilities include data and drawing management, display, editing, I/O, and numerical machining tape (NC) production (see Chapter 18). These capabilities make this system ideal for processing quantities of large-size drawings and NC machine tool tapes. The hardware components of this type of plotter-digitizer include

- CPU capable of controlling the unit while drafting or digitizing,
- A punched tape reader to read existing NC tapes,
- A tape punch to generate new NC machining tapes,
- A keyboard for entering data and instructions,
- An operator control console,
- A plotting and digitizing table,
- An optical line follower.

The *electrostatic printer/plotter*, shown in Fig. 2.50, plots data many times faster than a conventional pen plotter. Using an electronic matrix, an electrostatic printer prints dots on charge-sensitive paper. Special paper and chemicals are required with this process. It produces prints up

FIG. 2.48   Flatbed plotter. (*Courtesy CalComp*)

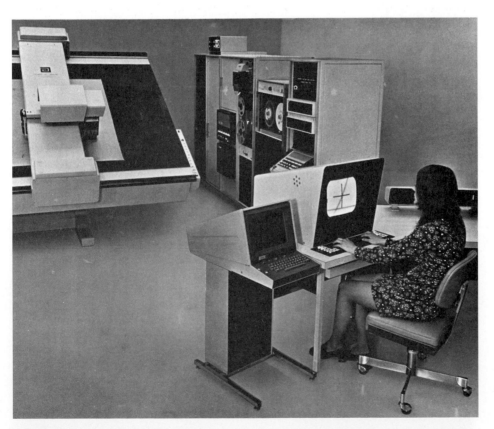

FIG. 2.49   Automatic drafting/digitizing system. (*Courtesy The Gerber Scientific Instrument Company*)

FIG. 2.50 Electrostatic printer/plotter. (*Courtesy CalComp*)

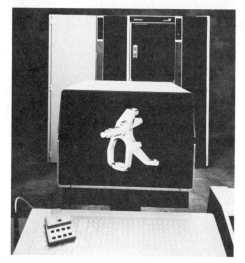

FIG. 2.51 Digitizing tablet. (*Courtesy Computervision Corporation*)

to 72 inches wide and functions as a large-scale version of a laser printer or rasterizer.

By using a MODEM (MOdulator-DEModulator), a graphics computer may be connected to other computers using phone lines, dedicated wires, or microwave. With a MODEM, a small computer can gain access to a large data bank of drawings in another computer system.

2.31 *Digitizing*

If plotting is a process of taking digital or numeric information and turning it into a drawing, *digitizing* is the process that takes a drawing and turns it into a digital data base or *file*. With any CADD system, there are always some drawings that exist on paper—sketches or engineering drawings made prior to computerization or received from noncomputerized vendors. Graphic data needed for lines, points, planes, arcs, circles, and special curves must be input to the computer. The input process for taking X (width), Y (height), and Z (depth) coordinates off a drawing is known as *digitizing*.

The digitizing process requires the use of a *digitizer*, which consists of either a digitizing tablet or a board similar to a drawing board (see Figs. 2.51 and 2.52). In a way, digitizing is much like tracing with a tracer. The tracer in this case is an input device called a *mouse*, *puck*, or a *stylus* which, when traced over a drawing, feeds the coordinates of the image directly into the computer.

A *digitizing tablet* is well suited for interactive design since the designer can work naturally and freely and at the same time see results on the screen. At the user's command, the computer can straighten lines and present an accurate drawing. Figure 2.53 shows the use of a digitizing tablet.

Since designers easily think in three-dimensions, 3-D digitizing opens many new possibilities for CAD/CAM. A 3-D digitizer (Fig. 2.54) provides Z-axis information in addition to X and Y. It does so without complicated operations, making it extremely easy to manipulate three-dimensional data when digitizing complex and irregular

FIG. 2.52 Digitizing. (*Courtesy Computervision Corporation*)

FIG. 2.53 **Microcomputer drafting system.** The drafter is using the digitizing tablet. (*Courtesy T&W Systems, Inc.*)

FIG. 2.54 **Digitizing an object in three dimensions.** (*Courtesy Micro Control Systems, Inc.*)

FIG. 2.55 Optical Line Follower (see Fig. 17.22). (*Courtesy The Gerber Scientific Instrument Company*)

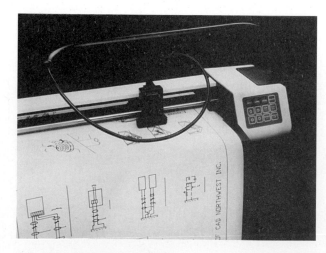

FIG. 2.56 Scanning a drawing.

shapes. The three-dimensional display shown in the monitor in Fig. 2.54 was created by tracing the physical object.

Commercial digitizing equipment may contain an *optical line follower*, as shown in Fig. 2.55. The line follower contains a light source for illuminating the digitizing surface, a light sensing device, a television camera, and a three-station tool holder. While the light-sensing device transmits signals to the digitizer's computer, the television camera transmits a magnified view of the area being digitized to the operator.

Another form of digitizing is called *scanning*. This uses a *raster-to-vector converter* to scan a drawing and translate the visual picture into geometric data (vectors) understood by the CADD system (Fig. 2.56). This process can also go the other way, in a vector-to-raster conversion. This is done to send vector data stored in a CADD system to a raster printer, such as a laser printer, capable of printing dot resolution.

FIG. 2.57   Magnetic tape unit.

2.32 *Storing CADD Drawings*

A CADD drawing is called a *file*. These files are arranged in hierarchial *directories* and given very specific and descriptive names. A certain number of files are stored permanently in the computer for daily use. But most drawing files are stored *off line* on specialized electronic storage devices like *magnetic tape* (Fig. 2.57) or *removable disks* (Fig. 2.58).

Drawing files stored in CPU memory are temporary and vanish when the computer is turned off or if the operator terminates a work session. This fact should never be forgotten. Information that needs to be retained indefinitely must be stored on disk or tape.

2.33 *CADD Software*

The necessary sequential instructions to run a CADD system are provided by *software* programs. For discussion

FIG. 2.58   Disk storage.

purposes it can be said that software falls into four categories:

1. Software for the operation of the system,
2. Graphics instructions,
3. Applications software,
4. User-developed software.

The first three types can be purchased from producers of CADD software or original equipment manufacturers.

Operating system (OS) software makes possible the general operation of the CADD system. It provides for such tasks as memory allocation, the scheduling of the CPU, and the operation of the input and output devices, and the OS prioritizes operations and interruptions.

Graphic software enables the computer to process graphic data and prepare drawings and graphs complete with notations. To meet the requirements of designers and drafting technicians, graphic software must be designed to permit creating, editing, revising, correcting, and storing data for graphic primitives such as points, lines, planes, arcs, and circles in a way that is familiar and easy to learn. The way in which this is done is called the *user interface*.

CADD *applications software* is created for specific tasks in technical fields relating to design. Such software permits drafting technicians to perform drafting tasks like the preparation of layouts, shop drawings, and piping schematics. Applications software is available for electronic, structural, pictorial, highway, and map drafting. Some applications software *postprocesses* the data base after creation on a CADD system.

Although CADD operators rarely develop their own software, they should be familiar with the several types of software required for the operation of the total system. User software is task-specific for applications that are unique to a company. *User software* is usually prepared by the company programmer, possibly assisted by the drafting technician. One common example of user software is the creation of specialized menus for input. Menu-driven programs may also be purchased from producers of software or the makers of CADD systems as part of the total system. *Macro commands*, another form of user software, are a way of executing commonly used steps in sequence with a single command.

A CADD menu is a common input device having a pattern of squares on the digitizing surface to which CADD functions are assigned (see Fig. 2.59). When a square is touched by an electronic stylus, the command assigned

FIG. 2.59 Menu form used in conjunction with a digitizing board as an input device. (*Courtesy Computervision Corporation*)

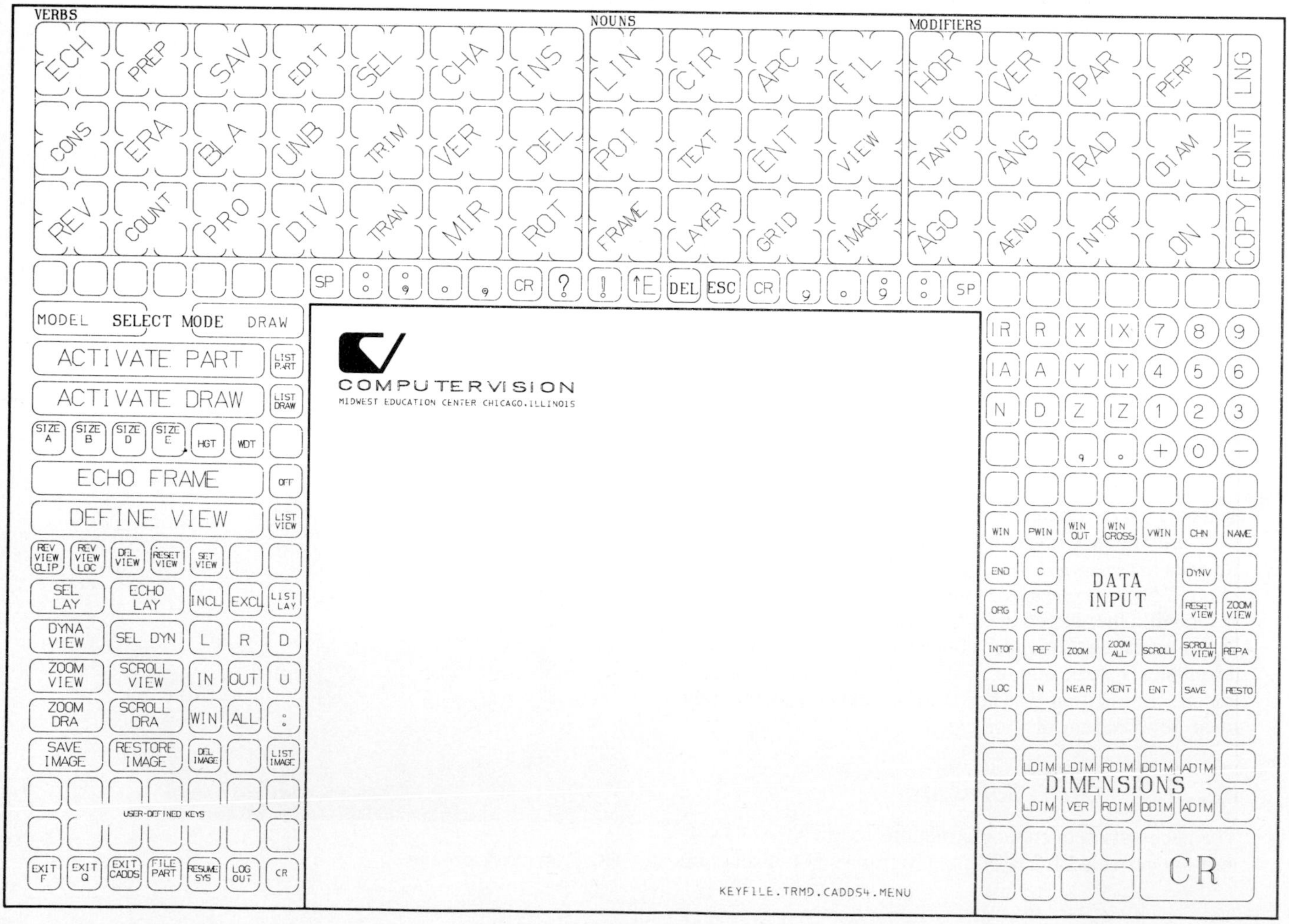

to that square is executed. The command may be represented in the box by a word, number, shape, value, or symbol. Because of the menu, the CADD operator finds it easy to prepare a drawing with frequently used symbols and then add all the notations to complete it. The menu is one example of user interface.

C / TECHNICAL LETTERING— MANUAL, MECHANICAL, AND CADD

2.34 Introduction to Manual Lettering

To present a complete *shape description* of a machine or structure, the drawing must be accompanied by *size description*—dimensions, notes, and text (Fig. 2.60). On traditional engineering drawings, dimensions and notes are lettered in a plain, legible style that can be rapidly executed. Poor lettering detracts from the appearance of a drawing and often impairs its usefulness, regardless of the quality of the line work.

2.35 Technical Lettering, Single-Stroke Letters

Single-stroke letters are used universally for technical drawings. This style is suitable for most purposes because it is legible and can be written quickly. On commercial drawings it appears in slightly variant forms since each person develops a unique style.

Single stroke means that the straight and curved lines that form the letters are the same width as the stroke of the pen or pencil.

2.36 General Proportions of Letters

Although there is no fixed standard for the proportions of hand lettering, certain rules must be observed if the lettering is to appear neat, readable, and pleasing. The best way to acquire lettering ability is to develop good lettering practices. Study the example of letter form in Fig. 2.60 and try to make your own lettering appear the same. Master the form first and then develop speed. It is advisable for the beginner, instead of relying on his untrained eye for proportions, to follow the fixed proportions shown in Figs. 2.67–2.72. Otherwise, the lettering will probably be dis-

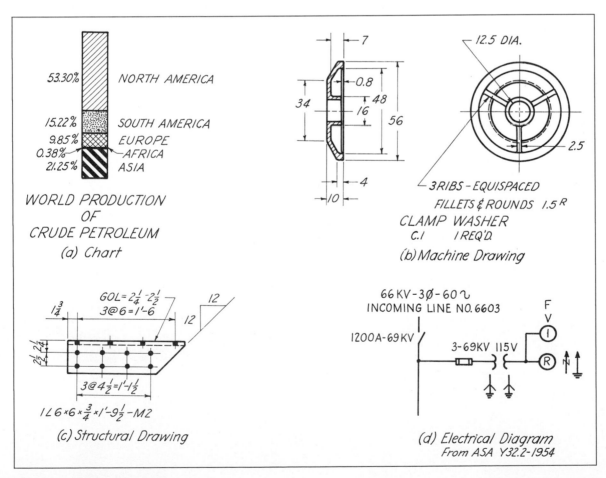

FIG. 2.60 Technical drawings.

pleasing to the trained eye of a professional. Later, after thoroughly mastering the art of lettering, individuality will be revealed naturally by light variations in the shapes and proportions of some of the letters.

It is often desirable to increase or decrease the width of letters in order to make a word or group of words fill a certain space. Letters narrower than normal are called *compressed letters*; those that are wider are called *extended letters* (Fig. 2.61).

2.37 Uniformity and Composition in Lettering

Uniformity in height, inclination, spacing, and strength of line are essential for good lettering (Fig. 2.62). Professional appearance depends as much on uniformity as on the correctness of the proportion and shape of individual letters. Uniformity in height and inclination is assured by the use of guide and slope lines (Fig. 2.63). Uniformity of weight and darkness is controlled by the type of pencil used and the pressure of its point on the paper.

In combining letters into words, the spaces for the various combinations of letters are arranged so that the areas

NORMAL LETTERS

COMPRESSED LETTERS

EXTENDED LETTERS

FIG. 2.61   Compressed and extended letters.

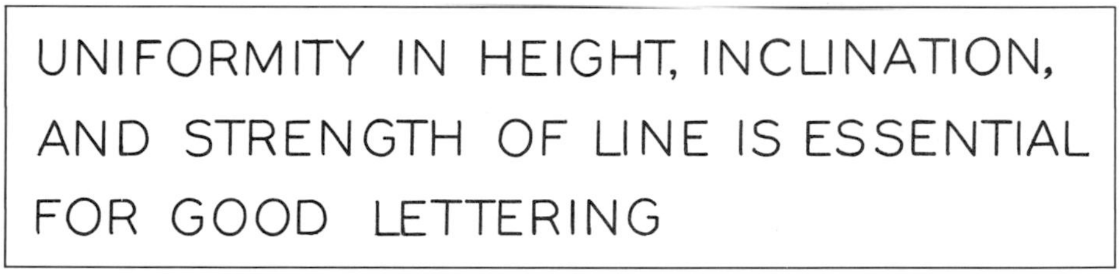

appear to be equal (Fig. 2.64). For standard lettering, this area should be about equal to one-half the area of the letter M.

The space between words should be equal to or greater than the height of a letter but not more than twice the height. The space between sentences should be somewhat greater.

Devices for drawing guide lines are available in a variety of forms. The two most popular are the *Braddock lettering triangle* (Fig. 2.65) and the *Ames lettering guide* (Fig. 2.66). Under no circumstance should technical lettering be attempted without the use of guide lines.

The Braddock lettering triangle is provided with sets of grouped holes that may be used to draw lines by inserting a sharp pointed pencil (4H or 6H lead) in the holes and sliding the triangle back and forth along a straight edge (Fig. 2.65). The holes are grouped to give guide lines for capitals and lower-case letters. The numbers below each set indicates the height of the capitals in thirty-seconds of an inch. For example, the number 3 set of holes is for capitals $\frac{3}{32}$ inch high, the number 4 set, for capitals $\frac{1}{8}$ inch high, and so on. For metric drawing, dimension letters should be 3.5 to 5mm high and the drawing number and title should be 7mm in height.

2.38 The Technique of Freehand Lettering

Any prospective engineer or technologist can learn to letter by practicing intelligently and maintaining a persistent desire to improve. The necessary muscular control, which

UNIFORMITY IN HEIGHT, INCLINATION, AND STRENGTH OF LINE IS ESSENTIAL FOR GOOD LETTERING

FIG. 2.62   Uniformity in lettering.

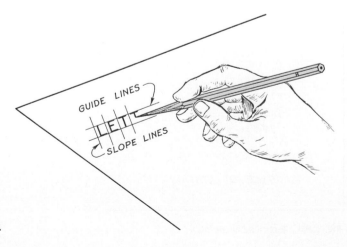

FIG. 2.63   Guide lines and slope lines.

FIG. 2.64   Letter areas.

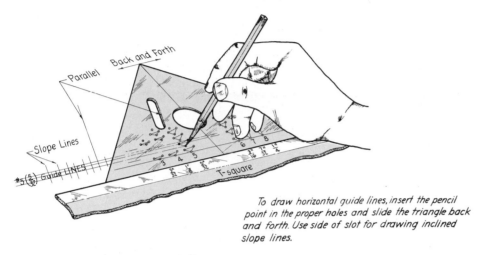

To draw horizontal guide lines, insert the pencil point in the proper holes and slide the triangle back and forth. Use side of slot for drawing inclined slope lines.

FIG. 2.65   Braddock lettering triangle.

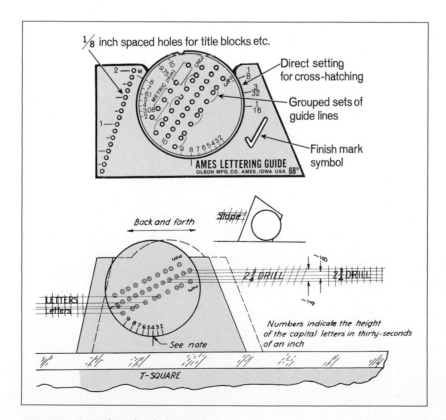

FIG. 2.66   Ames lettering instrument.

must accompany the knowledge of lettering, can only be developed through constant repetition.

Pencil letters should be formed with strokes that are dark and sharp, never with strokes that are grey and indistinct. Beginners should avoid the tendency to form letters by sketching, as strokes made in this manner vary in darkness and width.

2.39 Inclined and Vertical Letters and Numerals

The letters shown in Figs. 2.67 and 2.68 have been arranged in related groups. In laying out the characters, the number of widths and strokes have been reduced to the smallest number consistent with good appearance, similarities of shape have been emphasized, and minute differences have been eliminated. Arrows with numbers indicate the direction and order of the strokes. The curves of the inclined letters are portions of ellipses, while the curves of vertical letters are parts of circles.

The numerals shown in Figs. 2.69 and 2.70 have been arranged in related groups in accordance with the common characteristics that can be recognized in their construction.

Study the letterforms, noting the placement of cross bars and the proportions of upper and lower parts of each letter. Try to make your own lettering look like the example in Fig. 2.62.

Single-stroke lower case letters (Figs. 2.71 and 2.72), either vertical or inclined, are commonly used on map drawings, topographic drawings, structural drawings, and in survey field books. They are particularly suited for long notes and statements because they can be executed much faster than capitals and words and statements formed with them can be read more easily.

2.40 Large and Small Caps in Combination

Many commercial draftsmen use a combination of large and small capital letters in forming words, as illustrated in Fig. 2.73. When this style is used, the height of the small caps should be approximately three-fifths the height of the standard capital letter.

2.41

The height of each of the numbers in the numerator and denominator is equal to three-fourths the height of a non-fractional number, and the total height of the fraction is twice the height of this number. The division bar should

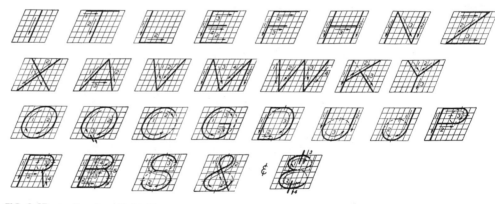

FIG. 2.67 Inclined capital letters.

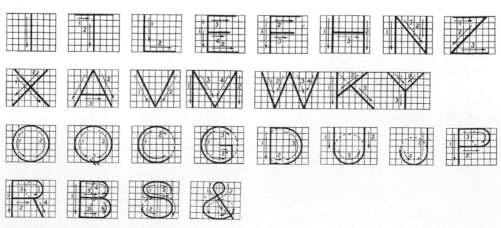

FIG. 2.68 Vertical capital letters.

FIG. 2.69 Inclined numerals.

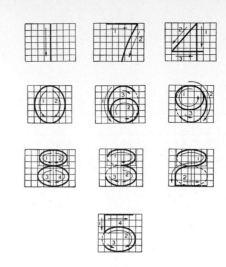

FIG. 2.70 Vertical numerals.

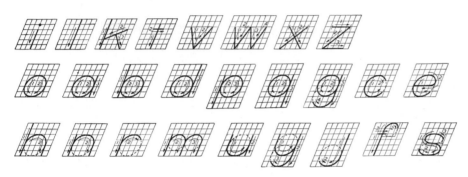

FIG. 2.71 Inclined lowercase letters.

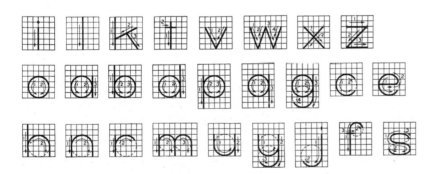

FIG. 2.72 Vertical lowercase letters.

be horizontal and centered between the fraction numerals, as shown in Fig. 2.74. It should be noted that the axis of the fraction bisects both numerator and denominator and is parallel to the axis of the whole number.

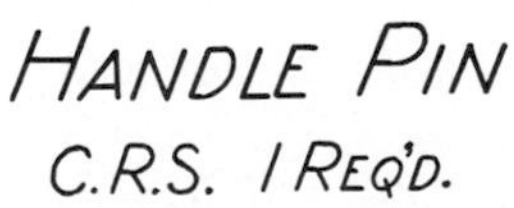

2.42 *Titles*

Every drawing, sketch, graph, chart, or diagram has some form of descriptive title to impart certain necessary information and to identify it. On machine drawings, where speed and legibility are prime requirements, titles are usually single stroke. On display drawings, maps, and the like, which call for an artistic effect, the titles are usually composed of "built-up" ornate letters.

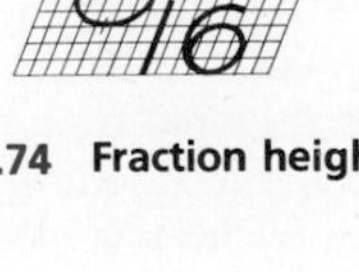

HANDLE PIN
C.R.S. 1 REQ'D.

FIG. 2.73 Use of large and small caps.

FIG. 2.74 Fraction height.

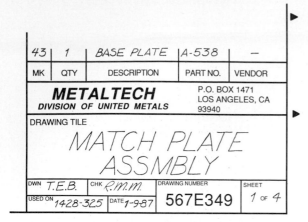

MK	QTY	DESCRIPTION	PART NO.	VENDOR
43	1	BASE PLATE	A-538	—

METALTECH
DIVISION OF UNITED METALS

P.O. BOX 1471
LOS ANGELES, CA
93940

DRAWING TILE

MATCH PLATE ASSMBLY

DWN T.E.B.	CHK R.M.M.	DRAWING NUMBER	SHEET
USED ON 1428-325	DATE 1-9-87	567E349	1 OF 4

FIG. 2.75   Title block.

FIG. 2.76   Use of a technical fountain pen and Rapido lettering guide. (*Courtesy Koh-I-Noor, Inc.*)

FIG. 2.77   Leroy lettering device.

Figure 2.75 shows a title block that might be used on a machine drawing. It should be noted that the important items are made more prominent by the use of larger letters formed with heavier lines. Less important information, such as the scale, date, drafting information, and so on, are given less prominence.

To be pleasing in appearance, a title should have some simple geometric form. An easy way to ensure the symmetry of a title is to count the letters and spaces, and then, working first toward the right from the middle, sketch the title lightly in pencil before lettering the final form. An alternative is to letter a trial title along the edge of a scrap piece of paper and place it in a balanced position just above the location of the line to be lettered.

2.43 Mechanical Lettering Devices and Templates

Up to this point our discussion of lettering has been confined to *freehand lettering*, done with a pencil or pen with only the help of guidelines. There are a number of mechanical aids available which improve the legibility of manually-produced letters.

Although mechanical lettering devices produce letters that may appear stiff to an expert, they are used in many drafting rooms for the simple reason that they enable even the most unskilled letterers to produce satisfactory results. These mechanical aids fall into three categories.

1. Lettering templates,
2. Mechanical guides with scribes,
3. Electronic lettering guides.

Lettering templates are an inexpensive alternative to freehand lettering. These plastic templates are held in place and moved along a straight edge while the inside of each letterform is traced (Fig. 2.76). These plastic guides are available to produce letters of $\frac{3}{32}$ to $\frac{1}{2}$ inch in height.

The *Leroy* lettering guide (Fig. 2.77) uses a tracer and precision-cut guide to form each letter. As with lettering templates, the user must space each letter, but generally the results are better.

Electronic lettering guides work very much like the Leroy guide except that text is lettered as a complete note, and the machine takes care of letter spacing and alignment (Fig. 2.78).

2.44 Introduction to CADD Lettering

The purpose behind lettering an engineering drawing is the same for both manual and CADD drawings. However, terminology and practices are different when using a computer to add text. A prospective draftsman or engineer should be familiar with this automated method of adding text to a drawing.

The obvious difference between manual and computer lettering is that manual lettering requires some measure of artistic skill where compuer lettering requires typing skill and a knowledge of computer commands. The better the typing skill of the CADD operator, the more easily text is entered with fewer mistakes.

Text is created in two distinctly different ways in CADD. Text blocks, called *strings* can be identified by the computer as entities in the data base (like circles, lines, or arcs). As an entity, a block of text can be identified by its origin and scaled, rotated, inclined, moved, made bolder, and so on. However, this CADD text is distinctly *different from word-processed text*. In the case of word-processed text, letters may be made larger or smaller, but they cannot be inclined, rotated, or moved independently of the text around them. There are times that word-processed text is more efficient for an engineering drawing, as for example when there are extensive bills of materials

FIG. 2.78 Electronic lettering guides.

ABCDEFGHIJKLMNOPQRSTUVWXYZ
abcdefghijklmnopqrstuvwxyz
1234567890
-_=+!@#$%^&*()[]{};:'",<.>/?`~\|

TEXT FONT(ITALIC) HEIGHT (.5 INCH)

ABCDEFGHIJKLMNOPQRSTUVWXYZ
abcdefghijklmnopqrstuvwxyz
1234567890
-_=+!@#$%^&*()[]{};:'",<.>/?`~\|

TEXT FONT(PLAIN) HEIGHT (.25 INCH)

FIG. 2.79 Text fonts.

and specifications. Word-processed text requires considerably less space inside the computer than does CADD text.

2.45 *Text Fonts*

A *text font* is a computer file of all of the characters in a particular lettering style and size (Fig. 2.79). The total number of characters in text fonts may vary due to specialized characters. Many CADD systems have a Leroy font that matches the lettering of the mechanical lettering guide previously described in Sec. 2.43. Which font to use is generally a company standard.

One factor governing the choice of text font is the method by which CADD drawings will be reproduced. Digital plotters (see Sec. 2.30) produce curved letters as a series of short vector lines or strokes. The smoother the letters, the smaller and more numerous the strokes and the greater the amount of memory required to store the text. If the CADD drawing contains a considerable amount of text, it may take longer than feasible to plot. A solution to this is the choice of a *fast font* (Fig. 2.80) where the minimum

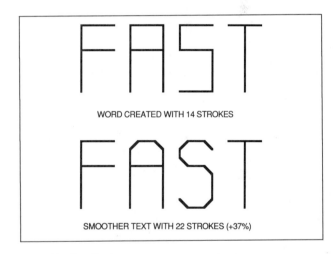

WORD CREATED WITH 14 STROKES

SMOOTHER TEXT WITH 22 STROKES (+37%)

FIG. 2.80 Fast font.

FIG. 2.81 Stroke and raster text.

number of strokes are used to form each letter. For a text-intensive drawing, plotting times may be lowered by 75%. CADD drawings, electrostatically printed (like a photo copy) or laser printed, are converted from strokes or vectors to a pattern of dots called a *rasterized image*. This process, called a vector-to-raster conversion, removes the plotting time benefit of a fast font (Fig. 2.81).

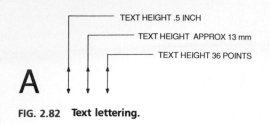

FIG. 2.82 Text lettering.

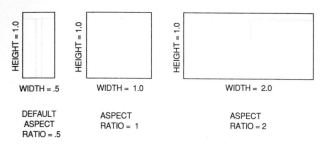

FIG. 2.83 Results of changing text aspect ratio.

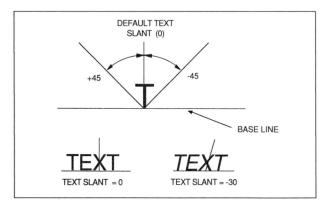

FIG. 2.84 Text slants.

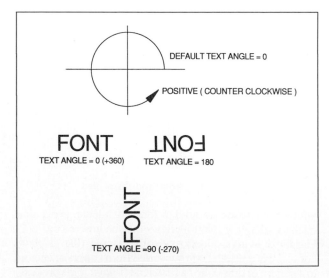

FIG. 2.85 Text angle.

2.46 Text Attributes

All CADD text has assigned to it certain attributes, often called *parameters*. These attributes control how the text appears and where it is located on the drawing. All attributes have standard settings called *default parameters* which are in effect when the CADD system is started. If you want larger text with fatter lines, you have to change the default parameters. Any attribute may be changed for a particular instance of text, or every instance of text on a drawing may be changed at once. These text attributes or parameters include

1. Font
2. Size
3. Aspect Ratio
4. Slant
5. Angle
6. Line font
7. Justification.

The *text size* attribute can be specified in a number of ways, but generally the height of the letter is what is important. Height can be specified as a fraction of an inch ($\frac{1}{4}$, $\frac{1}{2}$), a decimal fraction of an inch (.25, .50), in millimeters (25, 50), or by points (12, 14). The point system of text measurement is the standard used by commercial printers where 1 inch = 72 points (Fig. 2.82). Since CADD computers often send their drawings to typesetting computers for the production of engineering documents, many CADD systems provide for this specification of text. The default text height may be .25 inch, 12 millimeters, or 12 points.

The *aspect ratio* attribute controls the relationship of width to height. Text width may be 80% of its height by default. An aspect ratio of .5 would mean that each letter is 40% of its height (Fig. 2.83). As with hand lettering, text that is narrower than normal is called condensed and that which is wider is called expanded. The default aspect ratio is 1, meaning that regardless of the text size, the width is automatically adjusted to 80% of its height.

The text attribute *slant* refers to the deviation of the text axis from vertical (Fig. 2.84). The range of inclination can be from −45 to 45 degrees. The default inclination is zero degrees, or vertical.

Text angle describes the deviation from horizontal of the text's *baseline* (Fig. 2.85). This allows text to be placed at any angle—horizontal, vertical, even upside down. The default text angle is zero degrees. This means that unless you specify otherwise, text is entered on a horizontal baseline.

Line font is a number (1, 2, 3, etc.) that represents the type and thickness of the line in which the lettering is executed. Each letter stroke may be thin or fat, solid or dashed, black or in color. The default text line font is the solid black single stroke line (Fig. 2.86).

Text justification determines where text will be aligned and the point on the text string that acts as its handle for identification. This identification handle is referred to as

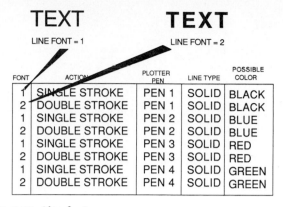

FIG. 2.86   Line font.

FONT	ACTION	PLOTTER PEN	LINE TYPE	POSSIBLE COLOR
1	SINGLE STROKE	PEN 1	SOLID	BLACK
2	DOUBLE STROKE	PEN 1	SOLID	BLACK
1	SINGLE STROKE	PEN 2	SOLID	BLUE
2	DOUBLE STROKE	PEN 2	SOLID	BLUE
1	SINGLE STROKE	PEN 3	SOLID	RED
2	DOUBLE STROKE	PEN 3	SOLID	RED
1	SINGLE STROKE	PEN 4	SOLID	GREEN
2	DOUBLE STROKE	PEN 4	SOLID	GREEN

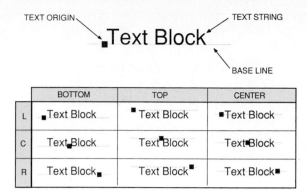

FIG. 2.87   Text justification.

	BOTTOM	TOP	CENTER
L	■Text Block	■ Text Block	■Text Block
C	Text■Block	Text■Block	Text■Block
R	Text Block■	Text Block■	Text Block■

the "origin" of the text and must be used to identify the piece of text for repositioning. Text may be left, right, or center justified at the base line, middle, or top (Fig. 2.87). The default text justification is generally at the left base line in most CADD systems. The location of the handle, or origin, is critical because identification of the origin of a block of lettering is necessary for any or all of its attributes to be changed.

EXERCISES IN INSTRUMENTAL DRAWING

The following elementary exercises have been designed to offer experience in the use of the drafting instruments. The designs should be drawn *lightly* with a hard pencil. After making certain that all constructions shown on a drawing are correct, the lines forming the designs should be heavied with a medium-hard pencil. The light construction lines need not be erased if the drawing is relatively clean.

1. (Fig. 2.88). On a sheet of drawing paper reproduce the line formations shown. If the principal border lines have not been printed on the sheet, they may be drawn first so that the large 130 × 195 mm rectangle can be balanced horizontally and vertically within the border. To draw the inclined lines, first draw the indicated measuring lines through the lettered points at the correct angle, and mark off the specified spacing. These division points establish the locations of the required lines of the formation. The six squares of the formation are equal in size.

2. (Fig. 2.89). Reproduce the line formations shown.

3. (Fig. 2.90). This exercise is designed to give the student practice with the bow pencil and compass by drawing some simple geometric figures. The line work within each large circle may be reproduced with only the knowledge that the diameter is $3\frac{1}{4}$ in. (82 mm). All circles and circle arcs are to be made finished-weight when they are first drawn, since retracing often produces a double line. Do not "overrun" the straight lines or stop them too short.

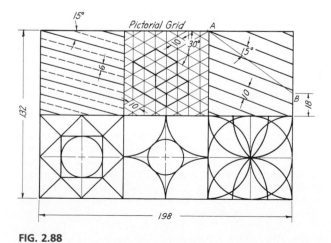

FIG. 2.88

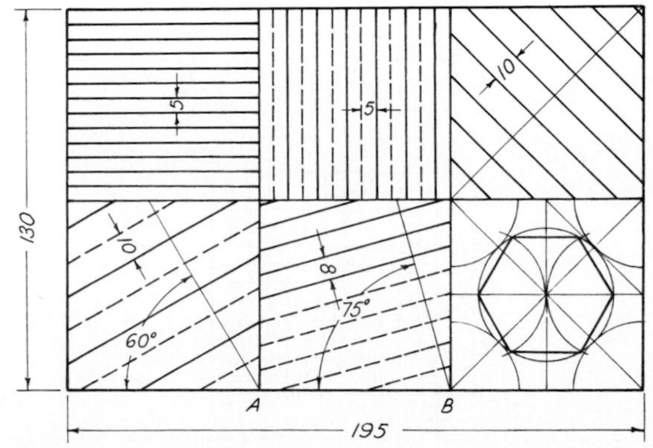

FIG. 2.89

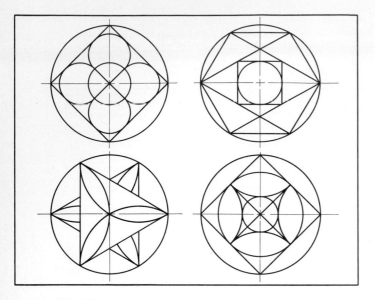

FIG. 2.90

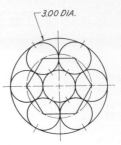

3.00 DIA.

FIG. 2.91

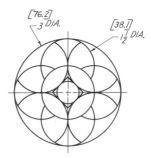

Dimensions in [ ] are in millimeters

FIG. 2.93

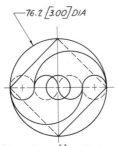

76.2 [3.00] DIA

Dimension in [ ] is in inches

FIG. 2.92

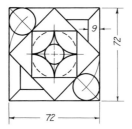

9

72

72

FIG. 2.94

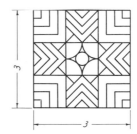

3

3

FIG. 2.95

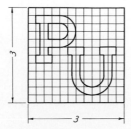

3

3

FIG. 2.96

4–6. (Figs. 2.91–2.93). Reproduce the following designs according to the instructions given for problem 3, making the dashes of the arcs approximately $\frac{1}{8}$ in. (3 mm) long.

7–9. (Figs. 2.94–2.96). Reproduce the line work within each square, using the dimensions given. (The dimensions shown, however, are for the student's use only and should not appear on the finished drawing). Arcs should be made finished-weight when first drawn. The straight lines of each design may be drawn with a hard pencil and later heavied with a softer pencil. Do not erase the construction lines.

10–13. (Figs. 2.97–2.100). Reproduce the geometric shapes.

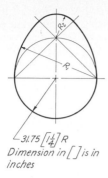

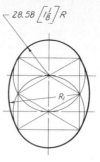

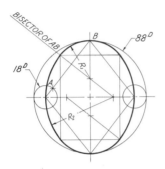

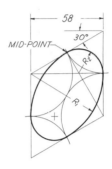

R2
R1
31.75 [1 1/4] R
Dimension in [ ] is in inches

FIG. 2.97 **Oval.**

28.58 [1 1/8] R
R1
Dimension in [ ] is in inches

FIG. 2.98 **Ellipse (approximate).**

BISECTOR OF AB
B
88D
18D
A
R1
R2

FIG. 2.99 **Ellipse (approximate).**

58
30°
MID-POINT
R2
R1

FIG. 2.100 **Ellipse (pictorial).**

LETTERING EXERCISES

These exercises are offered to give the student practice in letterforms and word composition. Each lettering exercise should be submitted to the instructor for severe criticism before the student proceeds to the next. Section 2.36 should be reread before starting the first exercise.

1. Letter the statement given in Fig. 2.101 in $\frac{5}{32}$-in. (4 mm) capital letters using an appropriate pencil that is suited to the type of paper being used and that will produce uniform opaque lines. The necessary guide lines should be drawn with a hard pencil.

2, 3. Letter the statements given in Figs. 2.102 and 2.103 $\frac{1}{8}$-in. (3 mm) capital letters using an appropriate pencil that is suited to the type of paper being used and that will produce uniform opaque lines. The necessary guide lines should be drawn with a hard pencil.

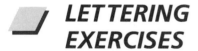

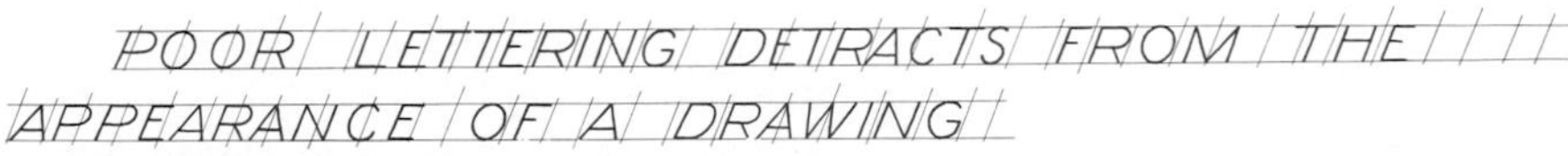

POOR LETTERING DETRACTS FROM THE
APPEARANCE OF A DRAWING

FIG. 2.101

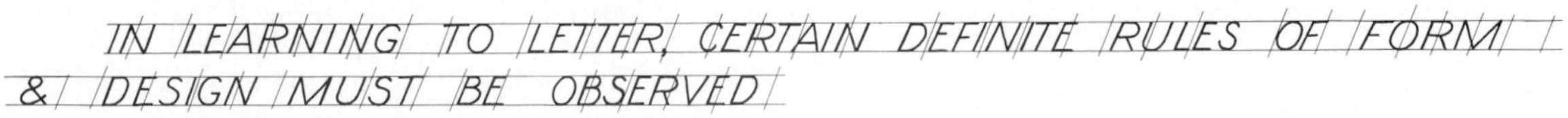

IN LEARNING TO LETTER, CERTAIN DEFINITE RULES OF FORM
& DESIGN MUST BE OBSERVED

FIG. 2.102

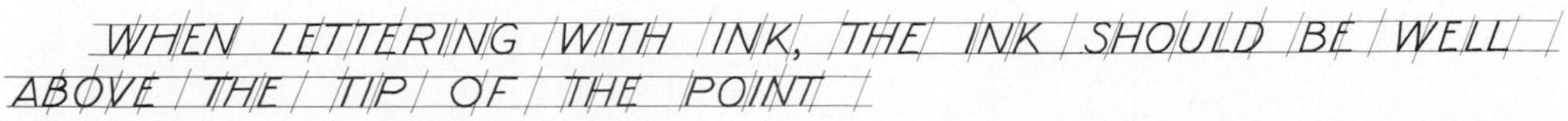

WHEN LETTERING WITH INK, THE INK SHOULD BE WELL
ABOVE THE TIP OF THE POINT

FIG. 2.103

4. Letter the statement given in Fig. 2.104 in $\frac{5}{32}$-in. (4 mm) capital letters using an appropriate pencil that is suited to the type of paper being used and that will produce uniform opaque lines. The necessary guide lines should be drawn with a hard pencil.

5–7. Letter the statement given in Figs. 2.105 to 2.107 in $\frac{1}{8}$-in. (3 mm) capital letters using an appropriate pencil that is suited to the type of paper being used and that will produce uniform opaque lines. The necessary guide lines should be drawn with a hard pencil.

#26 (.1470) DRILL AND REAM FOR #1×1 TAPER PIN
WITH PC #41 IN POSITION

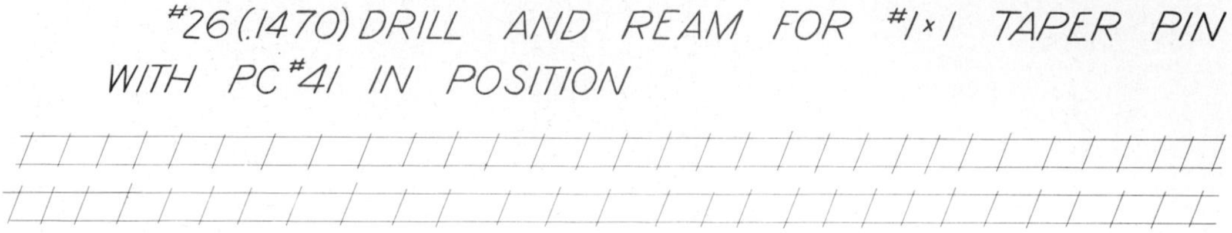

FIG. 2.104 Lettered statement. Draw guide lines as shown.

S.A.E. 1020 – COLD DRAWN STEEL BAR
1-12UNF-2B   1-8UNC-2A   1-5 SQUARE
BREAK ALL SHARP CORNERS UNLESS OTHERWISE SPECIFIED

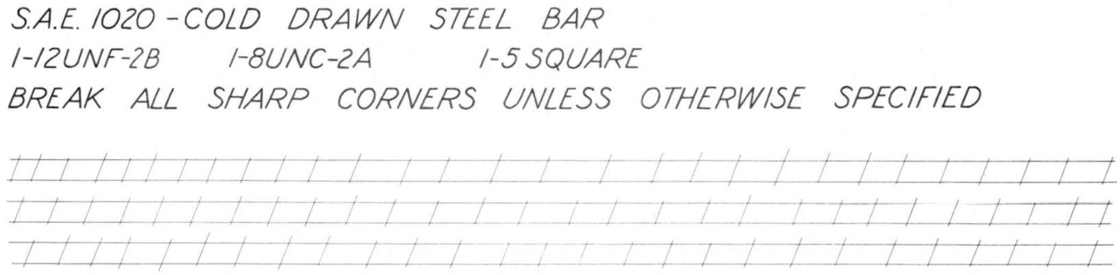

FIG. 2.105 Lettered statement. Draw guide lines as shown.

METRIC NOTES: TRUE R 18±0.1   Ø10±0.1-18 DEEP
3×45° CHAMFER
8±0.1-6 SLOTS EQL SP

METRIC

FIG. 2.106 Lettered statement. Draw guide lines as shown.

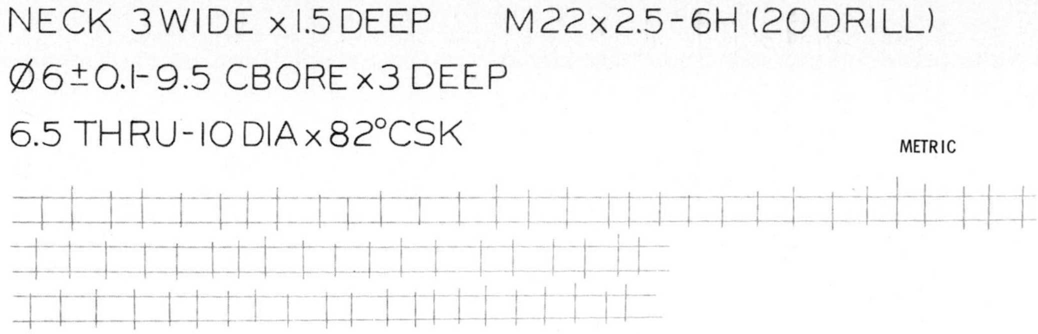

NECK 3 WIDE x 1.5 DEEP   M22 x 2.5 - 6H (20 DRILL)

Ø6 ± 0.1 - 9.5 CBORE x 3 DEEP

6.5 THRU - 10 DIA x 82°CSK

METRIC

FIG. 2.107 **Lettered statement. Draw guide lines as shown.**

The design of highway interchanges involves the application of the geometry of circle arcs. The up-and-down curves of the concrete roadway are parabolic. In the mechanical engineering field, geometry is used in the design of an endless number of machine parts as well as for gears, cams, and linkages. (*Courtesy Department of Public Works, State of California*)

Engineering Geometry

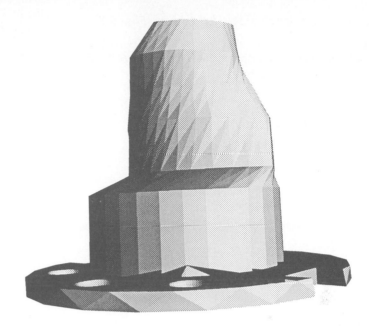

A / PLANE GEOMETRY— ENGINEERING CONSTRUCTIONS

3.1 Introduction

The simplified geometrical constructions presented in this chapter are those with which an engineer should be familiar, for they occur frequently in engineering drawing. The methods are applications of the principles found in textbooks on plane geometry. The constructions have been modified to take advantage of time-saving methods made possible by the use of drawing instruments.

Since a study of the subject of plane geometry should be a prerequisite for a course in engineering drawing, the mathematical proofs have been omitted intentionally. Geometric terms applying to lines, surfaces, and solids, are given in Figs. 3.53, 3.70 and 3.71.

3.2 To Bisect a Straight Line (Fig. 3.1)

(*a*) With *A* and *B* as centers, strike the intersecting arcs as shown using any radius greater than one-half of *AB*. A straight line through points *C* and *D* bisects *AB*.

(*b*) Draw either 60° or 45° lines through *E* and *F*. Through their intersection draw the perpendicular *GH* that will bisect *EF*.

The use of the dividers to divide or bisect a line by the trial method is explained in Sec. 2.14.

3.3 To Trisect a Straight Line (Fig. 3.2)

Given the line *AB*. Draw the lines *AO* and *OB* making 30° with *AB*. Similarly, draw *CO* and *OD* making 60° with *AB*. *AC* equals *CD* equals *DB*.

3.4 To Bisect an Angle (Fig. 3.3)

(*a*) Given the angle *BAC*. Use any radius with the vertex *A* as a center, and strike an arc that intersects the sides of the angle at *D* and *E*. With *D* and *E* as centers and a

radius larger than one-half of *DE*, draw intersecting arcs. Draw *AF*. Angle *BAF* equals angle *FAC*.

(*b*) Given an angle formed by the lines *KL* and *MN* having an inaccessible point of intersection. Draw *BA* parallel to *KL* and *CA* parallel to *MN* at the same distance from *MN* as *BA* is from *KL*. Bisect angle *BAC* using the method explained in part (*a*). The bisector *FA* of angle *BAC* bisects the angle between the lines *KL* and *MN*.

3.5 To Draw Parallel Curved Lines about a Curved Center Line (Fig. 3.4)

Draw a series of arcs having centers located at random along the given center line *AB*. Using the French curve, draw the required curved lines tangent to these arcs.

3.6 To Trisect an Angle (Fig. 3.5)

Given the angle *BAC*. Lay off along *AB* any convenient distnce *AD*. Draw *DE* perpendicular to *AC* and *DF* parallel to *AC*. Place the scale so that it passes through *A* with a distance equal to twice *AD* intercepted between the lines *DE* and *DF*. Angle *HAC* equals one-third of the angle *BAC*.

3.7 To Divide a Straight Line into a Given Number of Equal Parts (Fig. 3.6)

Given the line *LM*, which is to be divided into five equal parts.

(*a*) Step off, with the dividers, five equal divisions along a line making any convenient angle with *LM*. Connect the last point *P* and *M*, and through the remaining points draw lines parallel to *MP* intersecting the given line. These lines divide *LM* into five equal parts.

(*b*) Some commercial draftsmen prefer a modification of this construction known as the scale method. For the first step, draw a vertical *PM* through point *M*. Place the scale so that the first mark of five equal divisions is at *L* and the last mark falls on *PM*. Locate the four intervening division points, and through these draw verticals intersecting the given line. The verticals will divide *LM* into five equal parts.

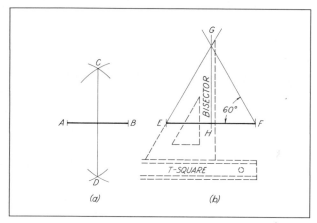

FIG. 3.1 To bisect a straight line.

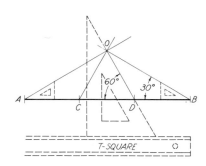

FIG. 3.2 To trisect a straight line.

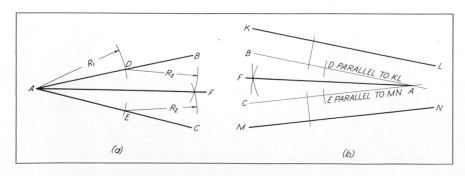

FIG. 3.3 To bisect an angle.

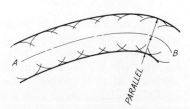

FIG. 3.4 To draw parallel curved lines.

3.8 To Divide a Line Proportionally (Fig. 3.7)

Given the line *AB*. Draw *BC* perpendicular to *AB*. Place the scale across *A* and *BC* so that the number of divisions intercepted is equal to the sum of the numbers representing the proportions. Mark off these proportions and draw lines parallel to *BC* to divide *AB* as required. The proportions in Fig. 3.7 are 1:2:3.

3.9 To Construct an Angle Equal to a Given Angle (Fig. 3.8)

Given the angle *BAC* and the line *A'C'* that forms one side of the transferred angle. Use any convenient radius with the vertex *A* as a center, and strike the arc that intersects the sides of the angle at *D* and *E*. With *A'* as a center, strike the arc intersecting *A'C'* at *E'*. With *E'* as a center and the chord distance *DE* as a radius, strike a short intersecting arc to locate *D'*. *A'B'* drawn through *D'* makes angle *B'A'C'* equal angle *BAC*.

3.10 To Draw a Line Through a Given Point and the Inaccessible Intersection of Two Given Lines (Fig. 3.9)

Given the lines *KL* and *MN*, and the point *P*. Construct any triangle, such as *PQR*, having its vertices falling on the given lines and the given point. At some convenient

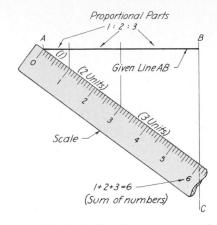

FIG. 3.7 **To divide a line proportionally.**

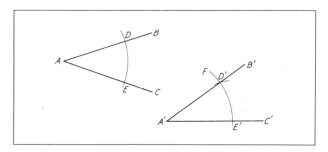

FIG. 3.8 **To construct an angle equal to a given angle.**

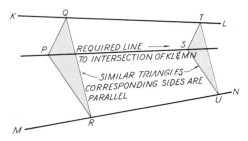

FIG. 3.9 **To draw a line through a given point and the inaccessible intersection of two given lines.**

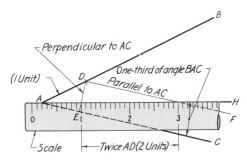

FIG. 3.5 **To trisect an angle.**

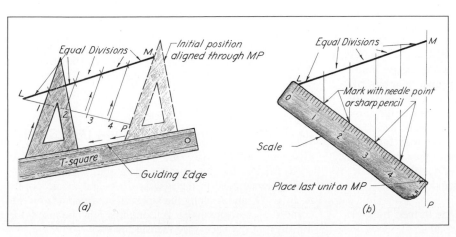

FIG. 3.6 **To divide a straight line into a number of equal parts.**

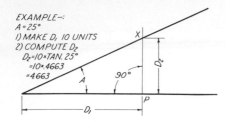

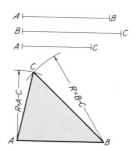

FIG. 3.11 To construct a triangle, given its three sides.

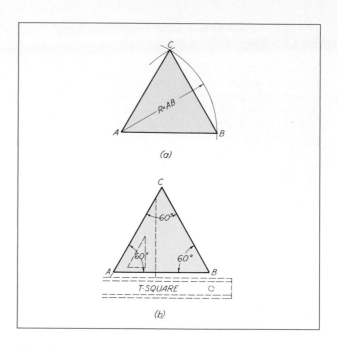

FIG. 3.12 To construct an equilateral triangle.

location construct triangle *STU* similar to *PQR*, by drawing *SU* parallel to *PR*, *TU* parallel to *QR*, and *ST* parallel to *PQ*. *PS* is the required line.

3.11 To Construct an Angle, Tangent Method (Fig. 3.10)

Draftsmen and designers often find it necessary to draw long lines having an angle between them that is not equal to an angle of a triangle. Such an angle may be laid off with a protractor, but it should be remembered that as the lines are extended any error is multiplied. To avoid this situation, the tangent method may be used. The tangent method involves trigonometry but, since it is frequently used, a discussion of it here is pertinent. (See Table 1 of the Appendix.)

In this method, a distance D_1 is laid off along a line that is to form one side of the angle, and a distance D_2, equal to D_1 times the natural tangent of the angle, is marked off along a perpendicular through point P. A line through point X is the required line, and angle A is the required angle. In laying off the distance D_1, unnecessary multiplication will be eliminated if the distance is arbitrarily made 10 full-size units. However, in order to make the construction large enough for accuracy and at the same time keep all the lines on the drawing, it may be necessary to lay off the 10 units to some scale other than full-size. When a decimal-inch scale is being used, the construction may be drawn using either a half-size or quarter-size scale. If a metric scale is used one might decide upon the 1:50 or some other selected scale.

This method is also used for angles formed by short lines whenever a protractor is not available.

3.12 To Construct a Triangle, Given Its Three Sides (Fig. 3.11)

Given the three sides *AB*, *AC*, and *BC*. Draw the side *AB* in its correct location. Using its end points *A* and *B* as centers and radii equal to *AC* and *BC*, respectively, strike the two intersecting arcs locating point *C*. *ABC* is the required triangle. This construction is particularly useful for developing the surface of a transition piece of triangulation.

3.13 To Construct an Equilateral Triangle (Fig. 3.12)

Given the side *AB*.

(*a*) Using the end points *A* and *B* as centers and a radius equal to the length of *AB*, strike two intersecting arcs to locate *C*. Draw lines from *A* to *C* and *C* to *B* to complete the required equilateral triangle.

(*b*) Using a 30° × 60° triangle, draw through *A* and *B* lines that make 60° with the given line. If the line *AB* is inclined, the 60° lines should be drawn as shown in Fig. 2.32.

3.14 To Transfer a Polygon (Fig. 3.13)

Given the polygon *ABCDE*.

(*a*) Enclose the polygon in a rectangle. Draw the "enclosing rectangle" in the new position and locate points *A*, *B*, *C*, *D*, and *E* along the sides by measuring from the corners of the rectangle. A compass may be used for transferring the necessary measurements.

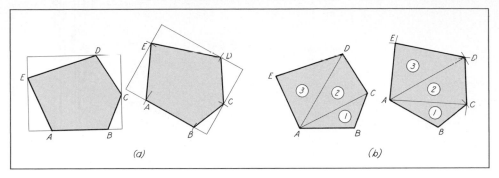

FIG. 3.13   To transfer a polygon.

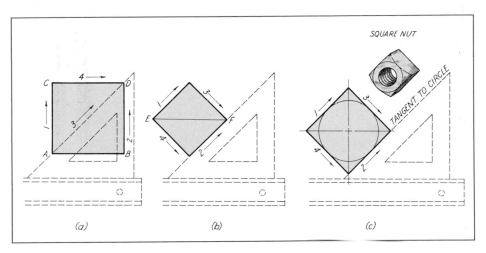

FIG. 3.14   To construct a square.

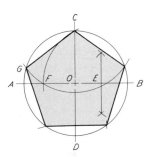

FIG. 3.15   To construct a regular pentagon.

(*b*) To transfer a polygon by the triangle method, divide the polygon into triangles and, using the construction explained in Sec. 3.12, reconstruct each triangle in its transferred position.

3.15 *To Construct a Square (Fig. 3.14)*

(*a*) Given the side *AB*. Using a T-square and a 45° triangle, draw perpendiculars to line *AB* through points *A* and *B*. Locate point *D* at the intersection of a 45° construction line through *A* and the perpendicular from *B*. Draw *CD* parallel to *AB* through *D* to complete the square. To eliminate unnecessary movements the lines should be drawn in the order indicated.

(*b*) Given the diagonal length *EF*. Using a T-square and a 45° triangle, construct the square by drawing lines through *E* and *F* at an angle of 45° with *EF* in the order indicated.

(*c*) The construction of an inscribed circle is the first step in one method for drawing a square when the location of the center and the length of one side are given.

Using a T-square and a 45° triangle, draw the sides of the square tangent to the circle. This construction is used in drawing square bolt heads and nuts.

3.16 *To Construct a Regular Pentagon (Fig. 3.15)*

Given the circumscribing circle. Draw the perpendicular diameters *AB* and *CD*. Bisect *OB* and, with its midpoint *E* as a center and *EC* as a radius, draw the arc *CF*. Using *C* as a center and *CF* as a radius, draw the arc *FG*. The line *CG* is one of the equal sides of the required pentagon. Locate the remaining vertices by striking off this distance around the circumference.

If the length of one side of a pentagon is given, the construction described in Sec. 3.19 should be used.

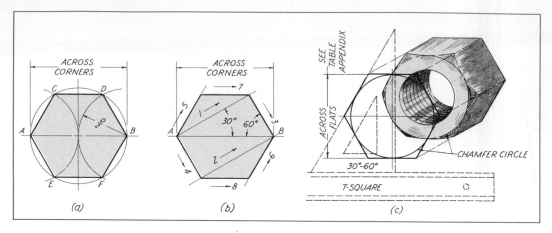

FIG. 3.16 To construct a regular hexagon.

3.17 To Construct a Regular Hexagon (Fig. 3.16)

(*a*) Given the distance *AB* across corners. Draw a circle having *AB* as a diameter. Using the same radius and with points *A* and *B* as centers, strike arcs intersecting the circumference. Join these points to complete the construction.

(*b*) Given the distance *AB* across corners. Using a 30° x 60° triangle and a T-square, draw the lines in the order indicated by the numbers on the figure.

(*c*) Given the distance across flats. Draw a circle whose diameter equals the distance across flats. Using a 30° × 60° triangle and a T-square, as shown, draw the tangents that establish the sides and vertices of the required hexagon.

This construction is used in drawing hexagonal bolt heads and nuts.

3.18 To Construct a Regular Octagon (Fig. 3.17)

(*a*) Given the distance across flats. Draw the circumscribed square and its diagonals. Using the corners as centers and one-half the diagonal as a radius, strike arcs across the sides of the square. Join these points to complete the required octagon.

(*b*) Given the distance across flats. Draw the inscribed circle; then, using a 45° triangle and T-square, draw the tangents that establish the sides and vertices of the required octagon.

3.19 To Construct Any Regular Polygon, Given One Side (Fig. 3.18)

Given the side *LM*. With *LM* as a radius, draw a semicircle and divide it into the same number of equal parts as the number of sides needed for the polygon. Suppose the polygon is to be seven-sided. Draw radial lines through points 2, 3, and so forth. Point 2 (the second division point) is always one of the vertices of the polygon, and line *L2* is a side. Using point *M* as a center and *LM* as a radius, strike an arc across the radial line *L6* to locate point *N*. Using the same radius with *N* as a center, strike another arc across *L5* to esablish *O* on *L5*. Although this procedure may be continued with point *O* as the next center, more accurate results will be obtained if point *R* is used as a center for the arc to locate *Q*, and *Q* as a center for *P*.

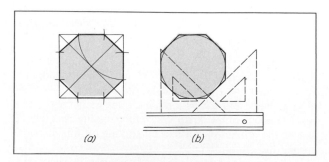

FIG. 3.17 To construct a regular octagon.

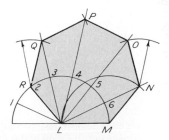

FIG. 3.18 To construct any regular polygon, given one side.

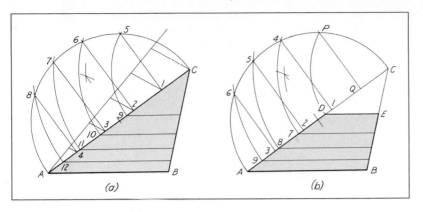

FIG. 3.19   To divide the area of a triangle or trapezoid into a given number of equal parts.

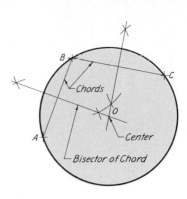

FIG. 3.20   To find the center of a circle through three points.

3.20 To Divide the Area of a Triangle or Trapezoid into a Given Number of Equal Parts (Fig. 3.19)

(a) Given the triangle ABC. Divide the side AC into (say, five) equal parts, and draw a semicircle having AC the diameter. Through the division points (1, 2, 3, and 4) draw perpendicular lines to points of intersection with the semicircle (5, 6, 7, and 8). Using C as a center, strike arcs through these points (5, 6, 7, and 8) that will cut AC. To complete the construction, draw lines parallel to AB through the points (9, 10, 11, and 12) at which the arcs intersect the side AC.

(b) Given the trapezoid DEBA. Extend the sides of the trapezoid to form the triangle ABC and draw a semicircle on AC with AC as a diameter. Using C as a center and CD as a radius, strike an arc cutting the semicircle at point P. Through P draw a perpendicular to AC to locate point Q. Divide QA into the same number of equal parts as the number of equal areas required (in this case, four), and proceed using the construction explained in (a) for dividing the area of a triangle into a given number of equal parts.

3.21 To Find the Center of a Circle Through Three Given Points Not in a Straight Line (Fig. 3.20)

Given the three points A, B, and C. Join the points with straight lines (which will be chords of the required circle), and draw the perpendicular bisectors. The point of intersection O of the bisectors is the center of the required circle, and OA, OB, or OC is its radius.

3.22 Tangent Circles and Arcs

Figure 3.21 illustrates the geometry of tangent circles. In (a) it can be noted that the locus of centers for circles of radius R tangent to AB is a line that is parallel to AB at a distance R from AB. The locus of centers for circles of the same radius tangent to CD is a line that is parallel to CD at distance R (radius) from CD. Since point O at which these lines intersect is distance R from both AB and CD, a circle of radius R with center at O must be tangent to both AB and CD.

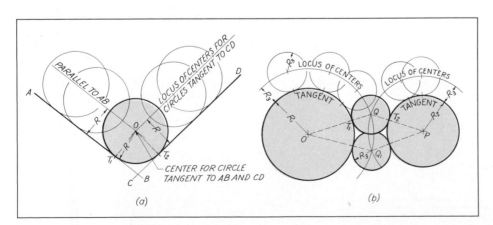

FIG. 3.21   Tangent circles.

In (b) the locus of centers for circles of radius R_3 that will be tangent to the circle with a center at O and having a radius R_1 is a circle that is concentric with the given circle at distance R_3. The radius of the locus of centers will be $R_1 + R_3$. In the case of the circle with center at point P, the radius of the locus of centers will be $R_2 + R_3$. Points Q and Q_1, where these arcs intersect, are points that are distance R_3 from both circles. Therefore, circles of radius R_3 that are centered at Q and Q_1 will be tangent to both circles with centers at O and P.

3.23 To Draw a Circular Arc of Radius R Tangent to Two Lines (Fig. 3.22)

(a) Given the two lines AB and CD at right angles to each other, and the radius of the required arc R. Using their point of intersection X as a center and R as a radius, strike an arc cutting the given lines at T_1 and T_2 (tangent points). With T_1 and T_2 as centers and the same radius, strike the intersecting arcs locating the center O of the required arc.

(b),(c) Given the two lines AB and CD, not at right angles, and the radius R. Draw lines EF and GH parallel to the given lines at a distance R. Since the point of intersection of these lines is distance R from both given lines, it will be the center O of the required arc. Mark the tangent points T_1 and T_2 that lie along perpendiculars to the given lines through O.

These constructions are useful for drawing fillets and rounds on views of machine parts.

3.24 To Draw a Circular Arc of Radius R₁ Tangent of a Given Circular Arc and a Given Straight Line (Fig. 3.23)

Given the line AB and the circular arc with center O.

(a),(b) Draw line CD parallel to AB at a distance R_1. Using the center O of the given arc and a radius equal to its radius plus or minus the required arc (R_2 plus or minus R_1), swing a parallel arc intersecting CD. Since the line CD and the intersecting arc will be the loci of centers for all circles of radius R_1, tangent respectively to the given line AB and the given arc, their point of intersection P will be the center of the required arc. Mark the points of tangency T_1 and T_2. T_1 lies along a perpendicular to AB through the center P, and T_2 along a line joining the centers of the two arcs.

This construction is also useful for drawing fillets and rounds on views of machine parts.

3.25 To Draw a Circular Arc of a Given Radius R₁ Tangent to Two Given Circular Arcs (Fig. 3.24)

Given the circular arcs AB and CD with centers O and P, and radii R_2 and R_3, respectively.

(a) Using O as a center and R_2 plus R_1 as a radius, strike an arc parallel to AB. Using P as a center and R_3 plus R_1 as a radius, strike an intersecting arc parallel to CD. Since each of these intersecting arcs of radius R_1 tangent to the given arc to which it is parallel, their point of intersection S will be the center for the required arc that is tangent to

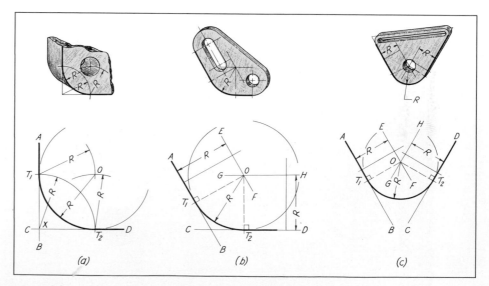

FIG. 3.22 To draw a circular arc tangent to two lines.

both. Mark the points of tangency T_1 and T_2 that lie on the lines of centers *PS* and *OS*.

(b) Using *O* as a center and R_2 plus R_1 as a radius, strike an arc parallel to *AB*. Using *P* as a center and R_3 minus R_1 as a radius, strike an intersecting arc parallel to *CD*. The point of intersection of these arcs is the center for the required arc.

3.26 To Draw a Reverse (Ogee) Curve (Fig. 3.25)

(a) *Reverse (ogee) curve connecting two parallel lines.* Given the two parallel lines *AB* and *CD*. At points *B* and *C*, the termini and tangent points of the reverse curve, erect perpendiculars. Join *B* and *C* with a straight line and

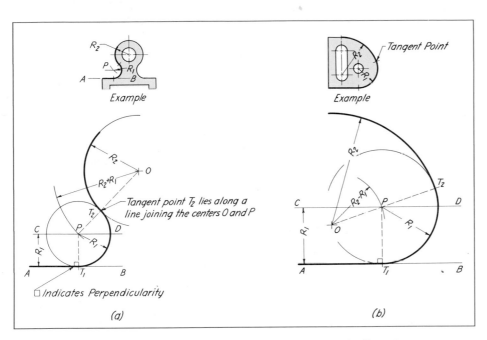

FIG. 3.23 **To draw a circular arc tangent to a given circular arc and a line.**

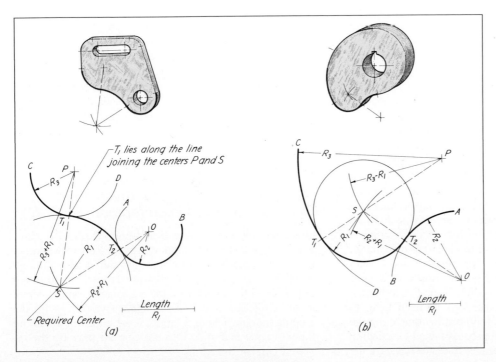

FIG. 3.24 **To draw a circular arc tangent to two given arcs.**

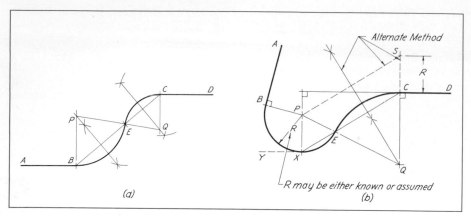

FIG. 3.25 To draw a reverse curve.

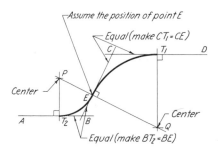

FIG. 3.26 To draw a reverse curve tangent to three lines.

assume a point E as the point at which the curves will be tangent to each other. Draw the perpendicular bisectors of BE and EC. Since an arc tangent to AB at B must have its center on the perpendicular BP, the point of intersection P of the bisector and the perpendicular is the center for the required arc that is to be tangent to the line at B and the other required arc at point E. For the same reason, point Q is the center for the other required arc.

This construction is useful to engineers in laying out center lines for railroad tracks, pipelines, and so forth.

(b) Reverse (ogee) curve connecting two nonparallel lines. Given the two nonparallel lines AB and CD. At points B and C, the termini and tangent points, erect perpendiculars. Along the perpendicular at B lay off the given (or selected) radius R and draw the arc having P as its center. Then draw a construction line through point P perpendicular to CD to establish the location of point X. With the position of X known, join points X and C with a straight line along which will lie the chords of the arcs forming the ogee curve between points X and C. The broken line XY (not a part of the construction) has been added to show that the procedure to be followed in completing the required curve will be as previously explained for drawing a reverse curve joining two parallel lines. In this case the parallel lines are XY and CD, instead of the lines AB and CD as in (a).

An alternative method for establishing the needed center for the required arc has been added to the illustration in (b). In this method the radius distance R is laid off upward along a perpendicular to CD through C. With point S established by this measurement, the line PS, as drawn, becomes the chord of an arc (not shown) that will have the same center as the required arc EC. The intersection of the perpendicular bisector of PS with the perpendicular erected downward from C will establish the position of point Q, the center of concentric arcs having chords PS and EC.

3.27 To Draw a Reverse Curve Tangent to Three Given Lines (Fig. 3.26)

Given the lines AB and CD that are intersected by a third line BC at points B and C. Assume the position of point E (point of tangency) along BC and locate the termini points T_1 and T_2 by making CT_1 equal to CE and BT_2 equal to BE. The intersections of the perpendiculars erected at points T_1, E, and T_2 establish the centers P and Q of the arcs that form the reversed curve.

3.28 To Draw a Line Tangent to a Circle at a Given Point on the Circumference (Fig. 3.27)

Given a circle with center O and point P on its circumference. Place a triangle supported by a T-square or another triangle in such a position that one side passes through the center O and point P. When using the method illustrated in (a), align the hypotenuse of one triangle on the center of the circle and the point of tangency; then, with the guiding triangle held in position, revolve the triangle about the 90° angle and slide into position for drawing the required tangent line.

Another procedure is shown in (b). To draw the tangent by this method, align one leg of a triangle, which is adjacent to the 90° angle, through the center of the circle and the point of tangency; then slide it along the edge of a guiding triangle into position.

This construction satisfies the geometric requirement that a tangent must be perpendicular to a radial line drawn to the point of tangency.

To Draw a Line Tangent to a Circle Through a Given Point Outside the Circle
3.29 (Fig. 3.28)

Given a circle with center O and an external point P. Join the point P and the center O with a straight line, and bisect it to locate point S. Using S as a center and SO (one-half PO) as a radius, strike an arc intersecting the circle at point T (point of tangency). Line PT is the required tangent.

To Draw a Tangent Through a Point P on a Circular Arc Having
3.30 an Inaccessible Center (Fig. 3.29)

Draw the chord PB; then erect a perpendicular bisector. With point P as a center swing an arc through point C where the perpendicular bisector cuts the given arc. With C as a center and a radius equal to the chord distance CE, draw an arc to establish the location of point F. A line drawn through points P and F is the required tangent.

To Draw a Line Tangent to a Circle Through a Given Point Outside the Circle
3.31 (Fig. 3.30)

Place a triangle supported by a T-square or another triangle in such a position that one leg passes through point P tangent to the circle, and draw the tangent. Slide the triangle along the guiding edge until the other leg coincides with the center O, and mark the point of tangency. Although this method is not as accurate as the geometric one explained in Sec. 3.30, it is frequently employed by commercial draftsmen.

3.32 To Draw a Line Tangent to Two Given Circles (Fig. 3.31)

Given two circles with centers O and P and radii R_1 and R.

(a) *Open belt.* Using P as a center and a radius equal to R minus R_1, draw an arc. Through O draw a tangent to this arc using the method explained in Sec. 3.30. With this location of tangent point T established, draw line PT and extend it to locate T_1. Draw OT_2 parallel to PT_1. The line from T_2 to T_1 is the required tangent to the given circles.

(b) *Crossed belt.* Using P as a center and a radius equal to R plus R_1, draw an arc. With the location of tangent point T determined through use of the method shown in Fig. 3.31, locate tangent point T_1 on line TP and draw OT_2 parallel to PT. The line T_1T_2, drawn parallel to OT, is the required tangent.

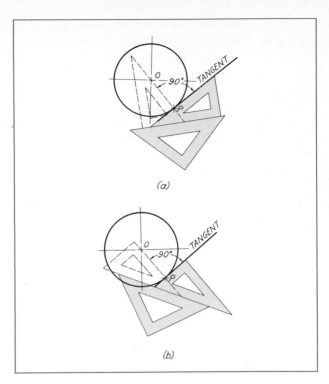

FIG. 3.27 **To draw a line tangent to a circle at a point on the circumference.**

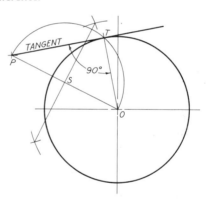

FIG. 3.28 **To draw a line tangent to a circle through a given point outside.**

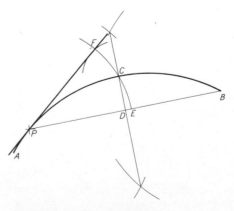

FIG. 3.29 **To draw a tangent to a circular arc having an inaccessible center.**

3.33 To Lay Off the Approximate Length of the Circumference of a Circle (Fig. 3.32)

Draw a line through point *A* tangent to the circle and lay off along it a distance *AB* equal to three times the diameter (3*D*). Using point *E* on the circumference as a center and a radius equal to the radius of the circle, strike an arc to establish the location of point *C*. Draw *CD* perpendicular to the vertical center line through point *A*. *DB* is the rectified length of the circumference; however, it is slightly longer than the true circumference by a negligible amount (approximate error 1/21,800).

3.34 To Lay Off the Approximate Length of a Circular Arc on Its Tangent (Fig. 3.33)

Given the arc *AB*.

(*a*) Draw the tangent through *A*, and extend the chord *BA*. Locate point *C* by laying off *AC* equal to one-half the length of the chord *AB*. With *C* as a center and a radius equal to *CB*, strike an arc intersecting the tangent at *D*. The length *AD* along the tangent is slightly shorter than the true length of the arc *AB* by an amount that is so minute it may be disregarded.

(*b*) Draw the tangent through *A*. Using the dividers,

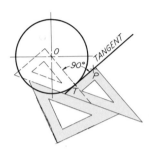

FIG. 3.30 **To draw a line tangent to a circle through a given point outside.**

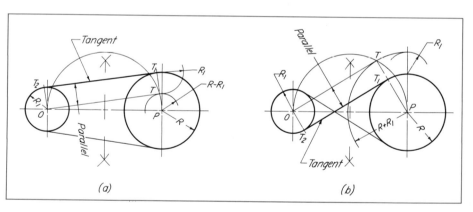

FIG. 3.31 **To draw a line tangent to two given circles.**

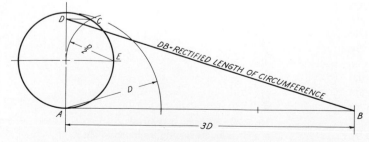

FIG. 3.32 **To lay off the approximate length of the circumference of a circle.**

start at B and step off equal chord distances around the arc until the point nearest A is reached. From this point (without raising the dividers) step off along the tangent an equal number of distances to locate point C. If the point nearest A is indented into the tangent instead of the arc, the almost negligible error in the length of AC will be still less.

Since the small distances stepped off are in reality the chords of small arcs, the length AC will be slightly less than the true length of the arc. For most practical purposes the difference may be disregarded.

When the central angle (Θ) and the radius of an arc are known, the length of the arc may be computed by the formula $L = 2\pi R(\Theta/360°) = 0.01745R\Theta$.

3.35 To Lay Off a Specified Length Along a Given Circle Arc (Fig. 3.34)

On the tangent to the arc, lay off the distance DE representing the specified length of arc. Divide DE into four equal parts. Then, using point 1 as a center and with the length 1-E as the radius R, strike an arc intersecting the given arc at F. The arc DF is approximately equal in length to the line DE. For large angles, it is advisable to make the construction for one-half of DE.

3.36 Conic Sections (Fig. 3.35)

When a right circular cone of revolution is cut by planes at different angles, four curves of intersection are obtained that are called *conic sections*.

When the intersecting plane is perpendicular to the axis, the resulting curve of intersection is a *circle*.

If the plane makes a greater angle with the axis than do the elements, the intersection is an *ellipse*.

If the plane makes the same angle with the axis as the elements, the resulting curve is a *parabola*.

Finally, if the plane makes a smaller angle with the axis than do the elements or is parallel to the axis, the curve of intersection is a *hyperbola*.

The geometric methods for constructing the ellipse, parabola, and hyperbola are discussed in succeeding sections.

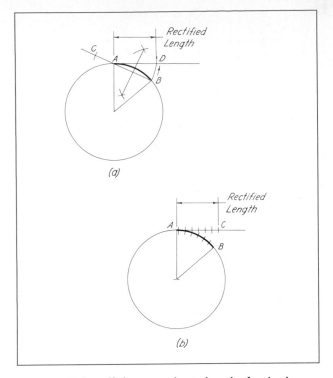

FIG. 3.33   To lay off the approximate length of a circular arc on its tangent.

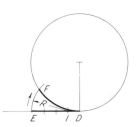

FIG. 3.34   To lay off a specified length along an arc.

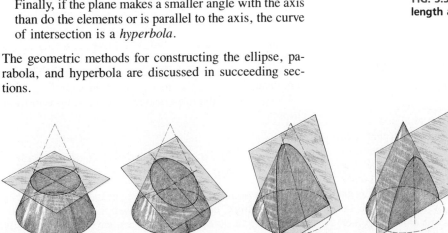

FIG. 3.35   Conic sections.

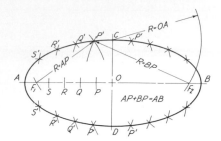

FIG. 3.36 To construct an ellipse, foci (definition) method.

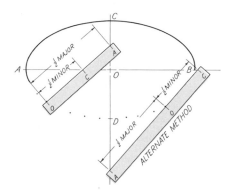

FIG. 3.37 To construct an ellipse, trammel method.

3.37 *Ellipse (Fig. 3.35)*

Mathematically the ellipse is a curve generated by a point moving so that at any position the sum of its distances from two fixed points (foci) is a constant (equal to the major diameter). It is encountered very frequently in orthographic drawing when holes and circular forms are viewed obliquely. Ordinarily, the major and minor diameters are known.

3.38 *To Construct an Ellipse, Foci Method (Fig. 3.36)*

Draw the major and minor axes (AB, CD) and locate the foci F_1 and F_2 by striking arcs, centered at C and having a radius equal to OA (one-half of the major diameter). The construction is as follows: Determine the number of points needed along the circumference of each quadrant of the ellipse for a relatively accurate layout (say, four) and mark off this number of division points (P, Q, R, and S) between O and F_1 on the major axis. In many cases it may be desirable to use additional points spaced closer together nearer F_1 in order to form accurately the sharp curvature at the end of the ellipse. Next, with F_1 and F_2 as centers and the distances AP and BP as radii, respectively, strike

intersecting arcs to locate P' on the circumference of the ellipse. Distances AQ and BQ are radii for locating points Q'. Locate points R' and S' in a similar manner and complete the ellipse using a French curve.

This method is sometimes known as the definition method, since it is based on the mathematical definition of the ellipse as given in Sec. 3.37.

3.39 *To Construct an Ellipse, Trammel Method (Fig. 3.37)*

Given the major axis AB and the minor axis CD. Along the straight edge of a strip of paper or cardboard, locate the points O, C, and A so that the distance OA is equal to one-half the length of the major axis, and the distance OC is equal to one-half the length of the minor axis. Place the marked edge across the axes so that point A is on the minor axis and point C is on the major axis. *Point O will fall on the circumference of the ellipse.* Move the strip, keeping A on the minor axis and C on the major axis, and mark at least five other positions of O on the ellipse in each quadrant. Using a French curve, complete the ellipse by drawing a smooth curve through the points. The ellipsograph, which draws ellipses mechanically, is based on this same principle. The trammel method is an accurate method.

An alternative method for marking off the location of points A, O, and C is given in Fig. 3.37.

3.40 *To Construct an Ellipse, Concentric Circle Method (Fig. 3.38)*

Given the major axis AB and the minor axis CD. Using the center of the ellipse (point O) as a center, describe circles having the major and minor axes as diameters. Divide the circles into equal central angles and draw diametrical lines such as P_1P_2. From point P_1 on the circumference of the larger circle, draw a line parallel to CD, the minor axis, and from point P_1' at which the diameter P_1P_2 intersects the inner circle, draw a line parallel to AB, the major axis. The point of intersection of these lines, point E, is on the required ellipse. At points P_2 and P_2' repeat the same procedure and locate point F. Thus, two points are established by the line P_1P_2. Locate at least five points in each of the four quadrants. The ellipse is completed by drawing a smooth curve through the points.

This is one of the most accurate methods used to form ellipses.

3.41 *To Draw a Tangent to an Ellipse (Fig. 3.39)*

Given any point, such as P, on the perimeter of the ellipse $ABCD$. Using C as a center and a radius equal to OA (one-half the major diameter), strike arcs across the major axis at F_1 and F_2. From these points, which are foci of the ellipse, draw F_1P and F_2G. The bisector of the angle GPF_1 is the required tangent to the ellipse.

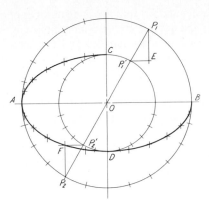

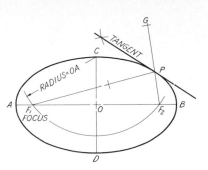

FIG. 3.39 **To draw a tangent to an ellipse.**

FIG. 3.38 **To construct an ellipse, concentric circle method.**

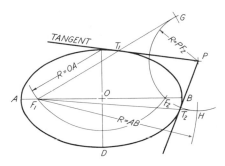

FIG. 3.40 **To draw a tangent to an ellipse through a point outside the ellipse.**

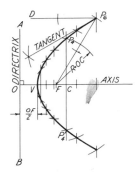

FIG. 3.41 **To construct a parabola.**

3.42 To Draw a Tangent to an Ellipse from Point P, Outside (Fig. 3.40)

With the end of the minor axis as a center and a radius R equal to one-half the length of the major axis, strike an arc to find the foci F_1 and F_2. With point P as a center and the distance PF_2 as a radius, draw an arc. Using F_1 as a center and the length AB as a radius, strike arcs cutting the arc with center P at points G and H. Draw lines GF_1 and HF_1 to establish the location of the tangent points T_1 and T_2. Draw the required tangent.

3.43 Parabola (Fig. 3.35)

Mathematically the parabola is a curve generated by a point moving so that at any position its distance from a fixed point (the focus) is always exactly equal to its distance to a fixed line (the directrix). The construction shown in Fig. 3.41 is based on this definition.

In engineering design, the parabola is used for parabolic sound and light reflectors, for vertical curves on highways, and for bridge arches.

3.44 To Construct a Parabola (Fig. 3.41)

Given the focus F and the directrix AB. Draw the axis of the parabola perpendicular to the directrix. Through any point on the axis, for example, point C, draw a line parallel to the directrix AB. Using F as a center and the distance OC as a radius, strike arcs intersecting the line at points P_4 and P'_4. Repeat this procedure until a sufficient number of additional points have been located to determine a smooth curve. The vertex V is located at a point midway between O and F.

To construct a tangent to a parabola, say, at point P_6, draw the line P_6D parallel to the axis; then bisect the angle DP_6F. The bisector of the angle is the required tangent.

3.45 To Construct a Parabola, Tangent Method (Fig. 3.42)

Given the points A and B and the distance CD from AB to the vertex. Extend the axis CD, and set off DE equal to CD. EA and EB are tangents to the parabola at A and B, respectively.

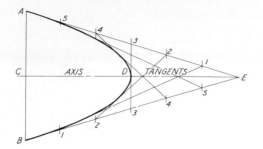

FIG. 3.42 To construct a parabola, tangent method.

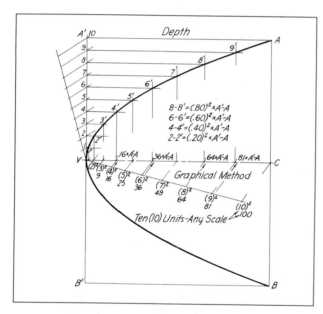

FIG. 3.43 To construct a parabola, offset method.

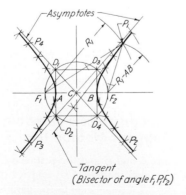

FIG. 3.44 To construct a hyperbola.

Divide *EA* and *EB* into the same number of equal parts (say, six) and number the division points as shown. Connect the corresponding points 1 and 1, 2, and 2, 3 and 3, and so forth. These lines, as tangents of the required parabola, form its envelope. Draw the tangent curve.

3.46 To Construct a Parabola, Offset Method (Fig. 3.43)

Given the enclosing rectangle *A'ABB'*. Divide *VA'* into any number of equal parts (say, ten) and draw from the division points the perpendiculars parallel to *VC*, along which offset distances are to be laid off. The offsets vary as the square of their distances from *V*. For example, since *V* to 2 is two-tenths of the distance from *V* to *A'*, 2–2' will be $(.2)^2$ or .04 of *A'A*. Similarly, 6–6' will be $(.6)^2$ or .36 of *A'A*; and 8–8' will be .64 of *A'A*. To complete the parabola, lay off the computed offset values along the perpendiculars and form the figure with a French curve.

The entire construction can be done graphically (as illustrated) by first calculating the values for the squared distances and then dividing the depth distance (along the axis) proportionally using these values. The graphical method shown in Fig. 3.7 was used.

The offset method is preferred by civil engineers for laying out parabolic arches and computing vertical curves for highways. The parabola shown in Fig. 3.43 could represent a parabolic reflector.

3.47 Hyperbola

Mathematically the hyperbola can be described as a curve generated by a point moving so that at any position the difference of its distances from two fixed points (foci) is a constant (equal to the transverse axis of the hyperbola). This definition is the basis for the construction shown in Fig. 3.44.

3.48 To Construct a Hyperbola (Fig. 3.44)

Given the foci F_1 and F_2, and the transverse axis *AB*. Using F_1 and F_2 as centers, and any radius R_1 greater than F_1B, strike arcs. With these same centers and a radius equal to $R_1–AB$, strike arcs intersecting the first arcs. These intersecting arcs establish the positions of four symmetrically located points (P_1, P_2, P_3, and P_4) using a single pair of radii. Additional sets of four points are obtained by assuming a different initial radius each time. Repeat this procedure, as outlined, until a sufficient number of points have been located to determine a smooth curve.

The tangent to the hyperbola at any point, such as P_1, is the bisector of the angle between the focal radii F_1P_1 and F_2P_1.

As hyperbolic curves extend toward infinity they gradually approach two straight lines known as *asymptotes*. These may be located by drawing a circle having the distance $F_1–F_2$ as a diameter and erecting perpendiculars to the transverse axis through points *A* and *B*. The points at which these perpendicular lines intersect the circle are points (D_1, D_2, D_3, and D_4) on the asymptotes.

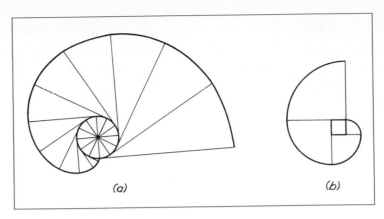

FIG. 3.45 **Involute.**

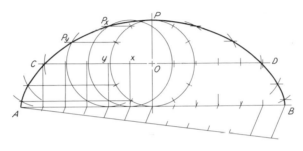

FIG. 3.46 **Cycloid.**

3.49 *Involute*

The spiral curve traced by a point on a cord as it unwinds from around a circle or a polygon is an *involute curve*. Figure 3.45(*a*) shows an involute of a circle, while (*b*) shows that of a square. The involute of a polygon is obtained by extending the sides and drawing arcs using the corners, in order, as centers. The circle in (*a*) may be considered to be a polygon having an infinite number of sides.

3.50 *To Draw an Involute of a Circle [Fig. 3.45(a)]*

Divide the circumference into a number of equal parts. Draw tangents through the division points. Then, along each tangent, lay off the rectified length of the corresponding circular arc, from the starting point to the point of tangency. The involute curve is a smooth curve through these points. The involute of a circle is used in the development of tooth profiles in gearing.

3.51 *To Draw the Involute of a Polygon [Fig. 3.45(b)]*

Extend the sides of the polygon as shown in (*b*). With the corners as centers, in order around the polygon, draw arcs terminating on the extended sides. The first radius is equal to the length of one side of the polygon. The radius of each successive arc is the distance from the center to the terminating point of the previous arc.

3.52 *Cycloid*

A cycloid is the curve generated by a point on the circumference of a moving circle when the circle rolls in a plane along a straight line, as shown in Fig. 3.46.

3.53 *To Draw a Cycloid (Fig. 3.46)*

Draw the generating circle and the line *AB* tangent to it. The length *AB* should be made equal to the circumference of the circle. Divide the circle and the line *AB* into the same number of equal parts. With this much of the construction completed, the next step is to draw the line of centers *CD* through point *O* and project the division points along *AB* to *CD* by drawing perpendiculars. Using these points as centers for the various positions of the moving circle, draw circle arcs. For the purpose of illustration, assume the circle is moving to the left. When the circle has moved along *CD* to *x*, point *P* will have moved to point P_x. Similarly, when the center is at *y*, *P* will be at P_y. To locate positions at *P* along the cycloidal curve, project the division points of the divided circle in their proper order, across to the position circles. A smooth curve through these points will be the required cycloid.

3.54 Epicycloid (Fig. 3.47)

An epicycloid is the curve generated by a point on the circumference of a circle that rolls in a plane on the outside of another circle. The method used in drawing an epicycloid is similar to the one used in drawing the cycloid.

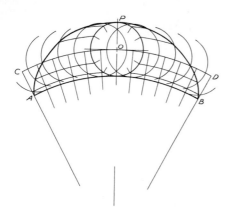

FIG. 3.47 **Epicycloid.**

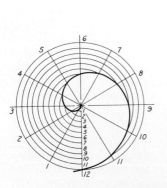

FIG. 3.48 **Hypocycloid.**

3.55 Hypocycloid (Fig. 3.48)

A hypocycloid is the curve generated by a point on the circumference of a circle that rolls in a plane on the inside of another circle. The method used to draw a hypocycloid is similar to the method used to draw the cycloid.

Additional information on the use of cycloidal curves to form the outlines of cycloidal gear teeth may be found in Chapter 21.

3.56 Spiral of Archimedes

Archimedes' spiral is a plane curve generated by a point moving uniformly around and away from a fixed point. In order to define this curve more specifically, it can be said that it is generated by a point moving uniformly along a straight line while the line revolves with uniform angular velocity about a fixed point.

The definition of the spiral of Archimedes is applied in drawing this curve as illustrated in Fig. 3.49. To find a sufficient number of points to allow the use of an irregular curve for drawing the spiral it is the practice to divide the given circle into a number of equal parts (say, 12) and draw radial lines to the division points. Next, divide a radial line into the same number of equal parts as the circle and number the division points on the circumference of the circle beginning with the radial line adjacent to the divided one. With the center of the circle as a center, draw concentric arcs that in each case will start at a numbered division point on the divided radial line and will end at an intersection with the radial line that is numbered correspondingly. The arc starting at point 1 gives a point on the curve at its intersection with radial line 1, the arc starting at 2 gives an intersection point on radial line 2, etc. The spiral is a smooth curve drawn through these intersection points.

3.57 Helix (Fig. 3.50)

The cylindrical helix is a space curve that is generated by a point moving uniformly on the surface of a cylinder. The point must travel parallel to the axis with uniform linear velocity while at the same time it is moving with

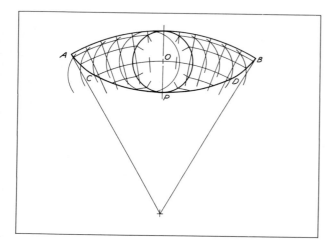

FIG. 3.49 **Spiral of Archimedes.**

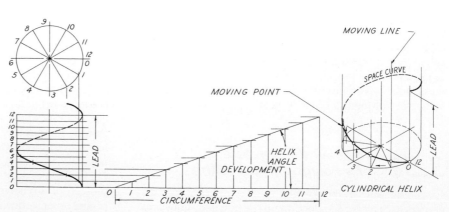

FIG. 3.50 **Helix.**

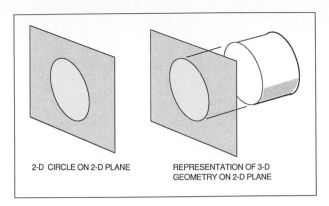

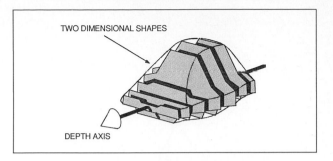

FIG. 3.52   **2-D Shapes arranged along depth axis to form 3-D shape.**

FIG. 3.51   **2-D Circle and representation of 3-D geometry.**

uniform angular velocity around the axis. The curve can be thought of as being generated by a point moving uniformly along a straight line while the line is revolving with uniform angular velocity around the axis of the given cylinder. Study the pictorial drawing, Fig. 3.50.

The first step in drawing a cylindrical helix is to lay out the two views of the cylinder. Next, the lead should be measured along a contour element and divided into a number of equal parts (say, 12). Divide the circular view of the cylinder into the same number of parts and number the division points.

The division lines of the lead represent the various positions of the moving point as it travels in a direction parallel to the axis of the cylinder along the moving line. The division points on the circular view are the related position of the moving line. For example, when the line has moved from the 0 to the 1 position, the point has traveled along the line a distance equal to one-twelfth of the lead; when the line is in the 2 position, the point has traveled one-sixth of the lead. (See pictorial drawing, Fig. 3.50.) In constructing the curve the necessary points are found by projecting from a numbered point on the circular view to the division line of the lead that is numbered similarly.

A helix may be either right-hand or left-hand. The one shown in Fig. 3.50 is a left-hand helix.

When the cylinder is developed, the helix becomes a straight line on the development, as shown. It is inclined to the base line at an angle known as the "helix angle." A screw thread is an example of a practical application of the cylindrical helix.

B / GEOMETRY AND THE COMPUTER—PLANE AND SOLID, TWO- AND THREE-DIMENSIONAL

3.58   *The Importance of Geometry*

For a designer to fully understand how a part functions in an assembly, how it operates, and how it interacts with other parts around it, the designer must visualize in three-dimensions. Before the advent of the computer, designers drew or *drafted* their ideas as two-dimensional diagrams. These geometric constructions have been covered in the first part of this chapter. If a real feeling was needed of how the design might appear, models or mock-ups were often crafted out of clay or wood. Then the designer could study the design, viewing it from different vantage points, checking how the design functioned in the space around it. (Refer to Chapter 12 for examples of engineering models.)

Using CADD, a three-dimensional *model* of the design is kept as a mathematical description in the memory of the computer; thus allowing the designer to see any and all views of the design and how the design operates. This is a fundamental change in the way a designer works. Previously, a designer had to translate the mental conception or mathematical model into a two-dimensional diagram before the object could be made. With CADD, it is the three-dimensional model that is translated as needed into two-dimensional diagrams for documentation and communication.

This necessitates thinking directly in three dimensions, always knowing the direction in space that you are viewing, and (the topic of this section) the characteristics of the geometry you are creating.

3.59   *Two- and Three-Dimensional Geometry*

The difference between 2-D and 3-D geometry can be seen in Fig. 3.51. A circle may be drawn on a two-dimensional surface and may represent either a circle or the 2-dimensional view of a three-dimensional form such as a sphere or a cylinder. A square may exist on a two-dimensional plane and may represent a quadrilateral (four-sided form) or may be the 2-dimensional view of three-dimensional geometry like cubes or cylinders.

Two-dimensional geometry may exist in three dimensional space as shown in Fig. 3.52 where 2-D profiles through a car body are arranged along the depth axis. Each profile is flat, but when taken together, they create a three-dimensional design.

The difference between two- and three-dimensional geometry is evident when an image is viewed from different

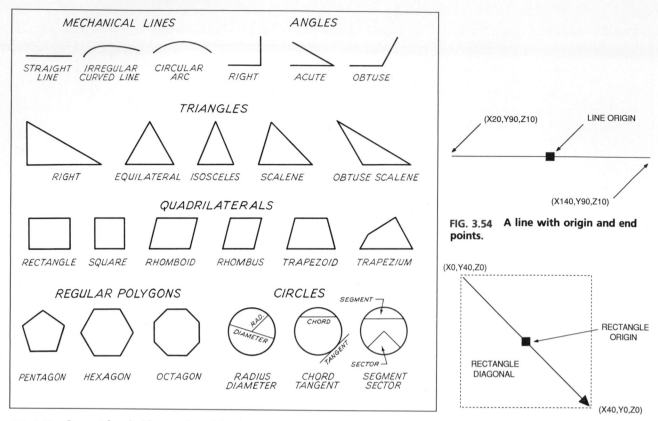

FIG. 3.53 **Geometric primitives and combinations.**

FIG. 3.54 **A line with origin and end points.**

FIG. 3.55 **Specification of a rectangle by its diagonal.**

vantage points: three-dimensional geometry yields additional correct geometric views. Two-dimensional geometries are really diagrams whose 3-D properties are derived from the arrangement of various flat views on the page.

3.60 Geometric Primitives (Fig. 3.53)

The plane geometry constructions presented in the first part of this chapter required basic geometric drawing. Lines, arcs, circles, points, and curves were used to construct useful forms (see Fig. 3.53). Many of the same constructions apply to computer drawing, except that the method of supplying the needed information is changed.

All CADD computers have basic two-dimensional geometric shapes called *primitives* that are available for the operator and from which more complex shapes can be made. Most CADD systems would include the following primitives.

• *Line* may be horizontal, vertical, oblique, parallel, perpendicular, or tangent, and of specific length.

• *Point* is a dimensionless location where two lines cross.

• *Rectangle* is a four-sided figure specified by its diagonal.

• *Polygon* is a multi-sided figure defined by a contiguous set of lines.

• *Circle* is a round form specified by center and diameter or radius, or by three points.

• *Arc* is a piece of the circumference of a circle specified by radius and starting and ending angle.

• *Irregular curve* passes through three or more specific points.

3.61 Geometric Lines

CADD computers store lines by the coordinates of their end points (Fig. 3.54). This allows lines, which make up the majority of CADD drawings, to be stored very efficiently. The *origin* of a line is located at its center, but the line may be identified at any point between its ends. A special kind of line is a *rectangle* (Fig. 3.55). By giving the computer the coordinates of the rectangle's diagonal, a box is formed. Each line segment of the box may then be treated separately.

3.62 Circles

To draw a circle, the CADD computer must be given the coordinates of its center and either the radius or diameter. This corresponds to the manual process of marking the center, setting the compass radius, and swinging the arc. Even when three points on the circumference of the circle are specified, the center is found in much the same way as in Fig. 3.20 where a circle was circumscribed around a triangle. *A CADD circle is drawn counterclockwise* from its origin at 0 degrees to 360 degrees (Fig. 3.56). A circle is identified by its origin or by any point on its circumference. The origin of a circle and its geometric center may be different points.

3.63 Arcs

Arcs are special cases of circles that are also drawn counterclockwise. Information necessary for drawing an arc is the beginning angle or point, the ending angle or point, and the radius. As an alternative method, an arc may be specified by the center, radius, beginning point, and sweep angle (Fig. 3.57). An arc whose beginning angle is zero degrees and whose ending angle is 360 degrees is an *arc circle*. The origin of an arc is at the center of the arc curve and the geometric center of the arc is at its usual position, equidistant from the curve (Fig. 3.58). CADD systems do not confuse arcs and circles because they are different objects.

3.64 Curves

A curve may be drawn through a number of points not on the same line by circular arc approximation (Fig. 3.59) or by a number of mathematical curves. An important curve to engineers is the *Bezier curve*, a curve that is *attracted* to points. The curve may actually touch a point and if it does it always maintains contact with the point. Or, a point may lie off the curve and attract the curve like a magnet. The closer the point is to the curve, the greater the attraction. A Bezier curve may be reshaped by moving these points called *control points* or *handles* and adjusting the curve as shown in Fig. 3.60. Note that the curve is not defined by circular arcs as was the case in Fig. 3.59. Rather, it behaves much like a spring, trying to connect smoothly the control points to which it is attached while also being attracted to other points.

3.65 Figures

A group of primitives that have been grouped together by a CADD operator is called a *figure* and may be identified and moved like one of the predefined CADD primitives.

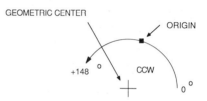

FIG. 3.58 Arc identification-positive movement.

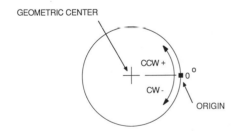

FIG. 3.56 Circle identification of clockwise (negative) and counterclockwise (positive) movement.

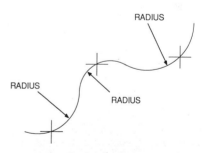

FIG. 3.59 Arc through three points —circular arc approximation.

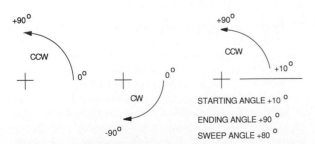

FIG. 3.57 Positive arc, negative arc, and arc by starting, ending, and sweep angles.

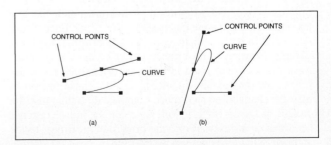

FIG. 3.60 Arc through three points—Bezier curve method.

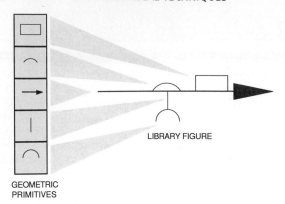

GEOMETRIC PRIMITIVES

LIBRARY FIGURE

FIG. 3.61 Library figure constructed from geometric primitives.

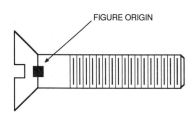

FIGURE ORIGIN

FIG. 3.62 Figure origin is located at a convenient position.

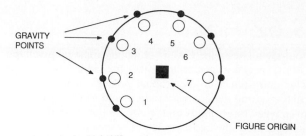

GRAVITY POINTS

FIGURE ORIGIN

FIG. 3.63 Gravity points allow the correct attachment of lines to a figure.

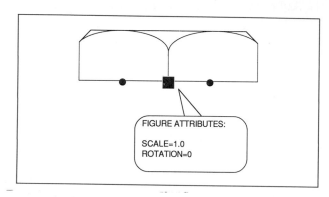

FIGURE ATTRIBUTES:

SCALE=1.0
ROTATION=0

FIG. 3.64 A square fastener head as a library figure with its figure attributes.

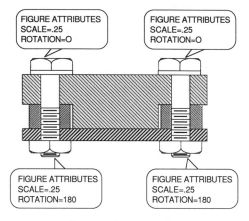

FIGURE ATTRIBUTES
SCALE=.25
ROTATION=O

FIGURE ATTRIBUTES
SCALE=.25
ROTATION=O

FIGURE ATTRIBUTES
SCALE=.25
ROTATION=180

FIGURE ATTRIBUTES
SCALE=.25
ROTATION=180

FIG. 3.65 The use of a library figure (square fastener head) in a sectional detail.

A graphic that can be identified as a unit is called an *entity*. Figures and primitives are both entities. Some CADD computers have *libraries* of figures, such as electronic, hydraulic, pneumatic, and mechanical parts, and of course these figures have been made from primitives (Fig. 3.61). The advantage of CADD is in having to create a figure only once, and then using that figure over and over again.

3.66 Figure Attributes

Each figure created on a CADD computer has assigned to it certain properties or attributes. The most important of these properties is the origin of the figure. As noted earlier, the origin is the handle that is used for identification and as the reference point for placing the figure on the drawing (Fig. 3.62). Figures may also have *gravity points*, spots on the figure that will attract other figures or primitives (Fig. 3.63). Figure attributes are much like the text attributes that were discussed in Sec. 2.47. Each figure is created in a convenient size, with its origin at the most popular position, with gravity points where the operator would expect them, and in a typical orientation (zero degrees rotation) as shown in Fig. 3.64. These attributes may be changed so that more complex drawings may be created as shown in Fig. 3.65. Note that the square fastener head has been inserted into the drawing four times, twice at the default (zero degree rotation) and twice at 180 degree rotation.

3.67 Geometric Modeling

Sophisticated computer programs allow a designer to *build* the three-dimensional mathematical description of a part just as one might sculpt the part out of wood or plastic. This is a fundamental change in how designers work and requires a solid understanding of the two-dimensional geometric principles covered in the first part of this chapter as well as three-dimensional geometry. Geometric modeling combines the power of the computer with the creativity of the designer, producing solutions to engineering problems that are more accurate, more standardized, and less time-consuming. Not all computers are capable of

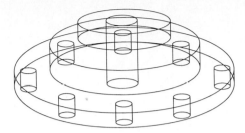

FIG. 3.66 Wire frame (edge) model.

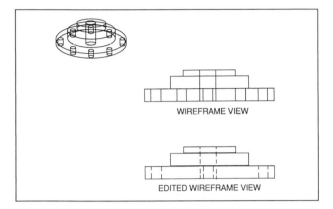

WIREFRAME VIEW

EDITED WIREFRAME VIEW

FIG. 3.67 Wire frame (edge) model, wire frame front view, and edited engineering drawing.

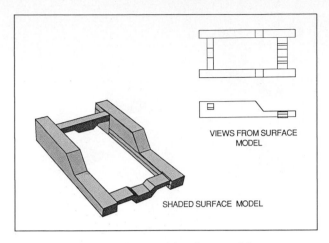

VIEWS FROM SURFACE MODEL

SHADED SURFACE MODEL

FIG. 3.68 Shaded surface model and top and front views of that model.

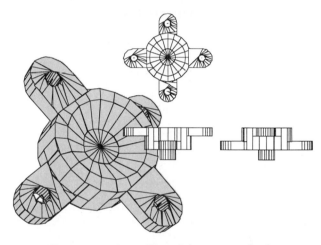

FIG. 3.69 Constructive solid model constructed using the Boolean operators difference and union.

geometric modeling and among those that are, the extent of capabilities varies.

3.68 *Wire Frame Modeling*

The most elementary modeling technique describes the object as a set of lines. Actual objects do not have lines, they have surfaces, intersections, and limits. Lines provide a technique to approximate the visual representation of solid geometry (Fig. 3.66). Yet with *wire frame modeling*, the computer knows nothing of the material inside the object; it recognizes only the artificial lines that connect the object's corners or vertices. This modeling technique is of little benefit to designers other than for getting a quick idea of the geometry and for producing views of the design as a drawing. The views must be changed or edited to reflect standard drafting practices as in Fig. 3.67, where solid lines have been changed to dashed lines.

3.69 *Surface Modeling*

Objects defined by their surfaces yield more information than do wire frame descriptions. The *surface model* may look like a wire frame, or the planes themselves may be shaded to yield a more realistic view (Fig. 3.68). However, the surface model is hollow, and nothing is known about the material inside the object. The surface model, like the wire frame model, can be turned into standard engineering drawings.

3.70 *Constructive Solid Geometry*

Constructive solid geometry (CSG) fully defines the object, both its planes and the material inside. This gives an engineer full freedom to carve or model the object and to combine objects into more sophisticated designs. This type of modeling is called *solid modeling* because of its similarity to traditional model making.

CSG models can be shaded or shown as wireframe representations. Since CSG models do not have the artificial outline of wire frames, these lines must be added so that an unshaded object can be seen (Fig. 3.69).

3.71 *Solid Modeling Primitives*

Solid models are constructed from three-dimensional primitive shapes (Fig. 3.70). These shapes may include

- prism
- cylinder
- sphere

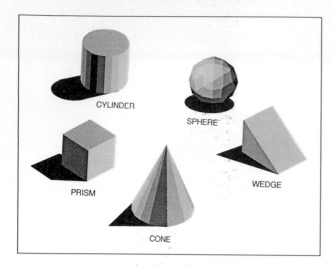

FIG. 3.70 Constructive solid geometry primitives: prism, wedge, cylinder, cone, and sphere.

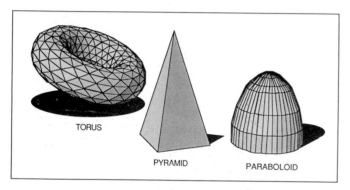

FIG. 3.71 Additional solids formed from the constructive primitives.

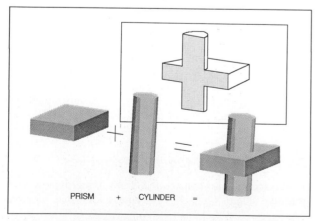

FIG. 3.72 Union (addition) operation of prism and cylinder primitives.

- wedge
- cone

Other basic shapes may be used to make additional primitive shapes (Fig. 3.71) such as

- torus
- pyramid
- paraboloid

3.72 Boolean Operators

Primitive shapes are combined and changed by mathematical operators called *Boolean operators*. These operations are *union* (addition), *difference* (subtraction), and *intersection* (multiplication). Solids may be added or subtracted just as numbers are added or subtracted.

■ UNION

Two solid forms may be joined together by the *union operator*. The forms may occupy the same space or they may be separated (Fig. 3.72). In either case, after the addition operation, the two forms are considered to be one shape. (They may just happen to be separated by empty space.) If the objects do occupy the same space, all intersections are calculated and visibility is determined. In Fig. 3.72, all points belonging to the prism are added to all points belonging to the cylinder, and the resulting shape is a new geometric form.

■ DIFFERENCE

The most common use of the *difference operation* is to subtract (or extract) a hole from solid material (Fig. 3.73). In this case the objects *must* occupy the same physical space for the subtraction operation to have any effect. In Fig. 3.73, the cylinder is subtracted from the prism, resulting in a hole in the prism. In this case, all points in the cylinder that are common with the prism are subtracted. The outcome of a subtraction operation depends on which form is being subtracted. Note the inset in Fig. 3.73 and the difference between subtracting the cylinder from the prism, and subtracting the prism from the cylinder.

■ INTERSECTION

The objects shown in the previous examples may also be intersected. In this case, all material *not common* to both shapes is removed. What remains is the intersection (Fig. 3.74). Note that this is different from union where no part of either object is removed. The intersection uses the multiplication operator ($\times$), and it is unimportant which object is specified first. Only points common to *both* the prism and cylinder are kept.

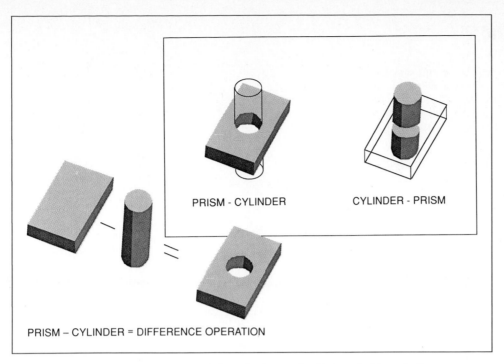

FIG. 3.73 Difference (subtraction) operation of prism and cylinder primitives: both cases.

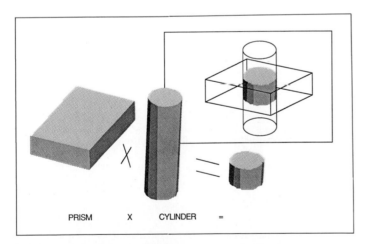

FIG. 3.74 Intersection operation of prism and cylinder primitives.

3.73 Other Modeling Techniques

Once the object has been described three-dimensionally, a number of additional operations may be performed on that object's data base. From the mathematical description of the object, *mass properties* may be extracted. These include the *center of mass*, the *geometric centroid*, *mo-*ments *of inertia*, *weight*, and *mass*. Even accounting information such as *unit cost* can be figured from the model. In each of these cases, the data base is acted upon after it is created. This is called *postprocessing*. Under no circumstances do any of these postprocessing techniques alter the actual geometric description or data base of the object.

PROBLEMS

The following exercises not only require the student to study and use certain common geometric constructions but also furnish additional practice in applying good line technique to the drawing of instrumental figures and practical designs. All work should be very accurately done. Tangent points should be indicated by a light, short dash across the line.

1. Draw a line 80 mm (3.15 in.) long. Divide it proportionally in the ratio 1:2:3. Use the method shown in Fig. 3.7.

2. Construct a regular hexagon having a 66 mm (2.6 in.) distance across flats. Select the most practical procedure.

3. Construct a regular hexagon having a 80 mm (3.15 in.) distance across corners. Select the most practical method.

4. Construct a regular pentagon having 30 mm (1.18 in.) sides. Use the method illustrated in Fig. 3.18.

5. Draw two horizontal lines 50 mm (1.97 in.) apart. Locate two points 75 mm (2.95 in.) apart horizontally, one on each line. Draw an ogee curve tangent to these lines. Study the procedure illustrated in Fig. 3.25(a).

6. Construct an ellipse having a major diameter of 108 mm (4.25 in.) and a minor diameter of 70 mm (2.76 in.). Use the trammel method illustrated in Fig. 3.37.

7. Construct an ellipse having a major diameter of 100 mm (3.94 in.) and a minor diameter of 70 mm (2.75 in.). Use the concentric circle method illustrated in Fig. 3.38. Find a sufficient number of points to obtain a smooth curve.

8. Construct a parabola with vertical axis. Make the focus 19 mm (.75 in.) from the directrix. Select a point on the curve and draw a line tangent to the parabola. Study Sec. 3.42 and Fig. 3.41.

9. Construct a hyperbola having a transverse axis of 25 mm (.98 in.) and foci 41 mm (1.62 in.) apart. Study Sec. 3.47 and Fig. 3.44.

10. Construct the involute of an equilateral triangle with 25 mm (.98 in.) sides. Study Secs. 3.48 and 3.50.

11. Construct the cycloid generated by a 38 mm (1.50 in.) circle. Study Sec. 3.51 and Fig. 3.52.

12. Construct the epicycloid generated by a 38 mm (1.50 in.) circle rolling on a 127 mm (5.00 in.) circle. Study Sec. 3.53 and Fig. 3.47.

13. Construct the hypocycloid generated by a 38 mm (1.50 in.) circle rolling on a 114 mm (4.50 in.) circle. Study Sec. 3.54 and Fig. 3.48.

14. Reconstruct the view of the wrench and hexagonal nut shown in Fig. 3.75. Mark all tangent points with short lines.

15. Construct the shape of the slotted guide shown in Fig. 3.76. Show all construction for locating centers, and mark points of tangency.

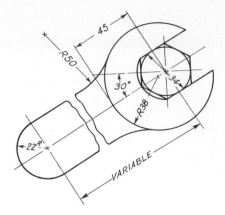

FIG. 3.75   Wrench.

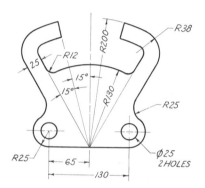

FIG. 3.76   Slotted guide.

16. Construct the adjustable Y-clamp shown in Fig. 3.77. Show all construction for locating centers and mark points of tangency.

17. Reconstruct the end view of the dolly block shown in Fig. 3.78.

18. Reconstruct the view of the electrode shown in Fig. 3.79.

19. Reconstruct the plat of a land survey shown in Fig. 3.80. Use the tangent method, as explained in Sec. 3.11 to determine the direction of the center line of State Road 26. The triangles used in combination will produce the other angles.

20. Reconstruct the view of the spline plate shown in Fig. 3.81.

22. (Fig. 3.83.) The design of a counterweight system is such that point *A* (pivot point) of the counterweight and point *C* (end view of the axis of the shaft and roller) are in line horizontally. A point *B* on the counterweight is 105 from *A* on a line making an angle of 41° with *AC* through *A*. Points *A* and *B* are the centers of R40 and R25 arcs, respectively.
   Draw an arc of R80 tangent to the 25 and 40 arcs to form the lower portion of the outline of the counterweight. Complete the upper part of the outline by drawing a reverse curve tangent to the top of the 25 and 40 arcs. The radius of the curve tangent to the 40 arc is to be 43% of the total length of the two chords of the reverse curve. Using geometry, determine the location of the point of tangency of the counterweight and roller when the counterweight swings clockwise into contact with the outside surface of the stop-roller.

23. Prepare a plan view drawing showing the center lines of the highway interchange (Fig. 3.84). Select a suitable scale.

24. It is desired to know the angular displacement of the center line of the cam (Fig. 3.85) when the follower moves to a position 18.8 below the position shown. Show the cam and follower in their new positions in phantom outline. Use only an approved geometrical method to find the location of the centers of the arcs. A trial-and-error method is not acceptable. Show all construction and mark all points of tangency.

Determine the angle through which the center line of the cam has moved. Dimension the angle between the center lines.

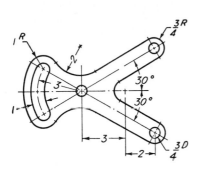

FIG. 3.77 Adjustable Y-clamp.

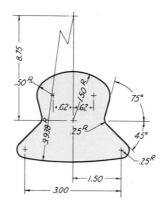

FIG. 3.78 End view, dolly

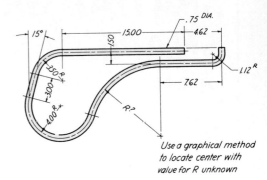

Use a graphical method to locate center with value for R unknown

FIG. 3.79 Electrode.

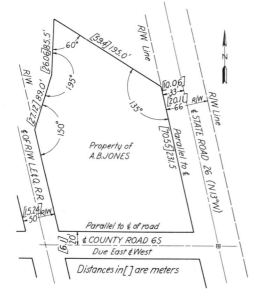

FIG. 3.80 Plat of a land survey.

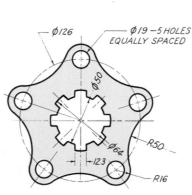

FIG. 3.81 Spline plate.

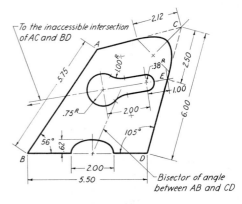

FIG. 3.82 Adjustment plate.

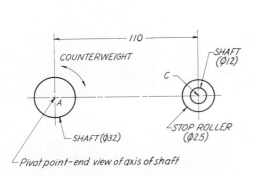

FIG. 3.83 Design of a counterweight

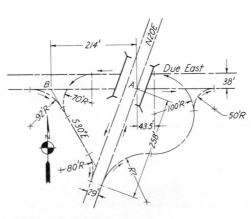

FIG. 3.84 Highway interchange.

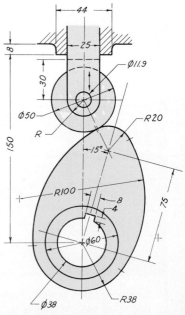

FIG. 3.85 Cam and follower.

SPATIAL GRAPHICS: SHAPE DESCRIPTION AND SPATIAL RELATIONSHIPS

Computer-aided drafting systems that have the capability of preparing both orthographic (multiview) and 3-D drawings are discussed in Chapters 1–5, 9, and 17. This capability extends to the solution of basic spatial geometry problems for design and analysis. See left-hand page facing the start of Chapter 9. Performing the drawing manipulations is Professor Jerry Smith. (*Photo courtesy of Department of Technical Graphics, Purdue University*)

CHAPTER · 4

The Representation of Space Relationships: Two- and Three-Dimensional

4.1 Introduction

Engineers record the shapes and sizes of three-dimensional objects on a sheet of drawing paper. This drawing paper is the engineer's *picture plane* (Fig. 4.1) on which the geometry is projected or drawn. *Size description* and *shape description* are equally important, but in order to simplify learning to make drawings and sketches, this chapter deals only with methods used to describe shape. A later chapter will discuss size description.

Three methods for representing shape are used by engineers and technologists and the theory governing each method should be understood thoroughly before it is used. These are

- orthographic projection
- axonometric projection
- oblique projection

FIG. 4.1 Perspective projection.

A/ ONE-PLANE PROJECTION— PICTORIAL

4.2 Perspective Projection (Convergent Projection)

Before discussing the three methods, let us analyze *perspective projection*, the view each of us is accustomed to seeing.

In perspective projection, the projecting lines or *visual rays* converge at a point, as shown in Figure 4.1. The representation on the transparent picture plane may be considered the view that would be seen by a single eye at a known point in space. The picture is formed on the picture plane by the *piercing points* of the projecting lines from the eye to the object. The size of the view depends on the distance from the observer to the plane and the distance from the plane to the object.

Perspective projections are *not* used by engineers for manufacturing and construction because the perspective view does not reveal exact size and shape. Perspectives may be used in marketing where a *natural* view of a product may be desirable.

4.3 Axonometric and Oblique Projection

If the object is turned and then tilted so that the three faces are inclined to the plane of projection, the resulting projection is a special type of *orthographic projection* (see Sect. 4.4) known as *axonometric projection*. Figure 4.2 illustrates an axonometric projection of a cube. Note that the projectors from the plane to the object are perpendicular to the plane. This axonometric or pictorial view shows three of the object's sides in one projection and therefore

is called a *one-plane projection*. The three subdivisions of axonometric projection are *isometric*, where the three sides are equally inclined, *dimetric*, where two of the three sides are equally inclined, and *trimetric*, where all three sides are inclined differently.

Another form of one-plane projection is *oblique projection*. This is not an orthographic projection because, although one face is imagined to be parallel to the plane of projection, the projectors are not perpendicular to it (Fig. 4.3). Oblique projection provides an easy way of turning an existing orthographic view into a pictorial view. Oblique projection and oblique drawing are covered in detail in Chapter 11.

Perspective projection, axonometric projection, and oblique projection may be classed together as *one-plane pictorial projections*.

B/ COORDINATE PLANES (2-D) PROJECTION

4.4 Orthographic Projection (Parallel Projection)

The projection system that engineers use for manufacturing and construction drawings is called *orthographic projection*. If the observer in Figure 4.1 moves straight back from the picture plane an infinite distance, the projecting lines (visual rays or projectors) from the eye to the object become parallel to each other and perpendicular to the picture plane. The resulting projection (Fig. 4.4) will be an accurate representation of the object's shape parallel to the picture plane. For convenience, the projection may be formed by extending perpendicular projectors from the object to the plane. This view is the orthographic projection.

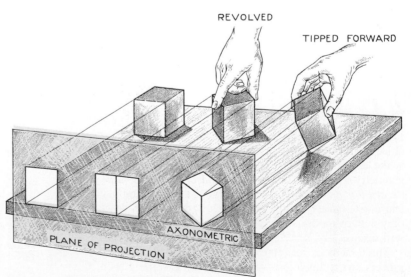

FIG. 4.2 Theory of axonometric projection.

Since the view shown in Fig. 4.4 does not reveal the thickness of the object or the shape on planes perpendicular to the first picture plane, one or more additional picture planes may be established (Fig. 4.5). Two projections are usually sufficient to describe most simple objects, but three or more may be needed for complicated geometry.

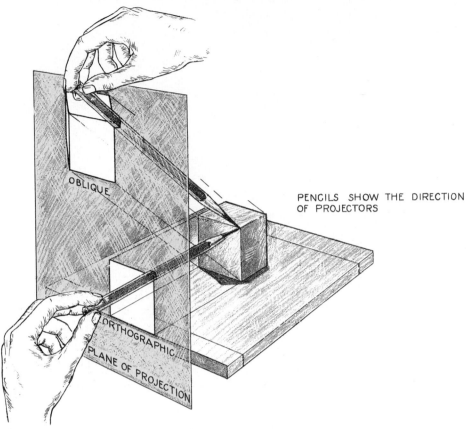

PENCILS SHOW THE DIRECTION OF PROJECTORS

FIG. 4.3   Oblique projection.

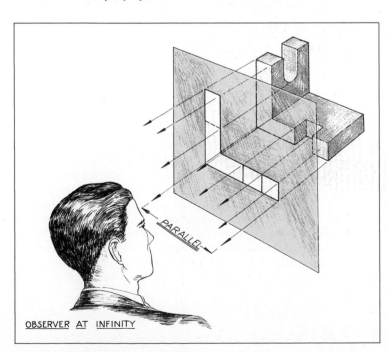

FIG. 4.4   Orthographic projection.

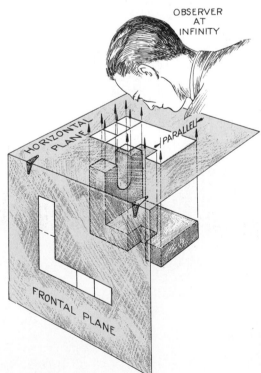

FIG. 4.5   Planes of projection.

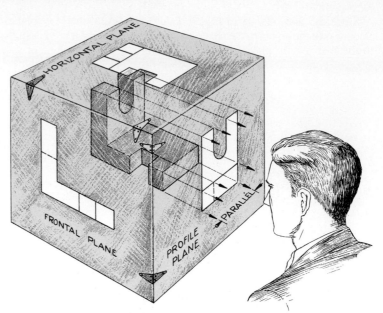

FIG. 4.6   Planes of projection.

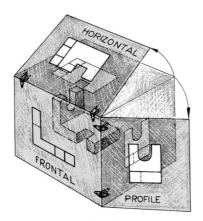

FIG. 4.7   Revolution of the planes of projection.

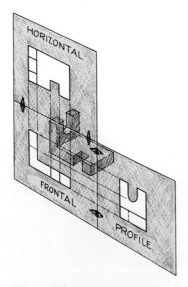

FIG. 4.8   Planes revolved into the plane of the paper.

The picture planes are customarily called the *principal* or *coordinate* planes of projection and the perpendiculars, *projectors*. There are three principal coordinate planes of projection: the *frontal* plane, the *horizontal* plane, and the *profile* plane. In engineering drawing the planes are usually arranged as shown in Fig. 4.6. All three are *mutually perpendicular*. Were all views perpendicular to the three coordinate planes drawn, a total of six views would be formed.

- Front view
- Rear view   ⎤ frontal projections

- Top View
- Bottom view   ⎤ horizontal projections

- Right side view
- Left side view   ⎤ profile projections

To maintain this mutual relationship when laying out the views, the frontal plane is usually considered to coincide with the plane of the paper and the horizontal (top) and profile (side) planes as revolving 90° into the position shown in Figs. 4.7 and 4.8. Note in Fig. 4.7 the manner in which the planes are revolved. This theoretical treatment of the coordinate planes establishes an absolute relationship between the views. Visualizing an object would be considerably more difficult than it is were it not for this fixed relationship, since it would be impossible to determine quickly the direction of sight for a particular view.

4.5   First- and Third-Angle Projection

If the frontal and horizontal projection planes are assumed to extend infinitely in space on one side of the profile

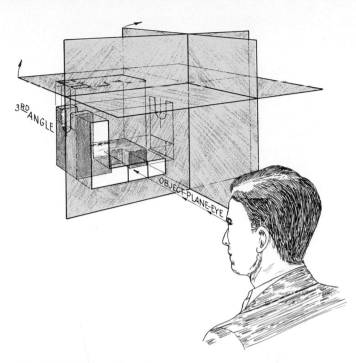

FIG. 4.10   **Third-angle projection.**

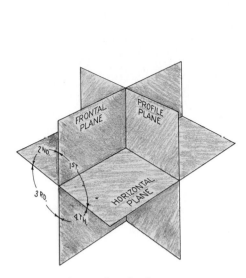

FIG. 4.9   **Planes of projection.**

plane, four *dihedral angles* (90 degree) are formed and are designated as the first, second, third, and fourth angles (Fig. 4.9). The lines of intersection between these planes are called *coordinate axes* and their point of intersection is called the *origin*.

Assume an object is placed so that its main faces are parallel to the frontal, horizontal, and profile planes (Fig. 4.10). The respective projections will show the true size and shape of all surfaces that are parallel to the planes. Theoretically, the object could be placed in any of the four quadrants. It has been placed in the third quadrant because engineering custom in the United States dictates the use of the third angle. This quadrant is used because the views, when revolved 90° into the plane of the front view, are in their natural positions. That is, the top view appears above the front view. The profile view showing the right side falls to the right of the front view, and so on.

In some countries, the first-angle projection is used for engineering drawings. Study the differences between Fig. 4.10 and Fig. 4.11. Observe that, when the planes are revolved, the top view will be below the front view and that the left side view will be to the right of the front view. Two views of a truncated cone (Fig. 4.12) are used to identify which angle of projection was used in a drawing.

4.6   *Systems of Projection*

As a review, the different systems of projection are shown diagramatically in Fig. 4.13.

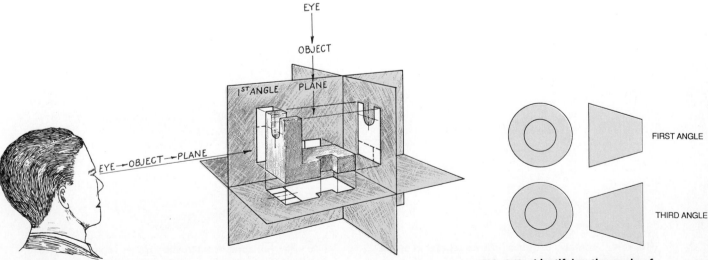

FIG. 4.11   **First-angle projection.**

FIG. 4.12   **Identifying the angle of projection.**

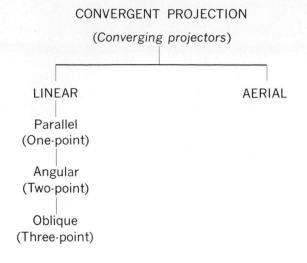

CONVERGENT PROJECTION

(*Converging projectors*)

LINEAR   AERIAL

Parallel
(One-point)

Angular
(Two-point)

Oblique
(Three-point)

PARALLEL PROJECTION

(*Parallel projectors*)

OBLIQUE   ORTHOGRAPHIC

General   MULTIVIEW   AXONOMETRIC

Cavalier   Two views   Isometric

Cabinet   Three views   Dimetric

Clinographic   Auxiliary   Trimetric
views

Sectional
views

FIG. 4.13   Systems of projection.

C / *CADD CONSTRUCTION PLANES*

4.7 *Construction Planes Used in CADD*

Construction planes are used in CADD to create 3-D geometry on 2-D surfaces which are parallel to coordinate planes. Construction planes are perpendicular or *normal* to the direction of sight for that view. For example, the front and rear of an object can be drawn on *frontal construction planes*, separated by the depth or thickness of the object (Fig. 4.14). Right and left side geometry can be drawn on *profile construction planes*, separated by the width of the object (Fig. 4.15) with the front and rear geometry placed in the proper relationship. Horizontal construction planes, positioned perpendicular to the frontal and profile planes, are used for required top or plan views (Fig. 4.16). The views made on these construction planes will define a 3-D object that is not too complex. The result

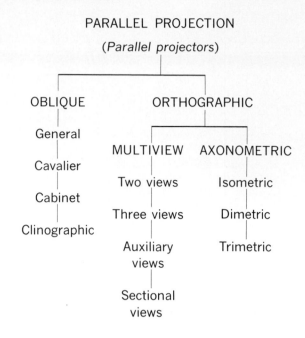

FIG. 4.15   Profile construction planes separated by the width of the object.

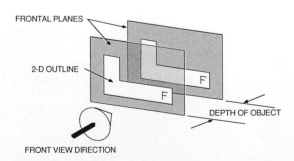

FIG. 4.14   Frontal construction planes separated by the depth of the object.

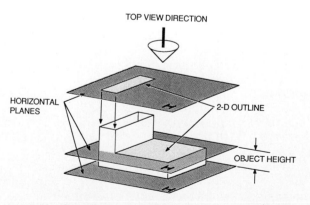

FIG. 4.16   Horizontal construction planes separated by the height of the object.

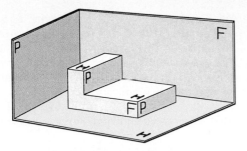

FIG. 4.17 Completed object and its relationship to principal construction planes.

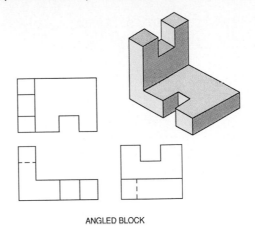

ANGLED BLOCK

FIG. 4.18 Angle block to be digitized.

is the representation of the object defined on construction planes set in three dimensions (Fig. 4.17).

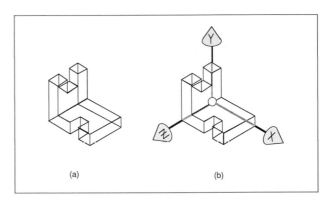

(a)   (b)

FIG. 4.19 Wire frame of angle block (a) and Cartesian axis system with origin installed at left, rear, and bottom corner.

D/ COORDINATE AXES

4.8 Creating a Three-Dimensional Data Base

The fundamental difference between projection onto coordinate planes and the use of coordinate axes is that with the axes, a three-dimensional numerical description of the object must be created first. This description or data base may be created by a variety of methods including two- and three-dimensional digitizing, and may be a full solid geometric model or a simple wire frame description. For example, an object may be manually digitized as a wire frame where the object is represented by vertices and boundary edges. The geometry is transparent and points that would normally be obscured can be seen. This description is the most elemental form of modeling. The student should realize, however, that no matter what modeling technique is used, the process is the same—describing the coordinates of the object in X-Y-Z space.

To demonstrate this process of building a data base, we will manually digitize a familiar object—the angle block (Fig. 4.18). The first step in manually digitizing an object is to describe the object as a picture [Fig. 4.19(a)]. All vertices and connections should be shown whether or not they can actually be seen. Next, locate the object in space relative to the origin. Placing the origin at the lower left rear of the object results in positive values along the X, Y, and Z axes [Fig. 4.19(b)]. Note that in Fig. 4.20 all vertices lying in the Z = 0 plane have been identified. Only their X and Y values change. Finally, assign values for the rest of the vertices relative to the origin as shown in Fig. 4.21. With this done the numeric data base can be built.

When the object is being drawn, the computer must know whether to "move" the pen without drawing a line or put the pen down and connect two points. This is called *pen control*. A pen control value of 0 may result in a move, a pen control value of 1 in a line being drawn.

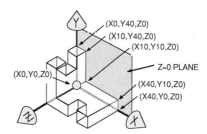

FIG. 4.20 Geometry on Z = 0 plane digitized relative to origin.

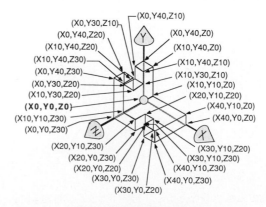

FIG. 4.21 Object completely digitized.

NUMERIC DATA BASE

Data Set	X	Y	Z	Pen	Data Set	X	Y	Z	Pen
01	00	00	00	1	29	20	00	30	2
02	00	40	00	2	30	20	00	20	2
03	10	40	00	2	31	30	00	20	2
04	10	10	00	2	32	30	00	30	2
05	40	10	00	2	33	40	00	30	2
06	40	00	00	2	34	40	00	00	2
07	00	00	00	2	35	40	00	30	1
08	00	00	30	2	36	40	10	30	2
09	00	40	30	2	37	30	00	30	1
10	00	40	20	2	38	30	10	30	2
11	00	30	20	2	39	30	00	20	1
12	00	30	10	2	40	30	10	20	2
13	00	40	10	2	41	20	00	20	1
14	00	40	00	2	42	20	10	20	2
15	10	40	00	1	43	20	00	30	1
16	10	40	10	2	44	20	10	30	2
17	10	30	10	2	45	00	40	30	2
18	10	30	20	2	46	10	40	30	1
19	10	40	20	2	47	10	40	20	1
20	10	40	30	2	48	00	40	20	2
21	10	10	30	2	49	10	30	20	1
22	20	10	30	2	50	00	30	20	2
23	20	10	20	2	51	10	30	10	1
24	30	10	20	2	52	00	30	10	2
25	30	10	30	2	53	10	40	10	1
26	40	10	30	2	54	00	40	10	2
27	40	10	00	2					
28	00	00	30	1					

FIG. 4.22 Numeric data base for angle block.

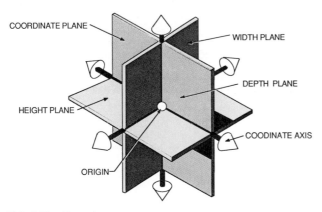

FIG. 4.23 Cartesian axes.

COORDINATE PLANE

WIDTH PLANE

DEPTH PLANE

HEIGHT PLANE

COODINATE AXIS

ORIGIN

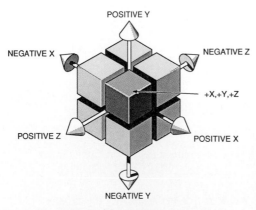

POSITIVE Y

NEGATIVE X

NEGATIVE Z

+X,+Y,+Z

POSITIVE Z

POSITIVE X

NEGATIVE Y

FIG. 4.24 Axes positive and negative values.

The data base is built by writing down the X, Y, and Z values in order as one moves from vertex to vertex, starting by moving to the origin without drawing a line (Fig. 4.22). An attempt should be made to complete the object without redrawing a line and with the fewest number of pen = 0 moves. If it appears necessary to redraw a line, the pen should be picked up and moved to another vertex. This is repeated until all edge boundaries are drawn.

A CADD computer program can use this data to display the wire frame from any orientation.

4.9 The Cartesian (World) Axis System

Once the object is described any and all views may be found by either of two methods. The axes and all space geometry may be revolved, or the object may be left alone and the viewing direction changed. Although this may seem to be a fine distinction, the distinction is important in determining one's position in relation to the object and from what direction the object is being viewed.

Figure 4.9 shows the coordinate planes and their intersection in space to form the angles of projection. Consider Fig. 4.23 where these planes are shown to form coordinate or Cartesian axes, also called *world axes*. The engineer designs in world units and in relation to the world axis system.

The horizontal plane is used to define all points of height in space where Y = 0 at the origin. The frontal plane defines all points of depth where Z = 0 at the origin. The profile plane defines all points of width where X = 0 at the origin. The intersection of these three planes is the location in space where X = 0, Y = 0, and Z = 0, the point known as the origin. The intersection of any two of the coordinate planes forms an axis. An object may have positive and negative values in its data base, depending on how it is positioned in relation to the origin (Fig. 4.24). At times, it may be convenient to position the object in the +X, +Y, +Z octant.

4.10 Right Axes

The description in the previous section is of a *right-hand axes system*. In a right-hand system, the system generally used in drawing, the front view contains normal width (X) and height (Y) dimensions. However, the fields of mathematics and physics, as well as the machining and aircraft industries, have adopted a modified right-handed system. Figure 4.25 shows the difference between these two methods of specifying directions in space. If the representation of the axes is included with the drawing, an engineer can always relate the object to special directions.

4.11 Device Axes

The world axes system establishes height (Y), width (X), and depth (Z) dimensions in space. However, a CADD

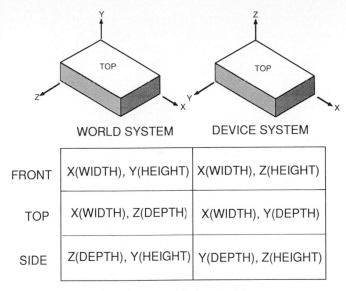

	WORLD SYSTEM	DEVICE SYSTEM
FRONT	X(WIDTH), Y(HEIGHT)	X(WIDTH), Z(HEIGHT)
TOP	X(WIDTH), Z(DEPTH)	X(WIDTH), Y(DEPTH)
SIDE	Z(DEPTH), Y(HEIGHT)	Y(DEPTH), Z(HEIGHT)

FIG. 4.25 Right and left-hand axes systems.

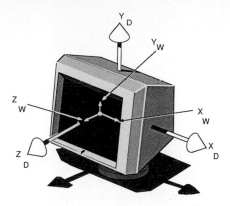

FIG. 4.26 Normally aligned world (W) and device (D) axes.

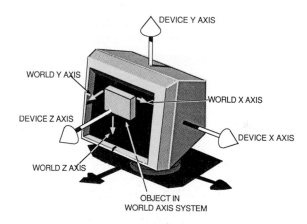

FIG. 4.27 Stationary device (D) axes and top view of world (W) axis system.

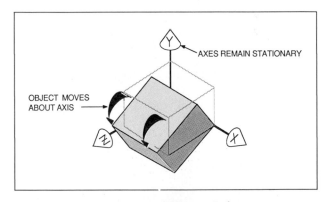

FIG. 4.28 Rotation about stationary cartesian axes.

system may choose to define space relative to a device, such as a milling machine or the terminal used by the designer. In general, it is best to keep device axes and world axes aligned. The more powerful the CADD computer, the greater the chance that device and world axes systems may be manipulated independent of one another. To be able to predict how a view will be altered by a rotate command, for example, this difference must be understood. In the case of device axes, the Z axis is considered to always come directly toward the operator.

Note, in Fig. 4.26, the axis markers attached to the terminal. These represent device axes and do not change. The world axes system that the operator sees inside the computer has not been rotated and is in alignment with the device axis system. CADD operators are advised to establish world axes in the workspace so that world and device axes may be aligned when needed.

Figure 4.27 shows the relationship between device and world axes systems. The operator is looking down the device Z axis but is seeing what is generally called the top view. Only the device and world X axes are aligned. This means that rotation or translation relative to the Y or Z axes must be done by specifying world or device axes. Otherwise, a completely different result than had been intended may result.

4.12 Rotation About Cartesian Axes

To change the orientation of the object relative to a stationary viewer, the object may be rotated about the Cartesian axes as shown in Fig. 4.28. Rotating the shape relative to stationary axes assumes a constant viewing direction. *The device and world axes remain aligned.* Rotation is positive in a counterclockwise direction when the axis is viewed as a point (Fig. 4.29).

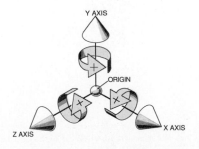

FIG. 4.29 Positive rotation.

FIG. 4.30 Specifying new view by a direction of sight vector. FIG. 4.31 View in direction of sight vector.

4.13 Rotation of Geometry by a Direction of Sight Vector

Another way of specifying axis orientation is through a *direction of sight vector* (Fig. 4.30). In this case, world and device axes become misaligned. In Fig. 4.30 this vector is normal to surface A, and when the vector is viewed as a point, a view *in the direction of the vector* results. This causes the direction of sight vector and the device Z axis to coincide (Fig. 4.31). Note that two of the three world axes are out of alignment with the matching device axes. Were the object to be subsequently rotated about the device Y axis, all three sets of axes would be misaligned.

4.14 2-D CADD Rotation

CADD computers which maintain a 2-D data base can only perform 2-D rotations. A command like ROTATE ALL 45 tells the computer to rotate the X-Y plane of the figure 45° counterclockwise around a point to be specified. Consider the shaded object shown in Fig. 4.32 to be the front view of the L-shaped bracket used to illustrate con-

cepts throughout this chapter. The command ROTATE ALL 45 has been entered and a Z axis of rotation identified. The rotation, counterclockwise from zero, is generally from the right horizontal (Fig. 4.33). Note that the computer understands about which axis the rotation will occur because, as was stated previously, a 2-D CADD computer can only rotate geometry about the device Z axis.

4.15 3-D CADD Rotation

Rotation of 3-D geometry is predicated on a computer's ability to manipulate 3-D data. This is definitely the future for all engineering drawing and design, since with 3-D capabilities any and all views of an object are available by either altering the orientation of the object relative to the Cartesian axes, or by taking an auxiliary viewing position (direction of sight vector). Also, designers work more effectively when they "model" in 3-D than when they draw on paper. To model in 3-D may require retraining and the aquisition of new design skills, but the benefits are well worth the effort.

To be able to use 3-D rotations a CADD operator must

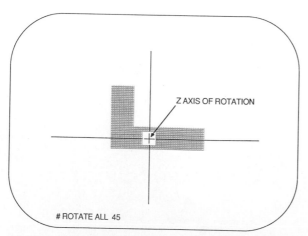

FIG. 4.32 2-D rotation-front view of L-shaped bracket.

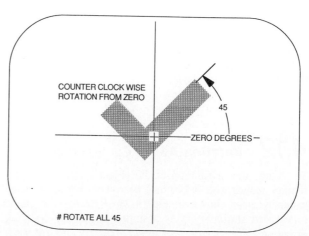

FIG. 4.33 2-D counterclockwise revolution.

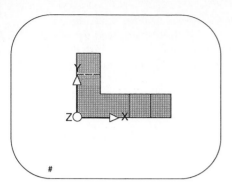

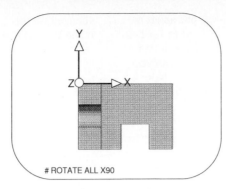

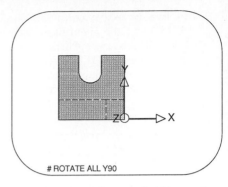

FIG. 4.34   3-D rotation; object is normal to the axis system.

FIG. 4.35   Revolved 90 degrees about the X axis to produce a top view.

FIG. 4.36   Front view revolved to produce a left-side view.

• know whether the computer rotates geometry relative to world or device axis systems.

• know the orientation of the object relative to that axis system.

• be able to specify axis direction (X-Y-Z, or user defined; positive or negative), and the angle in degrees through which the object will be revolved.

Figure 4.34 demonstrates 3-D rotation. The object appears normal to the axis system, not unlike the object in the 2-D CADD example in the previous section. In both cases the operator sees at first a negative Z or front view. The instructions to the computer, ROTATE ALL X90, has caused all geometry to be revolved $+90°$ about the X axis, resulting in a top view (Fig. 4.35). In 4.36, the original front view has been revolved with the command ROTATE ALL Y90. This results in a left-side view. These separate objects, when assembled together as shown in Fig. 4.37, form a three-view orthographic drawing. (See Chapter 5, Fig. 5.9.)

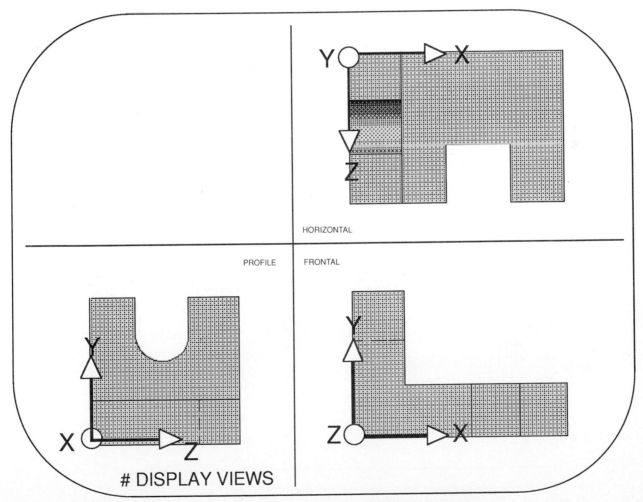

FIG. 4.37   Previously given rotation commands result in a three-view orthographic drawing.

Two successive rotations of single object, one about the Y axis and one about the X axis, can be used to further illustrate this concept of how rotation changes views. With the command ROTATE ALL Y45 the geometry has been rotated around the vertical Y axis [Fig. 4.38(*a*)]. Note in Fig. 4.38(*b*) how the front and profile views have been changed by this command. Next, in Fig. 4.39, the command ROTATE ALL X35 has further changed the object relative to the axes. The top and left side views are strange, but the front view should look pleasing and even a bit familiar. This front view is an *isometric view*. See Chapter 11 for an in-depth discussion of isometric presentation.

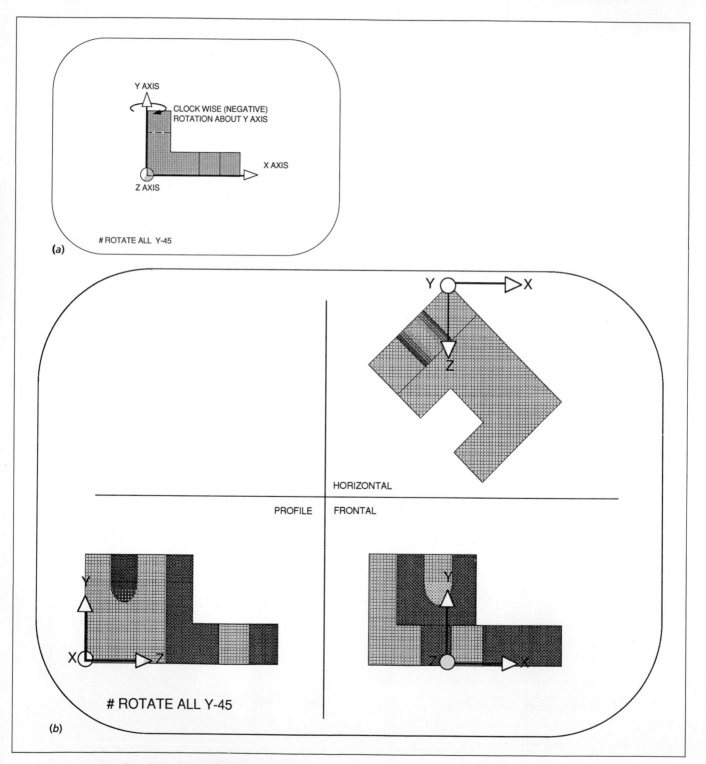

FIG. 4.38   Rotation about Y axis (ROTATE ALL Y45).

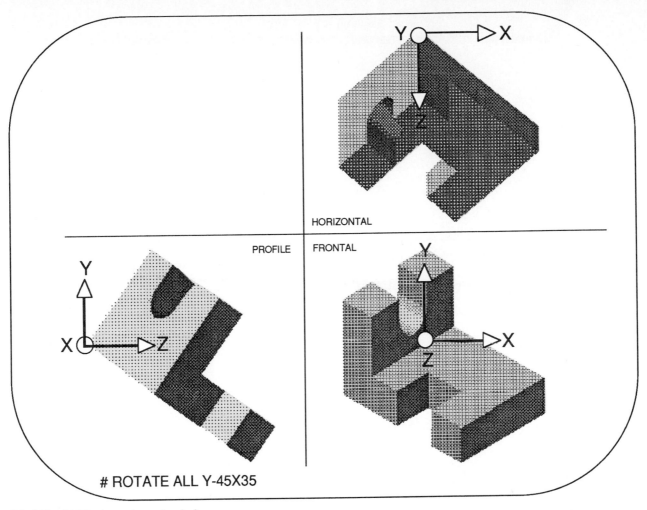

FIG. 4.39 CADD views through windows.

4.16 The CADD Approach to Orthographic Views

In the previous sections the manner in which CADD computers manipulate geometry through rotation has been discussed. We now know the difference between a device axis system and the world axis system and between 2-D and 3-D data. To complete this discussion, various strategies for displaying multiple views will be presented in Chapter 5.

CADD Computers offer a view of the world's geometry through windows. These windows may correspond to top, front, side, and axonometric views as shown in Fig. 4.39. Though the screen may be flat, the display of views is far from a flat drawing. Rather, each window provides the CADD operator a means of viewing and manipulating the geometric model.

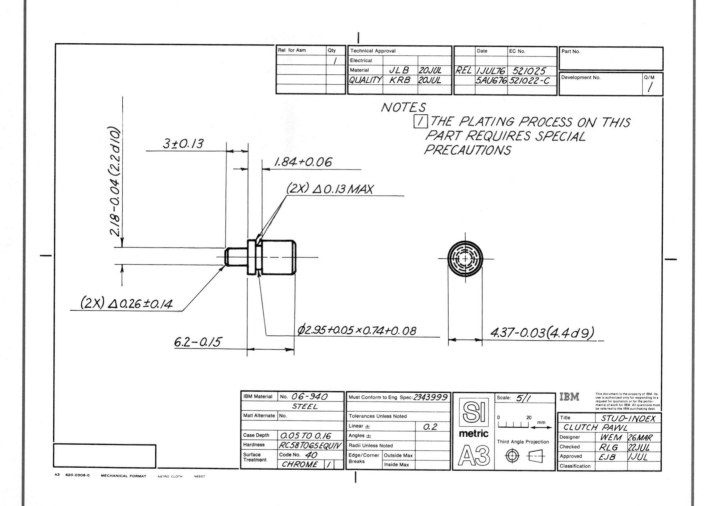

C H A P T E R · 5

Multiview Representation for Design and Product Development

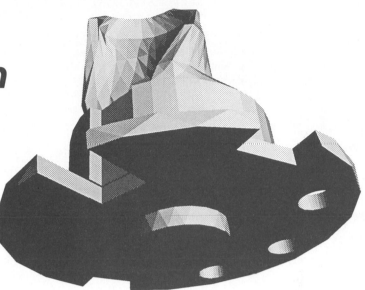

A | MULTIVIEW PROJECTION—COORDINATE PLANES METHOD

5.1 Introduction

Engineers use the orthographic system of projection for describing the shape of machine parts and structures (see facing page). Practical application of this method of describing an object results in a drawing consisting of a number of systematically arranged views that reproduce the object's exact shape. It was explained in Chapter 4,

Sec. 4.3 that a set of views, showing the object from different positions, is always taken. These views, positioned in strict accordance with the universally recognized arrangement, must show the three dimensions, width, height, and depth. Although three views (Fig. 5.1) are usually required to describe an ordinary object, only two may be needed for a particularly simple one. A very complicated object may require four or more views. A view projected on an auxiliary plane also may be desirable (Fig. 8.3). Such a view often makes possible the elimination of one of the principal views. Therefore, it is up to the individual to determine the number and type of views needed to produce a satisfactory drawing. You will soon develop a knack for this, if you bear in mind that the number of views required depends entirely on the complexity of the shape to be described.

5.2 Definition

Multiview (multiplanar) projection is a method by which the exact shape of an object can be represented by two or more separate views produced on projection planes that are at right angles to each other.

5.3 Methods of Obtaining the Views

The views of an object may be obtained by either of two methods:

1. The *natural* method.
2. The *glassbox* method.

Since the resulting views will be the same in either case, the beginner should adopt the method which is the easiest to understand. Both methods are explained here in detail.

5.4 Natural Method

In using this method, each of the necessary views is obtained by looking directly at the particular side of the object the view is to represent.

Figure 5.1 shows three of the principal views of an object: the front, top, and side views. They were obtained by looking directly at the front, top, and right side, respectively. In the application of this method, some consider the position of the object as fixed and the position of the observer as shifted for each view; others find it easier to

consider the observer's position as fixed and the position of the object as changed for each view (Fig. 5.1). Regardless of which procedure is followed, the top and side views must be arranged in their natural positions relative to the front view.

Figure 5.2 illustrates the natural relationship of views. Note that the top view is *vertically above* the front view, and the side view is *horizontally in line with* the front view. In both of these views *the front of the block is toward the front view.*

5.5 "Glass Box" Method

An imaginary glass box is used widely by instructors to explain the arrangement of orthographic views. An explanation of this scheme can best be made by reviewing the use of planes of projection (Chapter 4). It may be considered that planes of projection placed parallel to the six faces of an object form an enclosing glass box (Fig. 5.3). The observer views the enclosed object from the outside. The views are obtained by running projectors from points on the object to the planes. This procedure is in accordance with the theory of orthographic projection explained in Sec. 4.4, as well as the definition in Sec. 5.2. The top, front, and right side of the box represent the *H* (horizontal), *F* (frontal), and *P* (profile) projection planes.

Since the projections on the sides of the three-dimensional transparent box are to appear on a sheet of drawing paper, it must be assumed that the box is hinged (Fig. 5.4) so that, when it is opened outward into the plane of the paper, the planes assume the positions illustrated in Figs. 5.4 and 5.5. Note that all of the planes, except the back one, are hinged to the frontal plane. In accordance with this universally recognized assumption, the top projection must take a position directly above the front pro-

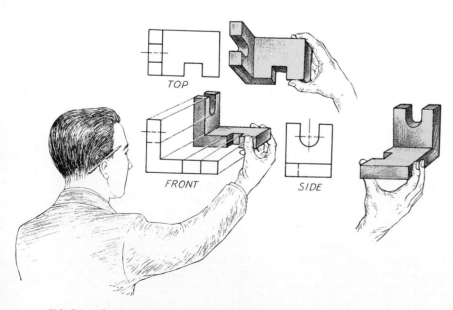

FIG. 5.1 Obtaining three views of an object.

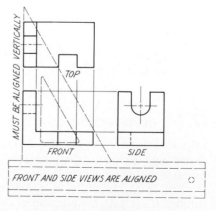

FIG. 5.2 Position of views.

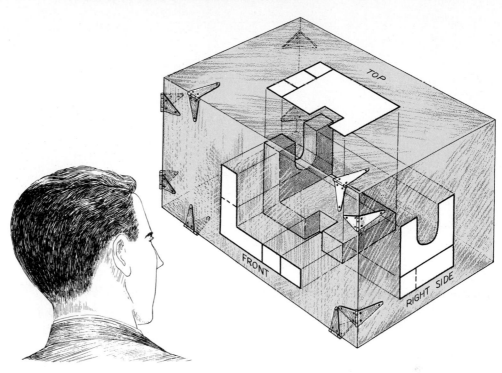

FIG. 5.3 "Glass box."

jection, and the right-side projection must lie horizontally to the right of the front projection. To identify the separate projections, engineers call the one on the frontal plane the *front view* or *front elevation*, the one on the horizontal plane the *top view* or *plan*, and the one on the side or profile plane the *side view*, *side elevation*, or *end view*. Figure 5.5 shows the six views of the same object as they would appear on a sheet of drawing paper. Ordinarily, only three of these views are necessary (front, top, and right side). A bottom or rear view will be required in comparatively few cases.

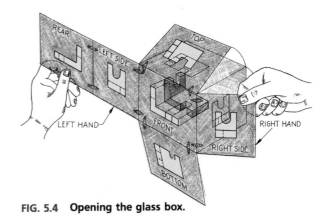

FIG. 5.4 **Opening the glass box.**

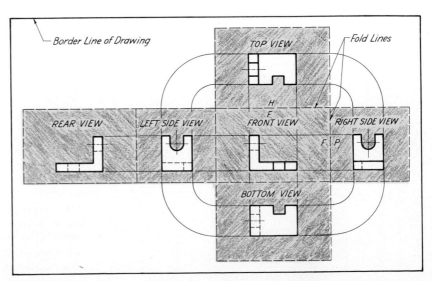

FIG. 5.5 **Six views of an object on a sheet of drawing paper.**

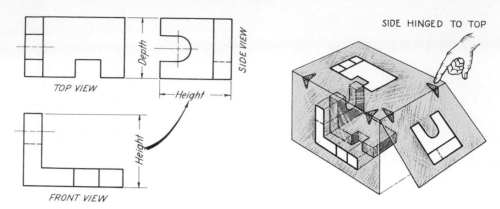

FIG. 5.6 "Second position" for the side view.

5.6 "Second Position"

Sometimes, especially in the case of a broad, flat object, it is desirable to hinge the sides of the box to the horizontal plane so that the side view will fall to the right of the top view, as illustrated in Fig. 5.6. This arrangement conserves space on the paper and gives the views better balance.

5.7 Principles of Multiview Drawing

The following principles should be studied carefully and understood thoroughly before any attempt is made to prepare an orthographic drawing:

1. The front and top views are *always* in line vertically (Fig. 5.2).

2. The front and side views are in line horizontally, except when the second position is used (Fig. 5.2).

3. The front of the object in the top view faces the front view (Fig. 5.4).

4. The front of the object in the side view faces the front view (Fig. 5.4).

5. The depth of the top view is the same as the depth of the side view (or views) (see Fig. 5.7).

6. The width of the top view is the same as the width of the front view (Fig. 5.7).

7. The height of the side view is the same as the height of the front view (Fig. 5.7).

8. A view taken from above is a top view and *must* be placed above the front view (Fig. 5.5).

9. A view taken from the right, in relation to the selected front, is a right-side view and *must* be placed to the right of the front view (Fig. 5.5).

10. A view taken from the left is a left-side view and *must* be placed to the left of the front view (Fig. 5.5).

11. A view taken from below is a bottom view and *must* be placed below the front view (Fig. 5.5).

B / CADD STRATEGIES FOR PRINCIPAL VIEWS

5.8 CADD Strategies for Displaying Multiple Views

CADD computers are capable of displaying multiple views of complex geometry by any of several methods. First, and most powerful, is the display of multiple views from a single data base (Fig. 5.8). This method's results are similar to those obtained by placing video cameras in specific positions relative to the object and projecting the image captured by each camera onto the appropriate location of the CADD screen. Here, *changes made to the object in one view alter the data base from which all other views are made*. This means that a change made in one view will automatically be made in the other views. The 3-D model can be finished as an engineering drawing by the operator, adding 2-D notes, symbols, and dimensions to the plane of the drawing (usually the X-Y plane).

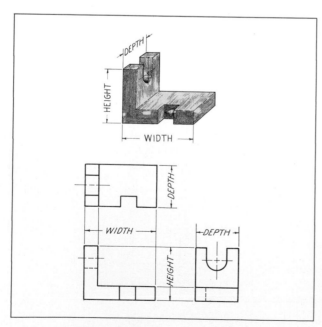

FIG. 5.7 View terminology.

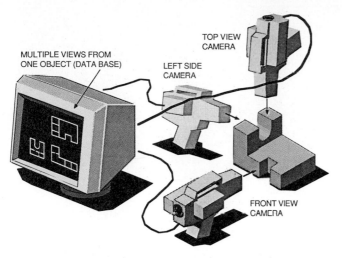

FIG. 5.8 **Multiple views from a single data base.***

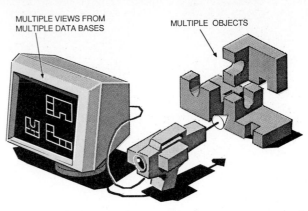

FIG. 5.9 **Single camera to view duplicate geometry rotated into positions of principle views.***

Next in terms of power is the use of a single camera to view *duplicate geometry* rotated into positions of principal views (Fig. 5.9). This was shown in Figs. 4.34–4.37. For three views, you have three identical objects, each in its unique orientation. Changes made to one view are not recorded in the other views because each view has its own data base. As with a single data base, a multiple data base can be finished as an engineering drawing.

Least powerful are *snapshots* of 2-D views placed in proper orientation on a 2-D construction plane (Fig. 5.10). This is very much like traditional paper and pencil drafting. Not only are the views not a display of the same data base where changes in one view will correctly change all views, but all three-dimensionality has been lost, and the views are literally graphic diagrams. The 2-D diagram of views, with the addition of dimensions, notes, and symbols, becomes an engineering drawing.

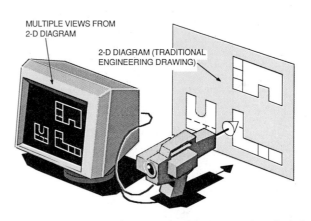

FIG. 5.10 **Snapshot views on a 2-D construction plane.***

C / PROJECTION OF POINTS, LINES, AND PLANES

5.9 *Projection of Lines*

A line may project either in *true length*, *foreshortened*, or as a *point* in a view depending on its relationship to the projection plane on which the view is projected (see Fig. 5.11). In the top view, the line projection $a^H b^H$ shows the true length of the edge AB (see pictorial) because AB is parallel to the horizontal plane of projection. Looking directly at the frontal plane, along the line, AB projects as a point ($a^F b^F$). Lines, such as CD, that are inclined to

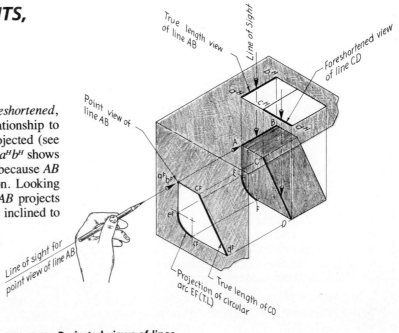

FIG. 5.11 **Projected views of lines.**

* (Computer generated.)

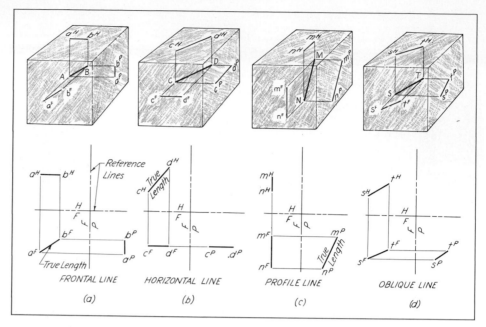

FIG. 5.12   Some typical line positions.

one of the planes of projection, will show a foreshortened projection in the view on the projection plane to which the line is inclined and true length in the view on the plane of projection to which the line is parallel. The curved line projection $e^F f^F$ shows the true length of the curved edge.

The student should study Fig. 5.12 and attempt to visualize the space position of each of the given lines. It is necessary both in preparing and reading graphical representations to recognize the position of a point, line, or plane and to know whether the projection of a line is true length or foreshortened, and whether the projection of a plane shows the true size and shape. The indicated reference lines may be thought of as representing the edges of the glass boxes shown. The projections of a line are identified as being on either a frontal, horizontal, or profile plane by the use of the letters F, H, or P with the lowercase letters that identify the end points of the line. For example, in Fig. 5.12(a), $a^H b^H$ is the horizontal projection of line AB, $a^F b^F$ is the frontal projection, and $a^P b^P$ is the profile projection.

It is suggested that the student hold a pencil and move it into the following typical line positions to observe the conditions under which the pencil representing a line, appears in true length.

1. *Vertical line.* The vertical line is perpendicular to the horizontal and will therefore appear as a point in the H (top) view. It will appear in true length in the F (frontal) view and in the P (profile) view.

2. *Horizontal line* [Fig. 5.12(b)]. The horizontal line will appear in true length when viewed from above because it is parallel to the H plane of projection and its end points are theoretically equidistant from an observer looking downward.

3. *Inclined line* [Fig. 5.12(c)]. The inclined line is any line not vertical or horizontal that is parallel to either

the frontal plane of the profile plane of projection. An inclined line will show true length in the F (frontal) view or P (profile) view.

4. *Oblique line* [Fig. 5.12(d)]. The oblique line will not appear in true length in any of the principal views because it is inclined to all of the principal planes of projection. It should be apparent in viewing the pencil alternately from the directions used to obtain the principal views, namely, from the front, above, and side, that one end of the pencil is always farther away from the observer than the other. Only when looking directly at the pencil from such a position that the end points are equidistant from the observer can the true length be seen. On a drawing, the true length projection of an oblique line will appear in a supplementary A (auxiliary) view projected on a plane that is parallel to the line (Sec. 9.7).

5.10 *Meaning of Lines*

On a multiview drawing a visible or invisible line may represent

1. The *intersection* of two surfaces,

2. The *edge view* of a surface,

3. The *limiting element* of a surface.

These three different meanings of a line are illustrated in Fig. 5.13. In the top view, the curved line is an edge view of surface C, while a straight line is the edge view of surface A. The full circle in the front view may be considered as the edge view of the cylindrical surface of the hole. In the side view, the top line, representing the limiting element of the cylindrical surface, indicates the limits for

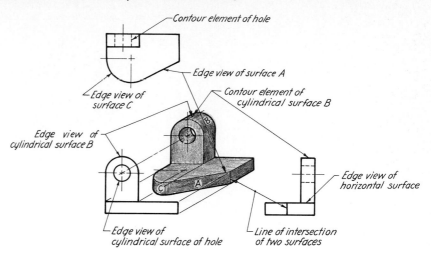

FIG. 5.13 **Meaning of lines.**

the surface and therefore can be thought of as being a surface limit line. The short vertical line in this same view represents the intersection of two surfaces. In reading a drawing, one can be sure of the meaning of a line on a view only after an analysis of the related view or views. All views must be studied carefully.

5.11 *Projection of Surfaces*

The components of most machine parts are bounded by either plane or single-curved surfaces. *Plane surfaces* bound cubes, prisms, and pyramids, while *single-curved surfaces*, ruled by a moving straight line, bound cylinders and cones. The projected representations (lines or areas) of both plane and single-curved surfaces are shown in Fig. 5.14. From this illustration the student should note that

1. when a surface is *parallel to a plane* of projection, it will appear in true size in the view on the plane of projection to which it is parallel,

2. when it is *perpendicular to the plane* of projection, it will project as a line in the view, and

3. when it is positioned at an *angle*, it will appear foreshortened.

A surface will always project either as a line or an area on a view. The area representing the surface may be either a full-size or foreshortened representation.

In Fig. 5.14 the cylindrical surface *A* appears as a line in the side (profile) view and as an area in the top and front views. Surface *B* shows true size in top view and as a line in both the front and side views. Surface *C*, a vertical surface, will appear as a line when observed from above.

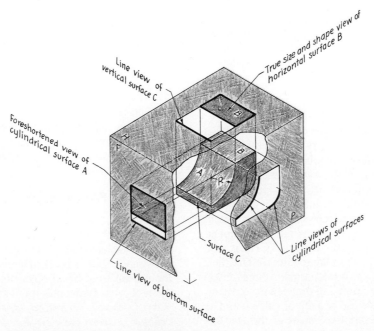

FIG. 5.14 **Projected views of surfaces.**

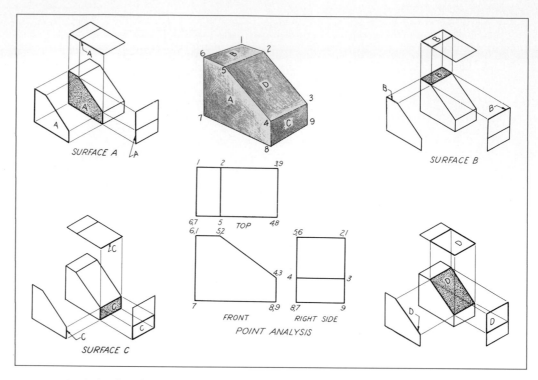

FIG. 5.15   Analysis of surfaces, lines, and points.

5.12   Analysis of Surfaces, Lines, and Points in Three Principal Views

An analysis of the representation of the surfaces of a mutilated block is given pictorially in Fig. 5.15. It can be noted that each of the surfaces *A*, *B*, and *C* appears in true size and shape in one view and as a line in each of the other two related views. Surface *D*, which is inclined, appears with foreshortened length in the top and side views and as an inclined line in the front view.

Three views of each of the visible points are shown on the multiview drawing. At the very beginning of an elementary course in drawing, a student will often find it helpful to number the corners of an object in all views.

5.13   Selection of Views

Careful study should be given to the outline of an object before the views are selected; otherwise there is no assurance that the object will be described completely from the reader's viewpoint (Fig. 5.16). Only those views that are necessary for a clear and complete description should be selected. Since the repetition of informaion only tends to confuse the reader, superfluous views should be avoided.

Although some objects, such as cylinders, bushings, bolts, and so forth, require only two views (front and side), more complicated pieces may require an auxiliary or sectional view in addition to the ordinary three views.

The space available for arranging the views often governs the choice between the use of a top or side view. The difference between the descriptive values of the two frequently is not great. For example, a draftsman often finds that the views of a long object will have better balance if

a top view is used [see Fig. 5.17(*a*)]; while in the case of a short object [see (*b*)]; the use of a side view may make possible a more pleasing arrangement. It should be remembered that the choice of views for many objects is definitely fixed by the contour alone, and no choice is offered as far as spacing is concerned. It is more important to have a set of views that describes an object clearly than one that is artistically balanced.

Often there is a choice between two equally important views, such as between a right-side and left-side view or between a top and bottom view (Fig. 5.5). In such cases, one should adhere to the following rule. *A right-side view should be used in preference to a left-side view and a top view in preference to a bottom view.* When this rule is applied to irregular objects, the front (contour) view should be drawn so that the most irregular outline is toward the top and right side.

Another rule, one that must be considered in selecting the front or profile view is as follows. *Place the object to obtain the smallest number of hidden lines.*

5.14   Principal (Front) View

The principal view is the one that shows the characteristic contour of the object [see Fig. 5.18(*a*) and (*b*)]. Good practice dictates that this be used as the front view on a drawing. It should be clearly understood that the view of the natural front of an object is not always the principal view, because frequently it fails to show the object's characteristic shape. Therefore, another rule to be followed is *ordinarily, select the view showing the characteristic contour shape as the front view, regardless of the normal or natural front of the object.*

* longer for top
* shorter for side

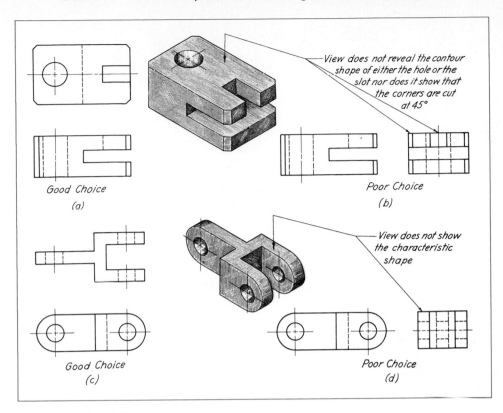

View does not reveal the contour shape of either the hole or the slot nor does it show that the corners are cut at 45°

Good Choice
(a)

Poor Choice
(b)

View does not show the characteristic shape

Good Choice
(c)

Poor Choice
(d)

FIG. 5.16 Choice of views.

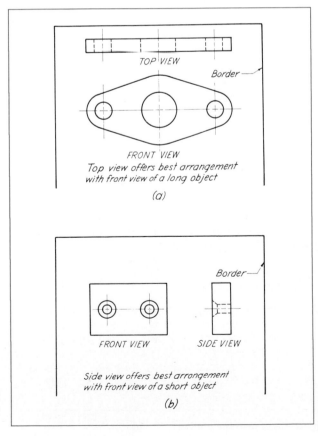

TOP VIEW

Border

FRONT VIEW
Top view offers best arrangement
with front view of a long object

(a)

Border

FRONT VIEW   SIDE VIEW

Side view offers best arrangement
with front view of a short object

(b)

FIG. 5.17 Choice of views.

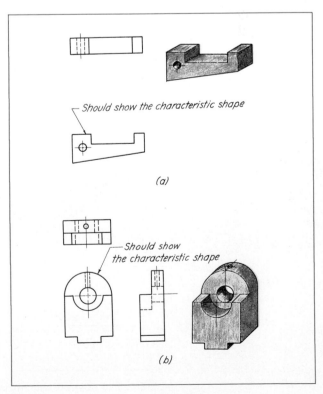

Should show the characteristic shape

(a)

Should show
the characteristic shape

(b)

FIG. 5.18 Principal view of an object.

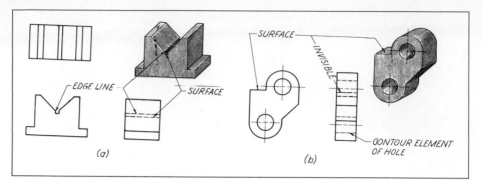

FIG. 5.19   Invisible lines.

When an object does have a definite normal position, however, the front view should be in agreement with it. In the case of most machine parts, the front view can assume any convenient position that is consistent with good balance.

5.15   Invisible Lines

Dotted or dashed lines are used on an external view of an object to represent surfaces, intersections, and limits invisible at the point from which the view is taken. In Fig. 5.19(*a*), one invisible line represents a line of intersection or edge line, while the other invisible line may be considered to represent either the surface or lines of intersection. On the side view in (*b*) there are invisible lines, which represent the contour elements of the cylindrical holes.

5.16   Treatment of Invisible Lines

The short dashes that form an invisible line should be drawn carefully in accordance with the recommendations in Sec. 5.25. An invisible line always starts with a dash in contact with the object line from which it starts, unless it forms a continuation of a visible line. In the latter case, it should start with a space, in order to establish at a glance the exact location of the end point of the visible line (see Fig. 5.20C). Note that the effect of definite corners is secured at points *A*, *B*, *E*, and *F*, where, in each case, the end dash touches the intersecting line. When the point of intersection of an invisible line and another object line does not represent an actual intersection on the object, the intersection should be open as at points *C* and *D*. An open intersection tends to make the lines appear to be at different distances from the observer.

Parallel invisible lines should have the breaks staggered.

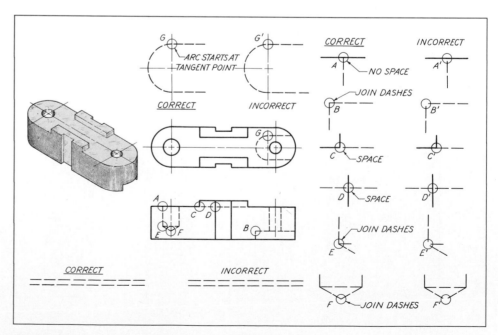

FIG. 5.20   Correct and incorrect junctures of invisible outlines.

The correct and incorrect treatment for starting invisible arcs is illustrated at *G* and *G′*. Note that an arc should start with a dash at the point of tangency. This treatment enables the reader to determine the exact end points of the curvature.

5.17 Omission of Invisible Lines

It is common practice for commercial drafters to omit hidden lines when their use tends to further confuse an already overburdened view or when the shape description of a feature is sufficiently clear in another view. It is not advisable for a beginning student to do so. The beginner, until the discrimination that comes with experience has more developed, will be wise to show *all* hidden lines.

5.18 Precedence of Lines

When one discovers in making a multiview drawing that two lines coincide, the question arises, which line should be shown or, in other words, which line will have precedence. For example, as revealed in Fig. 5.21, a solid line may have the same position as an invisible line representing the contour element of a hole, or an invisible line may occur at the same place as a center line for a hole. In these cases the decision rests on the relative importance of each of the two lines that can be shown. The precedence of lines is as follows:

1. *Solid lines* (visible object lines) take precedence over all other lines.

2. *Dashed lines* (invisible object lines) take precedence over center lines, although evidence of center lines may be indicated as shown in both the top and side views of Fig. 5.21.

3. *A cutting-plane line* takes precedence over a center line where it is necessary to indicate the position of a cutting plane.

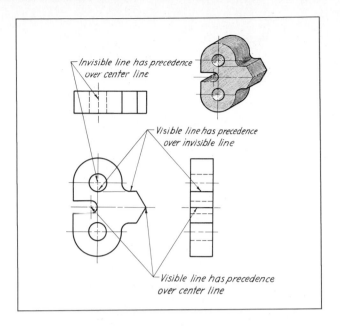

FIG. 5.21 Precedence of lines.

5.19 Projection of Angles

When an angle lies in a plane that is parallel to one of the planes of projection the angle will show in true size in the normal view of that particular plane. In Fig. 5.22 those angles indicated as actual show in their true size. The 60° angle, which lies in a surface that is not parallel to the *H* plane, appears at less than 60° in the top view. The 30° angle for the sloping line on the component portion that is inclined backward projects at greater than 30° in the front view. It may be said that, except for a 90° angle having one leg as a normal line, angles lying on inclined planes will project either larger or smaller than true size, depending on the position of the plane in which the angle lies. A 90° angle always projects in true size, even on an inclined plane, if the line forming one side of the angle is parallel to the plane of projection and a normal view

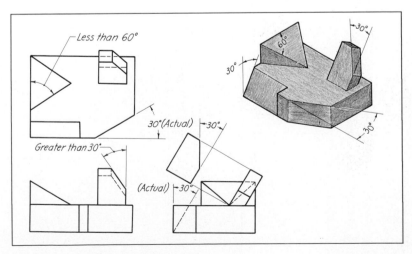

FIG. 5.22 Projection of angles.

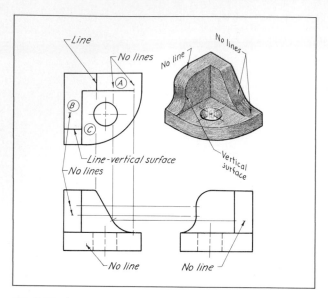

FIG. 5.23 Treatment of tangent surfaces.

of the line results. The normal view of a line is any view of a line that is obtained with the direction of sight perpendicular to the line.

What has been stated concerning angles on inclined surfaces can be easily verified by the student when observing what happens to the angles of a 30° × 60° triangle resting on the long leg, as it is revolved from a vertical position downward onto the surface of his desk top.

5.20 *Treatment of Tangent Surfaces*

When a curved surface is tangent to a plane surface, as illustrated in several ways on the pictorial drawing in Fig. 5.23, no line should be shown as indicated at A and B in

the top view and as noted for the front and side views. At C in the top view the line represents a small vertical surface that must be shown even though the upper and lower lines for this surface may be omitted in the front view, depending on the decision of the drafter. In the top view a line has been drawn to represent the intersection of the inclined and horizontal surfaces at the rear, even though they meet in a small round instead of a sharp edge. The presence of this line emphasizes the fact that there are two surfaces meeting here that are at a definite angle, one to the other. Several typical examples of tangencies and intersections have been illustrated in Fig. 5.24.

5.21 *Parallel Lines*

When parallel surfaces are cut by a plane, the resulting lines of intersection will be parallel. This is shown by the pictorial drawing in Fig. 5.25(*b*), where the near corner of the object has been removed by the oblique plane *ABC*. It can be observed from the multiview drawing in (*c*) that *when two lines are parallel in space, their projections will be parallel in all of the views*, even though at times both lines may appear as points on one view.

In Fig. 5.25, three views are to be drawn that show the block after the near front corner has been removed [see (*b*)]. Several of the required lines of intersection can be readily established through the given points *A*, *B*, and *C* that define the oblique plane. For example, $c^F b^F$ can be drawn in the front view and the line through a^F can be drawn parallel to it. In the top view, $a^H b^H$ should be drawn first and the intersection line through c^H should then be drawn parallel to this *H* view of *AB*. The drawing can now be completed by working back and forth from view to view while applying the rule that a plane intersects parallel planes along lines of intersection that are parallel. The remaining lines are thus drawn parallel to either *AB* or *CB* [see the pictorial in (*b*)].

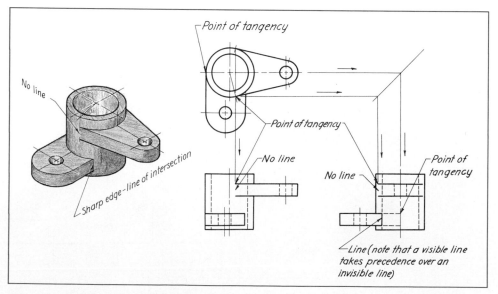

FIG. 5.24 Treatment of tangent surfaces.

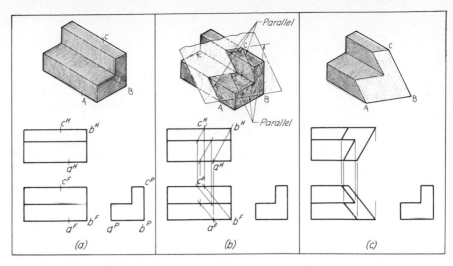

FIG. 5.25 Parallel lines.

5.22 *Plotting an Elliptical Boundary*

The actual intersection of a circular cylinder or cylindrical hole with a slanting surface (inclined plane) is an *ellipse* (Fig. 5.26). The elliptical boundary in (*a*) appears as an ellipse in the top view, as a line in the front view, and as a semicircle in the side view. The ellipse is plotted in the top view by projecting selected points (such as points *A* and *B*) from the circle arc in the side view, as shown. For example, point *A* was projected first to the inclined line in the front view and then to the top view. The mitre line shown was used to project the depth distance for *A* in the

top view for illustrative purposes only. Ordinarily, dividers should be used to transfer measurements to secure great accuracy.

In (*b*) the intersection of the hole with the sloping surface is represented by an ellipse in the side view. Points selected around the circle in the top view (such as points *C* and *D*) projected to the side view as shown permit the draftsman to form the elliptical outline. It is recommended that a smooth curve be sketched freehand through the projected points before a French curve or ellipse guide is applied to draw the finished ellipse, because it is easier to fit a curved ruling edge to a line than to scattered points.

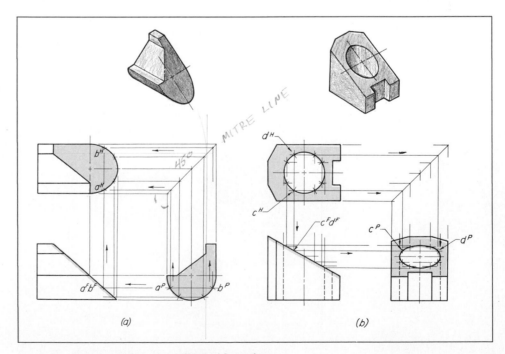

FIG. 5.26 Representation of an elliptical boundary.

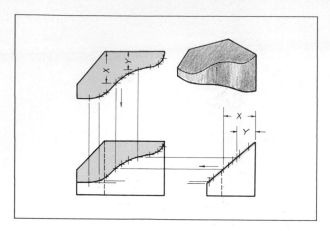

FIG. 5.27 Projecting a space curve.

5.23 Projecting a Curved Outline (Space Curve)

When a boundary curve lies in an inclined plane, the projection of the curve may be found in another view by projecting points along the curve, as illustrated in Fig. 5.27. In the example, points selected along the arcs forming the curve in the top view were first located in the side view, using distances taken from the top view, as shown by the X and Y measurements. Then the front view positions of these points, through which the front view of the curve must pass, were established by projecting horizontally from the side view and downward from the top view.

5.24 Treatment of Intersecting Finished and Unfinished Surfaces

Figure 5.28 illustrates the removal of material when machining surfaces, cutting a slot, and drilling a hole in a small part. The italic *f* on a surface of a pictorial drawing in this text indicates that the surface has been machined. The location of sharp and rounded corners, as illustrated in (*b*) and (*c*), are noted on the multiview drawing. A discussion covering rounded internal and external corners is given in Sec. 5.36. The *f* is not recommended for use on a detail drawing in any ANSI standard. See Sec. 13.7.

5.25 To Make an Orthographic Drawing

The location of all views should be determined before a drawing is begun. This will ensure balance in the appearance of the finished drawing. The contour view is usually started first. After the initial start, the drafter should construct views simultaneously by projecting back and forth from one to the other. It is poor practice to complete one view before starting the others, as much more time will be required to complete the drawing. Figure 5.29 shows the procedure for laying out a three-view drawing. The general outline of the views first should be drawn lightly with a hard pencil and then heavied with a medium-grade pencil. Although experienced persons sometimes deviate from this procedure in drawing in the lines of known length and location in finished weight while constructing the views, it is not recommended that beginners do so (see Fig. 5.29, step III).

Although a 45° mitre line is sometimes used for transferring depth dimensions from the top view to the side view, or vice versa, as shown in Fig. 5.30(*b*), it is better practice to use dividers, as in (*a*). Continuous lines need not be drawn between the views and the mitre line, as in the illustration, for one may project from short dashes across the mitre line. The location of the mitre line may be obtained by extending the construction lines representing the front edge of the top view and the front edge of the side view to an intersection.

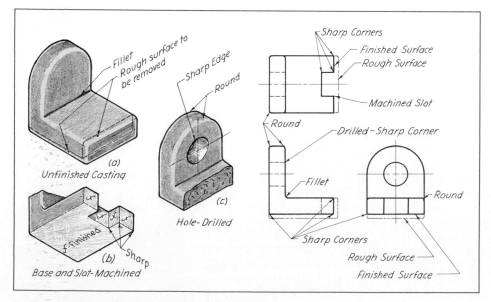

FIG. 5.28 Rough and finished surfaces on a casting.

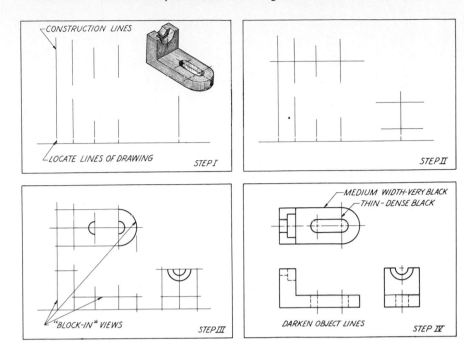

FIG. 5.29 Steps in making a three-view drawing of an object.

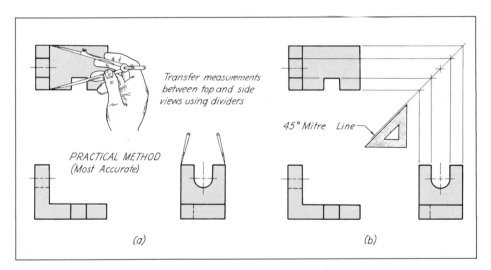

FIG. 5.30 Methods for transferring depth dimensions.

When making an orthographic drawing in pencil, the beginner should endeavor to use the line weights recommended in Sec. 2.17. The object lines should be made very dark and bright, to give snap to the drawing as well as to create the contrast necessary to cause the shape of the object to stand out. Special care should be taken to gauge the dashes and spaces in invisible object lines. On ordinary drawings, 3 mm (.12 in.) dashes and 0.8 mm (0.3 in.) spaces are recommended (Fig. 5.31).

Center lines consist of alternate long and short dashes. The long dashes are from 20 to 40 mm (.80 to 1.60 in.) long, the short dashes 3 mm (.12 in.), and the spaces 0.8 mm (.03 in.) (Fig. 5.31). The following technique is recommended in drawing center lines.

1. Where *center lines* cross, the short dashes should intersect symmetrically (Fig. 5.31). (In the case of very small circles the breaks may be omitted.)

2. The *breaks* should be so located that they will stand out and allow the center line to be recognized as such.

3. *Center lines* should extend approximately 3 mm (.12 in.) beyond the outline of the part whose symmetry they indicate (Fig. 5.31).

4. *Center lines* should not end at object lines.

5. *Center lines* that are aligned with object lines should have not less than a 1.5 mm (.06 in.) space between the end of the center line and the object line.

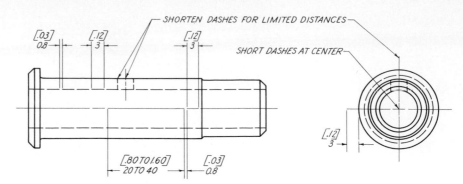

FIG. 5.31   Invisible lines and center lines.

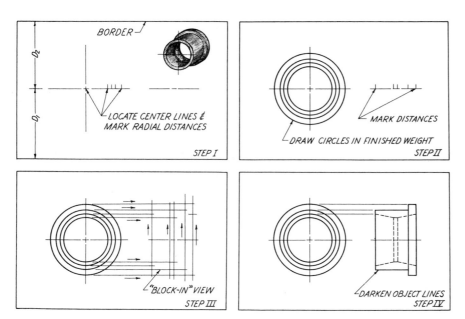

FIG. 5.32   Steps in making a two-view drawing of a circular object.

For a finished drawing to be pleasing in appearance, all lines of the same type must be uniform, and each type must have proper contrast with other symbolic types. The contrast between the types of pencil lines is similar to that of ink lines (Fig. 2.22), except that pencil lines are never as wide as ink lines (read Sec. 2.17). On commercial drawings, the usual practice is to ''burn in'' the object lines by applying heavy pressure.

If reasonable care is taken not to soil a drawing, it will not be necessary to clean any part of it with an eraser. Since the practice in engineering drawing is not to erase construction lines if they have been drawn lightly, the student, at the very beginning of the first course, should try to acquire habits that ensure cleanliness.

When constructing a two-view drawing of a circular object, the pencil work must start with the drawing of the center lines, as shown in Fig. 5.32. This is necessarily the first step, because the construction of the circular (contour) view is based on a horizontal and a vertical center line. The horizontal object lines of the rectangular view are projected from the circles.

5.26   *Visualizing an Object from Given Views*

Most students in elementary graphics courses find it difficult to visualize an object from two or more views. This trouble is largely due to the lack of systematic procedure for analyzing complex shapes.

The simplest method of determining shape is illustrated pictorially in Fig. 5.33. This method of ''breaking down'' may be applied to any object, since all objects may be thought of as consisting of elemental geometric forms, such as prisms, cylinders, cones, and so on. This is consistent with computer-aided geometric modeling. These imaginary component parts may be additions in the form of projections or subtractions in the form of cavities. Following such a detailed geometric analysis, a clear picture of an entire object can be obtained by mentally assembling a few easily visualized forms.

It should be realized, when analyzing component parts, that it is usually impossible to determine whether a form is an addition or a subtraction by looking at one view. For

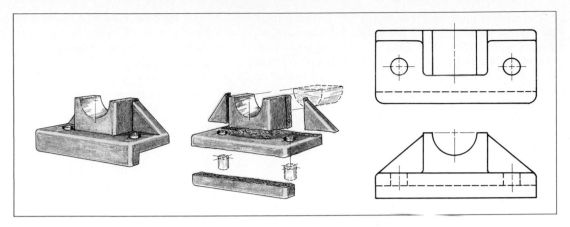

FIG. 5.33 "Breaking down" method.

example, the small circles in the top view in Fig. 5.33 indicate a cylindrical form, but they do not reveal whether the form is a hole or a projection. By consulting the front view, however, the form is shown to be a hole (subtracted cylinder).

The graphic language is similar to the written language in that neither can be read at a glance. A drawing must be read patiently by referring systematically back and forth from one view to another. At the same time the reader must imagine a three-dimensional object and not a two-dimensional flat projection.

A student usually will find that a pictorial sketch will clarify the shape of a part that is difficult to visualize. The method for preparing quick sketches in isometric projection is explained in Secs. 6.14 and 6.15.

5.27 Interpretation of Adjacent Areas of a View

To obtain a full understanding of the true geometric shape of a part, all the areas on a given view must be carefully analyzed, because each area represents a surface on the part. For example, in reading a drawing it must be determined whether a particular area in a top view represents a surface that is inclined or horizontal and whether the surface is higher or lower than adjacent ones. Five distinctly different objects are shown in Fig. 5.34, all having the same top view. In determining the actual shape of these objects, memory and previous experience can be a help, but one can easily be misled if the analysis is not approached with an open mind, for it is only by trial-and-error effort and by referring back and forth from view to view that a drawing can be read. In considering area *A* in (*a*) it might be thought that the triangular surface could be high and horizontal, which would be correct because of the arrangement of lines in the front view. However, in considering the top view alone, *A* could be either sloping, as in (*c*) and (*e*), or low and horizontal, as in (*b*). An analysis of the five parts reveals that the surface represented by area *B* can also be either sloping, high and horizontal, or low and horizontal. Area *C* offers even a wider variety of possibilities in that it may be either low and horizontal (*a*), sloping (*b*), cylindrical [(*c*) and (*d*)], or high and horizontal (*e*).

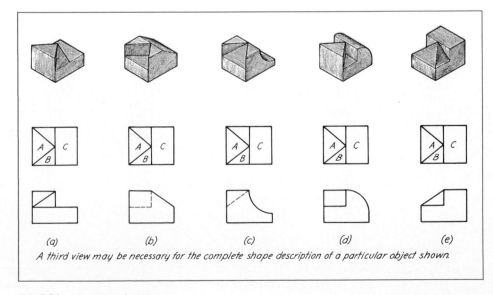

(a)   *(b)*   *(c)*   *(d)*   *(e)*

A third view may be necessary for the complete shape description of a particular object shown.

FIG. 5.34 Meaning of areas.

The student must realize at this point that since there are infinite possibilities for the shape, position, and arrangement of surfaces that form objects, one must learn to study tediously the views of any unfamiliar object until sure of the exact shape. Multiview drawings cannot be read with the ease of our written language, which lists all the components in a dictionary.

5.28 *True-length Lines*

Students who, lacking a thorough understanding of the principles of projection (Sec. 5.9), find it difficult to determine whether or not a projection of a line in one of the principal views shows the true length of the line, should study carefully the following facts.

1. If the principal projection of a line shows the true length of the line, one of the other projections must appear as a horizontal line, a vertical line, or a point, on one of the other views of the drawing.

2. If the front view of the line is horizontal, and the top view of the line is parallel to the frontal plane, both views show true length.

3. If the top view of a line is a point, the front and side views show the true length.

4. If the front view of a line is a point, the top and side views show the true length.

5. If the top and front views of a line are parallel to the profile plane, the side view shows true length.

6. If the side projection of a line is a point, the top and front views show the true length.

7. If the front view of a line is horizontal and the top view is inclined to the frontal plane, the top inclined view shows the true length.

8. If the top view of a line is horizontal and the front view is inclined to the horizontal plane, the front inclined view shows the true length.

5.29 *Representation of Holes*

In preparing drawings of parts of mechanisms, a drafter finds it necessary to represent machined holes, which most often are drilled, drilled and reamed, drilled and countersunk, drilled and counterbored, or drilled and spotfaced. Graphically, a hole is represented to conform with the finished form. The form may be completely specified by a note attached to the view showing the circular contour (Fig. 5.35). The shop note, as prepared by the drafter, usually specifies the several shop operations in the order that they are to be performed. For example, in (*d*) the hole, as specified, is drilled before it is counterbored. When depth has not been given in the note for a hole, it is understood to be a through hole; that is, the holes goes entirely through the piece [(*a*), (*c*), (*d*), and (*e*)]. A hole that does not go through is known as a *blind hole* (*b*). For such holes, depth is the length of the cylindrical portion. Drilled, bored, reamed, cored, or punched holes are always specified by giving their diameters, never their radii. Drill diameters for number- and letter-size drills may be found in Table 52 in the Appendix. Metric drills have been listed in Table 27.

In drawing the hole shown in (*a*), which must be drilled before it is reamed, the limits are ignored and the diameter

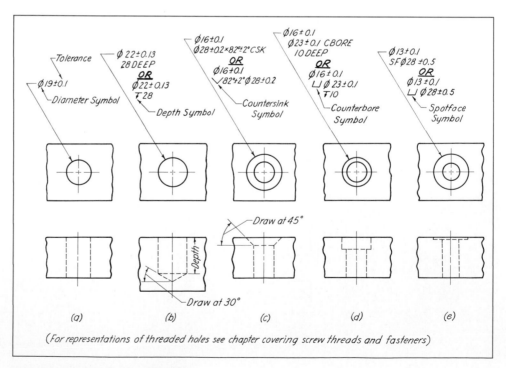

FIG. 5.35 Representation of holes.

is scaled to the nearest regular inch or millimeter size. In (b) the 30° × 60° triangle is used to draw the approximate representation of the conical hole formed by the drill point. In (c) a 45° triangle has been used to draw an approximate representation of the outline of the conical enlargement. The actual angle of 82° is ignored in order to save time in drawing. The spotface in (e) is most often cut to a depth of 1.5 mm (.06 in.); however, the depth is usually not specified.

The beginner should now scan the several sections in the chapter on shop processes to obtain some general information on the production of holes. Complete information on the preparation of shop notes for holes may be found in Chapters 13, 14, and 15.

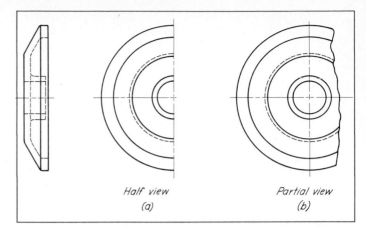

FIG. 5.36 Half views and partial views.

D/ CONVENTIONAL PRACTICES

5.30 Practices Defined

To reduce the high cost of preparing engineering drawings and at the same time to convey specific and concise information without a great expenditure of effort, some generally recognized systems of symbolic representation and conventional practices have been adopted by American industry.

A *standard symbol* or conventional representation can express information that might not be understood from a true-line representation unless accompanied by a lettered statement. In many cases, even though a true-line representation would convey exact information, very little more would be gained from the standpoint of better interpretation. Some conventional practices have been adopted for added clarity. For instance, they can eliminate awkward conditions that arise from strict adherence to the rules of projection.

These methods of drawing have slowly developed with the graphic language until at the present time they are universally recognized and observed and appear in the various standards of the American National Standards Institute.

Skilled professionals have learned to accept and respect the use of the symbols and conventional practices, for they can interpret these representations accurately and realize that their use saves valuable time in both the drawing room and the shop.

5.31 Half Views and Partial Views

When the available space is insufficient to allow a satisfactory scale to be used for the representation of a *symmetrical piece*, it is considered good practice to make one view either a half view or a partial view, as shown in Fig. 5.36. The half view, however, must be the top or side view and not the front view, which shows the characteristic contour. The half view should be the front half of the top or side view. In the case of the partial view shown in (b), a break line is used to limit the view.

5.32 Accepted Violations of True Projection in the Representation of Boltheads, Slots, and Holes for Pins

A departure from true projection is encountered in representing a *bolthead*. For example, on a working drawing, it is considered the best practice to show the head *across corners* in both views, regardless of the fact that in true projection one view would show *across flats*. This method of treatment eliminates the possibility of a reader's interpreting a hexagonal head to be a square head (Fig. 5.37).

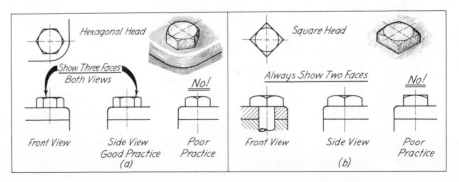

FIG. 5.37 Treatment of bolt heads.

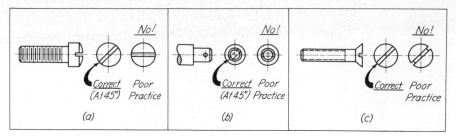

FIG. 5.38   Treatment of slots and holes in fasteners and pins.

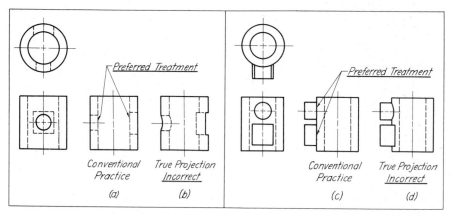

FIG. 5.39   Treatment of unimportant intersections.

Furthermore, the showing of a head across corners in both views clearly reveals the space needed for proper clearance.

In the case of the slotted head fasteners, the slots are shown at 45° in the end views in order to avoid placing a slot on a center line, where it is usually difficult to draw so that the center line passes accurately through the center (Fig. 5.38). This practice does not affect the descriptive value of the drawing, because the true size and shape of the slot is shown in the front view. The hole for a pin is shown at 45° for the same reason. In such a position it may be more quickly observed.

5.33   Treatment of Unimportant Intersections

The conventional methods of treating various unimportant intersections are shown in Fig. 5.39. To show the true line of intersection in each case would add little to the value of the drawing. Therefore, in the views designated as preferred, true projection has been ignored in the interest of simplicity. On the side views, in (a) and (b), for example, there is so little difference between the descriptive values of the true and approximate representations of the holes that the extra labor necessary to draw the true representation is unwarranted.

5.34   Aligned Views

Pieces that have arms, ribs, lugs, or other features at an angle are shown aligned or *straightened out* in one view,

as illustrated in Fig. 5.40. By this method, it is possible to show the true shape as well as the true position of such features. In Fig. 5.41, the front view has been drawn as though the slotted arm had been revolved into alignment with the element projecting outward to the left. This practice is followed to avoid drawing an element—that is at an angle—in a foreshortened position.

5.35   Conventional Treatment of Radially Arranged Features

Many objects that have radially arranged features may be shown more clearly if true projection is violated, as in Fig. 5.40(b). Violation of true projection in such cases consists of intentionally showing such features swung out of position in one view to present the idea of symmetry and show the true relationship of the features at the same time. For example, while the radially arranged holes in a flange (Fig. 5.42) should always be shown in their true position in the circular view, they should be shown in a revolved position in the other view in order to show their true relationship with the rim.

Radial ribs and radial spokes are similarly treated [Fig. 5.40(a)]. The true projection of such features may create representations that are unsymmetrical and misleading. The preferred conventional method of treatment, by preserving symmetry, produces representations that are more easily understood and that at the same time are much simpler to draw. Figure 5.43 illustrates the preferred treatment for radial ribs and holes.

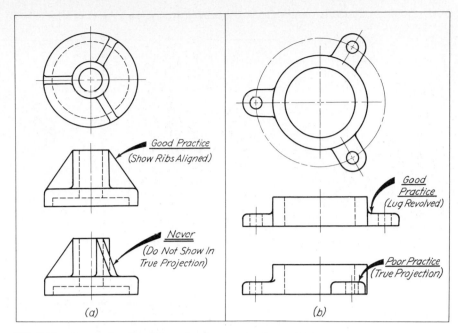

FIG. 5.40 **Conventional practice of representing ribs and lugs.**

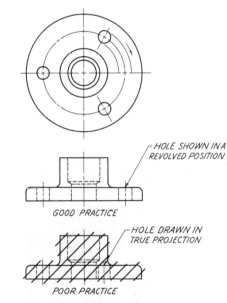

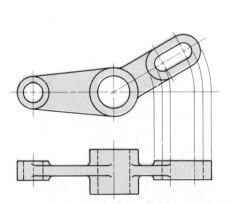

FIG. 5.41 **Aligned views.**

FIG. 5.42 **Radially arranged holes.**

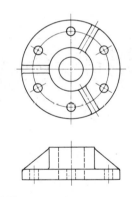

FIG. 5.43 **Conventional treatment of radially arranged ribs.**

5.36 *Representations of Fillets and Rounds*

Interior corners, which are formed on a casting by unfinished surfaces, are always *filleted* at the intersection in order to avoid possible fracture at that point. Sharp corners are also difficult to obtain and are avoided for this reason as well. Exterior corners are *rounded* for appearance and for the comfort of persons who must handle the part when assembling or repairing the machine on which the part is used. A rounded internal corner is known as a *fillet*; a rounded external corner is known as a *round* (Fig. 5.44).

When two intersecting surfaces are machined, however, their intersection will become a sharp corner. For this reason, all corners formed by unfinished surfaces should be shown "broken" by small rounds, and all corners

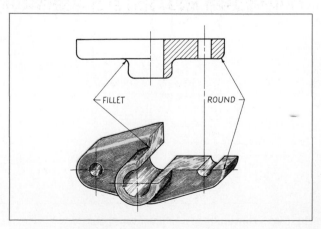

FIG. 5.44 **Fillets and rounds.**

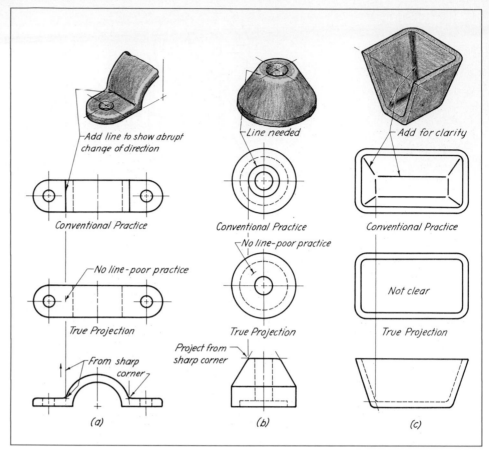

FIG. 5.45   Conventional practice of representing nonexisting lines of intersection.

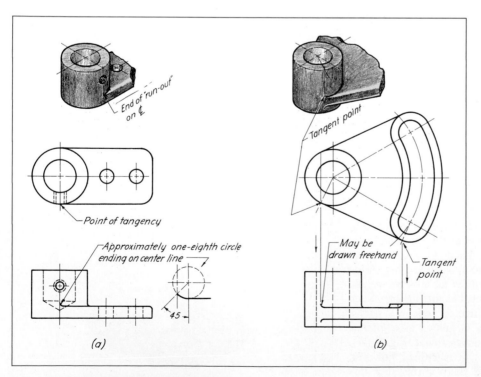

FIG. 5.46   Conventional treatment for fillets.

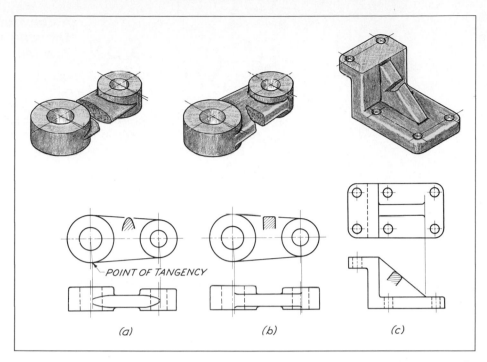

FIG. 5.47 Approximate methods of representing run-outs for intersecting fillets and rounds.

formed by two finished surfaces, or one finished surface and one unfinished surface, should be shown "sharp." Although in the past it has been the practice to allow patternmakers to use their judgment about the size of fillets and rounds, many present-day companies require their designers and draftsmen to specify their size even though their exact size may not be important.

Since fillets and rounds eliminate the intersection lines of intersecting surfaces, they create a special problem in orthographic representation. To treat them in the same manner as they would be treated if they had large radii results in views that are misleading. For example, the true-projection view in Fig. 5.45(c) confuses the reader, because at first glance it does not convey the idea that there are abrupt changes in direction. To prevent such a probable first impression and to improve the descriptive value of the view, it is necessary to represent these theoretically nonexisting lines. These characteristic lines are projected from the approximate intersections of the surfaces, with the fillets disregarded.

Figure 5.46 illustrates the accepted conventional method of representing the *run-out* intersection of a fillet in cases

where a plane surface is tangent to a cylindrical surface. Although run-out arcs such as these are usually drawn freehand, a French curve or a bow instrument may be used. If they are drawn with the bow instrument, a radius should be used that is equal to the radius of the fillet, and the completed arc should form approximately one-eighth of a circle.

The generally accepted methods of representing intersecting fillets and rounds are illustrated in Fig. 5.47. The treatment, in each of the cases shown, is determined by the relationship existing between the sizes of the intersecting fillets and rounds.

5.37 *Conventional Breaks*

A relatively long piece of uniform section may be shown to a larger scale, if a portion is broken out so that the ends can be drawn closer together (Fig. 5.48). When such a scheme is employed, a conventional break is used to indicate that the length of the representation is not to scale. The American National Standard conventional breaks,

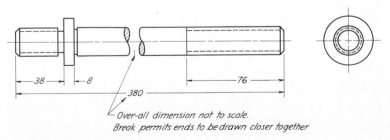

FIG. 5.48 Broken-out view.

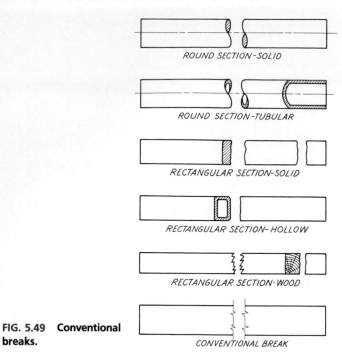

FIG. 5.49 Conventional breaks.

shown in Fig. 5.49, are used on either detail or assembly drawings. The break representations for indicating the broken ends of rods, shafts, tubes, and so forth, are designed to reveal the characteristic shape of the cross section in each case. Although break lines for round sections may be drawn freehand, particularly on small views, it is better to draw them with either an irregular curve or a bow instrument. The breaks for wood sections, however, always should be drawn freehand.

5.38 Ditto Lines

When it is desirable to minimize labor in order to save time, *ditto lines* may be used to indicate a series of identical features. For example, the threads on the shaft shown in Fig. 5.50 are just as effectively indicated by ditto lines as by a completed profile representation. When ditto lines are used, a long shaft of this type may be shortened without actually showing a conventional break.

FIG. 5.50 Ditto lines.

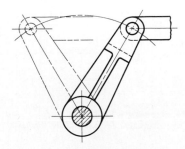

FIG. 5.51 Alternative positions.

5.39 A Conventional Method for Showing a Part in Alternative Positions

A method frequently used for indicating an alternative position of a part or a limiting position of a moving part is shown in Fig. 5.51. The dashes forming the object lines of the view showing the alternative position should be of medium weight. The *phantom line* shown in Fig. 5.51 is recommended for representing an alternative position.

5.40 Conventional Representation

Symbols are used on topographic drawings, architectural drawings, electrical drawings, and machine drawings. No engineer serving in a professional capacity can very well escape their use.

Most of the illustrations that are shown in Fig. 5.52 should be easily understood. However, the *crossed-lines* (diagonals) symbol has two distinct and different meanings. First, this symbol may be used on a drawing of a shaft to indicate the position of a surface for a bearing or, second, it may indicate that a surface perpendicular to the line of sight is flat. These usages are illustrated with separate examples.

◢ PROBLEMS

The problems that follow are intended primarily to furnish study in multiview projection through the preparation of either sketches or instrumental drawings. Many of the problems in this chapter, however, may be prepared in more complete form. Their views may be dimensioned as are the views of working drawings, if the student will study carefully the beginning of the chapter covering dimensioning before attempting to record size description (Chapter 13). All dimensions should be placed in accordance with the general rules of dimensioning. The problems given at the end of Chapter 6 offer further study in multiview representation.

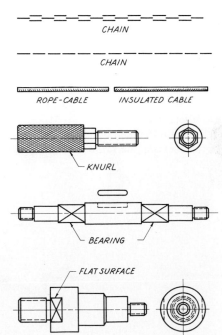

FIG. 5.52 Conventional symbols.

The views shown in a sketch or drawing should be spaced on the paper with aim for balance within the borderlines. Ample room should be allowed between the views for the necessary dimensions. If the views are not to be dimensioned, the distance between them may be made somewhat less than would be necessary otherwise.

Before starting to draw, the student should reread Sec. 5.25 and study Fig. 5.27, which shows the steps in making a multiview drawing. The preparation of a preliminary sketch proves helpful to the beginner.

All construction work should be done in light lines with a sharp hard pencil. A drawing should be checked by an instructor before the lines are "heavied in," unless the preliminary sketch was checked beforehand.

A part showing linear dimensions given in millimeters may be dimensioned in inches by converting the values using Table 4 in the Appendix.

1. (Fig. 5.53). Add the missing line or lines in one view of each of the three-view drawings. When the missing line or lines have been determined, the three views of each object will be consistent with one another.

FIG. 5.53 Missing-line (or lines) exercises.

2. (Fig. 5.54). Draw or sketch the third view for each of the given objects.

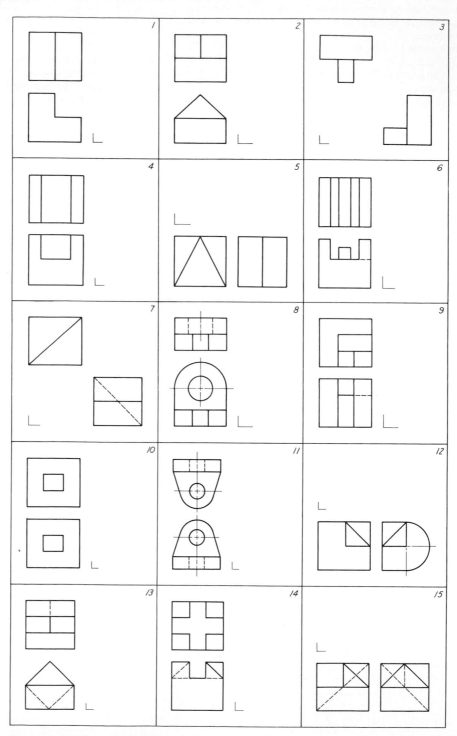

FIG. 5.54   Third-view problems.

3–6. (Figs. 5.55–5.58). Reproduce the given views and draw
the required view. Show all hidden lines.

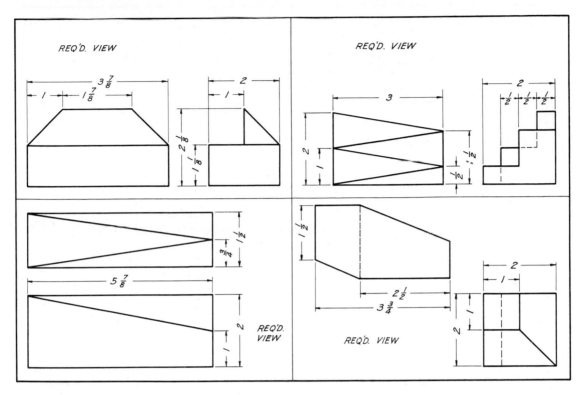

FIG. 5.55

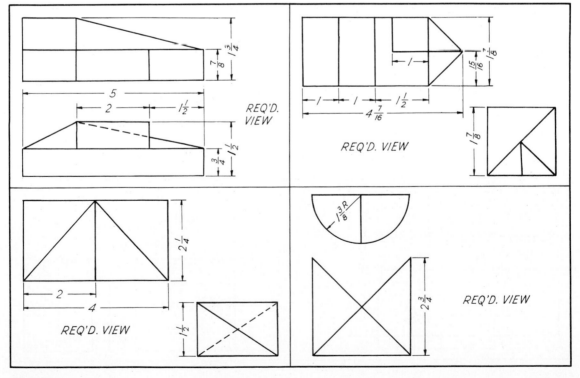

FIG. 5.56

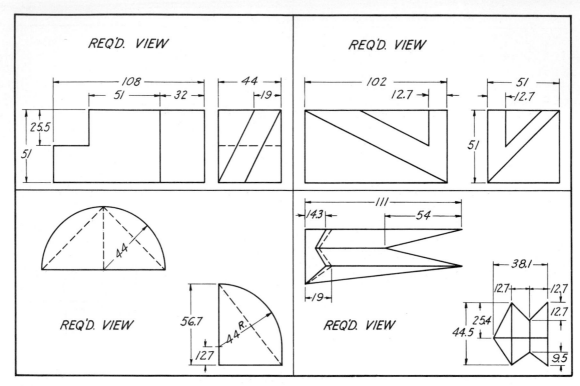

FIG. 5.57

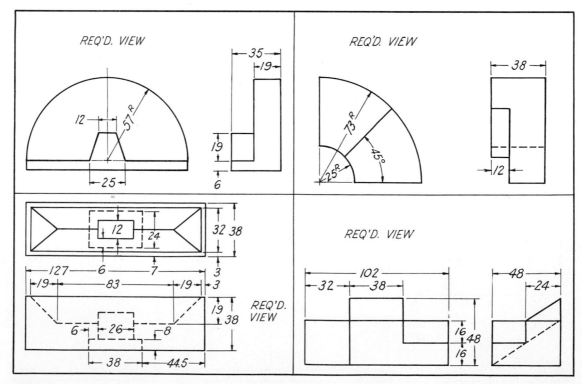

FIG. 5.58

7. (Fig. 5.59). Make an orthographic drawing or sketch of the bench stop. The views may be dimensioned. The shaft portion that fits into the hole in the bench top is 19 mm in diameter and 50 mm long.

8–39. (Figs. 5.60–5.91). Make multiview drawings of the given objects. The views of a drawing may or may not be dimensioned.

FIG. 5.59   Bench stop.

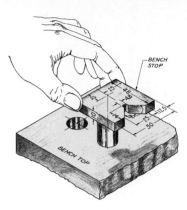

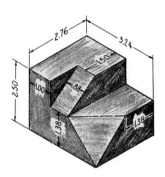

FIG. 5.60   Corner block.

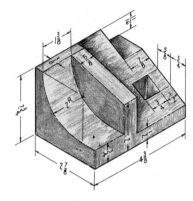

FIG. 5.61   Stop block.

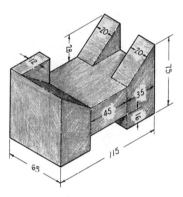

FIG. 5.62   Rest block.

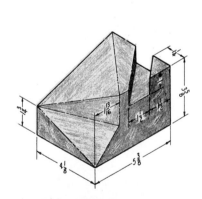

FIG. 5.63   Adjustment block.

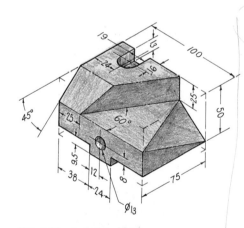

FIG. 5.64   Locating block.

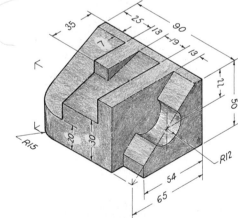

FIG. 5.65   Safety block.

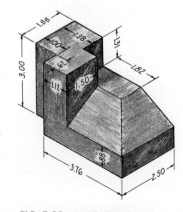

FIG. 5.66   Angle block.

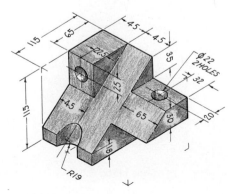

FIG. 5.67   Cross stop.

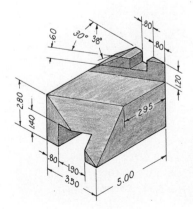

FIG. 5.68   End block.

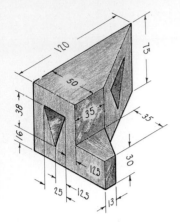

FIG. 5.69 Bevel block.

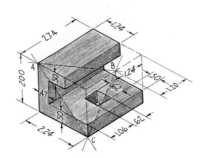

FIG. 5.70 Stabilizer block.

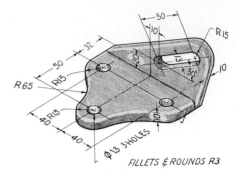

FIG. 5.71 Mounting bracket.

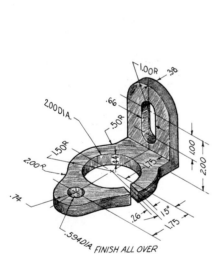

FIG. 5.72 Index guide.

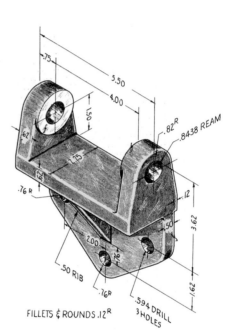

FIG. 5.73 Guide bracket.

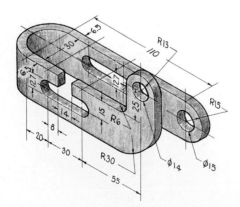

FIG. 5.74 Control guide.

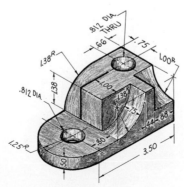

FIG. 5.75 Shoe block.

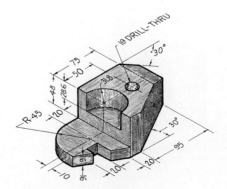

FIG. 5.76 Holder block.

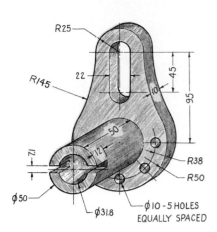

FIG. 5.77   Slotted guide.

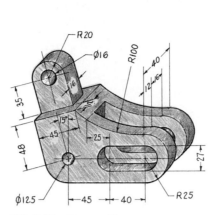

FIG. 5.78   Pivot guide.

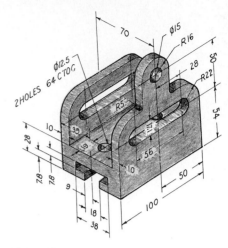

FIG. 5.79   Control guide.

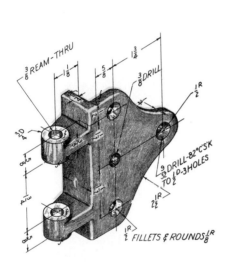

FIG. 5.80   Corner bracket.

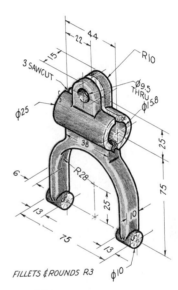

FIG. 5.81   Auxiliary fork.

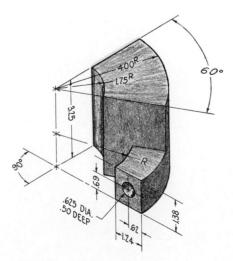

FIG. 5.82   Jaw block.

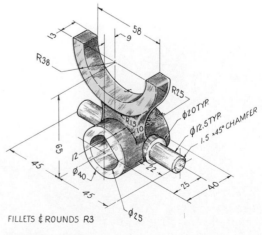

FILLETS & ROUNDS R3

FIG. 5.83   Shifter.

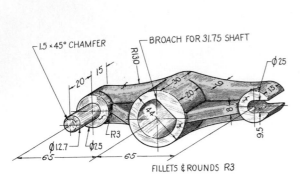

FIG. 5.84 Stud guide.

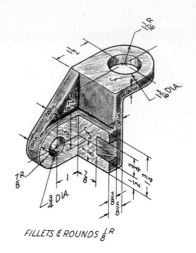

FIG. 5.85 Ejector bracket.

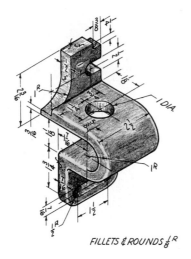

FIG. 5.86 Guide clip.

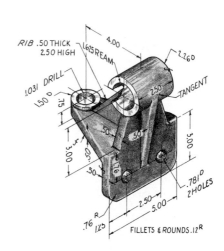

FIG. 5.87 Control rod guide.

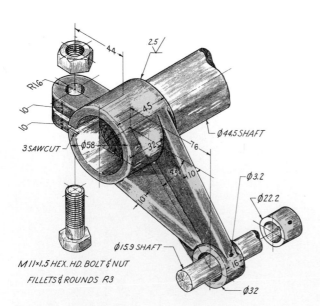

FIG. 5.88 Shaft bracket.

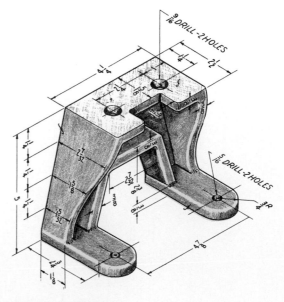

FIG. 5.89 Lathe leg.

40. (Fig. 5.92). Make a complete orthographic drawing of the anchor bracket.

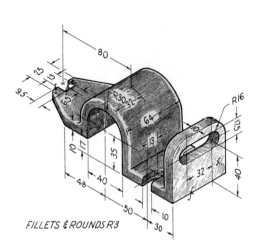

FILLETS & ROUNDS R3

FIG. 5.90 **Feed guide.**

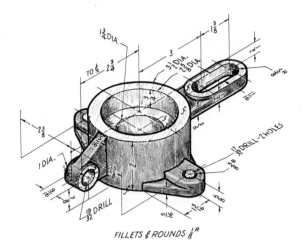

FILLETS & ROUNDS $\frac{1}{8}$R

FIG. 5.91 **Control bracket.**

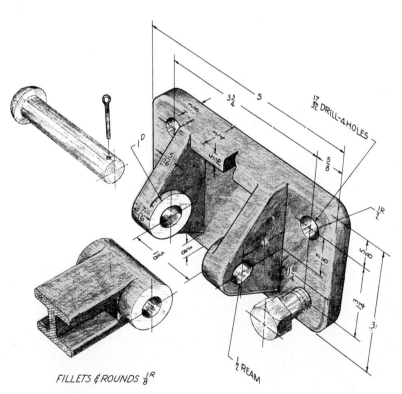

FILLETS & ROUNDS $\frac{1}{8}$R

FIG. 5.92 **Anchor bracket.**

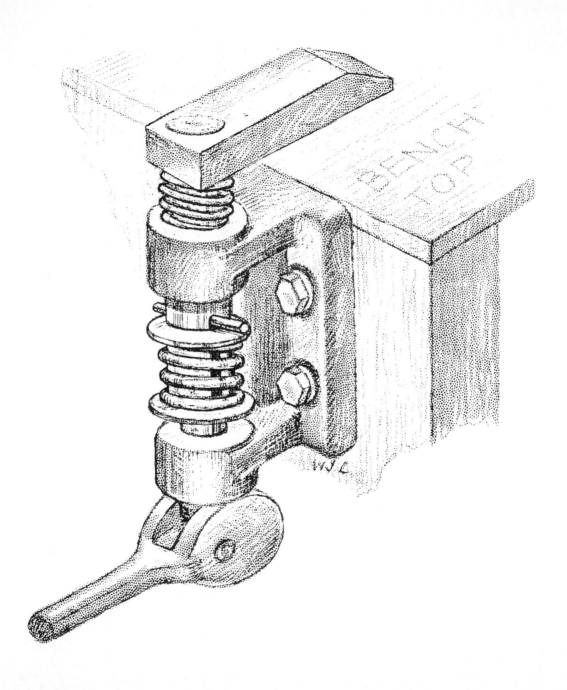

The ideas of a designer may be sketched in pictorial form as they are visualized. Later during a period of preliminary study, a combination of orthographic and pictorial sketches will be prepared as problems are recognized and possible solutions are considered. Sales engineers frequently include pictorial sketches along with orthographic sketches when preparing field reports.

Sketches, when used in combination with the written and spoken language, lead to a full understanding by all persons with whom one finds it necessary to communicate.

The pictorial sketch above shows a designer's idea for a quick acting bench clamp.

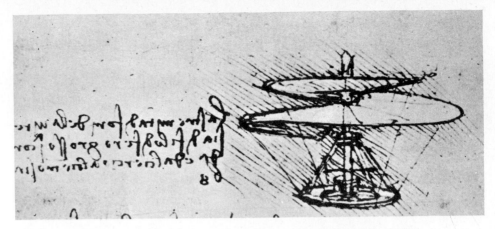

FIG. 6.3 **Idea sketch of a helicopter prepared by Leonardo da Vinci (1452–1519).** (*From Collections of Fine Arts Department, International Business Machines Corporation*)

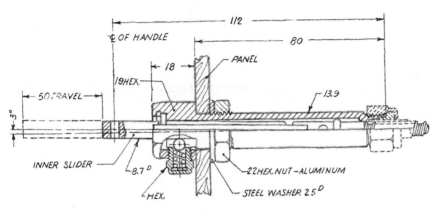

FIG. 6.4 **Design sketch for a connector of a remote control unit.** (*Courtesy Teleflex, Inc.*)

all stages in the development of a product problems must be solved and instructions clarified. Very often a pictorial sketch of some detail of construction will prove to be more intelligible and will convey the idea much better than an orthographic sketch, even when dealing with an experienced drafter or detailer (Fig. 6.22).

Design sketches may be done in the quiet of the designer's office or amid the confusion of the conference table. To meet the requirement of speed of preparation, one must resist all temptation to use instruments of any type and rely on the pencil alone, for the true measure of the quality of a finished sketch is neatness and good proportion rather than the straightness of the lines. *A pictorial sketch need not be an artistic masterpiece to be useful.*

Training for making pictorial sketches must include the presentation of basic fundamentals, as is done with other how-to-do-it subjects. *As learning the mechanics of English does not make one a creative writer, so training in sketching will not make one a creative engineer.* However, sketching is the means of recording creative thoughts.

Some design sketches drawn by an electrical engineer are shown in Fig. 12.19. These sketches were prepared in making a study of the wiring to the electronic control panel for an automatic machine.

6.2 *Thinking with a Pencil*

As an attempt is made to bring actuality to a plan, sketches undergo constant change as different ideas develop. An eraser may be in constant use or new starts may be made repeatedly, even though one should think in-depth and sketch only when it would appear to be worthwhile. Sketching should be done as easily and freely as writing, so that the mind is always centered on the idea and not on the technique of sketching. To reach the point where one can "think with the pencil" is not easy. Continued practice is necessary until one can sketch with as little thought about how it is done as is given to how a knife and fork are used at the dinner table.

6.3 *Value of Freehand Drawing*

Freehand technical drawing is primarily the language of those in charge of the development of technical designs and plans. Chief engineers, chief drafters, designers, and supervisors have found that the best way to present their ideas for either a simple or complex design is through the

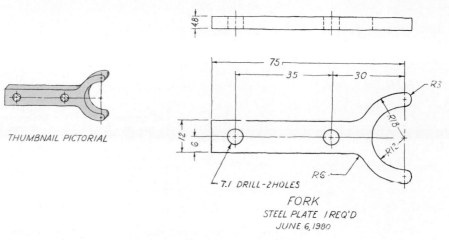

THUMBNAIL PICTORIAL

FORK
STEEL PLATE 1 REQ'D
JUNE 6,1980

FIG. 6.5 Freehand sketch for the manufacture of a part.

medium of sketches. Sketches may be *schematic*, as are those that are original expressions of new ideas (Fig. 6.3), or they may be *instructional*, their purpose being to convey ideas to drafters or technicians. Some sketches, especially those prepared for the manufacture of parts that are to replace worn or broken parts on existing machines, may resemble complete working drawings (Fig. 6.5).

tiview, pictorial, and the other divisions of mechanical drawing. For this reason, one must be thoroughly familiar with projection, in all its many forms, before one is adequately trained to prepare sketches.

B SKETCHING TECHNIQUES

6.4 Projections

Although freehand drawing lacks the refinement given by mechanical instruments, it is based on the same principles of projection and conventional practices that apply to mul-

6.5 Sketching Materials

For the type of sketching discussed here, the required materials are an F pencil, a soft eraser, and some paper. In the industrial field, workers who have been improperly

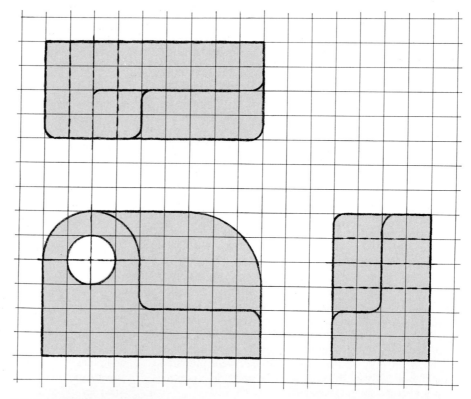

FIG. 6.6 Sketch on grid paper.

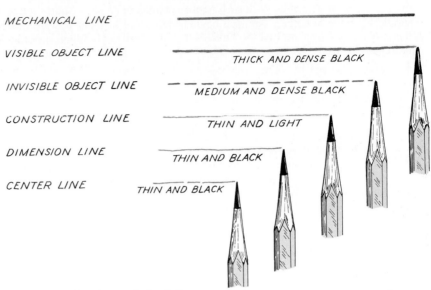

FIG. 6.7 **Pencil points and sketch lines.**

trained in sketching often use straightedges and cheap pocket compasses that they could dispense with if they would adopt the correct technique. Preparing sketches with instruments consumes much unnecessary time.

For the person who cannot produce a satisfactory sketch without guide lines, grid paper is helpful (Fig. 6.6).

6.6 *Technique of Lines*

Freehand lines quite naturally will differ in their appearance from mechanical ones. A well-executed freehand line will never be perfectly straight and absolutely uniform in weight, but an effort should be made to approach *exacting uniformity*. As in the case of mechanical lines, they should be black and clear, not broad and fuzzy (Fig. 6.7).

6.7 *Sharpening the Sketching Pencil*

A sketching pencil should be sharpened to a conical point. The point then should be rounded slightly, on the back of the sketch pad or on another sheet of paper, to the correct

degree of dullness. When rounding the point, rotate the pencil to prevent the formation of sharp edges.

6.8 *Straight Lines*

The pencil should rest on the second finger and be held loosely by the thumb and index finger about 25–40 mm. (1–1½ in.) above the point.

Horizontal lines are sketched from left to right, if you are right-handed, with an easy arm motion that is pivoted about the muscle of the forearm. The straight line thus becomes an arc of infinite radius. When sketching a straight line, it is advisable first to mark the end points with light dots or small crosses (Fig. 6.8).

The complete procedure for sketching a straight line is as follows.

1. Mark the end points.

2. Make a few trial motions between the marked points to adjust the eye and hand to the contemplated line.

3. Sketch a *very* light line between the points by moving the pencil in two or three sweeps. When sketching the trial line, the eye should be on the point toward which

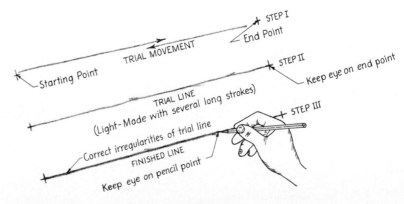

FIG. 6.8 **Steps in sketching a straight line.**

the movement is directed. With each stroke, an attempt should be made to correct the most obvious defects of the stroke preceding, so that the finished trial line will be relatively straight.

4. Darken the finished line, keeping the eye on the pencil point on the trial line. The final line, replacing the trial line, should be distinct, black, uniform, straight, and made in one stroke.

It is helpful to turn the paper to a *convenient angle* so that the horizontal and vertical lines assume a slight inclination (Fig. 6.9). A horizontal line, when the paper is in this position, is sketched to the right and upward, thus allowing the arm to be held slightly away from the body and making possible a free arm motion.

FIG. 6.9 **Sketching horizontal lines.**

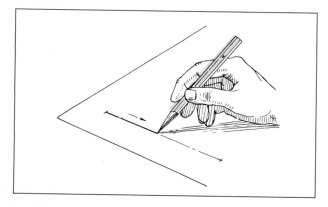

FIG. 6.10 **Sketching vertical lines.**

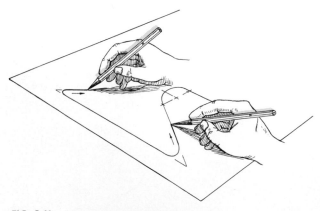

FIG. 6.11 **Sketching inclined lines.**

Short vertical lines may be sketched either downward or upward, without changing the position of the paper. When sketching downward, the arm is held slightly away from the body and the movement is toward the sketcher (Fig. 6.10). To sketch vertical lines upward, the arm is held well away from the body. By turning the paper, a long vertical line may be made to assume the position of a horizontal line and can be sketched with the same general movements used for the latter.

Inclined lines running upward from lower left to upper right may be sketched upward with the same movements used for horizontal lines, but those running downward from upper left to lower right are sketched with the general movements used for either horizontal or vertical lines, depending on their inclination (Fig. 6.11). Inclined lines may be more easily sketched by turning the paper to make them conform to the direction of horizontal lines. A left-handed person will sketch from right to left.

6.9 Circles

Small circles may be sketched by marking radial distances on perpendicular center lines. When additional points are needed, the distances can be marked off either by eye or by measuring with a marked strip of paper (Fig. 6.12). Larger circles may be constructed more accurately by sketching two or more diagonals, in addition to the center lines, and by sketching short construction lines perpendicular to each, equidistant from the center. Tangent to these lines, short arcs are drawn perpendicular to the radii. The circle is completed with a light construction line, and all defects are corrected before darkening (Fig. 6.13).

C MULTIVIEW SKETCHES

6.10 Making a Multiview Sketch (Fig. 6.14)

When making orthographic working sketches a systematic order should be followed, and all the rules and conventional practices used in making working drawings should be applied. The following procedure is recommended.

1. Examine the object, giving particular attention to detail.

2. Determine which views are necessary.

3. Block-in the views, using light construction lines.

4. Complete the detail and darken the object lines.

5. Sketch extension lines and dimension lines, including arrowheads.

6. Complete the sketch by adding dimensions, notes, title, date, sketcher's name or initials, and so on.

7. Check the entire sketch carefully to see that no dimensions have been omitted.

The beginning student should read Part A of Chapter 5 before attempting to make a multiview sketch.

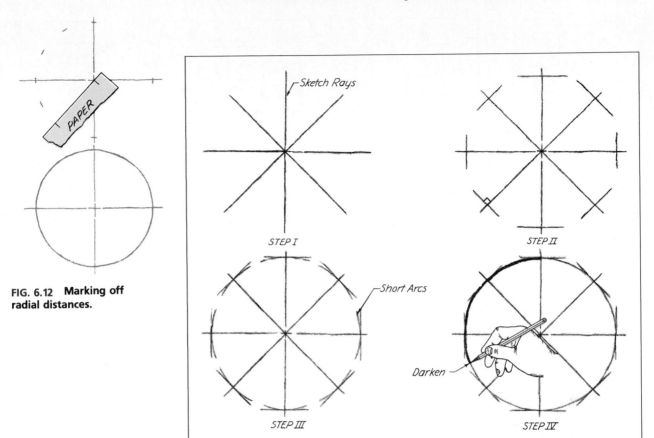

FIG. 6.12 **Marking off radial distances.**

FIG. 6.13 **Sketching large circles.**

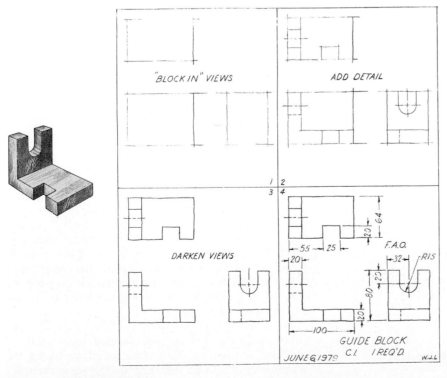

FIG. 6.14 **Steps in sketching.**

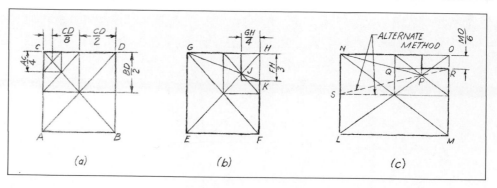

FIG. 6.15 **Methods of proportioning a rectangle representing the outline of a view.**

6.11 Proportions

The beginner must recognize the importance of being able to *estimate comparative relationships between the width, height, and depth of an object being sketched.* The complete problem of proportioning a sketch also involves relating the estimated dimensions for any component parts, such as slots, holes, and projections, to the over-all dimensions of the object. It is not the practice to attempt to estimate actual dimensions, for sketches are not usually made to scale. Rather one must decide, for example, that the width of the object is twice its height, that the width of a given slot is equal to one-half the width of the object, and that its depth is approximately one-fourth the overall height.

To become proficient at sketching one must learn to recognize proportions and be able to compare dimensions by eye. Until one is able to do this, one cannot really "think with a pencil." Some people can develop a keen eye for proportion with only a limited amount of practice and can maintain these estimated proportions when making the views of a sketch. Others have alternately discouraging and encouraging experiences. Discouragement comes when one's knowledge of sketching is ahead of ability, not having as much practice as needed. The many who find it difficult to make the proportions of the completed sketch agree with the estimated proportions of the object may begin by using the graphical method shown in Figs. 6.15(a), (b), and (c). This method is based on the fact that a rectangle (enclosing a view) may be divided to obtain intermediate distances along any side that are in proportion to the total length, such as one-half, one-fourth, one-third, and so on. Those who start with this rectangle method as an aid in proportioning should abandon its use when they have developed their eye and sketching skills so that it is no longer needed.

Sketching must be done rapidly, and the addition of unnecessary lines consumes much valuable time. Furthermore, the addition of construction lines distracts the reader, and it is certain that they do not contribute to the neatness of the sketch.

This midpoint of a rectangle is the point of intersection of the diagonals, as shown in Fig. 6.15(a). A line sketched through this point that is perpendicular to any side will establish the midpoint of that side. Should it be necessary to determine a distance that is equal to one-fourth the length of a side, (say, *AC*), the quarter-point may be located by repeating this procedure for the small rectangle representing the upper left-hand quarter of *ABCD*.

With the midpoint *J* located by the intersecting diagonals of the small rectangle [representing one-fourth of the larger rectangle *EFGH*, as in (b)], the one-third point along *FH* may be located by sketching a line from point *G* through *J* and extending it to line *FH*. The point *K* at the intersection of these lines establishes the needed one-third distance.

To determine one-sixth the length of a side of a rectangle, as in (c), sketch a line from *N* through point *P*, as was done in (b), to determine a one-third distance. Point *Q* at which the line *NP* crosses the center line of the rectangle establishes the one-sixth distance along the center line.

Figure 6.16 shows how this method for dividing the sides of a rectangle might be used to proportion an orthographic sketch.

The square may be used to proportion a view after one dimension for the view has been assumed. In this method, additional squares are added to the initial one having the assumed length as one side (Fig. 6.17). As an example,

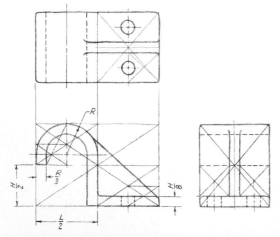

FIG. 6.16 **Rectangle method applied in making an orthographic sketch.**

suppose that it has been estimated that the front view of an object should be three times as wide as it is high. In Fig. 6.17 the height of the view has been represented by the line AB sketched to an assumed length. The first step in making the construction is to sketch the initial square $ABCD$ and extend AC and BD to indefinite length, being certain that the overall length from A and B will be slightly greater than three times the length of AB. Then the center line must be sketched through the intersection of AD and BC. Now BX extended to E locates EF to form the second square, and DY extended to point G locates the line GH. Line AG will be three times the length of AB.

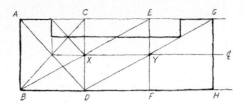

FIG. 6.17   Build-up method.

persons lack only the necessary confidence to start making pictorial sketches.

D / SKETCHING IN ISOMETRIC, OBLIQUE, AND PERSPECTIVE

6.12 *Pictorial Sketching*

Students may employ pictorial sketches to advantage as an aid in visualizing and organizing problems. Sales engineers may frequently include pictorial sketches with orthographic sketches when preparing field reports on the needs and suggestions of the firm's customers.

With some training anyone can prepare pictorial sketches that will be satisfactory for most purposes. *Artistic ability is not needed.* This fact is important, for many

6.13 *Mechanical Methods of Sketching*

Many engineers have found that they can produce satisfactory pictorial sketches by using one of the so-called mechanical methods. They rely on these methods because of their familiarity with the procedures used in making pictorial drawings with instruments.

The practices presented in Chapter 11 for the mechanical methods, axonometric, oblique, and perspective, are followed generally in pictorial sketching, except that angles are assumed and lengths are estimated. For this reason, an eye for good proportion must be developed before one can create a satisfactory pictorial sketch that will be in no way misleading.

A student having difficulty in interpreting a multiview drawing usually will find that a pictorial sketch, prepared as illustrated in Fig. 6.18, will clarify the form even before the last lines of the sketch have been drawn.

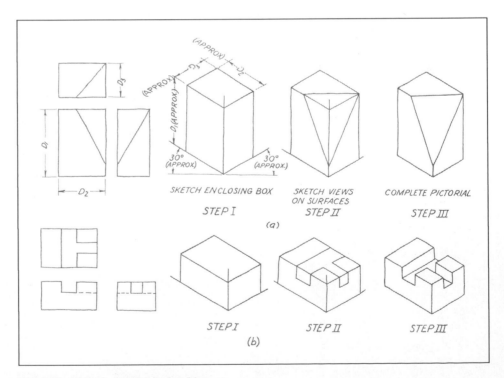

FIG. 6.18   Steps in isometric sketching.

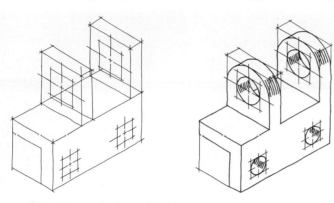

FIG. 6.19 Blocking in an isometric sketch.

6.14 Isometric Sketching

Isometric sketching starts with three *isometric lines*, called axes, which represent three mutually perpendicular lines. One of these axes is sketched vertically, the other two at 30° with the horizontal. In Fig. 6.18 (step I), the near front corner of the enclosing box lies along the vertical axis, while the two visible receding edges of the base lie along the axes receding to the left and to the right.

If the object is of simple rectangular form, as in Fig. 6.18, it may be sketched by drawing an enclosing isometric box (step I) on the surfaces of which the orthographic views may be sketched (step II). Care must be taken in assuming lengths and distances so that the finished view (step III) will have relatively correct proportions. In constructing the enclosing box (step I), the vertical edges are parallel to the vertical axis, and edges receding to the right and to the left are parallel to the right and left axes, respectively.

Objects of more complicated construction may be "blocked in," as shown in Fig. 6.19. Note that the projecting cylindrical features are enclosed in isometric prisms and that the circles are sketched within isometric squares. The procedure in Fig. 6.19 is the same as in Fig. 6.18, except that three enclosing isometric boxes are needed in the formation of the final representation instead of one.

In sketching an ellipse to represent a circle pictorially, an enclosing isometric square or rhombus is drawn having sides approximately equal to the diameter of the true circle (step I, Fig. 6.20). The ellipse is formed by first drawing arcs tangent to the midpoints of the sides of the isometric square in light pencil lines (step II). In finishing the ellipse (step III) with a dark heavy line, care must be taken to obtain a nearly elliptical shape.

Figure 6.21 shows the three positions for an isometric circle. Note that the major axis is horizontal for an ellipse on a horizontal plane (I).

An idea sketch prepared in isometric is shown in Fig. 6.22.

6.15 Proportioning

As stated in Sec. 6.11, one should eventually be able to judge lengths and recognize proportions. Until this ultimate goal has been reached, the graphical method presented in Fig. 6.15 may be used with pictorial sketching (Fig. 6.23). The procedures for each face are identical; the only recognizable difference being that the rectangle in the first case now becomes a rhomboid. Figure 6.24 illustrates how the method might be applied in making a sketch of a simple object. The enclosing box was sketched

SKETCH "ISOMETRIC SQUARE"
STEP I

SKETCH SHORT ARCS
STEP II

COMPLETE ELLIPSE
STEP III

FIG. 6.20 Isometric circles.

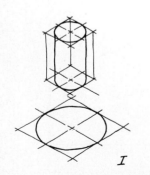
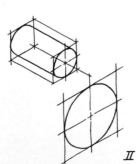
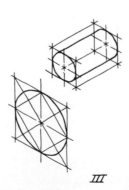

I   II   III

FIG. 6.21 Isometric circles.

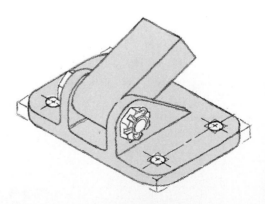

FIG. 6.22 Idea sketch in isometric.

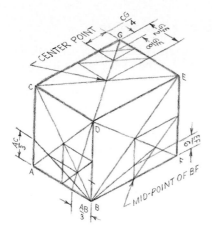

FIG. 6.23 Method for proportioning a rhomboid.

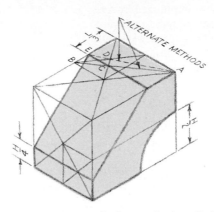

FIG. 6.24 Proportioning method applied.

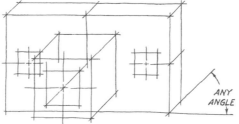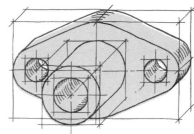

FIG. 6.25 Blocking in an oblique sketch.

first with light lines, and then the graphical method was applied as shown to locate the points at one-quarter and one-half of the height. To establish the line of the top surface that is at a distance equal to one-third of the length from the end, a construction line was sketched from *A* to the midpoint *B* to locate *C* at the point of intersection of *AB* with the diagonal. Point *C* will fall on the required line.

6.16 *Sketches in Oblique*

A sketch in oblique shows the front face without distortion, in its true shape. It has this one advantage over a representation prepared in isometric, even though the final result usually will not present so pleasing an appearance. *It is not recommended for objects having circular or irregularly curved features on any but the front plane or in a plane parallel to it.*

The beginner who is familiar with isometric sketching will have very little difficulty in preparing a sketch in oblique, for, in general, the methods of preparation presented in the previous sections apply to both. The principal difference between these two forms of sketching is in the position of the axes. Oblique sketching is unlike isometric in that two of the axes are at right angles to each other. The third axis may be at any convenient angle, as indicated in Fig. 6.25.

Figure 6.26 shows the steps in making an oblique sketch using the proportioning methods previously explained for dividing a rectangle and a rhomboid. The receding lines are made parallel when a sketch is made in oblique projection.

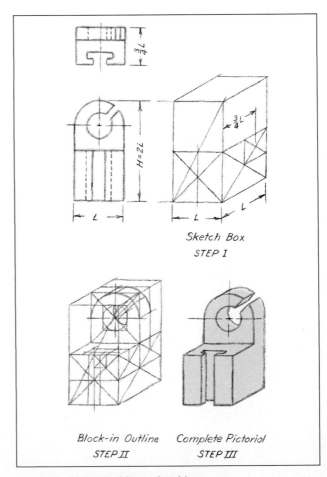

FIG. 6.26 Steps in oblique sketching.

The distortion and illusion of extreme elongation in the direction of the receding axis may be minimized by fore-shortening to obtain proportions that are more realistic to the eye and by making the receding lines converge slightly. The resulting sketch will then be in a form of pseudo-perspective, which resembles parallel perspective to some extent.

FIG. 6.27 **Sketch in parallel perspective.**

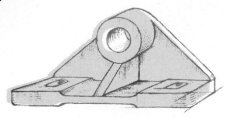

FIG. 6.28 **Sketch in angular perspective.**

6.17 *Perspective Sketching*

A sketch that has been prepared in accordance with the concepts of perspective will present a somewhat more pleasing and realistic effect than one in oblique or iso-metric. A perspective sketch actually presents an object as it would appear naturally. The recognition of this fact, along with an understanding of the concepts that an object will appear smaller at a distance than when it is close and that horizontal lines converge as they recede until they meet at a vanishing point, should enable one to produce sketches having a perspective appearance. In sketching an actual object, a position should be selected that will show it to the best advantage. When the object exists only in one's mind or on paper in orthographic form, then the object must be visualized and the viewing position as-sumed.

At the start, the principal lines should be sketched in lightly, each line extending for some length toward its vanishing point. After this has been accomplished, the enclosing perspective squares for circles should be blocked in and the outline for minor details added. When the object lines have been darkened, the construction lines extending beyond the figure may be erased.

Figure 6.27 shows a *parallel* or *one-point perspective* that bears some resemblance to an oblique sketch. All faces in planes parallel to the front show their true shape. All receding lines should meet at a single vanishing point. Figure 6.28 is an angular or two-point perspective.

As stated previously, an *inclined* or *two-point perspec-tive* sketch shows an object as it would appear to the human eye at a fixed point in space and not as it actually exists.

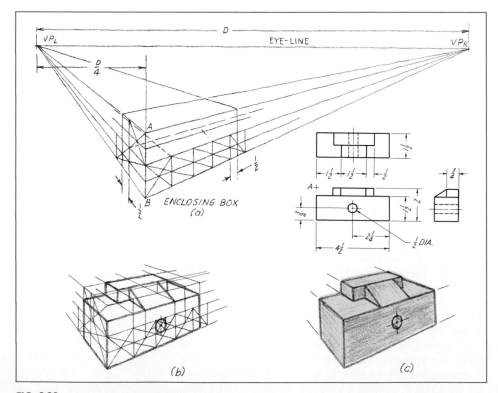

FIG. 6.29 **Preparing a perspective sketch.**

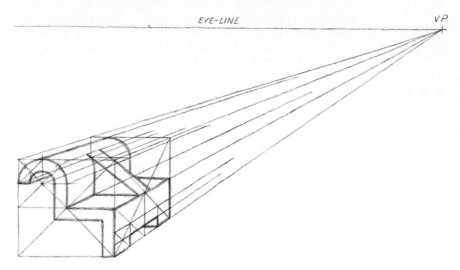

FIG. 6.30 **Sketch made in one-point perspective.**

All parallel receding lines converge (Fig. 6.29). Should these receding lines be horizontal, they will converge at a vanishing point on the eyeline. Those lines extending toward the right converge to a vanishing point to the right (VP_R), and those to the left converge to the left (VP_L). These vanishing points are at the level of the observer's eye (Fig. 6.29). A system of lines that is neither perpendicular nor horizontal will converge to a *VP* for inclined lines.

In one-point perspective one of the principal faces is parallel to the picture plane. All of the vertical lines will appear as vertical, and the receding horizontal lines will converge to a single vanishing point (Fig. 6.30).

Those interested in a complete discussion of the geometry of perspective drawing should read Secs. 11.27–11.40. The beginner should make two or three mechanically drawn perspectives at the start to fix the fundamentals of the methods of perspective projection in his mind, even though there is some difference between sketching what one sees or imagines and true geometric perspective.

In making a sketch in perspective, several fundamental concepts must be recognized.

1. A circle sketched in perspective will resemble an ellipse (Fig. 6.29). The long diameter of the representation of a circle on a horizontal plane is always in a horizontal direction.

2. If an object or a component part of an object is above the eyeline, it will be seen from below. Should the object be below the eyeline, it will be seen from above (Fig. 6.29). The farther the object is removed below or above the eyeline, the more one can see of the top or the bottom surface.

3. The nearest vertical edge of an object will be the longest vertical line of the view, as shown in Fig. 6.29. When two or more objects of the same actual height appear in a perspective sketch, their represented heights will decrease in the view as they near the vanishing point.

6.18 *Preparing a Perspective Sketch*

The application of proportioning methods to the construction and division of an enclosing box is shown in Fig. 6.29. Read Secs. 6.11 and 6.15. The construction of a required perspective by steps is as follows.

STEP 1. Sketch the eyeline. This line should be well toward the top of the sheet of sketch paper.

STEP 2. Locate VP_L and VP_R on the eyeline. These vanishing points should be placed as far apart as possible.

STEP 3. Assume the position and height for the near front edge *AB*. The height of this line, along with the spacing of the vanishing points, establishes the size of the finished sketch. The position of *AB* determines how the visible surfaces are to appear. For instance, if the line *AB* had been moved downward from the position shown in (*a*), much more of the top surface would be seen in (*b*) and (*c*). Should *AB* have been moved to the right from its position shown in (*a*), the left side would have become more prominent. If *AB* were placed midway between the two vanishing points, then both the front and left-side surfaces would be at 45° with the picture plane for the perspective. As it has been placed in (*a*), the side face is at 60° to the picture plane, while the front face is at 30°. Before establishing the position for the near front edge, one must decide which surfaces are the most important surfaces and how they may best be displayed in the sketch.

STEP 4. Sketch light construction lines from points *A* and *B* to each vanishing point.

STEP 5. Determine the proportions for the enclosing box, in this case 4½, 2, and 1½, and mark off two equal units along *AB*.

STEP 6. Sketch perspective squares, representing the faces of 1-unit cubes, starting at *AB* and working toward each vanishing point. In cases where an overall length must be completed with a partial unit, a full-perspective square must be sketched at the end.

STEP 7. Subdivide any of the end squares if necessary and sketch in the enclosing box (*a*).

STEP 8. Locate and block in the details, subdividing the perspective squares as required to establish the location of any detail. When circles are to be sketched in perspective by a beginner, it is advisable to sketch the enclosing box first using light lines (*b*).

STEP 9. Darken the object lines of the sketch. Construction lines may be removed and some shading added to the surfaces as shown in (*c*), if desired.

A sketch in one-point perspective might be made as shown in Fig. 6.30. For this particular sketch the enclosing box was made to the assumed overall proportions for the part. Then the location of the details was established by subdividing the regular rectangle of the front face of the enclosing box and the perspective rectangle of the right side.

6.19 Pencil Shading

The addition of some shading to the surfaces of a part will force its form to stand out against the white surface of the sketching paper and will increase the effect of depth in a view that might otherwise appear to be somewhat flat.

Technologists and drafters should be able to do creditable work in artistic shading, with shadows included. Within the scope of this chapter, written for beginning students, it will only be possible to present a few simple rules as a guide for those making a first attempt at surface shading. However, continued practice and some thought should lead one to the point where a creditable job of shading can be done and a pictorial sketch improved.

When shading, a designer may consider the source of the light to be located in a position to the left, above, and in front of the object. Of course, if the part actually exists and is being sketched by viewing it, then the sketcher should attempt to duplicate the degrees of shade and shadows as they are observed.

With the light source considered to be to the left, above, and in front of the object, a rectangular part would be shaded as shown in Fig. 6.31(*a*). The use of gradation of tone on the surfaces gives additional emphasis to the depth. To secure this added effect by shading, the darkest tone on the surface that is away from the light must be closest to the eye. As the surface recedes, the tone must be made lighter in value with a faint trace of reflected light showing along the entire length of the back edge. On the lighted side, the lightest area must be closest to the eye, as indicated by the letter L_1 in (*a*). To make this lighted face appear to go into the distance, it is made darker as it recedes, but it should never be made as dark as the lightest of the dark tones on the dark surface.

Shading a cylindrical part is not as easy to do as shading a rectangular part but, if it is realized that practically half of the cylinder is in the light and half in the dark and that the lightest light and the darkest dark fall along the elements at the quarter-points of each half, then one should not find the task too difficult (*b*). The two extremes are separated by lighter values of shade. The first quarter on the lighted side must be made lighter than the last quarter on the dark side. In starting at the left and going counterclockwise there is a dark shade of light blending into the full light at the first quarter-point. From this point and passing the center to the dark line, the tone should become gradually darker. If vertical lines are used for shading, they should be spaced closer and closer together as they approach the dark line. The extreme-right-hand quarter should show the tones of reflected light.

There are two ways that pencil shading can be applied. If the paper has a medium-rough surface, solid tone shading may be used with one shade blending into the other. For the best results, the light tones are put on first over all areas to be shaded. The darker tones are then added by building up lighter tones to the desired intensity for a particular area. For this form of shading, a pencil with flattened point is used.

The other form of shading, and the one that is best suited for quick sketches, is produced with lines of varied spacing and weight. Light lines with wide spacing are used on the light areas and heavy lines that are closely spaced give the tone for the darkest areas. No lines are needed for the lightest of the light areas.

6.20 Conventional Treatment of Fillets, Rounds, and Screw Threads

Sketches that are not given full pencil shading may be given a more or less realistic appearance by representing the fillets and rounds of the unfinished surfaces as shown

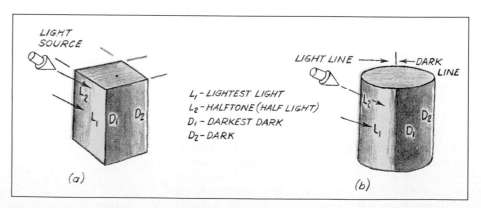

FIG. 6.31 **Shading rectangular and cylindrical parts.**

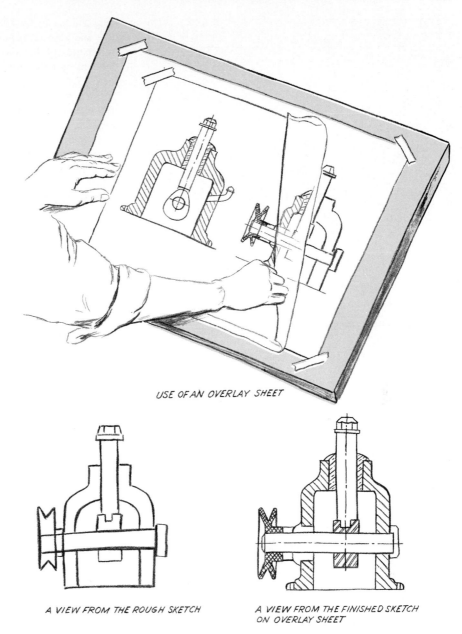

USE OF AN OVERLAY SHEET

A VIEW FROM THE ROUGH SKETCH   A VIEW FROM THE FINISHED SKETCH
                                    ON OVERLAY SHEET

FIG. 6.32 Use of an overlay sheet for creating a final and complete sketch of a mechanism.

in Fig. 11.41. The conventional treatment for screw threads is shown in (*b*) and (*c*) of the same illustration.

6.21 *Use of an Overlay Sheet*

In preparing a design sketch, an *overlay sheet* may be used to advantage in making a sketch that is complicated by many details (Fig. 6.32). In this case, a quick sketch showing the general outline of the principal parts is made first in a rather rough form. Then an overlay sheet is placed over this outline sketch and the lines are retraced. In doing so, slight corrections can be made for any errors existing in the proportions of the parts or in the position of any of the lines of the original rough sketch. When this has been

done, the representation of the related minor parts are added. If at any time one becomes discouraged with a sketch (multiview or pictorial) and wants to make a new start, an overlay sheet should be used for there are usually many features on the existing sketch that may be retraced with a great saving of time.

6.22 *Illustration Sketches Showing Mechanisms Exploded*

A sketch of a mechanism showing the parts in exploded positions along the principal axes is shown in Fig. 6.33. Through the use of such sketches, those who have not

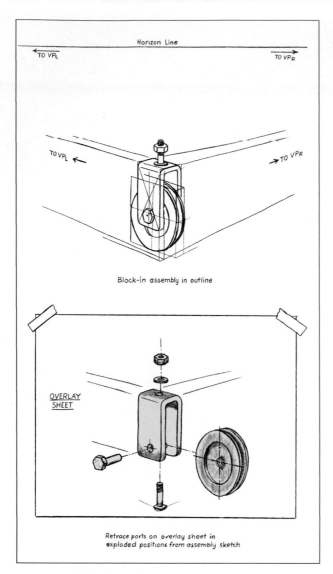

FIG. 6.33 Sketch showing the parts of a mechanism in exploded positions.

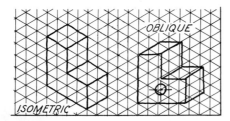

FIG. 6.34 Sketches on isometric paper.

been trained to read multiview drawings can readily understand how a mechanism should be assembled, for both the shapes of the parts and their order of assembly, as denoted by their space relationship, is shown in pictorial form.

Illustration sketches may be made for discussions dealing with ideas for a design, but more frequently they are prepared for explanatory purposes to clarify instructions for preparing illustration drawings in a more finished form as assembly illustrations, advertising illustrations, catalogue illustrations, and illustrations for service and repair charts.

Many persons find that it is desirable, when preparing sketches of exploded mechanisms, to first block in the complete mechanism with all of the parts in position. At this initial stage of construction, the parts are sketched in perspective in rough outline and the principal axes and object lines are extended partway toward their vanishing points.

When the rough layout has been completed to the satisfaction of the person preparing the sketch, an overlay sheet is placed over the original sketch and the parts are traced directly from the sketch beneath in exploded positions along the principal axes and along the axes of holes. Frequently some beginners make a traced sketch of each individual part and place the sketches in exploded positions before preparing the finished sketch. Others with more experience accomplish the same results by first tracing the major part along with the principal axes and then moving the overlay sheet as required to trace off the remaining parts in their correct positions along the axes.

It should be recognized that in preparing sketches of exploded mechanisms in this manner, the parts are not shown perspectively reduced although they are removed from their original places in the pictorial assembly, outward and toward the vanishing points. To prepare a sketch of this type with all of the parts shown in true geometrical perspective would result in a general picture that would be misleading and one that would be apt to confuse the nontechnical person because some parts would appear much too large or too small to be mating parts.

6.23 Pictorial Sketching on Ruled Paper

Although one should become proficient in sketching on plain white bond paper, a specially ruled paper, shown in Fig. 6.34, can be used by those who need the help of guidelines.

PROBLEMS

The problems presented with this chapter have been selected to furnish practice in freehand drawing. The individual pieces that appear in pictorial form have been taken from a wide variety of mechanisms used in different fields of engineering. The student may be required to prepare complete working sketches of these parts if his instructor desires. Such an assignment, however, would presuppose an understanding of the fundamentals of dimensioning as they are presented in the beginning sections of Chapter 13.

The problems presented in Figs. 6.51–6.58 were selected to give practice in preparing pictorial sketches—isometric, oblique, and perspective. In addition to developing proficiency in sketching, these problems offer the student further opportunity to gain experience in reading drawings. Additional problems that are suitable for pictorial sketching may be found in Chapter 11.

1. (Fig. 6.35). Reproduce the one-view sketch on a sheet of sketching paper.

2–4. (Figs. 6.36–6.38). Sketch, freehand, the necessary views of the given objects as assigned. The selected length for the unit

will determine the size of the views. Assume any needed dimensions that are not given in units.

5–14. (Figs. 6.39–6.48). These problems are designed to give the student further study in multiview representation and, at the same time, offer him the opportunity to apply good line technique to the preparation of sketches.

Only the necessary views on which all of the hidden lines are to be shown should be drawn.

If dimensions are to be given, ample space must be allowed between the views for their placement. The beginning sections of Chapter 13 present the basic principles of size description.

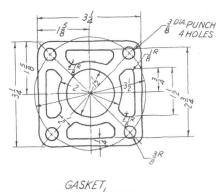

GASKET,
PUNCH FROM 1/16 CORK

FIG. 6.35

FIG. 6.36 **Wedge block.**

FIG. 6.37 **End block.**

FIG. 6.38 **Corner block.**

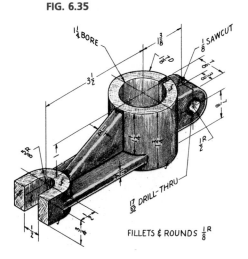

FILLETS & ROUNDS 1/8 R

FIG. 6.39 **Shifter.**

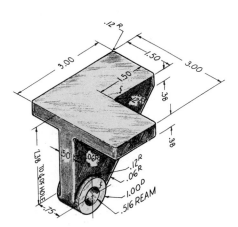

FIG. 6.40 **Tool rest.**

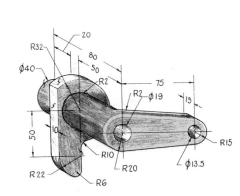

FIG. 6.41 **Offset trip**

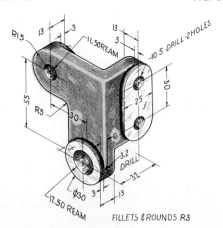

FILLETS & ROUNDS R3

FIG. 6.42 **Guide link.**

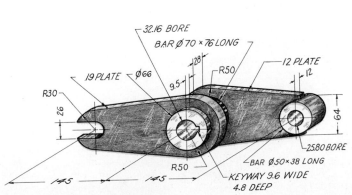

FIG. 6.43 **Idler lever (weldment).**

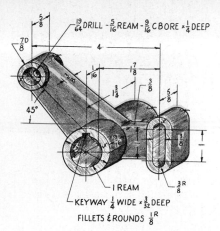

FIG. 6.44 **Control bracket.**

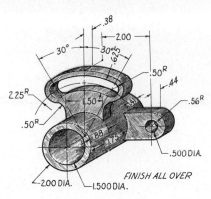

FIG. 6.45 **Index guide.**

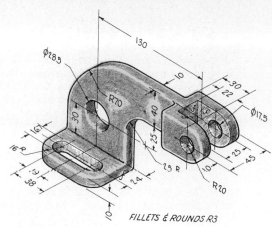

FIG. 6.46 **Arm bracket.**

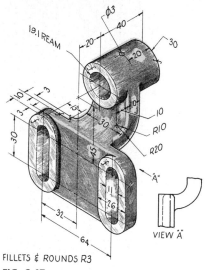

FIG. 6.47 **Bearing bracket.**

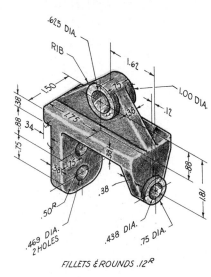

FIG. 6.48 **Rear support bracket.**

15. (Fig. 6.49). Make a complete three-view sketch of the motor base. The ribs are 10 mm. thick. At points *A*, four holes are to be drilled for 12 mm bolts that are to be 64 mm. center to center in one direction and 80 mm. in the other. At points *B* four holes are to be drilled for 12 mm. bolts that fasten the motor base to a steel column. Fillets and rounds are *R3*.

16. (Fig. 6.50). Make a three-view orthographic sketch of the motor bracket.

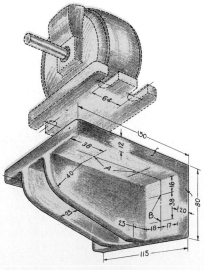

FIG. 6.49 **Motor base.**

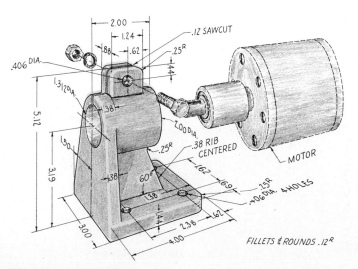

FIG. 6.50 **Motor bracket.**

17–22. (Figs. 6.51–6.56). Make freehand isometric sketches of the objects as assigned.

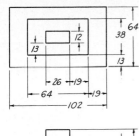
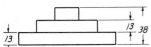

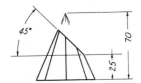

FIG. 6.51

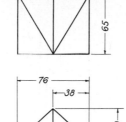
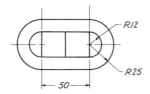

FIG. 6.52

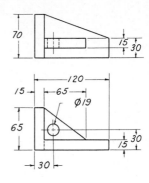
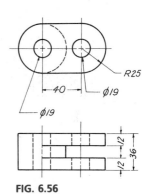

FIG. 6.53

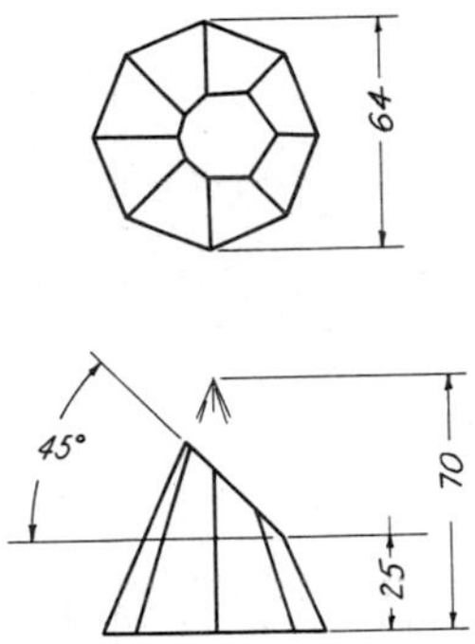

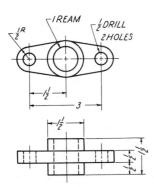

FIG. 6.54

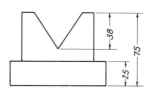
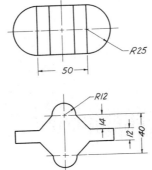

FIG. 6.55

FIG. 6.56

23–26. (Figs. 6.57–6.58). Make freehand oblique sketches of the objects as assigned.

FIG. 6.57

FIG. 6.58

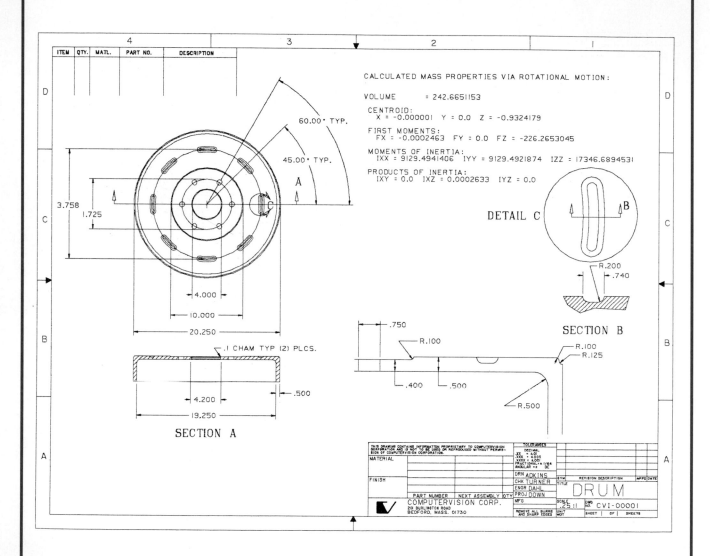

DETAIL C

SECTION B

60.00° TYP.

45.00° TYP.

3.758

1.725

4.000

10.000

20.250

.I CHAM TYP (2) PLCS.

4.200

.500

19.250

SECTION A

.750

R.100

.400

.500

R.500

R.100

R.125

R.200

.740

The drawing shown was generated and drawn by an interactive computer graphics system. Section A and Section B emphasize and clarify design details. The full section view was obtained by imagining the drum to have been cut by an imaginary plane passing entirely through the part as indicated by the cutting plane line labeled A. See circular view. (*Courtesy Computervision, Inc.*)

Sectional Views

7.1 Sectional Views (Fig. 7.1)

Although the interior features of a simple object usually may be described on an exterior view by the use of hidden lines, it is unwise to depend on a perplexing mass of such lines to describe the interior of a complicated object or an assembled mechanism. Whenever a representation becomes so confused that it is difficult to read, it is customary to make one or more of the views "in section" (Fig. 7.2). A view "in section" is one obtained by imagining the object to have been cut by a cutting plane, the front portion being removed to reveal clearly the interior features. Figure 7.3 illustrates the use of an imaginary cutting plane. At this point it should be understood that a portion is shown

removed only in a sectional view, not in any of the other views (Fig. 7.3).

When the cutting plane cuts an object lengthwise, the section obtained is commonly called a *longitudinal section*; when crosswise, it is called a *cross section*. It is designated as being either a *full section*, a *half section*, or a *broken section*. If the plane cuts entirely across the object, the section represented is known as a full section. If it cuts only halfway across a symmetrical object, the section is a half section. A broken section is a partial one, used when less than a half section is needed (Fig. 7.3).

On a completed sectional view, fine *section lines* are drawn across the surface cut by the imaginary plane, to emphasize the contour of the interior (see Sec. 7.8).

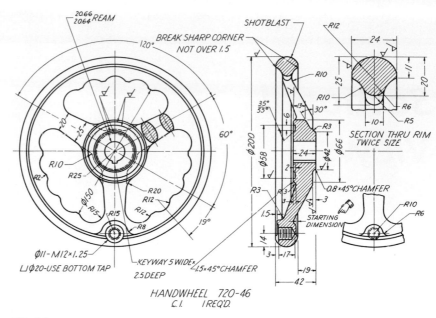

FIG. 7.1 **Working drawing with sectional views.** (*Courtesy Warner and Swasey Co.*)

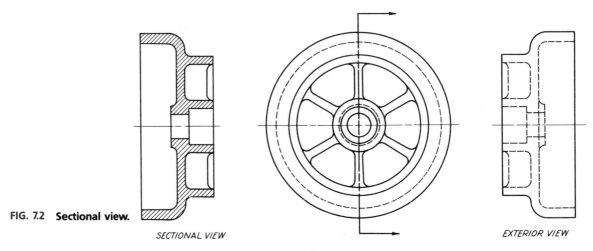

FIG. 7.2 **Sectional view.**

SECTIONAL VIEW

EXTERIOR VIEW

7.2 Full Section

Since a cutting plane that cuts a *full section* passes entirely through an object, the resulting view will appear as illustrated in Fig. 7.3. Although the plane usually passes along the main axis, it may be *offset* (Fig. 7.4) to reveal important features.

A full-sectional view, showing an object's characteristic shape, usually replaces an exterior front view; however, one of the other principal views, side or top, may be converted to a sectional view if some interior feature thus can be shown to better advantage or if such a view is needed in addition to a sectioned front view.

The procedure in making a full-sectional view is simple, in that the sectional view is an orthographic one. The imaginary cut face of the object simply is shown as it would appear to an observer looking directly at it from a point an infinite distance away. In any sectional view, *it is considered good practice to omit all invisible lines unless such lines are necessary to clarify the representation*. Even then they should be used sparingly.

7.3 Half Section

The cutting plane for a *half section* removes one-quarter of an object. The plane cuts halfway through to the axis or center line so that *half the finished sectional view appears in section and half appears as an external view* (Fig. 7.3). This type of sectional view is used when a view is needed showing both the exterior and interior construction of a symmetrical object. Good practice dictates that hidden lines be omitted from both halves of the view unless they are absolutely necessary for dimensioning purposes or for explaining the construction. Although the use of a solid object line to separate the two halves of a half section has been approved by the Society of Automotive Engineers and has been accepted by the American National Standards Institute [Fig. 7.5(*a*)], many draftsmen prefer to use a center line, as shown in Fig. 7.5(*b*). They reason that the removal of a quarter of the object is theoretical and imaginary and that an actual edge, which would be implied by a solid line, does not exist. The center line is taken as denoting a theoretical edge.

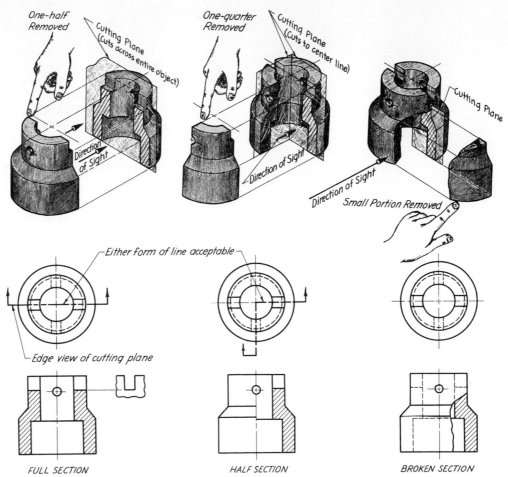

One-half Removed

Cutting Plane (Cuts across entire object)

Direction of Sight

One-quarter Removed

Cutting Plane (Cuts to center line)

Direction of Sight

Cutting Plane

Direction of Sight

Small Portion Removed

Either form of line acceptable

Edge view of cutting plane

FULL SECTION

HALF SECTION

BROKEN SECTION

FIG. 7.3 **Types of sectional views.**

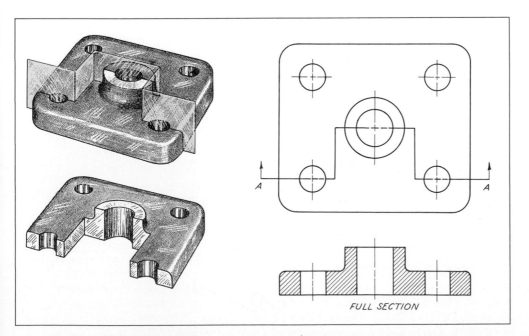

A   A

FULL SECTION

FIG. 7.4 **Offset cutting plane.**

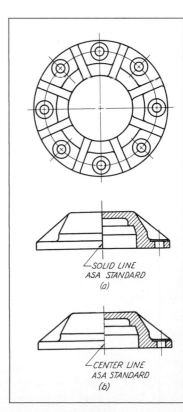

SOLID LINE
ASA STANDARD
(a)

CENTER LINE
ASA STANDARD
(b)

FIG. 7.5 **Half section.**

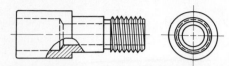

FIG. 7.6   Broken section.

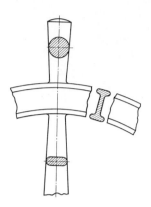

**FIG. 7.7   Revolved
section.***

7.4 Broken Section

A *broken section* is used mainly to expose the interior of objects so constructed that less than a half section is required for a satisfactory description (Fig. 7.6). The object theoretically is cut by a cutting plane and the front portion is removed by breaking it away. The "breaking away" gives an irregular boundary line to the section.

7.5 Revolved Section

A revolved section is useful for showing the true shape of the *cross section* of an object like a bar or some feature of an object, such as an arm, spoke, or rib (Figs. 7.1 and 7.7).

To obtain such a cross section, an imaginary cutting plane is passed through the member perpendicular to the longitudinal axis and then is revolved through 90° to bring the resulting view into the plane of the paper (Fig. 7.8). When revolved, the section should show in its true shape and in its true revolved position, regardless of the location

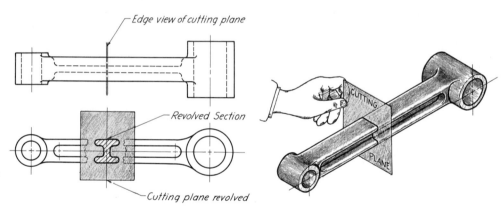

FIG. 7.8   Revolved section and cutting plane.

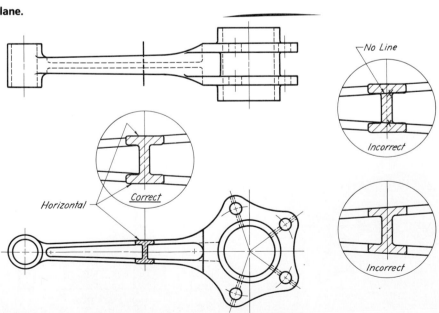

FIG. 7.9   Correct and incorrect treatment of a revolved section.

* ANSI Y14.2

of the lines of the exterior view. If any lines of the view interfere with the revolved section, they should be omitted (Fig. 7.9). It is sometimes advisable to provide an open space for the section by making a break in the object (Fig. 7.7).

7.6 *Removed (Detail) Sections*

A *removed section* is similar to a revolved section, except that it does not appear on an external view but instead is drawn "out of place" and appears adjacent to it (Fig. 7.10). There are two good reasons why detail sections

frequently are desirable. First, their use may prevent a principal view of an object, the cross section of which is not uniform, from being cluttered with numerous revolved sections (Fig. 7.11). Second, they may be drawn to an enlarged scale in order to emphasize detail and allow for adequate dimensioning.

Whenever a detail section is used, there must be some means of identifying it. Usually this is accomplished by showing the cutting plane on the principal view and then labeling both the plane and the resulting view, as shown in Fig. 7.11.

7.7 *Phantom Sections*

A *phantom section* is a regular exterior view on which the interior construction is emphasized by crosshatching an imaginary cut surface with dashed section lines (Fig. 7.12). This type of section is used only when a regular section or a broken section would remove some important exterior detail or, in some instances, to show an accompanying part in its relative position with regard to a particular part (Fig. 7.13). Instead of using a broken line with dashes of equal length, the phantom line shown in Fig. 2.59 could have been used to represent the outline of the adjacent parts shown in Fig. 7.13.

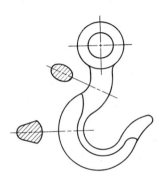

FIG. 7.10 Removed sections.*

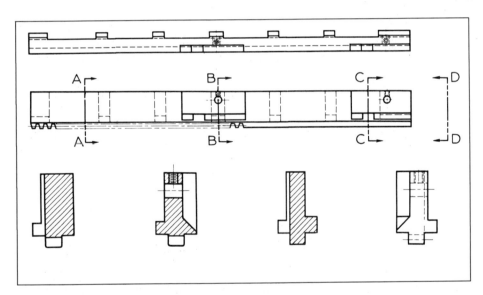

FIG. 7.11 Removed (detail) sections.*

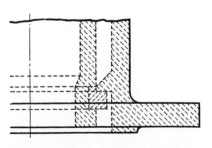

FIG. 7.12 Phantom section.

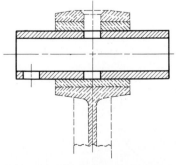

FIG. 7.13 Phantom sectioning—adjacent parts.

* ANSI Y14.2

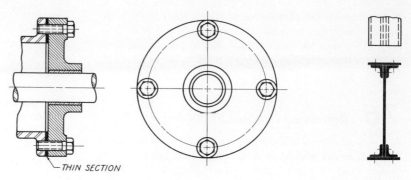

FIG. 7.14   Thin sections.

7.8 *Section Lining*

Section lines are light continuous lines drawn across the imaginary cut surface of an object for the purpose of emphasizing the contour of its interior. Usually they are drawn at an angle of 45° except in cases where a number of adjacent parts are shown assembled (Fig. 7.17).

To be pleasing in appearance, these lines must be correctly executed. While on ordinary work they are spaced about 2 mm (.09 in.) apart, there is no set rule governing their spacing. They simply should be spaced to suit the drawing and the size of the areas to crosshatched. For example, on small views having small areas, the section lines may be as close as 0.8 mm (.03 in.), while on large views having large areas they may be as far apart as 3 mm (.12 in.). In the case of very thin plates, the cross section is shown solid black (Fig. 7.14).

The usual mistake of the beginning student is to draw the lines too close together. This, plus the unavoidable slight variations, causes the section lining to appear streaked. Several forms of mechanical section liners are available. The student must be careful to see that the initial pitch as set by the first few lines is maintained across the area. To accomplish this, check back from time to time to make sure there has been no slight general increase or decrease in the spacing. An example of correct section lining is shown in Fig. 7.15(*a*), and, for comparison, examples of faulty practice may be seen in Fig. 7.15(*b*), (*c*), and (*d*). Experienced draftsmen realize that nothing will do more to ruin the appearance of a drawing than carelessly executed section lines.

As shown in Fig. 7.16, the section lines on two adjacent pieces should slope at 45° in opposite directions. If a third piece adjoins the two other pieces, as in Fig. 7.17(*a*), it ordinarily is section-lined at 30°. An alternative treatment that might be used would be to vary the spacing without changing the angle. On a sectional view showing an assembly of related parts, *all portions of the cut surface of any part must be section-lined in the same direction, for a change would lead the reader to consider the portions*

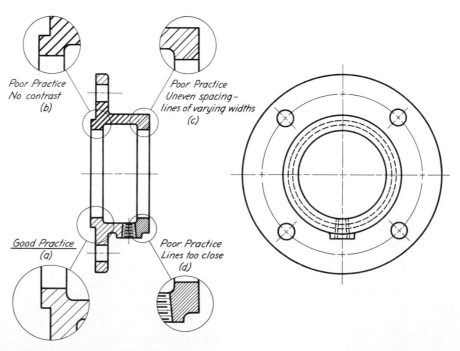

FIG. 7.15   Faults in section lining.

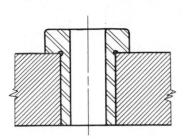

FIG. 7.16   Two adjacent pieces.

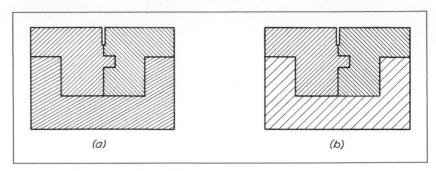

FIG. 7.17   Three adjacent pieces.

as belonging to different parts. Furthermore, to allow quick identification, each piece (and all identical pieces) in every view of the assembly drawing should be section-lined in the same direction.

Shafts, bolts, rivets, balls, and the like, whose axes lie in the plane of section, are not treated the same as ordinary parts. Having no interior detail to be shown, they are drawn in full and thus tend to make the adjacent sectioned parts stand out to better advantage (Fig. 7.18).

Whenever section lines drawn at 45° with the horizontal are parallel to part of the outline of the section (see Fig. 7.19), it is advisable to draw them at some other angle (say, 30° or 60°). Those drawn as in (*a*) and (*c*) produce an unusual appearance that is contrary to what is expected. Note the more natural effect obtained in (*b*) and (*d*) by sloping the lines at 30° and 75°.

7.9   *Outline Sectioning*

Very large surfaces may be section-lined around the bounding outline only, as illustrated in Fig. 7.20.

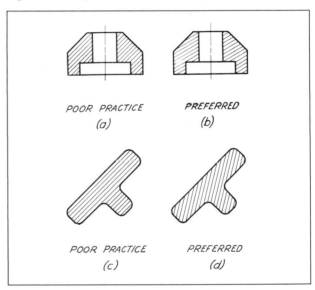

FIG. 7.19   Section lining at 30°, 60°, or 75°.

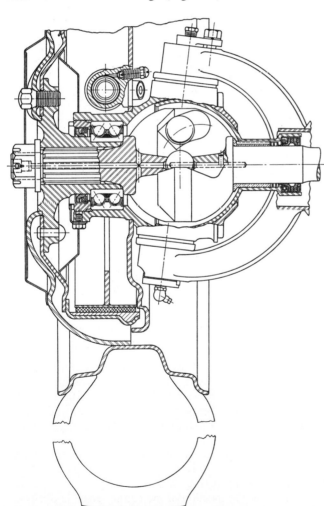

FIG. 7.18   Treatment of shafts, fasteners, ball bearings, and other parts. (*Courtesy New Departure Division, General Motors Corporation*)

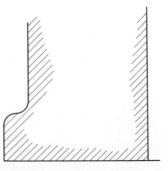

FIG. 7.20   Outline sectioning.

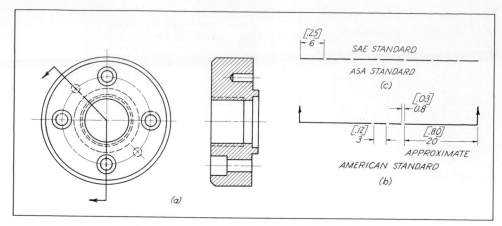

FIG. 7.21 Cutting plane lines (AN Standard).

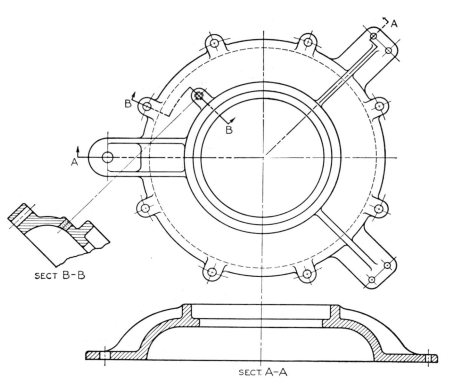

SECT B-B

SECT. A-A

FIG. 7.22 Sectional view.*

7.10 Symbolic Representation for a Cutting Plane

The symbolic lines that are used to represent the edge view of a *cutting plane* are shown in Fig. 7.21. The line is as heavy as an object line and is composed of either alternate long and short dashes or a series of dashes of equal length. The latter form is used in the automobile industry and has been approved by the SAE (Society of Automotive Engineers) and the American National Standards Institute. On drawings of ordinary size, when al-

ternate long and short dashes are used for the cutting-plane line, the long dashes are 20 mm (.80 in.) long, the short dashes 3 mm (.12 in.) long, and the spaces 0.8 mm (.03 in.) wide, depending on the size of the drawing. When drawn in pencil on manila paper, they are made with a medium pencil.

Arrowheads are used to show the direction in which the imaginary cut surface is viewed, and reference letters are added to identify it (Fig. 7.22).

Whenever the location of the cutting plane is obvious, it is common practice to omit the edge-view representation, particularly in the case of symmetrical objects. But if it is shown, and coincides with a center line, it takes precedence over the center line.

* ANSI Y14.2

7.11 Summary of the Practices of Sectioning

1. A cutting plane may be offset in order to cut the object in such a manner as to reveal an important detail that would not be shown if the cutting plane were continuous (Fig. 7.4).

2. All visible lines beyond the cutting plane for the section are usually shown.

3. Invisible lines beyond the cutting plane for the section are usually not shown, unless they are absolutely necessary to clarify the construction of the piece. In a half section, they are omitted in the unsectioned half, and either a center line or a solid line is used to separate the two halves of the view (Figs. 7.3 and 7.5).

4. On a view showing assembled parts, the section lines on adjacent pieces are drawn in opposite directions at an angle of 45° (Fig. 7.16).

5. On an assembly drawing, the portions of the cut surface of a single piece in the same view or different views always should be section-lined in the same direction, with the same spacing (Fig. 7.18).

6. The symbolic line indicating the location of the cutting plane may be omitted if the location of the plane is obvious (Fig. 7.1).

7. On a sectioned view showing assembled pieces, an exterior view is preferred for shafts, rods, bolts, nuts, and the like, which have no interior detail (Fig. 7.18).

7.12 Auxiliary Sections

A sectional view, projected on an auxiliary plane, is sometimes necessary to show the shape of a surface cut by a plane or to show the cross-sectional shape of an arm or rib (Fig. 7.23). When cutting an object, as in Fig. 7.23, arrows should show the direction in which the cut surface is viewed. *Auxiliary sections* are drawn by the usual method for drawing auxiliary views. When the bounding edge of the section is a curve, it is necessary to plot enough points to obtain a smooth one. Section 8.11 explains in detail the method for constructing the required view. A

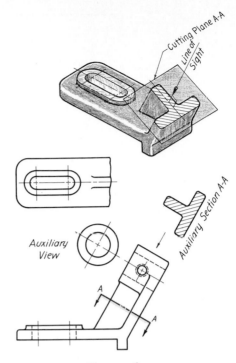

FIG. 7.23 Auxiliary section.

section view of this type usually shows only the inclined cut surface and is called a *partial section*.

7.13 Conventional Sections

Sometimes a less confusing sectioned representation is obtained if certain of the strict rules of projection as explained in Chapter 5 are violated. For example, an unbalanced and confused view results when the sectioned view of the pulley shown in Fig. 7.24 is drawn in the true projection, as in (*a*). It is better practice to preserve symmetry by showing the spokes as if they were aligned into one plane, as in (*c*). Such treatment of unsymmetrical features is not misleading, since their actual arrangement is revealed in the circular view. The spokes are not shown in the preferred sectional view. If they were, the first impression would be that the wheel had a solid web (*b*). See also Fig. 7.25.

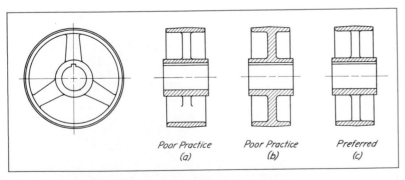

Poor Practice (*a*) *Poor Practice* (*b*) *Preferred* (*c*)

FIG. 7.24 Conventional treatment of spokes in section.

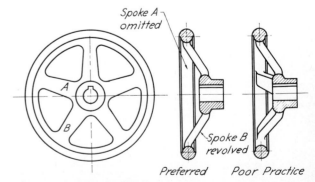

Spoke A omitted

Spoke B revolved

Preferred *Poor Practice*

FIG. 7.25 Spokes in section.*

* ANSI Y14.2

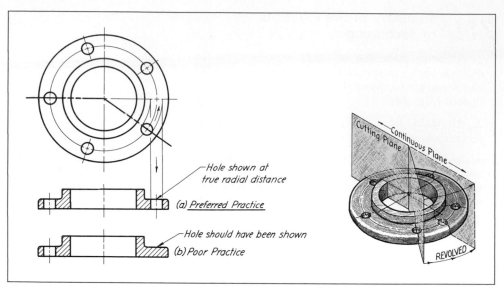

Hole shown at true radial distance

(a) Preferred Practice

Hole should have been shown

(b) Poor Practice

FIG. 7.26 Drilled flanges.

When there are an odd number of holes in a flange, as is the case with the part in Fig. 7.26, two should be shown aligned in the sectioned view to reveal their true location with reference to the rim and the axis of the piece. To secure the so-called aligned section, one usually considers the cutting plane to be bent to pass through the angled hole, as shown in the pictorial drawing. Then, the bent portion of the plane (with the hole) is imagined to be revolved until it is aligned with the other portion of the cutting plane. Straightened out, the imaginary continuous plane produces the preferred section view shown in (a).

Figure 7.27 shows another example of conventional representation. The sectional view is drawn as though the rear projecting lug had been swung forward until the portion of the cutting plane through it forms a continuous plane with the other lug (Sec. 5.42). It should be noted that the hidden lines in the sectioned view are necessary for a complete description of the construction of the lugs.

7.14 Ribs in Section

When a machine part has a rib cut by a plane of section (Fig. 7.28), a "true" sectional view taken through the rib would prove to be false and misleading because the cross-hatching on the rib would cause the object to appear "solid." The preferred treatment is to omit the section lines from the rib, as illustrated by Fig. 7.28(a). The resulting sectional view may be considered the view that would be obtained if the plane were offset to pass just in front of the rib (b).

An alternative conventional method, approved but not as frequently used, is illustrated in Fig. 7.29. This practice of omitting alternate section lines sometimes is adopted

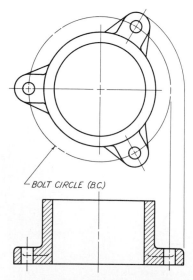

BOLT CIRCLE (B.C.)

FIG. 7.27 Revolution of a portion of an object.

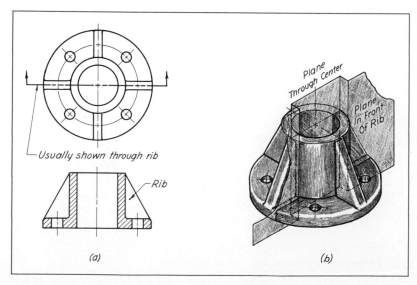

Usually shown through rib

Rib

(a)

(b)

FIG. 7.28 Conventional treatment of ribs in section.

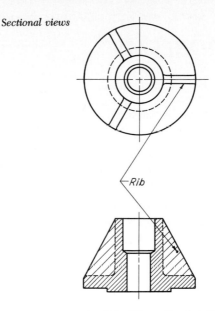

when it is necessary to emphasize a rib that might otherwise be overlooked. Note the use of invisible lines in Fig. 7.29.

7.15 Half Views

When the space available is insufficient to allow a satisfactory scale to be used for the representation of a symmetrical piece, it is considered good practice to make one view a *half view*, as shown in Fig. 7.30. The half view, however, must be the top or side view and not the front view, which shows the characteristic contour. The half view should be the rear half.

7.16 Material Symbols

The *section-line symbols* recommended by the American National Standards Institute for indicating various materials are shown in Fig. 7.31. Symbolic section lining is not usually used on a working (detail) drawing of a separate part. It is considered unnecessary to indicate a material symbolically when its exact specification must be given as a note. For this reason, and in order to save time as well, the easily drawn symbol for cast iron is commonly used on detail drawings for all materials. Contrary to this general practice, however, some insist that symbolic section lining be used on all detail drawings.

Symbolic section lining is usually employed on an assembly section showing the various parts of a unit in position, because a distinction between the materials causes the parts to "stand out" to better advantage. Furthermore, a knowledge of the type of material of which an individual part is composed often helps the reader to identify it more quickly and understand its function.

FIG. 7.29 Alternative treatment of ribs in section.

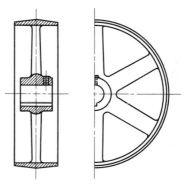

FIG. 7.30 Half view.

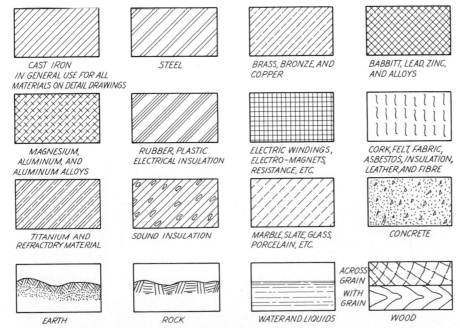

FOR OTHER SYMBOLS FOR MATERIALS IN SECTION AND ELEVATION SEE ARCHITECTURAL DRAWING CHAPTER

FIG. 7.31 Material symbols (AN Standard).

 **PROBLEMS**

The following problems were designed to emphasize the principles of sectioning. Those drawings that are prepared from the pictorials of objects may be dimensioned if the elementary principles of dimensioning (Chapter 13) are carefully studied.

1. (Fig. 7.32). Reproduce the top view and change the front view to a full section view in accordance with the indicated cutting plane.

2. (Fig. 7.33). Reproduce the top view and change the front and side views to sectional views that will be in accordance with the indicated cutting planes.

3. (Fig. 7.34). Draw a front view of the pulley (circular view) and a side view in full section.

4. (Fig. 7.35). Reproduce the top view of the rod support and draw the front view in full section. Read Sec. 5.43 before starting to draw.

5. (Fig. 7.36). Draw a front view of the V pulley (circular view) and a side view in full section.

6. (Fig. 7.37). Reproduce the two views of the hand wheel and change the right-side view to a full section.

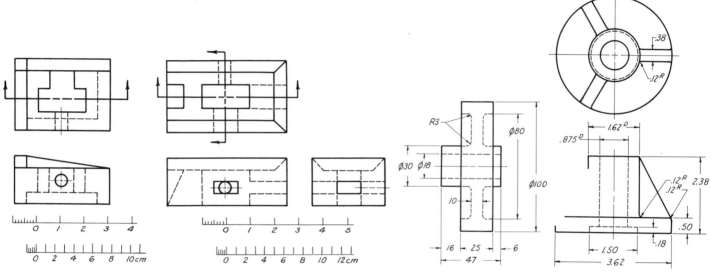

FIG. 7.32   Mutilated block.   **FIG. 7.33   Mutilated block.**   **FIG. 7.34   Pulley.**   **FIG. 7.35   Rod support.**

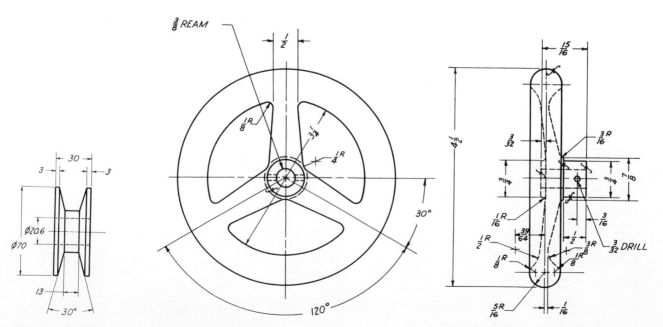

FIG. 7.36   V pulley.   **FIG. 7.37   Hand wheel.**

7. (Fig. 7.38). Reproduce the top view of the control housing cover and convert the front view to a full section.

8. (Fig. 7.39). Reproduce the circular front view of the pump cover and convert the right-side view to a full section.

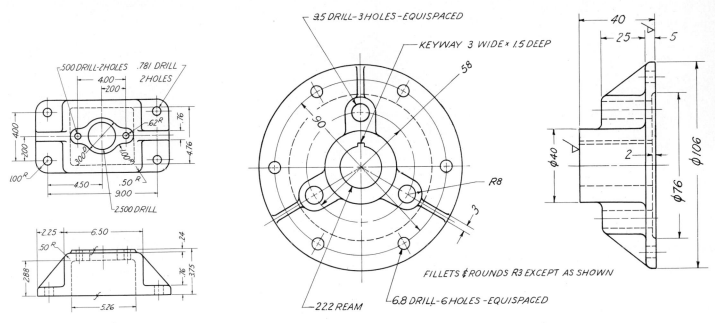

FIG. 7.38 Control housing cover.

FIG. 7.39 Pump cover.

9. (Fig. 7.40). Reproduce the front and top views of the steady brace. Complete the top view and draw the required side view and auxiliary section. Since this is a structural drawing, the figure giving the value of a distance appears above an unbroken dimension line in accordance with the custom in this field of engineering. The slope (45°) of the inclined member to which the plates are welded is indicated by a slope triangle with 12-in. legs.

This problem has been designed to test a student's ability to visualize the shape of the inclined structural member. Good judgment must be exercised in determining the location of the third hole in each plate. All hidden object lines should be shown.

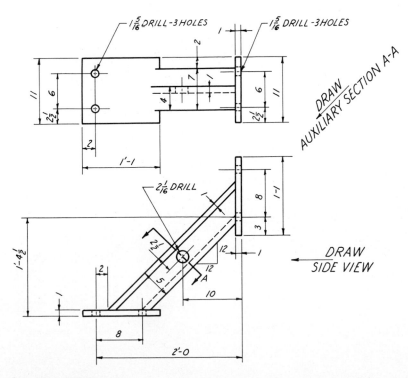

FIG. 7.40 Steady brace.

10–19. (Figs. 7.41–7.50). These problems may be dimensioned, as are working drawings. For each object, the student should draw all the views necessary for a working drawing of the part.

Good judgment should be exercised in deciding whether the sectional view should be a full section or a half section. After making this decision, the student should consult the instructor.

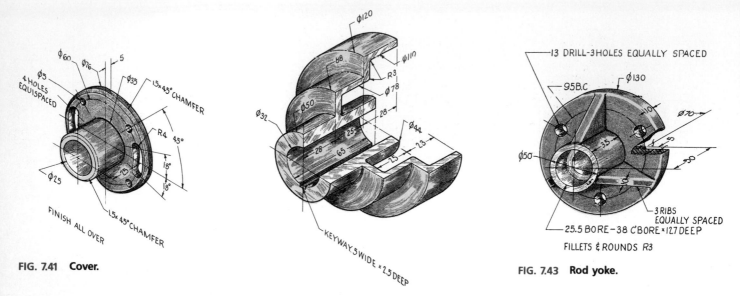

FIG. 7.41 Cover.

FIG. 7.42 Cone pulley.

FIG. 7.43 Rod yoke.

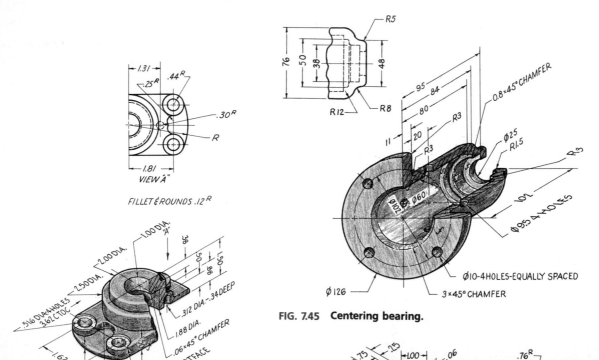

FIG. 7.44 Control housing cover.

FIG. 7.45 Centering bearing.

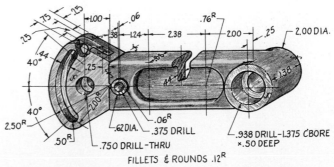

FIG. 7.46 Slotted guide link.

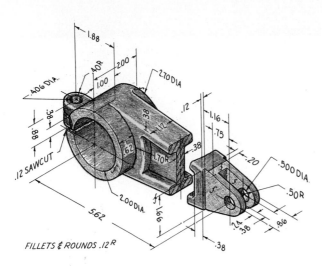

FILLETS & ROUNDS .12 R

FIG. 7.47   Shifter link.

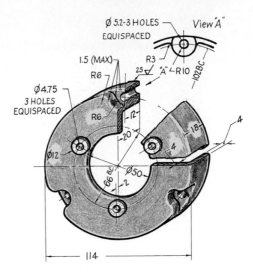

FIG. 7.48   Cover.

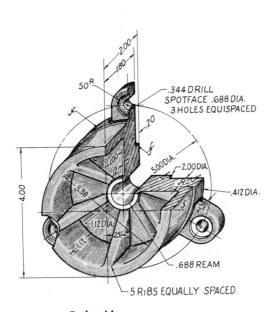

.344 DRILL
SPOTFACE .688 DIA.
3 HOLES EQUISPACED

5 RIBS EQUALLY SPACED

FIG. 7.49   End guide.

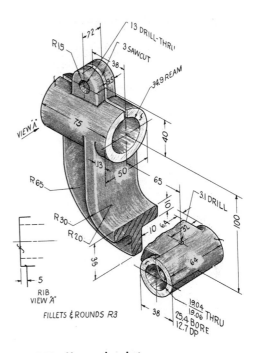

FILLETS & ROUNDS R3

FIG. 7.50   Hanger bracket.

An artist's concept of NASA's Space Shuttle making a landing at the Flight Research Center with one of the F-104 chase air- craft alongside. (*Courtesy National Aeronautics Space Admin- istration*)

CHAPTER · 8

Auxiliary Views

A PRIMARY (INCLINED) AUXILIARY VIEWS

8.1 Introduction

When it is desirable to show the *true size and shape* of an irregular surface, which is inclined to two or more of the coordinate planes of projection, a view of the surface must be projected on a plane parallel to it. This imaginary projection plane is called an *auxiliary plane*, and the view obtained is called an *auxiliary view* (Fig. 8.1).

The theory underlying the method of projecting principal views applies also to auxiliary views. In other words,

an auxiliary view shows an inclined surface of an object as it would appear to an observer stationed an infinite distance away and viewing the plane perpendicularly (Fig. 8.2).

8.2 Use of Auxiliary Views

In commercial drafting, an auxiliary view ordinarily is a partial view showing only an inclined surface. The reason for this is that a projection showing the entire object adds very little to the shape description. The added lines are likely to defeat the intended purpose of an auxiliary view. For example, a complete drawing of the casting in Fig. 8.3 must include an auxiliary view of the inclined surface

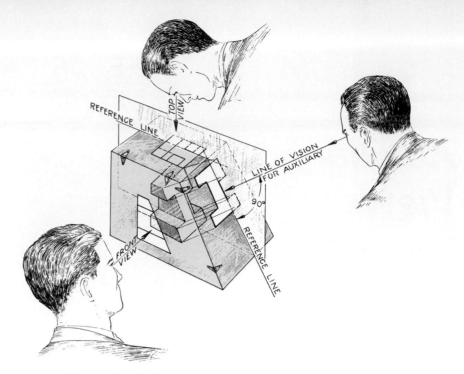

FIG. 8.1 Theory of projecting an auxiliary view.

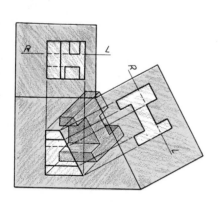

FIG. 8.2 Auxiliary view.

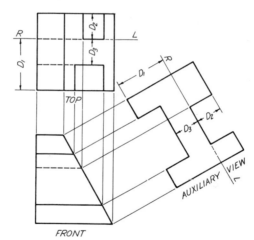

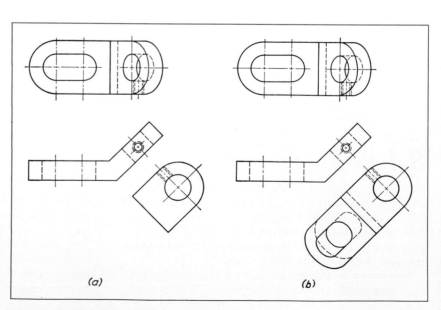

(a)　　　　　　　　　(b)

FIG. 8.3 Partial and complete auxiliary views.

in order to show the true shape of the surface and the location of the holes. Compare the views in (*a*) and (*b*) and note the confused appearance of the view in (*b*). Some instructors may require that an auxiliary view show the entire object, including all invisible lines. Such a requirement, though impractical commercially, is justified in the classroom, for the construction of a complete auxiliary view furnishes excellent practice in projection.

A partial auxiliary view often is needed to complete the projection of a foreshortened feature in a principal view. This second important function of auxiliary views is illustrated in Fig. 8.16 and explained in Sec. 8.13.

8.3 Types of Auxiliary Views

Although auxiliary views may have an infinite number of positions in relation to the three principal planes of projection, primary auxiliary views may be classified into three general types in accordance with position relative to the principal planes.

1. Adjacent to the front view.
2. Adjacent to the top view.
3. Adjacent to the side view.

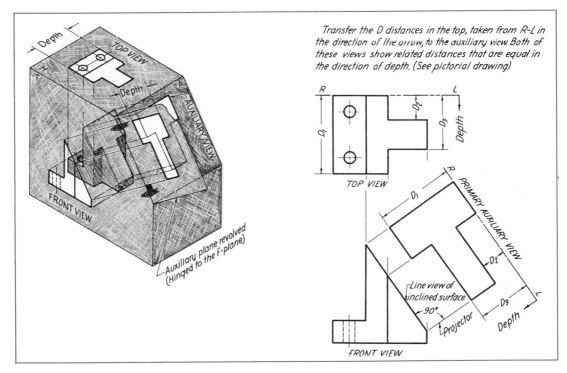

FIG. 8.4 Auxiliary view projected from front view.

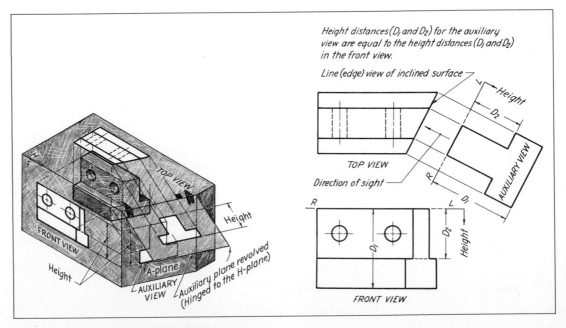

FIG. 8.5 Auxiliary view projected from top view.

Figure 8.4 shows the first type, where the auxiliary plane is *perpendicular to the frontal plane* and inclined to the horizontal plane of projection. Here the auxiliary view and top view have one dimension that is common to both: the depth. Note that the auxiliary plane is adjacent to the frontal plane and that the auxiliary view is projected from the front view.

In Fig. 8.5 the auxiliary plane is *perpendicular to the horizontal plane* and inclined to the frontal and profile planes of projection. The auxiliary view is projected from the top view, and its height is the same as the height of the front view.

The third type of auxiliary view, as shown in Fig. 8.6, is *perpendicular* to the side view and has a common dimension with both the front and top views. To construct it, distances may be taken from either the front or top view.

All three types of auxiliary views are constructed similarly. Each is projected from the view that shows the slanting surface as an edge, and the distances for the view are taken from the other principal view that has a common dimension with the auxiliary. A careful study of the three illustrations will reveal the fact that the inclined auxiliary plane is always adjacent to the principal plane to which it is perpendicular.

8.4 Symmetrical and Unsymmetrical Auxiliary Views

Since auxiliary views are either symmetrical or unsymmetrical about an axis, they may be termed (1) *symmetrical*, (2) *unilateral*, or (3) *bilateral*, according to the degree of symmetry. A symmetrical view is drawn symmetrically about a center line, the unilateral view entirely on one side of a reference line, and the bilateral view on both sides of a reference line.

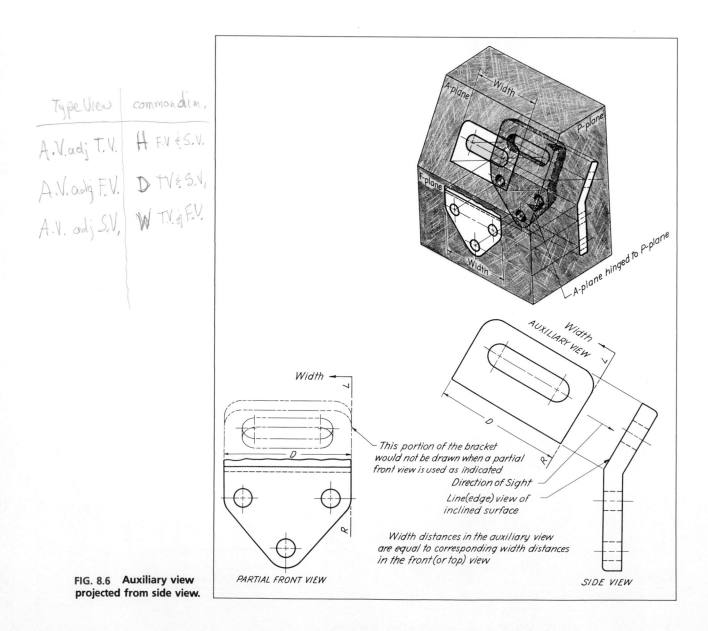

FIG. 8.6 **Auxiliary view projected from side view.**

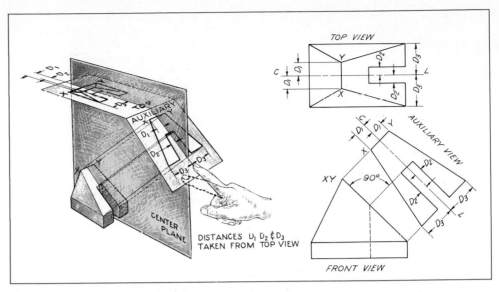

FIG. 8.7 Symmetrical auxiliary view of an inclined surface.

8.5 To Draw a Symmetrical Auxiliary View

When an inclined surface is symmetrical, the auxiliary view is "worked" from a center line (Fig. 8.7). The first step in constructing such a view is to draw a center line parallel to the inclined line that represents an edge view of the surface. If the object is assumed to be enclosed in a glass box, this center line may be considered the line of intersection of the auxiliary plane and an imaginary vertical center plane.

Although, theoretically, this working center line may be drawn at any distance from the principal view, it should be located to give the whole drawing a balanced appearance. If not already shown, it also must be drawn in the principal view showing the true width of the inclined surface.

The next step is to draw projection lines from each point of the sloping face, remembering that the projectors make an angle of 90° with the inclined line representing the surface. With the projectors drawn, the location of each point in the auxiliary can be established *by setting the dividers to each point's distance from the center line in the principal view and transferring the distance to the auxiliary view.* For example, point *X* is projected to the auxiliary by drawing a projector from point *X* in the front view perpendicular to the center line. Since its distance from the center line in the top view is the same as it is from the center line in the auxiliary view, the point's location along the projector may be established by using the distance taken from the top view. In the case of point *X*, the distance is set off from the center line toward the front view. Point *Y* is set off from the center line away from the front view. A careful study of Fig. 8.7 reveals the fact that if a point lies between the front view and the center line of the top view, it will lie between the front view and the center line of the auxiliary view, and, con-

versely, if it lies away from the front view with reference to the center line of the top view, it will lie away from the front view with reference to the center line of the auxiliary view.

8.6 Unilateral Auxiliary Views

When constructing a *unilateral auxiliary view*, it is necessary to work from a reference line that is drawn in a manner similar to the working center line of a symmetrical view. The reference line for the auxiliary view may be considered to represent the line of intersection of a reference plane, coinciding with an outer face, and the auxiliary plane (Fig. 8.8). The intersection of this plane with the top plane establishes the reference line in the top view. All the points are projected from the edge view of the surface, as in a symmetrical view, and it should be noted in setting them off that they all fall on the same side of the reference line.

Figure 8.9 shows an auxiliary view of an entire object. In constructing such a view, it should be remembered that the projectors from all points of the object are perpendicular to the auxiliary plane, since the observer views the entire figure by looking directly at the inclined surface. The distances perpendicular to the auxiliary reference line were taken from the front view.

8.7 Bilateral Auxiliary Views

The method of drawing a *bilateral view* is similar to that of drawing a unilateral view, the only difference being that in a bilateral view the inclined face lies partly on both sides of the reference plane, as shown in Fig. 8.10.

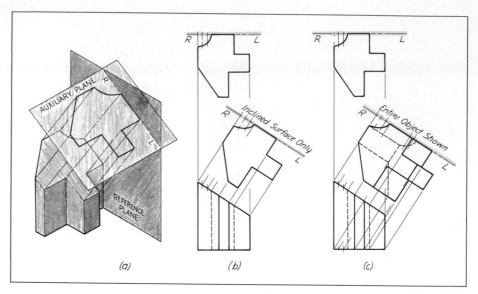

FIG. 8.8 Unilateral auxiliary view.

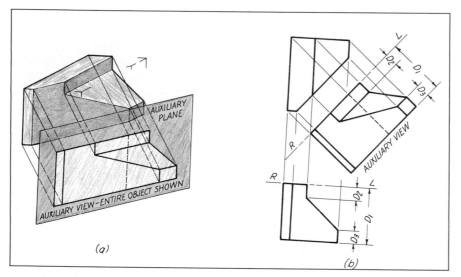

FIG. 8.9 Auxiliary view of an object.

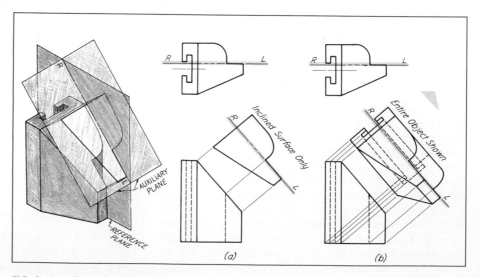

FIG. 8.10 Bilateral auxiliary view.

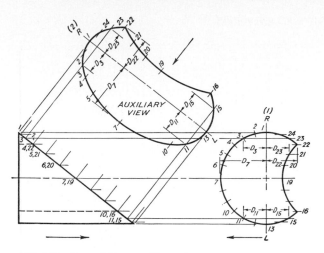

FIG. 8.11 Curved-line auxiliary view.

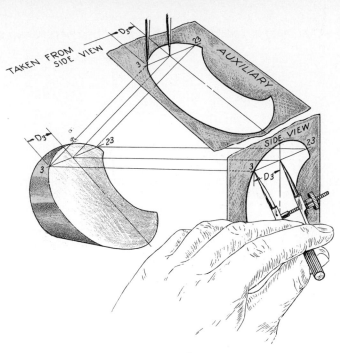

8.8 Curved Lines in Auxiliary Views

To draw a *curve* in an auxiliary view, a drafter must plot a sufficient number of points to ensure a smooth curve (Fig. 8.11). The points are projected first to the edge view of the surface in the front view and then to the auxiliary view. The distance of any point from the center line in the auxiliary view is the same as its distance from the center line in the end view.

8.9 Projection of a Curved Boundary

In Fig. 8.11, the procedure is illustrated for plotting the true size and shape of an inclined surface bounded by a *curved outline*. A similar procedure can be followed to plot a curve, such as the one on the left end of the object shown in Fig. 8.12, in an auxiliary view. Since the points on the curved outline of the vertical surface are being viewed from the same direction as those on the curved boundary of the inclined surface, the projectors from points on both surfaces will be parallel. From the pictorial drawing, it can be observed that points A and A' and B and B' lie on elements of the cylindrical surface. Also it should be noted that A and A' are the same distance from the reference plane as are B and B'. It is for this reason that point A' in the auxiliary view is at the intersection of the projector from A' in the front view and a line through A in the auxiliary view, drawn parallel to the reference line RL.

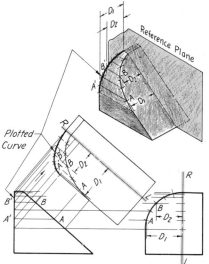

FIG. 8.12 Plotted boundary curves.

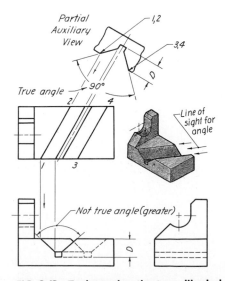

FIG. 8.13 To determine the true dihedral angle between inclined surfaces.

8.10 Dihedral Angles

Frequently, an auxiliary view may be needed to show the true size of a *dihedral angle*—that is, the *true size of the angle between two planes*. In Fig. 8.13, it is desirable to show the true size of the angle between the planes forming the V slot by means of a partial auxiliary view, as shown.

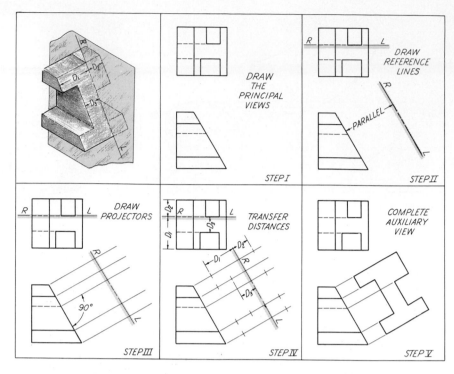

FIG. 8.14 Steps in constructing an auxiliary view.

The direction of sight (see pictorial) must be taken parallel to the edge planes and lines 1–2 and 3–4 so that these lines will appear as points and the surfaces forming the dihedral angle will project as edge views in the auxiliary view. The reference line for the partial auxiliary view would necessarily be drawn perpendicular to the lines 1–2 and 3–4 in the top view. Since the plane on which the auxiliary view is projected is a vertical one, height dimensions were used—that is, distances in the direction of the dimension *D* in the auxiliary view were taken from the front view.

8.11 To Construct an Auxiliary View, Practical Method

The usual steps in constructing an auxiliary view are shown in Fig. 8.14. The illustration should be studied carefully, as each step is explained in the drawing.

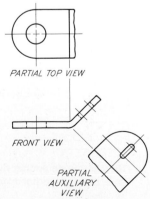

FIG. 8.15 Partial views.

8.12 Auxiliary and Partial Views

Often the use of an auxiliary view allows the elimination of one of the principal views (top or side) or makes possible the use of a *partial principal view*. The shape description furnished by the partial views shown in Fig. 8.15 is sufficient for a complete understanding of the shape of the part. The use of partial views simplifies the drawing, saves valuable drafting time, and tends to make the drawing easier to read.

A break line is used at a convenient location to indicate an imaginary break for a partial view.

8.13 Use of an Auxiliary View to Complete a Principal View

As previously stated, it is frequently necessary to project a foreshortened feature in one of the principal views from an auxiliary view. In the case of the object shown in Fig. 8.16, the foreshortened projection of the inclined face in the top view can be projected from the auxiliary view. The elliptical curves are plotted by projecting points from the auxiliary view to the front view and from there to the top view. The location of these points in the top view with respect to the center line is the same as their location in the auxiliary view with respect to the auxiliary center line. For example, the distance D_1 from the center line in the top view is the same as the distance D_1 from the auxiliary center line in the auxiliary view.

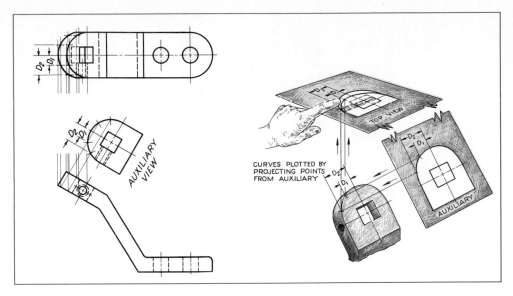

FIG. 8.16   Use of auxiliary to complete a principal view.

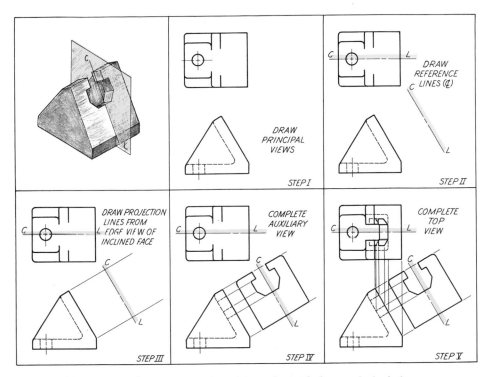

FIG. 8.17   Steps in preparing an auxiliary view and completing a principal view.

The steps in preparing an auxiliary view and using it to complete a principal view are shown in Fig. 8.17.

8.14   *Line of Intersection*

It is frequently necessary to represent a *line of intersection* between two surfaces when making a multiview drawing involving an auxiliary view. Figure 8.18 shows a method for drawing the line of intersection on a principal view. In this case the scheme commonly used for determining the intersection involves the use of elements drawn on the surface of the cylindrical portion of the part, as shown on the pictorial drawing. These elements, such as *AB*, are common to the cylindrical surface. Point *B*, where the

element pierces the flat surface, is a point that is common to both surfaces and therefore lies on the line of intersection.

On the orthographic views, element *AB* appears as a point on the auxiliary view and as a line on the front view. The location of the projection of the piercing point on the front view is visible upon inspection. Point *B* is found in the other principal view by projecting from the front view and setting off the distance *D* taken from the auxiliary view. The distance *D* of point *B* from the center line is a true distance for both views. The center line in the auxiliary view and side view can be considered as the edge view of a reference plane or datum plane from which measurements can be made.

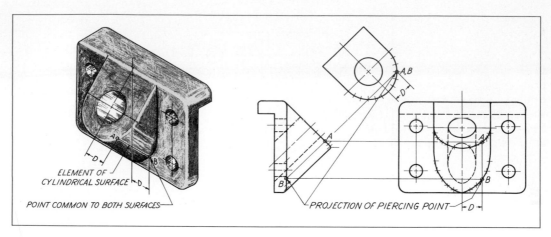

FIG. 8.18 **Line of intersection.**

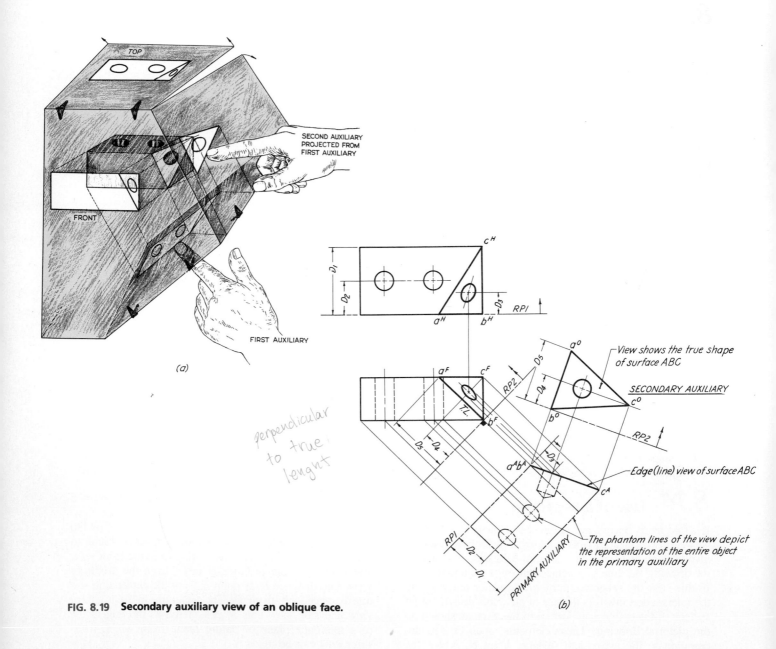

FIG. 8.19 **Secondary auxiliary view of an oblique face.**

8.15 *True Length of a Line*

The true length of an oblique line may be determined by means of an auxiliary view or by revolution. Separate discussions of the procedure to be followed in the application of these methods are given in Chapters 5 and 9. To determine the true length of a line by revolution, see Sec. 5.32. To find the true length through the use of an auxiliary view, read Sec. 9.7.

B / SECONDARY AUXILIARY VIEWS

go back two views for common dimension

8.16 *Secondary (Oblique) Auxiliary Views*

Frequently an object will have a face that is not perpendicular to any one of the principal planes of projection. In such cases it is necessary to draw a primary auxiliary view and a secondary auxiliary or oblique view (Fig. 8.19). The primary auxiliary view is constructed by projecting the figure on a primary auxiliary plane that is perpendicular to the inclined surface and one of the principal planes. This plane may be at any convenient distance. In the illustration, the primary auxiliary plane is perpendicular to the frontal plane. Note that the oblique plane appears as an edge in the primary auxiliary view. Using this view as a regular view, the secondary auxiliary view may be projected on a plane parallel to the oblique face. Figure 8.19(*b*) shows a practical application of the theoretical principles shown pictorially in (*a*).

It is suggested that the student read Sec. 9.14 in which the procedure for drawing the normal (true-shape) view of an oblique surface is presented step by step.

Figure 8.20 shows the progressive steps in preparing and using a secondary auxiliary view of an oblique face to complete a principal view. Reference planes have been used as datum planes from which to take the necessary measurements.

STEP I. Draw partial front and top views.

STEP II. Show the partial construction of the primary auxiliary view in which the inclined surface appears as a line.

STEP III. Show the secondary auxiliary view projected and completed from the primary view, using the known measurements of the lug. The primary auxiliary view is finished by projecting from the secondary auxiliary view.

STEP IV. Project the top view from the secondary auxiliary view through the primary auxiliary view in order to complete the foreshortened view of the lug. It should be noted that distance D_1 taken from reference R_2P_2 in the secondary auxiliary is transferred to the top view because both views show the same width distances in true length. A sufficient number of points should be obtained to allow the use of an irregular curve.

STEP V. Project these points on the curve to the front view. In this case the measurements are taken from the primary auxiliary view because the height distances from reference plane R_1P_1 are the same in both views.

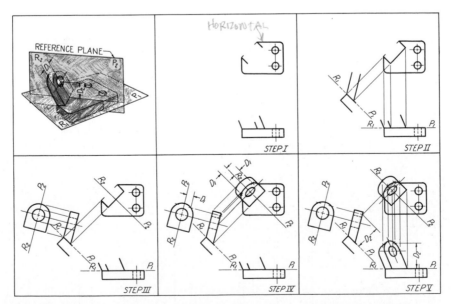

FIG. 8.20   Steps in drawing a secondary auxiliary view and using it to complete a principal view.

PROBLEMS

The problems shown in Fig. 8.21 are designed to give the student practice in constructing auxiliary views of the inclined surfaces of simple objects formed mainly by straight lines. They will provide needed drill in projection if, for each of the objects in Fig. 8.21, an auxiliary is drawn showing the entire object. Com-

plete drawings may be made of the objects shown in Figs. 8.22–8.30. If the views are to be dimensioned, the student should adhere to the rules of dimensioning given in Chapter 13 and should not take too seriously the locations for the dimensions on the pictorial representations.

1. (Fig. 8.21). Using instruments, reproduce the given views of an assigned object and draw an auxiliary view of its inclined surface.

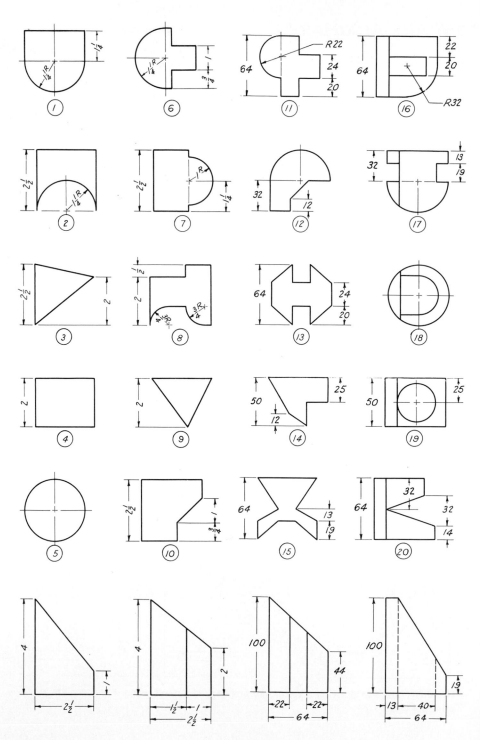

FIG. 8.21

2. (Fig. 8.22). Draw the views that would be necessary on a working drawing of the dovetail bracket.

3. (Fig. 8.23). Draw the necessary views of the anchor bracket. Make partial views for the top and end views.

4. (Fig. 8.24). Draw the views that would be necessary on a working drawing of the feeder bracket.

5. (Fig. 8.25). Draw the necessary views of the anchor clip. It is suggested that the top view be a partial one and that the auxiliary view show only the inclined surface.

6. (Fig. 8.26). Draw the views that would be necessary on a working drawing of the angle bracket. Note that two auxiliary views will be required.

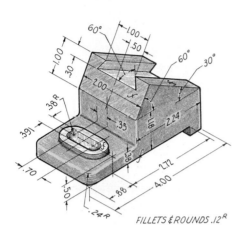

FIG. 8.22 Dovetail bracket.

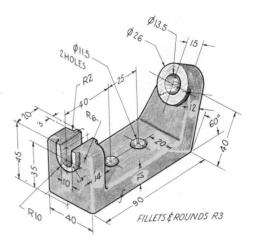

FIG. 8.23 Anchor bracket.

FIG. 8.24 Feeder bracket.

FIG. 8.25 Anchor clip.

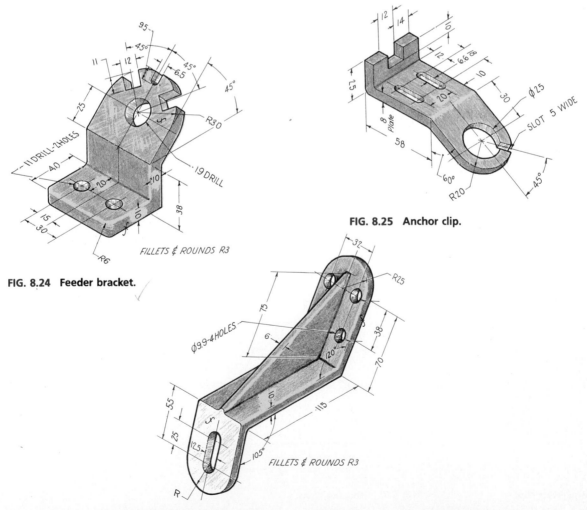

FIG. 8.26 Angle bracket. (Dimension values shown in [ ] are in millimeters.)

7. (Fig. 8.27). Draw the necessary views of the offset guide. It is suggested that partial views be used, except in the view where the inclined surface appears as a line.

8. (Fig. 8.28). Draw the views that would be needed on a working drawing of the gear cover. The opening on the inclined face is circular.

9. (Fig. 8.29). Draw the necessary views of the cutoff clip.

10. (Fig. 8.30). Draw the necessary views of the ejector clip.

11. (Fig. 8.31). Draw the views as given. Complete the top view.

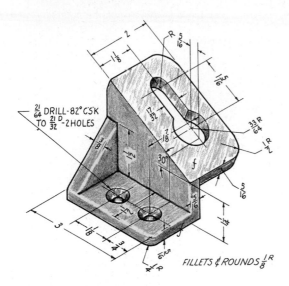

FIG. 8.27   Offset guide.

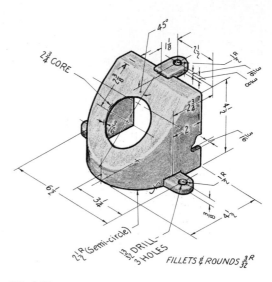

FIG. 8.28   Gear cover.

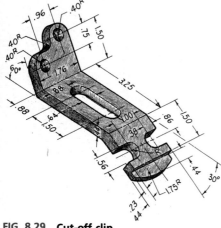

FIG. 8.29   Cut-off clip.

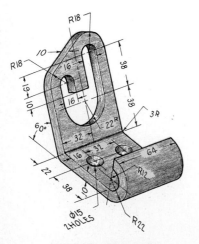

FIG. 8.30   Ejector clip.

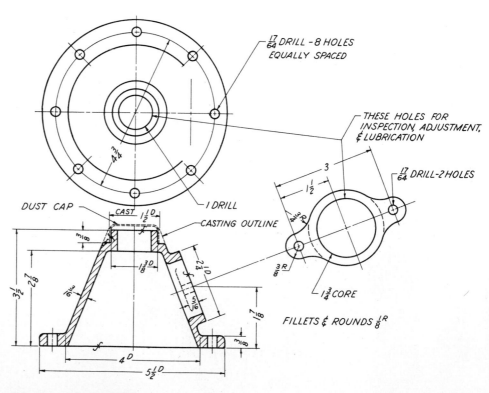

FIG. 8.31   Housing cover.

12. (Fig. 8.32). Draw the views as given. Complete the auxiliary view and the front view.

13. (Fig. 8.33). Draw the views that would be necessary on a working drawing of the 45° elbow.

14. Make a multiview drawing of the airplane engine mount shown in Fig. 8.34. The engine mount is formed of three pieces of steel plate welded to a piece of steel tubing. The completed drawing is to consist of four views. It is suggested that the front view be the view obtained by looking along and parallel to the axis of the tube. The remaining views that are needed are an auxiliary view showing only the inclined lug, a side view that should be complete with all hidden lines shown, and a partial top view with the inclined lug omitted.

15. (Fig. 8.35). Draw the views given and add the required primary and secondary auxiliary views.

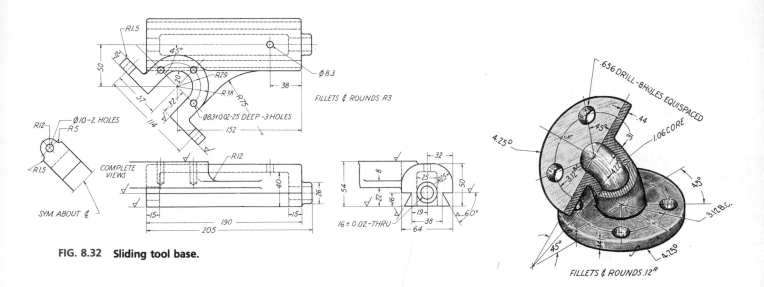

FIG. 8.32 Sliding tool base.

FIG. 8.33 45° elbow.

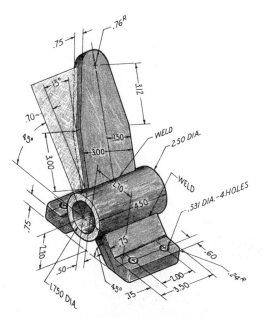

FIG. 8.34 Airplane engine mount.

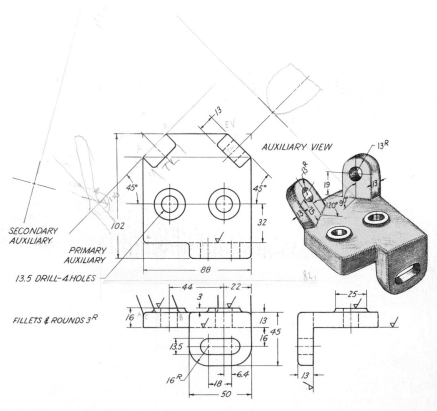

FIG. 8.35 Cross anchor.

16. (Fig. 8.36). Draw the necessary views of the tool holder.

17. (Fig. 8.37). Draw the necessary views of the angle block. One view should be drawn to show the true size and shape of the oblique surface.

18. (Fig. 8.38). Draw the necessary views of the locating slide.

19. (Fig. 8.39). Using instruments, draw a secondary auxiliary view that will show the true size and shape of the inclined surface of an assigned object. The drawing must also show the given principal views.

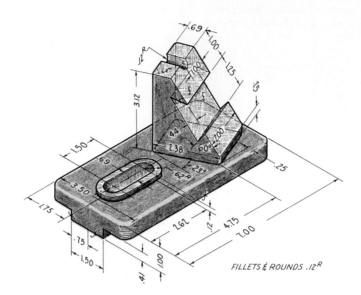

FIG. 8.36 Tool holder.

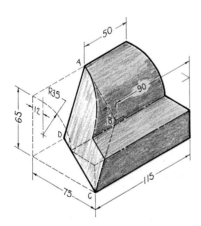

FIG. 8.37 Angle block.

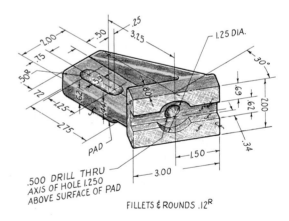

FIG. 8.38 Locating slide.

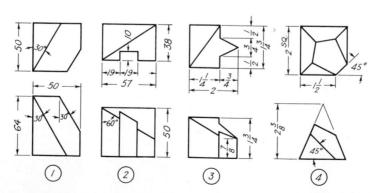

FIG. 8.39

20. (Fig. 8.40). Draw the layout for the support anchor as given and then, using the double-auxiliary-view method, complete the views as required.

The plate and cylinder are to be welded. Since the faces of the plate show as oblique surfaces in the front and top views, double auxiliary views are necessary to show the thickness and the true shape.

Start the drawing with the auxiliary views that are arranged horizontally on the paper, then complete the principal views. The inclined face of the cylinder will show as an ellipse in top and front views; but do not show this in the auxiliary view that shows the true shape of the square plate.

How would you find the view that shows the true angle between the inclined face and the axis of the cylinder?

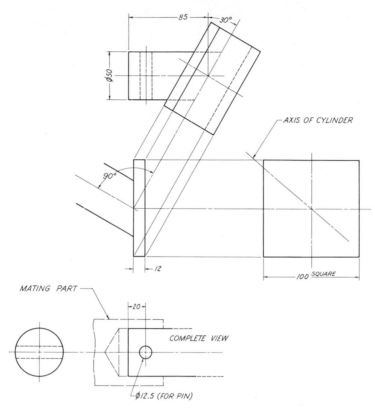

FIG. 8.40   **Support anchor.**

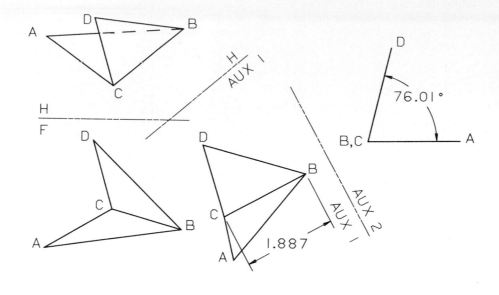

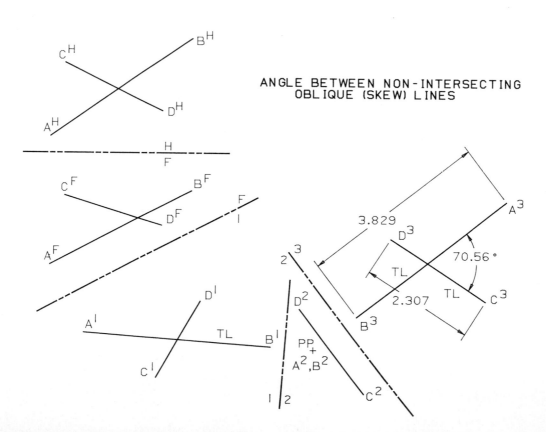

ANGLE BETWEEN NON-INTERSECTING
OBLIQUE (SKEW) LINES

Descriptive geometry is based on the theory of orthographic projection. The basic principles, as set forth in the chapter that follows, are applied by designers and drafters to the graphic representation and solution of more advanced space problems such as determining the true size and shape of an oblique plane, the true size of the dihedral angle between planes (*see above-top*), the true length of an oblique line, the distance from a point to an oblique line, the shortest distance between two skew lines, the angle between two intersecting lines, and the angle between two skew lines (*see above*). (*Drawings furnished by Professor Jerry Smith and Dennis Short of the School of Technology, Purdue University. Computer generated drawings.*)

Basic Spatial Geometry for Design and Analysis

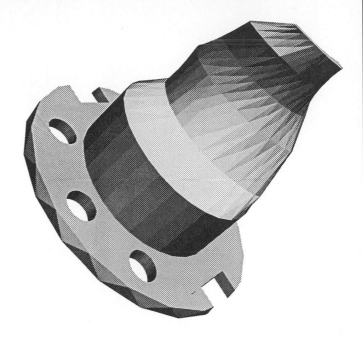

visualize
3D

A / BASIC DESCRIPTIVE GEOMETRY

9.1 Introduction

On many occasions, problems arise in engineering design that may be solved quickly by applying the basic principles of orthographic projection. If one thoroughly understands the solution for each of the problems presented, it should be easy, at a later time, to analyze and solve almost any of the practical problems encountered.

It should be pointed out at the very beginning that to solve most types of problems one must apply the principles

and methods used to solve a few basic problems, such as how (1) to find the true length of a line, (2) to find the point projection of a line, and (3) to find the true size and shape of a surface (Fig. 9.1). To find information such as the angle between surfaces, the angle between lines, or the clearance between members of a structure, one must use, in proper combination, the methods of solving these basic problems. Success in solving problems by projection depends largely on the complete understanding of the principles of projection, the ability to visualize space conditions, and the ability to analyze a given situation. Since the ability to analyze and to visualize are of utmost importance in engineering design, the student is urged to develop these abilities by resisting the temptation to memorize step procedures.

FIG. 9.1   The frame of this satellite could not have been designed without the application of descriptive geometry methods. (*Courtesy TRW Systems Group***)**

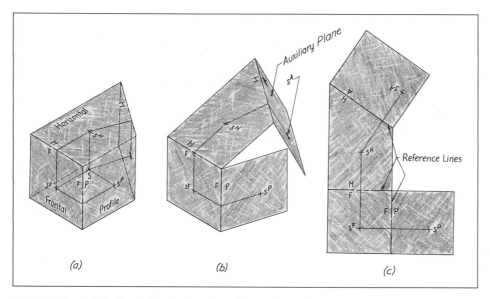

FIG. 9.2   Frontal, horizontal, profile, and auxiliary views of a point S.

9.2 Projection of a Point

Figure 9.2(*a*) shows the projection of point *S* on the three principal planes of projection and an auxiliary plane *A*. The notation used is as explained in Sec. 5.8. Point s^F is the view of point *S* on the frontal plane, s^H is the view of *S* on the horizontal plane, and s^P is its view on the profile plane. For convenience and ease in recognizing the projected view of a point on a supplementary plane, the supplementary planes are designated as *A* planes and *O* planes. *A* (auxiliary) planes are always perpendicular to one of the principal planes. Point s^A is the view of *S* on the *A* plane. The view of *S* on an *O* (oblique) plane would be designated s^O. Additional *O* planes are identified as O_1, O_2, O_3, etc., in the order that they follow the first *O* plane.

Since it is necessary to represent on one plane (the working surface of our drawing paper) the views of point *S* that lie on mutually perpendicular planes of projection, the planes are assumed to be hinged so that they can be revolved, as shown in Fig. 9.2(*b*), until they are in a single plane, as in (*c*). The lines about which the planes of projection are hinged are called *reference lines*. A ref-

erence line is identified by the use of capital letters representing the adjacent planes, as *FH, FP, FA, HA, AO,* and so forth (see Fig. 9.2).

It is important to note in (*c*) that the projections s^F and s^H fall on a vertical line, s^F and s^P lie on a horizontal line, and s^H and s^A lie on a line perpendicular to the reference line *HA*. In each case this results from the fact that point *S* and its projections on adjacent planes lie in a plane perpendicular to the reference line for those planes [Fig. 9.2(*a*)]. This important principal of projection determines the location of views when the relationship of lines and planes form the problem.

9.3 Projection of a Straight Line

Capital letters are used for designating the end points of the actual line in space. In the projected views, these points are identified as shown in Fig. 9.3. The student should read Sec. 5.8, which presents the principles of multiview drawing. In particular study the related illustration, which shows some typical line positions.

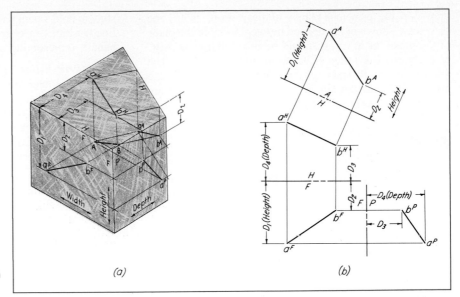

(a)   (b)

FIG. 9.3 The line in space and in successive views.

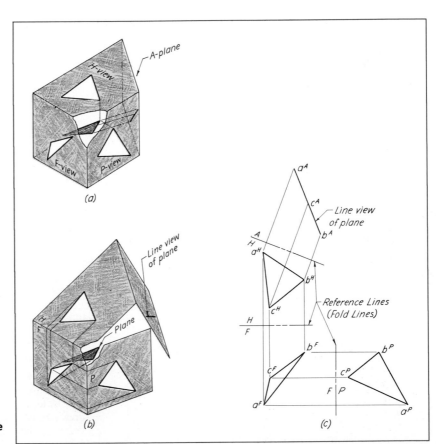

FIG. 9.4 The plane in space and in successive views.

9.4 *Projection of a Plane in Space*

Theoretically, a plane is considered to be flat and unlimited in extent. A plane can be delineated graphically by

1. Two intersecting lines,
2. A line and a point not on the line,
3. Two parallel lines,
4. Three points not on a straight line.

For graphical purposes and to facilitate the solution of space problems as presented in this chapter, planes will be bounded and usually triangular. The space picture and multiview representation of a plane *ABC* are given in Fig. 9.4. The pictorial at the left in (*a*) shows the projected views on the principal planes of projection. In (*b*) the planes are shown being opened outward to be in the plane of the paper, as in (*c*). It should be noted that plane *ABC* projects as a line (edge view) on the auxiliary plane. The three points *A*, *B*, and *C* of the plane are projected in the same manner as the single point *S* in Fig. 9.2 and are identified similarly (a^F, a^H, a^P, a^A, etc.).

9.5 Parallel Lines

Any two lines in space must be parallel, intersecting, or nonintersecting and nonparallel (called *skew lines*). Figure 9.18 shows intersecting lines, while Fig. 9.17 shows skew lines. Parallel lines are shown in Fig. 9.5.

It might be stated as a rule of projection, with one exception, that *when two lines are parallel their projections will be parallel in every view* (Fig. 9.5). In other words the lines will appear to be parallel in every view in which both appear. This is true even though in specific views they may appear as points or their projections may coincide. In either case they are still parallel because both conditions indicate that the lines have the same direction. The exception that has been mentioned occurs when the *F* and *H* projections of two inclined profile lines are shown. For proof of parallelism a supplementary view should be drawn, which may or may not be the profile view.

The true or shortest distance between two parallel lines can be determined on the view that will show these lines as points. The true distance can be measured between the points (Fig. 9.5).

9.6 To Determine the True Length of a Line

An observer can see the true length of a line when looking in a direction perpendicular to it. It is suggested that the student hold a pencil and move it into the following typical line positions to observe the conditions under which the pencil, representing a line, appears in true length.

1. *Vertical line.* The vertical line is perpendicular to the horizontal and will therefore appear as a point in the *H* (top) view. It will appear in true length in the *F* (frontal) view, in true length in the *P* (profile) view, and in true length in any auxiliary view that is projected on an auxiliary plane that is perpendicular to the horizontal plane of projection.

2. *Horizontal line.* The horizontal line will appear in true length when viewed from above because it is parallel to the *H* plane of projection and its end points are theoretically equidistant from an observer looking downward.

3. *Inclined line.* The inclined line will show true length in the *F* view or *P* view, for by definition (Sec. 5.8) an inclined line is one that is parallel to either the *F* plane or the *P* plane of projection. However, it cannot be parallel to both planes of projection at the same time.

4. *Oblique line.* The oblique line will not appear in true length in any of the principal views because it is inclined to all of the principal planes of projection. It should be apparent, in viewing the pencil alternately from the directions used to obtain the principal views, namely, from the front, above, and side, that one end of the pencil is always farther away from the observer than the other. Only when looking directly at the pencil from such a position that the end points are equidistant from the observer can the true length be seen. On a drawing, the true length projection of an oblique line will appear in a supplementary *A* (auxiliary) view on a plane that is parallel to the line.

9.7 To Determine the True Length of an Oblique Line

In order to find the true length of an oblique line, it is necessary to select an auxiliary plane of projection that will be parallel to the line (Figs. 9.6 and 9.7).

GIVEN: The *F* (frontal) view $a^F b^F$ and the *H* (top) view $a^H b^H$ of the oblique line *AB* (Fig. 9.6).

SOLUTION: (1) Draw the reference line *HA* parallel to the projection $a^H b^H$. The *A* plane for this reference line will be parallel to *AB* and perpendicular to the *H* plane (see pictorial drawing). (2) Draw lines of projection from points a^H and b^H perpendicular to the reference line. (3) Transfer height measurements from the *F* view to the *A* view to locate a^A and b^A. In making this transfer of measurements, the students should attempt to visualize the space condition for the line and understand that, since the *F* view and *A* view both show height and because the planes of projection for these views are perpendicular to the *H* plane, the perpendicular distance D_1 from the reference line *HF* to point a^F must be the same as the distance D_1 from the reference line *HA* to point a^A.

The projection $a^A b^A$ shows the true length of the line *AB*.

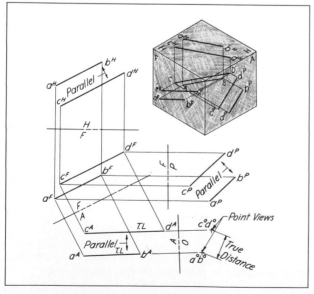

FIG. 9.5 Parallel lines.

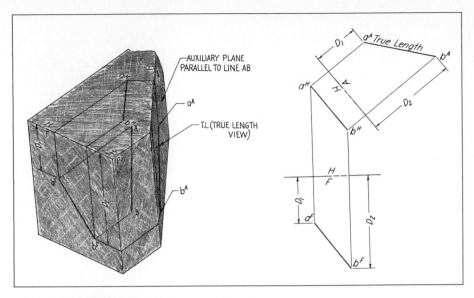

FIG. 9.6 To find the true length of an oblique line.

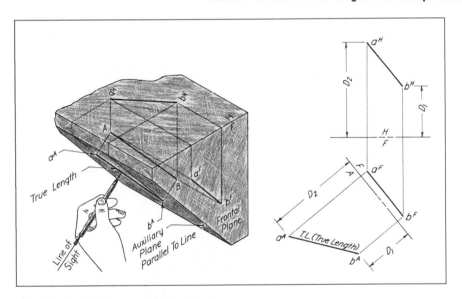

FIG. 9.7 To find the true length of a line.

It was not necessary to use an auxiliary plane perpendicular to the *H* plane to find the true length of line *AB* in Fig. 9.6. The auxiliary plane could just as well have been perpendicular to either the *F* or *P* planes. Figure 9.7 shows the use of an auxiliary plane perpendicular to the frontal plane to find the true length of the line. In this case the auxiliary view has depth distances in common with top view, as indicated.

9.8 *Perpendicular Lines*

Lines that are perpendicular in space will have their projections perpendicular in any view that shows either or both of the lines in true length. A second rule of perpendicularity might be that, when a line is perpendicular to a plane, it will be perpendicular to every line in that plane. A careful study of Fig. 9.8 will verify these rules.

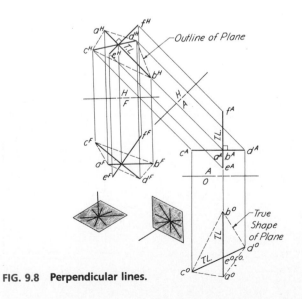

FIG. 9.8 Perpendicular lines.

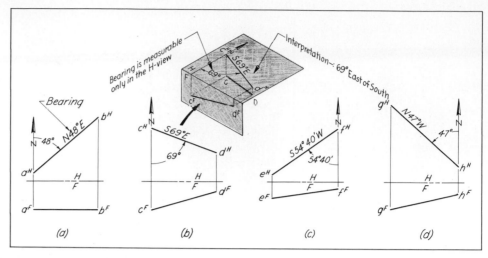

FIG. 9.9 Bearing of a line.

For instance, it should be noted that the lines *AB* and *CD* lie in a plane that is outlined with broken lines and that $e^H f^H$ is perpendicular to $a^H b^H$ because $a^H b^H$ shows the true length of the line *AB*. In the *A* view, we see that $e^A f^A$ is perpendicular to the line view of the plane and is therefore perpendicular to both *AB* and *CD*. The *O* view shows the true shape (TSP) of the plane and the line *EF* as a point. This again verifies the fact that *EF* is perpendicular to the plane and to lines *AB* and *CD*. Otherwise line *EF* would not appear as a point. Note also that the *O* view shows the true length (TL) of *AB* and *CD*.

9.9 Bearing of a Line

The bearing of a line is the horizontal angle between the line and a north-south line. A bearing is given in degrees with respect to the meridian and is measured from 0° to 90° from either north (N) or south (S). The bearing reading indicates the quadrant in which the line is located by use of the letters N and E, S and E, S and W, or N and W, as N 48° E or S 54° 40′ W. The bearing of a line is measured in the *H* view (Fig. 9.9).

9.10 Point View of a Line (Fig. 9.10)

It was pointed out in the first section of this chapter that the solutions of many types of problems depend on an understanding of a few basic constructions. One of these basic constructions involves the finding of the view showing the *point view* or point projection of a line. For instance, this construction is followed when it is necessary to determine the dihedral angle between two planes, for the true size of the angle will appear in the view that shows the line common to the two planes as a point.

A line will show as a point on a projection plane that is perpendicular to the line. The observer's direction of sight must be along and parallel to the line. When a line appears in true length on one of the principal planes of

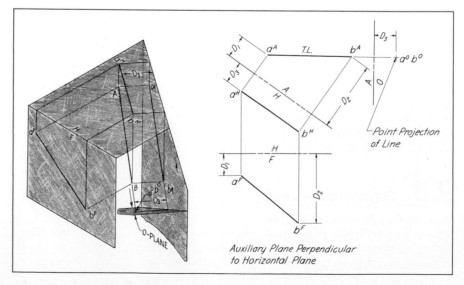

FIG. 9.10 Point view of a line.

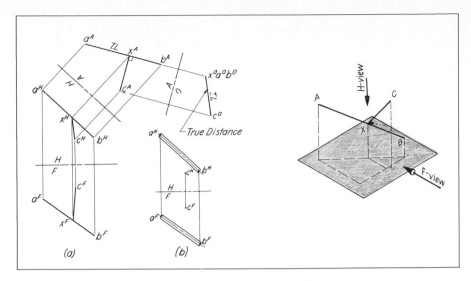

FIG. 9.11 To find the shortest distance from a point to a line.

projection, only an auxiliary view is needed to show the line as a point. However, in the case of an oblique line both an inclined auxiliary and an oblique auxiliary view are required, for a point view must always follow a true-length view. In other words, the plane of projection for the view showing the line as a point must be adjacent to the plane for the true view and be perpendicular to it.

GIVEN: The F view $a^F b^F$ and the H view $a^H b^H$ of the oblique line AB (Fig. 9.10).

SOLUTION: (1) Draw the view showing the TL (true length) of AB. This is an auxiliary view drawn, as explained in Sec. 9.7. (2) Draw reference line AO perpendicular to the true-length projection $a^A b^A$. This reference line is for an O plane that is perpendicular to the A plane. (3) Draw a projection line from $a^A b^A$ and transfer the distance D_3 from the H view to the O view. It should be noted from the pictorial drawing that the distance D_3 is common to both of these views, and that points a^o and b^o coincide to give a point or end view of line AB.

9.11 To Find the Shortest Distance from a Point to a Line

The shortest distance between a given point and a given straight line must be measured along a perpendicular drawn from the point to the line. Since lines that are perpendicular will have their projections shown perpendicular in any view showing either or both lines in true length (Sec. 9.8), the perpendicular must be drawn in the view showing the given line in true length, or as a point.

GIVEN: The F and H views of the line AB and point C (Fig. 9.11).

SOLUTION: (1) Draw the A (auxiliary) view showing the true-length view $a^A b^A$ of line AB and view c^A of point C. (2) Draw $c^A x^A$ perpendicular to $a^A b^A$. Line $c^A x^A$ is a view of the required perpendicular from point C to its juncture with line AB at point X. (3) Draw reference line AO parallel to $c^A x^A$.

This reference line locates an O plane, which will be parallel to the perpendicular CX and perpendicular to the A plane. The O view will show the true length of CX. Line CX does not show true length in any of the other views.

9.12 Principal Lines of a Plane (Fig. 9.12)

Those lines that are parallel to the principal planes of projection are the *principal lines of a plane*. A principal line may be either a horizontal line, a frontal line, or a profile line. Principal lines are true-length lines and one such line may be drawn in any plane to appear true length in any one of the principal views, as desired. This is an important principle that is the basis for the solution of many problems involving lines and planes.

9.13 To Obtain the Edge View of a Plane

When a plane is vertical, an edge view of it will be seen from above and it will be represented by a line in the top view. Should a plane be horizontal, it will appear as an edge in the frontal view. However, planes are not always vertical or horizontal; frequently they are inclined or oblique to the principal planes of projection.

Finding the edge view of a plane is a basic construction that is used to determine the slope of a plane (dip), to determine clearance, and to establish perpendicularity. The method presented here is part of the construction used to obtain the true size and shape of a plane, to determine the angle between a line and plane, and to establish the location at which a line pierces a plane.

The edge view of an oblique plane can be obtained by viewing the plane with direction of sight parallel to it. The edge view will then appear in an auxiliary view. When the auxiliary view shows height, the slope of the plane is shown (Fig. 9.13).

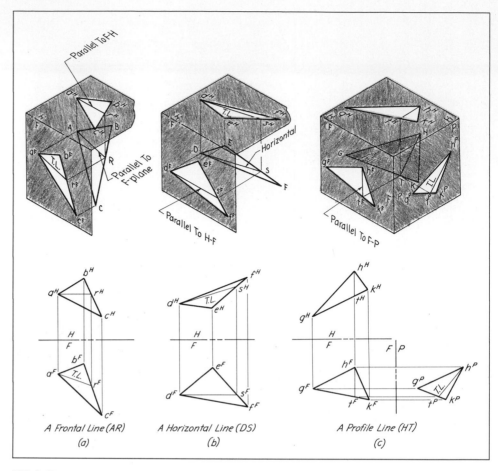

FIG. 9.12 Location of a principal line in a plane.

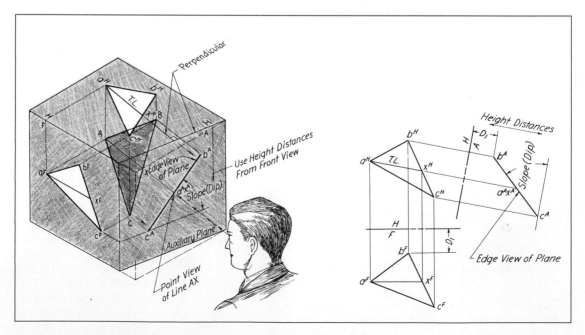

FIG. 9.13 To find the edge view of a plane.

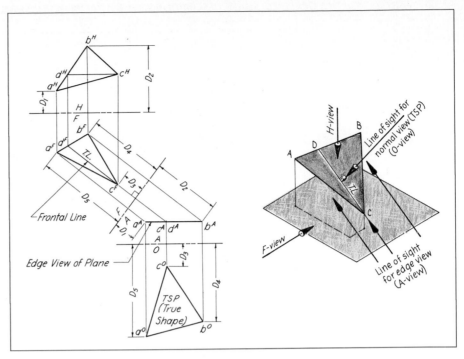

FIG. 9.14   To find the true shape of an oblique plane.

GIVEN: The *F* and *H* views of plane *ABC*.

SOLUTION: (1) Draw the horizontal line *AX* in the plane. Because *AX* is parallel to the horizontal, $a^F x^F$ will be horizontal and must be drawn before the position of $a^H x^H$ can be established. (2) Draw reference line *HA* perpendicular to $a^H x^H$. (3) Construct the *A* view, which will show the plane *ABC* as a straight line. Since the *F* view and *A* view have height as a common dimension, the distances used in constructing the *A* view were taken from the *F* view. It should be noted by the reader that an edge view found by projecting from the front view will show the angle that the plane makes with the *F* plane. Similarly, the edge view found by projecting from the side view will show the angle with the *P* plane.

9.14 To Find the True Shape (TSP) of an Oblique Plane

Finding the true shape of a plane by projection is another of the basic constructions that the student must understand, for it is used to determine the solution of two of the problems that are to follow. In a way, the construction shown in Fig. 9.14 is a repetition of that shown in Fig. 8.19. However, repetition in the form of another presentation should help even those students who feel they understand the method for finding the true shape of an oblique surface of an object.

To see the true size and shape of an oblique plane an observer must view it with a line of sight perpendicular to it. To do this one must, as the first step in the construction, obtain an edge view of the plane. An O view taken from the A view will then show the true shape of the plane.

GIVEN: The *F* and *H* views of plane *ABC*.

SOLUTION: (1) Draw a frontal line *CD* in plane *ABC*. (2) Draw reference line *FA* perpendicular to $c^F d^F$ and construct the *A* view showing the edge view of the plane. The auxiliary view has depth in common with the *H* view. (3) Draw reference line *AO* parallel to the edge view in the *A* view and construct the *O* view, which will show the true size and shape of plane *ABC*. The needed distances from the reference line to points in the *O* view are found in the *F* view.

9.15 To Find the Piercing Point of a Line and a Plane

Determining the location of the point where a line pierces a plane is another fundamental operation with which one must be familiar. A line, if it is not parallel to a plane, will intersect the plane at a point that is common to both. *In a view showing the plane as an edge, the piercing point appears where the line intersects (cuts) the edge view.* This method is known as the *edge-view method* to distinguish it from the cutting-plane method, which may be found in Chapter 10.

The simple cases occur when the plane appears as an edge in one of the principal views. The general case, for which we use an auxiliary plane, is given in Fig. 9.15.

GIVEN: Plane *ABC* and line *ST*.

SOLUTION: (1) Draw the horizontal line *BX* in *ABC*. (2) Draw reference line *HA* perpendicular to $b^H x^H$ and construct the *A* view showing an edge view of the plane as line $a^A b^A c^A$ and the view of the line $s^A t^A$. Point p^A where the line cuts the edge view of the plane is the *A* view of the piercing point. (3) Project point *P* back from the *A* view first to the *H* view and then to the *F* view.

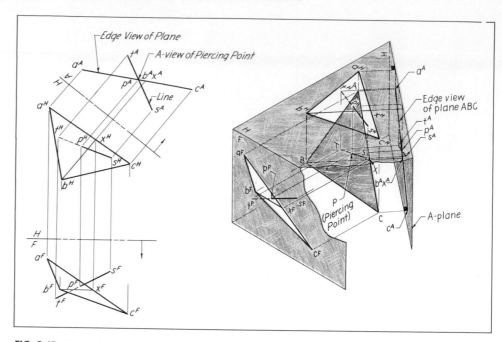

FIG. 9.15 To find the piercing point of a line and a plane.

9.16 To Determine the Angle Between a Line and a Given Plane

The true angle between a given line and a given plane will be seen in the view that shows the plane as an edge (a line) and the line in true length. The solution shown in Fig. 9.16 is based on this premise. The solution as presented might be called the *edge-view method*.

GIVEN: The *F* and *H* views of plane *ABC* and line *ST*.

SOLUTION: (1) Draw the frontal line *BX* in the plane *ABC*. (2) Draw reference line *FA* perpendicular to $x^F b^F$ and construct the *A* view. This view will show plane *ABC* and line $a^A b^A c^A$; however, since this view does not show *ST* in true length, the true angle is not shown. (3) Draw reference line *AO* parallel to the edge view of the plane and construct the *O* view that

will show line *ST* viewed obliquely and plane *ABC* in its true size and shape. (4) Draw reference line OO_1 parallel to $s^o t^o$ and construct the second oblique view.

Line $s_1^o t_1^o$ will show the true length of *ST* in this second oblique view. Plane *ABC* will be seen again as an edge (line), for it now appears on an adjacent view taken perpendicular to the view showing true shape (TSP). The required angle can now be measured between the true-length view of line ST and the edge view of plane *ABC*.

In the illustration of Fig. 9.16 three supplementary views were required to obtain the true angle. If the plane had appeared as an edge in one of the given views, only two supplementary views would have been needed; if it had appeared in true shape in a given view only one additional view, properly selected to show the plane as an edge and the line in true length, would have been needed.

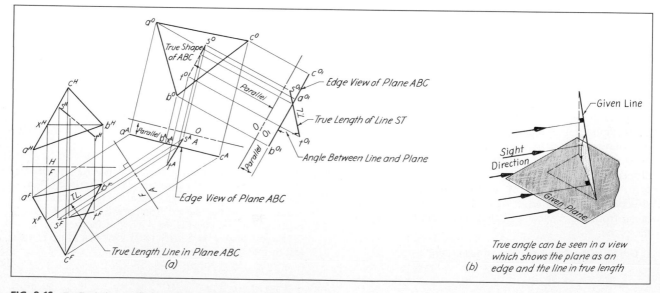

FIG. 9.16 To find the angle between a line and a plane.

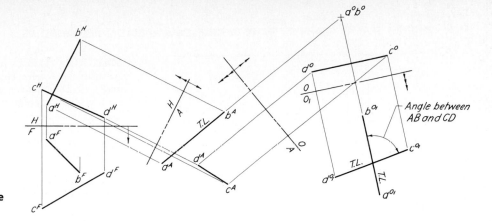

FIG. 9.17 **To find the angle between two skew lines.**

9.17 To Determine the Angle Between Two Nonintersecting (Skew) Lines

The angle between two nonintersecting lines is measurable in a view that shows both lines in true length. The analysis and construction that might be followed to obtain the needed view are shown in Fig. 9.17.

GIVEN: The two nonintersecting lines *AB* and *CD*.

SOLUTION: (1) Draw the reference line *HA* parallel to $a^H b^H$ and construct the *A* view that shows *AB* in true length and the view of the line *CD*. (2) Draw reference line *AO* perpendicular to $a^A b^A$ and draw the *O* view in which the line *AB* will appear as a point ($a^o b^o$). (3) Draw reference line OO_1, parallel to $c^o d^o$ and construct the *O* view showing the true length of *CD*. Line *AB* must also show true length in this view since *AB*, showing as a point in the *O* view, is parallel to the O_1 plane. With both lines shown in true length in the O_1 view, the required angle may be measured in this view.

9.18 To Determine the True Angle Between Two Intersecting Oblique Lines

Since, as previously stated, two intersecting lines establish a plane, *the true angle between the intersecting lines may be seen in a true-shape view of a plane containing the lines* (Fig. 9.14). In Fig. 9.18 line *AC* completes plane *ABC* containing the given line *AB* and *BC*. It is necessary to find the true angle between *AB* and *BC*.

SOLUTION: (1) Draw the frontal line *XC*. (2) Draw reference line *FA* perpendicular to $c^F x^F$ and construct the *A* view showing an edge view of plane *ABC*. (3) Draw reference line *OA* parallel to the edge view $a^A c^A b^A$ and construct the *O* view, which will show the TSP of plane *ABC*. In this view it is desirable to show only the given lines. The true angle between *AB* and *BC* is shown by $a^o b^o c^o$. As a practical application, this method might be used to determine the angle between two adjacent sections of bent rod, as shown in (*b*).

9.19 Distance Between Two Parallel Planes

When two planes are parallel their edge views will appear as parallel lines in the same view. The clearance or perpendicular distance between them can be measured in this view (Fig. 9.19). The existence of planes as parallel lines in a view is another proof that they are parallel.

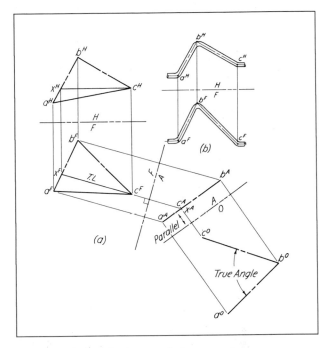

FIG. 9.18 **To find the true angle between two intersecting lines.**

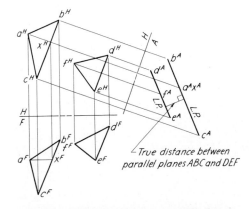

FIG. 9.19 **Distance between parallel planes.**

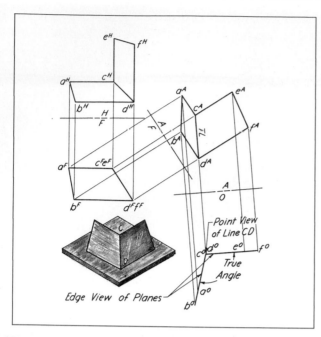

FIG. 9.20 **To find the dihedral angle between two planes.**

9.20 To Find the Dihedral Angle Between Two Planes

The angle between two planes is known as a *dihedral angle*. *The true size of this angle between intersecting planes may be seen in a plane that is perpendicular to both*. For this condition as set forth, the intersecting planes will appear as edges and the line of intersection of the two planes as a point. The true angle may be measured between the edge views of the planes (Fig. 9.20).

GIVEN: The intersecting planes *ABCD* and *CDEF*. The line of intersection is line *CD*, as shown in the pictorial drawing.

SOLUTION: (1) Draw reference line *FA* parallel to $c^F d^F$ and construct the *A* view. This view will show *CD* in true length ($c^A d^A$). (2) Draw reference line *AO* perpendicular to $c^A d^A$ and construct the adjacent *O* view. Since this view was taken looking along line *CD*, points *C* and *D* are coincident and appear as a single point identified as $c^O d^O$. The intersecting planes show as edge views and the true angle between the given planes may be measured between these edge-view lines. When two planes are given that do not intersect, the dihedral angle may be found after the line of intersection has been determined.

9.21 To Find the Shortest Distance Between Two Skew Lines

As was stated in Sec. 9.5, any two lines that are not parallel and do not intersect are called *skew lines. The shortest distance between any such lines must be measured along one line and only one line that can be drawn perpendicular to both.* This common perpendicular can be drawn in a view that is taken to show one line as a point. Its projection will be perpendicular to the view of the other line and will show in true length (Fig. 9.21).

GIVEN: The *F* and *H* views of two skew lines *AB* and *CD*.

SOLUTION: (1) Draw an *A* view adjacent to the *H* view to show line *AB* in true length ($a^A b^A$). Line *CD* should also be shown in this same view ($c^A d^A$). (2) Draw reference line *AO* perpendicular to $a^A b^A$ and draw the *O* view in which line *AB* will appear as a point ($a^O b^O$). It is in this view that the exact location of the required perpendicular can be established. (3) Draw the line $e^O f^O$ through point $a^O b^O$ perpendicular to $c^O d^O$. The shortest distance between the skew lines now appears in true distance as the length of $e^O f^O$. Although *CD* does not appear in true length in the *O* view, $e^O f^O$ does, and hence $c^O d^O$ and $e^O f^O$ will appear perpendicular. (4) Complete the *A* view by first locating point e^A on $c^A d^A$ and then draw $e^A f^A$ parallel to reference line *AO*. (5) Locate points *E* and *F* in the *H* and *F* views remembering that point *E* is located on line *CD* and point *F* on line *AB*.

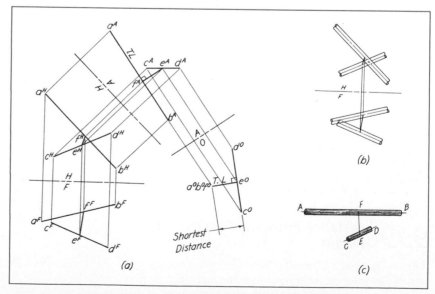

FIG. 9.21 **To find the shortest distance between two skew lines.**

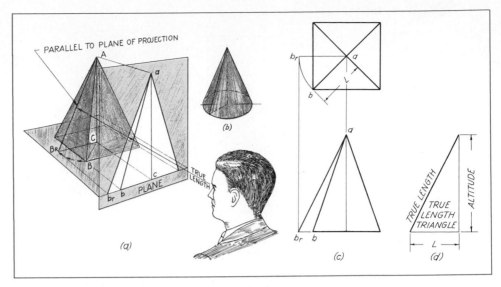

FIG. 9.22 **True length of a line, revolution method.**

In engineering design an engineer frequently has to locate and find the length of the shortest line between two skewed members in order to determine clearance or the length of a connecting member. In underground construction work one might use this method to locate a connecting tunnel.

In Fig. 9.21(b) and (c) we see the two rods for which the clearance distance was determined in (a). Lines *AB* and *CD* represent the center lines of the rods.

B REVOLUTION— COORDINATE PLANES

9.22 To Find the True Length of a Line by Revolution

In engineering drawing, it is frequently necessary to determine the true length of a line when constructing the development of a surface (Chapter 10). The true lengths must be found of those lines that are not parallel to any coordinate plane and therefore appear foreshortened in all the principal views. (see Sec. 5.9). The practical as well as theoretical procedure is to revolve any such oblique line into a position parallel to a coordinate plane such that its projection on that particular plane will be the same length as the line. In Fig. 9.22(a), this is illustrated by the edge *AB* on the pyramid. *AB* is oblique to the coordinate planes, and its projections are foreshortened. If this edge line is imagined to be revolved until it becomes parallel to the frontal plane, then the projection ab_r in the front view will be the same length as the true length of *AB*.

A practical application of this method is shown in Fig. 9.22(c). The true length of the edge *AB* in Fig. 9.22(a) would be found by revolving its top projection into the position ab_r, representing *AB* revolved parallel to the frontal plane, and then projecting the end point b_r down into

its new position along a horizontal line through *b*. The horizontal line represents the horizontal plane of the base, in which the point *B* travels as the line *AB* is revolved.

Commercial draftsmen who are unfamiliar with the theory of coordinate planes find the true-length projection of a line by visualizing the line's revolution. They think of an edge as being revolved until it is in a plane perpendicular to the line of sight of an observer stationed an infinite distance away (Figs. 10.10, 10.11, and 10.12). The process corresponds to that used in drawing regular orthographic views (Sec. 5.4). Usually this method is more easily understood by a student.

Note in Fig. 9.22(a) and (b) that the true length of a line is equal to the hypotenuse of a right triangle whose altitude is equal to the difference in the elevation of the end points and whose base is equal to the top projection of the line. With this fact in mind, many draftsmen determine the true length of a line by constructing a true-length triangle similar to the one illustrated in Fig. 9.22(d).

9.23 Revolution of an Object

Although in general the views on a working drawing represent a machine part satisfactorily when shown in a natural position, it is sometimes desirable to revolve an element until it is parallel to a coordinate plane in order to improve the representation or to reveal the true size and shape of a principal surface, or true length of a line.

The distinguishing difference between this method and the method of auxiliary projection (Chapter 8) is that, *in the procedure of revolution, the observer turns (revolves) the object with respect to the customary planes of projection, instead of shifting viewing position with respect to either an inclined or an oblique surface of the object.*

Despite the fact that the revolution of an entire object, as illustrated in Fig. 9.23, rarely has a practical application in industry, the making of such a drawing provides excellent drill in projection. Therefore, since the several

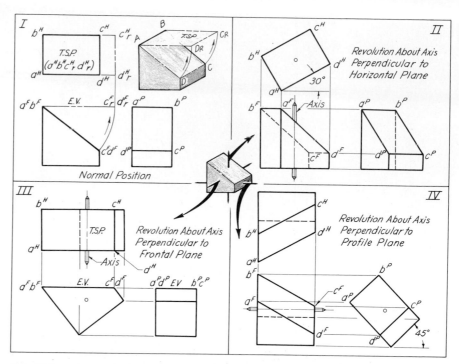

FIG. 9.23   Single revolution about an axis perpendicular to a principal (horizontal, frontal, or profile) plane.

articles that follow are intended primarily for training students, the practical applications have been omitted, while the procedures for revolving simple objects are explained in detail.

9.24   Simple (Single) Revolution

When the regular views are given, an object may be shown in another position, as may be required, by imagining it to be revolved about an axis perpendicular to one of the principal (horizontal, frontal, or profile) planes. A single revolution about such an axis is known as a "simple revolution." The three general cases are shown in Fig. 9.23, spaces II, III, and IV.

9.25   To Determine the True-Shape Projection (Normal View) of a Plane Surface—Revolution Method

If a surface is parallel to one of the principal planes of projection, its projection on that plane will show its true shape, as has been explained in Sec. 5.11. In Fig. 9.23, space I, the inclined surface $ABCD$ appears as a line $(a^F b^F c^F d^F)$ in the front view and is shown foreshortened in the top and side views. Line A–B was taken as the axis of rotation and the surface was revolved into a position parallel to the H plane, as shown by the pictorial drawing $(ABC_R D_R)$. First, the edge view of the surface, as represented by the line $a^F b^F c^F d^F$, was revolved about $a^F b^F$ into a horizontal position, as shown by the line

$a^F b^F c_r^F d_r^F$. The complete H view $(a^H b^H c_r^H d_r^H)$ of the surface as revolved shows the true shape of $ABCD$.

9.26   Revolution About a Vertical Axis Perpendicular to the Horizontal Plane

A simple revolution about an axis perpendicular to the horizontal plane is illustrated in Fig. 9.23, space II. The object is first revolved about the assumed imaginary axis until it is in the desired position (see pictorial). The views of the part in its revolved position are then obtained by orthographic projection, as in the case of any ordinary multiview drawing. Since the top view will not be changed in shape by the revolution, it must be drawn first in its revolved position (at 30° in this case) and the front and side views should be projected from it. Since the heights to all of the points on the object also remain unchanged by the revolution, height distances could be conveniently projected from the initial views in I.

The top view may be drawn directly in revolved position, without first drawing the usual orthographic views. If this procedure is followed, the height distances for the front and side views may be set off to known dimensions.

9.27   Revolution about a Horizontal Axis Perpendicular to the Frontal Plane

If an object is revolved about an imaginary axis perpendicular to the frontal plane, as shown in Fig. 9.23, space

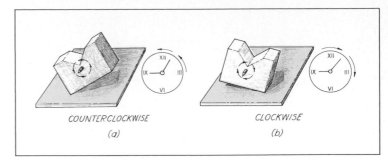

COUNTERCLOCKWISE
(a)

CLOCKWISE
(b)

FIG. 9.24 **Direction of revolution.**

III, the front view changes in position but not in shape. The front view, therefore, should be drawn first in its revolved position, and the top and side views should be projected from it. The depth of the top view and the side view remain unchanged since the depth distance is parallel to the axis. If the usual unrevolved views are not drawn first, the front view may be drawn directly in its revolved position and depth distances can then be laid off to known dimensions.

9.28 *Revolution about a Horizontal Axis Perpendicular to the Profile Plane*

A single revolution of an object about an axis perpendicular to the profile plane is illustrated in Fig. 9.23, space IV. Since in this case it is the side view that is perpendicular to the axis and revolves parallel to the coordinate plane of projection, it is the side view that remains unchanged in its shape. The width of the top and front views is not affected by the revolution. Therefore, horizontal dimensions for these views may be set off by using known measurements.

From these general cases of simple revolution, two principles have emerged that can be stated as follows.

1. The view that is perpendicular to the axis of revolution changes only in position.

2. The lengths of the lines parallel to the axis do not change during the revolution and, therefore, may be

either laid off to known measurements or projected from the usual orthographic views of the object.

9.29 *Clockwise and Counterclockwise Revolution*

An object may be revolved either clockwise or counterclockwise about an axis of revolution. The direction is indicated by the view to which the axis is perpendicular. For example, front views, when revolved as in Fig. 9.24(b), show a clockwise revolution. When revolved as in (a), their revolution is counterclockwise. Top views show a clockwise revolution when revolved to the right. Right-side views indicate a clockwise direction of revolution when they have been revolved to the right and a counterclockwise direction when revolved to the left. Remember that counterclockwise revolution is specified in positive degrees. Clockwise revolution is specified in negative degrees.

9.30 *Successive (Multiple) Revolution*

Since it is possible to show an object in any position relative to the coordinate planes of projection, it can be drawn as may be required by making a series of successive simple revolutions. Usually such a series is limited to three or four stages. Figure 9.25 shows an object revolved successively about two separate axes. The usual orthographic

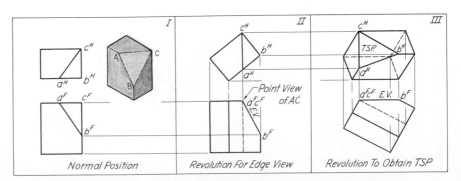

Normal Position   *Revolution For Edge View*   *Revolution To Obtain TSP*

FIG. 9.25 **Successive revolution to show the true shape of an oblique surface.**

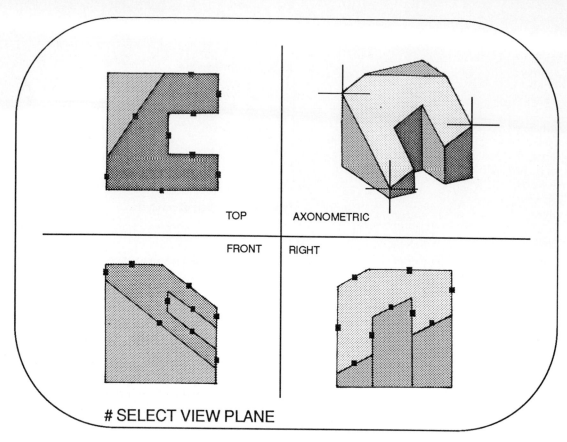

TOP   AXONOMETRIC

FRONT   RIGHT

SELECT VIEW PLANE

FIG. 9.26   **Select view plane. (Computer generated)**

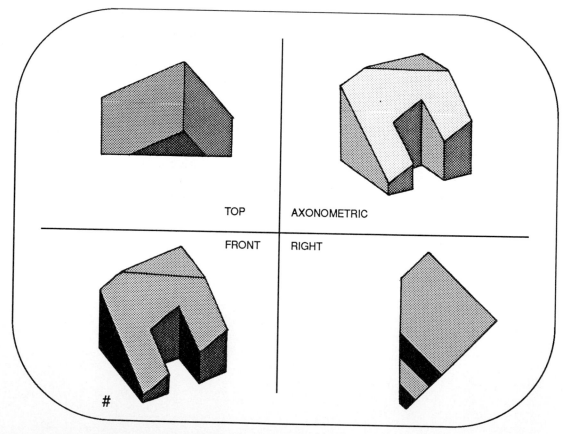

TOP   AXONOMETRIC

FRONT   RIGHT

#

FIG. 9.27   **Normal view of an oblique surface. (Computer generated)**

views of the given object are shown at the left in space I. In space II, the object has been revolved (counterclockwise) about a vertical axis to obtain an edge view (EV) of the oblique surface *ABC* (see pictorial). Note that the line *AC* projects as a point ($a^F c^F$) in the front view. The identification of each corner may not be necessary in the case of a simple object. However, if the object is in the least complex, possible confusion is avoided if each corner point is either numbered or identified by a letter, as was done for the oblique surface in the illustration. In space III at the right, the object is represented after it has been revolved from its previous position in space II. In this case, since the edge view (EV) of surface *ABC* is now horizontal, a true-shape projection (TSP) of the surface will appear in the top view.

A convenient method of copying a view in a new revolved position is first to trace it on a small piece of tracing paper and then to place the traced view in correct position and draw the new view. An alternative method is to place the traced view in correct position and indent the corner points, using the dividers opened out. With the traced view removed, the revolved view on the final drawing may be completed by joining the indentations.

C/ CADD STRATEGIES FOR DISPLAYING THE NORMAL VIEWS OF OBLIQUE SURFACES

9.31 CADD Strategies for Displaying Normal Views

CADD computers are capable of displaying geometry in any position if the operator can define the desired view direction in terms the computer understands. One way of doing this is to define a view direction vector normal to a surface. In Fig. 9.26 three principal views plus an axonometric view are displayed. A normal view of the oblique surface on the object is desired. The operator has defined a view plane in the axonometric view by identifying three points on the oblique surface. This has, within the computer, created a view vector perpendicular to the surface. The boundary lines on the oblique surface have been identified at line origins in the other views by small boxes. This is a check for the operator that the correct plane has been identified.

Upon command, the CADD computer aligns the normal vector with the device Z axis, presenting the normal view of the oblique surface in the front view (Fig. 9.27). Because the surface appears normally in the front view, one would expect edge views of the surface in adjacent views. This is the case as top and right-side views have been updated.

D/ VECTOR GEOMETRY

9.32 Vector Methods

In order to be successful in solving some types of problems that arise in design, a well-trained technologist should have a working knowledge of *vector geometry*. The methods presented in this chapter should furnish the student with some background knowledge for solving force problems as they appear in the study of mechanics, strength of materials, and design. Through discreet use of the methods of vector geometry, as well as mathematical methods, it is possible to solve engineering problems quickly within a fully acceptable range of accuracy. Since any quantity having both magnitude and direction may be represented by a fixed or rotating vector, vector operations are commonly used for problems in the design of frame structures, problems dealing with velocities in mechanisms, and for problems arising in the study of electrical properties. Because a student in a beginning course in engineering graphics should understand basic principles rather than specialized problems, the methods given in this chapter for solving both two-dimensional and three-dimensional force problems deal mainly with static structures or, in other words, structures with forces acting so as to be in equilibrium. In a study of physics, graphical methods are useful for the composition and resolution of forces.

It is hoped that as a student progresses through other undergraduate courses, a desire to learn more about the use of vector methods for solving problems will grow, and that the cases when there is a choice between a graphical and an algebraic method will be recognized. The graphical method is better for many cases because it is much quicker and can be checked more easily.

An example of a vector addition is shown in Fig. 9.28. An airplane is flying north with a cross wind from the west. If the speed of the plane is 250 kph (250 kilometers per hour) and the wind is blowing toward the east at 98 kph, the plane will be flying NE (northeast) at 270 kph. Vectors can be used for problems of this type because forces acting on a body have both magnitude and direction.

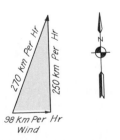

FIG. 9.28 Vector problem.

9.33 Force

In our study of vector methods, a *force* may be defined as *a cause which tends to produce motion in an object.*

A force has four characteristics which determine it. First, a force has *magnitude*. The value of this magnitude is expressed in terms of some standard unit. Second, a force has a *line of action*. This is the line along which the force acts. Third, a force has *direction*. This is the direction in which it tends to move the object upon which it acts. Fourth, and last, a force has a *point of application*. This is the place at which it acts upon the object, often assumed to be a point at the center of gravity.

When values are given for magnitude, it is essential that consistent units be used. The English gravitational system uses feet, pounds, and seconds as basic units while the metric(SI) absolute system uses the kilogram, meter, and second as units of mass, length, and time. Since the student will find it necessary to be familiar with both systems, both systems have been used in this chapter.

The SI unit of force (Fig. 9.29), called a newton (N), is the force that must be applied to a mass of one kilogram (kg) to produce an acceleration of one meter per second per second. The weight of a body in newtons is equal to its mass in kilograms multiplied by 9.81 m/s². See Appendix Table 9.

Conversion Factors

English to SI	SI to English
Force 1 lb = 4.448 N	1 N = 0.2248 lb
= 4.448 kg · m/s²	

9.34 Vector

A force can be represented graphically by a straight line segment with an arrowhead at one end. Such an arrow when used for this purpose is known as a *vector* (Fig. 9.29). The position of the body of the arrow represents the line of action of the force while the arrowhead points out the direction. The magnitude is represented to some selected scale by the overall length of the arrow itself.

When a force acts in a two-dimensional plane, only one view of the vector is needed. However, if the force is in space, two views of the vector must be given.

9.35 Addition of Vector Forces—Two Forces

For a thorough understanding of the principles of vector addition, two simple examples will be considered.

If one of two men who find it necessary to move a supply cabinet pushes on it with 60 lb of force while the other pushes in the same direction with 40 lb of force, the total force exerted to move the cabinet is 100 lb. The representation of two or more such forces in the manner shown in Fig. 9.30 amounts to a *vector addition*. Should these men be in a prankish mood and decide to push in opposite directions, as illustrated in Fig. 9.31, the cabinet

might move provided the 20-lb resultant force were sufficient to overcome friction. The 20-lb resultant comes from a graphical addition.

Now let it be supposed that force A represented by F_A and force B represented by F_B in Fig. 9.32 act from a point P, the point of application. The resultant force on the body will not now be the sum of forces A and B, but instead will be the graphical addition of these forces as represented by the diagonal of a parallelogram having sides equal to the scaled length of the given forces. This single force R of 470 N would produce an effect upon the body that would be equivalent to the combined forces F_A and F_B. The single force which could replace any given force system, is known as the *resultant* (R) for the force system.

Figure 9.32 shows that the resultant R divides the parallelogram into two equal triangles. Therefore, R could have been found just as well by constructing a single triangle as shown in Fig. 9.33 provided that the vector F_B

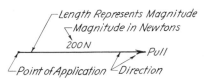

FIG. 9.29 Vector.

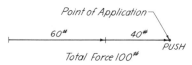

FIG. 9.30 Vector addition (English Gravitational System).

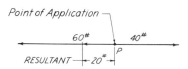

FIG. 9.31 Forces in opposite directions (English Gravitational System).

FIG. 9.32 Parallelogram of forces.

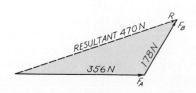

FIG. 9.33 Vector triangle.

is drawn so that its tail-end touches the tip-end of F_A, and R is drawn with its arrow-end of to the tip-end of F_B. Since either of the triangles shown in Fig. 9.33 could have been drawn to determine R, it should be obvious the resultant is the same regardless of the order in which the vectors are added. However, it is important that they be added tip-end to tail-end and that the vector arrows show the true direction for the action of the concurrent forces in the given system.

To find the resultant of two forces, which are applied as shown in Fig. 9.34, it is first necessary to move the vector arrows along their lines of action to the intersection point P before one can apply the parallelogram method.

The forces of a system whose lines of action all lie in one plane are called *coplanar forces*. Should the lines of action pass through a common point, the point of application, the forces are said to be *concurrent*. Figure 9.35 shows a system of forces that are both concurrent and coplanar.

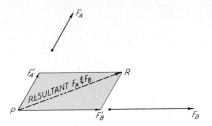

FIG. 9.34 Resultant of two forces.

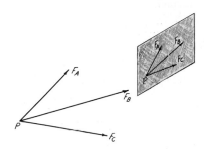

FIG. 9.35 Coplanar, concurrent forces.

9.36 Addition of Vectors—Three or More Forces

The parallelogram method may be used to determine the resultant for a system of three or more forces that are concurrent and coplanar. In applying this method to three or more forces, it is necessary to draw a series of parallelograms, the number depending upon the number of vector quantities that are to be added graphically. For example, in Fig. 9.36 two parallelograms are required to determine the resultant R for the system. The resultant R_1 for forces F_A and F_B is determined by the first parallelogram to be drawn, and then R_1 is combined in turn with F_C by forming the second and larger parallelogram. By combining the forces in this way R becomes the resultant for the complete system.

Where a considerable number of vectors form a system, a somewhat less complicated diagram results, and less work is required when the triangle method is extended and applied to the formation of a vector diagram such as the one shown in Fig. 9.37. In this case the diagram is formed by three vector triangles, one adjacent to the other, and the resultant of forces F_A and F_B is combined with F_C to form the second triangle. Finally, by combining the resultant of the three forces F_A, F_B, and F_C with F_D, the vector R is obtained, which represents the magnitude and direction of the resultant of the four forces. In the construction F_B, F_C, and F_D in the diagram must be drawn so as to be parallel respectively to their lines of action in the system. However, the order in which they are placed in the diagram is optional as long as one vector joins another tip to tail.

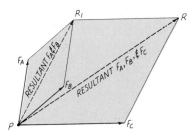

FIG. 9.36 Resultant of three or more forces with a common point of application (parallelogram method).

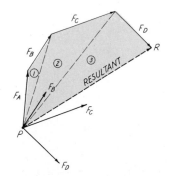

FIG. 9.37 Resultant of forces (polygon of forces).

9.37 Vector Components

A component may be defined as one of two or more forces into which an original force may be resolved. The components, which together have the same action as the single force, are determined by a reversal of the process for vector addition, that is, the original force is resolved using the parallelogram method (see Fig. 9.38). The resolution of a plane vector into two components, horizontal and vertical, is illustrated in (*a*). In (*b*), the resolution of a force into components of specified direction is shown.

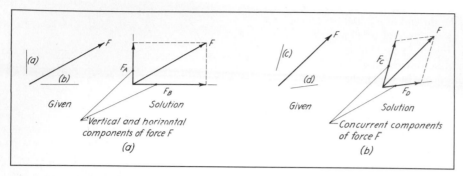

FIG. 9.38   Components.

9.38   *Forces in Equilibrium*

A body is said to be in equilibrium when the opposing forces acting upon it are in balance. In such a state the resultant of the force system will be zero. The concurrent and coplanar force system shown in Fig. 9.39 is in a state

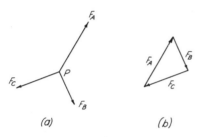

FIG. 9.39   Forces in equilibrium.

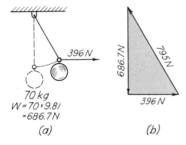

FIG. 9.40   Determination of forces (graphically).

of equilibrium, for the vector triangle closes and each vector follows the other tip to tail.

An *equilibrant* is the force that will balance two or more forces and produce equilibrium. It is a force that would equal the resultant of the system but would necessarily have to act in an opposite direction.

Figure 9.40 shows a weight supported by a short steel cable. The force to be determined is that needed to hold the weight in a state of equilibrium when it is swung from the position indicated by the broken lines into the position shown by solid lines.

This may be done by drawing a vector triangle with the forces in order from tip to tail. The 396 N force vector represents the equilibrant, the force that will balance the 686.7 N force and the 795 N tension force now in the cable. The reader may wonder at the increase in the tension force in the cable from a 686.7 N force when hanging straight down to a 795 N force when the cable is at an angle of 30° with the vertical. It might help to realize that as the weight is swung outward toward a position where the cable will be horizontal, both the tension force and the equilibrant will increase. Theoretically it would require forces infinitely large to hold the system in equilibrium with the cable in a horizontal position.

In solving a force system graphically it is possible to determine two unknowns in a coplanar system.

Now suppose that it is desired to determine the forces acting in the members of a simple truss as shown in Fig. 9.41(*a*). To determine these forces graphically, one should isolate the joint supporting the weight and draw a diagram,

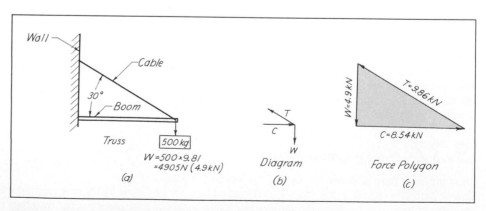

FIG. 9.41   Determination of the forces at the joint of a simple truss.

known as a free-body diagram to show the forces acting at the joint (*b*). Although the lines of this diagram may have any length, they must be parallel to the lines in the space diagram in (*a*). Since the boom will be in compression, a capital letter *C* has been placed along the line that represents the boom in the diagram. A letter *T* has been placed along the line for the cable because it will be in tension. Although the diagram may not have been essential in this particular case, such a diagram does play an important part in solving more complex systems.

In constructing the force polygon, it is necessary to start by drawing the vertical vector, for the load is the only force having a known magnitude and direction. After this vertical vector has been drawn to a length representing 4.9 kN, using a selected scale, the force polygon (triangle) may be completed by drawing the remaining lines representing the unknown forces parallel to their known lines of action as shown in (*a*). The force polygon will close since the force system is in equilibrium.

The magnitude of the unknown forces in the members of the truss can now be determined by measuring the lines of the diagram using the same scale selected to lay out the length of the vertical vector. This method might be used to determine the forces acting in the members at any point in a truss.

9.39 Coplanar, Nonconcurrent Force Systems

Forces in one plane having lines of action that do not pass through a common point are said to be *coplanar, nonconcurrent forces* (Fig. 9.42).

9.40 Two Parallel Forces

When two forces are parallel and act in the same direction, their resultant will have a line of action that is parallel to the lines of action of the given forces and it *will be located between them*. The magnitude of the resultant will be equal to the sum of the two forces [Fig. 9.43(*a*)], and it will act through a point that divides any perpendicular line joining the lines of action of the given forces inversely as the forces.

Should the two forces act in opposite directions, as shown in (*b*), the resultant will be located outside of them and will have the same direction as the greater force. Its magnitude will be equal to the difference between the two given forces. The proportion shown with the illustration in (*b*) may be used to determine the location of the point of application of the resultant. Those who prefer to determine graphically the location of the line of action for the resultant may use the method illustrated in Fig. 9.44. This method is based on well-known principles of geometry.

With the two forces F_A and F_B given, any line 1–2 is drawn joining their lines of action. From this line two distances must be laid off along the lines of action of the given forces. If the given forces act in the same direction,

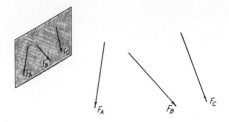

FIG. 9.42   Coplanar, nonconcurrent forces.

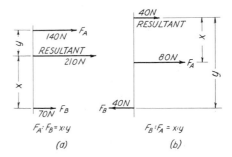

FIG. 9.43   Parallel forces.

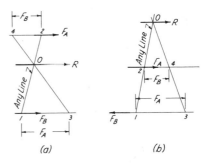

FIG. 9.44   Determination of the position of the resultant of parallel forces (graphical method).

then the distances are laid off in opposite directions from the line 1–2, (*a*). If they act in opposite directions, the distances must be laid off on the same side of line 1–2, (*b*). In Fig. 9.44(*a*), a length equal by scale to F_A was laid off from point 1 on the line of action of F_B. Then from point 2 a length equal to F_B was marked off in an opposite direction. These measurements located points 3 and 4, the end points of the line intersecting line 1–2 at point *O*. Point *O* is on the line of action of the resultant *R*. In Fig. 9.44(*b*) this method has been applied to establish the location of the resultant for two forces acting in opposite directions.

9.41 Moment of a Force

The *moment of a force* with respect to a point is the product of the force and the perpendicular distance from the given point to the line of action of the force. In the illustration,

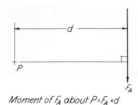

Moment of F_A about $P = F_A \times d$

FIG. 9.45   Moment of a force.

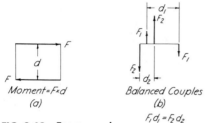

Moment $= F \times d$
(a)

Balanced Couples
(b)

$F_1 d_1 = F_2 d_2$

FIG. 9.46   Force couples.

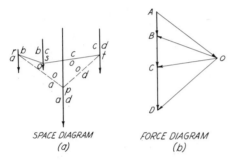

SPACE DIAGRAM
(a)

FORCE DIAGRAM
(b)

FIG. 9.47   Resultant of parallel forces— Bow's notation.

Fig. 9.45, the moment of the force F_A about point P is Mom. $= F_A \times d$. The perpendicular distance d is known as the lever arm of the force.

9.42   Force Couples (Fig. 9.46)

Two equal forces that act in opposite directions are known as a *couple*. A couple does not have a resultant, and no single force can counteract the tendency to produce rotation. The measurement of this tendency is the moment of the couple that is the product of one of the forces and the perpendicular distance between them.

To prevent the rotation of a body that is acted upon by a couple, it is necessary to use two other forces that will form a second couple. The body acted upon by these couples will be in equilibrium if each couple tends to rotate the body in opposite directions and the moment of one couple is equal to the other.

9.43   String Polygon—Bow's Notation

A system for lettering space and force diagrams, known as *Bow's notation*, is widely used by technical authors.

Its use in this chapter will tend to simplify the discussions which follow.

In the space diagram, shown in Fig. 9.47(a), each space from the line of action of one force to the line of action of the next one is given a lowercase letter such as a, b, c, and d in alphabetical order. Thus the line of action for any particular force can be designated by the letters of the areas on each side of it. For example, in Fig. 9.49 the line of action for the 1080-lb force, acting downward on the beam, would be designated as line of action bc. On the force diagram, corresponding capital letters are used at the ends of the vectors. In Fig. 9.47(b), AB represents the magnitude of ab in the space diagram and BC represents the magnitude of bc.

To find the resultant of three or more parallel forces graphically, the *funicular* or *string polygon* is used. The magnitude and direction of the required resultant for the system shown in Fig. 9.47 are known. The magnitude, representing the algebraic sum of the given forces, appears as the heavy line AD of the force polygon. It is required to determine the location of its line of action. With the forces located in the space diagram and the force polygon drawn, the steps for the solution are as follows.

1. Assume a pole point O and draw the rays OA, OB, OC, and OD. Each of the triangles formed is regarded as a vector triangle with one side representing the resultant of the forces represented by the other two sides. For example: If we consider AB to be a resultant force, then OA and OB are two component forces that could replace AB. For the second vector triangle, OB and BC have OC as their resultant. OC, when combined with CD, will have OD as the resultant. OA and OD combine with AD, the final resultant of the system.

2. Draw directly on the space diagram the corresponding strings of the funicular polygon. The funicular polygon may be started at any selected point r along the line of action ab. The string ob will then be parallel to OB of the force polygon. From point s, where ob intersects bc, draw oc parallel to OC. The line oc extended to cd establishes the location of point t. Line od drawn parallel to OD and line oa drawn parallel to OA intersect at point p. Point p is a point on the line of action of force ad, the resultant force AD for the given force system.

When one or more forces of a parallel system are directed oppositely from the others, the magnitude and direction of the resultant will be equal to the algebraic sum of the original forces.

9.44   Coplanar, Nonconcurrent, Nonparallel Forces

In further study of coplanar and nonconcurrent forces it might be supposed that it is necessary to determine the magnitude, direction, and line of action of the one force that will establish a state of equilibrium when combined with the given forces AB, BC, and CD of the force system shown in Fig. 9.48. The direction and line of action of the original forces are given in both the space diagram in (*a*) and the force polygon in (*b*). The magnitude and direction of the force that will produce equilibrium is represented by *DA*, the force needed to close the force polygon. With the force polygon completed, the next step is to assume a pole point *O* and draw the rays *OA*, *OB*, *OC*, and *OD*. Now *OA* and *OB* are component forces of *AB*, and *AB* might be replaced by these forces. To clarify this statement: each of the four triangles may be considered to be a vector triangle, and in the case of vector triangle *OAB*, *AB* can be regarded as the resultant for the other two forces *OA* and *OB*. It should be noted that component force *OB* of the vector triangle *OBC* must be equal and opposite in direction to component force *OB* of *OAB*.

All that remains to be done is to determine the line of action of the required force *DA* by drawing the string diagram as explained in Sec. 9.43, remembering that point *r* may be any point along the line of action of *ab*. The intersection point *p* for strings *oa* and *od* is a point along the line of action *da* of force *DA*. Although lines of action *ab*, *bc*, *cd*, and *da* were drawn to a length representing their exact magnitude in Fig. 9.48(*a*), they could have been drawn to a convenient length to allow for the construction of the string polygon, for these lines merely represent lines of action for the forces *AB*, *BC*, *CD*, and *DA*. The lines were presented in scaled length for illustrative purposes.

9.45 Equilibrium of Three or More Coplanar Parallel Forces

When a given system, consisting of three or more coplanar forces, is in equilibrium, both the force polygon and the funicular polygon must close. If the force polygon should close and the funicular polygon not close, the resultant of the given system will be found to be a force couple.

Two unknown forces of a parallel coplanar force system may be determined graphically by drawing the force and funicular polygons as shown in Fig. 9.49, since the forces are known to be in equilibrium, and all are vertical. Although one may be aware that the sum of the two reaction forces R_1 and R_2 is equal to the sum of forces *AB*, *BC*, *CD*, and *DE*, the magnitude of R_1 and R_2 as single forces is unknown. The location of point *F* in the force polygon, which is needed if one is to determine the magnitudes of R_1 and R_2, may be found by fulfilling the requirement that the funicular polygon be closed.

The funicular polygon is started at any convenient point along the known line of action of R_1, and successive strings are drawn parallel to corresponding rays of the force polygon. In area *b* the string will be parallel to *OB*, in area *c* the string will be parallel to *OC*, and so on, until the string that is parallel to *OE* has been drawn. String *of* may then be added to close the funicular polygon. This closing line from point *x* to the starting point *y* determines the position of *OF* in the force polygon, for ray *OF* must be parallel to the string *of*. The magnitude of the reaction R_1 is represented to scale by vector *FA*, and R_2 is represented by the vector *FE*.

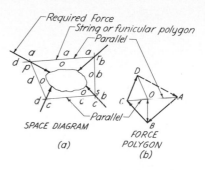

FIG. 9.48 Funicular or string polygon.

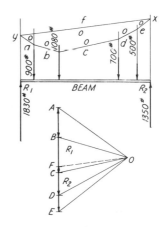

FIG. 9.49 To determine the reaction forces of a loaded beam (English Gravitational System).

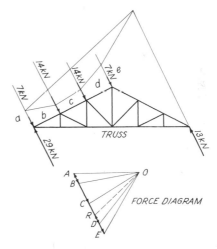

FIG. 9.50 Determination of the reactions for wind loads.

The graphical method for determining the values of wind load reactions for a roof truss having both ends fixed is shown in Fig. 9.50. The solution given is practically identical with the solution applied to the beam in Fig. 9.49.

9.46 Concurrent, Noncoplanar Force Systems

Up to this point in our study of force systems, the student's attention has been directed solely to systems lying in one

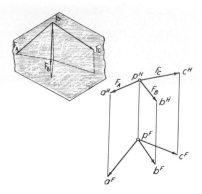

FIG. 9.51 Concurrent, noncoplanar forces.

1. in corresponding views (*H* view and *H* view, or *F* view and *F* view), each vector in the vector diagram will be parallel to its corresponding representation in the space diagram;

2. if a system of concurrent, noncoplanar forces are in equilibrium, the force polygon in space closes, and the projection on each plane will close;

3. the true magnitude of a force can be measured only in the vector diagram when it appears in true length or is made to do so.

Determination of the Resultant of a Force System of Concurrent, 9.47 Noncoplanar Forces

The parallelogram method for determining the resultant of concurrent forces as explained in Sec. 9.36 may be employed to find the resultant of the three forces *OA*, *OB*, and *OC* in Fig. 9.52. Any number of given concurrent, noncoplanar forces can be combined into their resultant by this method. In the illustration, forces *OA* and *OB* were combined into their resultant, which is the diagonal of the smaller parallelogram; then this resultant in turn was combined with the third force *OC* to obtain the final resultant *R* for the given system. Since the true magnitude of *R* can be scaled only in a view showing its true length, an auxiliary view was projected from the front view. The true length of *R* could also have been determined by revolution.

Since the single force needed to hold a force system in balance, known as the *equilibrant*, is equal to the resultant in magnitude but is opposite in direction, this method might be used to determine the equilibrant for a system of concurrent, noncoplanar forces.

In presenting this problem and the two problems that follow, it has been assumed that the student has read the previous sections of this chapter and that his knowledge of the principles of projection is sufficient for him to find the true length of a line, having two views given, and to draw the view of a plane so that it will appear as an edge.

plane in order that the graphical methods dealing with the composition and resolution of forces could be presented in a clear and simple manner, free from the thinking needed for understanding force systems involving the third dimension.

In dealing with noncoplanar forces it is necessary to use at least two views to represent a structure in space. Although the methods as applied to coplanar force systems for solving problems may be extended to noncoplanar systems, the vector diagram for noncoplanar forces must have two views instead of one view as in the case of coplanar forces. To understand the discussions that are to follow, the student must grasp the idea that for the composition and resolution of noncoplanar forces two distinct and separate space representations must be used: the space diagram for the given structure and the related vector diagram (force polygon). Figure 9.51 shows two views of a concurrent, noncoplanar force system not in equilibrium.

There are a few basic relationships that exist between a space diagram and its related vector diagram, which must be kept in mind when solving noncoplanar force problems. These relationships are:

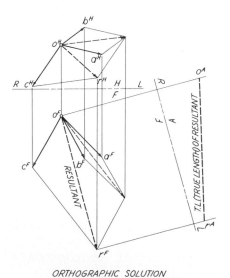

ORTHOGRAPHIC SOLUTION

FIG. 9.52 Determination of the resultant of concurrent, noncoplanar forces.

To Find the Three Unknown Forces of a Simple Load-bearing 9.48 Frame—Special Case

In dealing with the simple load-bearing frame in Fig. 9.53, it should be realized that this is a special case rather than a general one, for two of the truss members appear as a single line in the frontal view of the space diagram. This condition considerably simplifies the task of finding the unknown forces acting in the members, and it is this particular spatial situation that makes this a special case. However, it should be pointed out now that this condition must exist or be set up in a projected view when a vector solution is to be applied to any problem involving a system of concurrent, noncoplanar forces. More will be said about the necessity for having two unknown forces coincide in one of the views in the discussion in Sec. 9.50.

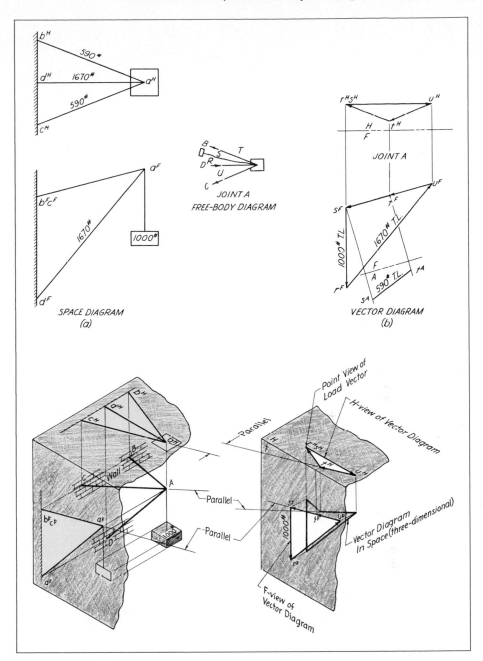

FIG. 9.53 Solution of a concurrent, noncoplanar force system—special case (English Gravitational System)

After the space diagram has been drawn to scale, the steps toward the final solution are as follows.

1. Draw a free-body diagram showing the joint at *A*, the joint at which the load is applied. The lines of this diagram may be of any length, but each line in it must be parallel respectively to a corresponding line in the view of the space diagram to which the free-body diagram is related. In this case, it is the horizontal (top) view. A modified form of Bow's notation was used for convenience in identifying the forces. The vertical load has been shown pulled to one side in order that this force can be made to fall within the range of the notation. This diagram is important to the solution of this prob-

lem, for it enables one to see and note the direction of all of the forces that the members exert upon the joint (note arrowheads). Capital letters were used on the free-body diagram to identify the spaces between the forces, rather than lowercase letters as is customary, so that lowercase letters could be used for the ends of the views of the vectors in the vector diagram.

2. Using a selected scale, start the two views of the vector diagram (*b*) by laying out vector *RS* representing the only known force, in this case the 1000-lb load. Since *RS* is a force acting in a vertical direction $r^F s^F$ will be in true length in the *F* view and will appear as point $r^H s^H$ in the *H* view.

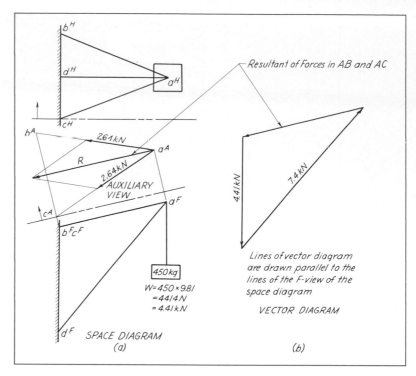

FIG. 9.54   To determine the three unknown forces of a simple load-bearing frame—composition method.

3. Complete the *H* and *F* views of the vector diagram. Since each vector line in the top view must be parallel to a corresponding line in the top view of the space diagram, $s^H t^H$ must be drawn parallel to $a^H b^H$, $t^H u^H$ parallel to $a^H c^H$, and $u^H r^H$ parallel to $a^H d^H$. Since the forces acting at joint *A* are in equilibrium, the vector triangle will close and the vectors will appear tip to tail.

In the frontal view of the vector diagram, $s^F t^F$ will be parallel to $a^F b^F$, $t^F u^F$ will be parallel to $a^F c^F$, and $r^F u^F$ will be parallel to $d^F a^F$.

4. Determine the magnitude of the forces acting on joint *A*. Since vector *RU* shows its true length in the *F* view, the true magnitude of the force represented may be determined by scaling $r^F u^F$ using the same scale used to lay out the length of $r^F s^F$. Although it is known that vectors *ST* and *TU* are equal in magnitude, it is necessary to find the true length representation of one or the other of these vectors by some approved method force scaling to determine the true value of the force.

An arrowhead may now be added to the line of action of each force in the free-body diagram to indicate the direction of the action. Since the free-body diagram was related to the top view of the space diagram, the arrowhead for each force will point in the same direction as does the arrowhead on the corresponding vector in the *H* view of the vector diagram. These arrowheads

show that the forces in members *AB* and *AC* are acting away from joint *A* and are therefore *tension forces*. The force in *AD* acts toward *A* and thus is a *compression force*.

To Find the Three Unknown Forces of a Simple Load-bearing Frame—Composition Method

9.49 (Fig. 9.54)

Since the forces in *AB* and *AC* lie in an inclined plane that appears as an edge in the *F* view, they may be composed into a single force that will have the same effect in the force system as the forces it replaces. This replacement force along with the force in *AD* and the load now become the forces that would be acting in a simple load-bearing truss (see Fig. 9.41). After the vertical vector of this concurrent coplanar force system has been drawn to a selected scale, the triangular force polygon in (*b*) may be completed by drawing the remaining lines representing the forces in *AD* and the resultant *R*. These lines are drawn parallel to their known lines of action as shown in the *F* view of the space diagram in (*a*). Finally, the true length view of *R* must be transferred to the auxiliary view in (*a*) and resolved into its component forces, acting in *AB* and *AC*, that collectively have the same action as the resultant *R*.

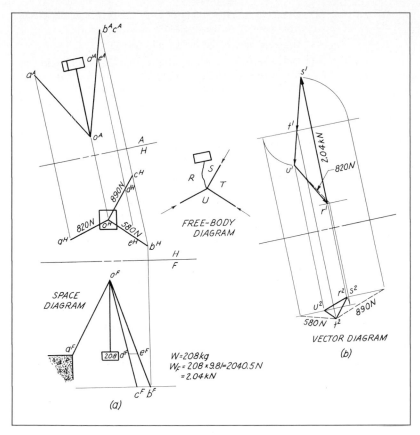

FIG. 9.55 **Solution of a concurrent, noncoplanar force system—general case.**

9.50 To Find the Three Unknown Forces of a Simple Load-bearing Truss—General Case

For the general case shown in Fig. 9.55, the known force is in a vertical position as in the previous problem, but no two of the three unknown forces appear coincident in either of the two given views. For this reason, it is necessary at the very start to add a complete auxiliary view to the space diagram that will combine with the existing top view to give a point view of one member and a line view of two of the three unknown forces. To obtain this desired situation, one should start with the following steps, which will transform the general case into the special case with which one should now be familiar.

STEP 1. Draw a true length line in the plane of two of the members. In Fig. 9.55(*a*) this line is *DE*, which appears in true length (TL) in the *H*-view.

STEP 2. Draw the needed auxiliary view, taken so that *DE* will appear as a point ($d^A e^A$) and *OB* and *OC* will be coincident (line $o^A b^A c^A$). This construction involves finding the edge view of a plane (see Sec. 9.13). In this particular case, the auxiliary view has height in common with the frontal view.

STEP 3. Draw the two views of the vector diagram by assuming the *H*-view and the *A*-view to be the given views of the special case. Proceed by the steps set forth for the special case in Sec. 9.48.

STEP 4. Determine the magnitude of the forces and add arrowheads to the free-body diagram to show the direction of action of the forces acting on point *O*.

9.51

In practice, engineers and technologists find wide use for methods that solve problems through the use of three-dimensional vector diagrams, for any quantity having both magnitude and direction may be represented by a vector. And, although the examples used in the chapter dealt with static structures, which are in the field of the structural engineer, vector diagram methods are used frequently by the electrical engineer for solving electrical problems and by the mechanical engineer for problems dealing with bodies in motion. The student will encounter some of these methods in the textbook for a later course or when presented by an instructor.

 # PROBLEMS

Descriptive Geometry

Problems 1 through 5 have been selected and arranged to offer the student an opportunity to apply basic principles of descriptive geometry.

The problems can be reproduced to a suitable size by transferring the needed distances from the drawing to one or the other of the scales that have been provided for each group of problems. If the inch scale is used, metric values in problem instructions should be converted to decimal inches.

1. (Fig. 9.56). Reproduce the given views of the line or lines of a problem as assigned and determine the true length or lengths using the auxiliary view method explained in Sec. 9.7. A problem may be reproduced to a suitable size by transferring the needed distances from the drawing to one of the given scales to determine values. The distances, as they are determined, should be laid off on the drawing paper using a full-size scale.

2. (Fig. 9.57). Reproduce the given views of the plane of a problem as assigned and draw the view showing the true size and shape. Use the method explained in Sec. 9.14. Determine needed distances by transferring them from the drawing to one of the given scales, by means of the dividers.

3. (Fig. 9.58). These problems are intended to give some needed practice in manipulating views to obtain certain relationships of points and lines. Determine needed distances by transferring them from the drawing to one of the accompanying scales, by means of the dividers.

1. Determine the distance between points *A* and *B*.

2. Draw the *H* and *F* views of a 13 mm perpendicular erected from point *N* of the line *MN*.

3. Draw the *H* view of the 95 mm line *ST*.

4. Draw the *H* and *F* views of a plane represented by an equilateral triangle and containing the *AB* as one of the edges. The added plane *ABC* is to be at an angle of 30° with plane *ABDE*.

5. If the figure *MNOP* is a plane surface, an edge view of the surface would appear as a line. Draw such a view to determine whether or not *MNOP* is a plane.

6. A vertical pole with top *O* is held in place by three guy wires. Determine the slope in tangent value of the angle for the guy wire that has a bearing of N 23° W.

4. (Fig. 9.59). In this group of problems it is required to determine the shortest distance between skew lines and the angle formed by intersecting lines.

1. Show proof that the plane *ABCD* is an oblique plane.

2. Determine the shortest distance between the lines *MN* and *ST*.

3. Through point *K* on line *GH* draw the *F* and *H* views of a line that will be perpendicular to line *EF*.

4. Determine the angle between line *AB* and a line intersecting *AB* and *CD* at the level of point *E*.

5. Through a point on line *MN* that is 32 mm from point *N*, draw the *F* and *H* views of a line that will be perpendicular to line *ST*.

6. Erect a 25 mm perpendicular at point *K* in the plane *EFGH*. Connect the outer end point *L* of the perpendicular with *F*. Determine the angle between *LF* and *KF*.

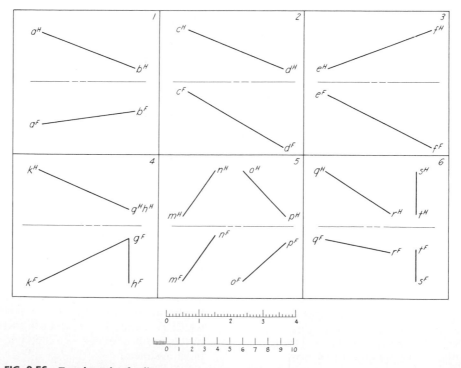

FIG. 9.56   True length of a line.

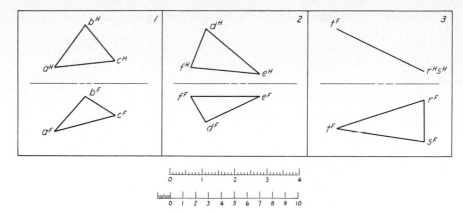

FIG. 9.57 True size and shape of a plane.

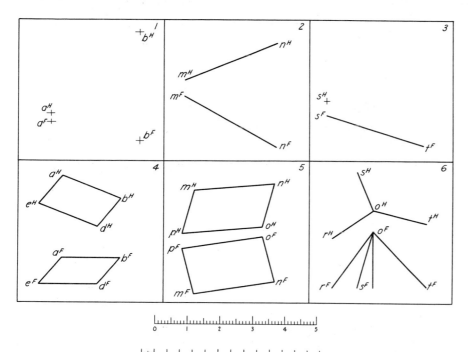

FIG. 9.58 Relationships of points and lines.

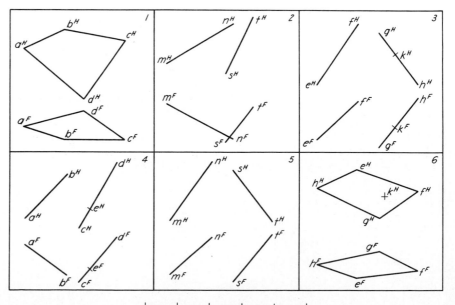

FIG. 9.59 Shortest distance between skew lines and the angle formed by intersecting lines.

5. (Fig. 9.60). These problems require that the student determine the angle between a line and a plane and the angle between two given planes.

1. Determine the angle between the planes *MNQP* and *RST*.

2. Determine the angle between the line *ST* and
   (a) The *H* plane.
   (b) The *F* plane.

3. The line *EF* has a bearing of N 53° E. What angle does this line make with the *P* plane?

4. Draw the *F* and *H* views of a line through point *K* that forms an angle of 35° with plane *MNQR*.

5. The top and front views of planes *ABC* and *RST* are partially drawn.
   (a) Complete the views including the line of intersection.
   (b) Determine the angle between the line of intersection and the *H* plane of projection.

6. Two views of a plane *RST* and the top view of an arrow are shown. The arrow, pointing downward and toward the left, is in a plane that forms an angle of 68° with plane *RST*. The arrow point is 6 mm from the plane *RST*.
   (a) Draw the front view of the arrow.
   (b) Draw the top and front views of the line of intersection of the 68° plane and plane *RST*.

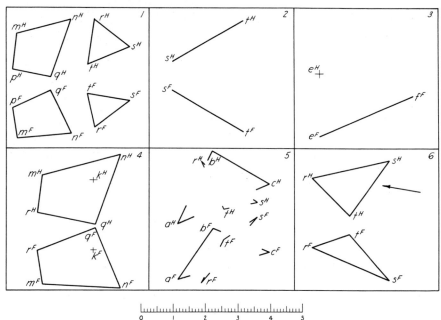

FIG. 9.60   Angle between a line and a plane and the angle between two planes.

Vector Geometry

The following problems have been selected to emphasize the basic principles underlying vector geometry. By solving a limited number of the problems presented, the student should find that he has a working knowledge of some vector methods that are useful for solving problems in design that involve the determination of the magnitude of forces as well as their composition and resolution. The student is to select his own scale remembering that a drawing made to a large scale usually assures more accurate results.

1. (Fig. 9.61). A force of 408 N acts downward at an angle of 60° with the horizontal. Determine the vertical and horizontal components of this force.

2. (Fig. 9.62). Determine the resultant force for the given coplanar, concurrent force system.

3. (Fig. 9.63). Determine the magnitude of the force F_C and the angle that the resultant force *R* makes with the horizontal for the given coplanar, concurrent force system.

4. (Fig. 9.64). Determine the magnitude and direction of the equilibrant for the given coplanar, concurrent force system.

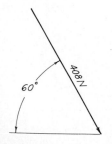

FIG. 9.61

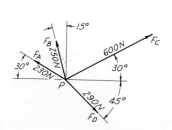

FIG. 9.62   Force system.

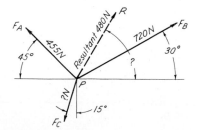

FIG. 9.63

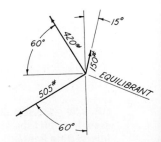

FIG. 9.64

5. (Fig. 9.65). A block weighing 45 lb is to be pulled up an inclined plane sloping at an angle of 30° with the horizontal. If the frictional resistance is 16 lb, what is the magnitude of the force F_M that is required to move the block uniformly up the plane?

6. (Fig. 9.66). A horizontal beam AB is hinged at B as shown. The end of the beam at A is connected by a cable to a hook in the wall at C. The load at A is 110 kg. Using the dimensions as given, determine the tension force in the cable and the reaction on the hinge at B. The weight of the beam is to be neglected.

7. (Fig. 9.67). A 270 kg load is supported by cables as shown. Determine the magnitude of the tension in the cables.

8. (Fig. 9.68). A ship that is being pulled through the entrance of a harbor is headed due east through a cross current moving at 4 knots as shown. If the ship is moving at 12 knots, what is the speed of the tugboat?

9. (Fig. 9.69). Determine the magnitude of the reactions R_1 and R_2 of the beam with loads as shown.

10. (Fig. 9.70). Determine the magnitude of the reactions R_1 and R_2 of the beam.

11. (Fig. 9.71). Determine the magnitude of the reactions R_1 and R_2 for the roof truss shown. Each of the six panels is of the same length.

12. (Fig. 9.72). Determine the magnitude of the reactions R_1 and R_2 to the wind loads acting on the roof truss as shown.

13. (Fig. 9.73). A tripod with a 40 kg load is set up on a level floor as shown. Determine the stresses in the three legs due to the vertical load on the top.

14–15. (Figs. 9.74 and 9.75). Determine the stresses in the members of the space frame shown.

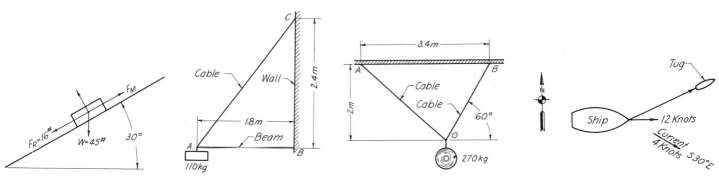

FIG. 9.65   **FIG. 9.66**   **FIG. 9.67 Cable support system.**   **FIG. 9.68**
(Values shown in [ ] are in the metric system.)

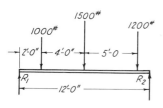

FIG. 9.69

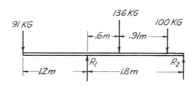

FIG. 9.70 Simple beam.

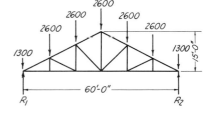

FIG. 9.71

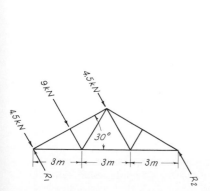

FIG. 9.72

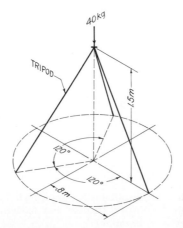

FIG. 9.73

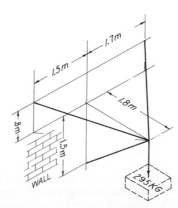

FIG. 9.74 Space frame.

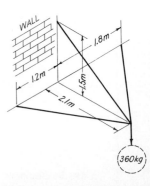

FIG. 9.75 Space frame.
(Values shown in [ ] are in the metric system.)

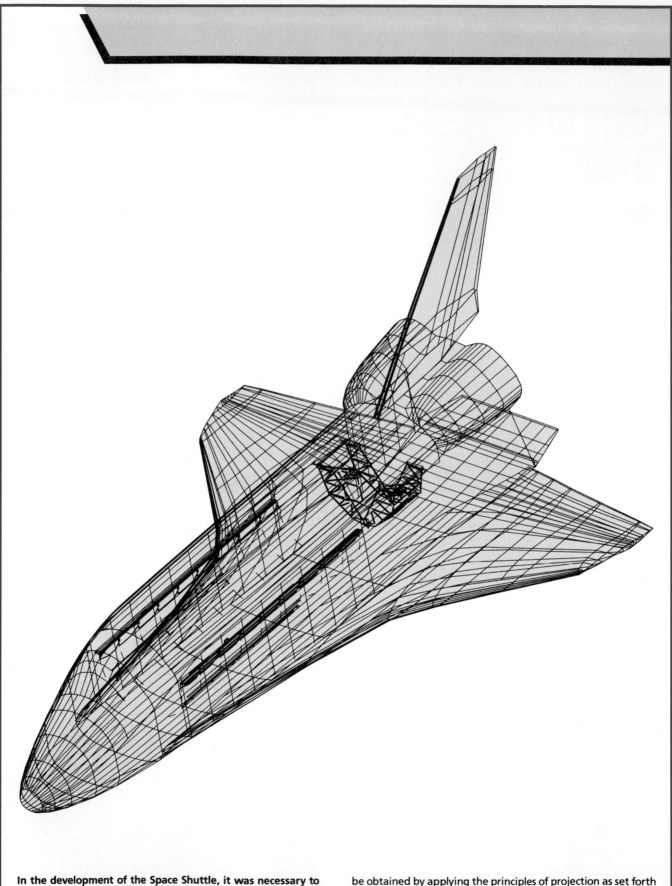

In the development of the Space Shuttle, it was necessary to solve many problems involving the intersection and development of surfaces. The intersection of geometric surfaces may be obtained by applying the principles of projection as set forth in the chapter that follows. (*Courtesy Computervision Corporation*)

Developments and Intersections

10.1 Introduction

Intersections and developments are logically a part of the subject of *descriptive geometry*. A few of the many applications that can be handled without advanced study in projection, however, are presented in this chapter.

Desired lines of intersection between geometric surfaces may be obtained by applying the principles of projection with which the student is already familiar. Although developments are laid out and are not drawn by actual projection in the manner of exterior views, their construction nevertheless requires the application of orthographic projection in finding the true lengths of elements and edges.

10.2 Geometric Surfaces

A geometric surface is generated by the motion of a geometric line, either straight or curved. Surfaces that are generated by a moving straight line are known as *ruled surfaces*, and those generated by a curved line are known as *double-curved surfaces*. Any position of the generating line, known as a *generatrix*, is called an *element of the surface*.

Ruled surfaces include planes, single-curved surfaces, and warped surfaces.

A plane is generated by a straight line moving in such

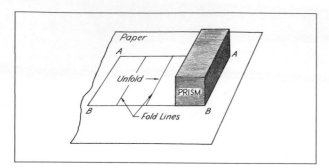

FIG. 10.1 **Development of a prism.**

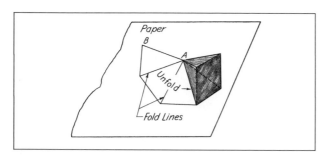

FIG. 10.2 **Development of a pyramid.**

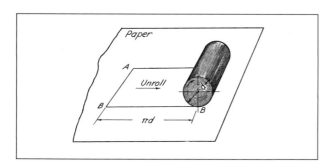

FIG. 10.3 **Development of a cylinder.**

sector of circle

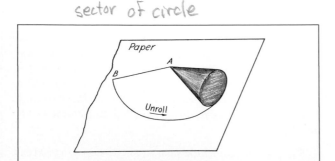

FIG. 10.4 **Development of a cone.**

a manner that one point touches another straight line as it moves parallel to its original position.

A *single-curved surface* is generated by a straight line moving so that in any two of its near positions it is in the same plane.

A *warped surface* is generated by a straight line moving so that it does not lie in the same plane in any two near positions.

Double-curved surfaces include surfaces that are generated by a curved line moving in accordance with some mathematical law.

10.3 *Geometric Objects*

Geometric solids are bounded by geometric surfaces. They may be classified as follows.

1. Solids bounded by plane surfaces: Tetrahedron, cube, prism, pyramid, and others.

2. Solids bounded by single-curved surfaces: Cone and cylinder (generated by a moving straight line).

3. Solids bounded by warped surfaces: Conoid, cylindroid, hyperboloid of one nappe, and warped cone.

4. Solids bounded by double-curved surfaces: Sphere, spheroid, torus, paraboloid, hyperboloid, and so on (surfaces of revolution generated by curved lines).

A/ DEVELOPMENTS

10.4 *Introduction*

A layout of the complete surface of an object is called a *development* or *pattern*. The development of an object bounded by plane surfaces may be thought of as being obtained by turning the object, as illustrated in Figs. 10.1 and 10.2, to unroll the imaginary enclosing surface on a plane. Practically, the drawing operation consists of drawing the successive surfaces in their true sizes with their common edges joined.

The surfaces of cones and cylinders also may be unrolled on a plane. The development of a right cylinder (Fig. 10.3) is a rectangle having a width equal to the altitude of the cylinder and a length equal to the cylinder's computed circumference (πd). The development of a right circular cone (Fig. 10.4) is a sector of a circle having a radius equal to the slant height of the cone and an arc length equal to the circumference of its base.

Warped and double-curved surfaces cannot be developed accurately, but they may be developed by some approximate method. Ordinarily, an approximate pattern will prove to be sufficiently accurate for practical purposes if the material of which the piece is to be made is somewhat flexible.

Plane and single-curved surfaces (prisms, pyramids, cylinders, and cones), which can be accurately developed, are said to be developable. Warped and double-curved surfaces, which can be only approximately developed, are said to be nondevelopable.

10.5 *Practical Developments*

On many industrial drawings, a development must be shown to furnish the necessary information for making a pattern to facilitate the cutting of a desired shape from sheet metal. Because of the rapid advance of the art of manufacturing an ever-increasing number of pieces by folding, rolling, or pressing cut sheet-metal shapes, one must have a broad knowledge of the methods of constructing varied types of developments. Patterns also are used in stonecutting as guides for shaping irregular faces.

A development of a surface should be drawn with the inside face up, as it theoretically would be if the surface were unrolled or unfolded, as illustrated in Figs. 10.1–10.4. This practice is further justified because sheet-metal workers must make the necessary punch marks for folding on the inside surface.

Although in actual sheet-metal work extra metal must be allowed for lap at seams, no allowance will be shown on the developments in this chapter. Many other practical considerations have been purposely ignored, as well, in order to avoid confusing the beginner.

10.6 *To Develop a Right Truncated Prism*

Before the development of the lateral surface of a prism can be drawn, the true lengths of its lateral edges and the true size of the right section must be determined. On the right truncated prism, shown in Fig. 10.5, the true lengths

of the lateral edges are shown in the front view and the true size of the right section is shown in the top view.

The lateral surface is "unfolded" by first drawing a "stretch-out line" and marking off the widths of the faces (distances 1–2, 2–3, 3–4, and so on, from the top view) along it in succession. Through these points light construction lines are then drawn perpendicular to the line 1_D1_D, and the length of the respective edge is set off on each by projecting from the front view. When projecting right section circumference lengths into the development, the points should be taken in a clockwise order around the perimeter, as indicated by the order of the numbers in the top view. The outline of the development is completed by joining these points. Thus far, nothing has been said about the lower base or the inclined upper face. These may be joined to the development of the lateral surface, if so desired.

In sheet-metal work, it is usual practice to make the seam on the shortest element in order to save time and conserve solder or rivets.

10.7 *To Develop an Oblique Prism*

The lateral surface of an *oblique prism*, such as the one shown in Fig. 10.6, is developed by the same general method used for a right prism. Similarly, the true lengths of the lateral edges are shown in the front view, but it is necessary to find the true size of the right section by auxiliary plane construction. The widths of the faces, as taken from the auxiliary right section, are set off along the stretch-out line, and perpendicular construction lines representing the edges are drawn through the division points. The lengths of the portions of each respective edge, above and below plane *XX*, are transferred to the corre-

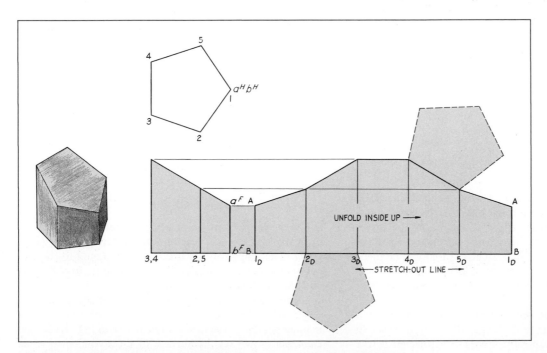

FIG. 10.5 Standard method of developing the lateral surface of a right prism.

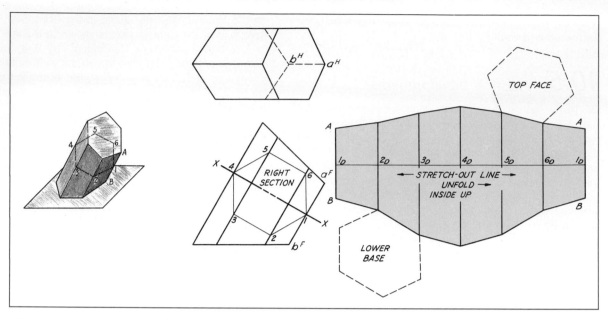

FIG. 10.6 Development of an oblique prism.

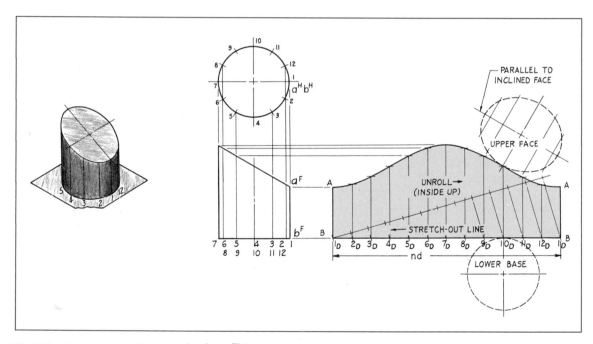

FIG. 10.7 Development of a right circular cylinder.

sponding line in development. Distances above plane *XX* are laid off above the stretch-out line, and distances below *XX* are laid off below it. The development of the lateral surface is then completed by joining the end points of the edges by straight lines. Since an actual fold will be made at each edge line when the prism is formed, it is the usual practice to heavy these edge (fold) lines on the development.

The stretch-out line might well have been drawn in a position perpendicular to the edges of the front view, so that the length of each edge might be projected to the development (as in the case of the right prism).

10.8 *To Develop a Right Cylinder*

When the lateral surface of a *right cylinder* is rolled out on a plane, the base develops into a straight line (Fig. 10.7). The length of this line, which is equal to the circumference of a right section ($\pi \times$ diam), may be calculated and laid off as the stretch-out line 1_D1_D.

Since the cylinder can be thought of as being a *many-sided prism*, the development may be constructed in a manner similar to the method illustrated in Fig. 10.5. The elements drawn on the surface of the cylinder serve as lateral edges of the many-sided prism. Twelve or 24 of these elements ordinarily are used, the number depending on the size of the cylinder. Usually they are spaced by dividing the circumference of the base, as shown by the circle in the top view, into an equal number of parts. The stretch-out line is divided into the same number of equal parts, and perpendicular elements are drawn through each division point. Then the true length of each element is projected to its respective representation on the development, and the development is completed by joining the points with a smooth curve. In joining the points, it is advisable to sketch the curve in lightly, freehand, before using the French curve. Since the surface of the finished cylindrical piece forms a continuous curve, the elements on the development are not heavied. When the development is symmetrical, as in this case, only one-half need be drawn.

A piece of this type might form a part of a two-piece, three-piece, or four-piece elbow. The pieces are usually developed as illustrated in Fig. 10.8. The stretch-out line of each section is equal in length to the computed perimeter of a right section.

10.9 *To Develop an Oblique Cylinder*

Since an *oblique cylinder* theoretically may be thought of as enclosing a regular oblique prism having an infinite number of sides, the development of the lateral surface of the cylinder shown in Fig. 10.9 may be constructed by using a method similar to the method illustrated in Fig. 10.6. The circumference of the right section becomes stretch-out line 1_D1_D for the development.

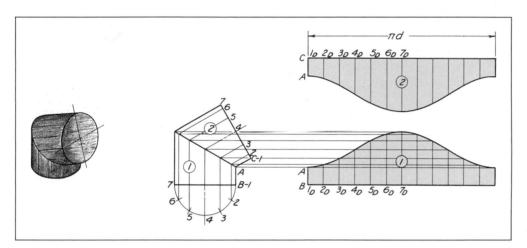

FIG. 10.8 Two-piece elbow.

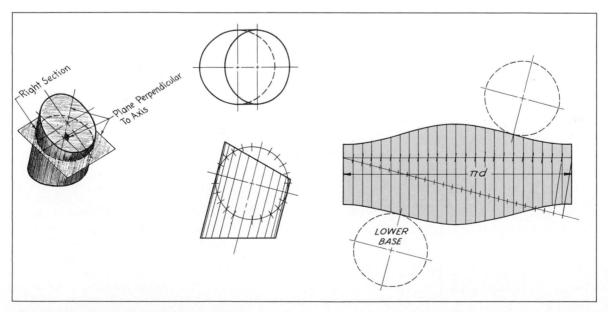

FIG. 10.9 Development of an oblique cylinder.

10.10 To Determine the True Length of a Line

In order to construct the development of the lateral surface of some objects, it frequently is necessary to determine the true lengths of oblique lines that represent the lateral edges. The general method for determining the true lengths of lines inclined to all of the coordinate planes of projection has been explained in detail in Chapter 5.

10.11 True-length Diagrams

When it is necessary in developing a surface to find the true lengths of a number of lateral edges or elements, some confusion may be avoided by constructing a *true-length diagram* adjacent to the orthographic view, as shown in Fig. 10.10. The elements were revolved into a position parallel to the *F* (frontal) plane so that their true lengths show in the diagram. This practice prevents the front view in the illustration from being cluttered with lines, some of which would represent elements and others their true lengths.

Figure 10.12 shows a diagram that gives the true lengths of the lateral edges of the pyramid. Each line representing the true length of an edge is the hypotenuse of a right triangle whose altitude is the altitude of the edge in the front view and whose base is equal to the length of the projection of the edge in the top view. The top view distances from the vertex to the base corners are shown horizontally from a vertical reference. Since all the edges have the same altitude, this line is a common vertical leg for all the right triangles in the diagram. The true-length diagram shown in Fig. 10.10 could very well have been constructed by this method.

10.12 To Develop a Right Pyramid

To develop (unfold) the lateral surface of a right pyramid, it is first necessary to determine the true lengths of the lateral edges and the true size of the base. With this information, the development can be constructed by laying out the faces in successive order with their common edges joined. If the surface is imagined to be unfolded by turning the pyramid, as shown in Fig. 10.2, each triangular face is revolved into the plane of the paper about the edge that is common to it and the preceding face.

Since the edges of the pyramid shown in Fig. 10.11 are all equal in length, it is necessary only to find the length of the one edge A1 by revolving it into the position a^F1r. The edges of the base, 1–2, 2–3, and so on, are parallel to the horizontal plane of projection and consequently show

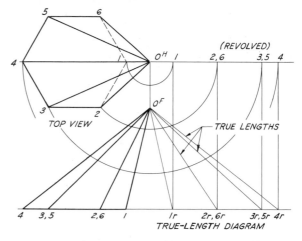

FIG. 10.10 True-length diagram (the revolution method).

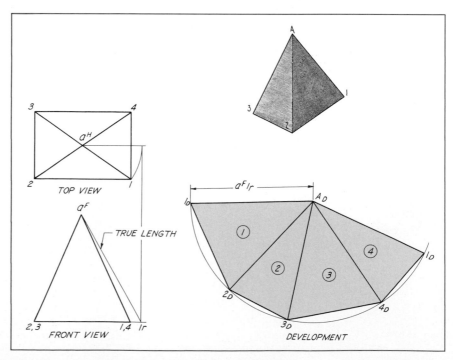

FIG. 10.11 Development of a rectangular right pyramid.

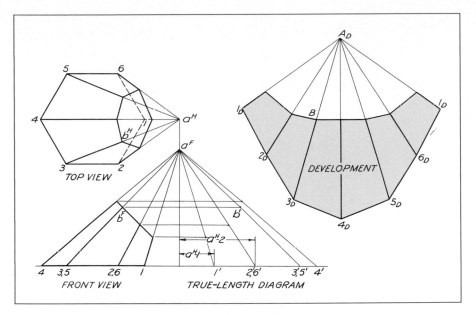

FIG. 10.12 Development of the frustum of a pyramid.

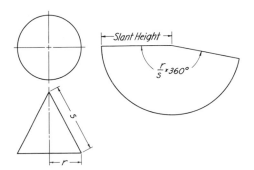

FIG. 10.13 Development of a right cone.

in their true length in the top view. With this information, the development is easily completed by constructing the four triangular surfaces.

10.13 To Develop the Surface of a Frustum of a Pyramid

To develop the lateral surface of the *frustum of a pyramid* (Fig. 10.12), it is necessary to determine the true lengths of edges of the complete pyramid as well as the true lengths of edges of the frustum. The desired development is obtained by first constructing the development of the complete pyramid and then laying off the true lengths of the edges of the frustum on the corresponding lines of the development.

It may be noted with interest that the true length of the edge $B3$ is equal to the length $b'3'$ on the true-length line a^F3' and that the location of point b' can be established by the shortcut method of projecting horizontally from point b^F. Point b' on a^F3' is the true revolved position of point B, because the path of point B is in a horizontal plane that projects as a line in the front view.

10.14 To Develop a Right Cone

As previously explained in Sec. 10.4, the development of a regular right circular cone is a sector of a circle. The development will have a radius equal to the *slant height* of the cone and an included angle at the center equal to $(r/s) \times 360°$ (Fig. 10.13). In this equation, r is the radius of the base and s is the slant height.

10.15 To Develop a Right Truncated Cone

The development of a right truncated cone must be constructed by a modified method of *triangulation*, in order to develop the outline of the elliptical inclined surface. This commonly used method is based on the theoretical assumption that *a cone is a pyramid having an infinite number of sides*. The development of the incomplete right cone shown in Fig. 10.14 is constructed on a layout of the whole cone by a method similar to the standard method illustrated for the frustum of a pyramid in Fig. 10.12.

Elements are drawn on the surface of the cone to serve as lateral edges of the many-sided pyramid. Either 12 or 24 are used, depending on the size of the cone. Their location is established on the developed sector by dividing the arc representing the unrolled base into the same number of equal divisions, into which the top view of the base has been divided. At this point in the procedure, it is necessary to determine the true lengths of the elements of the frustum in the same manner that the true lengths of the edges of the frustum of a pyramid were obtained in Fig. 10.12. With this information, the desired development can be completed by setting off the true lengths on the corresponding lines of the development and joining the points thus obtained with a smooth curve.

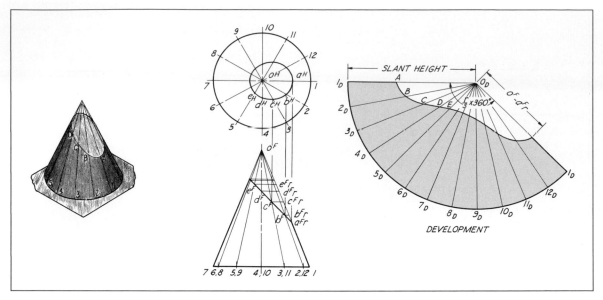

FIG. 10.14   **Development of a truncated cone.**

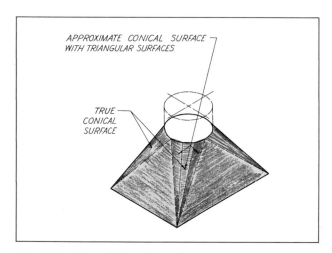

FIG. 10.15   **Triangulation of a surface.**

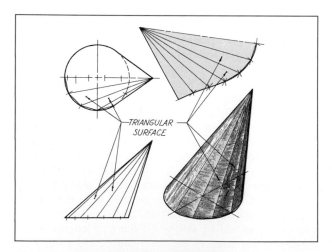

FIG. 10.16   **Triangulation of an oblique cone.**

10.16 *Triangulation Method of Developing Approximately Developable Surfaces*

A nondevelopable surface may be developed approximately if the surface is assumed to be composed of a number of small developable surfaces (Fig. 10.15). The particular method ordinarily used for warped surfaces and the surfaces of oblique cones is known as the *triangulation method*. The procedure consists of covering the lateral surface with numerous small triangles that will lie approximately on the surface (Fig. 10.16). These triangles, when laid out in their true size with their common edges joined, produce an approximate development that is accurate enough for most practical purposes.

Although this method of triangulation is sometimes used to develop the lateral surface of a right circular cone, it is not recommended for such a purpose. The resulting development is not as accurate as it would be if constructed by one of the standard methods (Secs. 10.14 and 10.15).

10.17 *To Develop an Oblique Cone Using the Triangulation Method*

A development of the lateral surface of an oblique cone is constructed by a method similar to that used for an oblique pyramid. The surface is divided into a number of unequal triangles having sides that are elements on the cone and bases that are the chords of short arcs of the base.

The first step in developing an oblique cone (Fig. 10.17) is to divide the circle representing the base into a convenient number of equal parts and draw elements on the surface of the cone through the division points (1, 2, 3, 4, 5, and so on). To construct the triangles forming the development, it is necessary to know the true lengths of the elements (sides of the triangles) and chords. In the illustration, all the chords are equal. Their true lengths are shown in the top view. The true lengths of the oblique elements may be determined by one of the standard methods explained in Sec. 10.11.

Since the seam should be made along the shortest element, $A1$ will lie on the selected starting line for the development and $A7$ will be on the center line. To obtain the development, the triangles are constructed in order, starting with the triangle $A–1–2$ and proceeding around the cone in a clockwise direction (as shown by the arrow in the top view). The first step in constructing triangle $A–1–2$ is to set off the true length a^F1' along the starting line. With point A_D of the development as a center, and

with a radius equal to a^F2', strike an arc; then, with point 1_D as a center, and with a radius equal to the chord $1–2$, strike an arc across the first arc to locate point 2_D. The triangle $A_D2_D3_D$ and the remaining triangles are formed in exactly the same manner. When all the triangles have been laid out, the development of the complete conical surface is completed by drawing a smooth curve through the end points of the elements.

10.18 *Transition Pieces*

A few of the many types of transition pieces used for connecting pipes and openings of different shapes and sizes are illustrated pictorially in Fig. 10.18.

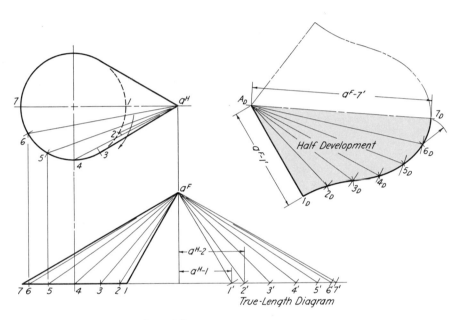

FIG. 10.17 **Development of an oblique cone.**

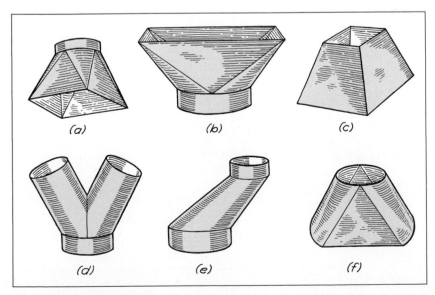

FIG. 10.18 **Transition pieces.**

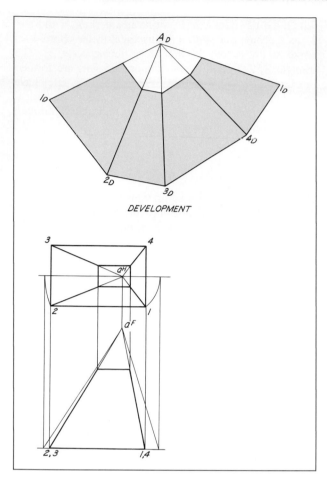

FIG. 10.19 Transition piece.

10.19 To Develop a Transition Piece Connecting Rectangular Pipes

The transition piece shown in Fig. 10.19 is designed to connect two rectangular pipes of different sizes on different axes. Since the piece is a frustum of a pyramid, it can be accurately developed by the method explained in Sec. 10.13.

10.20 To Develop a Transition Piece Connecting Two Circular Pipes

The transition piece shown in Fig. 10.20 connects two circular pipes on different axes. Since the piece is a frustum of an oblique cone, the surface must be triangulated, as explained in Sec. 10.17, and the development must be constructed by laying out the triangles in their true size in regular order. The general procedure is the same as that illustrated in Fig. 10.17. In this case, however, since the true size of the base is not shown in the top view, it is necessary to construct a partial auxiliary view to find the true lengths of chords between the end points of the elements.

10.21 To Develop a Transition Piece Connecting a Circular and a Square Pipe

A detailed analysis of the transition piece shown in Fig. 10.21 reveals that it is composed of four isosceles triangles whose bases form the square base of the piece and four partial oblique cones. It is not difficult to develop this type

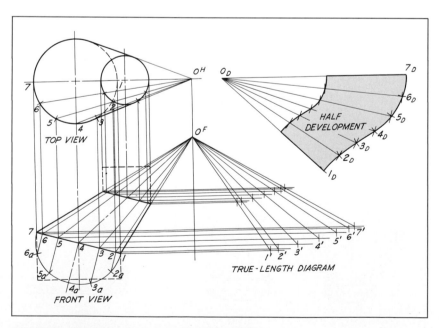

FIG. 10.20 Transition piece connecting two pipes.

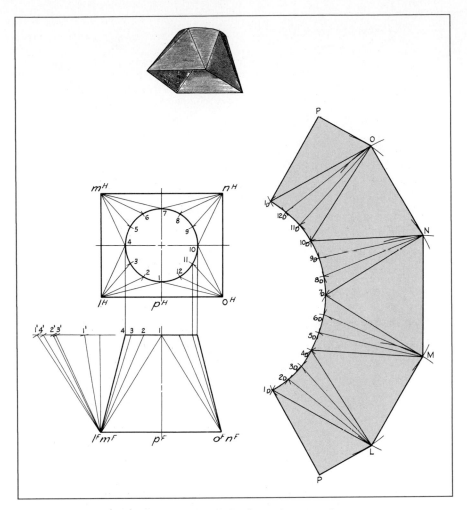

FIG. 10.21 Transition piece connecting a circular and square pipe.

of transition piece because, since the whole surface may be "broken up" into component surfaces, the development may be constructed by developing the first and then each succeeding component surface separately (Fig. 10.15). The surfaces are developed around the piece in a clockwise direction, in such a manner that each successive surface is joined to the preceding surface at their common element. In the illustration, the triangles $1LO$, $4LM$, $7MN$, and $10NO$ are clearly shown in top view. Two of these, $1LO$ and $10NO$, are visible on the pictorial drawing. The apexes of the conical surfaces are located at the corners of the base.

Before starting the development, it is necessary to determine the true lengths of the elements by constructing a true-length diagram, as explained in Sec. 10.11. The true lengths of the edges of the lower base (LM, MN, NO, and OL) and the true lengths of the chords (1–2, 2–3, 3–4, and so on) of the short arcs of the upper base are shown in the top view. The development is constructed in the following manner. First, the triangle 1_DPL is constructed, using the length p^Hl^H taken from the top view

and true lengths from the diagram. Next, using the method explained in Sec. 10.17, the conical surface, whose apex is at L, is developed in an attached position. Triangle 4_DLM is then added, and so on, until all component surfaces have been drawn.

10.22 To Develop a Transition Piece Having an Approximately Developable Surface by the Triangulation Method

Figure 10.22 shows a half development of a transition piece that has a *warped surface* instead of a partially conical one, such as that discussed in Sec. 10.21. The method of constructing the development is somewhat similar, however, in that it is formed by laying out, in true size, a number of small triangles that approximate the surface. The true size of the circular intersection is shown in the top view, and the true size of the elliptical intersection is shown in the auxiliary view, which was constructed for that purpose.

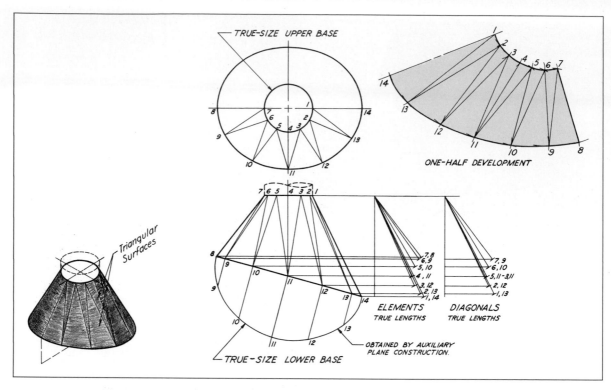

FIG. 10.22   Development of transition piece by triangulation.

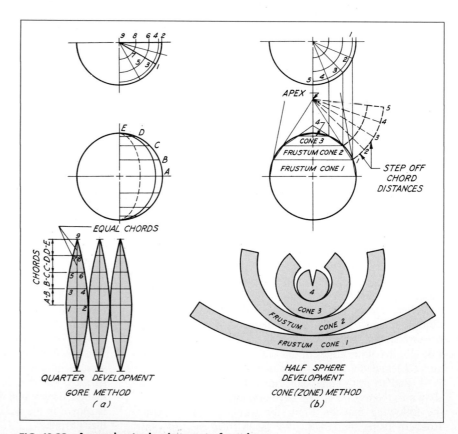

FIG. 10.23   Approximate development of a sphere.

The front half of the circle in the top view should be divided into the same number of equal parts as the half-auxiliary view. By joining the division points, the lateral surface may be initially divided into narrow quadrilaterals. These in turn may be subdivided into triangles by drawing diagonals, which, though theoretically curved lines, are assumed to be straight. The true lengths of the elements and the diagonals are found by constructing two separate true-length diagrams by the method illustrated in Fig. 10.12.

10.23   To Develop a Sphere

The surface of a sphere is a double-curved surface that can be developed only by some approximate method. The standard methods commonly used are illustrated in Fig. 10.23.

In (*a*) the surface is divided into a number of equal *meridian sections* of cylinders. The developed surfaces of these form an approximate development of the sphere. In drawing the development it is necessary to develop the surface of only one section, for this can be used as a pattern for the developed surface of each of the others.

In (*b*) the sphere is cut by parallel planes, which divide it into a number of *partial right circular cones*, the surfaces of which approximate the surface of the sphere. Each of these sections may be considered the frustum of a right cone whose apex is located at the intersection of the chords extended.

B / INTERSECTIONS

10.24   Lines of Intersection of Geometric Surfaces

The line of intersection of two surfaces is a line that is common to both. It may be considered the line that would contain the points in which the elements of one surface would pierce the other. Many lines on a orthographic representation are lines of intersection; therefore, the following discussion may be deemed an extended study of the same subject. The methods presented in this chapter are the recognized easy procedures for finding the more complicated lines of intersection created by intersecting geometric surfaces.

In order to complete a view of a working drawing or a view necessary for developing the surfaces of intersecting geometric shapes, one frequently must find the line of intersection between surfaces. On an ordinary working drawing the line of intersection may be "faked in" through a few critical points, in particular, the beginning, end, and limits of the line. On a sheet-metal drawing, however, a sufficient number of points must be located to obtain an accurate line of intersection and an ultimately accurate development.

The line of intersection of two surfaces is found by determining a number of points common to both surfaces

and drawing a line or lines through these points in correct order. The resulting line of intersection may be straight, curved, or straight and curved. The problem of finding such a line may be solved by one of two general methods, depending on the type of surfaces involved.

For the purpose of simplifying this discussion of intersections, it should be assumed that all problems are divided into these two general groups:

Group I. Problems involving two geometric figures, both of which are composed of *plane surfaces*.

Group II. Problems involving geometric figures which have either *single- or double-curved surfaces*.

For instance, the procedure for finding the line of intersection of two prisms is the same as that for finding the line of intersection of a prism and a pyramid; hence, both problems belong in the same group (Group I). Since the problem of finding the line of intersection of two cylinders and the problem of finding the line of intersection of a cylinder and a cone both involve single-curved surfaces, these two also belong in the same group (Group II).

Problems of the first group are solved by locating the points through which the edges of each of two geometric shapes pierce the other. These points are vertices of the line of intersection. Whenever one of two intersecting plane surfaces appears as an edge in one view, the points through which the lines of the other surfaces penetrate it usually may be found by inspecting that view.

Problems of the second group may be solved by drawing elements on the lateral surface of one geometric shape in the region of the line of intersection. The points at which these elements intersect the edge view of the other surface are points that are common to both surfaces and consequently lie on their line of intersection. A curve, traced through these points, will be a representation of the required intersection. To obtain accurate results, some of the elements must be drawn through certain critical points at which the curve changes sharply in direction. These points usually are located on contour elements. Hence, the usual practice is to space the elements equally around the surface, starting with a contour element.

10.25   Determination of a Piercing Point by Inspection (Fig. 10.24)

It is easy to determine where a given line pierces a surface when the surface appears as an edge view in one of the given views. For example, when the given line *AB* is extended as shown in (*a*), the *F* view of the *piercing point* *C* is observed to be at c^F, where the frontal view of the line *AB* extended intersects the line view of the surface. With the position of c^F known, the *H* view of point *C* can be quickly found by projecting upward to the *H* view of *AB* extended.

In (*b*) the *H* view (f^H) of the piercing point *F* is found first by extending $d^H e^H$ to intersect the edge view of the surface pierced by the line. By projecting downward, f^F is located on $d^F e^F$ extended.

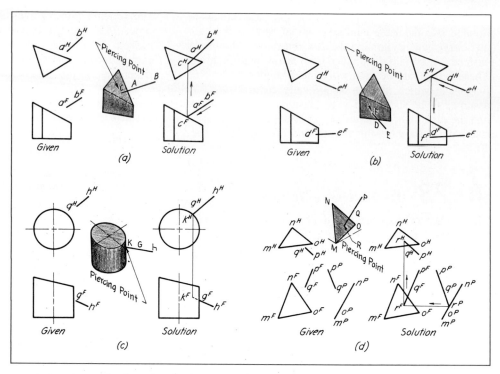

FIG. 10.24   Determination of a piercing point by inspection.

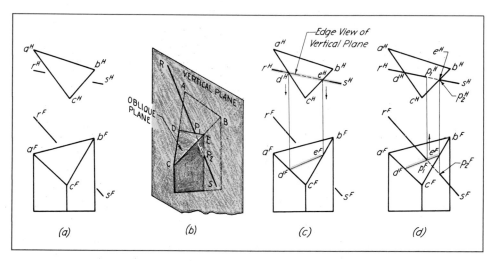

FIG. 10.25   Use of a line-projecting plane.

In (c) the views of the piercing point K are found in the same manner as in (b), the only difference being that the edge view of the surface pierced by the line appears as a circle arc in the H view instead of a straight line. It should be noted that a part of the line is invisible in the F view because the piercing point is on the rear side of the cylinder.

The F and H views of the piercing point R in (d) may be found easily by projection after the P view (r^P) of R has been once established by extending p^P q^P to intersect the line view of the surface.

10.26 Determination of a Piercing Point Using a Line-projecting Plane

When a line pierces a given oblique plane and an edge view is not given, as in Fig. 10.25, a line-projecting plane (cutting plane) may be used to establish a line of intersection that will contain the piercing point. In the illustration, a vertical projecting plane was selected that would contain the given line RS and intersect the given plane ABC along line DE, as illustrated by the pictorial drawing.

SOLUTION: Draw the *H* view of the projecting plane through $r^H s^H$ to establish $d^H e^H$, as shown in the *H* view in (c). Locate $d^F e^F$ and draw the *F* view of the line of intersection. Then, complete the line view $r^F s^F$ to establish p_1^F at the point of intersection of $r^F s^F$ and $d^F e^F$. Finally, locate p_1^H on $f^H s^H$ by projecting upward from p_1^F, as shown in (d).

To Find Where a Line Pierces a Geometric Solid-cylinder–cone–sphere Using Cutting Planes (Fig. 10.26).

10.27

The points where a line pierces a cylinder, cone, or sphere may be found easily through the use of a cutting plane that contains the given line, as illustrated in (a), (b), and (c).

In (a) the intersections of the cutting plane and the cylinder are straight-line elements because the projecting plane used is parallel to the axis of the cylinder. The use of planes parallel to the axis permits the rapid solution of this type of problem. As shown by the pictorial drawing, the vertical projecting plane cuts elements on both the right and left sides of the cylinder. The line *AB* intersects the element *RS* at *C* and the other element at *D*. Points *C* and *D* are the piercing points.

The piercing points of a line and a cone are the points of intersection of the line and the two specific elements of the cone that lie in the cutting plane containing the line as shown in (b). The vertex of the cone and the given line fix the position of the projecting plane, the plane of the elements. In the illustration, the vertical projecting plane,

taken through the line *EF* and the vertex of the cone *T*, cuts the base of the cone at *U* and *V*, the points needed to establish the *F* views of the elements lying in the plane. The points of intersection of the given line *EF* and these elements are points *G* and *H*, the points where the line pierces the cone. If the given line had not been in a position to intersect the axis of the cone, it would have been necessary to use an oblique cutting plane through the apex.

A cutting plane that contains a line piercing a sphere will cut a circle on the surface of the sphere; therefore, points where the given line intersects the circle will be points where the line pierces the sphere. [See the pictorial drawing in (c).] In the illustrations, a vertical projecting plane was used containing the given line *JK*. The *F* views of the piercing points *M* and *N* ($m^F n^F$) were found first at the points of intersection of the line and the circle. The *H* views ($m^H n^H$) of the piercing points were found by projecting upward from m^F and n^F in the *F* view.

To Determine the Points Where a Line Pierces a Cone, General Case

10.28

In Sec. 10.27, the statement was made that the piercing points of a line and a cone are the points of intersection of the line and the two specific elements of the cone that lie in the cutting plane containing the line and the apex of the cone [Fig. 10.26(b)]. This pertained to a special condition for which a cutting plane could be used. For other cases, the following general statement applies: *The piercing points of a line and any surface must lie on the*

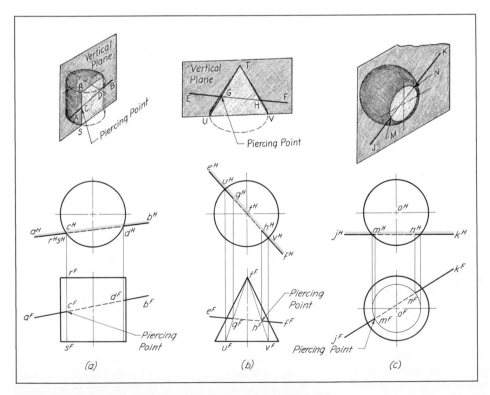

FIG. 10.26 To determine where a line pierces a geometric solid.

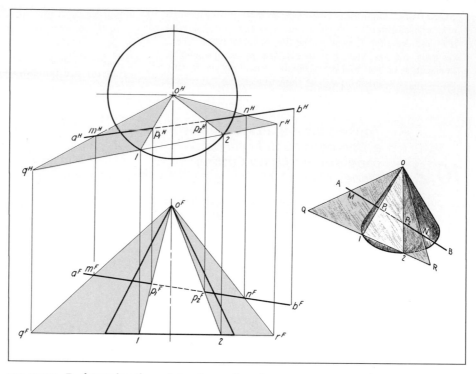

FIG. 10.27   To determine the points where a line pierces a cone—general case.

lines of intersection of the given surface and a cutting plane that contains the line. Obviously, an infinite number of cutting planes could have been assumed that would have contained the line *AB* in Fig. 10.27, but all would have resulted in curved lines of intersection, except in the case of the one plane that was selected to pass through the apex *O* of the cone. As can be noted by observing the pictorial drawing, this choice gives a plane that intersects the cone along two straight-line elements.

SOLUTION: (1) Form a cutting plane containing the line *AB* and the apex *O* by drawing a line from *O* to an assumed point *M* on the line *AB*. Lines *AB* and *OM* define the cutting plane. (2) Extend the cutting plane until it intersects the plane of the base of the cone. In the pictorial illustration it should be noted that *OM* and *ON* extended establish the line of intersection *QR* between the cutting plane and the plane of the base of the cone. Line *ON*, drawn from *O* to an assumed point *N*, is an additional line in the plane. (3) Project points 1 and 2, where the line $q^H r^H$ intersects the curve of the base of the cone in the *H* view, to the *F* view, and draw the views of the elements of intersection $o^F 1$ and $o^F 2$ in the *F* view. Points p_1^F and p_2^F, where the *F* views of these elements intersect $a^F b^F$, are the *F* views of the piercing points of the line *AB* and the given cone. (4) Project the *F* views of the piercing points to the *H* view to locate p_1^H and p_2^H.

When the line and cone are both oblique an added view (auxiliary view), showing the base as an edge, may be needed to obtain a quick and accurate solution employing the steps illustrated in Fig. 10.27.

To Find the Intersection of Two Planes, Cutting Plane
10.29   Method

The intersection of two oblique planes may be determined by finding where two of the lines of one plane pierce the other plane, as illustrated by the pictorial drawing in Fig. 10.28. The procedure that is illustrated employs cutting planes to find the piercing points of the lines *XY* and *XZ* and the oblique plane *RST*. Therefore, it might be said that the solution requires the determination of the piercing point of a line and an oblique plane, as explained in Sec. 10.26.

GIVEN: The oblique planes *RST* and *XYZ*.

SOLUTION: Since the vertical cutting plane $C_1 P_1$ is to contain the line *XY* of the plane *XYZ*, draw the line-view representation of this projecting plane to coincide with $x^H y^H$. Next, project the line of intersection *AB* between the line-projecting plane $C_1 P_1$ and plane *RST* from the top view, where it appears as $a^H b^H$ in the edge-view representation of $C_1 P_1$ to the front view. Then, since it is evident that the line *AB* is not parallel to *XY*, which lies in the projecting plane (see *F* view), the line *XY* intersects *AB*. The location of this intersection at *E* is established first in the *F* view, where the line $x^F y^F$ intersects $a^F b^F$ at e^F. The *H* view of *E*, that is, e^H, is found by projecting upward from e^F in the *F* view to the line view $x^H y^H$. The other end of the line of intersection between the two given planes at *F* is found by using the cutting plane $C_2 P_2$ and following the same procedure as for determining the location of point *E*.

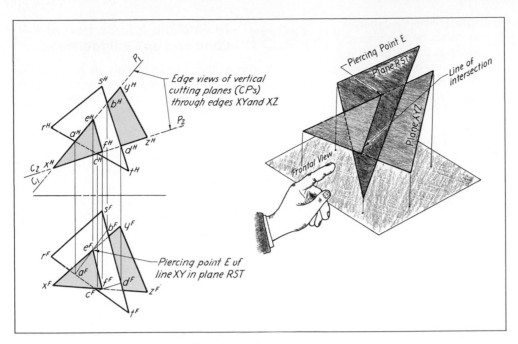

FIG. 10.28 To find the intersection line of two planes.

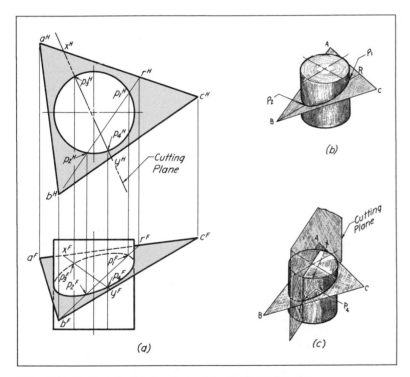

FIG. 10.29 To find the intersection of a cylinder and a plane.

10.30 To Find the Intersection of a Cylinder and an Oblique Plane

There are two distinct and separate methods for finding the line of intersection of an oblique plane and a cylinder. Both methods are shown in Fig. 10.29. The selected line method is illustrated pictorially in (*b*), while the cutting-plane method is shown in (*c*).

In the application of the selected line method, any line of the given plane, such as line *BR*, is drawn in the *F* and *H* views (*a*). It can be noted by observing, in the pictorial drawing in (*b*), that this particular line pierces the cylinder at points P_1 and P_2 to give two points on the line of intersection. On the multiview drawing, the locations of the *H* views of points P_1 and P_2 can readily be recognized as being at the points labeled p_1^H and p_2^H, where the line view $b^H r^H$ intersects the edge view of the surface of the cylinder. The *F* views (p_1^F and p_2^F) of points P_1 and P_2 were found by projecting downward from p_1^H and p_2^H to the line view $b^F r^F$. Additional points along the line of intersection, as needed, can be obtained by using other lines of the plane.

The line of intersection of this same plane and cylinder could have been almost as easily determined through the use of a series of cutting planes passed parallel to the axis of the cylinder, as illustrated in (*c*). It should be noted that the vertical cutting plane shown cuts elements on the cylinder that intersect *XY*, the line of intersection of the cutting plane and the given plane, at points P_3 and P_4. Points P_3 [not visible in (*c*)] and P_4 are two points on the line of intersection, for they lie in both the cutting plane and the given plane and are on the surface of the cylinder. After the position of the cutting plane has been established in (*a*) by drawing the line representation in the *H* view, x^H and y^H must be projected downward to the corresponding lines of the plane in the *F* view. Line $x^F y^F$ as then drawn is the *F* view of the line of intersection of the cutting plane and given plane. Finally, as the last step, the intersection elements that appear as points in the *H* view at p_3^H and p_4^H must be drawn in the *F* view. The *F* views of points P_3 and P_4 (p_3^F and p_4^F) are at the intersection of the *F* views of the elements and the line $x^F y^F$.

A series of selected planes will give the points needed to complete the *F* view of the intersection.

10.31 To Find the Intersection of a Cone and an Oblique Plane

When the intersecting plane is oblique, as is true in Fig. 10.30, it is usually desirable to employ the cutting-plane method shown in (*a*) rather than to resort to the use of an additional view, as in (*b*). Since the given cone is a right cone, any one vertical cutting plane, passing through the apex *O*, will simultaneously cut straight lines on the conical surface and across the given oblique plane. The pictorial illustration shows the cutting plane *XY* intersects the cone along the two straight-line elements *O*–2 and *O*–8 and the plane along the line *RS*. Points *G* and *H*, where the elements intersect the line *RS*, are points along the required line of intersection because both points lie in the given plane *ABCD* and are on the surface of the cone. In (*a*), the line representation of the cutting plane *XY* in the *H* view ($r^H s^H$) of the line *RS*. The *H* views of elements *O*–2 and *O*–8 also lie in the edge view of the cutting plane *XY*. With this much known, the *F* views of the two elements and line *RS* may be drawn. Points g^F and h^F are at the intersection of $r^F s^F$ and o^F–2 and o^F–8, respectively. A series of cutting planes passed similarly furnishes the points needed to complete the solution.

At times one might prefer to determine the line of intersection through the use of a constructed auxiliary view that shows the given plane as an edge. In this case, when selected elements may be seen to intersect the line view of the plane in the auxiliary view, the solution becomes quite simple, because all that is required is to project the point of intersection of an element and the line view of the plane to the *F* and *H* views of the same element. For example, the *A* view (o^A–8) of the element *O*–8 can be

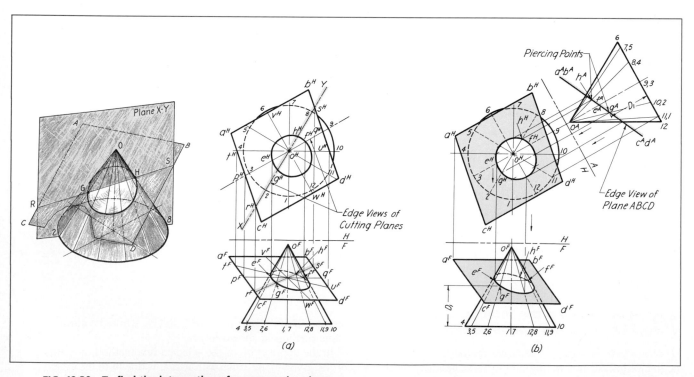

FIG. 10.30 To find the intersection of a cone and a plane.

seen to intersect the edge line $a^A b^A c^A d^A$ at h^A, the A view of point H. By projecting to line o^H–8, the H view of H (h^H) may be easily established. The F view of H (h^F) lies directly below h^H on o^F–8. Through the projected views of other points, located similarly, smooth curves may be drawn to form the F and H view representations, as shown.

10.32 To Find the Intersection of a Sphere and an Oblique Plane

Horizontal cutting planes have been used to find the line of intersection of the sphere and oblique plane shown in Fig. 10.31. Two approaches to the solution have been given on the line drawing. *The horizontal cutting planes, as selected, cut circles from the sphere and straight lines from the given oblique plane.* For example, the cutting plane CP_3 in the F view cuts the horizontal line 3–3 from plane $ABCD$ and a circle from the sphere. This circle appears as an edge in the F view and shows in its true diameter. In the H view it will show in true shape. The line and circle intersect at two points that also have been identified by the number 3, the number assigned to the cutting plane in which these two points lie. These points now located in the H view are projected to the CP_3 line in the F view. The curved line through points, that have been determined by a series of planes, is a line common to both surfaces and is therefore the line of intersection.

This problem could also have been solved by using an auxiliary view showing the plane $ABCD$ as an edge. As before, the horizontal cutting planes will cut circles from the sphere and lines from the plane. However, in this case both the lines and the intersections show as points in the A view. For each CP, these points must be projected from the auxiliary view to the corresponding circle in the H view. The F views of these points on the line of intersection may be found by projection and by using measurements taken from the A view.

10.33 To Find the Intersection of Two Prisms

In Fig. 10.32, points A, C, and D, through which the edges of the horizontal prism pierce the vertical prism, are first found in the top view (a^H, c^H, and d^H) and are then projected downward to the corresponding edges in the front view. Point B, through which the edge of the vertical prism pierces the near face of the triangular prism, cannot be found in this manner because the side view from which it could be projected to the front view is not shown. Its location, however, can be established in the front view without even drawing a partial side view, if some scheme like the one illustrated in the pictorial drawing is used. In this scheme, the intersection line AB, whose direction is shown in the top view as line $a^H b^H$, is extended on the triangular face to point X on the top edge. This amounts to passing a cutting plane in the face of the vertical prism, causing line abx to be formed. Point x^H is projected to the corresponding edge in the front view and a light con-

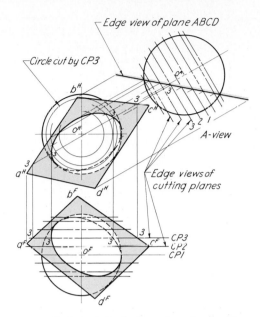

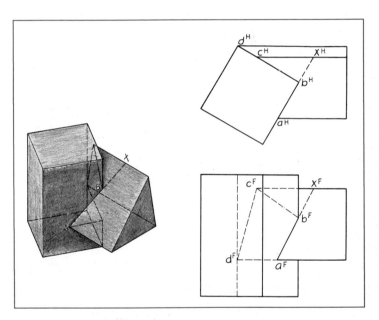

FIG. 10.31   To find the intersection of a sphere and an oblique plane.

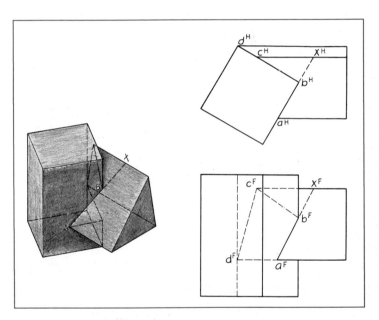

FIG. 10.32   Intersecting prisms.

struction line is drawn between the points a^F and x^F. Since point B is located on line AX (see pictorial) at the point where the edge of the prism pierces the line, its location in the front view is at point b^F where the edge cuts the line $a^F x^F$.

10.34 To Find the Intersection of a Pyramid and a Prism (Fig. 10.33)

The intersection of a right pyramid and a prism may be found by the same general method used for finding the intersection of two prisms (Sec. 10.33).

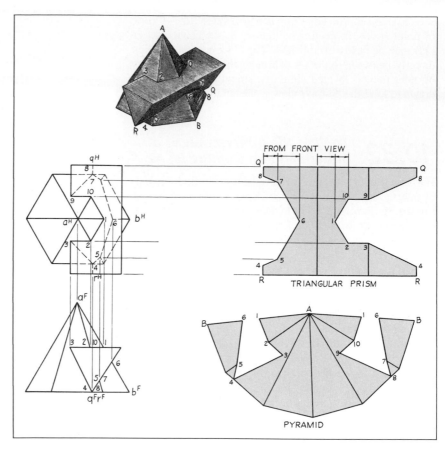

FIG. 10.33 Intersecting pyramid and prism.

10.35 To Determine the Intersection of a Prism and a Pyramid Using Cutting Planes

Frequently, it becomes necessary to draw the line of intersection between two geometric shapes so positioned that the piercing points of edges cannot be found by inspection if only the principal views are to be used. In this case, one must resort to the method discussed in Sec. 10.26 to determine where a line, such as the edge line *GD* of the prism shown in Fig. 10.34, pierces a surface. As illustrated by the pictorial drawing, a vertical plane passed through the edge *DG* of the prism, intersects the surface *ABC* of the pyramid along line *MN* that contains point *D*, the piercing point of *DG*. In (*a*), the *H*-view ($m^H n^H$) of the line *MN* lies along $d^H g^H$ extended to m^H on the edge of the pyramid, because the *H* view of the cutting plane appears as an edge that coincides with $d^H g^H$. With the *F* view of *MN* established by projecting downward from $m^H n^H$ in the *H* view, the frontal view of the piercing point *D* is at d^F where the view of the edge line *DG* of the prism intersects $m^F n^F$. The *H* view of *D* is found by projecting upward from d^F. The two other piercing points, at *E* and *F*, are found in the same manner using two other cutting planes.

FIG. 10.34 Intersecting pyramid and prism.

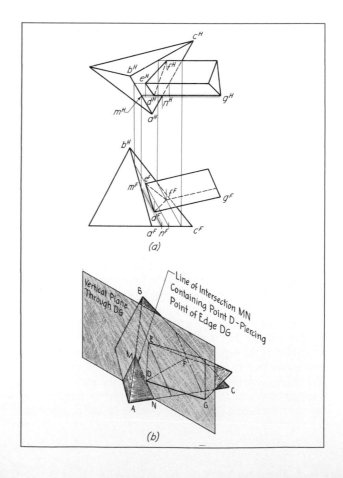

(*a*)

(*b*)

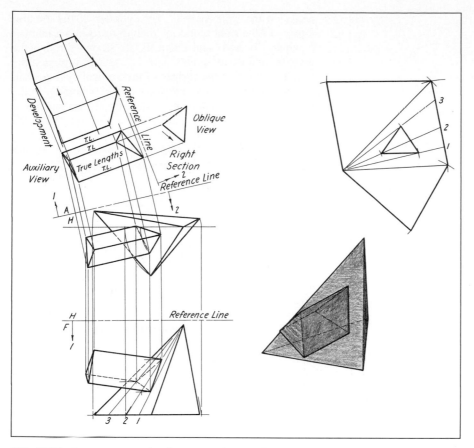

FIG. 10.35 To construct a development using auxiliary views.

10.36 To Construct a Development Using Auxiliary Views

When one of the components is oblique to the principal planes of projection, as is the prism in Fig. 10.35, the construction needed for the development can be simplified somewhat through the use of an auxiliary view to find the true lengths of the edges and an oblique (secondary auxiliary) view to show a right section. Since the plane on which the auxiliary view is projected is a vertical one that is parallel to the edges of the prism, the distances perpendicular to the *AH* reference line are height distances. In making the construction, distances are taken from the *F* view in the direction of the single-headed arrow for use in the auxiliary view in the direction indicated by a similar arrow. For the oblique view, projected on an *O* plane that is perpendicular to both the *A* plane and the edges of the prism, distances are taken from the *AH* reference line in the direction of the arrow numbered 2 to be laid out from the reference line for the *O* view. Since this is the second auxiliary, the arrows indicating the direction for equal distances have been given two heads.

If there is sufficient space available, the true-length measurements for the edges in the development may be projected directly from the auxiliary view showing the true lengths. The true distances between these edges, taken from the right section in the direction of the arrow, are laid off along the stretch-out line. The arrow on the development indicates the direction in which the successive faces are laid down when the prism is turned to unroll the lateral surface inside-up.

10.37 To Find the Intersection of Two Cylinders

If a series of elements are drawn on the surface of the small horizontal cylinder, as in Fig. 10.36, the points *A*, *B*, *C*, and *D* in which they intersect the vertical cylinder will be points on the line of intersection (see pictorial). These points, which are shown as a^H, b^H, c^H, and d^H in the top view, may be located in the front view by projecting them downward to the corresponding elements in the front view, where they are shown as points a^F, b^F, c^F, and d^F. The desired intersection is represented by a smooth curve drawn through these points.

10.38 To Find the Intersection of Two Circular Cylinders Oblique to Each Other

The first step in finding the line of intersection of two cylinders that are oblique to each other (Fig. 10.37) is to draw a revolved right section of the oblique cylinder di-

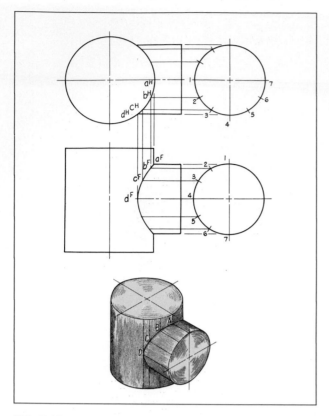

FIG. 10.36 **Intersecting cylinders.**

rectly on the front view of that cylinder. If the circumference of the right section is then divided into a number of equal divisions and elements are drawn through the division points, the points *A*, *B*, *C*, and *D* in which the elements intersect the surface of the vertical cylinder will be points on the line of intersection (see pictorial). In the case of the illustration shown, these points are found first in the top view and then are projected downward to the corresponding elements in the front view. The line of intersection in the front view is represented by a smooth curve drawn through these points.

10.39 To Find the Intersection of Two Circular Cylinders Using Cutting Planes

The line of intersection of the two cylinders shown in Fig. 10.37 could have been determined through the use of a series of parallel cutting planes passed parallel to their axes (Fig. 10.38). The straight-line elements cut on the cylinders by any one cutting plane, such as *C*, intersect on the line of intersection of the cylinders. As many line-projecting planes as are needed to obtain a smooth curve should be used and they should be placed rather close together where a curve changes sharply.

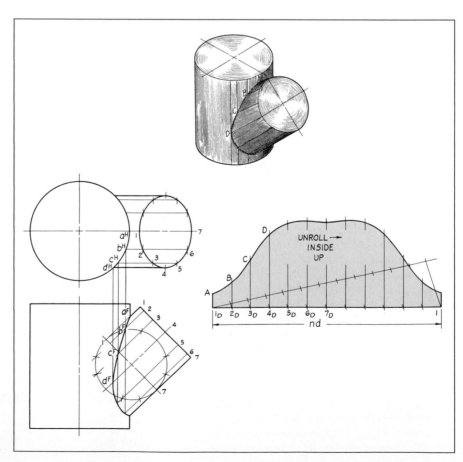

FIG. 10.37 **Intersecting cylinders.**

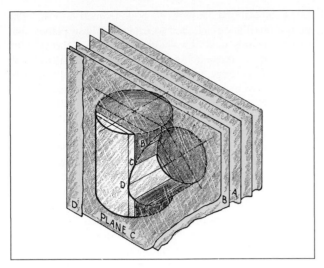

FIG. 10.38 To find the intersection of two cylinders using line-projecting planes.

10.40 To Find the Intersection of a Cylinder and a Cone

The intersection of a cylinder and a cone may be found by assuming a number of elements on the surface of the cone. The points at which these elements cut the cylinder are on the line of intersection (see Figs. 10.39 and 10.40). In selecting the elements, it is the usual practice to divide the circumference of the base into a number of equal parts and draw elements through the division points. To obtain needed points at locations where the intersection line will change suddenly in curvature, however, there should be additional elements.

In Fig. 10.39, the points at which the elements pierce the cylinder are first found in the top view and are then projected to the corresponding elements in the front view. A smooth curve through these points forms the figure of the intersection.

To find the intersection of the cone and cylinder combination shown in Fig. 10.40, cutting planes were passed through the vertex O parallel to the axis of the cylinder

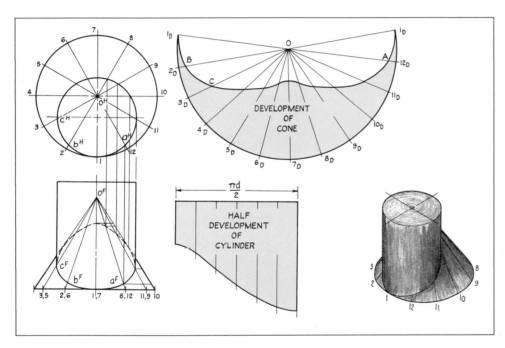

FIG. 10.39 Intersecting cylinder and cone.

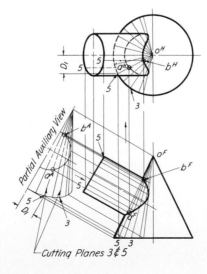

FIG. 10.40 To find the intersection of a cone and a cylinder.

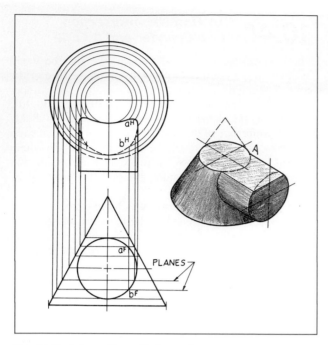

FIG. 10.41   Intersecting cylinder and cone.

to cut intersecting elements on both geometric forms. The partial auxiliary is needed to establish these planes, because it is only in a view showing the axis of the cylinder as a point that these planes and the surface of the cylinder will show as edge views. Each cutting plane cuts one needed straight-line element from the conical surface and two straight-line elements from the cylindrical surface. Cutting plane 5, for example, cuts one element (numbered 5) from the near side of the cone and an upper and lower element from the surface of the cylinder. The intersection of these three elements, all numbered 5, establishes the location of points *A* and *B* on the line of intersection.

Another method for finding the line of intersection of a cylinder and a right cone is illustrated in Fig. 10.41. Here horizontal cutting planes are passed through both geometric shapes in the region of their line of intersection. In each cutting plane, the circle cut on the surface of the cone will intersect elements cut on the cylinder at two points common to both surfaces (see pictorial). A curved line traced through a number of such points in different planes is a line common to both surfaces and is therefore the line of intersection. This second method presents a more direct solution.

10.41 To Find the Intersection of an Oblique Cone and a Paraboloid

The method that is illustrated in Fig. 10.42 for finding the line of intersection of a cone and a paraboloid may be used effectively for the solution of many other intersection problems involving different geometric forms, such as two cones or a cone and a cylinder having parallel bases. For the arrangement as shown, a horizontal cutting plane will intersect each geometric form in a circle. When a number of horizontal cutting planes are passed through both geometric shapes in the region of their line of intersection, each plane cuts a circle on the cone that intersects a circle cut on the paraboloid at two points common to both surfaces. The curved line traced through the points that have been determined by the several planes is a line common to both surfaces and is therefore the line of intersection.

10.42 To Find the Intersection of a Prism and a Cylinder

In Fig. 10.43 it is required to find the intersection between the cylinder and two of the plane surfaces of the prism so that a pattern for the hole can be cut in the prism to match the cylinder. Although the intersection could have been secured merely by determining, through inspection, the piercing points of elements drawn arbitrarily on the cylinder, vertical cutting planes were used to illustrate a common approach to this type of problem. The cutting planes are located in the view showing the right section of the cylinder. From this first step, the positions for the lines and elements cut on the prism and cylinder, respectively,

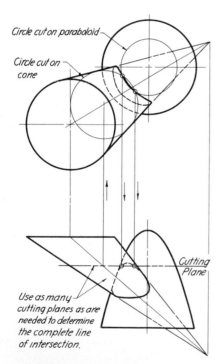

FIG. 10.42   Intersecting cone and paraboloid.

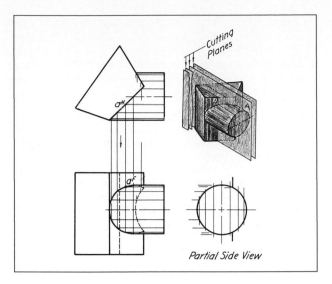

FIG. 10.43 **To find the intersection of a prism and a cylinder.**

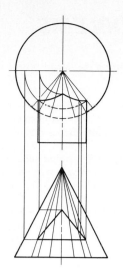

FIG. 10.44 **Intersecting cone and prism.**

can be determined by projection. Each cutting plane cuts a line on a plane surface of the prism and related elements on the cylinder that intersect on the line of intersection. A sufficient number of cutting planes should be used to enable one to draw a smooth curved-line representation of the intersection. If more points are desired, at a location where a curve changes sharply, additional cutting planes may be added.

10.43 To Find the Intersection of a Prism and a Cone

The complete line of intersection may be found by drawing elements on the surface of the cone (Fig. 10.44) to locate points on the intersection as explained in Sec. 10.40. To obtain an accurate curve, however, some thought must be given to the placing of these elements. For instance, although most of the elements may be equally spaced on the cone to facilitate the construction of its development, additional ones should be drawn through the critical points and in regions where the line of intersection changes sharply in curvature. The elements are drawn on the view that will reveal points on the intersection, then the determined points are projected to the corresponding elements in the other view or views. In this particular illustration a part of the line of intersection in the top view is a portion of the arc of a circle that would be cut by a horizontal plane containing the bottom surface of the prism. This solution relies on a combination of inspection of elements and the use of cutting planes.

If the surfaces of the prism are parallel to the axis of

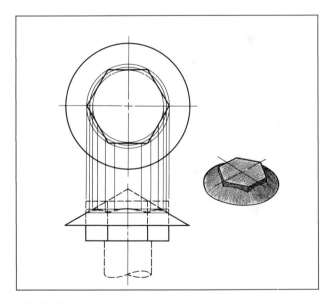

FIG. 10.45 **Intersecting cone and hexagonal prism.**

the cone, as in Fig. 10.45, the line of intersection will be made up of the tips of a series of hyperbolas. The intersection may be found by using planes that will cut circles on the surface of the cone. The points at which these cutting circles pierce the faces of the prism are points common to the lateral surfaces of both shapes and are therefore points on the required line of intersection. It should be noted that the resulting solution represents a chamfered bolthead.

PROBLEMS

1. (Fig. 10.46). Develop the lateral surface of one or more of the prisms as assigned.

2. (Fig. 10.47). Develop the lateral surface of one or more of the pyramids as assigned. Make construction lines light. Show construction for finding the true lengths of the lines.

3. (Fig. 10.48). Develop the lateral surface of one or more of the cylinders as assigned. Use a hard pencil for construction lines and make them light.

4. (Fig. 10.49). Develop the lateral surface of one or more of the cones as assigned. Show all construction. Use a hard pencil for construction lines and make them light. In each case start with the shortest element and unroll, inside-up. It is suggested that 12 elements be used, in order to secure a reasonably accurate development.

5. (Fig. 10.50). Develop the lateral surface of one or more of the transition pieces as assigned. Show all construction lines in light, sharp pencil lines. Use a sufficient number of elements on the curved surfaces to ensure an accurate development.

6. (Fig. 10.51). Develop the sheet-metal connections. On pieces 3 and 4, use a sufficient number of elements to obtain a smooth curve and an accurate development.

7–8. (Figs. 10.52 and 10.53). Draw the line of intersection of the intersecting geometric shapes as assigned. Show the invisible portions of the lines of intersection as well as the visible. Consider that the interior is open.

9–10. (Figs. 10.54 and 10.55). Draw the line of intersection of the intersecting geometric shapes as assigned. It is suggested that the elements used to find points along the intersection be spaced 15° apart. Do not erase the construction lines. One shape does not pass through the other.

11. (Fig. 10.56). Draw the line of intersection of the intersecting geometric shapes as assigned. Show the invisible portions of the line of intersection as well as the visible. The interior of the combination is hollow. One shape does not pass through the other. Show construction with light, sharp lines drawn with a hard pencil.

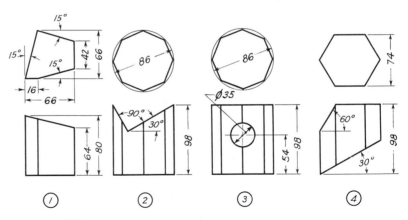

FIG. 10.46 Prisms.

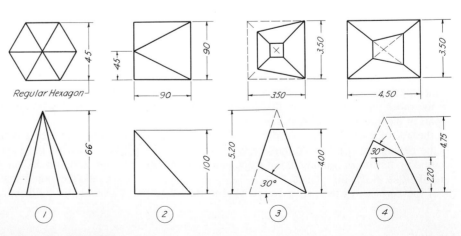

FIG. 10.47 Pyramids.

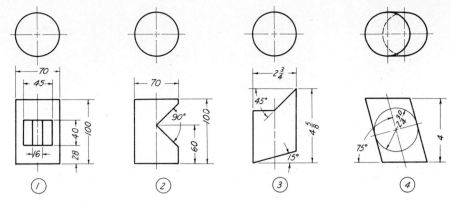

FIG. 10.48 Cylinders.

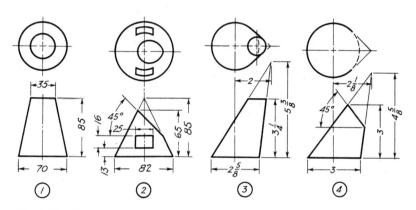

FIG. 10.49 Cones.

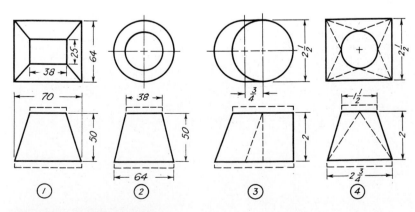

FIG. 10.50 Transition pieces.

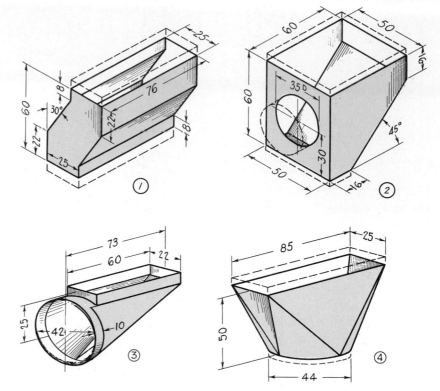

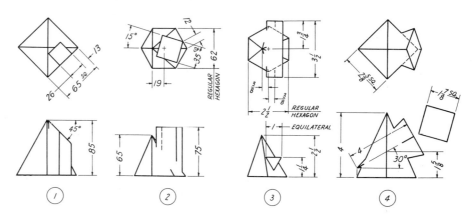

FIG. 10.51   Sheet-metal connections (transitions).

FIG. 10.52   Intersecting surfaces.

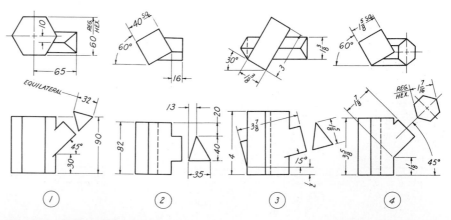

FIG. 10.53   Intersecting surfaces.

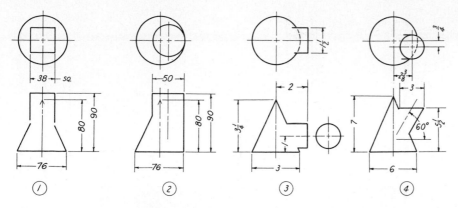

FIG. 10.54 Intersecting surfaces.

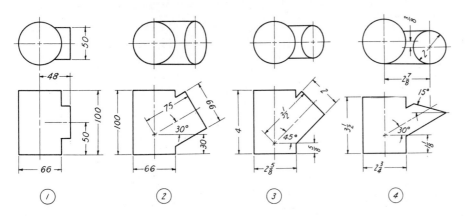

FIG. 10.55 Intersecting surfaces.

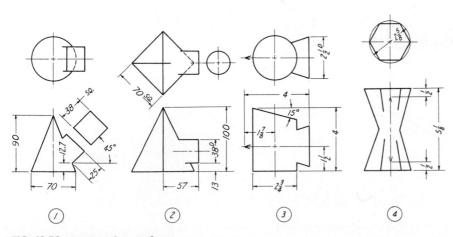

FIG. 10.56 Intersecting surfaces.

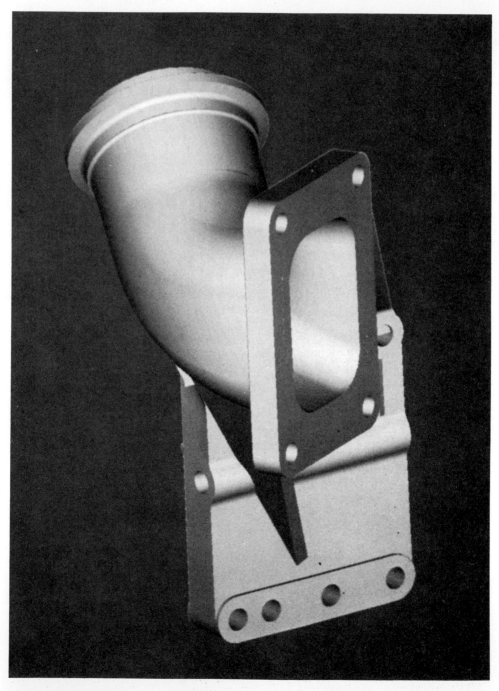

This 3-D drawing was prepared using an interactive computer-aided drafting system.
(*Courtesy Computervision Corporation*)

Pictorial
Presentation

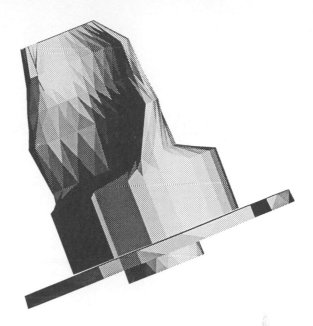

11.1 Introduction

An orthographic drawing of two or more views describes an object accurately in form and size, but, since each of the views shows only two dimensions without any suggestion of the third, such a drawing can convey information only to those who are familiar with graphic representation. For this reason, multiview drawings are used mainly by engineers, designers, and technologists.

Frequently, however, engineers and technologists find they must use conventional picture drawings to convey specific information to persons who do not possess the training necessary to construct an object mentally from views. To make such drawings, several special schemes of *one-plane pictorial drawing* have been devised that combine the pictorial effect of perspective with the advantage of having the principal dimensions to scale. But pictorial drawings, in spite of certain advantages, have disadvantages that limit their use. A few of these are as follows.

1. Some drawings frequently have a distorted, unreal appearance that is disagreeable.

2. The time required for execution is, in many cases, greater than for an orthographic drawing.

3. They are difficult to dimension.

4. Some of the lines cannot be measured.

Even with these limitations, pictorial drawings are used extensively for technical publications, Patent Office rec-

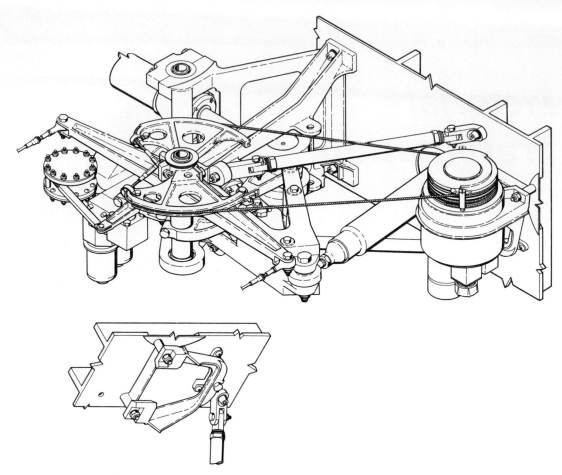

FIG. 11.1   **Pictorial illustration.** (*Courtesy Lockheed Aircraft Corp.*)

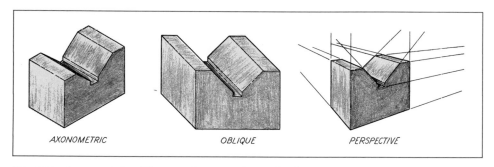

FIG. 11.2   **Axonometric, oblique, and perspective projection.**

ords, piping diagrams, and furniture designs. Occasionally they are used, in one form or another, to supplement and clarify machine and structural details that would be difficult to visualize (Fig. 11.1).

11.2   *Divisions of Pictorial Drawing*

Single-plane pictorial drawings are classified in three general divisions as shown in Fig. 11.2:

   1. Axonometric projection,

   2. Oblique projection,

   3. Perspective projection.

Perspective methods produce the most realistic drawings, but the necessary construction is more difficult and tedious than the construction required for the conventional methods classified under the other two divisions. For this reason, engineers customarily use some form of either axonometric or oblique projection. Modified methods, which are not theoretically correct, are often used to produce desired effects.

A/ AXONOMETRIC PROJECTION

11.3 Divisions of Axonometric Projection

Theoretically, axonometric projection is a form of orthographic projection. The distinguishing difference is that only one plane is used instead of two or more, and the object is turned from its customary position so that three faces are displayed (Fig. 11.3). Since an object may be placed in a countless number of positions relative to the picture plane, *an infinite number of views may be drawn*, which will vary in general proportions, lengths of edges, and sizes of angles. For practical reasons, a few of these possible positions have been classified in such a manner as to give the recognized divisions of axonometric projection:

1. isometric,
2. dimetric,
3. trimetric.

Isometric projection is the simplest of these, because the principal axes make equal angles with the plane of projection and the edges are therefore foreshortened equally.

11.4 Isometric Projection

If the cube in Fig. 11.3 is revolved through an angle of 45° about the vertical axis, as shown in II, and then tilted forward until its body diagonal is perpendicular to the vertical plane, the edges will be foreshortened equally and the cube will be in the correct position to produce an isometric projection.

The three front edges, called *isometric axes*, make angles of approximately 35° 16′ with the vertical plane of projection, or picture plane. In this form of pictorial, the visual angles between the projections of these axes are 120°, and the projected lengths of the edges of an object, along and parallel to these axes, are approximately 81% of their true lengths. It should be observed that the 90° angles of the cube appear in the isometric projection as either 120° or 60°.

Now, if instead of turning and tilting the object in relation to a principal plane of projection, an auxiliary plane is used that will be perpendicular to the body diagonal, the view projected on the plane will be an axonometric projection. Since the auxiliary plane will be inclined to the principal planes on which the front, top, and side views would be projected, the auxiliary view, taken in a position perpendicular to the body diagonal, will be a secondary auxiliary view, as shown in Fig. 11.4.

11.5 Isometric Drawing

Objects are seldom drawn in true isometric projection, since the use of an isometric scale is inconvenient and impractical. Instead, a conventional method is used in which all foreshortening is ignored, and actual true lengths are laid off along isometric axes and isometric lines. To avoid confusion and to set this method apart from true isometric projection, it is called *isometric drawing*.

The isometric drawing of a figure is slightly larger (approximately 22½%) than the isometric projection, but, since the proportions are the same, the increased size does not affect the pictorial value of the representation (see Fig. 11.4). The use of a regular scale makes it possible to

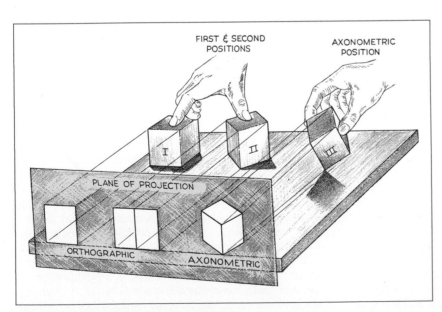

FIG. 11.3 **Theory of axonometric projection.**

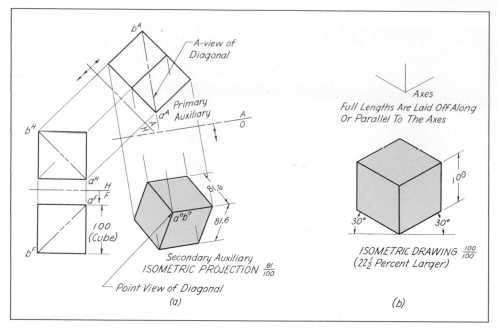

FIG. 11.4 **Comparison of isometric projection and isometric drawing.**

produce a satisfactory drawing with a minimum expenditure of time and effort.

In isometric drawing, lines that are parallel to the isometric axes are called *isometric lines* and can be directly measured.

11.6 To Make an Isometric Drawing of a Rectangular Object

The procedure followed in making an isometric drawing of a rectangular block is illustrated in Fig. 11.5. The three axes that establish the front edges, as shown in (*b*), should be drawn through point *A* so that one extends vertically downward and the other two upward to the right and left at an angle of 30° from the horizontal. Then the actual lengths of the edges may be set off, as shown in (*c*) and (*d*), and the remainder of the view completed by drawing lines parallel to the axes through the corners thus located, as in (*e*) and (*f*).

Hidden lines, unless absolutely necessary for clearness, always should be omitted on a pictorial representation.

11.7 Nonisometric Lines

Those lines that are inclined and are not parallel to the isometric axes are called *nonisometric lines*. Since a line of this type does not appear in its true length and cannot be measured directly, its position and projected length must be established by locating its extremities. In Fig. 11.6, *AB* and *CD*, which represent the edges of the block, are nonisometric lines. The location of *AB* is established in the pictorial view by locating points *A* and *B*. Point *A* is on the top edge, *X* distance from the left-side surface.

Point *B* is on the upper edge of the base, *Y* distance from the right-side surface. All other lines coincide with or are parallel to the axes and, therefore, may be measured off with the scale.

The pictorial representation of an irregular solid containing a number of nonisometric lines may be conveniently constructed by the *box method*; that is, the object may be enclosed in a rectangular box so that both isometric and nonisometric lines may be located by points of contact with its surfaces and edges (see Fig. 11.7).

A study of Figs. 11.6 and 11.7 reveals the important fact that *lines that are parallel on an object are parallel in the pictorial view*, and, conversely, *lines that are not parallel on the object are not parallel on the view*. It is often possible to eliminate much tedious construction work by the practical application of the principle of parallel lines.

11.8 Coordinate Construction Method

When an object contains a number of inclined surfaces, such as the one shown in Fig. 11.8, the use of the coordinate construction method is desirable. In this method, the end points of the edges are located in relation to an assumed isometric base line located on an *isometric reference plane*. For example, the line *RL* is used as a base line from which measurements are made along isometric lines, as shown. The distances required to locate point *A* are taken directly from the orthographic views.

Irregular curved edges are most easily drawn in isometric by the *offset method*, which is a modification of the coordinate construction method (Fig. 11.9). The position of the curve can be readily established by plotted points located by measuring along isometric lines.

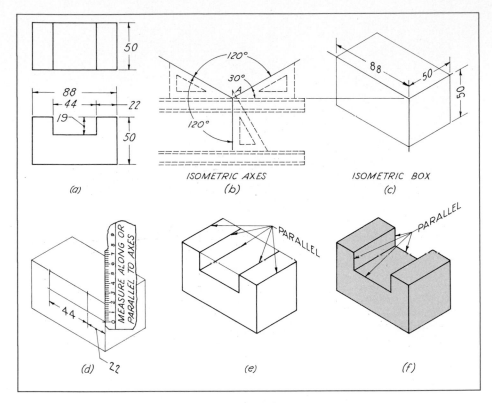

FIG. 11.5   Procedure for constructing an isometric drawing.

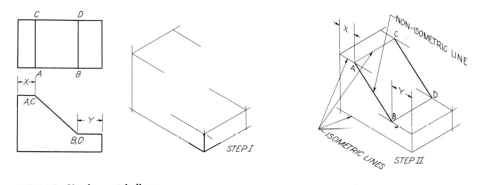

FIG. 11.6   Nonisometric lines.

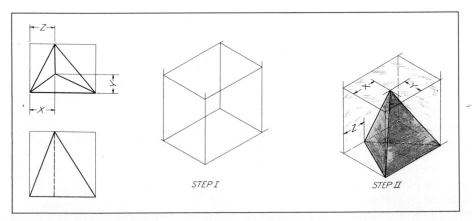

FIG. 11.7   Box construction.

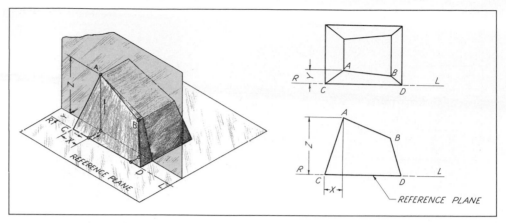

FIG. 11.8   Coordinate construction.

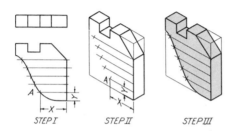

FIG. 11.9   Offset construction.

11.9   Angles in Isometric Drawing

Since angles specified in degrees do not appear in true size on an isometric drawing, angular measurements must be converted in some manner to linear measurements that can be laid off along isometric lines. Usually, one or two measurements taken from an orthographic view may be laid off along isometric lines on the pictorial drawing to locate an inclined edge that has been specified by an angular dimension.

In Fig. 11.10(*a*), the position of the inclined line *AB* was established on the isometric drawing by using the distance *X* taken from the front view of the orthographic drawing. When an orthographic drawing has already been prepared to a different scale than the scale being used for the pictorial representation, one can draw a partial orthographic view and take off the needed dimensions. A practical application of this idea is shown in (*b*). By making the construction of a partial view at the place where the angle is to appear on the isometric drawing, the position of the required line can be obtained graphically.

If desired, the tangent method, as explained in Sec. 3.11, may be used, as shown in (*c*). In using this method, a length equal to 10 units (any scale) is laid off along an isometric line that is to form one side of the angle. Then, a distance equal to 10 times the tangent of the angle is set off along a second isometric line that represents the second leg of the right triangle in pictorial. A line drawn through the end points of these lines will be the required line at the specified angle.

11.10   Circle and Circle Arcs in Isometric Drawing

In isometric drawing, a circle appears as an ellipse. The tedious construction required for plotting an ellipse accurately (Figs. 11.11 and 11.12) often is avoided by using some approximate method of drawing. The representation thus obtained is accurate enough for most work, although the true ellipse, which is slightly narrower and longer, is more pleasing in shape (Fig. 11.11). For an approximate construction, a four-center method is generally used.

To draw a circle in pictorial representation, a square is conceived to be circumscribed about the circle in the orthographic projection. When transferred to the isometric plane, the square becomes a rhombus (isometric square) and the circle an ellipse tangent to the rhombus at the midpoints of its sides. If the ellipse is to be drawn by the four-center method (Fig. 11.13), the points of intersection of the perpendicular bisectors of the sides of the rhombus will be centers for the four arcs forming the approximate ellipse. The two intersections that lie on the corners of the rhombus are centers for the two large arcs, while the remaining intersections are centers for the two small arcs. Furthermore, the length along the perpendicular from the center of each arc to the point at which the arc is tangent to the rhombus (midpoint) will be the radius.

The amount of work may be still further shortened and the accuracy of the construction improved by following the procedure shown in Fig. 11.14. The steps in this method are as follows.

STEP 1.   Draw the isometric center lines of the required circle.

STEP 2.   Using a radius equal to the radius of the circle, strike arcs across the isometric center lines.

STEPS 3 AND 4.   Through each of these points of intersection erect a perpendicular to the other isometric center line.

STEPS 5 AND 6.   Using the intersection points of the perpendiculars as centers and lengths along the perpendiculars as radii, draw the four arcs that form the ellipse (Fig. 11.15).

A circle arc will appear in pictorial representation as a segment of an ellipse. Therefore, it may be drawn by using

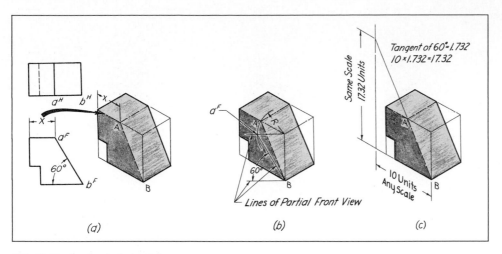

FIG. 11.10 Angles in isometric.

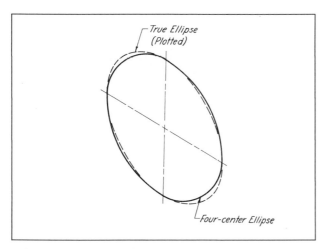

FIG. 11.11 Pictorial ellipses.

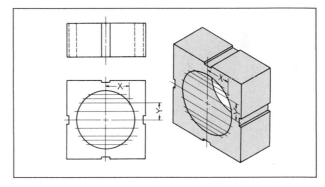

FIG. 11.12 To plot an isometric circle.

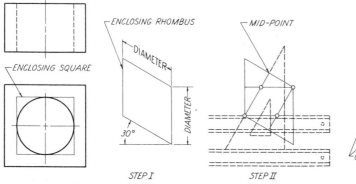
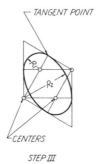

FIG. 11.13 Four-center approximation.

as much of the four-center method as is required to locate the needed centers (Fig. 11.16). For example, to draw a quarter circle, it is only necessary to lay off the true radius of the arc along isometric lines drawn through the center and to draw intersecting perpendiculars through these points.

To draw isometric concentric circles by the four-center method, a set of centers must be located for each circle (Fig. 11.17).

When several circles of the same diameter occur in parallel planes, the construction may be simplified. Figure 11.18 shows two views of an object and its corresponding

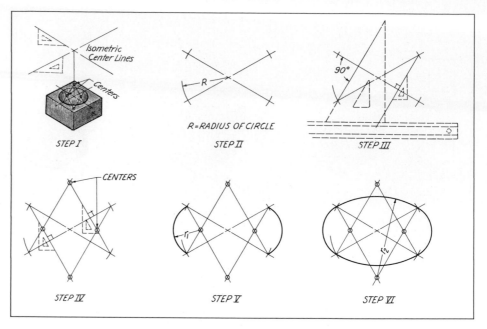

FIG. 11.14   Steps in drawing a four-center isometric circle (ellipse).

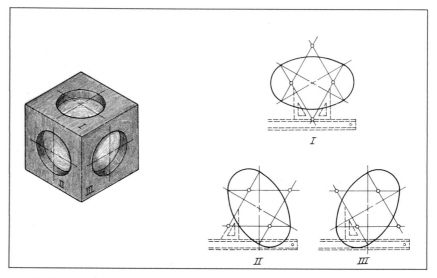

FIG. 11.15   Isometric circles.

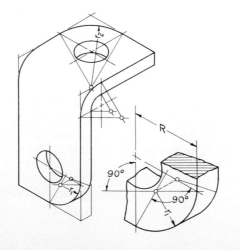

FIG. 11.16   Isometric circle arcs.

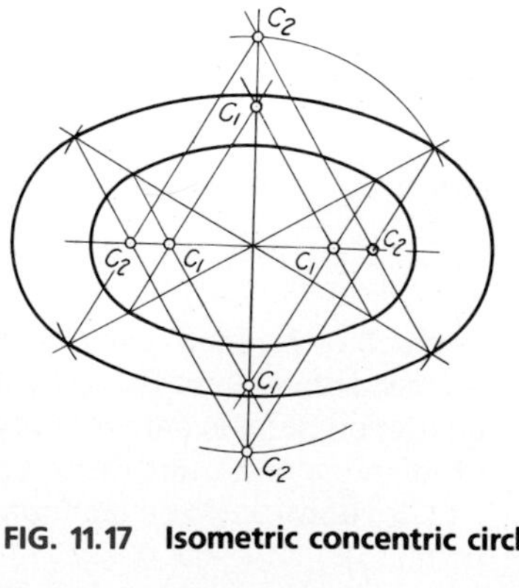

FIG. 11.17   Isometric concentric circles.

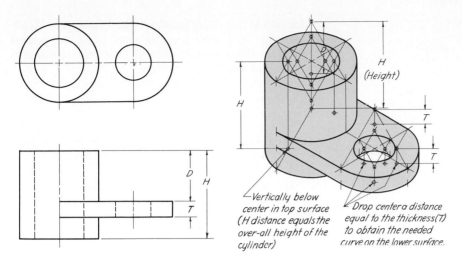

FIG. 11.18 **Isometric parallel circles.**

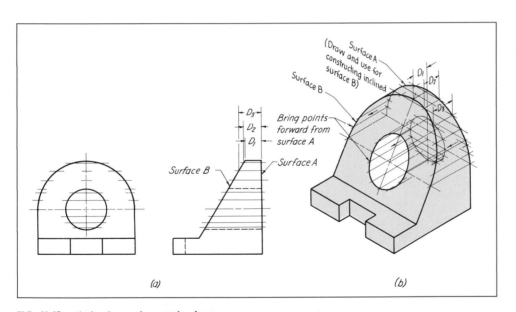

FIG. 11.19 **Circles in nonisometric planes.**

isometric drawing. In Fig. 11.18, the centers for the ellipse representing the upper base of the large cylinder are found in the usual way, while the centers for the lower base are located by moving the centers for the upper base downward a distance equal to the height of the cylinder. By observing that portion of the object projecting to the right, it can be noted that corresponding centers lie along an isometric line parallel to the axis of the cylinder.

Circles and circle arcs in nonisometric planes may be plotted by using the offset or coordinate method. Sufficient points for establishing a curve must be located by transferring measurements from the orthographic views to isometric lines in the pictorial view. There is a rapid and easy way for drawing the cylindrical portion of the object shown in Fig. 11.19. The semicircular arc must be plotted on the rear surface as the first step. Then, after this has been done, each point is brought forward to the inclined face.

The offset distances (D_1, D_2, D_3, etc.) at each level are taken from the side view in (*a*).

The pictorial representation of a sphere is the envelope of all of the great circles that could be drawn on the surface. In isometric drawing, the great circles appear as ellipses and a circle is their envelope. In practice it is necessary to draw only one ellipse, using the true radius of the sphere and the four-center method of construction. The diameter of the circle is the major axis of the ellipse (Fig. 11.20).

11.11 *Positions of Isometric Axes*

It is sometimes desirable to place the principal isometric axes so that an object will be in position to reveal certain faces to a better advantage (Fig. 11.21).

The difference in direction should cause no confusion, since the angle between the axes and the procedure fol-

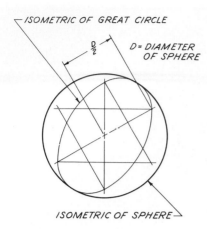

FIG. 11.20 Isometric drawing of a sphere.

lowed in constructing the view are the same for any position. The choice of the direction may depend on the construction of the object, but usually this is determined by the position from which the object is ordinarily viewed.

Reversed axes (b) are used in architectural work to show a feature as it would be seen from a natural position below.

Sometimes long objects are drawn with the long axis horizontal, as shown in Fig. 11.22.

11.12 Isometric Sectional Views (Fig. 11.23)

Generally, an isometric sectional view is used for showing the inner construction of an object when there is a complicated interior to be explained or when it is desirable to emphasize features that would not appear in a usual outside view. Sectioning in isometric drawing is based on the same

principles as sectioning in orthographic drawing. Isometric planes are used for cutting an object, and the general procedure followed in constructing the representation is the same as for an exterior view.

Figure 11.23 shows an isometric half section. It is easier, in this case, to outline the outside view of the object in full and then remove a front quarter with isometric planes.

Figure 11.24 illustrates a full section in isometric. The accepted procedure for constructing this form of sectional view is to draw the cut face and then add the portion that lies behind.

Section lines should be sloped at an angle that produces the best effect, but they should never be drawn parallel to object lines. In Fig. 11.25, (a) illustrates the slope that is correct for most drawings, while (b), (c), and (d) show the poor effect produced when this phase of section lining is ignored. Ordinarily, isometric section lines are drawn at 60°.

B / OBLIQUE PROJECTION

11.13 Oblique Projection

In oblique projection, the view is produced by using parallel projectors that make some angle other than 90° with the plane of projection. Thus, an oblique projection is *not* an orthographic or axonometric view. Rather, it is a contrived view without orthogonal basis. Generally, one face is placed parallel to the picture plane and the projection lines are taken at 45°. This gives a view that is pictorial in appearance, as it shows the front and one or more additional faces of an object. In Fig. 11.26, the orthographic and oblique projections of a cube are shown. When

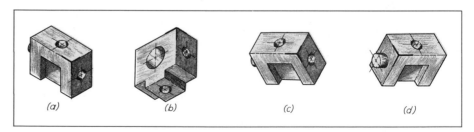

FIG. 11.21 Convenient positions of axes.

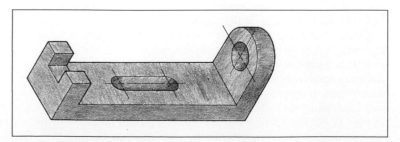

FIG. 11.22 Main axis horizontal—long objects.

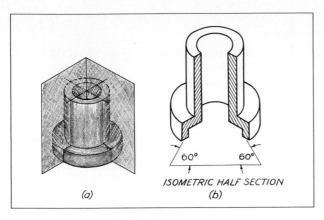

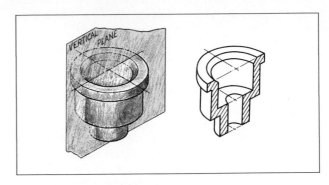

FIG. 11.23 **Isometric half section.**

FIG. 11.24 **Isometric full section.**

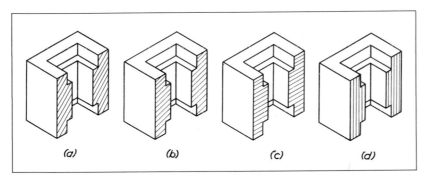

FIG. 11.25 **Section lining.**

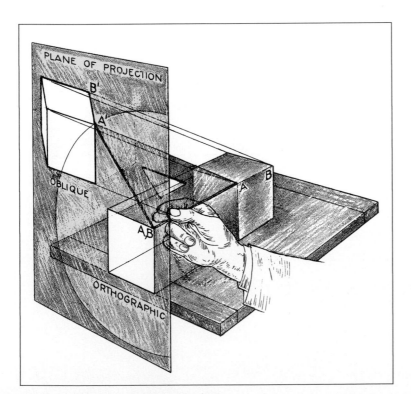

FIG. 11.26 **Theory of oblique projection.**

the angle is 45°, as in this illustration, the representation is sometimes called *cavalier projection*. It is generally known, however, as an *oblique projection* or an *oblique drawing*.

11.14 Oblique Drawing

This form of drawing is based on three mutually perpendicular axes along which, or parallel to which, the necessary measurements are made for constructing the representation. Oblique drawing differs from isometric drawing principally in that two axes are always perpendicular to each other, while the third (receding axis) is at some convenient angle, such as 30°, 45°, or 60° with the horizontal (Fig. 11.28). It is somewhat more flexible and has the following advantages over isometric drawing:

1. circular or irregular outlines on the front face show in their true shape;

2. distortion can be reduced by foreshortening along the receding axis;

3. a greater choice is permitted in the selection of the positions of the axes.

A few of the various views that can be obtained by varying the inclination of the receding axis are illustrated in Fig. 11.27. Usually, the selection of the position is governed by the character of the object.

11.15 To Make an Oblique Drawing

The procedure to be followed in constructing an oblique drawing is illustrated in Fig. 11.28. The three axes that establish the perpendicular edges in (b) are drawn through

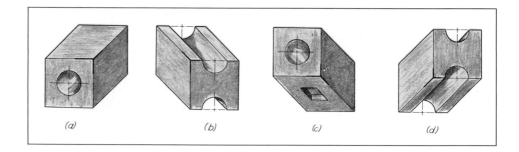

FIG. 11.27 **Various positions of the receding axis.**

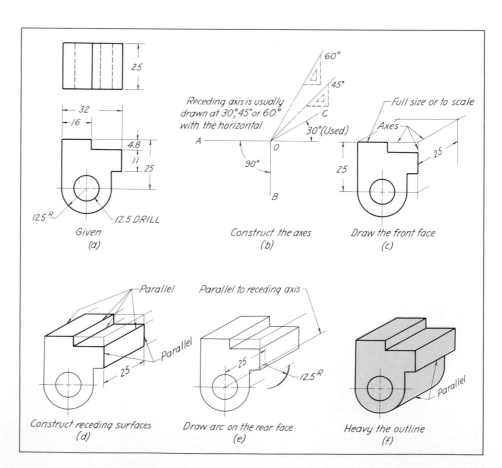

FIG. 11.28 **Procedure for constructing an oblique drawing.**

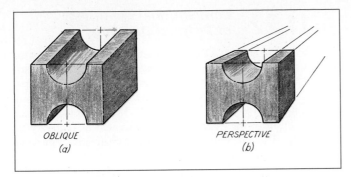

FIG. 11.34 **Comparison of oblique and perspective.**

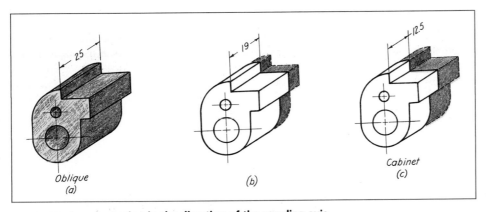

FIG. 11.35 **Foreshortening in the direction of the receding axis.**

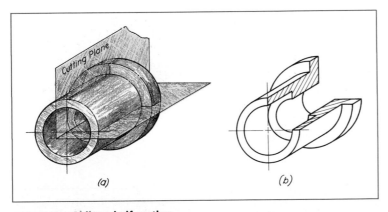

FIG. 11.36 **Oblique half section.**

For practical purposes, measurements are usually reduced one-half, but any scale of reduction may be arbitrarily adopted if the view obtained will be more realistic in appearance. When the receding lines are drawn one-half their actual length, the resulting pictorial view is called a *cabinet drawing*. Figure 11.35 shows an oblique drawing (*a*) and a cabinet drawing (*c*) of the same object, for the purpose of comparison.

11.20 *Oblique Sectional Views*

Oblique sectional views are drawn to show the interior construction of objects. The construction procedure is the same as for an isometric sectional view, except that oblique planes are used for cutting the object. An oblique half section is illustrated in Fig. 11.36.

11.21 Pictorial Dimensioning

The dimensioning of isometric and other forms of pictorial working drawings is done in accordance with the following rules.

1. Draw extension and dimension lines (except those dimension lines applying to cylindrical features) parallel to the pictorial axes in the plane of the surface to which they apply (Fig. 11.37).

2. If possible, apply dimensions to visible surfaces.

3. Place dimensions on the object, if, by so doing, better appearance, added clearness, and easy readings result.

4. Notes may be lettered either in pictorial or as on ordinary drawings. When lettered as on ordinary drawings the difficulties encountered in forming pictorial letters are avoided (Fig. 11.37).

5. Make the figures of a dimension appear to be lying in the plane of the surface whose dimension it indicates, by using vertical figures drawn in pictorial (Fig. 11.37). (*Note*: Guide lines and slope lines are drawn parallel to the pictorial axes.)

11.22 Conventional Treatment of Pictorial Drawings

When it is desirable for an isometric or an oblique drawing of a casting to present a somewhat more or less realistic appearance, it becomes necessary to represent the fillets and rounds on the unfinished surfaces. One method commonly used is shown in Figure 11.38(*a*). On the drawing in (*b*) all of the edges have been treated as if they were sharp. The conventional treatment for threads in pictorial is illustrated in (*b*) and (*c*).

C / PERSPECTIVE PROJECTION

11.23 Perspective

In perspective projection an object is shown much as the human eye or camera would see it at a particular point. Actually, it is a geometric method by which a picture can be projected on a picture plane in much the same way as in photography. Perspective drawing differs from the methods previously discussed in that the projectors or visual

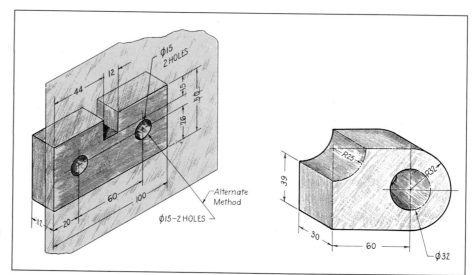

FIG. 11.37 Extension and dimension lines in isometric (left); numerals, fractions, and notes in oblique (right).

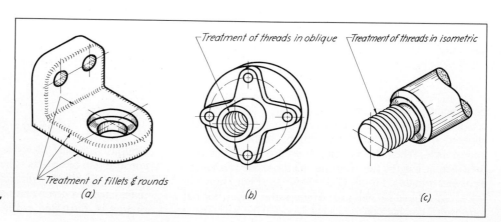

FIG. 11.38 Conventional treatment of fillets, rounds, and threads in pictorial.

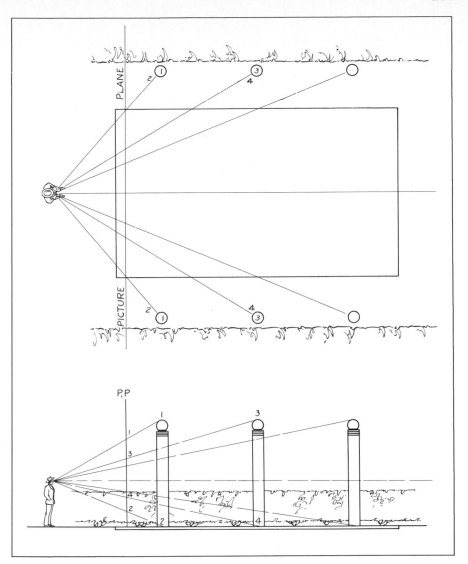

FIG. 11.39 Picture plane.

rays intersect at a common point known as the *station point* (Fig 11.41).

Since the perspective shows an object as it appears instead of showing its true shape and size, it is rarely used by engineers. It is more extensively employed by architects to show the appearance of proposed buildings, by artist-draftsmen for production illustrations, and by illustrators in preparing advertising drawings.

Figure 11.1 shows a type of production illustration that has been widely used in assembly departments as an aid to those persons who find it difficult to read an orthographic assembly. This form of presentation, which may show a mechanism both exploded and assembled, has made it possible for industrial concerns to employ semitrained personnel. Figure 11.52 shows a type of industrial drawing made in perspective that has proved useful in aircraft plants. Because of the growing importance of this type of drawing, and also because engineers frequently will find perspective desirable for other purposes, its elementary principles should be discussed logically in this text. Other books on the subject, some of which are listed in

the bibliography, should be studied by architectural students and those interested in a more thorough discussion of the various methods.

The fundamental concepts of perspective can be explained best if the reader will imagine looking through a picture plane at a formal garden with a small pool flanked by lampposts, as shown in Fig. 11.39. As noted above, the point of observation, at which the rays from the eye to the objects in the scene meet, is called the station point, and the plane on which the view is formed by the piercing points of the visual rays is known as the *picture plane* (*PP*). The piercing points reproduce the scene, the size which depends on the location of the picture plane.

It should be noted that objects of the same height intercept a greater distance on the picture plane when close to it than when farther away. For example, rays from the lamppost at 2 intercept a distance 1–2 on the picture plane, while the rays from the pole at 4, which actually is the same height, intercept the lesser distance 3–4. From this fact it should be observed that the farther away an object is, the smaller it will appear, until a point is reached at

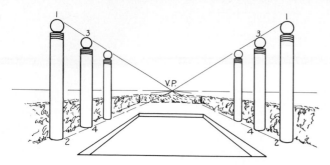

FIG. 11.40   The picture (perspective).

front of the eye in the plane of the horizon on the horizon line.

11.25   Location of Picture Plane

The picture plane is usually placed between the object and the *SP* (station point). In parallel perspective (Sec. 11.31) it may be passed through a face of the object in order to show the true size and shape of the face.

11.26   Location of the Station Point

Care must be exercised in selecting the location for the station point, for its position has much to do with the appearance of the finished perspective drawing. A poor choice of position may result in an overly distorted perspective that will be decidedly displeasing to the eye.

In general, the station point should be offset slightly to one side and should be located above or below the exact center of the object. However, it must be remembered that the center of vision must be near the center of interest for the viewer.

One should always think of the station point as the viewing point, and its location should be where the object can be viewed to the best advantage. It is desirable that it be at a distance from the picture plane equal to at least twice the maximum dimension (width, height, or depth) of the object, for at such a distance, or greater, the entire object can be viewed naturally, as a whole, without turning the head.

A wide angle of view is to be avoided in the interest of good picturization. It has been determined that best results are obtained when the visual rays from the station

which there will be no distance intercepted at all. This happens at the horizon.

Figure 11.40 shows the scene observed by the man in Fig. 11.39 as it would be formed on the picture plane. The posts farther from the picture plane diminish in height, as each one has a height on the picture plane equal to the distance it intercepts, (Fig. 11.39). The lines of the pool and hedge converge to the center of vision or vanishing point, which is located directly in front of the observer on the horizon.

11.24   Perspective Nomenclature

Figure 11.41 illustrates pictorially the accepted nomenclature of perspective drawing. The *horizon line* is the line of intersection of the horizontal plane through the observation point (eye of the observer) and the picture plane. The horizontal plane is known as the *plane of the horizon*. The *ground line* is the line of intersection of the ground plane and the picture plane. The *CV point* is the center of vision of the observer. It is located directly in

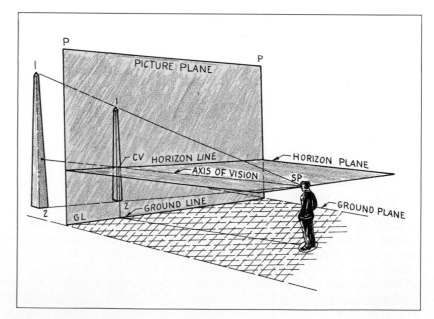

FIG. 11.41   Nomenclature.

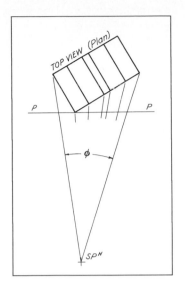

FIG. 11.42 **Angle of vision.**

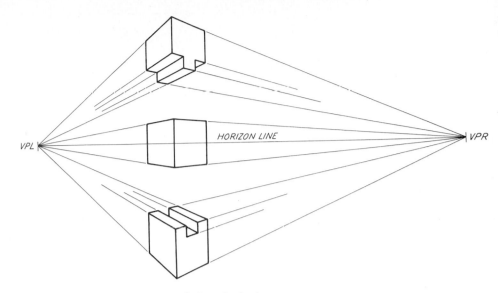

FIG. 11.43 **Objects on, above, or below the horizon.**

point (*SP*) to the object are kept within a cone having an angle of not more than 30° between diametrically opposing elements (see Fig. 11.42).

In locating an object in relation to the picture plane, it is advisable to place it so that both of the side faces do not make the same angle with the picture plane and thus will not be equally visible. It is common practice to choose angles of 30° and 60° for rectangular objects.

11.27 *Position of the Object in Relation to the Horizon*

When making a perspective of a tall object, such as a building, the horizon usually is assumed to be at a height above the ground plane equal to the standing height of a viewer's eye, normally about 5 ft 6 in.

A small object may be placed either above or below the horizon (eye level), depending on the view desired. If an object is above the horizon, it will be seen looking up from below, as shown in Fig. 11.43. Should the object be below the horizon line, it will be seen from above.

11.28 *Lines*

The following facts should be recognized concerning the perspective of lines.

1. Parallel horizontal lines vanish at a single *VP* (vanishing point). Usually the *VP* is at the point where a line parallel to the system through the *SP* pierces the *PP* (picture plane).

2. A group of horizontal lines has its *VP* on the horizon.

3. Vertical lines, since they pierce the picture plane at infinity, will appear vertical in perspective.

4. When a line lies in the picture plane, it will show its true length.

5. When a line lies behind the picture plane, its perspective will be shorter than the line.

6. When a line lies in front of the picture plane, its perspective will be longer than the line.

11.29 *Perspective by Multiview Projection*

Adhering to the theory that a perspective drawing is formed on a picture plane by visual rays from the eye to the object, as illustrated in Fig. 11.39, a perspective can be drawn by using multiview projection (see Fig. 11.44). The multiview method may be the easiest for a student to understand but it is not often used by an experienced person because considerably more line work is required if the scene or object to be represented is at all complicated. The top and side views are drawn in multiview projection and the picture plane (as an edge view) and the station point are shown in each case. SP^H and its related views of the rays are in the top view, while SP^P and its projections of the rays belong to the side view. The front view of the picture plane is in the plane of the paper.

After the preliminary layout has been completed point *A* may be located in the perspective by the following procedure.

STEP 1. Draw the top view ($SP^H a^H$) and the side view ($SP^P a^P$) of the visual ray from the eye to point *A*.

STEP 2. From a' (the top view of the piercing point of the ray) draw a projection line downward.

STEP 3. From a'' (the side view of the piercing point) draw a horizontal projection line to intersect the one drawn from a'.

Point *A* of the perspective is at this intersection.

Point *B* and the other points that are needed for the perspective representation are found in the same manner as point *A*.

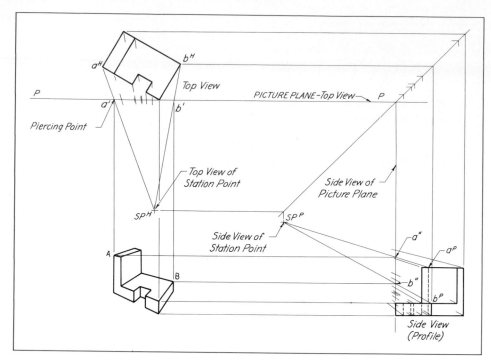

FIG. 11.44   Perspective drawing—orthographic method.

11.30 Types of Perspective

In general, there are two types of perspective: *parallel perspective* and *angular perspective*. In parallel perspective, one of the principal faces is parallel to the picture plane. All vertical lines are vertical, and the receding horizontal lines converge to a single vanishing point. In angular perspective, the object is placed so that the principal faces are at an angle with the picture plane. The horizontal lines converge at two vanishing points.

11.31 Parallel Perspective

Figure 11.45 shows the parallel perspective of a rectangular block. The *PP* line is the top view of the picture plane, SP_H is the top view of the station point, and *CV* is the center of vision. The receding horizontal lines vanish at *CV*. The front face, since it lies in the picture plane, is shown in true size. The lines representing the edges back of the picture plane are found by projecting downward from the points at which the visual rays pierce the picture plane, as shown by the top views of the rays. Figure 11.46 shows a parallel perspective of a cylindrical machine part.

11.32 Angular Perspective

Figure 11.47 shows pictorially the graphical method for the preparation of an angular perspective drawing of a cube. To visualize the true layout on the surface of a sheet of drawing paper, it is necessary to mentally revolve the

horizontal plane downward into the vertical or picture plane. On completion of Sec. 11.32, it is suggested that the reader turn back and endeavor to associate the development of the perspective in Fig. 11.48 with the pictorial presentation in Fig. 11.47. For a full understanding of the construction in Fig. 11.48, it is necessary to differentiate between the lines that belong to the horizontal plane and those that are on the vertical or picture plane. In addition, it must be fully realized that there is a top view for the perspective that is a line and that in this line view lie the points that must be projected downward to the perspective representation (front view).

Figure 11.48 shows an angular perspective of a block. The block has been placed so that one vertical edge lies in the picture plane. The other vertical edges are parallel to the plane, while all of the horizontal lines are inclined to it so that they vanish at the two vanishing points, *VPL* and *VPR*, respectively.

In constructing the perspective shown in this illustration, an orthographic top view was drawn in such a position that the visible vertical faces made angles of 30° and 60° with the picture plane. Next, the location of the observer was assumed and the horizon line was established. The vanishing points *VPL* and *VPR* were found by drawing a 30° line and a 60° line through the *SP*. Since these lines are parallel to the two systems of receding horizontal lines, each will establish a required vanishing point at its intersection with the picture plane. The vertical line located in the picture plane, which is its own perspective, was selected as a measuring line on which to project vertical measurements from the orthographic front view. The lines shown from these division points along this line to the

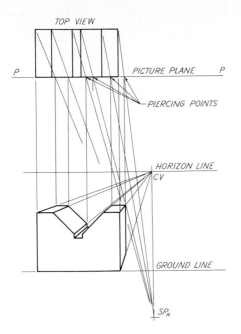

FIG. 11.45 Parallel perspective.

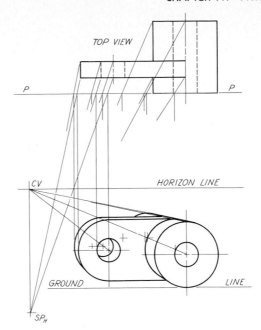

FIG. 11.46 Circles in parallel perspective.

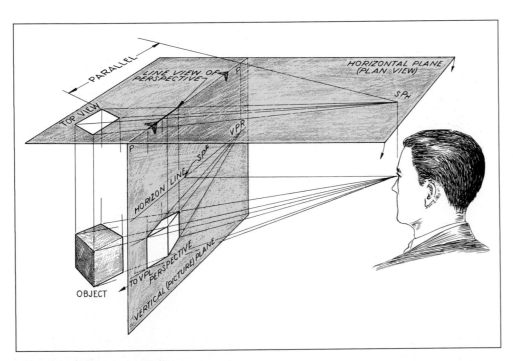

FIG. 11.47 Angular perspective.

vanishing points (*VPL-VPR*) established the direction of the receding horizontal edge lines in the perspective. The positions of the back edges were determined by projecting downward from the points at which the projectors from the station point (*SP*) to the corners of the object pierced the picture plane, as shown by the top view of the object and projectors.

11.33 *Use of Measuring Lines*

Whenever the vertical front edge of an object lies in the picture plane, it can be laid off full length in the perspective, because theoretically, it will be in true length in the picture formed on the plane by the visual rays (see Fig. 11.48). Should the near vertical edge lie behind the

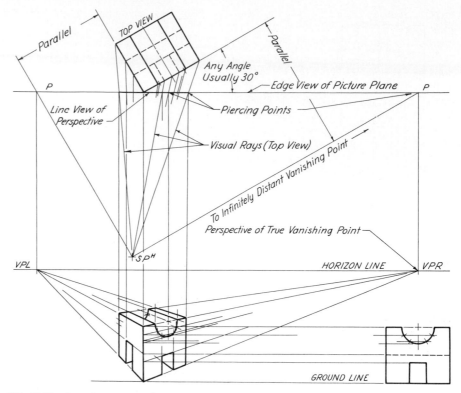

FIG. 11.48   Angular perspective.

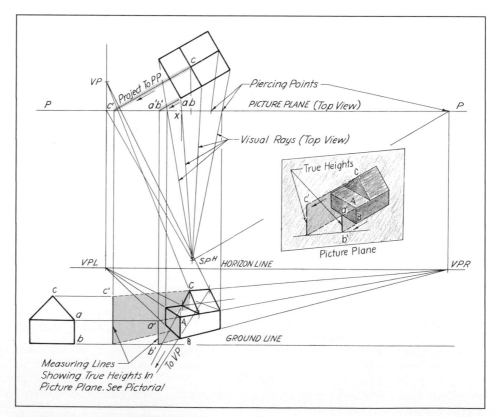

FIG. 11.49   Use of measuring lines.

picture plane, as is the case with the edge line *AB* in Fig. 11.49, the use of a *measuring line* becomes desirable. The measuring line *a′b′* is the vertical edge *AB* moved forward to the picture plane, where it will appear in its true height. Some prefer to think of the vertical side as being extended to the picture plane so that the true height of the side is revealed. The length and position of *AB* is established in the perspective picture by first drawing vanishing lines from *a′* and *b′* to *VPR*; then the top view of the edge in the picture plane (point *X*) is projected downward to the front-view picturization. Points *A* and *B* must fall on the vanishing lines from *a′* and *b′* to *VPR* respectively.

A measuring line may be used to establish the "picture height" of any feature of an object. For example, in Fig. 11.49, the vertical measuring line through *c′* was used as needed to locate point *C* and the top line of the object in the perspective.

11.34 Vanishing Points for Inclined Lines

In general, the perspective of any inclined straight line can be found by locating the end points of the line perspectively. By this method, an end point may be located by drawing the perspective representations of any two horizontal lines intersecting at the end point. Where several parallel inclined lines vanish to the same vanishing point, it may prove to be worthwhile to locate the *VP* for the group in order to conserve time and achieve a higher degree of accuracy.

Just as in the case of other lines, the vanishing point for any inclined line may be established by finding the piercing point in the picture plane (*PP*) of a line through *SP* parallel to the given inclined line.

In Fig. 11.50 the inclined lines are *AB* and *CD*. If one is to understand the construction shown for locating vanishing point *VPI*, one must recognize that vertical planes vanish in vertical lines and that a line in a vertical plane will vanish at a point on the vanishing line of the plane. The preceding statements being true, the vanishing point *VPI* for line *CD* must lie at some point on a vertical line through *VPR*, since *CD* and *CE* lie in the same vertical plane. *VPI* is the point at which a line drawn parallel (in space) to *CD* through *SP* pierces the picture plane.

On the drawing, the distance *D* that *VPI* is above *VPR* may be found easily through the construction of a right triangle with *SP^H Q* as a base. Angle β is the slope angle for line *CD* (see side view). The needed distance *D* is the line *QR*, or the short leg of the triangle.

11.35 Circles in Perspective

If a circle is on a surface that is inclined to the picture plane (*PP*), its perspective will resemble an ellipse. It is the usual practice to construct the representation within an enclosing square by finding selected points along the curve in the perspective, as shown in Fig. 11.51(*a*). Any points might be used, but it is recommended that points be located on 30° and 60° lines.

In (*a*), the perspective representation of the circle was found by using visual rays and parallel horizontal lines in

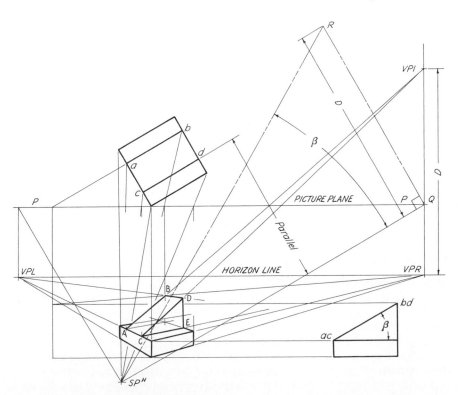

FIG. 11.50 Vanishing point for inclined lines.

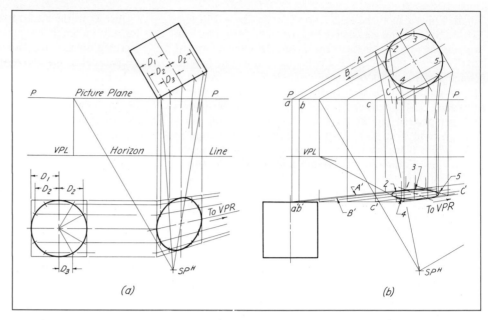

FIG. 11.51 Circles in perspective.

combination. In starting the construction, the positions of several selected points, located on the circumference and lying on horizontal lines in the plane of the circle, were established in both views. After these lines had been drawn in the perspective in the usual manner, the locations of the points along them were determined through the use of visual rays, as shown. Specifically, the position of a point in the perspective was found by projecting downward from the piercing point of the ray from the point and the picture plane (see line view) to the perspective view of the line on which the point must lie. In (*b*), the same method was applied to construct the perspective view of a circle in a horizontal plane. It should be noted in this case that the horizontal lines, as established in the top view, were extended to cut the picture plane so that the true height of these lines at the plane could be used in the perspective view for locating the end points of the perspective representations (at the left).

D/ INDUSTRIAL ILLUSTRATIONS

11.36 Technical Illustration Drawings

The design and manufacturing procedures of present-day mass production require various types of pictorial illustrations to communicate ideas and concepts to large numbers of persons who are all working toward a common objective. These illustration drawings are of value at all stages of a project from the design phase, where they may be only pencil-shaped freehand sketches, through all of the stages of production that we may consider to include not only the assembly but final installation of the systems

as well. They are used in operation and maintenance manuals to make complex and difficult tasks understandable to those persons who may be unable to interpret conventional drawings (Fig. 11.52). Pictorial illustrations range from simple types of line drawings (Figs. 11.53 and 11.54) that have already been discussed to artists' renderings that have the realism of photographs. An artist's rendering, such as the one shown in Fig. 11.56, is usually prepared to reinforce an oral or written report that is to be presented before a decision-making group that consists to some extent of persons who would otherwise be unable to understand construction details. Drawings of this type are depended on to sell a project (Fig. 11.57).

Illustration drawings are used for many purposes in every field of engineering and technology (Fig. 11.55). They appear in advertising literature, operation and service manuals, patent applications (Fig. 11.59), and textbooks (Fig. 11.66). Illustration drawings may be working drawings, assembly drawings, piping and wiring diagrams, and architectural and engineering renderings that are almost true-to-life. Typical examples of pictorial drawings that were prepared to facilitate assembly are shown in Figs. 11.1 and 16.11. Figure 22.1 shows an electronic diagram that appeared in a service manual.

11.37 Design Illustrations

Design illustrations are prepared to clarify conventional engineering design and production drawings and written specifications. These are used for the communication of ideas and concepts concerning the details of complicated designs. Properly prepared, drawings of this type reveal the relationship of the components of a system so clearly that the principles of operation of a unit can be understood by almost everyone, even by persons who may be relatively

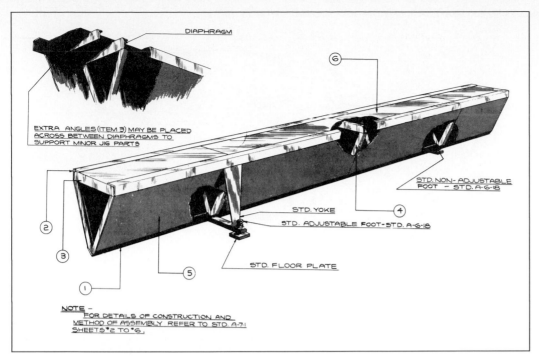

DIAPHRAGM

EXTRA ANGLES (ITEM 3) MAY BE PLACED
ACROSS BETWEEN DIAPHRAGMS TO
SUPPORT MINOR JIG PARTS

STD. NON-ADJUSTABLE
FOOT – STD. A-6-18

STD. YOKE

STD. ADJUSTABLE FOOT-STD. A-6-18

STD. FLOOR PLATE

NOTE –
FOR DETAILS OF CONSTRUCTION AND
METHOD OF ASSEMBLY REFER TO STD. A-7-1
SHEETS *2 TO *6.

FIG. 11.52 **Production illustration prepared in perspective.** (*Courtesy Craftint Mfg. Co.*)

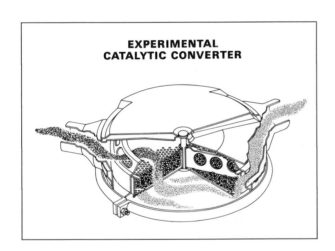

EXPERIMENTAL
CATALYTIC CONVERTER

FIG. 11.53 **This pictorial representation of a catalytic converter appeared in a technical report prepared by General Motors.** Pictorial drawings such as this one are often needed in final design reports. Some engineers and technologists think that the utilization of a complex combination of components along with a catalytic converter in the exhaust line will prove to be the most effective method of treating hydrocarbons and carbon monoxide. (*Courtesy General Motors Corporation*)

FIG. 11.54 **Items other than the catalytic converter that are a part of the total package of emission control are shown in the pictorial representation of the advanced emission control system under development at General Motors.** This presentation was extracted from a technical report prepared by GM engineers. (*Courtesy General Motors Corporation*)

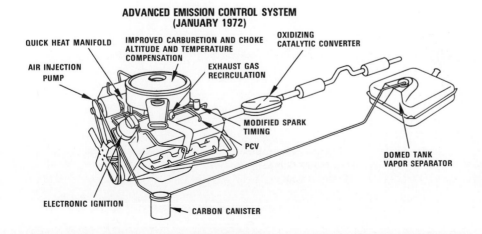

ADVANCED EMISSION CONTROL SYSTEM
(JANUARY 1972)

QUICK HEAT MANIFOLD

IMPROVED CARBURETION AND CHOKE
ALTITUDE AND TEMPERATURE
COMPENSATION

OXIDIZING
CATALYTIC CONVERTER

AIR INJECTION
PUMP

EXHAUST GAS
RECIRCULATION

MODIFIED SPARK
TIMING

PCV

DOMED TANK
VAPOR SEPARATOR

ELECTRONIC IGNITION

CARBON CANISTER

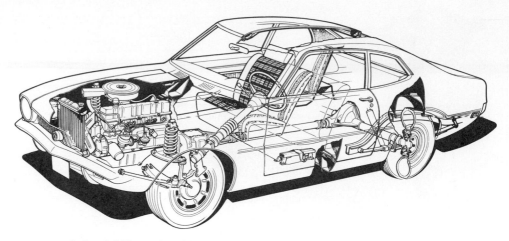

FIG. 11.55 **Industrial illustration.** (*Courtesy Ford Motor Company*)

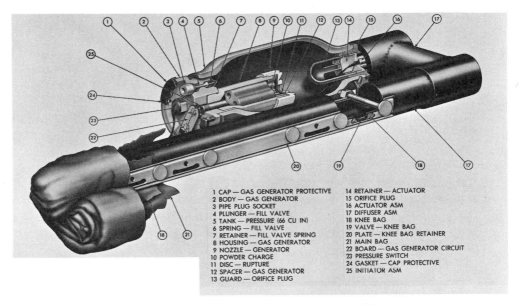

1 CAP — GAS GENERATOR PROTECTIVE	14 RETAINER — ACTUATOR
2 BODY — GAS GENERATOR	15 ORIFICE PLUG
3 PIPE PLUG SOCKET	16 ACTUATOR ASM
4 PLUNGER — FILL VALVE	17 DIFFUSER ASM
5 TANK — PRESSURE (66 CU IN)	18 KNEE BAG
6 SPRING — FILL VALVE	19 VALVE — KNEE BAG
7 RETAINER — FILL VALVE SPRING	20 PLATE — KNEE BAG RETAINER
8 HOUSING — GAS GENERATOR	21 MAIN BAG
9 NOZZLE — GENERATOR	22 BOARD — GAS GENERATOR CIRCUIT
10 POWDER CHARGE	23 PRESSURE SWITCH
11 DISC — RUPTURE	24 GASKET — CAP PROTECTIVE
12 SPACER — GAS GENERATOR	25 INITIATOR ASM
13 GUARD — ORIFICE PLUG	

FIG. 11.56 **Components of the Allied Chemical Corporation air-bag modules to be used in the air-bag— seat-belt restraint system under development by the Ford Motor Company.** The system has a main bag (cushion) and a small knee bag. (*Courtesy Ford Motor Company*)

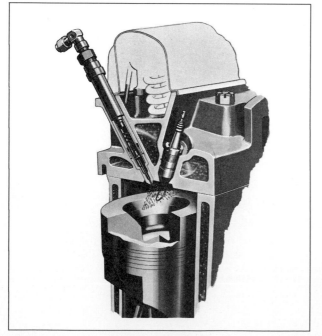

FIG. 11.57 **New combustion process developed by the Ford Motor Company, cutaway view.** (*Courtesy Ford Motor Company*)

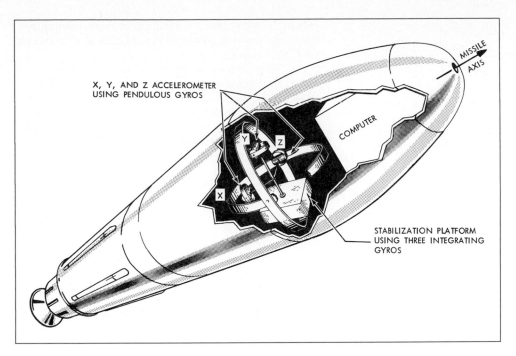

FIG. 11.58 **Pictorial representation of the principal elements of the inertial guidance system in a missile vehicle.** (*Courtesy General Motors Corporation*)

unfamiliar with graphic methods. It is common practice to prepare a series of such drawings to clarify complex details of construction, to indicate the function of closely related parts, and reveal structural features. A pictorial illustration of this type is shown in Fig. 11.57. It was used to supplement a technical paper presented at a Society of Automotive Engineers Congress.

11.38 Shading Methods

A pictorial illustration can be improved and given a more realistic appearance by shading to produce the effect of surface texture. To this shades and shadows may be added to give additional realism. The use of pencil shading is most common; however, ink shading (see Fig. 11.66) produces clean illustrations of high quality that are well suited for reproductions in texts and brochures. The techniques of surface shading by means of ink lines will be discussed in Sec. 11.39.

Some of the other more basic methods of representing surface textures under light and shade involve the use of Rossboard, Craftint paper, Zip-a-tone overlay film, and the airbrush.

Most of the shaded pictorial drawings in this text were drawn on Rossboard. This popular drawing paper, with its rough plaster-type surface, is available in many textured patterns. Surface shading is done with a very soft pencil.

Craftint papers, single-tone or double-tone, have the pattern in the paper. Drawings are prepared on these papers in the usual manner and regular black waterproof drawing ink is used for the finished lines. Solid black areas are filled in with ink before the areas where shading is desired are brushed with a developer that brings out the surface

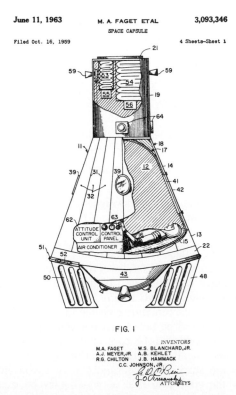

FIG. 11.59 **Patent drawing with surface shading.**

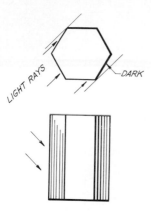

FIG. 11.60   **Surface shading on a prism.**

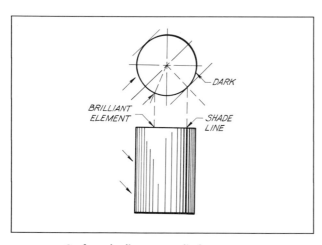

FIG. 11.61   **Surface shading on a cylinder.**

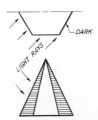

FIG. 11.62   **Pyramid.**

pattern. These papers are available in many shading patterns. The drawing shown in Fig. 11.52 was prepared on Craftint paper.

Zip-a-tone overlay screens with printed shading patterns of dots and lines provide an easy method of surface shading suitable for high-quality printed reproductions (Fig. 11.58). The screen, backed with a clear adhesive, is applied as a sheet or partial sheet to the areas to be shaded. Unwanted portions are removed with a sharp knife before the screen is rubbed down firmly to complete the bond between screen and paper.

A high degree of realism can be achieved using a small spray gun that in the language of the artist is known as an *airbrush*. This delicate instrument sprays a fine mist of diluted ink over the surface of the drawing to produce variations of tone. A capable artist can produce a representation that has the realism of a photograph. The pictorial illustration in Fig. 11.56 is an excellent example of airbrush rendering by a commercial illustrator. The airbrush is sometimes used to retouch photographs to improve their appearance for reproduction.

11.39   *Surface Shading by Means of Lines (Fig. 11.59)*

Line shading is a conventional method of representing, by ruled lines, the varying degrees of illumination on the surfaces of an object. It is a means of giving clearer definition to the shapes of objects and a finished appearance to certain types of drawings. In practice, line shading is used on Patent Office drawings, display drawings, and on some illustrations prepared for publications. It is never used on engineering drawings, and for this reason few draftsmen ever gain the experience necessary to enable them to employ it effectively.

In shading surfaces, the bright areas are left white and the dark areas are represented by parallel shade lines (Figs. 11.60 and 11.61). Varying degrees of shade may be represented in one of the following ways:

1. By varying the weight of the lines while keeping the spacing uniform, as in Fig. 11.60.

2. By using uniform straight lines and varying the spacing.

3. By varying both the weight of the lines and the spacing, as in Fig. 11.61.

The rays of light are assumed to be parallel and coming from the left, over the shoulder of the drafter (Fig. 11.60). In accordance with this, two of the visible faces of the hexagonal prism, shown in Fig. 11.60, would be illuminated, while the remaining visible inclined face would be dark. It should be noted that the general principle of shading is modified in the case of flat inclined surfaces, which, theoretically, would be uniformly lighted. Such surfaces are shaded in accordance with a conventional scheme, the governing rule of which may be stated as follows: *The portion of an illuminated inclined surface nearest the eye is the lightest, while the portion of a shaded inclined surface nearest the eye is the darkest.* In the application of this rule, the shading on an inclined illuminated face will increase in density as the face recedes, while on an unilluminated inclined surface it will decrease in density as the face recedes, as shown in Fig. 11.60.

A cylinder would be shaded as shown in Fig. 11.61. The lightest area on the surface is at the brilliant line, where light strikes it and is reflected directly to the eye; the darkest is along the shade line, where the light rays are tangent to the surface. The brilliant line passes through the point at which the bisector for the angle formed by a light ray to the center and the visual ray from the center line to the eye would pierce the external surface. In shading a cylinder, the density of the shading is increased in both directions, from the bright area along the brilliant line to the contour element on the left and the shade line on the right. The density of the shading is slightly decreased on the shaded portion beyond the shade line. Since one would expect this area to be even darker, some drafters extend the dark portion to the contour element. On very small cylinders, the bright side (left side) usually is not shaded.

A pyramid may be shaded as shown in Fig. 11.62.

The surface of a cone will be the darkest at the element where the light rays are tangent to the surface (Fig. 11.63).

A sphere is shaded by drawing concentric circles having either the geometric center of the view or the "brilliant point" as a center (Fig. 11.64). The darkest portion of the surface is along the shade line, where the parallel rays of light are tangent to the sphere and the lightest portion is around the *brilliant point*. The *brilliant point* is where the bisector of the angle between a light ray to the center and the visual ray from the center to the eye pierces the external surface. The construction necessary to determine the dark line and the brilliant point is shown in Fig. 11.64(*b*).

Although good line shading requires much practice and some artistic sense, skillful drafters should not avoid shading the surfaces of an object simply because they have never before attempted to do so. After careful study, one should be able to produce fairly satisfactory results. Often the shading of a view makes it possible to eliminate another view that otherwise would be necessary.

Figures 11.65 and 11.66 show applications of line shading on technical illustrations.

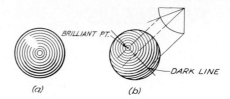

FIG. 11.64 Surface shading on a sphere.

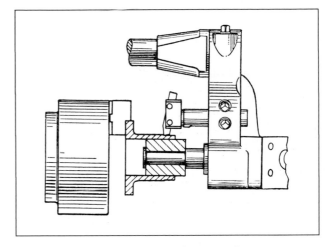

FIG. 11.65 Line-shaded drawing. (*From Doyle, Tool Engineering, Prentice-Hall, Inc.*)

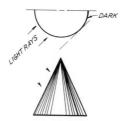

FIG. 11.63 Cone.

FIG. 11.66 Line-shaded pictorial drawing. (*Courtesy Socony-Vacuum Oil Co.*)

PROBLEMS

The student will find that a preliminary sketch will facilitate the preparation of isometric and oblique drawings of the problems of this chapter. On such a sketch he may plan the procedure of construction. Since many technologists and designers frequently find it necessary to prepare pictorial sketches during discussions with untrained persons who cannot read orthographic views, it is recommended that some problems be sketched freehand on either plain or pictorial grid paper (see Fig. 6.34). Additional problems may be found at the end of Chapter 6.

1–9. (Figs. 11.67–11.75). Prepare instrumental isometric drawings or freehand sketches of the objects as assigned.

10–14. (Figs. 11.71–11.75). Prepare instrumental oblique drawings or freehand sketches of the objects as assigned.

15. (Fig. 11.76). Make an isometric drawing of the differential spider.

16. (Fig. 11.77). Make an isometric drawing of the stepladder. Select a suitable scale.

17. (Fig. 11.78). Make an isometric drawing of the sawhorse. Select a suitable scale.

18. (Fig. 11.79). Make an oblique drawing of the locomotive driver nut.

19. (Fig. 11.80). Make an oblique drawing of the adjustment cone.

20. (Fig. 11.81). Make an oblique drawing of the fork.

21. (Fig. 11.82). Make an oblique drawing of the feeder guide.

22. (Fig. 11.83). Make an isometric drawing of the hinge bracket.

23. (Fig. 11.84). Make an isometric drawing of the alignment bracket.

24. (Fig. 11.85). Make an isometric drawing of the stop block.

25. (Fig. 11.80–11.82). Make a parallel perspective drawing as assigned.

26. (Fig. 11.83–11.85). Make an angular perspective drawing as assigned.

27. (Fig. 11.86). Make a pictorial drawing (oblique, isometric, or perspective) of the slotted bell crank.

28. (Fig. 11.87). Make a pictorial drawing (oblique, isometric, or perspective) of the control guide.

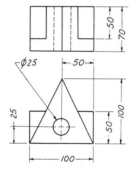

FIG. 11.67

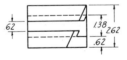

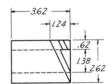

FIG. 11.69

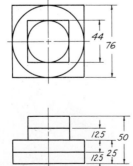

FIG. 11.68

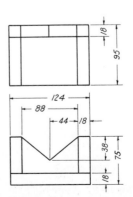

FIG. 11.70

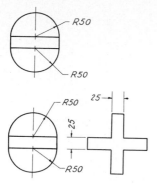

FIG. 11.71

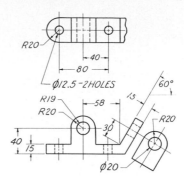

FIG. 11.74

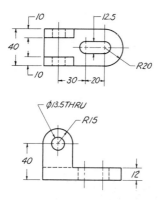

FIG. 11.72

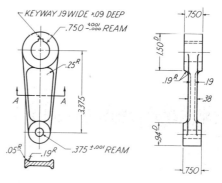

FIG. 11.75

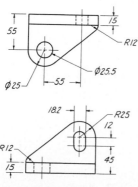

FIG. 11.73

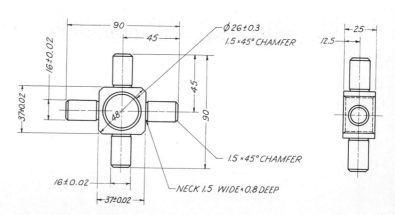

FIG. 11.76 Differential spider.

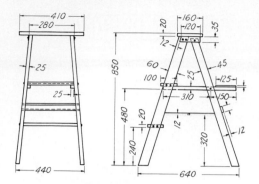

FIG. 11.77   Stepladder.

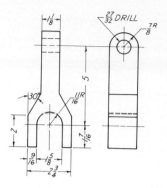

FIG. 11.81   Fork.

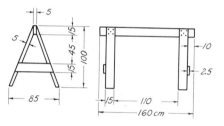

FIG. 11.78   Sawhorse.

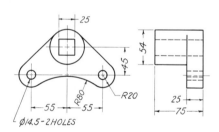

FIG. 11.82   Feeder guide.

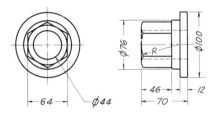

FIG. 11.79   Locomotive driver nut.

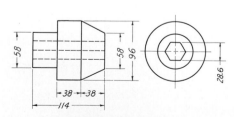

Dimensions are in millimeters
FIG. 11.80   Adjustment cone.

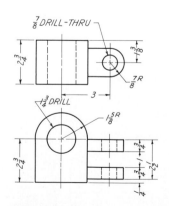

FIG. 11.83   Hinge bracket.

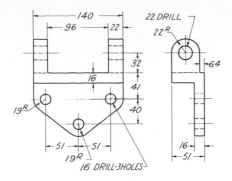

FIG. 11.84   Alignment bracket.

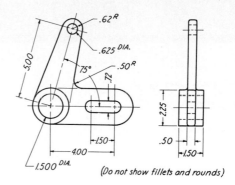

(Do not show fillets and rounds)

FIG. 11.86   Slotted bell crank.

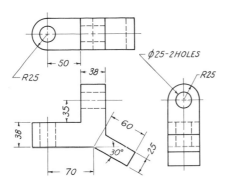

FIG. 11.85   Stop block.

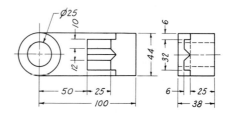

FIG. 11.87   Control guide.

PART · 3

DESIGN

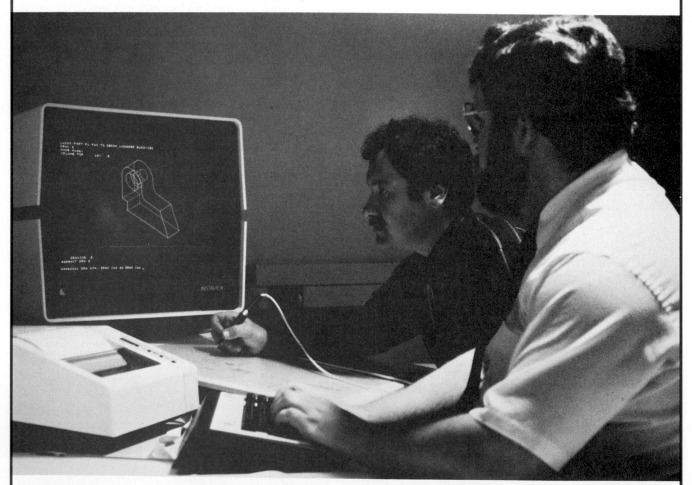

In computer-aided design, man and electronic machine are blended together to form a problem-solving team. The computer has speed, works accurately and has almost unlimited storage and rapid-recall capabilities. Its use makes possible a direct pass-through arrangement from design to fabrication without shop drawings. See Chapter 17. Shown at the workstation are Stephen Faulkner and Dennis Short of the Purdue staff. (*Courtesy Department of Technical Graphics, School of Technology, Purdue University*)

CHAPTER · 12

The Design Process and Graphics

A / THE DESIGN PROCESS

12.1 Design

In the dictionary, design is defined as follows: (1) to form or conceive in the mind, (2) to contrive a plan, (3) to plan and fashion the form of a system (structure), and (4) to prepare the preliminary sketches and/or plans for a system that is to be produced. *In engineering, design is a decision-making process used for the development of engineering systems for which there is human need* (Fig. 12.1). To design is to conceive, to innovate, to create. One may design an entirely new system or modify and rearrange existing things in a new way for improved usefulness or

performance. Engineering design begins with the recognition of a social or economic need (Fig. 12.2). The need must first be translated into an acceptable idea by conceptualization and decision making. Then the idea must be tested against the physical laws of nature before one can be certain that it is workable. This requires that the designer have a full knowledge of the fundamental physical laws of the basic sciences, a working knowledge of the engineering sciences, and the ability to communicate ideas both graphically (Fig. 12.19) and orally. The designer should be well grounded in economics, have some knowledge of engineering materials, and be familiar with manufacturing methods. In addition, some knowledge of both marketing and advertising will prove worthwhile, since usually what is produced must be distributed at a profit.

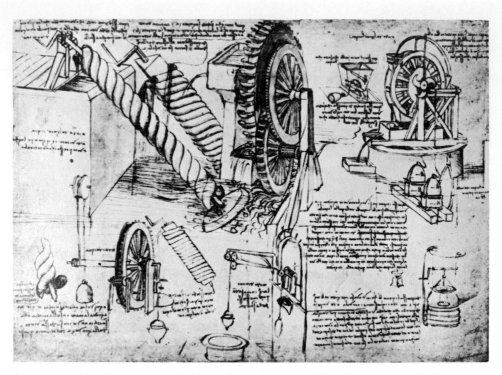

FIG. 12.1   **Archimedean screw and wheel.** Sketch showing Archimedean screw and wheel by Leonardo da Vinci (1452–1519), engineer, scientist, and painter. (*From Collection of Fine Arts Department, IBM Corporation*)

Proficiency in designing can be attained only through total involvement, since it is only through practice that the designer acquires the art of continually providing new and novel ideas. In developing the design, the engineer or engineering technologist must apply his knowledge of engineering and material sciences while taking into account related human factors, reliability, visual appearance, manufacturing methods, and sale price. It may therefore be said that *the ability to design is both an art and a science*.

A creative person will almost never follow a set pattern of action in developing an idea. To do so would tend to structure his thinking and might limit the creation of possible solutions. The design process calls for unrestrained creative ingenuity and continual decision making by a free-wheeling mind. However, the total development of an idea, from recognition of a need to the final product, does appear to proceed loosely in stages that are recognized by authors and educators (Fig. 12.2).

Creative thinking usually begins when a design team, headed by a project leader, has been given an assignment to develop something that will satisfy a particular need. The need may have been suggested by a salesperson, a consumer, or even an engineer from another company now using a product or a machine produced by the design team's own company. Most often the directive will come down from top management, as was the case with the development of the electric knife some years ago. Although it is always pleasant for an individual to think about the

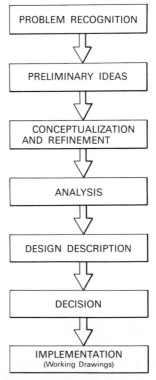

FIG. 12.2   **The design process.**

careers of famous inventors of the past and dream of the fame and fortune that might await the development and marketing of an idea, the fact is that almost all new and improved products, from food choppers to aircraft engines, represent a team effort.

12.2 Design Synthesis

The process of combining constituent elements in a new or altered arrangement to achieve a unified entity is known as *design synthesis*. It is a process that involves reasoning from assumed propositions and known principles to arrive at comparatively new design solutions to recognized problems. The synthesis of systems for simple combinations as well as for complicated assemblies requires creative ability of the very highest order. The synthesis of both parts and systems usually requires successive trials to create new arrangements of old components and new features.

Proof of this point is the Land camera, once considered by many people to be an entirely new product. In reality, the camera represents a combination of features and principles common to existing cameras to which Mr. Land added several new ideas of his own, including a new type of film and film pack that for the first time made it possible to develop film and pictures within the camera itself. Land's design activity no doubt started with a recognition of the need and desire that people had to take pictures that could be seen almost immediately. The early automobile is another example that bears out the fact that old established products have features that are used as a starting basis for a new product. For example, at the turn of the century the automobile looked like a horseless carriage; the horse was taken to the barn and a motor was added in its place.

12.3 Design of Systems and Products

In general most design problems may be classified as being either a *systems design* or a *product design*, even though it may be quite difficult in many instances to recognize a problem as belonging entirely to one classification or the other. This is due to the fact that there often will be an overlap of identifying characteristics.

A *systems design* problem involves the interaction of numerous components that together form an operating unit. A complex system such as the climatic-control system for an automobile (heating and cooling), a movie projector, a stadium, an office building, or even a parking lot represents a composition of several component systems that together form the complete composite system (Fig. 12.3). Some of the component systems of an office building, aside from the structural system itself, are: the electrical system, the plumbing system (including sewers), the heating and cooling systems, the elevator system, and the parking facilities system. All of these component systems, when combined, will meet the needs of a total system; but usually the total system design will involve more than just a technological approach. In the design of composite systems for use by the public, as in the design of a highway, an office building, or a stadium, the designer must adhere to the availability of funds. One must build into the design the safety features that are required by law and, at the same time, give due consideration to human factors, trends, and even present-day social problems. In many cases, a designer will find that there are political and special-interest groups that limit the freedom of decision making. Under such conditions the designer must be willing to compromise in the best interests of all concerned. This means that a successful designer must have experience in dealing with people. Social studies courses taken along with and in addition to the scientific courses offered in

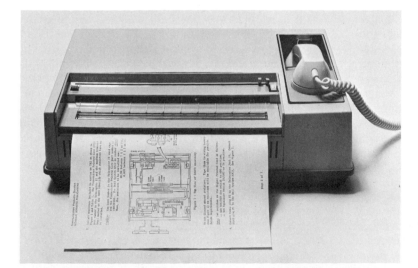

FIG. 12.3 Telecopier Transceiver. This transceiver marks the beginning of a new era in facsimile transmission. It sends and receives, over regular telephones, a letter-size document or drawing. (*Courtesy Business Products Group, Xerox Corporation*)

engineering colleges and technical schools will prove to be helpful.

Product design is concerned with the design of some appliances, systems components, and other similar unit-type items for which there appears to be a market. Such a product may be an electric lock, a power car-jack, a lawn sprinkler, a food grinder, a toy, a piece of specialized furniture, an electrical component, a valve, or other item that can be readily marketed as a commercial unit. Such products are designed to perform a specific function and to satisfy a particular need. Total product design includes not only the design of an item, but the testing, manufacture, and distribution of the product as well. Design does not end with the solution of a problem through creative thinking. The design phase for a product to be sold on the open market ends only when the item has received wide acceptance by the public. Of course, if the item being designed is not for the consumer market, the designer will have less reason to be concerned with details of marketing. This would be true in the case of components for systems and for items to be used in products produced in finished form by other companies.

After initially recognizing a need or desire for a new product, the creative individual enters the next step in the design procedure—the *research and exploration* phase. During this phase, the possible ways and the feasibility of fulfilling the need are investigated (Fig. 12.4). That is, the question is raised at this point as to whether the contemplated product can be marketed at a competitive price or at a price the public will be willing to pay. Except in the case of the public sector, where a system is being designed under the direction of a government unit (Fig. 12.5), profit is the name of the game. A search of literature and patent records will often show that there have been previous developments with the possibility of infringements on the design and patents of other companies. The discovery that others already may have legal protection for one or more of the possible solutions will tend to control the paths that innovative thinking may take in seeking an acceptable design solution. Although freewheeling innovative thinking is the basic element of successful design, the experienced engineer never fails to consider known restraints when he is brainstorming possible solutions. In addition to avoiding patent infringements, the designer must be fully aware of possible changes in desires and needs and give full consideration to the production processes. Also not to be overlooked in the decision-making stage are the visual design (styling aspects) and the price range within which the product must be sold.

The ultimate success of a design is judged on the basis of its acceptability in the marketplace and on how well it satisfies the needs of a particular culture. In these latter decades of the twentieth century, we have added to these judgments a requirement that what has been created must not damage the physical environment around us, an environment which has been rapidly deteriorating because of some past technological advances. New requirements are being prepared and approved in rapid order that set standards to reduce if not eliminate pollution. These restrictions were almost unknown to the creative designers of the first half of this century. Today, it has become almost mandatory for a designer selecting a material or an energy source to ask if the material can be recycled and whether any residue discharged into the atmosphere is damaging to life.

12.4 Innovative Design—Individuals and Groups

In the past there have been a number of creative individuals, working almost alone, who have created products that have advanced our culture, products for which people still have a great desire. Examples of such products are the printing press, the steam engine, the gasoline engine, the automobile, the telephone, the phonograph, the motion picture camera and projector, the radio, the television, and many more. These individuals were keen observers of their culture and they possessed inquisitive attitudes that led to experimentation. Some called them dreamers. If so, their eyes were wide open, they worked long hours, and they persisted in making a new try after each failure. Their fame and their fortune, although rightly due them, resulted from the fact that each brought forth a new product or system at a time when it would be readily accepted and when it could be more or less mass-produced. These people possessed a good sense of timing and they were not afraid of failure or ridicule. They belonged to the age in which they lived.

At present, it is the practice of large industrial organizations to use group procedures in order to stimulate the imaginations of the individual members of the group and thereby benefit from their combined thinking about a specific problem (Fig. 12.6). Within the group, the innovative idea of one individual stimulates another individual to present an alternative suggestion. Each idea forms the basis for still other ideas until a great number have been listed that hopefully will lead to a workable solution to the problem at hand. This group attack on a problem produces a long list of ideas that would be difficult for a single individual to assemble in so short a time (Fig. 12.7). A single person attempting to solve the problem alone would find it necessary to conduct extensive searches of current literature and patent records and to hold numerous discussions with knowledgeable experts in seeking the guidance needed for making necessary decisions.

The most widely used of the group-related procedures is known as *brainstorming*. In applying this procedure to seek the solution to a design problem, a group of optimal size meets in a room where they will be relatively free of interruptions and distractions. The people selected should be knowledgeable and there should be no one assigned to the group who might take a strong negative attitude toward the design problem to be considered. Usually, group size ranges from six to fifteen persons. If the group consists of less than six people, the back-and-forth interchange of ideas is reduced and the length of the list of ideas, which will be the basis for eventual decision making, is shorter.

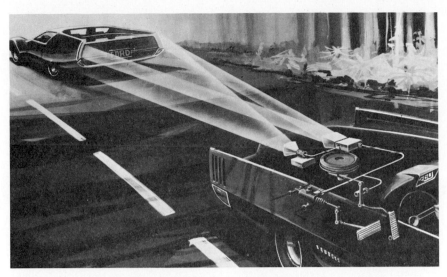

FIG. 12.4 An entirely new concept for the electronic control of automobiles is illustrated above by an artist's rendering. Many additional drawings of a more conventional type will be needed to make the idea a reality. The trailing car is equipped with a transmitter (at left) which projects an invisible beam at the car ahead. The taillights of the car in front, which does not have to be especially equipped, reflect the beam back to a receiver (right). A computer "reads" the signal and adjusts brakes and accelerator automatically so that a preset safe following distance will be maintained. (*Courtesy Ford Motor Company*)

FIG. 12.5 NEMO on the ocean floor. NEMO is a 168-cm-diameter sphere of acrylic plastic. The sphere is constructed from 12 identical curved pentagons. The capsule, with 63.5-mm walls, is bonded with acrylic adhesive. One of the first uses of NEMO will be as a diver control center at points of Seabee underwater construction sites. In operation, the manned observatory is lowered into the sea from a Navy support vessel, with the crewmen flooding the ballast tanks for the descent. NEMO operates independently, controlled by an array of pushbuttons (41) linked to a solid-state circuitry control system. NEMO has been designed for a normal stay underwater of 8 hours. Under emergency conditions it may remain submerged for as long as 24 hours. (*Official Navy photograph. Courtesy* The Military Engineer Magazine)

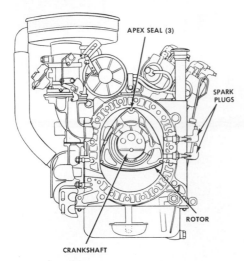

FIG. 12.6 **Wankel (rotary) engine.** Because of its different mechanical configuration, a variety of new and different mechanical problems have been encountered that have been subject to extensive research and development. Emissions are very similar to those of the piston engine. (*Courtesy General Motors Corporation*)

FIG. 12.7 **A Picturephone is shown in use.** (*Courtesy American Telephone and Telegraph Company*)

On the other hand, should the group be larger than fifteen, some individuals who may be capable of making excellent suggestions will have little chance to talk. Also, very large groups tend to be dominated by a few individuals.

During the brainstorming phase of design, no appraisals or judgments should be made nor should criticisms or ridicule of any nature be permitted. The group leader, who should be a capable person, will encourage positive thinking and stimulating comments, and will discourage those who may want to dominate the discussion.

The ideas and suggestions of the members of the group should be listed on a chalk-board and, no matter how long the list, none should be omitted. All ideas should be welcomed and recorded. After the meeting, all the ideas persented may be typed and reproduced for the information of the group and for reference at later discussions.

As the list of suggestions lengthens, several possible solution patterns usually emerge (Fig. 12.8). These in turn lead to still more suggestions for other possible combinations and for improvements to likely solutions to the design problem.

One or two lengthy brainstorming sessions can result in several acceptable design solutions that represent the combined suggestions of the individuals in the group. However, making a group evaluation of the ideas that have evolved from brainstorming is another matter, since no one member feels completely responsible for the results. The major weakness of any group procedure, such as brainstorming, is that individual motivation is dampened to some extent. However, since group procedures are productive and have become widely used in industry, there have been studies made that will hopefully lead to changes that will minimize this recognized weakness and improve group effectiveness. Means must be found to raise the motivation of each participant to the highest possible level.

12.5 *Creation of New Products*

A product or a system may be said to have been produced either through evolutionary change or by what appears to be pure innovation. The word *appears* has been used appropriately, since few, if any, products are ever entirely new in every respect. Most products that appear to have drawn heavily on innovation usually combine both old and new ideas in a new and more workable arrangement. A product of evolutionary change, however, develops rather slowly, over a long period of time, and with slight improvements being made only now and then. Such a product may be reliable and virtually free of design and production errors, but the small amount of design work involved, done at infrequent intervals, will never really challenge a creative person.

In today's competitive world, when products are produced for world-wide consumption, evolutionary change is hardly sufficient to ensure either the economic well-being or even the survival of those companies that seem to be willing to let well enough alone. Rapid technological changes coupled with new scientific discoveries have increased the emphasis on the importance of new and marketable products that can gain a greater share of the total market than is possible with the product the company may now be promoting. In meeting this need for new and marketable ideals, the designer will find that one's innovative ability and one's experience and knowledge are being taxed to the limit and, since one may in a sense be stepping into the unknown, some risks must be taken.

The unusual characteristics that seem to be a part of the general makeup of every outstanding designer are

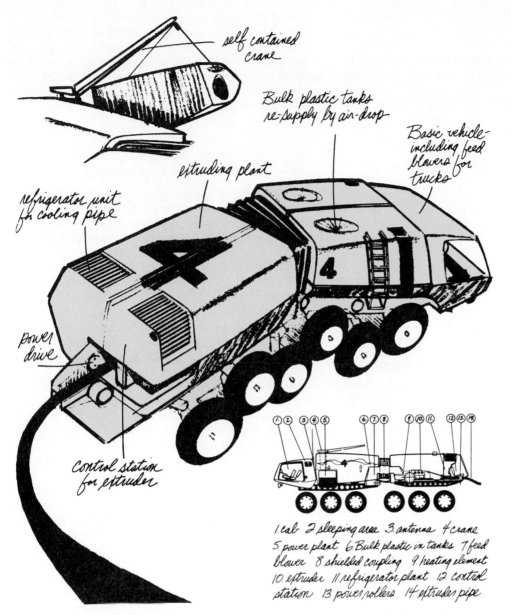

Labels on drawing:
self contained crane
Bulk plastic tanks re-supply by air-drop
Basic vehicle-including feed blowers for trucks
extruding plant
refrigerator unit for cooling pipe
power drive
control station for extruder

1.cab 2.sleeping area 3.antenna 4.crane 5.power plant 6.Bulk plastic in tanks 7.feed blower 8.shielded coupling 9.heating element 10.extruder 11.refrigerator plant 12.control station 13.power rollers 14.extruder pipe

FIG. 12.8 Design of a self-contained pipe-layer. It is intended to facilitate irrigation of large tracts of desert land. The equipment as designed is capable of transporting sufficient bulk plastic to lay approximately 2 miles of plastic pipe from each pair of storage tanks. Tanks are to be discarded when empty and replaced by air drop. (*Courtesy Donald Desky Associates, Inc., and Charles Bruning Company*)

1. the ability to recognize a problem,

2. the ability to take a questioning approach toward all possible solutions,

3. the possession of an active curiosity about a problem at hand,

4. the innate willingness to take responsibility for what one has done or may do,

5. the ability to make needed decisions and to defend those decisions in writing and orally,

6. the possession of intellectual integrity.

William Lear, one of the most prominent designers of the last three decades, spent his entire working life discovering needs and then finding ways to fulfill them. In the case of the development of the eight-track stereo tape, he was working from economic considerations that required more repertoire on tape without adding more tape. This meant either running the tape slower or adding more tracks. The practical answer from Lear's viewpoint was to add more tracks. In addition to the Learoscope, an automatic direction-finder for use on airplanes, Lear is responsible for the development of the car radio, the automatic pilot, the Lear jet plane, and more than 150 other inventions. A few years ago, he began development work on the problem of steam-powered automobile engines. Lear's inventing was done when surrounded by many people with considerable knowhow. His work involved the gathering of a maze of information and ideas from which he could pick out the salient facts and discard the unimportant ones, while always keeping the goal in mind and solving the problem at the least possible cost.

12.6 Background for Innovative Designing

Designing should be done by people who have a diversified background and who are not entirely unfamiliar with the problem at hand. As an example, even though the design of a product may be thought of as being in the field of mechanical engineering, a designer with a knowledge of electrical applications and controls will find it a distinct advantage since many of our present-day products use electric current as an energy source. When a design group is involved, it is important that the mechanical engineer and others should have at least some understanding of the electrical engineer's suggestions. This added knowledge will enable them to modify their thinking about a product that is largely a mechanical device. Examples of these products include electric locks, electric food choppers, and electric typewriters.

The background required will vary considerably depending upon the field in which the individual works. For example, a person who may be designing small household appliances would probably never need more than the knowledge acquired from his basic engineering courses, while a designer in the aerospace field would need a background based upon advanced study in chemistry, physics, and mathematics. On the other hand, there are respected and competent designers in industry who have had as little as two years of tehcnical education. With this limited training and several years of on-the-job experience, these men and women have become able to design complex solutions.

Because of the increasing complexity of engineering, the rapid development of new materials, and the accumulation of new knowledge at an almost unbelievable rate, it has become absolutely necessary for engineering design to become a team effort in some fields. Under such conditions, the design effort becomes the responsibility of highly qualified specialists. A project requiring designers of varied specialized backgrounds might need, for example, people with experience in mechanical, electrical, and structural design and persons with considerable knowledge of materials and chemical processes (Fig. 12.6). If it is decided that styling is important, then one or more stylists must be added to the team. A complete design group for a major project could include pure scientists, metallurgists, technicians, technologists, and stylists in addition to the designers.

Finally, graphics must not be overlooked when one considers the background needed to become a successful designer. Anyone who hopes to enter the field of design, other than as a specialist, must have a thorough training in this area. A working knowledge of all of the forms of graphical expression that are presented in this text is required along with the ability to explain the design orally and in writing in preliminary and final reports. The methods used for the preparation or oral and written reports are discussed in Sec. 12.24.

12.7 History and Background of Human Engineering

Human engineering is a relatively new area in the field of design. In order to simplify matters we might define human engineering as adapting design to the needs of humans; that is, we engineer our designs to suit human behavior, human motor activities, and human physical and mental characteristics. The applications of human engineering apply not only to machine systems and consumer products but to work methods and to work environments as well. In the early years of its development, human engineering was concerned mainly with the working environment and with the comfort and general welfare of all human beings. As time passed, designers came to realize that more than just safety and comfort should be considered, and that almost all the designs with which they were concerned were in some way related to the general physical characteristics, behavior, and attitudes of people. At this point, designers began to include these added human factors in their approach to the solution of a design problem so as to secure the most satisfying and efficient relationship to machines.

It is in the area of human engineering that an engineering designer must adhere to the input of specialists who may be involved in a wide range of disciplines. These disciplines may be industrial engineering, industrial psychology, medicine, physiology, climatology, and statistics. Stimulated to a great extent by the space program, scientists from all of these disciplines have, together and separately, become deeply involved in basic research and laboratory experimentation, which has led to a continuous input of new man-machine information into design. Although in the past human engineering has been associated mainly with industrial engineering, industrial psychology, and industrial design, designers in all fields of engineering must now not only be knowledgeable about the principles of human engineering but they must be capable of utilizing these principles and related information whenever they are developing a product that involves human relationships.

Typical body dimensions, representing an average-size person, are used when a product is being designed for general use. The measurements of typical adult males and females, as determined from studies made by Henry Dryfuss, may be used for most designs requiring close adaptation to human physical characteristics. Since many designs involve both foot and hand movements these average body dimensions, as tabulated, include arms, hands, legs, and feet along with other parts of the human body. Data relating to body proportions and dimensions are known as anthropometric data. Information relating to body proportions may be obtained from *The Measure of Man*, Whitney Library of Design.

The first known serious studies of the human body were made by Leonardo da Vinci. To record his studies for his own use and for the use of others, he made some of the finest and most accurate detail sketches and drawings of the human body known to man. These drawings show even the intricate details of muscle formation. His work, done in defiance of the laws of his time, is still used today in several textbooks. Da Vinci's studies of the human body mark the beginning of the science of biomechanics. See Fig. 12.9.

From the time of Leonardo da Vinci until early in the twentieth century, very little work was done toward the development of this science. This lack of interest in people and their relationship with equipment and tasks to be performed was due largely to the fact that from the beginning of the Industrial Revolution (which many people say began with invention of the steam engine patented by James Watt in 1769) until the early 1900s, the interest of designers was centered mainly on the creation of new products and the raising of production efficiency to the high levels needed to compete in world markets. Until about 1900, revolutionary change was the order of the day and there was very little time available for consideration of the human anatomical, physiological, behavior, and attitude factors that have now become the basis of our present human-machine-task systems. Even though our computer-programmed numerically controlled machine tools permit us to do almost any task with only minimal human intervention, people are still needed in most of our human-machine-task systems and they must be taken largely as they are. In our designs we must not overlook even the possibility of boredom, for a person who become sufficiently bored might just "pull the plug" and everything would stop.

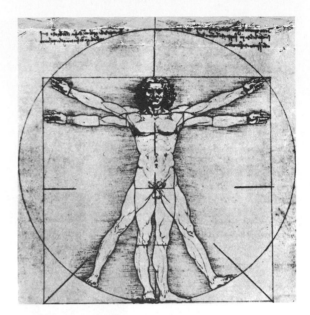

FIG. 12.9 Human proportions and body dimensions as illustrated by Leonardo da Vinci (1452–1519).

12.8 *Human Engineering in Design*

There are a number of factors in human engineering, other than human anthropometric measurements, that must be dealt with in design. These factors include motor activities (Fig. 12.10) and body orientation; the five human senses (sight, hearing, touch, smell, and sometimes even taste); atmospheric environment, temperature, humidity, and light; and, finally, accelerative forces if they are exceptional and are likely to cause undue physical discomfort.

During the first half of this century most of the research done in human engineering, aside from the anthropometric studies already mentioned, was directed toward: work areas and the position of controls; physical effort and fatigue; and the speed and accuracy to be expected in the performance of particular tasks. Finally, when it became evident that this was not enough, industrial engineers and industrial psychologists turned their attention to the more complex activities of the average human being. These new studies dealt primarily with receiving information through sight and sound, the making of decisions in response to

stimuli, and, finally, the performance in direct response to these decisions.

The study of body motion deals primarily with the effective range of operation of parts of the body, usually the arms and legs, and the amount of body force a human being may reasonably be expected to exert in the performance of an assigned task. For example, many time and motion studies have been made of the range of operation of persons performing given tasks while seated at assigned work areas. At the same time, in many of these studies attention has been given to the location and the amount of force required to operate levers and controls in relation to the size and strength of the operator.

Vision is an important factor in all designs where visual gauges or colored lights on control panels are a part of a control system that involves manual operation or, in the case of numerically controlled machines, monitoring for manual intervention at specified times. Control panels for such equipment must be designed to be within the visual range of the average person, and distinctive colors must be selected and used for the colored lights. The colors selected must be easily recognizable and must be capable of quickly attracting the operator's attention. Not to be overlooked is the fact that vision studies have produced much new information for use in highway design. From these same studies have come new ideas for our highway warning and information signs along with suggestions for their placement.

When a designer must consider the working environment as a part of the total design, that environment includes

1. temperature,
2. humidity,
3. lighting,

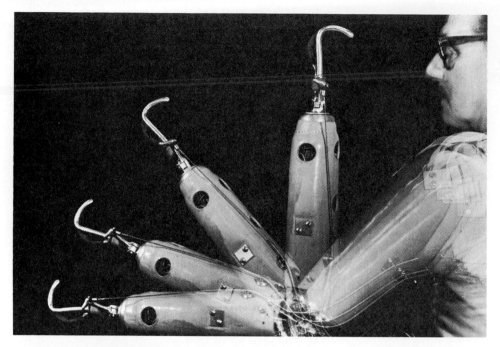

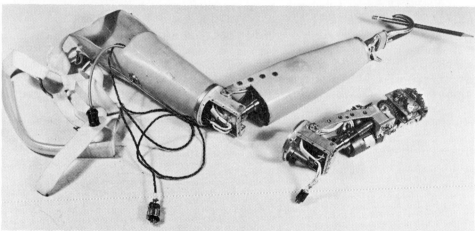

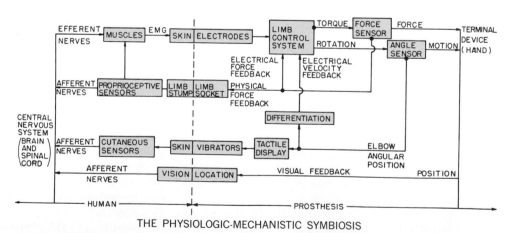

THE PHYSIOLOGIC-MECHANISTIC SYMBIOSIS

FIG. 12.10 Boston Arm. The range of movement of the Boston Arm is demonstrated in the upper photograph. The arm, developed as a joint project of Liberty Mutual Insurance Companies, Harvard Medical School, Massachusetts General Hospital, and the Massachusetts Institute of Technology, acts as does a normal arm through thought impulses transmitted from the brain to existing arm muscles. The design of a product for the handicapped can provide much satisfaction to the designer. (*Courtesy Liberty Mutual Insurance Companies*)

FIG. 12.11   Artist's conception of the Atlantis undersea habitat. The project was to be designed and developed as a joint project of the University of Miami and the Space Division of Chrysler Corporation. The project goal was to explore the continental shelf along our coasts. (*Courtesy Space Division, Chrysler Corporation*)

4. color schemes,

5. sound.

These are a few of the factors that also deserve full consideration in the design of a large industrial plant, a particular work area in a plant, the cockpit of an airplane, or the cabin of a space vehicle.

The overall environment and the design of the working areas and the living quarters of the undersea laboratory shown in Fig. 12.11 were based on human requirements. The design of any undersea craft or laboratory involves problems that are similar in many ways to those encountered in the design of space vehicles, in that an artificial living environment must be created and maintained for extended periods of time. This requires a self-contained atmosphere. The members of the crew also must have ample space in which to work and live under climatic conditions that duplicate those on land. Crew members must be able to perform their tasks under normal lighting conditions and they must be able to see to the outside. Aside from the design of features and components that are related directly to the performance of the research assignments, the overall development of the undersea craft can be said to be based on the physical needs and the psychological attitudes and reactions of human beings.

Human engineering is applied to a wide range of consumer items. Automobiles (Fig. 12.12), refrigerators, furniture, office equipment, lawn mowers, hand tools, and

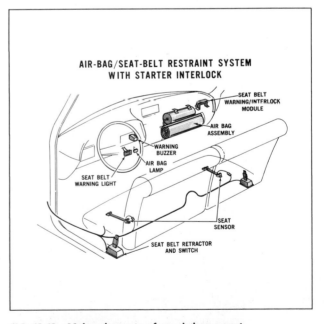

FIG. 12.12   Major elements of an air-bag—seatbelt restraint system now under development by the Ford Motor Company. The air-bag assembly and the seat-belt starter-interlock components are identified. It should be noted that there are separate signal lights for the seat-belt and airbag systems. (*Courtesy Ford Motor Company*)

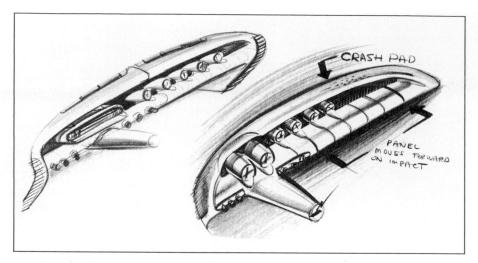

FIG. 12.13 Design sketches for a dashboard panel. (*Courtesy General Motors Corporation*)

FIG. 12.14 Several styling design sketches prepared for the U.S. supersonic transport (SST). Visual appeal is important to the success of the total design project. (*Courtesy the Boeing Company*)

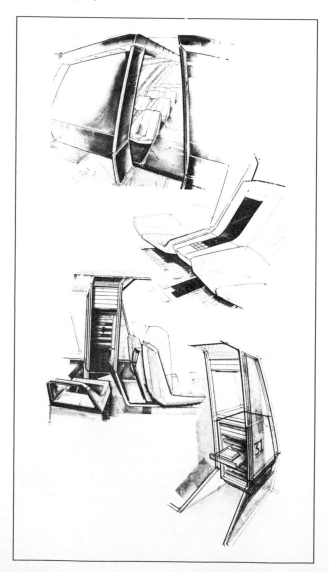

like items have long been designed with human factors in mind.

The automotive industry, as it designs and redesigns cars with human factors in mind, is producing cars in accordance with stricter government safety regulations.

Under pressure from consumer groups and organizations interested in safety and in protecting our natural environment, there has been an increase in governmental laws and regulations. Many of these laws and directives provide controls in the field of human engineering. In the case of environmental pollution, automobile companies, acting to meet regulations set forth by the Environmental Protection Agency, developed several devices to reduce undesirable pollutants at a scheduled percentage rate to produce an almost pollution-free car. In the interest of safety, some type of passive protection system, such as the air-bag device shown in Fig. 12.12, will probably be added.

These government laws and regulations have come into existence because of a growing interest on the part of the general public in human engineering. Designers in the years ahead must be fully cognizant of all such regulations and must be willing to abide by them or seek to have them changed should they appear to be unreasonable or impractical.

12.9 *Visual Design*

Visual design includes the use of line, form, proportion, texture, and color to produce the eye-pleasing appearance needed to bring about the acceptance of a consumer item. Without this acceptance there would be no profit, and even though the item might otherwise have been carefully engineered, it would soon disappear from the marketplace. The sketches shown in Fig. 12.13, for the dash panel of an automobile, tastefully combine these visual elements into an attractive design. An illusion of depth has been obtained by means of pencil shading.

Since styling is now recognized as being one of the

most important factors in sales, many engineers have come to accept the role of industrial designers in the development of a consumer product, particularly when they are employed by a company that is small and cannot afford to employ one or more trained stylists. Large companies, such as the Ford Motor Company, General Motors, and the Boeing Company (Fig. 12.14), having styling divisions. Medium-size companies often turn to nationally known organizations to get needed help; the industrial design is then done under contract agreement. Many books have been written about aesthetics that have proved to be helpful to engineers (see Appendix F). At present, design engineers, who have been trained largely to solve technical problems, are reading more about styling and they are considering the eye-appeal and the overall appearance of products as part of their engineering interests.

12.10 *Constructive Criticism*

There are people who seem to find it easier to criticize than to mix praise with alternative suggestions. In any group meeting a critic should show respect for good ideas and be able to offer constructive suggestions. If there is to be feedback, which will hopefully lead to the introduction of more ideas, the discussion must be free of any harsh criticism. Harsh criticism may cause a sensitive person to assume a defensive position or to withdraw almost entirely from participation in group action. It is the responsibility of the project leader to prevent this from happening.

12.11 *Recognition of a Need*

A design project usually begins with the recognition of a need and with the willingness of a company to enter the market with a new product. At other times an idea may be initiated and developed by an individual who either seeks personal economic benefits or who seeks to solve some social or environmental problem. In either case, the identification of the need in itself represents a high order of creative thinking and the search for a solution to the need requires considerable self-confidence and inner courage. As can be easily observed from reviewing the achievements of our distinguished inventors of the past, those who are closely attuned to life around them become aware of needs or less-than-ideal situations that are worthy of their attention.

It is important that a proposed design project have clear and definite objectives that will justify the money and effort to be expended in product design and development. The statement defining the objectives should identify the need and state the function the product is to perform in satisfying this need. The identification of the need may be based on the designer's personal observations, suggestions from sales people in the field, opinion surveys, or on new scientific concepts. The identification of the design problems involved in creating the needed product comes later in the design process.

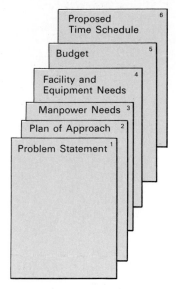

FIG. 12.15 **Elements of a proposal.**

12.12 *Formal Proposal*

The statement covering the recognition of need can be used as a basis for a formal proposal that may be either a few short paragraphs or several typewritten pages. A complete proposal may include supporting data in addition to the description of the plan of action that is to be taken to solve the problem as identified (Fig. 12.15). The report should have the same general form as other technical reports and might include a listing of requirements and possible limitations as then recognized. In preparing the report one should keep in mind that the proposal, when approved, gives the broad general parameters of an agreement under which the project will be developed to its conclusion.

12.13 *Phases of the Design Process*

Many people in the past have prepared outlines of the steps that can be followed in the process of design (Fig. 12.16). They have prepared these outlines in order to give some semblance of order to the total design process from the point of recognition of the need to the point of marketing the product. One must recognize, however, that there are actually many combinations of steps in the overall procedure, with no single pattern either the best or the only combination. The design procedure required in many cases can be very complex and successful designers have found different ways to achieve their goals. However, the phases of design, as recognized by these authors, have been listed here in sequential order to provide some degree of direction to the student who is making his first attempt to design a product under a contrived classroom situation. More experienced persons may find it desirable to alter this outline to make it more suitable to their own method of designing.

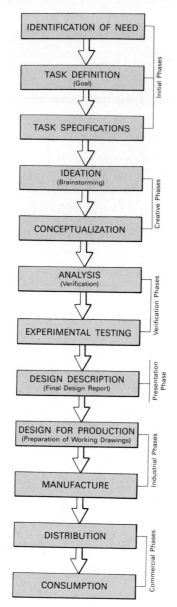

FIG. 12.16 Basic phases of total design.

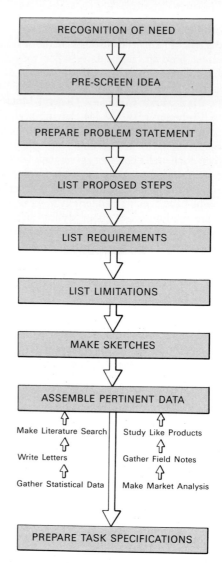

FIG. 12.17 Initial steps of the design process.

The basic phases in the design process are

1. identification of need,
2. task definition (goal),
3. task specifications,
4. ideation,
5. conceptualization,
6. analysis,
7. experimental testing (Fig. 12.23),
8. design (solution) description,
9. design for production.

Phase 9 (implementation) is not usually a primary concern of a designer. Also, the designer may or may not give some thought to manufacture, distribution, and consump-

tion of the product. Consideration of these factors may be thought of as being the tenth, eleventh, and twelfth stages of total design.

12.14 Task Definition—Definition of Goals

Briefly stated, *task definition* is the expression of a commitment to produce either a product or a system that will satisfy the need as identified. By means of broad statements, the product and the goals of the project are identified. The statements, as written, must be clear and concise to avoid at least some of the difficulties (often encountered in design) that can be traced directly back to poorly defined goals.

Even though it is probable that the person who has initiated the project has already gathered some pertinent information and has some preconceived ideas, it is desirable that this material not be included (Fig. 12.17). It

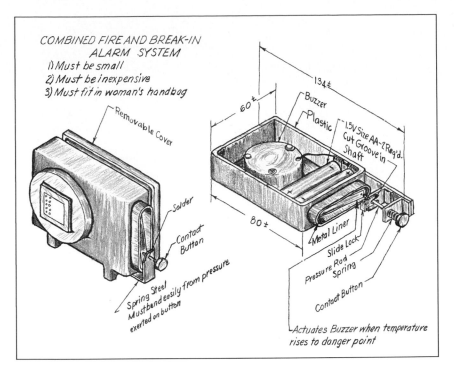

COMBINED FIRE AND BREAK-IN
ALARM SYSTEM
1) Must be small
2) Must be inexpensive
3) Must fit in woman's handbag

Removable Cover

Solder

Contact Button

Spring Steel
Must bend easily from pressure
exerted on button

134±

60±

80±

Buzzer

Plastic

1.5V Size AA-2 Req'd.
Cut Groove in Shaft

Metal Liner

Slide Lock

Pressure Rod Spring

Contact Button

Actuates Buzzer when temperature
rises to danger point

**FIG. 12.18   Idea sketches for a small portable safety alarm that will give
warning for both fire and an attempted break-in. (Dimension values are in
millimeters.)**

is better to present the goals in terms of objectives and then allow the designers to pursue the project in their own way, as free of restrictions as possible. The task definition should be included in the proposal.

12.15 Task Specifications

This is a listing of parameters and data that will serve to control the design. This stage will ordinarily be preceded by some preliminary research to collect information related to the goal as defined. In preparing the task specifications, the designer or design group lists all the pertinent data that can be gathered from research reports, trade journals, patent records, catalogues, and other sources that possess information relating to the project at hand. Included in this listing should be the parameters that will tend to control the design. Other factors that deserve consideration, such as materials to be used, maintenance, and cost may also be noted.

12.16 Ideation

The ideation phase of design has been discussed for group procedure situations in Sec. 12.4. It is recommended that the reader review this discussion to refresh his memory regarding brainstorming procedures. It is well to remember that often a lasting solution to a problem has resulted from a creative idea selected from a number of alternative ones (Fig. 12.18). The likelihood of finding an optimum solution

is greatly enhanced as the list of possible alternatives grows longer. Truly great creations are possible when someone's imagination is allowed to soar with little restraint, either when working alone or with a group. If engineers on a design project can set aside their engineering know-how and blind themselves, at least temporarily, they will be in the right frames of mind to meet almost any challenge. This is the mood and the mental approach for great discoveries. Some call it ideation; others have called it *imagineering*. If we can learn to open up our engineering minds to new approaches to our technical problems as well as to our existing problems of air pollution, industrial waste, transportation, and even unemployment, there is no limit to what we can accomplish.

This open-minded *imagineering* approach to problems must be tempered with a sense of professional responsibility. It is no longer acceptable to solve an immediate problem using a solution that in years to come can endanger the environment. Imagineering is engineering for the total well-being of mankind and all other living things.

12.17 Conceptualization

Conceptualization follows the preliminary idea (ideation) stage when all the rough sketches (Figs. 12.19–12.21) and notes have been assembled and reviewed to determine the one or more apparent solutions that seem to be worthy of further consideration. In evaluating alternative solutions, consideration must be given to any restrictions that have been placed on the final design. It is at this stage that the

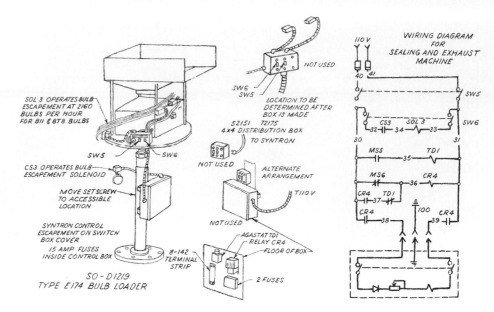

FIG. 12.19   Portion of a design sketch. (*Courtesy General Electric Company*)

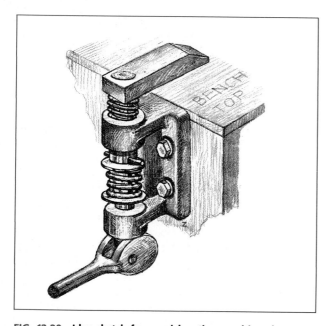

FIG. 12.20   Idea sketch for a quick-acting machine clamp.

FIG. 12.21   Idea sketch showing remote control system for a motor boat.

preliminary sketches should be restudied to see that all worthwhile ideas are being included and that none have been inadvertently overlooked. At no time during this phase should designers become so set in their thinking that they do not feel free to develop still another and almost entirely new and different concept, if necessary. They should realize that it is more sensible to alter or even abandon a concept at this stage than later, when considerable money and time will have been invested in the project. *The conceptualization stage of design is that stage where alternative solutions are developed and evaluated in the form of concepts.* Considerable research may be necessary and task specifications must be continually reviewed. As activity progresses, many idea sketches are made as alternative approaches are worked out; these approaches are evaluated for the best possible chance of product success. It is not necessary at this stage of the design procedure for any of the alternative solutions to be worked out in any great detail.

12.18   *Selection of Optimum Concept*

As the design of a product or system progresses, a point is reached in the procedure when it becomes necessary to select the best design concept to be presented to the administrators in the form of a proposal. In making this final selection, a more-or-less complete design evaluation is made for each of the alternative concepts under consideration. These evaluations may reveal ways that costs can be reduced and value improved; means of simplifying the design to reduce costs may also become apparent.

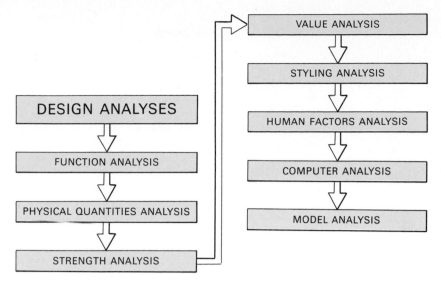

FIG. 12.22 **Analysis phase of design.**

12.19 *Design Analysis*

After a design concept has been chosen as the best possible solution to the problem at hand, it must be subjected to a design analysis; that is, it must be tested against physical laws and evaluated in terms of certain design factors that are almost certain to be present (Fig. 12.22).

Total analysis of a proposed design will include a review of the engineering principles involved and a study of the materials to be used. In addition, there should also be an evaluation of such design considerations as

1. the environmental conditions under which the device will operate,

2. human factors,

3. possible production methods and production problems,

4. assembly methods,

5. maintenance requirements,

6. cost,

7. styling and market appeal.

If the design is based on newly discovered scientific principles, some research may be in order before a final decision can be reached.

It is at this stage in the design procedure that physics, chemistry, and the engineering sciences are employed most fully. In making the usual design analysis, the engineers or engineering technologists must depend on the formal training that they received in school, and although considerable mathematics may be needed in making most of the necessary calculations, they will find it convenient at times to resort to graphical methods. Over the years, graphical methods have proved to be most helpful in evaluating and developing a design. For example, descriptive geometry methods can be employed for making spatial analyses and critical information can be obtained by scaling accurate drawings.

If design analysis proves that the design as proposed is inadequate and does not meet requirements, the designer may either make certain modifications or incorporate into the designs some new concept that might well be a modification of an earlier idea that was abandoned along the way.

12.20 *Experimental Testing*

The experimental testing phase of the design process ranges from the testing of a single piece of software or hardware to verify its workability, durability, and operational characteristics through the construction and testing of a full-size prototype of the complete physical system.

A component of a product should be tested in such a way that the designer can predict its durability and performance under the conditions that will be encountered in its actual use. Needed tests may be performed using standard test apparatus or with special devices that have been produced for a particular test.

There are three types of models that may be constructed for the purpose of testing and evaluating a product. These are (1) the mockup, (2) the scale model, and (3) the prototype.

12.21 *Mockups*

A mockup is a full-sized model constructed primarily to show the size, shape, component relationships, and styling of the finished design. At this point the designer's conception begins to take shape for the first time. Automobile manufacturers customarily produce mockups to evaluate proposed changes in the styling of automobile bodies for new models. Needed modifications in size and body configuration can be determined by studying the mockup and

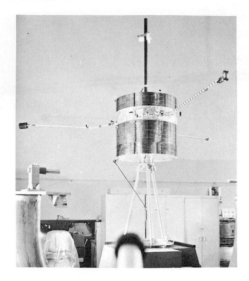

FIG. 12.23 Full-size prototype of the Pioneer Satellite A-B-C-D series. A prototype provides the best means for an evaluation of a complex design. (*Courtesy Systems Group, TRW, Inc.*)

analyzing its overall appearance. Since interior styling is important also, interiors are modeled in clay to reveal the aesthetic appearance the stylist had in mind. In the automobile industry, mockups are made to secure early approval of management for model change. A mockup is more meaningful than a sketch to those whose support and approval is needed. One must realize, however, that numerous sketches and artistic renderings (Fig. 12.14) are made before any work is started on a mockup. These sketches are used as guides. Mockups may be made of clay, wood, plastic, and so forth.

12.22 Preliminary and Scale Models

Models may be made at almost any stage in the design process to assist the designer in evaluating and analyzing the design. Models are made to strengthen three-dimensional visualization, to check the motion and clearance of parts, and to make necessary tests to clear up questions that have arisen in the designer's own mind or in the mind of a colleague.

The designer may prepare a preliminary model to understand more fully what the shape of a component should be, how well it may be expected to operate, and how it might be fabricated most economically. In some cases, the model might be so simple that it could be made of paper, wood, or clay.

Scale and test models may be constructed either for analysis and evaluation or for the purpose of presenting the design in a more or less refined stage for approval. Scale models may be made of balsa wood, plastic, aluminum, wire, steel, or any other material that can be used to a good advantage. The designer should select a scale that will make the model large enough to permit the movement of parts should be demonstration of movement be desirable.

12.23 Prototypes

A prototype is the most expensive form of model that can be constructed for experimental purposes. Yet, since it will yield valuable information that is difficult to obtain in any other way, its cost is usually justified. A prototype is a full-size working model of a physical system built in accordance with final specifications, and it represents the final step of the experimental stage (Fig. 12.23). In it the designers and stylists see their ideas come to life. From a prototype the designers can gain information needed for mass-production procedures that are to come later. Much can be learned at this point about workability, durability, production techniques, assembly procedures, and, most important of all, performance under actual operating conditions. Since prototype testing offers the last chance for modification of the design, possible changes to improve the design should not be overlooked, nor should a designer ever be reluctant to make a desirable change.

Since a prototype is a one-of-a-kind working model, it is made by hand using general purpose machine tools. Although it might be best to use the same materials that will be used for the mass-produced product, this is not always done. Materials that are easier to fabricate by hand are often substituted for hard-to-work materials.

In the development of a design, designers deal first with the mockup; next they work out specific problems relating to single features with preliminary models; then they evaluate the design with scale models; and, finally, they may, if desirable, test the whole concept using a prototype. This order in the use of models maintains a desirable relationship between concept and analysis and represents a logical procedure for the total design process.

12.24 Design (Solution) Description— Final Report

In the solution description the designers describe their designs on paper, to communicate their thinking to others. Although the purpose of the final report may be to sell the idea to upper management, it may also be used to instruct the production division on how to construct the product. It usually will contain specific information relating to the product or system. In some cases a process will be described in considerable detail.

A complete design description (see Fig. 12.24), prepared as the main part of a formal report, should include

1. a detailed description of the device or system,

2. a statement of how the device or system satisfies the need,

3. an explanation of how the device operates,

4. a full set of layout drawings, sketches, and graphs,

5. pictorial renderings, if needed,

6. a list of parts,

7. a breakdown of costs,

8. special instructions to ensure that the intent of the designer is followed in the production stage.

After the design description is accepted and approved, there remain only the commercial stages of the total design process, namely, implementation, manufacture, distribution (sales), and consumption.

12.25 Implementation—Design for Production

Implementation is that phase of the total design process when working drawings are prepared for those who must fabricate the nonstandard parts and assemble the product. A complete set of working drawings, both detail and assembly, are needed to permit the manufacture of a product.

Even though in theory the production design phase directly follows the preliminary design phase and the acceptance of the design recommended in the final report, in actual practice there is often no clear dividing line between these two phases. This is because detailed drawings of some components may have been started and to some extent completed back in the preliminary design stage, along with one or more design layout assemblies (see Fig. 12.25). Furthermore, it is not unusual for detail design drawings to be made for two or more likely solutions in the design-for-production stage and the final selection of the one best solution delayed until the information derived from these detail design drawings and related assemblies can be used in making the final decision. Such a delay in decision making tends to cause the preliminary design and the production design phases to appear to melt into one another. However, there is still a division line even though the designers themselves may not recognize this fact.

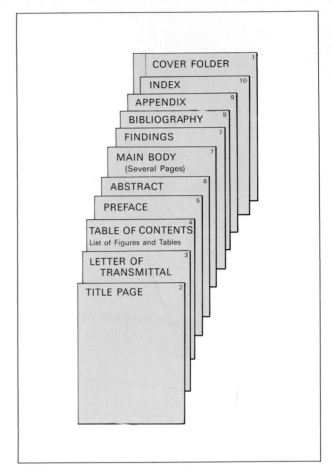

FIG. 12.24 Contents of a design (final) report.

12.26 Manufacture (Fig. 12.26)

From the time the task specifications are written and through all the stages up to the manufacture, the designer works closely with a production engineer who is familiar with the available shop facilities, production methods, inspection procedures, quality control, and the assembly line. If this is done, the problems encountered in the manufacturing stage will be few in number.

12.27 Distribution

Since a designer usually has little expertise in the area, task problems relating to distribution are passed along to *marketing specialists*. These specialists have the knowledge and the supporting staff required to decide on the proper release date and to set a competitive price based upon market testing and cost and profit studies. Included among these specialists, will be experienced advertising personnel who prepare the needed advertising and promotional literature. However, these specialists do consult the designer frequently during this stage since no one else knows more about the product. The designer is the one

person who can be depended upon to supply needed technical data and information concerning the capabilities and limitations of the product. Furthermore, the sales promotion people expect ideas from the designer that will lead to a wide and favorable distribution.

12.28 Consumption

There should hopefully be *consumer feedback* from this last stage of the design process that will prove useful to the designer when it becomes necessary to alter and improve the product at the time of the next model change. Most of this feedback information will come from the sales force, from distributors, and from service departments. Some of this information will be in the form of complaints made by irate users, but much of it will be received as constructive suggestions. These constructive suggestions take the designer back to the need stage for the next model, and a new design cycle starts.

The consumption stage represents the goal sought by the designer, and this is where the design has its ultimate test. At this stage the design will be pronounced a success or a failure by the users and consumers who are and always will be the final judges.

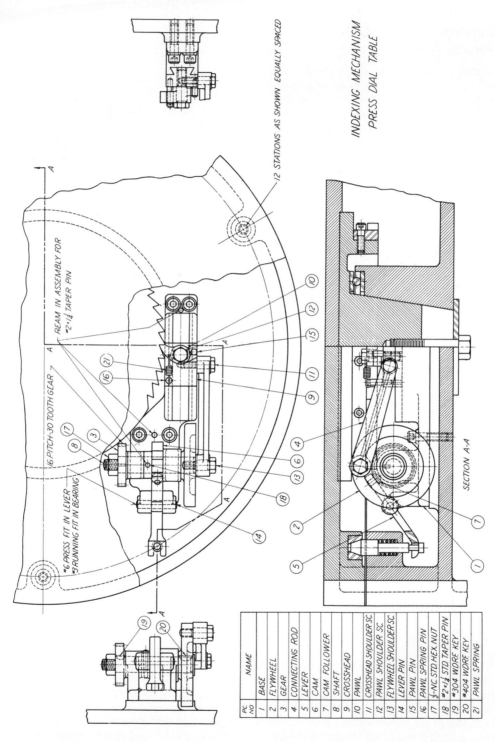

INDEXING MECHANISM
PRESS DIAL TABLE

12 STATIONS AS SHOWN EQUALLY SPACED

REAM IN ASSEMBLY FOR
#2 x ¼ TAPER PIN

16 PITCH-30 TOOTH GEAR

#6 PRESS FIT IN LEVER
#3 RUNNING FIT IN BEARING

SECTION A-A

PC NO	NAME
1	BASE
2	FLYWHEEL
3	GEAR
4	CONNECTING ROD
5	LEVER
6	CAM
7	CAM FOLLOWER
8	SHAFT
9	CROSSHEAD
10	PAWL
11	CROSSHEAD SHOULDER SC.
12	PAWL SHOULDER SC.
13	FLYWHEEL SHOULDER SC.
14	LEVER PIN
15	PAWL PIN
16	PAWL SPRING PIN
17	½-NC STD.HEX NUT
18	#2 x 1½ STD TAPER PIN
19	#304 WDRF KEY
20	#404 WDRF KEY
21	PAWL SPRING

FIG. 12.25   Design layout drawing.

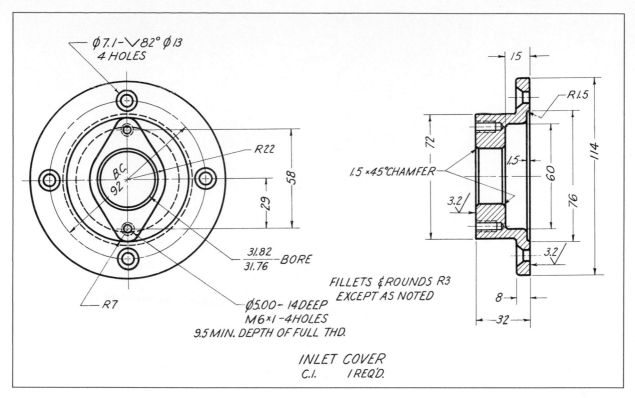

FIG. 12.26   Detail drawing.

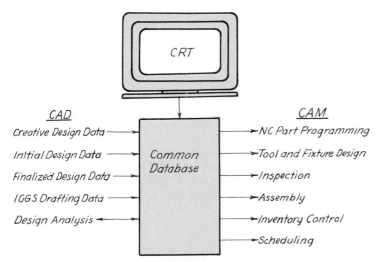

FIG. 12.27   CAD/CAM flowchart—from design through production.

B / IMPLICATIONS OF THE COMPUTER IN THE DESIGN AND PRODUCTION PROCESSES

12.29 CAD/CAM

The progressive steps of CAD/CAM fit well into the general areas of design: design drafting, production planning, and fabrication (machining) (Fig. 12.27). In following the steps and activities, we find that CAD/CAM tends to integrate subsystems and erase the distinction that has existed in the past between engineering and the production department. With CAD/CAM, marketing may be closely linked to design and production through the sharing of a common data base. The ultimate goal of CAD/CAM is complete integration of the design and manufacturing processes such that an entire firm acts and reacts as a single entity controlled by a network of computers.

FIG. 12.28 Designer at the workstation of CADD system. (*Courtesy ComputerVision Corporation*)

12.30 The Computer and the Design Process (CAD)

In computer-aided design, the engineer and the computer work together to form a problem-solving team. The combination of human being and machine produces better results in much less time than if a person performed alone. The human being and the electronic machine complement one another, with each having characteristics superior to the other. In comparing the capabilities of people and computers, it will be found that people can think and make decisions using intuitive and analytical thought. The computer has speed, work accurately, and has almost unlimited storage and rapid recall capabilities.

As a part of the CAD team, the functions of the computer are to provide an extension of the designer's memory, enhance the analytical powers of the designer, and perform the repetitious tasks of design to free its human partner for other work. With the computer performing its functions, the designer will be left free to control the design process; apply powers of intuition, creativity, imagination and judgment to the development of the design; and finally, apply experience to the analysis of significant information.

12.31 CAD

Computers are used by architects, design and production engineers, and drafting technicians in nearly all of the basic phases of total design from the identification of need

through the production stages (see Fig. 12.28). A CAD system provides the designer with a calculator of exceptional capabilities, a storage bank for design data, and a drafting aid for the development of a final design from the designer or designers' preliminary rough sketches. Furthermore, since a considerable portion of design work involves changes to an existing design, a designer can recall and reuse data from the earlier design, modifying it as needed and adding any new parts as may be required. The computer's memory capabilities may be relied upon in the first stages of design when a data bank is needed for the collection and storage of raw data for quick recall. Graphical data together with alphanumeric data can provide a data bank from which information can be quickly retrieved by the designer. The computer's use continues through the ideation and conceptualization stages. Concepts and brainstorming ideas are recorded to be recalled at a later time to stimulate new ideas or for reevaluation of past thoughts. With a graphics display in use, large amounts of information can be presented quickly for review in decision making (Fig. 12.29). The computer is needed early in the design process for making calculations relating to the design specifications and again at the point of selecting the optimum concept that will give maximum performance at minimum cost.

It is in the final experimental phases, where the workability, durability, and operational characteristics are tested with expensive scale models and working prototypes that the computer proves its worth in the design process by

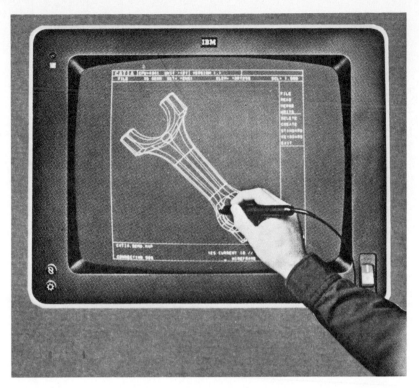

FIG. 12.29 Design displayed on a CRT screen. (*Courtesy IBM Corporation*)

eliminating the need for expensive physical models and testing facilities. This has all been made possible by CAD/CAM computer programs that simulate these models under operating conditions. Programs are available or may be prepared that simulate the effects of structural, thermal, and kinematic conditions. When necessary a designer can evaluate the electrical properties of a circuit quickly and easily.

Through simulation a new design may be displayed on a CRT screen to be evaluated under operating conditions. The model, as displayed, under program control, is continually updated to simulate dynamic motion under load conditions. Different designs can be tested without the expense of constructing models.

12.32 *Computer-aided Drafting (Fig. 12.30)*

In a plant where numerically controlled machines are widely employed, it is possible to fabricate parts without any hard-copy drawings being made for the shop. This direct pass through arrangement from design to fabrication without shop drawings is true CAD/CAM, the ultimate goal of high-technology industry. However, many companies still hold to their old practice of using hard-copy drawings in the production shops (Fig. 12.30). This may be due either to the nature of their products or to the fact that they have too few NC machines. In any case, drawings for fabrication no longer need be made at the drafting

board using drafting instruments to convert the engineer's preliminary sketches into useful plans. Now, a computer-aided drafting system can be called upon to take over this tedious and time-consuming hand-drawing work. The drawing shown in Fig. 12.30 was prepared at the Ross Gear Division of TRW, Inc., using an IBM Graphics System.

Finally, as has already been pointed out, much drafting work involves only the modification of an existing design. Should this be the case and graphical data for the original design have been stored in the computer, the design can be modified as required, parts can be added, and new drawings produced using computer-aided drafting. Computer-aided design and drafting systems have found wide acceptance for this purpose and the future appears bright.

C/ PATENTS AND DESIGN RECORDS

12.33 *Patent*

A patent, when granted to an inventor, excludes others from manufacturing, using, or selling the device or system covered by the patent anywhere in the country for a period of seventeen years. In the patent document, in which the invention is fully described, the rights and privileges of

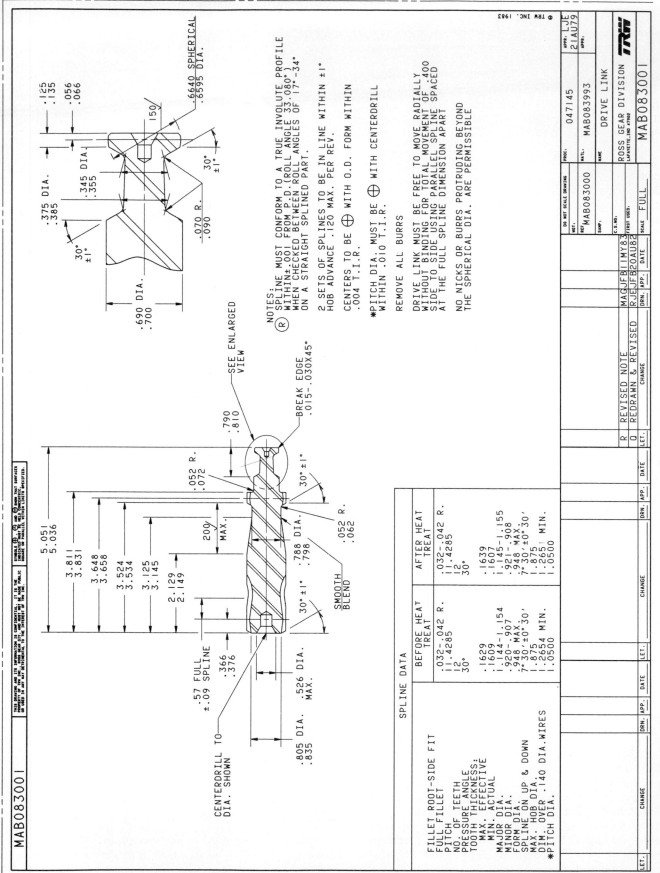

FIG. 12.30   A computer-plotter-prepared drawing. *(Courtesy of Ross Gear Division, TRW, Inc.)*

the inventor are set forth and defined. Upon the issuance of the patent, the inventor has the right to either manufacture and sell the product with a protected market or assign rights to others and collect fees for the manufacture and sale or use of the invention. After the seventeen-year period has expired, the inventor no longer has any protection and the invention becomes public property and may be produced, sold, and used by anyone for the good of all.

To ensure full protection of the patent laws, patented products must be marked *Patent* with the patent number following. Even though an invention is not legally protected until a patent actually has been issued, a product often bears a statement that reads *Patent Pending*.

In accordance with patent law "any person who has invented any new and useful process, machine, manufacture or composition of matter, or any new and useful improvement thereof, may obtain a patent," subject to restrictions, conditions, and requirements imposed by patent law. To be patentable an invention must be new, original, and uniquely different; it must perform a useful function; and the invention must not have been previously described in any publication anywhere nor have been sold or in general use in the United States before the applicant made the invention. It should be noted that an idea in itself is not patentable, since it is a requirement that a specific design and design description of a device accompany an application for a patent.

12.34 *Patent Attorney*

When an inventor has a patentable device, a patent attorney should be engaged to help prepare the necessary application for a patent. The inventor should depend upon the attorney for advice concerning whether or not the product or process may infringe upon the rights of others. Since many patent attorneys are also graduate engineers, the inventor will find the attorney to be a person who can guide the application through the searching, investigating, and processing that takes place before a patent is granted. Generally, an application remains pending for as long as four years or more before a patent is finally granted. In this period the pending patent application may well be amended to include engineering changes that have been made in the device or system.

If, after some investigation, the attorney-engineer thinks that the device is novel and therefore patentable, an application for a patent should be prepared and filed. A patent application includes a formal portion consisting of the petition, a power of attorney, and an oath or declaration. This is followed by a description of the invention, called the specifications, and a list of claims relating to it. If the device can be illustrated, one or more drawings should be included (Fig. 12.31).

In selecting a patent attorney, the inventor should remember that this professional can be of service for many years, going well beyond the time when the patent is issued. The attorney or legal firm may be retained to prepare all agreements covering the sale and leasing of patent rights, to assist in negotiations relating to these rights, and if the need should arise, to handle charges of alleged infringements.

Many large corporations have patent divisions that operate as a section of the home office legal staff. An additional office to deal with patent matters may also be maintained in Washington, D.C. In this case, both offices, staffed with patent attorneys, will have all activities coordinated by a director of patents.

12.35 *Role of the Inventor in Obtaining a Patent*

Even though an inventor must rely almost entirely upon a patent attorney to take the final steps to secure a patent, the inventor plays a vital role up to the time of filing the application. Very often what the inventor has or has not done determines whether rights to the invention can be safeguarded by the attorney.

Since court decisions in patent suits usually depend upon the inventor's ability to prove that certain design events happened on a specific date, it is always advisable to keep the design records in a hardcover, permanently bound notebook so that there will be no question about whether or not pages have been added (as could be charged if a loose leaf notebook were used). A well-kept patent notebook is a complete file of information covering a design. It can serve as the basis of project reports; it can spare the designer the unnecessary expense of repeating portions of the experimental work; and it furnishes indisputable proof of dates of conception and development.

To prepare a legally effective patent notebook the designer should

1. Use a bound notebook having printed page numbers.

2. Make entries directly using either a black ink pen or an indelible pencil.

3. Keep entries in a chronological order. Do not add retroactive entries.

4. Add references to sources of information.

5. Describe all procedures, equipment, and instruments used for developmental work.

6. Insert photos of instrument and equipment setups. Add other photos showing models, mockups, and so forth.

7. Sign and date all photos.

8. Have a qualified witness sign and date each completed page. The witness cannot be a coinventor and cannot have a financial interest in the development of the device. The witness must be a person who is capable of understanding the construction and operation of the device or system, who is also experienced in reading drawings and understanding the specifications.

9. Have the witness write the words "Witnessed and Understood," sign it, and date the signature.

10. Have two or more witnesses sign and date every page after reading it.

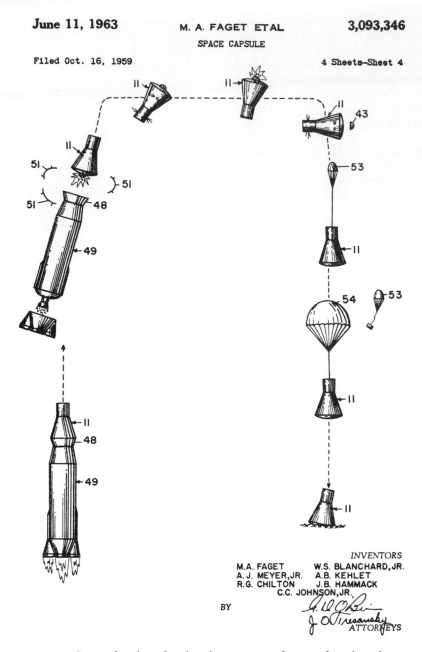

June 11, 1963   M. A. FAGET ETAL   3,093,346
SPACE CAPSULE

Filed Oct. 16, 1959   4 Sheets—Sheet 4

INVENTORS
M.A. FAGET   W.S. BLANCHARD, JR.
A. J. MEYER, JR.   A.B. KEHLET
R.G. CHILTON   J.B. HAMMACK
C.C. JOHNSON, JR.
BY
ATTORNEYS

FIG. 12.31 **Patent drawing, showing the sequence of events from launch to landing.**

11. Avoid having blank spaces on any page.

12. Have all new entries witnessed at least once a week.

13. Have the notebook evaluated periodically by a patent attorney.

It takes a considerable time and effort to adhere to these thirteen recommendations for the preparation of an effective patent notebook. However, an inventor entangled in an infringement lawsuit realizes that a little extra effort pays off.

12.36 *Patent Drawings*

A person who has invented a new machine or device, or an improvement for an existing machine, and who applies for a patent is required by law to submit a drawing showing every important feature of the invention. When the invention is an improvement, the drawing must contain one or more views of the new invention alone, and a separate view showing it attached to the portion of the machine for which it is intended.

United States Patent Office

3,465,153
Patented Sept. 2, 1969

1

3,465,153
RADIATION PROTECTION SYSTEM AND
APPARATUS
Willard F. Libby, Los Angeles, Calif., assignor to
McDonnell Douglas Corporation, Santa Monica,
Calif., a corporation of Maryland
Filed Aug. 14, 1964, Ser. No. 389,734
Int. Cl. G21f *1/12, 3/02, 7/00*
U.S. Cl. 250—108   **16 Claims**

My present invention relates generally to astronautics,
the science of space flight, and more particularly to a
system and apparatus and method for the protection of
astronauts from the hazards of suddenly encountered
radiation fields of extreme intensity in space.

Manned space flights have now been successfully
achieved by both the United States and the Soviet Union.
Such flights will be followed by manned space probes including lunar and interplanetary missions for the manned
exploration of the Moon, Mars and Venus. The manned
space program of the United States is directed towards
manned exploration first of the moon and then initially
only of the two planets Mars and Venus of the solar
system since all of its other planets appear to be barren
and lifeless. These and other probes will, of course,
eventually lead to interstellar journeys over vast distances to other stellar systems for the purpose of con-
ducting explorations aimed at discovering new worlds
which are susceptible to colonization by the human race.

2

by using a greater number of stages are offset by the additional complexity involved, and it is very difficult to
increase the propellant-weight ratio much beyond a certain value in the present concepts of vehicle systems.

The only feasible alternative remaining is to reduce
the inert weight which is not useful for propulsion in the
vehicle system so that a greater payload weight can be
obtained without the need to increase initial launch weight
of the vehicle system. This is an important consideration since any unnecessary inert weight in the various
stages of the vehicle system imposes a heavy, additional
demand on required engine thrust which is functionally
related to launch weight. In a large, three stage booster
system to be used on a lunar flight, for example, any
change in weight of the final stage will be reflected in
a similar change in the total launch weight multiplied,
however, by a growth factor which may easily number
in the hundreds.

In undertaking manned, space exploration missions, the
astronauts may be exposed to radiation fields of high intensity in space. Biological damage is done by the ion-
ization produced by radiation and high energy charged
particles which pass through the tissues of the astronauts.
A lethal action arises when the radiation dosage is excessive such that changes in living cells result in their
death when they attempt division. Of course, extremely
high radiation dose rates which may be lethal to an
astronaut after a relatively short exposure period are

FIG. 12.32   Portion of the first of six sheets of specifications (petition) for the radiation protection system illustrated in Figs. 12.33 and 12.34.

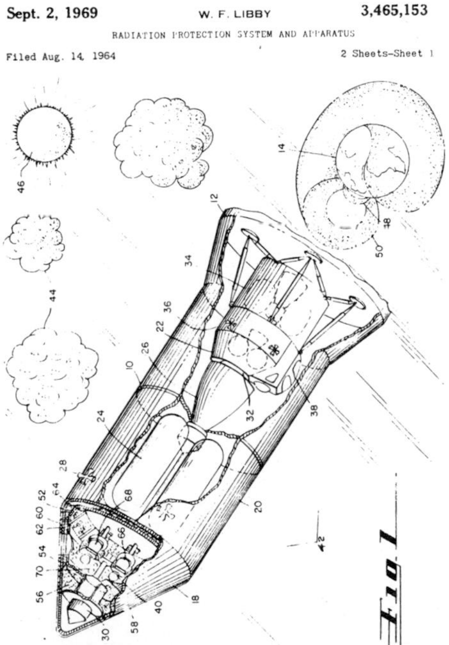

Sept. 2, 1969   W. F. LIBBY   3,465,153
RADIATION PROTECTION SYSTEM AND APPARATUS
Filed Aug. 14, 1964   2 Sheets-Sheet 1

INVENTOR.
WILLARD F. LIBBY
BY
AGENT

FIG. 12.33   Patent drawing showing a radiation protection system. See also Fig. 12.34.

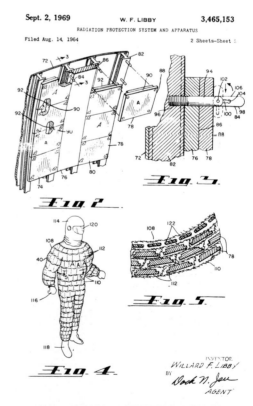

Sept. 2, 1969   W. F. LIBBY   3,465,153
RADIATION PROTECTION SYSTEM AND APPARATUS
Filed Aug. 14, 1964   2 Sheets-Sheet 2

INVENTOR.
WILLARD F. LIBBY
BY
AGENT

FIG. 12.34   Drawings showing the details that are related to the claims.

Patent drawings must be carefully prepared in accordance with the strict rules of the U.S. Patent Office. These rules are published in a pamphlet entitled *Rules of Practice in the United States Patent Office*, which may be obtained, without charge, by writing to the Commissioner of Patents, Washington, D.C.

In the case of a machine or mechanical device, the complete application for a patent will consist of a petition, a "specification" (written description), and a drawing. Ordinary drafters lack the skill and experience necessary to produce such drawings. The services of a patent drafter should be secured, along with those of a patent attorney. Three sheets of drawings for a patent for a radiation protection system are shown in Figs. 12.32–12.34.

Several U.S. Patent Office publications that are available to the general public have been listed in Appendix G.

 # DESIGN PROJECTS

Note to Instructor: Design projects may be included within traditional courses covering engineering drawing. Either two or three weeks of a semester may be set aside for design projects, or the design tasks may be spread out over a much longer time to parallel the work on the regularly assigned classroom drawing problems. In developing the design of a simple product, a student working in a group can acquire much of the basic knowledge deemed essential. The following problems are offered as suggestions to stimulate creativity and give some additional experience in both pictorial and multiview sketching. Students who have an inclination to design useful mechanisms should be encouraged to select a problem for themselves, for the creative mind works best when directed to a task in which it already has some interest. However, the young beginner should be limited to ideas for simple mechanisms that do not require extensive training in machine design and the engineering sciences.

It is suggested that the group leader for the design project prepare a design event and activities schedule using a form that is similar to the one shown in Fig. 12.35. A carefully prepared record should be kept of the progress of the design work.

DESIGN SCHEDULE & PROGRESS REPORT

Project:_____

Design Team:_____

Leader:_____

Member:_____

_____

_____

_____

Work Periods:_____

Due Dates

Investigation Report:_____

Proposal:_____

Preliminary Report:_____

Final Report:_____

ASSIGNMENT		EST. HRS	START DATE	PERCENT COMPLETE				ACTUAL HOURS
	Members			25%	50%	75%	100%	
1								
2								
3								
4								
5								

FIG. 12.35 **Design schedule and progress record.**

1. Prepare design sketches (both pictorial and multiview) for an open-end wrench to fit the head of a bolt having a body diameter of 24 mm. Give dimensions on the multiview sketch and specify the material. See Table 12 in the Appendix.

2. Prepare design sketches (both pictorial and multiview) for a wrench having a head with four or more fixed openings to fit the heads of bolts having nominal diameters of $\frac{5}{8}$, $\frac{3}{4}$, $\frac{7}{8}$, and 1 in. Dimension the multiview sketch and specify the material.

3. Prepare a series of design sketches for a bumper hitch (curved bumper) for attaching a light two-wheel trailer to a passenger automobile. Weight of trailer is 135 kg.

4. Prepare sketches for a hanger bracket to support a 24 mm control rod. The bracket must be attached to a vertical surface to which the control rod is parallel. The distance between the vertical surface and the center line of the control rod is 100 mm.

5. Prepare sketches for a mechanism to be attached to a two-wheel hand truck to make it easy to move the truck up and down stairs with a heavy load.

6. Prepare sketches for a quick-acting clamp that can be used to hold steel plates in position for making a lap weld.

7. Prepare a pictorial sketch of a bracket that will support an instrument panel at an angle of 45° with a vertical bulkhead to which the bracket will be attached. The bracket should be designed to permit the panel to be raised or lowered a height distance of 100 mm as desired.

8. Prepare sketches for an adjustable pipe support for a $1\frac{1}{2}$-in. pipe that is to carry a chemical mixture in a factory manufacturing paint. The pipe is overhead and is to be supported at 10-ft intervals where the adjustable supports can be attached to the lower chords of the roof trusses. The lower chord of a roof truss is formed by two angles $2\frac{1}{2} \times 2\frac{1}{2} \times \frac{5}{16}$ that are separated by $\frac{3}{8}$-in.-thick washers.

9. Prepare sketches and working drawings for an easy-to-operate, fast-release glider hitch. It is suggested that the student talk with several members of a local glider club to determine the requirements for the hitch and what improvements can be made for the hitch that is being used. Follow the stages of design listed in Sec. 12.13.

10. Prepare design sketches for a camera mounting that may be quickly attached and removed from any selected surface on an automobile, boat, or other type of moving vehicle. The device is to sell for not more than $9.95. Standard parts are to be used if possible. Follow through the several stages of design listed in Sec. 12.13 as required by the instructor. Make either a complete set of working drawings or drawings of selected parts.

11. Design a tire pump that will be more efficient and easier to operate than the ones now on the market.

12. Design an electric door lock that can be opened by pushing buttons in a special pattern.

13. Design a thermostatically controlled, electrically heated sidewalk upon which snow and ice will not accumulate.

14. Design an automatic automobile theft alarm.

15. Design a weather-controlled window-closing system.

16. Design a child's toy.

17. Design an automatic pet food dispenser.

18. *Class Projects.* Each student in the class is to prepare a description (on a single sheet of paper) of an innovational idea of a design suitable for a group project. From among the ideas collected several of the most worthy will be assigned to the class for development, one idea to each group. Each group of students is to be considered as a project group and be headed by one student member who may be thought of as the project engineer. The instructor will assume the role of coach for all of the groups. About midway in the total time period assigned for the development of the design, each group is to submit a written preliminary design report accompanied by sketches. A final design report (see Sec. 12.24) will be due when the project has been completed. Each report will be judged on the following: (1) evidence of good group organization, (2) quality of technical work, (3) function and appearance, (4) economic analysis, (5) manufacturing methods and requirements, and (6) over-all effectiveness of communication (written and/or oral). (*Note:* The instructor will decide whether or not finished shop drawings are to be made for all of the parts. Information needed for the preparation of finished drawings may be found in Chapters 13–16.)

PART 4

GRAPHICS FOR DESIGN AND COMMUNICATION

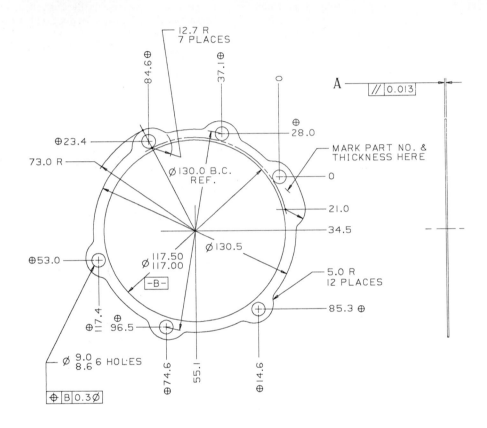

SYMBOL CODE	
PARALLELISM	//
TRUE POSITION	⊕
DENOTES BASIC DIMENSIONS	⊕

CHANGE IN DESIGN, COMPOSITION, PROCESSING OR
IN-PROCESS GAGING FROM PART PREVIOUSLY APPROVED
FOR PART PRODUCTION REQUIRES PRIOR PRODUCT
ENGINEERING APPROVAL.

ENGINEERING APPROVAL & TEST OF SAMPLES FROM
EACH SUPPLIER IS REQUIRED PRIOR TO FIRST
PRODUCTION SHIPMENT OF PARTS.

FOR ENGINEERING APPROVED SOURCE SEE
ENGINEERING NOTICE.

Fully dimensioned detail drawing computer-generated and produced by a plotter. (*Courtesy Ford Motor Company*)

Dimensions, Notes, Limits, and Geometric Tolerances

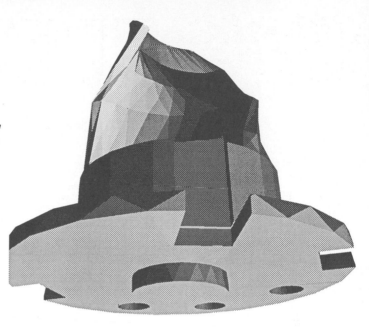

A **FUNDAMENTALS AND TECHNIQUES**

13.1 Introduction

A detail drawing, in addition to giving the shape of a part, must furnish information such as the distances between surfaces, locations of holes, kind of finish, type of material, number required, and so forth. The expression of this information on a drawing by the use of lines, symbols, figures, and notes is known as *dimensioning*.

Intelligent dimensioning requires engineering judgment and a thorough knowledge of the practices and require-

ments of the production departments. See Chapters 15 and 18.

13.2 Theory of Dimensioning

Any part may be dimensioned easily and systematically by dividing it into simple geometric solids. Even complicated parts, when analyzed, are usually found to be composed principally of cylinders and prisms and, frequently, frustums of pyramids and cones (Fig. 13.1). The dimensioning of an object may be accomplished by dimensioning each elemental form to indicate its size and relative location from a center line, base line, or finished surface. A machine drawing requires two types of dimensions: *size dimensions and location dimensions*.

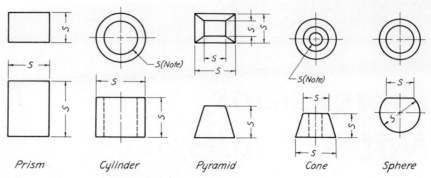

Prism   Cylinder   Pyramid   Cone   Sphere

FIG. 13.1 Dimensioning geometric shapes.

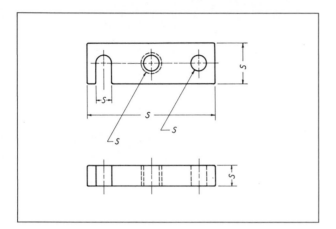

FIG. 13.2 **Size dimensions.**

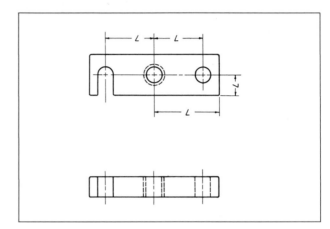

FIG. 13.3 **Location dimensions.**

13.3 *Size Dimensions (Fig. 13.2)*

Size dimensions give the size of a piece, component part, hole, or slot.

Figure 13.1 should be carefully analyzed, as the placement of dimensions shown is applicable to the elemental parts of almost every piece.

The rule for placing the three principal dimensions (width, height, and depth) on the drawing of a prism or modification of a prism is as follows: *Give two dimensions on the principal view and one dimension on one of the other views.*

The circular cylinder, which appears as a boss or shaft, requires only *the diameter and length, both of which are shown preferably on the rectangular view.* It is better practice to dimension a hole (negative cylinder) by giving the diameter and operation as a note on the contour view with a leader to the circle (Figs. 13.1 and 13.59).

Cones are dimensioned by giving *the diameter of the base and the altitude on the same view.* A taper is one example of a conical shape found on machine parts (Figs. 13.51 and 13.52).

Pyramids, which frequently form a part of a structure, are dimensioned by giving *two dimensions on the view showing the shape of the base.*

A sphere requires only the diameter.

13.4 *Location Dimensions*

Location dimensions fix the relationship of the component parts (projections, holes, slots, and other significant forms) of a piece or structure (Fig. 13.3). Particular care must be exercised in the selection and placing of location dimensions because on them depends the accuracy of the operations in making a piece and the proper mating of the piece with other parts. To select location dimensions intelligently, one must first determine the contact surfaces, finished surfaces, and center lines of the elementary geometric forms and, with the accuracy demanded and the method of production in mind, decide from what other surface or center line each should be located. Mating locating dimensions must be given from the same center line or finished surface on both pieces.

Location dimensions may be from center to center, surface to center, or surface to surface (Fig. 13.4).

13.5 *Procedure in Dimensioning*

The theory of dimensioning may be applied using the following six steps.

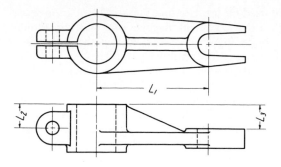

FIG. 13.4   Types of location dimensions.

L_1 – Center to Center
L_2 – Surface to Center
L_3 Surface to Surface

1. Mentally divide the object into its component geometrical shapes.

2. Place the size dimensions on each form.

3. Select the locating center lines and surfaces after giving careful consideration to mating parts and to the processes of manufacture.

4. Place the location dimensions so that each geometrical form is located from a center line or finished surface.

5. Add the overall dimensions.

6. Complete the dimensioning by adding the necessary notes.

13.6   *Placing Dimensions*

Dimensions must be placed where they will be most easily understood—in the locations where the reader will expect to find them. They generally are attached to the view that shows the contour of the features to which they apply, and a majority of them usually will appear on the principal view (Fig. 13.14). Except in cases where special convenience and ease in reading are desired, or when a dimension would be so far from the form to which it referred that it might be misinterpreted, dimensions should be placed outside a view. They should appear directly on a view only when clarity demands.

All extension and dimension lines should be drawn before the arrowheads have been filled in or the dimensions, notes, and titles have been lettered. Placing dimension lines not less than 15 mm (.62 in.) from the view and at least 10 mm (.38 in.) from each other will provide spacing ample to satisfy the one rule to which there is no exception: *Never crowd dimensions*. If the location of a dimension forces a poor location on other dimensions, its shifting may allow all to be placed more advantageously without sacrificing clarity. Important location dimensions should be given where they will be conspicuous, even if a size dimension must be moved.

The person dimensioning a drawing must make certain that every feature has been completely dimensioned and that no dimension has been repeated in a second view.

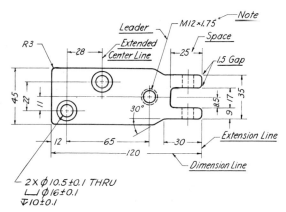

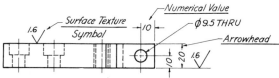

FIG. 13.5   Terms and dimensioning notation. (Dimension values are in millimeters.)

13.7   *Dimensioning Practices*

A generally recognized system of lines, symbols, figures, and notes is used to indicate size and location. Figure 13.5 illustrates dimensioning terms and notation.

A *dimension line* is a lightweight line that is terminated at each end by an arrowhead. A numerical value, given along the dimension line, specifies the number of units for the measurement that is indicated (Fig. 13.6). When the numerals are in a single line, the dimension line is broken near the center, as shown in (*a*) and (*b*). Under no circumstances should the line pass through the numerals.

Extension lines are light, continuous lines extending from a view to indicate the extent of a measurement given by a dimension line that is located outside a view. They start 1.5 mm (.06 in.) from the view and extend 3 mm (.12 in.) beyond the dimension line (Fig. 13.5).

Arrowheads are drawn for each dimension line, before the figures are lettered. They are made with the same pen

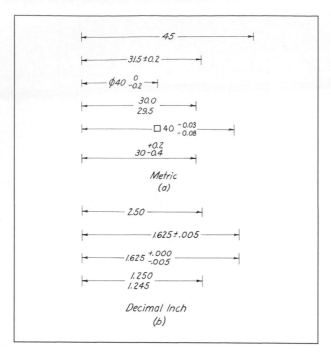

FIG. 13.6 Dimension line.

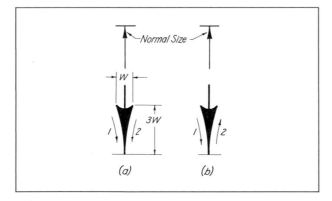

FIG. 13.7 Formation of arrowheads.

or pencil used for the lettering. The size of an arrowhead, although it may vary with the size of a drawing, should be uniform on any one drawing. To have the proper proportions, the length of an arrowhead must be approximately three times its spread. This length for average work is usually 3 mm (.12 in.). Figure 13.7 shows enlarged drawings of arrowheads of correct proportions. Although many draftsmen draw an arrowhead with one stroke, the beginner will get better results by using two slightly concave strokes drawn toward the point (*a*) or, as shown in (*b*), one stroke drawn to the point and one away from it.

A *leader* or *pointer* is a light, continuous line (terminated by an arrowhead) that extends from a note to the feature of a piece to which the note applies (Fig. 13.5). It should be made with a straightedge and should not be curved or made freehand.

A leader pointing to a curve should be radial, and the first 3 mm (.12 in.) of it should be in line with the note (Fig. 13.5).

Finish marks indicate the particular surfaces of a rough casting or forging that are to be machined or "finished." They are placed in all views, touching the visible or invisible lines that are the edge views of surfaces to be machined (Fig. 13.8). When a surface is to be finished by the removal of material and it is not necessary to indicate surface quality, a bar is added to the check-mark ($\sqrt{}$) portion of the texture symbol at the top of the short leg. When material removal is prohibited, a small circle is added to the vee in place of the horizontal bar. See Sec. 13.38.

It is not necessary to show finish marks on holes. They are also omitted, and a title note, "finish all over," is substituted, if the piece is to be completely machined. Finish marks are not required when limit dimensions are used.

Dimension figures should be lettered either horizontally or vertically with the whole numbers equal in height to the capital letters in the notes and guidelines and slope lines must be used. The numerals must be legible; otherwise, they might be misinterpreted in the shop and cause errors.

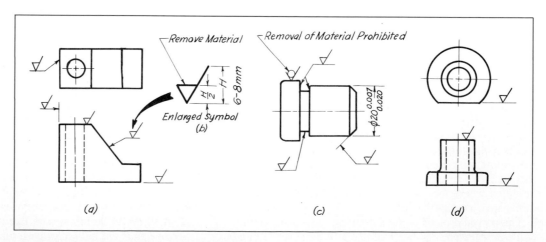

FIG. 13.8 Material removal symbol (finish mark).

13.8 *Fractional Dimensioning*

For ordinary work, where accuracy is relatively unimportant, technicians work to nominal dimensions given as common fractions of an inch, such as $\frac{1}{2}$, $\frac{1}{4}$, $\frac{1}{8}$, $\frac{1}{16}$, $\frac{1}{32}$, and $\frac{1}{64}$. When dimensions are given in this way, many large corporations specify the required accuracy through a note on the drawing that reads as follows: *Permissible variations on common fraction dimensions to machined surfaces to be $\pm.010$ unless otherwise specified.* It should be understood that the allowable variations will differ among manufacturing concerns because of the varying degree of accuracy required for different types of work.

13.9 *Decimal System*

A student should first be interested in selecting and placing dimensions. Later it will become easy to use fractional or decimal inch or the metric system as required by the needs of the design. To assist you in using any standard system, a standard conversion table has been provided in the Appendix (Table 3).

In Fig. 13.9 a drawing is shown that illustrates decimal-inch dimensioning.

The following practices are recommended for decimal-inch dimensioning.

1. Two-place decimals should be used for dimensions where tolerance limits of $\pm.01$ or more can be allowed (Fig. 13.9).

2. Decimals to three or more places should be used for tolerance limits less than $\pm.010$ [see Fig. 13.6(*b*)].

3. In the case of a two-place decimal, the second decimal place should preferably be an even digit such as .02, .04, and .06 so that when it is divided by 2, the result will remain a two-place decimal. Odd two-place decimals may be used where necessary for design reasons.

4. Common fractions may be used to indicate standard nominal sizes for materials and for features produced by standard tools as in the case of drilled holes, threads, keyways, and so forth.

5. When desired, decimal equivalents of nominal commercial sizes may be used for materials and for features such as drilled holes, threads, and so forth that are produced by standard tools.

13.10 *Use of the Metric System (Fig. 13.10)*

Under the metric system, drawings are prepared to scales based on divisions of 10, such as 1 to 2, 1 to 5, 1 to 10, and so forth. A millimeter is one-thousandth part of a meter, which was established as being 39.37 in. in length.

Most of the illustrations in this chapter have been dimensioned using millimeters.

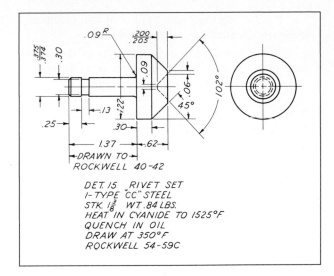

FIG. 13.9 Decimal-inch dimensioning. (*Courtesy Ford Motor Company*)

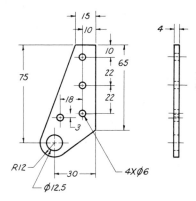

FIG. 13.10 Metric dimensioning. (Dimensions are in millimeters.)

B / GENERAL DIMENSIONING PRACTICES

13.11 *Selection and Placement of Dimensions and Notes*

The reasonable application of the selected dimensioning practices that follow should enable a student to dimension acceptably. The practices in **boldface** type should never be violated. In fact, these have been so definitely established by practice that they might be called rules.

When applying the dimensioning practices that follow, the student should realize that each company has its own drafting standards and thus the practices set forth in this chapter may not be exactly the same as the practices followed in all corporations throughout the United States. However, the practices given are in general agreement with recommendations set forth in ANSI standards.

1. Place dimensions using either of two recognized methods—aligned or unidirectional.

a. *Aligned method.* Place the numerals for the dimension values so that they are readable from the bottom and right side of the drawing. An aligned expression is placed along and in the direction of the dimension line (Fig. 13.11). Make the values for oblique dimensions readable from the directions shown in Fig. 13.32.

b. *Unidirectional method.* Place the numerals for the dimension values so that they can be read from the bottom of the drawing (see Fig. 13.12). The fraction bar for a common fraction should be parallel to the bottom of the drawing.

2. Place dimensions outside a view, unless they will be more easily and quickly understood if shown on the view (Figs. 13.12 and 13.13).

3. Place dimensions between views unless the rules, such as the contour rule, the rule against crowding, and so forth, prevent their being so placed.

4. Do not use an object line or a center line as a dimension line.

5. Locate dimension lines so that they will not cross extension lines.

6. If possible, avoid crossing two dimension lines.

7. A center line may be extended to serve as an extension line (Fig. 13.14).

8. Keep parallel dimensions equally spaced [usually 10 mm (.38 in.) apart] and the figures staggered (Fig. 13.15).

9. Always give locating dimensions to the centers of circles that represent holes, cylindrical projections, or bosses (Figs. 13.5 and 13.14).

10. If possible, attach the location dimensions for holes to the view on which they appear as circles (Fig. 13.16).

11. Group related dimensions on the view showing the contour of a feature (Fig. 13.14).

12. Arrange a series of dimensions in a continuous line (Fig. 13.17).

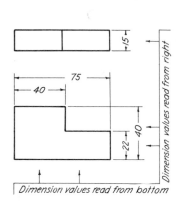

FIG. 13.11 Reading dimensions— aligned method.

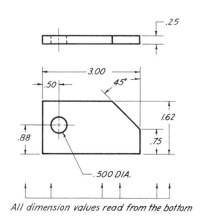

FIG. 13.12 Reading dimensions— unidirectional method.

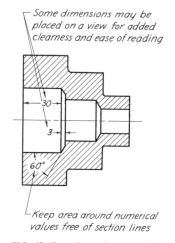

FIG. 13.13 Dimensions on the view.

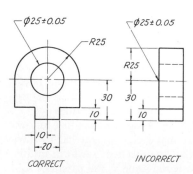

FIG. 13.14 Contour principle of dimensioning.

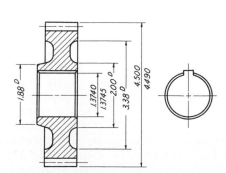

FIG. 13.15 Parallel dimensions.

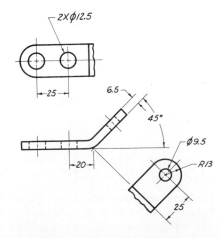

FIG. 13.16 Dimensioning an angle bracket. (Dimension values are in millimeters.)

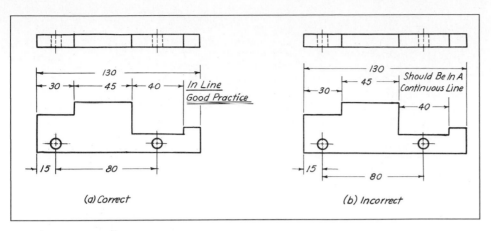

FIG. 13.17 Consecutive dimensions.

13. Dimension from a finished surface, center line, or base line that can be readily established (Figs. 13.40 and 13.54).

14. Stagger the figures in a series of parallel dimension lines to allow sufficient space for the figures and to prevent confusion (Fig. 13.15).

15. Place longer dimensions outside shorter ones so that extension lines will not cross dimension lines.

16. Give three overall dimensions located outside any other dimensions (unless the piece has cylindrical ends—see practice 43 and Fig. 13.44).

17. When an overall is given, one intermediate distance should be omitted unless given for reference. Reference dimensions are to be identified by enclosure within parentheses (Figs. 13.18 and 13.19).

18. Do not repeat a dimension. One of the duplicated dimensions may be missed if a change is made. Give only those dimensions that are necessary to produce or inspect the part.

19. Make decimal points of a sufficient size so that dimensions cannot be misread.

20. When dimension figures appear on a sectional view, show them in a small uncrosshatched portion so that they may be easily read. This may be accomplished by doing the section lining after the dimensioning has been completed (Fig. 13.20).

21. When an arc is used as a dimension line for an angular measurement, use the vertex of the angle as the center [Fig. 13.21(a)]. It is usually undesirable to terminate the dimension line for an angle at lines that represent surfaces. It is better practice to use an extension line [Fig. 13.21(b)].

22. Place the figures of angular dimensions so they will read from the bottom of a drawing, except in the case of large angles (Fig. 13.22).

23. Dimension an arc by giving its radius preceded by the abbreviation R, and indicate the center with a small cross. [Locate the center by dimensions. (Fig. 13.23)].

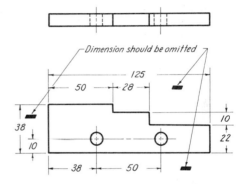

FIG. 13.18 Omit unnecessary dimensions.

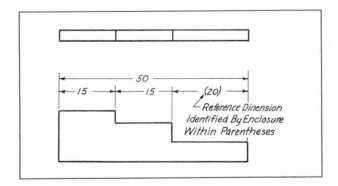

FIG. 13.19 Place reference dimensions in parentheses.

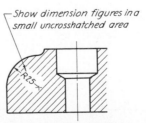

FIG. 13.20 Dimension figures on a section view.

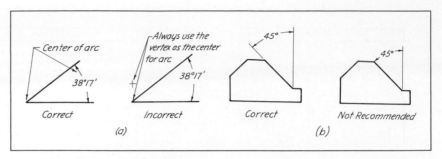

FIG. 13.21   Dimensioning an angle.

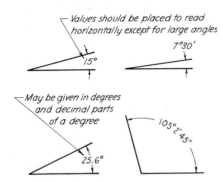

FIG. 13.22   Angular dimensions.

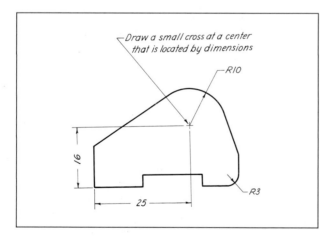

FIG. 13.23   Dimensioning a circular arc. (Dimension values are in millimeters.)

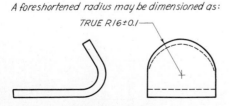

FIG. 13.24   Dimensioning a circular arc— true R.

24. TRUE R precedes the radius value, where the radius is dimensioned in a view that does not show the true shape of the arc (Fig. 13.24).

25. Show the diameter of a circle, never the radius. If it is not clear that the dimension is a diameter, the figures should be preceded by the diameter symbol φ (Figs. 13.25 and 13.26). Often this will allow the elimination of one view.

26. When dimensioning a portion of a sphere with a radius the term SPHER R or SR is added (Fig. 13.27).

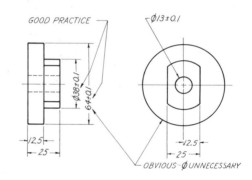

FIG. 13.25   Dimensioning a cylindrical piece.

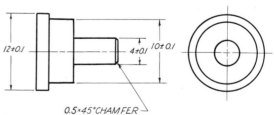

Note:
Although it is better practice to use a minimum of two views, a cylindrical part may be completely described in one view (no end view) by using the φ symbol with the value of the diameter.

FIG. 13.26   Dimensioning machined cylinders.

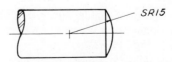

FIG. 13.27   Dimensioning a piece with a spherical end.

27. Letter all notes horizontally (Fig. 13.54).

28. Make dimensioning complete, so that it will not be necessary for a worker to add or subtract to obtain a desired dimension or to scale the drawing.

29. Give the diameter of a circular hole, never the radius, because all hole-forming tools are specified by diameter. If the hole does not go through the piece, the depth may be given as a note (Fig. 13.28).

30. Never crowd dimensions into small spaces. Use the practical methods suggested in Fig. 13.29.

31. When space is limited and a radius dimension cannot be placed between the arrowhead and the center as shown in Fig. 13.43, the radius line can be extended and the dimensional value placed outside the arc with a leader as shown in Fig. 13.30(a). Two other methods of dimensioning small arcs are shown in (b) and (c). Where the center of a radius is not dimensionally located, the center should not be indicated.

32. Avoid placing inclined dimensions in the shaded areas shown in Fig. 13.31. Place them so that they may be conveniently read from the right side of the drawing. If this is not desirable, make the figures read from the left in the direction of the dimension line [Fig. 13.32(b)]. The unidirectional method is shown in (a).

33. Omit superfluous dimensions. Do not supply dimensional information for the same feature in two different ways.

34. Give dimensions up to 72 in. in inches, except on structural and architectural drawings (Fig. 13.33). Omit the inch marks when all dimensions are in inches.

35. Show dimensions in feet and inches as illustrated in Fig. 13.34. Note that the use of the hyphen in (a) and (b) and the cipher in (b) eliminates any chance of uncertainty and misinterpretation.

36. If feasible, design a piece and its elemental parts to such dimensions as .10, .40 and .50 in. or 4, 10, 15, and 20 mm. Except for critical dimensions and hole sizes, dimension values in millimeters should be given in a full number of millimeters, preferably to an even number (12, 16, 20, 40, etc.) so that, when divided by 2, they will remain a full number. Drill sizes are given in Appendix Tables 28 and 53.

37. Chamfers may be dimensioned either by an angle and a linear dimension or by two linear dimensions (Fig. 13.35). When an angle and a linear dimension are given, the dimension specifies the distance from the indicated surface to the start of the chamfer. A note may be used for 45° chamfers. Where the edge of a rounded

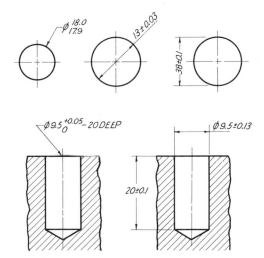

FIG. 13.28 Dimensioning holes.

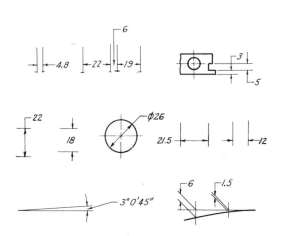

FIG. 13.29 Dimensioning in limited spaces.

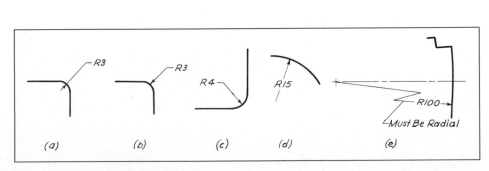

FIG. 13.30 Dimensioning radii.

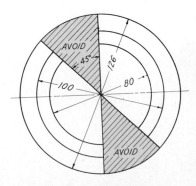

FIG. 13.31 Areas to avoid.

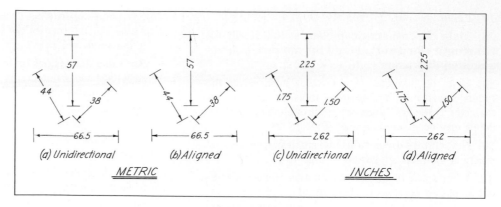

FIG. 13.32   Reading horizontal, vertical, and oblique dimensions.

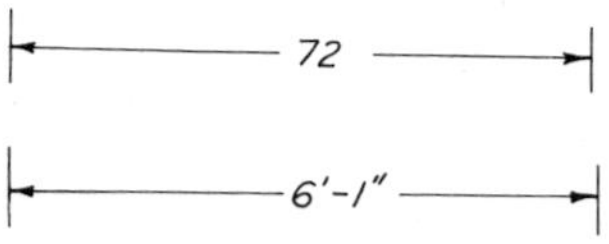

FIG. 13.33   Dimension values.

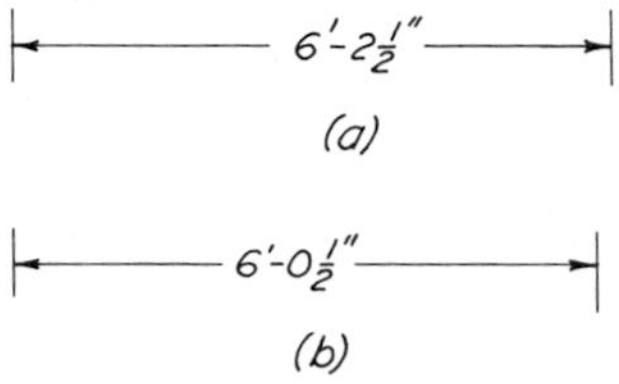

FIG. 13.34   Feet and inches.

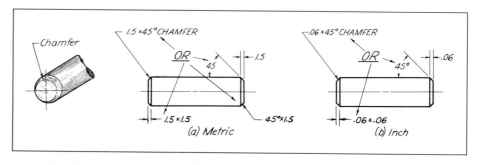

FIG. 13.35   Dimensioning an external chamfer.

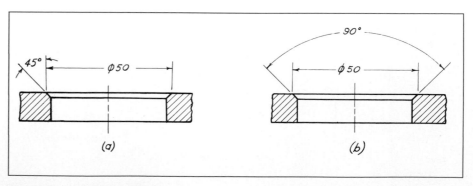

FIG. 13.36   Internal chamfers.

hole is to be chamfered, the practices shown in Fig. 13.36 are followed

38. Equally spaced holes in a circular flange may be dimensioned by giving the diameter of the bolt circle, across the circular center line, and the size and number of holes, in a note (Fig. 13.37).

39. When holes are unequally spaced on a circular center line, give the angles as illustrated in Fig. 13.38.

40. Holes that must be accurately located should have their location established by the coordinate method. Holes arranged in a circle may be located as shown in Fig. 13.39 rather than through the use of angular measurements. Figure 13.40 shows the application of the coordinate method to the location of holes arranged in a general rectangular form. The method with all dimensions referred to datum lines is sometimes called *base-line dimensioning*.

41. Dimension a curved line by giving offsets or radii.
a. A noncircular curve may be dimensioned by the coordinate method illustrated in Fig. 13.41. Offset measurements are given from datum lines.
b. A curved line, which is composed of circular arcs, should be dimensioned by giving the radii and locations of either the centers or points of tangency (Figs. 13.42 and 13.43).

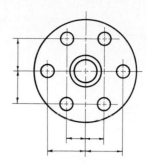

FIG. 13.39 Accurate location dimensioning of holes.

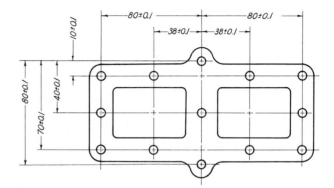

FIG. 13.40 Location dimensioning of holes.

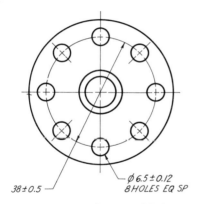

FIG. 13.37 Equally spaced holes.

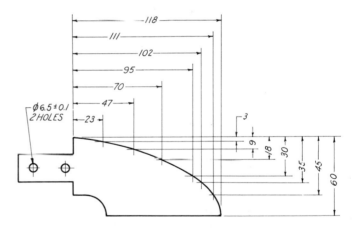

FIG. 13.41 Dimensioning curves by offsets.

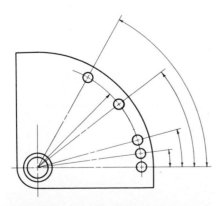

FIG. 13.38 Locating holes on a circle by polar coordinates.

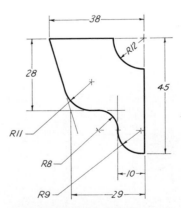

FIG. 13.42 Dimensioning curves consisting of circular arcs.

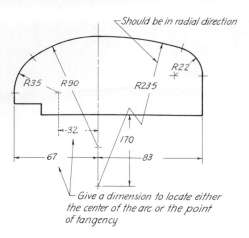

FIG. 13.43 **Dimensioning curves by radii.**

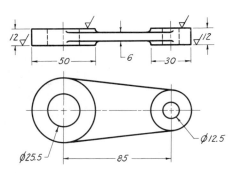

FIG. 13.44 **Dimensioning a part with cylindrical ends—link.**

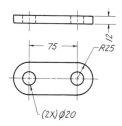

FIG. 13.45 **Link with rounded ends.**

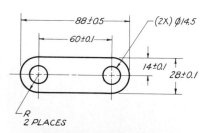

FIG. 13.46 **Dimensioning a part with rounded ends.**

42. Show an offset dimension line for an arc having an inaccessible center (Fig. 13.43). Locate with true dimensions the point placed in a convenient location that represents the true center.

43. Dimension, as required by the method of production, a piece with cylindrical ends. Give the diameters and center-to-center distance (Fig. 13.44). No overall is required.

44. The method to be used for dimensioning a piece with rounded ends is determined by the degree of accuracy required and the method of production (Figs. 13.45–13.47).

a. It has been customary to give the radii and center-to-center distance for parts and contours that would be laid out and/or machined using centers and radii. A link (Fig. 13.45) or a pad with a slot is dimensioned in this manner to satisfy the requirements of the patternmaker and machinist. An overall dimension is not needed.

b. Overall dimensions are recommended for parts having rounded ends when considerable accuracy is required. The radius is indicated but is not dimensioned when the ends are fully rounded. In Fig. 13.46, the center-to-center hole distance has been given because the hole location is critical.

c. Slots are dimensioned by giving length and width dimensions. They are located by dimensions given to their longitudinal center line and to either one end or a center line (Fig. 13.47).

d. When the location of a hole is more critical than the location of a radius from the same center, the radius and the hole should be dimensioned separately. (Fig. 13.48).

45. A keyway on a shaft or hub should be dimensioned as shown in Fig. 13.49.

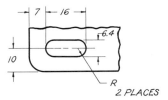

FIG. 13.47 **Dimensioning a slot.**

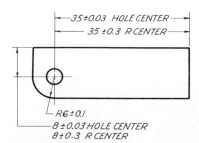

FIG. 13.48 **Dimensioning a piece with a hole and rounded end.**

46. When knurls are to provide a rough surface and control is not important, it is necessary to specify only the pitch and kind of knurl, as shown in Fig. 13.50(a) and (c). When specifying knurling for a press fit, it is best practice to give the diameter before knurling with a tolerance and include the minimum diameter after

knurling in the note that gives the pitch and type of knurl, as shown in (b).

47. Conical tapers can be specified in one of the following ways.

a. Basic taper and basic diameter (Fig. 13.51).

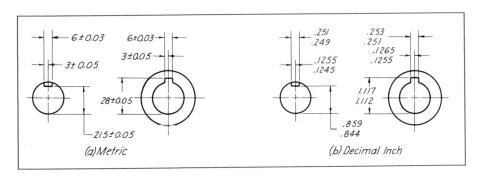

FIG. 13.49 Dimensioning keyways and keyslots.

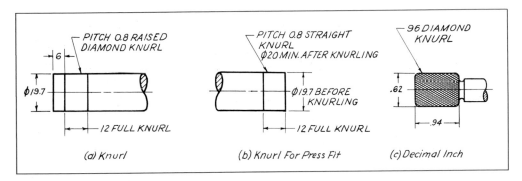

FIG. 13.50 Dimensioning knurls.

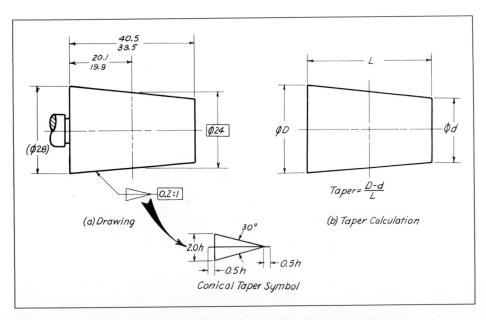

FIG. 13.51 Specifying a conical taper using a basic taper and a basic diameter.

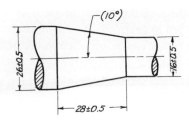

FIG. 13.52 Specifying a noncritical taper.

b. A toleranced diameter at each end and a toleranced length (Fig. 13.52). This method is used mainly for noncritical tapers such as a transition taper between the diameters of a shaft.

Flat tapers may be specified by a toleranced slope and a toleranced height at one end as shown in Fig. 13.53.

48. A half section may be dimensioned through the use of hidden lines on the external portion of the view (Fig. 13.54).

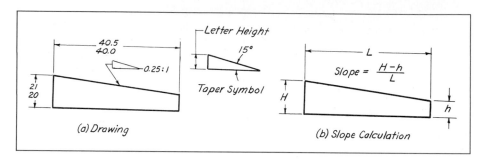

FIG. 13.53 Specifying a flat taper.

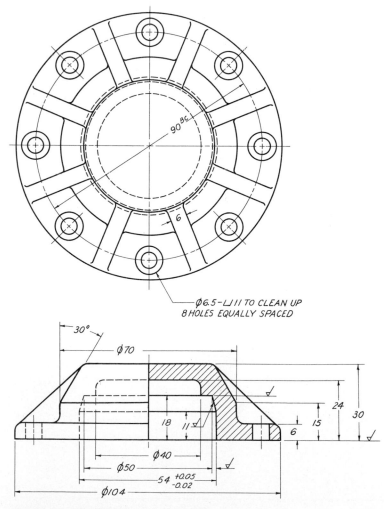

FIG. 13.54 Dimensioning a half section.

49. The fact that a dimension is out of scale may be indicated by a straight line placed underneath the dimension value (Fig. 13.55).

50. In sheet-metal work, mold lines are used in dimensioning instead of the centers of the arcs (see Fig. 13.56). A mold line (construction line) is the line at the intersection of the plane surfaces adjoining a bend.

51. Features such as holes and slots which may be repeated in a series or in the form of a pattern may be specified by indicating the number of the like features followed by an X and the size dimension of the feature. As indicated in Fig. 13.57, a space is shown between the X and the size dimension of the feature.

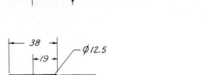
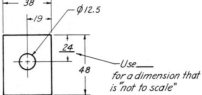

FIG. 13.55 Dimension out of scale.

13.12 Dimensions from Datum

When it is necessary to locate the holes and surfaces of a part with a considerable degree of accuracy, it is the usual practice to specify their positions by dimensions taken from a datum (Fig. 13.58) in order to avoid the buildup of error. By this method the different features of a part are located with respect to carefully selected bits of data and not with respect to each other. Lines and surfaces that are selected to serve as datums must be easily recognizable and accessible during production. Corresponding datum points, lines, or surfaces must be used as datums on mating parts.

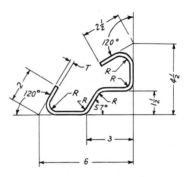

FIG. 13.56 Profile dimensioning.

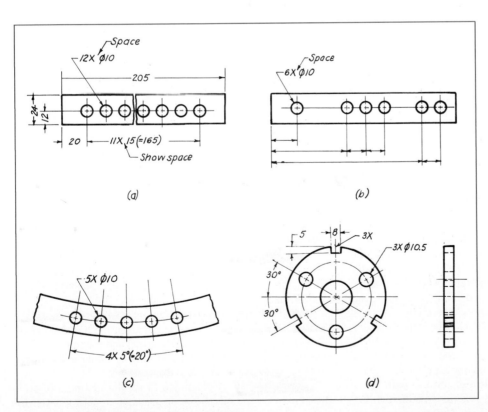

FIG. 13.57 Dimensioning repetitive features.

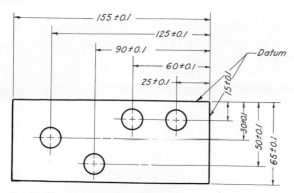

FIG. 13.58 **Dimensions from datum lines.**

13.13 Notes (Fig. 13.59)

The use of properly composed notes often adds clarity to the presentation of dimensional information involving specific operations. Notes also are used to convey supplementary instructions about the kind of material, kind of fit, degree of finish, and so forth. Brevity in form is desirable for notes of general information or specific instruction. In the case of threaded parts one should use the terminology recommended in Chapter 14.

13.14 Drawing Symbols

The symbols shown in Fig. 13.60 are used with dimensions and notes. A symbol must be shown before its associated linear dimension value in a dimension line or in place with a note as shown by the examples. The symbols have been shown at the left in the illustration and their interpretations directly follow.

C / LIMIT DIMENSIONING AND CYLINDRICAL FITS: CYLINDRICAL FITS—INCHES

13.15 Limit Dimensions

Present-day competitive manufacturing requires quantity production and interchangeability for many parts. The production of each of these parts to an exact decimal dimension, although theoretically possible, is economically unfeasible, since the cost of a part rapidly increases as an absolute correct size is approached. For this reason, the commercial drafter specifies an allowable error (tolerance) between decimal limits (Fig. 13.61). The determination of these limits depends on the accuracy and clearance required for the moving parts to function satisfactorily in the machine. Although manufacturing experience is often used to determine the proper limits for the parts of a mechanism, it is better and safer practice to adhere to the fits recommended by the American National Standards

Institute in ANSI B4.1 and ANSI B4.2. These standards apply to fits among plain cylindrical parts. Recommendations are made for preferred sizes, allowances, tolerances, and fits for use where applicable.

There are many factors that a designer must take into consideration when selecting fits for a particular application. These factors might be the bearing load, speed, lubrication, materials, and length of engagement. Frequently temperature and humidity must be taken into account. Considerable practical experience is necessary to make a selection of fits or to make the subsequent adjustments that might be needed to satisfy critical functional requirements. In addition, manufacturing economy must never be overlooked.

Those interested in the selection of fits should consult texts on machine design and technical publications, for coverage of this phase of the dimensioning of cylindrical parts is not within the scope of this book. However, since it is desirable to be able to determine limits of size following the selection of a fit, attention in this section will be directed to the use of Table 52 in the Appendix. Whenever the fit to be used for a particular application has not been specified in the instructions for a problem or has not been given on the drawing, the student should consult the instructor after a tentative choice has been made based on the brief descriptions of fits as given in this section.

To compute limit dimensions it is necessary to understand the following associated terms.

Nominal size. Nominal size is the approximate size used for the purpose of identification. In Fig. 13.62, the nominal size of the hole and shaft is $\frac{1}{2}$, or .50 in.

Basic size. Basic size is the theoretical size from which the limits of size are determined by the application of allowances and tolerances. Values to be used for plus-and-minus tolerancing are given in Tables 52A, B, C, D, and E in the Appendix.

Allowance. An allowance is the intentional difference between the maximum material limits (clearance or maximum interference) of mating parts. For clearance fits the allowance is positive. For interference fits it will be negative. In Fig. 13.62, the allowance is positive and is equal to the difference between the smallest hole and the largest shaft (.5000 − .4988 = .0012).

Tolerance. Tolerance is the permissible variation of a size that has been prescribed for a part. It is the difference between the specified limits. In Fig. 13.62, the tolerance for the shaft is .0010, the difference between the upper limit (.4988) and the lower limit (.4978).

Limits of size. The limits of size are the extreme maximum and minimum sizes specified by a toleranced dimension. In Fig. 13.62, the limits of size for the hole are .5016 and .5000.

Fit. Fit is the term commonly used to signify the tightness or looseness that may result from the application of a specific combination of allowances and tolerances in the design of mating parts. The fit may be either a clearance fit, an interference fit, or a transition fit.

Clearance fit. A clearance fit results in limits of size that assure clearance between assembled mating parts. In Fig. 13.62, the RC6 fit provides clearance under maximum

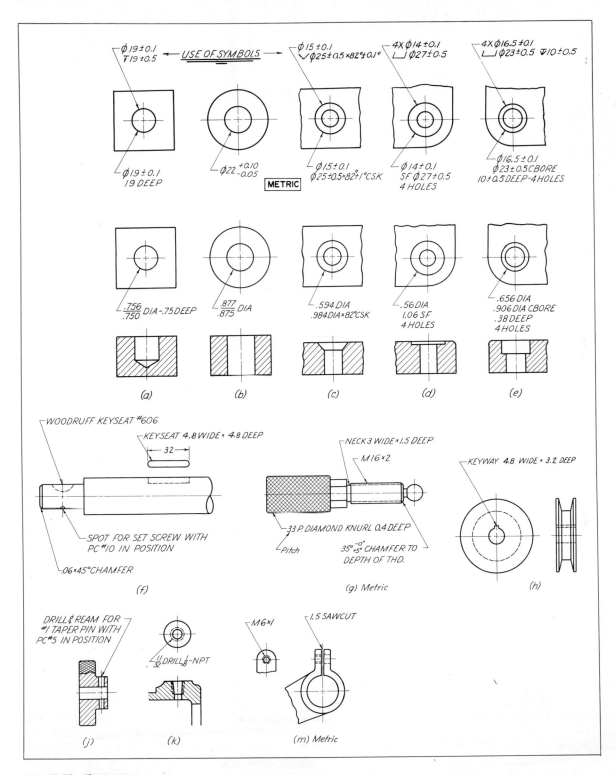

FIG. 13.59 Shop notes.

ANSI Y14.5M-1982 AND ISO			
SYMBOL FOR	ANSI	ISO	EXAMPLES
Radius	R	R	R100
Spherical Radius	SR	None	SR 24 —Space
Diameter	φ	φ	φ10±0.1
Spherical Diameter	Sφ	None	Sφ36 $_{-0.4}^{0}$
Arc Length	⌒	None	40
Square (Shape)	□	□	φ24±0.1 □10
Dimension Origin	⊕→	None	26±0.4
Dimension Not To Scale	18	18	95
Number Of Times/Places	X	X	5X φ8 Space
Reference Dimension	( )	None	(50)
Counterbore/Spotface	⊔	None	φ6±0.1 THRU ⊔φ10±0.1 ↧3±0.1
Countersink	∨	None	φ6±0.1 THRU ∨φ10±0.1×82°
Depth/Deep	↧	None	↧3±0.1
Slope	◁	◁	◁ 0.25:1
Conical Taper	▷	▷	▷ 0.25:1

FIG. 13.60 **General drawing symbols.**

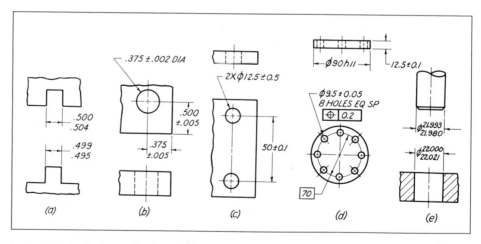

FIG. 13.61 **Limit dimensioning for the production of interchangeable parts.**

material conditions. The allowance or minimum clearance is .0012.

Interference fit. An interference fit has limits of size that result in an interference between two mating parts. In the case of a hole and shaft, the shaft will be larger than the hole, to give an actual interference of metal that will result in either a force or a press fit. This has the effect of producing an almost permanent assembly for two assembled parts.

Transition fit. A transition fit has limits of size that can lead to either clearance or interference. A shaft may be either larger or smaller than the hole in a mating part.

Basic hole system. For the widely used basic hole system, the design size of the hole (smallest size) is the basic size. The allowance (if any) is applied to the shaft. In Fig. 13.62, the basic hole size is .5000, the smallest diameter. The allowance of .0012 can be subtracted from this diameter to obtain the diameter of the largest shaft

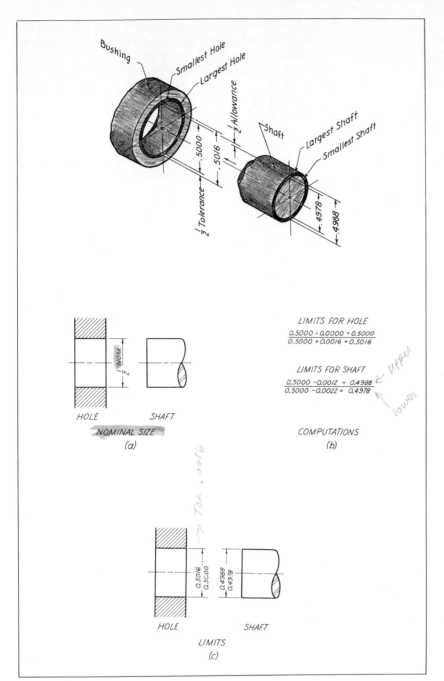

FIG. 13.62 Computation of limits (basic hole system).

(.4988). In design, the basic hole system is used whenever conditions permit because holes are usually formed and checked using standard tools and gauges while cold-rolled shafting can be much more easily machined to any specific size.

Basic shaft system. For the basic shaft system the design size (maximum diameter) of the shaft is the basic size. The allowance, if any, is applied to the hole. In Fig. 13.63, the basic shaft size is .5000. The basic shaft system should be used only when several parts having different fits are to be mounted on standard-size cold-finished shafting of fixed diameter. For each part to be assembled on

the shaft the allowance can be applied to the basic shaft size to obtain the smallest diameter of the hole.

Tables 52A, B, C, D, and E in the Appendix cover three general types of fits: running fits, locational fits, and force fits. For educational purposes standard fits may be designated by means of letter symbols, as follows:

RC—Running or Sliding Clearance Fit

LC—Locational Clearance Fit

LT—Transition Clearance or Interference Fit

LN—Locational Interference Fit

FN—Force or Shrink Fit

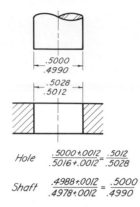

Hole $\dfrac{.5000 + .0012}{.5016 + .0012} = \dfrac{.5012}{.5028}$

Shaft $\dfrac{.4988 + .0012}{.4978 + .0012} = \dfrac{.5000}{.4990}$

FIG. 13.63 Computation of limits (basic shaft system).

It should be understood that these letters are not to appear on working drawings. Only the limits for sizes are shown.

When a number is added to these letter symbols a complete fit is represented. For example, FN4 specifies, symbolically, a class 4 force fit for which the limits of size for mating parts may be determined from use of Table 52E. The minimum and maximum limits of clearance or interference for a particular application may be read directly from this table.

Classes of fits given in these tables are as follows:

Running and Sliding Fits—Classes RC1–RC9

Locational Clearance Fits—Classes LC1–LC11

Locational Transition Fits—Classes LT1–LT6

Locational Interference Fits—Classes LN1–LN3

Force and Shrink Fits—Classes FN1–FN5

Running and sliding fits. Running and sliding fits (RC1–RC9) provide a similar running performance, with suitable lubrication allowance, throughout the range of sizes. The clearances for the RC1 and RC2 classes, which are used mainly for slide fits, increase more slowly with diameter than the other classes, so that accurate location can be maintained even at the expense of free relative motion.

Locational fits. These fits are intended for mating parts requiring some degree of accuracy of location. Locational fits provide for parts requiring rigidity and accurate alignment (interference fits—LN) as well as fits for mating parts where some freedom of location is permissible (locational clearance—LC). Locational fits are divided into three groups: locational clearance fits—LC, locational transition fits—LT, and locational interference fits—LN. Locational transition fits offer a compromise between LC and LN fits and are used where either a small amount of clearance or interference can be permitted.

Force fits. Force fits and shrink fits constitute a special type of interference fit that is normally characterized by maintenance of constant bore pressures throughout the range of sizes. Force fits may be light drive fits (FN1),

medium drive fits (FN2), heavy drive fits (FN3), and force fits (FN4) and (FN5). Medium drive fits are suitable for ordinary steel parts. FN4 and FN5 force fits are for parts that can be highly stressed. Normally, FN2 fits are used for shrink fits on light sections, while FN3 is applicable for shrink fits in medium sections.

13.16 Computation of Limits of Size for Cylindrical Parts

To obtain the correct fit between two engaging parts, compute limit dimensions that modify the nominal size of both. Numerical values of the modifications necessary to obtain the proper allowance and tolerances for various diameters for all fits mentioned previously are given in Tables 52A, B, C, D, and E in the Appendix.

The two systems in common use for computing limit dimensions are (1) the basic hole system, and (2) the basic shaft system. The same ANSI tables may be used conveniently for both systems.

13.17 Basic Hole System

Because most limit dimensions are computed on the basic hole system, the illustrated example shown in Fig. 13.62 involves the use of this system. If, as is the usual case, the nominal size is known, all that is necessary to determine the limits is to convert the nominal size to the basic hole size and apply the figures given under "standard limits," adding or subtracting (according to their signs) to or from the basic size to obtain the limits for both the hole and the shaft.

EXAMPLE Suppose that a $\frac{1}{2}$-in. shaft is to have a class RC6 fit in a $\frac{1}{2}$-in. hole [Fig. 13.62(*a*)]. The nominal size of the hole is $\frac{1}{2}$ in. The basic hole size is the exact theoretical size .5000.

From Table 52A it is found that the hole may vary between +.0000 and +.0016, and the shaft between −.0012 and −.0022. As can be readily observed, these values result in a variation (tolerance) of .0016 between the upper and lower limits of the hole, while the variation (tolerance) for the shaft will be .0010. The allowance (minimum clearance) is .0012, as given in the table.

The limits of the hole are

$$\frac{.5000 + .0000}{.5000 + .0016} = \frac{.5000}{.5016}$$

The limits on the shaft are

$$\frac{.5000 - .0012}{.5000 - .0022} = \frac{.4988}{.4978}$$

It is recommended that the maximum limit always be placed directly above the minimum limit where dimensions are associated with dimension lines (ANSI Y14.5). However, when both limits are given in a single line in association with a leader or note, the minimum limit is to be given first.

13.18 Basic Shaft System

When a number of parts requiring different fits but having the same nominal size must be mounted on a shaft, the basic shaft system is used because it is much easier to adjust the limits for the holes than to machine a shaft of one nominal diameter to a number of different sets of limits required by different fits.

For basic shaft fits the maximum size of the shaft is basic. The limits of clearance or interference are the same as those shown in Tables 52A, B, C, D, and E for the corresponding fits. The symbols for basic shaft fits are identical with those used for the standard fits with a letter S added. For example, LC4S specifies a locational clearance fit, class 4, as determined on a basic shaft basis.

■ BASIC SHAFT SYSTEM—CLEARANCE FITS

To determine the needed limits, increase each of the limits obtained, using the basic hole system, by the value given for the upper shaft limit. For example, if the same supposition (nominal size and fit requirement) is made as for the preceding illustration, the limits shown in Fig. 13.63 can be most easily obtained by adding .0012 to each of the limits shown for the hole and shaft in Fig. 13.62.

■ BASIC SHAFT SYSTEM—INTERFERENCE AND TRANSITION FITS

To determine the needed limits, subtract the value shown for the upper shaft limit from the basic hole limits.

EXAMPLE The basic shaft limits are to be determined using an FN2 fit and the same nominal diameter as for the two previous illustrations.
For the hole:

$$\frac{.5000 - .0016}{.5007 - .0016} = \frac{.4984}{.4991}$$

For the shaft:

$$\frac{.5016 - .0016}{.5012 - .0016} = \frac{.5000 \text{ (basic size)}}{.4996}$$

In brief, it can be stated that the limits for hole and shaft, as given in Tables 52A–E, are increased for clearance fits or decreased for transition or interference fits by the value shown for the upper shaft limit, which is the amount required to change the maximum shaft to basic size.

D/ LIMITS AND FITS: SI METRIC SYSTEM

13.19 SI Metric—Limits and Fits

Representing approximately 60 nations, the ISO (International Organization for Standardization) has developed a universally recognized system of limits and fits for com-

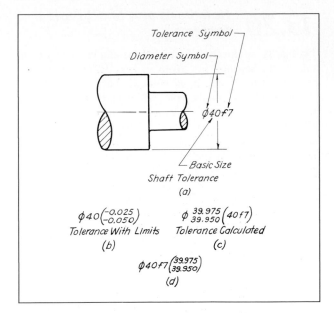

FIG. 13.64 Shaft with a toleranced dimension.

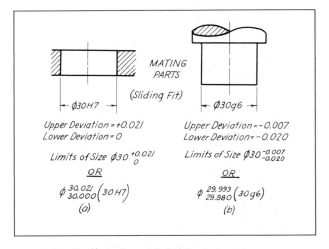

FIG. 13.65 Computation of limits for hole and mating shaft.

ponent parts. The related symbologies used ensure ready recognition of nominal size and tolerance.

The components of a toleranced dimension are given in the following order: (1) diameter symbol, if appropriate, (2) the basic size, and (3) the tolerance symbol (Fig. 13.64). Upper-case letters are used for hole tolerances and lower-case letters are used for shaft tolerances (Fig. 13.65).

Even though it is common practice to present a tolerance as given in Fig. 13.64(*a*), the upper and lower deviations may also be placed in parentheses as in (*b*). Still a third alternative is to present toleranced dimensions as shown in (*c*) in order to avoid further calculations.

Figure 13.65 shows the limits for a H7/g6 mating hole and shaft. The basic size is 30 mm.

13.20 Hole Basis and Shaft Basis Systems

Ordinarily, engineering requirements can be met by using one of the selected fits given in Fig. 13.66. Hole basis fits, listed in the left-hand column, have a fundamental deviation of "H" on the hole, while shaft basis fits, listed in the second column, have a fundamental deviation of "h" on the shaft. As described, related types of fits provide the same relative fit condition. For example, both H8/f7 and F8/h7 provide the same fit condition when a close running fit is needed. Normally, the hole basis system is used. However, when there are several parts to be mounted that have different fits with a common shaft, the fits are calculated using the shaft basis system. It should be noted from the table shown in Fig. 13.66 that the locational clearance fit H7/h6 is included in both systems.

When a designer encounters special conditions that may not be met by one of the selected fits, he or she should consult the standard ANSI B4.2, Preferred Metric Limits and Fits.

With the hole basis system, the selection of preferred fits produces the results that follow when shafts are assembled with mating H holes.

1. Shafts c through h—clearance fits
2. Shafts k and n—transition fits
3. Shafts p through u—interference fits

Under the shaft basis system, all shafts are h6 except for the first three.

Figure 13.67 shows three examples of fits using the hole basis system for the calculation of the limits for mating H holes and shafts.

To calculate the limits of size for the hole and shaft designated H9/d9, one must determine both the upper and lower deviations for the hole and shaft from Table 23 in the appendix.

For a 30 mm hole the deviations for the hole are +0.052 and 0.000. Therefore, the limits of size of the hole designated ϕ30H9 are 30.052 and 30.000. The deviations

Selected ISO fits

		Hole basis	Shaft basis	Type of fit
Clearance fits		H11/c11	C11/h11	Loose running fit
		H9/d9	D9/h9	Free running fit
		H8/f7	F8/h7	Close running fit
		H7/g6	G7/h6	Sliding fit—intended to move freely
		H7/h6	H7/h6	Locational clearance fit
Transition fits		H7/k6	K7/h6	Locational transition fit
		H7/n6	N7/h6	Locational transition fit
Interference fits		H7/p6	P7/h6	Locational interference fit—press fit
		H7/s6	S7/h6	Medium drive fit or shrink fit
		H7/u6	U7/h6	Force fit

FIG. 13.66 Preferred fits—basic hole and basic shaft.

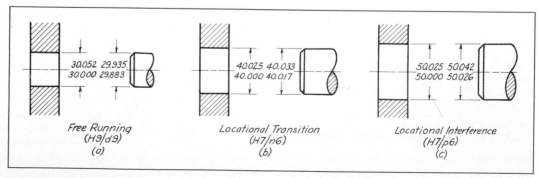

FIG. 13.67 Hole basis system.

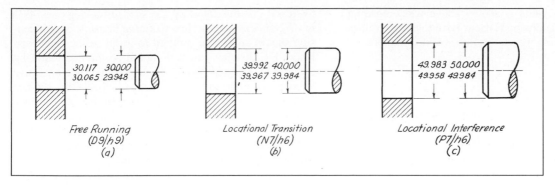

FIG. 13.68 Shaft basis system.

for the φ30d9 shaft that appear in the same table are −0.065 and −0.117.

The limits of size for the mating shaft are 29.935 and 29.883. The limits of size for the other examples in (b) and (c) were derived using deviations found in Table 23 under the column headed by the specified fit symbol.

Tables 24A and 24B must be used to determine the upper and lower deviations for the mating shaft and hole, shown in Fig. 13.68, using the shaft basis system. For a free running fit a D9/h9 fit is to be used as called for in the table. See Fig. 13.66. The deviations given in Table 24A for a φ30h9 shaft are −0.052 and 0.000. The deviation found in Table 24B for the mating hole are +0.117 and +0.065. Using these deviations, the limits of size for the shaft are 30.000 (basic) and 29.948. The limits for the mating hole are 30.117 and 30.065.

13.21 *Tolerances*

Necessary tolerances may be expressed by general notes printed on a drawing form or they may be given with definite values for specific dimensions as in the case shown in Fig. 13.73, where the part has been dimensioned using millimeters. When expressed in the form of a printed note, the wording might be as follows: ALLOWABLE VARIATION ON ALL FRACTIONAL DIMENSIONS IS ±.010 UNLESS OTHERWISE SPECIFIED. A general note for tolerance on decimal dimensions might read: ALLOWABLE VARIATION ON DECIMAL DIMENSIONS IS ±.001. This general note would apply to all decimal dimensions where limits were not given.

The general notes on tolerances should be allowed to apply to all dimensions where it is not necessary to use specific tolerances.

13.22 *Tolerancing Methods*

A tolerance applied directly to a dimension may be expressed by giving limits (maximum and minimum values) or by means of plus-and-minus tolerancing. When limit dimensioning is used, the high limit is given above the low limit, as: .750 .742. When given in a single line, the low

---

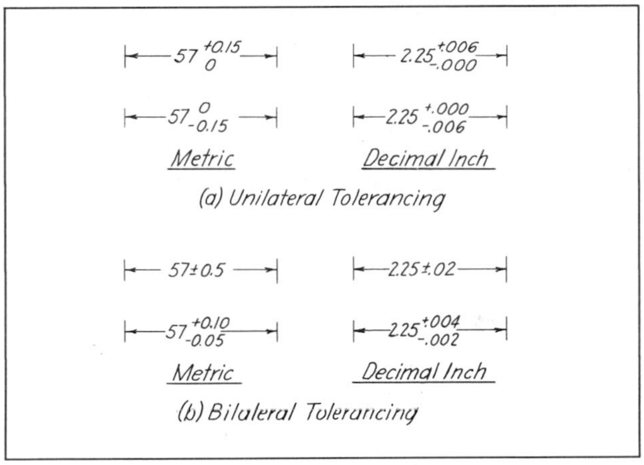

FIG. 13.69 Plus-and-minus tolerancing.

limit precedes the high limit, with a dash separating the values (.742–.750).

When the metric system is used, the same practices are followed. It should be noted that when the high and low limits are placed one above the other, they are not separated by a line.

EXAMPLES

31.8 31.5 mm   31.5–31.8 mm

13.23 *Plus-and-Minus Tolerancing*

When plus-and-minus tolerancing is used, the specified size is given first, followed by a plus-and-minus expression of the tolerance as shown in Fig. 13.69.

Unilateral plus-and-minus *tolerancing* is illustrated in (a) for both the metric and English systems of measurement. It should be noted for metric dimensioning that the nil value is shown by a zero with no plus-or-minus symbol being given. When the tolerance is a plus value, it is placed above the zero. When the tolerance has a minus value, it is placed below the zero.

Bilateral tolerancing is illustrated in (b). For equal plus-and-minus values, the mean value is preceded by a ±

symbol. The plus value is shown above when unequal plus-and-minus tolerancing is used. Both tolerance values should be given using the same number of decimal places.

13.24 Cumulative Tolerances

An undesirable condition may result when either the location of a surface or an overall dimension is affected by more than one tolerance dimension. When this condition exists, as illustrated in Fig. 13.70(a), the tolerances are said to be cumulative. In (a) surface B is located from surface A and surface C is related in turn to surface B. With the tolerances being additive, the tolerance on C is the sum of the separate tolerances ($\pm .002$). In respect to A, the position of C may vary from 1.998 to 2.002. This

tolerance, as illustrated by the shaded rectangle, is .004 in. When consecutive dimensioning is used, one dimension should always be omitted to avoid serious inconsistency. The distance omitted should be the one requiring the least accuracy. To avoid the inconsistency of cumulative tolerances it is the preferred practice to locate the surfaces from a datum plane, as shown in (b), so that each surface is affected by only one dimension. The use of a datum plane makes it possible to take full advantage of permissible variations in size and still satisfy all requirements for the proper functioning of the part.

13.25 Specification of Angular Tolerances

Angular tolerances may be expressed in degrees, minutes, or seconds (see Fig. 13.71). If desired, an angle may be given in degrees and decimal parts of a degree with the tolerance in decimal parts of a degree.

E / TOLERANCES OF LOCATION, TOLERANCES OF FORM, PROFILE, ORIENTATION, AND RUNOUT

13.26 Geometric Tolerancing

Geometric tolerances specify the maximum variation that can be allowed in form or position from true geometry. Actually, a geometrical tolerance is *either the width or diameter of a tolerance zone within which a surface or axis of a hole or cylinder can lie with the resulting part satisfying the necessary standards of accuracy for proper functioning and interchangeability.* Whenever tolerances of form are not specified on a drawing for a part, it is understood that the part as produced will be acceptable regardless of form variations. Expressions of tolerances of form control straightness, flatness, parallelism, squareness, concentricity, roundness, angular displacement, and so forth.

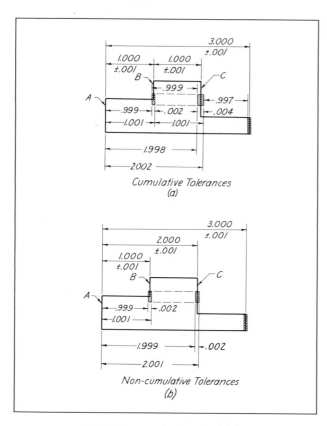

FIG. 13.70 Cumulative tolerances.

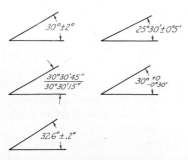

FIG. 13.71 Angular tolerances.

13.27 Symbols for Tolerances of Position and Form

The characteristic symbols shown in Fig. 13.72 have been adopted for use in lieu of notes to express positional and form tolerances. In general, these symbols are the same as those given in Mil. Std. 8C for use by the armed services. Fig. 13.73 shows typical feature control symbols applied to a drawing. After making a careful study of this drawing, the reader is urged to relate and compare the symbolic callouts used with callouts and their illustrated significance as given in Figs. 13.74–13.86 inclusive.

USE OF SYMBOLS

Geometric characteristic symbols. These symbols denoting geometric characteristics are shown in Fig. 13.72. Figs. 13.73–13.86 show their use.

Datum identifying symbol. The datum identifying symbol consists of a frame (box) containing the datum reference letter. It may be associated with the datum feature in one of the ways shown in Figs. 13.73–13.86.

Basic dimension symbol. In order that a basic dimension may be identified, it is enclosed in a frame as shown in Fig. 13.73.

Supplementary symbols—M, MC, RFS, and diameter. The symbols ⓜ and ⓢ are used to designate "maximum material condition" and "regardless of feature size" (Fig. 13.72). In notes, the abbreviations MMC and RFS are used.

The symbol φ is used to designate a diameter (Figs. 13.72 and 13.73). On a drawing it is used to replace DIA or D. It always precedes the associated dimension on the drawing and the specified tolerance in a feature control symbol.

Combined symbols. Individual symbols (characteristic and supplementary), datum reference letters, and the needed tolerance may be combined in one frame, as shown in the lower portion of Fig. 13.72, to express a callout for a tolerance symbolically.

In using the universally recognized symbols given in Fig. 13.72, a position or form tolerance is expressed by means of a feature control symbol consisting of a frame that contains the appropriate geometric characteristic symbol followed by the allowable tolerance. It should be noted in Fig. 13.73 that a vertical line separates the symbol from the tolerance. Where applicable, the tolerance should be preceded by the diameter symbol and followed by the symbol ⓜ or ⓢ. See Fig. 13.73.

Where a tolerance of position of form is related to a datum, this relationship shall be indicated in the feature control symbol by placing the datum reference letter(s) following the tolerance. As can be noted in the several examples shown in Fig. 13.72, vertical lines separate these entries. Each datum reference letter entered is supplemented by the symbol for MMC or RFS when applicable (Fig. 13.73).

Placement of feature control symbols. A symbol is related to the feature to which it applies by one of the four methods listed.

1. By adding the symbol to a note (Fig. 13.73).

2. By a leader from the feature to the symbol (Fig. 13.73).

3. By attaching the symbol frame to an extension line from the feature (Fig. 13.73).

4. By attaching a side or end of the symbol frame to a dimension line (Fig. 13.73).

The symbols shown in Fig. 13.72 provide the best means for specifying geometric characteristics and tolerances. Notes have been extensively used in the past but notes can be inconsistent and require more space. In addition, these symbols are understood internationally and surmount language barriers. This has been the principal reason for their quick adoption by multinational companies for use on drawings.

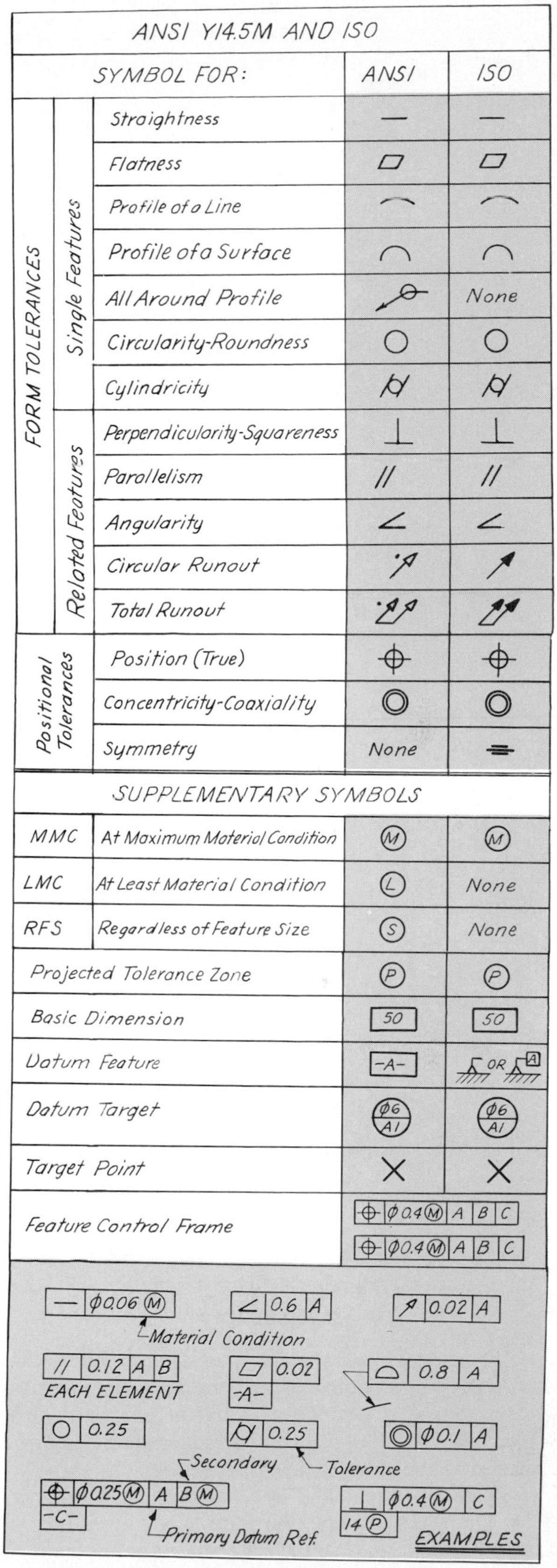

FIG. 13.72 Geometric characteristic symbols.

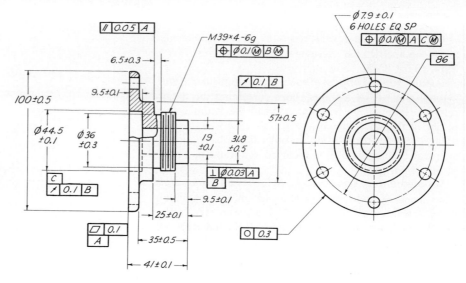

FIG. 13.73   Use of symbols in specifying positional and form tolerances.

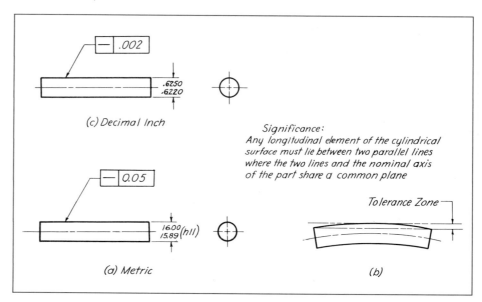

FIG. 13.74   Specifying straightness of surface elements.

13.28   Tolerance Specification for Straightness

A straightness tolerance specifies the tolerance zone within which an axis or all points of the considered element must lie. Straightness is a condition where an axis or element of a surface is a straight line. The straightness tolerance is applied in the straight-line view.

Straightness is specified by symbol as shown in Fig. 13.74. The given example illustrates straightness control of individual longitudinal surface elements of a cylindrical part. The symbol meaning is set forth in (*b*). In (*a*) the tolerance zone has been given as 0.05 mm, while in (*c*) it is .002 in. These values are equivalent.

In Fig. 13.75 the derived axis of the actual feature must lie within a cylindrical tolerance zone of 0.04 diameter regardless of feature size. In addition, each circular element must be within the specified limits of size.

13.29   Tolerance Specification for Flatness

The callout used to control flatness specifies that all points of the actual surface must lie between two parallel planes that are a distance apart equal to the specified tolerance (see Fig. 13.76). The symbol in (*a*) is interpreted to read: The entire actual surface shall be flat within 0.1 (mm)

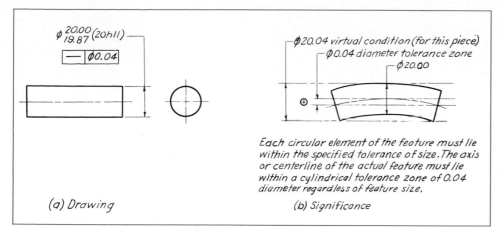

FIG. 13.75 Specifying straightness (RFS).

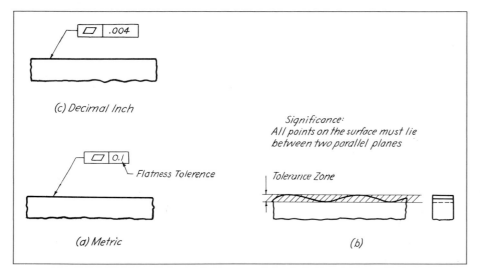

FIG. 13.76 Specification for flatness.

total tolerance zone. When necessary, the expressions MUST NOT BE CONCAVE or MUST NOT BE CONVEX may be added below the feature control symbol.

13.30 *Tolerance Specification for Perpendicularity*

Perpendicularity is the condition of a surface, median plane, or axis that is at a 90° angle to a datum plane or axis. A perpendicular tolerance relates to one of the following:

1. A tolerance zone defined by two parallel planes perpendicular to a datum plane or axis within which the surface must lie. See Figs. 13.77 and 13.78.

2. A tolerance zone defined by two parallel planes which are perpendicular to a datum axis within which the feature axis must lie.

3. A cylindrical tolerance zone perpendicular to a da-

tum plane within which the feature axis must lie. See Figs. 13.79 and 13.80.

4. A tolerance zone defined by two parallel lines perpendicular to a datum plane or axis within which each radial element of the surface must lie.

13.31 *Tolerance Specification for Parallelism*

Parallelism is the condition of a surface or axis equidistant at all points from a datum plane or axis. For the example in Fig. 13.81, the datum is considered as a plane established by the high points of surface *A*. All points of the other surface must lie between two planes that are parallel to the datum. In (*a*) the two planes are 0.05 (mm) apart. In Fig. 13.82 the feature axis must lie between two planes that are parallel to datum plane A. The two planes are 0.12 apart.

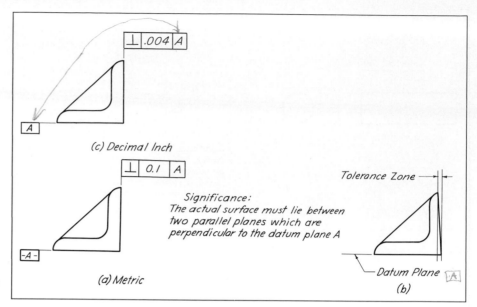

FIG. 13.77 Specification for perpendicularity.

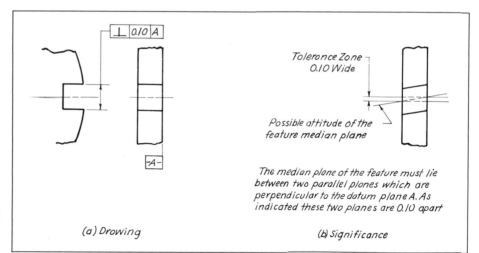

FIG. 13.78 Specifying perpendicularity for a median plane (feature RFS). (See Fig. 13.88(b).)

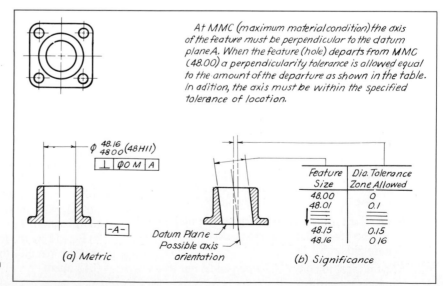

FIG. 13.79 Specification of perpendicularity.

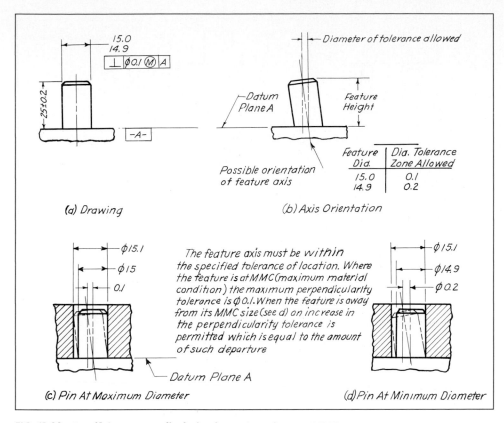

FIG. 13.80 Specifying perpendicularity for a pin or boss at MMC.

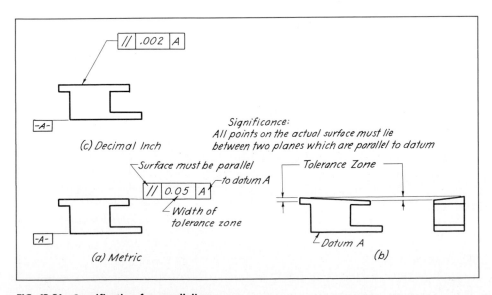

FIG. 13.81 Specification for parallelism.

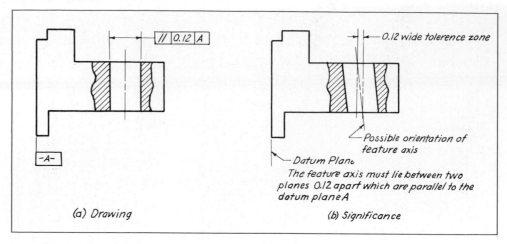

FIG. 13.82 **Specifying parallelism for an axis (feature RFS).**

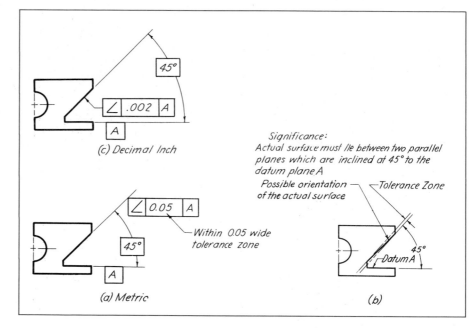

FIG. 13.83 **Specification for angularity.**

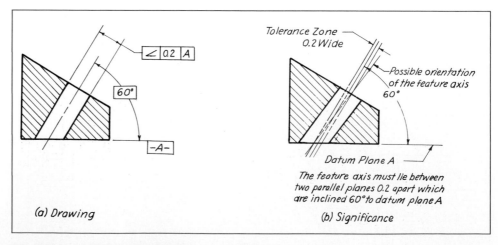

FIG. 13.84 **Specifying angularity for an axis (RFS).**

13.32 Tolerance Specification for Angularity

The feature control symbol that is commonly used to specify the tolerance for control of angularity is shown in Fig. 13.83. Angularity is the condition of a surface or axis that is at some specified angle (other than 90°) from a datum plane or axis. An angularity tolerance relates to one of the following:

1. A tolerance zone defined by two parallel planes at a specified basic angle to a datum plane or axis within which the surface of the feature must lie. See Fig. 13.83.

2. A tolerance zone defined by two parallel planes at a specified basic angle to a datum plane or axis within which an axis of a feature must lie. See Fig. 13.84.

13.33 Tolerance Specification for Concentricity (Fig. 13.85)

When the cylindrical or conical features of a part must be basically concentric, it is the practice to specify permissible eccentricity in terms of the maximum permissible deviation from concentricity. Concentricity tolerance, as illustrated in Fig. 13.85(*b*), is the diameter of the cylindrical tolerance zone within which the axis of the feature must lie. The axis of this tolerance zone must coincide with the axis of the datum feature that has been indicated.

13.34 Profile Tolerancing

Where a uniform amount of variation may be permitted along a profile, a zone tolerance may be specified. The zone is indicated at a conspicuous location by two phantom lines drawn parallel to the profile when the zone is bilateral,

that is, symmetrical about the contour line, as shown in Fig. 13.86. Only one phantom line is needed for a unilateral zone that may lie on either side of the true profile. As can be observed from the illustration, the finished surface must lie within the specified tolerance zone. The variation to be permitted and the extent of the tolerance zone must be specified. On the drawing, the applicable feature control symbol should appear with the view where the surface is represented in profile.

13.35 True-position Dimensioning

It is the usual practice to locate points by means of rectangular dimensions given with tolerances. A point located in this manner will lie within a square tolerance zone when the positioning dimensions are at right angles to each other, as in Fig. 13.87. Where features are located by radial and angular dimensions with tolerances, wedge-shaped tolerance zones result.

In making a comparison of coordinate tolerancing and true-position tolerancing of circular features, it can be noted in the case of coordinate tolerancing, as illustrated in Fig. 13.87, that the actual position of the feature can be anywhere within the 0.1 square and that the maximum allowable variation from the desired position occurs along the diagonal of the square. With this allowable variation along the diagonal being 1.4 times the specified tolerance, the diameter of the cylinder for true-position tolerancing of the same feature could be 1.4 times the tolerance that would be used in coordinate tolerancing without any increase in the maximum allowable variation. True-position tolerancing increases the permissible tolerance in all directions, without detrimental effect on the location of the feature. True-position dimensioning takes into full account the relations that must be maintained for the interchange-

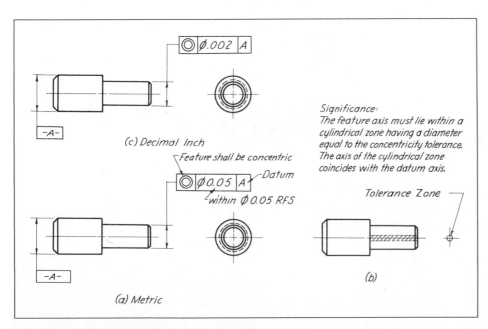

(c) Decimal Inch

⌖ Feature shall be concentric

⌖ Datum

within ⌀0.05 RFS

Significance:
The feature axis must lie within a cylindrical zone having a diameter equal to the concentricity tolerance. The axis of the cylindrical zone coincides with the datum axis.

Tolerance Zone

(b)

(a) Metric

FIG. 13.85 Concentricity tolerancing for coaxiality.

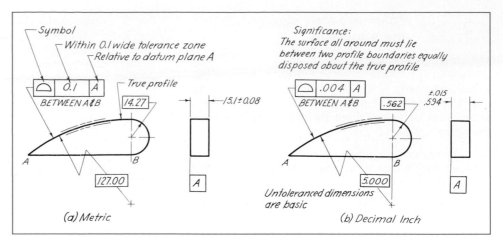

FIG. 13.86 **Profile (zone) tolerancing between points.**

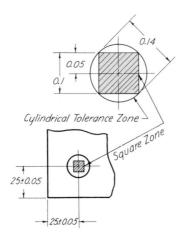

FIG. 13.87 **Comparison between coordinate tolerancing and true-position tolerancing.**

able assembly of mating parts and it permits the design intent to be expressed more simply and precisely. Furthermore, the true-position approach to dimensioning corresponds to the control furnished by position and receiver gauges with round pins. Such gauges are commonly used for the inspection of patterns of holes in parts that are being mass produced.

Positional tolerancing can be used for specific features of a machine part. When the part contains a number of features arranged in groups, positional tolerances can be used to relate each of the groups to one another as necessary and to tolerance the position of the features within a group independently of the features of the other groups.

The term *true position* denotes the theoretically exact position for a feature. In practice, the basic (exact) location is given with untoleranced dimensions that are excluded from the general tolerances, usually specified near the title block. This is done by enclosing each of the true-position-locating dimensions in a rectangular box to identify them. See Fig. 13.89.

When the alignment between mating parts depends on

some functional surface, this surface is selected as a datum for dimensioning, and the datum is identified.

The requirement of true-position dimensioning for a cylindrical feature is illustrated in Fig. 13.88(*a*). It must be understood that the axis of the hole at all points must lie within the specified cylindrical tolerance zone having its center located at true position. This cylindrical tolerance zone also defines the limits within which variations in the squareness of the axis of the hole in relation to the flat surface must be confined.

For noncircular features, such as slots and tabs, the positional tolerance is usually applied only to surfaces related to the center plane of the feature. In applying true-position dimensioning to such features, it will be found that the principal difference is in the geometric form of the tolerance zone within which the center plane of the feature must be contained. See Fig. 13.88(*b*). The center plane of the tolerance zone must be located at true position. It should be noted that this tolerance zone also defines the limits within which variations in the squareness of the center plane of the slot must be contained.

13.36 True-position Dimensioning—Application of the MMC Principle

The least desirable situation exists for the assembly of mating parts when both parts are at their maximum metal condition (designated MMC). The expression *maximum metal condition* by itself, as applied to an internal feature of a finished part, means that the internal feature (hole, slot, etc.) is at its minimum allowable size. In the case of external features (shafts, lugs, tabs, etc.), a maximum material condition exists when these features are at their maximum allowable sizes. MMC occurs for mating parts, say, for a hole and a shaft, when the shaft is at its maximum size and the hole is at its smallest size. Thus, at MMC, there is least clearance between these parts. In general, tolerance of position and the MMC of mating features are

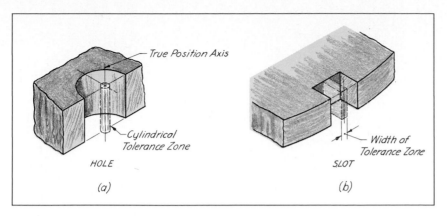

FIG. 13.88 **Meaning of true-position dimensioning.**

considered together in relationship to each other. This leads to the situation where the specified limits of location frequently may be exceeded and acceptable parts produced when the mating features are away from their maximum material limits of size. With this latter condition permissible, it becomes desirable to indicate the fact that specified limits of location need be observed only under MMC.

True-position tolerancing on the MMC basis is both practical and economical for the mass production of interchangeable parts. However, the MMC basis should not be applied where it would be inconsistent with functional requirements.

In those relatively few cases where the more economical MMC basis is not applicable and the positional tolerance must be stated without reference to MMC, the more restrictive ''regardless of feature size'' (RFS) basis is specified. This is accomplished by adding the abbreviation RFS (or S, its symbolic equivalent), to the true-position callout.

Additional information concerning the meaning of MMC as related to positional tolerances may be found in ANSI Y14.5.

The 0.14 position tolerance, illustrated in Fig. 13.87, would usually be based on the MMC size of the hole. Then as the hole size deviates from the MMC size, the position of the hole is permitted to shift off its true position somewhat beyond the specified tolerance zone to offer a realistic bonus to the extent of that departure. In Fig. 13.89, the position tolerance zone would show considerable enlargement as the hole size departs from MMC (17.5 mm) to its high limit size of 17.6 mm. Thus, position tolerancing provides greater production tolerances while meeting fully all design requirements.

Datum planes (surfaces), as the basis for position relationships, may be either specified on a drawing or implied. They have been specified for the part shown in Fig. 13.90. Datums must be specified on a drawing when feature interrelationships, involving form and position tolerances, must be directly and accurately controlled.

A circular pattern of six holes located by position tolerancing with respect to the center hole is shown in Fig. 13.91. In (*a*) the symbol specifies the six holes are to be

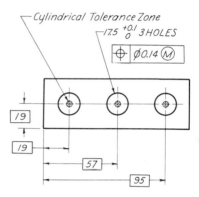

FIG. 13.89 **Position tolerancing and use of feature control symbol.**

located at position within ϕ 0.25 at 9.5 MMC size of the holes with respect to datum *A* and datum hole *B* at 24 MMC.

Tabs and slots are dimensioned as shown in Fig. 13.92. For such features of size a positional tolerance is used to locate the center plane established by parallel surfaces of the feature. The 0.5 tolerance value is the distance between two parallel planes at MMC.

F / DESIGNATION OF SURFACE TEXTURE

13.37 *Surface Quality*

The improvement in machining methods within recent years coupled with a strong demand for increased life for machined parts has caused engineers to give more attention to the quality of the surface finish. Not only the service life but also the proper functioning of the part as well may depend on obtaining the needed smoothness quality for contact surfaces.

On an engineering drawing a surface may be represented

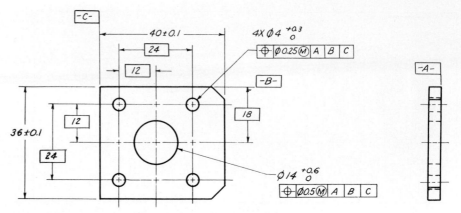

FIG. 13.90   Hole pattern with datum surfaces specified.

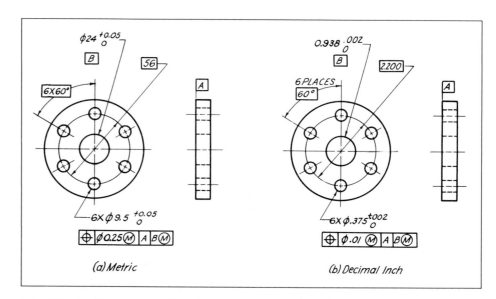

FIG. 13.91   Positional tolerancing with datum references.

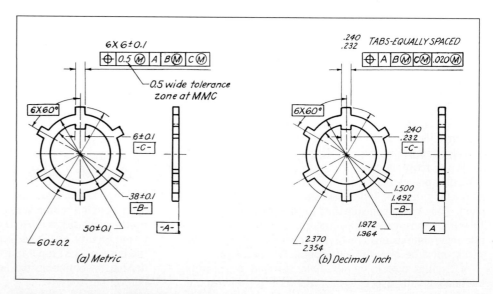

FIG. 13.92   Positional tolerancing of tabs.

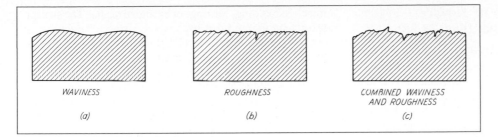

FIG. 13.93 Surface definitions illustrated.

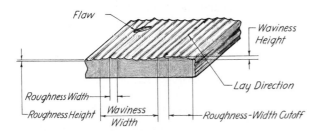

FIG. 13.94 Surface texture definitions.

TABLE 1   Preferred Roughness Average Values (R_a)

μm.*	μin.†	μm.	μin.
0.025	1	1.60	63
0.050	2	3.2	125
0.10	4	6.3	250
0.20	8	12.5	500
0.40	16	25	1,000
0.80	32		

* Micrometers—μm.
† Microinches—μin.

TABLE 2   Preferred Series Maximum Waviness Height Values

mm*	in.†	mm	in.	mm	in.
0.0005	0.00002	0.008	0.0003	0.12	0.005
0.0008	0.00003	0.012	0.0005	0.20	0.008
0.0012	0.00005	0.020	0.0008	0.25	0.010
0.0020	0.00008	0.025	0.001	0.38	0.015
0.0025	0.0001	0.05	0.002	0.50	0.020
0.005	0.0002	0.08	0.003	0.80	0.030

* Millimeters.
† Inches.

by line if shown in profile or it may appear as a bounded area in a related view. Machined and ground surfaces, however, do not have the perfect smoothness represented on a drawing. Actually a surface has three dimensions—length, breadth, and curvature (waviness)—as illustrated in Fig. 13.93(*a*). In addition, there will be innumerable peaks and valleys of differing lengths, widths, and heights. An exaggerated profile of surface roughness is shown in (*b*). Combined waviness and roughness are illustrated in (*c*).

The following terms must be understood before the surface symbol shown in Fig. 13.98 can be properly applied:

Surface texture. This term refers to repetitive or random deviations from the nominal surface, which form the pattern on the surface. Included are roughness, waviness, lay, and flaws (Fig. 13.94).

Roughness. Roughness is the relatively finely spaced surface irregularities that are produced by the cutting action of tool edges and abrasive grains on surfaces that are machined.

Roughness height. Roughness height is the average (arithmetical) deviation from the mean line of the profile. It is expressed in micrometers (microinches) (Fig. 13.94). See Table 1.

Roughness width. Roughness width is the distance between successive peaks or ridges, which constitute the predominant pattern of roughness. Roughness width is measured in millimeters (inches) (Fig. 13.94).

Roughness width cutoff. This term indicates the greatest spacing of repetitive surface irregularities to be included in the measurement of average roughness height. It is measured in inches or millimeters (Fig. 13.94).

Waviness. Waviness is the surface undulations that are of much greater magnitude than the roughness irregularities. Waviness may result from machine or work deflec-

tions, vibrations, warping, strains, or similar causes. See Table 2.

Waviness height. Waviness height is the peak-to-valley distance (Fig. 13.94). It is rated in inches or millimeters.

Waviness width. Waviness width (rated in inches or millimeters) is the spacing of successive wave valleys or wave peaks (Fig. 13.94).

Flaws. Flaws are irregularities, such as cracks, checks, blowholes, scratches, and so forth, that occur at one place or at relatively infrequent or widely varying intervals on the surface (Fig. 13.94).

Lay. Lay is the predominant direction of the tool marks of the surface pattern (Fig. 13.99).

Microinch. A *microinch* is one-millionth (.000001) of an inch (μin.). A micrometer is one-millionth of a meter (μm).

13.38 Designation of Surface Characteristics (Fig. 13.98)

A surface whose finish is to be specified should be marked with the finish mark having the general form of a check mark (✔) so that the point of the symbol shall be on the line representing the surface, on the extension line, or on a leader pointing to the surface. Good practice dictates that the long leg and the extension shall be to the right as the drawing is read. Figure 13.95 illustrates the specification of roughness, waviness, and lay by listing rating values on the symbol.

Surface texture symbols used to control surface irregularities are shown in Fig. 13.96. The recommended proportions for the basic symbol are given in (*a*). When the removal of material by machining is necessary, a horizontal bar is added at the top of the short leg (*b*). In (*c*) the added value indicates the amount of stock to be removed. This value may be given either in millimeters or in inches. When material removal is prohibited, a small circle is added in the vee (*d*). This circle indicates that the surface is to be produced without removal of material by some production process such as die casting, powder metallurgy,

injection molding, forging, or extruded shape. The surface texture symbol in (*e*) is used when surface characteristics are to be specified above or below the horizontal line or to the right of the symbol as illustrated in Fig. 13.98.

Surface texture values are applied to the surface texture symbol as illustrated in Fig. 13.97. Roughness average rating (maximum value) is placed above the vee to the left of the long leg (*a*). See Table 1. Maximum waviness height rating is the first rating placed above the horizontal line (*c*). See Table 2. It is followed by the maximum waviness spacing rating. Lay designation is indicated by positioning the proper lay symbol to the right of the long leg as shown in (*f*) and (*h*). When it is necessary to specify maximum roughness spacing, the value is placed at the right of the lay symbol (*h*). The value for roughness sampling length (cutoff rating) is placed below the horizontal line as shown in (*g*).

If the nature of the preferred lay is to be shown, it will be indicated by the addition of a combination of lines, as shown in Fig. 13.98. Parallel or perpendicular lines indicate that the dominant lines on the surface are parallel or perpendicular to the boundary line of the surface in contact with the symbol (Fig. 13.99).

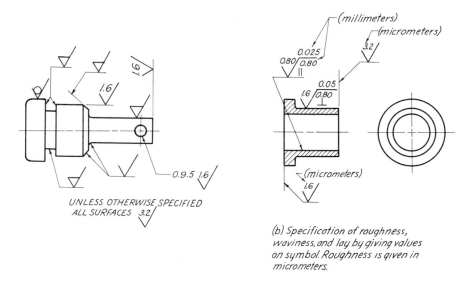

(b) Specification of roughness, waviness, and lay by giving values on symbol. Roughness is given in micrometers.

FIG. 13.95 Application of surface texture symbols to a drawing of a machine part.

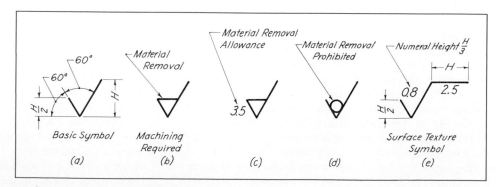

FIG. 13.96 Surface texture symbols.

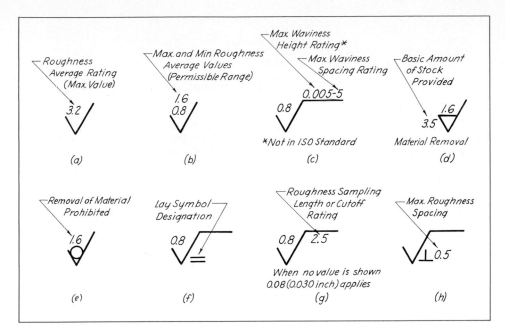

FIG. 13.97 Application of surface texture values to symbol.

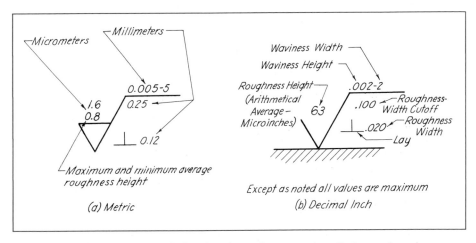

FIG. 13.98 Surface texture symbol. Values for surface control applied—metric and decimal inch.

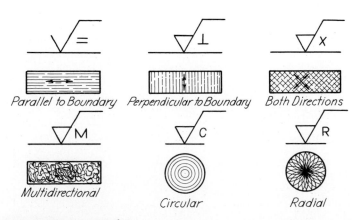

FIG. 13.99 Lay notations.

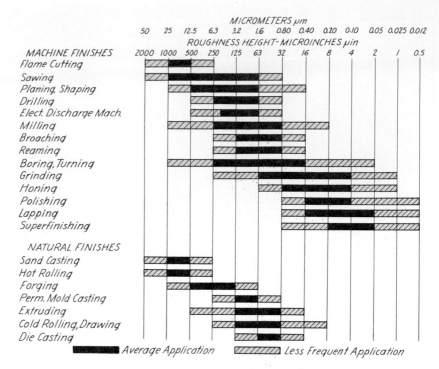

FIG. 13.100 **Surface finishes expected from common production methods.** (Micrometers, μm., and microinches, μin.)

The chart in Fig. 13.100 shows the expected surface roughness in microinches and micrometers for surfaces produced by common production methods.

The surface-quality symbol, which is used only when it is desirable to specify surface smoothness, should not be confused with a finish mark (√), which indicates the removal of material. A surface-quality symbol might be used for a surface on a die casting, forging, or extruded shape where the surface is to have a natural finish and no material is to be removed.

Surface finish should be specified only by experienced persons because the function of many parts does not depend on the smoothness quality of a surface or surfaces. In addition, surface quality need not be necessarily indicated for many parts that are produced to close dimensional tolerances because a satisfactory surface finish may result from the required machining processes. It should be remembered that the cost of producing a part will generally become progressively greater as the specification of surface finish becomes more exacting.

PROBLEMS

The following problems offer the student the opportunity to apply the rules of dimensioning given in this chapter. If it is desirable, either millimeters or decimals of an inch may be used in place of fractions. Use Table 3 in the Appendix.

1,2. (Figs. 13.101 and 13.102). Reproduce the given views of an assigned part. Determine the dimensions by transferring them from the drawing to one of the open-divided scales by means of the dividers.

3–9. (Figs. 13.103–13.109). Make a fully dimensioned multiview sketch or drawing of an assigned part. Draw all necessary views. Give a detail title with suitable notes concerning material, number required, etc. These parts have been selected from different fields of industry—automotive, aeronautical, chemical, electrical, etc.

10. (Fig. 13.110). Make a fully dimensioned multiview sketch or drawing of the rocker arm. Show a detail section taken through the ribs. *Supplementary information*: (1) The distance from the center of the shaft to the center of the hole for the pin in 4.00 in. The distance from the shaft to the threaded hole is 4.50 in. (2) The nominal diameter of the hole for the shaft is 1.875 in. The hole in the rocker arm is to be reamed for a definite fit. Consult your instructor. The diameter of the pin is .969 in. (3) The diameter of the threaded boss is 2.00 in. (4) The diameter of the roller is 2.25 in., and its length is 1.46 in. Total clearance between the roller and finished faces is to be .03 in. (5) The inside faces of the arms are to be milled in toward the hub far enough to accommodate the roller. (6) The rib is .62 in. thick. (7) The lock nut has $1\frac{1}{4}$-12 UNF thread. (8) Fillets and rounds .12 in. R except where otherwise noted.

11. (Fig. 13.111). Make a fully dimensioned drawing of an assigned part of the shaft support.

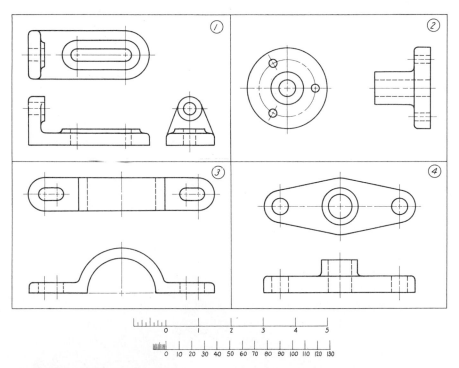

FIG. 13.101 Dimensioning problems.

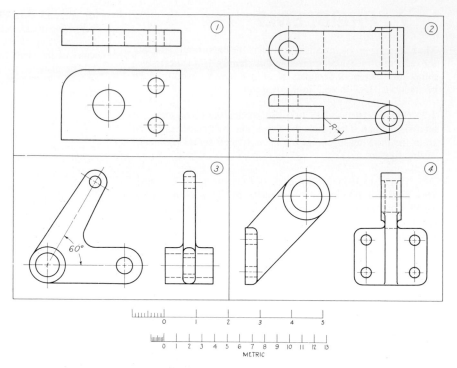

FIG. 13.102   Dimensioning problems.

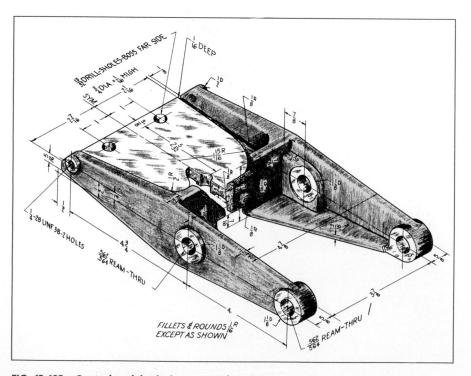

FIG. 13.103   Control pedal: airplane control system.

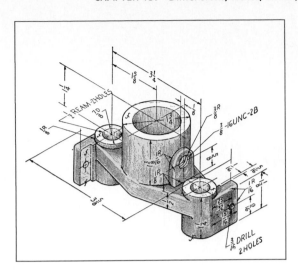

FIG. 13.104   Guide bracket.

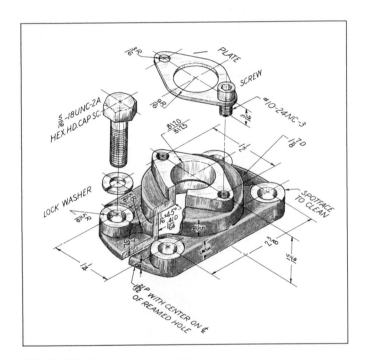

FIG. 13.105   Cover: mixing machine.

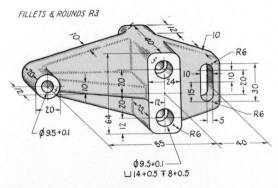

FIG. 13.106   Elevator bracket.

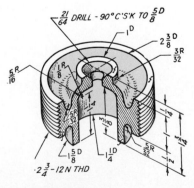

FIG. 13.107   Valve seat.

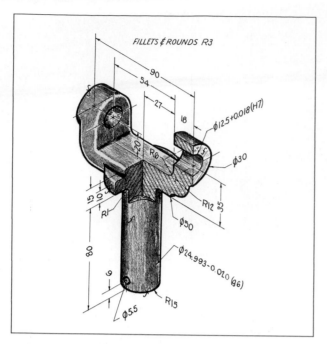

FIG. 13.108 **Yoke.**

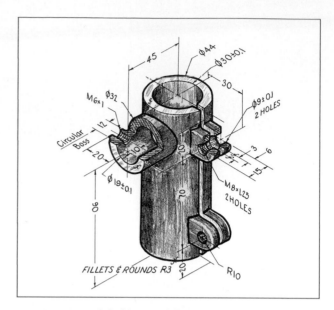

FIG. 13.109 **Torch holder—welding.**

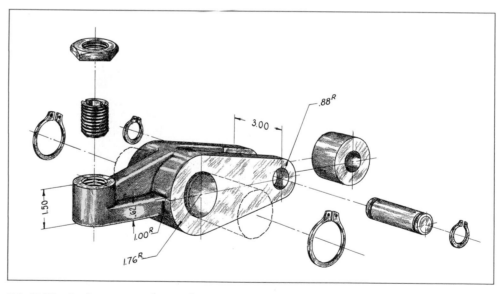

FIG. 13.110 **Rocker arm: marine engine.**

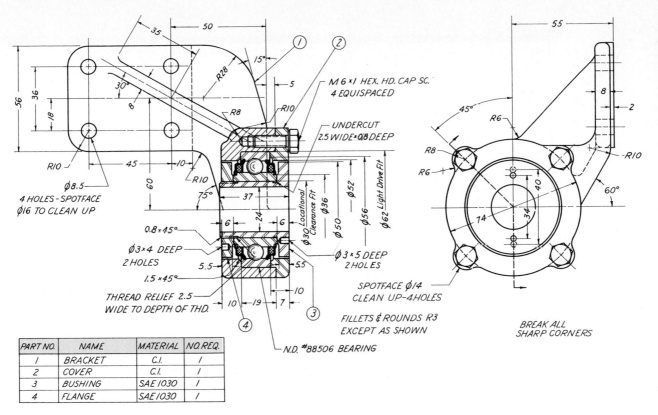

FIG. 13.111 **Shaft support.**

PART NO.	NAME	MATERIAL	NO. REQ.
1	BRACKET	C.I.	1
2	COVER	C.I.	1
3	BUSHING	SAE 1030	1
4	FLANGE	SAE 1030	1

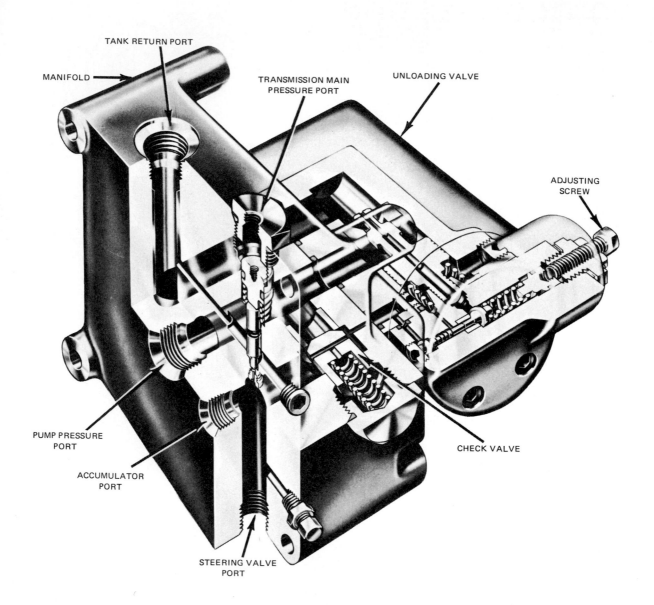

TANK RETURN PORT

MANIFOLD

TRANSMISSION MAIN
PRESSURE PORT

UNLOADING VALVE

ADJUSTING
SCREW

PUMP PRESSURE
PORT

ACCUMULATOR
PORT

CHECK VALVE

STEERING VALVE
PORT

This manifold and unloading control valve is part of the hydraulic steering and braking circuits. It maintains the hydraulic system at a constant pressure. Units of this type require many threaded parts and the use of springs, bearings, and fasteners. (*Courtesy General Motors Corporation*)

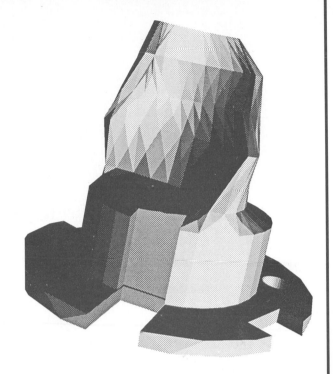

CHAPTER · 14

Fastening and Connecting Methods for Assembly

A / SCREW THREADS

14.1 Introduction

In the commercial field, where the practical application of engineering drawing takes the form of working drawings, knowledge of screw threads and fasteners is important. There is always the necessity for assembling parts either with permanent fastenings such as rivets, or with bolts, screws, and so forth, which may be removed easily.

Engineers, detailers, and draftsmen must be completely familiar with the common types of threads and fastenings, as well as with their use and the correct methods of rep-

resentation, because of the frequency of their occurrence in structures and machines. Information concerning special types of fasteners may be obtained from manufacturers' catalogues.

Young technologists in training should study Fig. 14.1 to acquaint themselves with the terms commonly associated with screw threads.

14.2 Threads

The principal uses of threads are (1) for fastening, (2) for adjusting, and (3) for transmitting power. To satisfy most of the requirements of the engineering profession, the var-

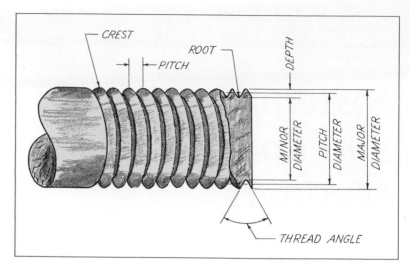

FIG. 14.1 Screw-thread nomenclature.

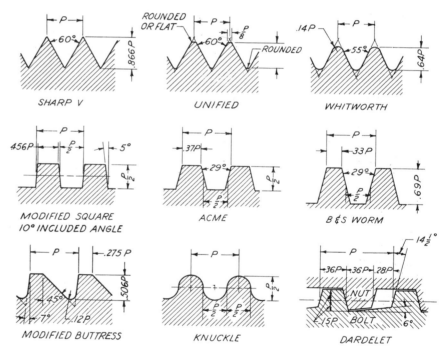

FIG. 14.2 Screw threads.

ious forms of threads shown in Fig. 14.2 are used.

The *Unified thread* has essentially the same basic profile as the ISO (International Organization for Standardization) thread form shown in Fig. 14.5. It is the thread form in general use in the United States.

The *sharp V thread* is used where adjustment and holding power are essential. For the transmission of power and motion, the modified square, Acme, and Brown and Sharpe *worm threads* have been adopted. The modified square thread, which is now rarely used, transmits power parallel to its axis. A further modification of the square thread is the stronger Acme, which is easier to cut and more readily disengages split nuts (as lead screws on lathes). The Brown and Sharpe worm thread, with similar

proportions but with longer teeth, is used for transmitting power to a worm wheel.

The *knuckle thread*, commonly found on incandescent lamps, plugs, and so on, can be cast or rolled.

The Whitworth and *buttress threads* are not often encountered by the average engineer. The former, which fulfills the same purpose as the American Standard thread, is used in England but is also frequently found in this country. The buttress or breech-block thread, which is designed to take pressure in one direction, is used for breech mechanisms of large guns and for airplane propeller hubs. The thread form has not been standardized and appears in different modified forms.

The *Dardelet thread* is self-locking in assembly.

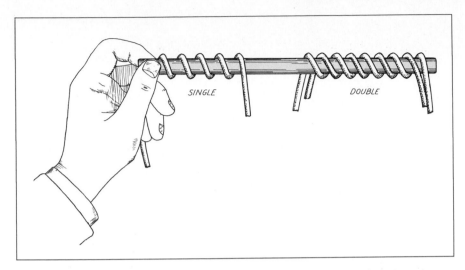

FIG. 14.3 Demonstration of a helix, related to a single-start and double (2-START) screw thread.

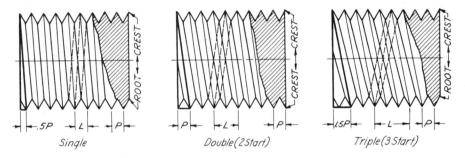

FIG. 14.4 Single-start and multiple-start threads.

14.3 *Multiple-start Threads*

Whenever a quick advance is desired, as on fountain pens, valves, and so on, two or more threads are cut side by side. Two threads form a double (2-START) thread; three, a triple (3-START) thread; and so on. A thread that is not otherwise designated is understood to be a single-start thread.

Fig. 14.3 shows wires wound around a rod for the purpose of demonstrating single-start and double (2-START) threads. The center line of the single line, representing the single thread, assumes the form of a helix. In the case of the double (2-START) thread, it should be noted that there are two wires side by side that are shaded differently for clarity. On a double (2-START) thread, each thread starts diametrically opposite the other one.

In drawing a single or an odd-number multiple-start thread, a crest is always diametrically opposite a root; in a double or other even-number multiple-start thread, a crest is opposite a crest and a root opposite a root (Fig. 14.4).

14.4 *Right-hand and Left-hand Threads*

A right-hand thread advances into a threaded hole when turned clockwise; a left-hand thread advances when turned counterclockwise. They can be easily distinguished by the thread slant. A right-hand thread on a horizontal shank always slants upward to the left (\) and a left-hand, upward to the right (/). A thread is always considered to be right-hand if it is not otherwise specified. A left-hand thread is always marked LH on a drawing.

14.5 *Pitch*

The pitch of a thread is the distance from any point on a thread to the corresponding point on the adjacent thread, measured parallel to the axis, as shown in Fig. 14.1.

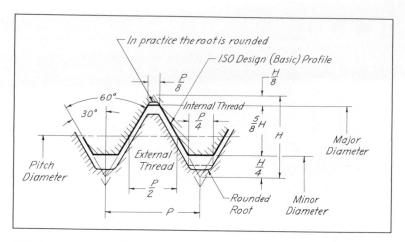

FIG. 14.5   ISO metric internal and external thread design profiles.

14.6   Lead

The lead of a screw may be defined as the distance advanced parallel to the axis when the screw is turned one revolution (Fig. 14.4). For a single thread, the lead is equal to the pitch; for a double (2-START) thread, the lead is twice the pitch; for a triple (3-START) thread, the lead is three times the pitch, and so on.

14.7   ISO Metric Screw Threads

Some years ago an ISO Committee was organized to investigate the possibility of establishing a single international system for screw threads.

Two systems were accepted, a new SI Metric Standard System and the Unified Inch System developed for use in the United States, Great Britain, and Canada. Fortunately for all concerned, the two thread standards are somewhat alike, with both using essentially the same basic thread profile. See Fig. 14.5. The principal differences between these two recognized standards are in the basic sizes, the magnitude and application of allowances and tolerances, and the method for designating and specifying threads. The ISO Metric Screw Thread Standard Series may be found in Table 10 in the Appendix.

14.8   Designation of ISO Metric Screw Threads (Fig. 14.6)

In general, ISO metric threads will be specified in this text using only the basic designation. The basic designation consists of the letter M followed by the nominal size (basic major diameter in millimeters) and then the pitch in millimeters (see Tables 10 and 11 in the Appendix), the nominal size and pitch being separated by the sign ×.

EXAMPLES

M8 × 1 (ISO designation)
M20 × 1.5

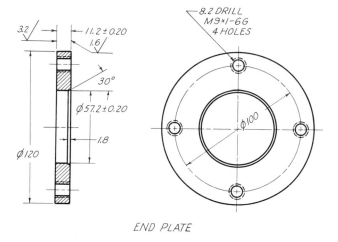

END PLATE

FIG. 14.6   Detail drawing with ISO (metric) thread specification.

For coarse series threads the indication of the pitch may be omitted. However, it is recommended in a number of company engineering standards that this option be disregarded. Therefore, the student should follow instructor's recommendations and specify, say, a 10-mm coarse thread as either M10 or M10 × 1.5, whichever is suggested.

The complete designation of an ISO metric thread includes the basic designation previously discussed, followed by an identification for the tolerance class. The tolerance-class designation includes the symbol for the pitch-diameter tolerance followed by the symbol for the crest-diameter tolerance. As can be observed from the examples that follow, each of these symbols consists of a number indicating the tolerance grade followed by a letter indicating the tolerance position. The tolerance-class designation is separated from the basic designation by a dash.

EXAMPLE

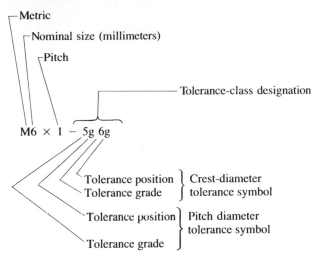

When the pitch-diameter and crest-diameter tolerance symbols are identical, the symbol is only given once. It is not repeated.

EXAMPLE

M6 × 1 — 6g

Pitch-diameter and crest-diameter tolerance symbols (equal)

For the principal thread elements (pitch diameter and crest diameter) the ISO standard provides a number of tolerance grades that reflect various magnitudes. Basically, three Metric Tolerance Grades are recommended by ISO. These are grades 4, 6, and 8. Grade 6 is commonly used for general-purpose threads with normal lengths of engagement. As might be expected, grade 6 is the closest ISO grade to our Unified 2A and 2B fits.

Under this system, the number of the tolerance grades reflect the size of the tolerance. Tolerances below grade 6 are smaller and are specified for fine-quality requirements or short lengths of engagement. Tolerances above grade 6 are larger and are therefore recommended for coarse quality or long lengths of engagement.

Tolerance positions establish the maximum material limits of the pitch and crest diameters for both internal and external threads. The series of tolerance-position symbols as established by ISO to reflect varying amounts of allowance are as follows:

For external threads (bolts, etc.):
Lowercase e—large allowance
Lowercase g—small allowance
Lowercase h—no allowance

For internal threads (nuts, etc.):
Uppercase G—small allowance
Uppercase H—no allowance

A desired fit between mating threads may be specified by giving the internal thread tolerance-class designation followed immediately by the external thread tolerance class. The two designations are separated by a slash (/), as shown by the example.

EXAMPLE

M6 × 1–6H/6g (or, since this is a coarse thread, as M6–6H/6g)

Metric screw thread designations in the United States are now based on ANSI B1.13M, ANSI B1.21M, and Federal Standard H28/21. The metric screw threads approved by ANSI are the already discussed M profile design that conforms to the ISO basic profile and the MJ profile metric screw thread.

The MJ metric screw thread has a modified ISO 68 profile design. This thread with the standard classes 4h6h compares with inch class 3A, while 4H6H (for sizes up to 5 mm diameter) and 4H5H (6 mm diameter and up) compare to the inch class 3B.

EXAMPLES

MJ8 × 1.25 4h6h
MJ8 × 1.25 4H5H

A detail drawing of a plug having a tapered metric thread is shown in Fig. 14.7. It should be noted that the taper has been specified as 1:16.

Unless otherwise designated metric threads are considered to be right-hand. Left-hand threads must bear the designation –LH.

EXAMPLE

M6 × 1–LH

The designation of multiple-start threads should have both the lead and pitch given with their values.

EXAMPLE

M24 × L6–P3–4h6h (TWO STARTS)

ANSI Y14.6aM–1981 should be consulted when preparing thread specifications for threads having a special length of engagement (LE), threads having modified crests or special rounded roots, and for both M and MJ metric threads that are to be either coated or plated.

14.9 Metric Threads for Production of Screws, Bolts, and Nuts

The standard for fastener threads recognizes only one series of diameter-pitch combinations, as listed here.

M1.6 × 0.35	M4 × 0.7
M2 × 0.4	M5 × 0.8
M2.5 × 0.45	M6 × 1
M3 × 0.5	M8 × 1.25
M3.5 × 0.6	M10 × 1.5
M12 × 1.75	M48 × 5
M14 × 2	M56 × 5.5
M16 × 2	M64 × 6
M20 × 2.5	M72 × 6
M24 × 3	M80 × 6

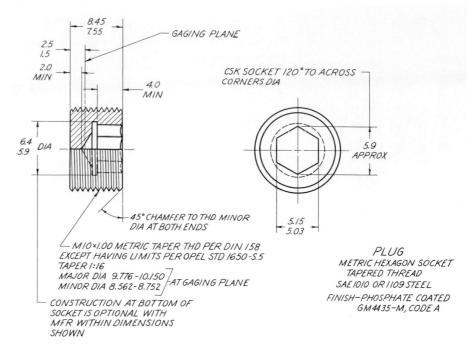

FIG. 14.7 **Part with a tapered metric thread.** (*Courtesy Standards Section, General Motors Corporation*)

M30 × 3.5   M90 × 6

M36 × 4   M100 × 6

M42 × 4.5

These 25 thread sizes, adopted for metric fasteners, cover the size range of the inch series from No. 0 (.060) through 4 inches.

14.10 Detailed Screw-thread Representation

The true representation of screw threads by helical curves, requiring unnecessary time and laborious drafting, is rarely used. The detailed representation, closely approx-imating the actual appearance, is preferred in commercial practice, for it is much easier to represent the helices with slanting lines and the truncated roots and crests with sharp Vs (Fig. 14.8). Since detailed rendering is also time consuming, its use is justified only in those few cases where appearance and permanency are important factors and when it is necessary to avoid the possibility that confusion might result from the use of one of the symbolic methods. The presentation of a detailed representation is a task that belongs primarily to a draftsman, the engineer being concerned only with specifying that this form be used.

The steps in drawing a metric or Unified thread are shown in Figs. 14.9 and 14.10.

The stages in drawing the detailed representation of modified square and Acme threads are shown in Figs.

FIG. 14.8 **Detailed representation.**

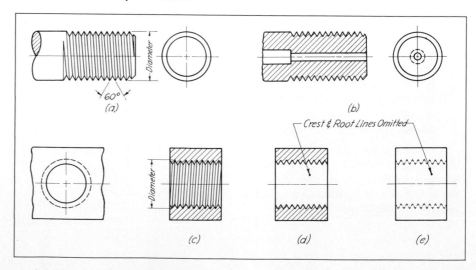

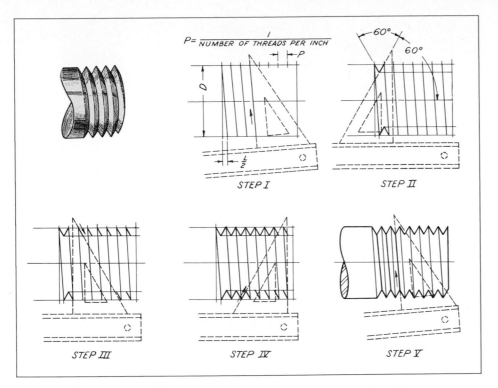

FIG. 14.9 **Detailed representation of Unified, ISO metric, and sharp V-threads (external).**

14.11, 14.12, and 14.13. All lines of the finished square thread are made the same weight. The root lines of the Acme thread may be made heavier than the other lines.

14.11 American National Standard Conventional Thread Symbols (Fig. 14.14)

To save valuable time and expense in the preparation of drawings, the American National Standards Institute has adopted the "schematic" and "simplified" series of thread symbols to represent threads having a diameter of 25 mm (1 in.) or less.

The root of the thread for the *simplified representation* of an external thread is shown by invisible lines drawn parallel to the axis [Fig. 14.15(*a*)].

The *schematic representation* consists of alternate long and short lines perpendicular to the axis. Although these lines, representing the crests and roots of the thread, are not spaced to actual pitch, their spacing should indicate noticeable differences in the number of threads per inch of different threads on the same working drawing or group of drawings (Fig. 14.16). The root lines are made heavier than the crest lines (Fig. 14.17).

Before a hole can be tapped (threaded), it must be drilled to permit the tap to enter. See Tables 10 and 29 in the Appendix for tap drill sizes for standard threads. Since the last of the thread cut is not well formed or usable, the hole must be shown drilled and tapped deeper than the screw will enter [Fig. 14.18(*d*), (*e*), and (*f*)]. To show the

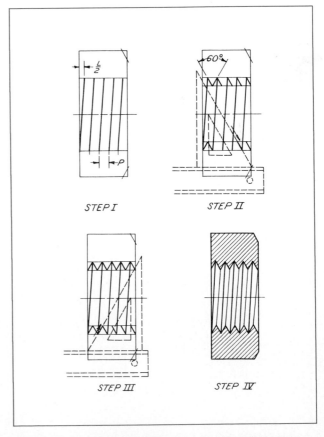

FIG. 14.10 **Detailed representation of Unified, ISO metric, and sharp V-threads (internal).**

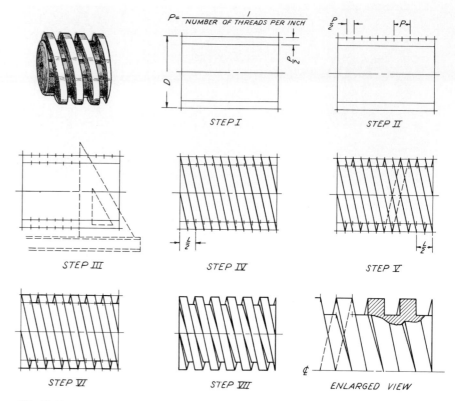

FIG. 14.11   Detailed representation of square threads (modified).

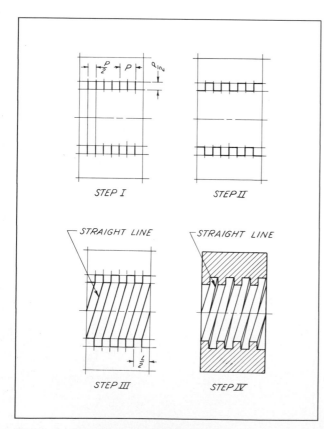

FIG. 14.12   Detailed representation of square threads (internal).

threaded portion extending to the bottom of the drilled hole indicates the use of a bottoming tap to cut full threads at the bottom. This is an extra and expensive operation not justified except in cases where the depth of the hole and the distance the screw must enter are limited [see Fig. 14.18(*g*), (*h*), and (*i*)].

Figure 14.19 shows a simplified method of representation for square threads.

14.12 *Threads in Section*

The detailed representation of threads in section, which is used for large diameters only, is shown in Fig. 14.10. Since the far side of an internal thread in section is visible, the crest and root lines incline in the opposite direction to those of an external thread having the same specifications.

The schematic and simplified representations for threads of small diameter are shown in Figs. 14.14 and 14.18.

A sectioned assembly drawing is shown in Fig. 14.20. When assembled pieces are both sectioned, the detailed representation is used, and the thread form is drawn. Schematic representation could have been used for the thread on the cap screw instead of the simplified representation shown (Fig. 13.111).

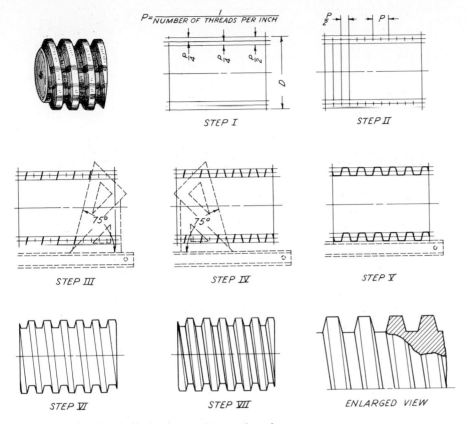

FIG. 14.13   **Detailed representation of Acme thread.**

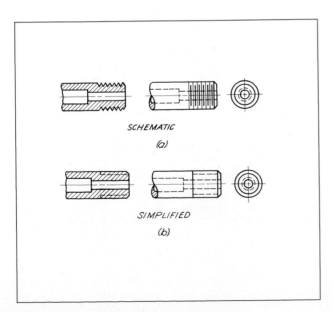

FIG. 14.14   **External thread representation.**

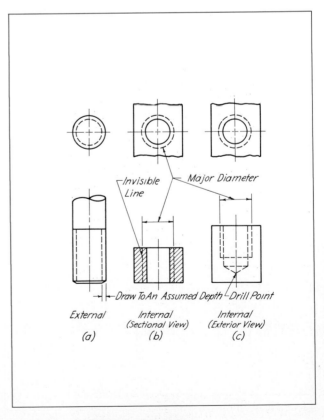

FIG. 14.15   **Simplified representation.**

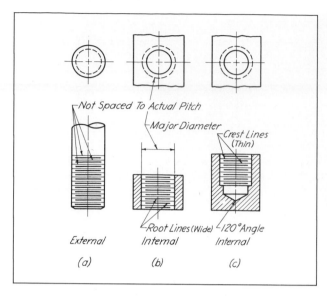

FIG. 14.16   Schematic representation.

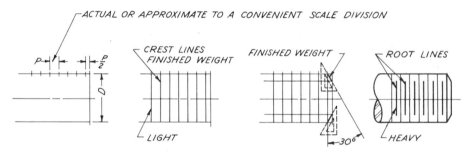

FIG. 14.17   Drawing conventional threads—schematic representation.

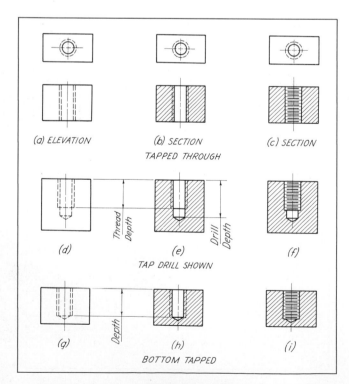

FIG. 14.18   Representation of internal threads.

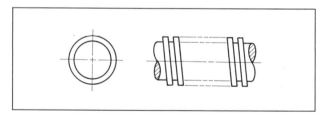

FIG. 14.19   Simplified representation of a square thread.

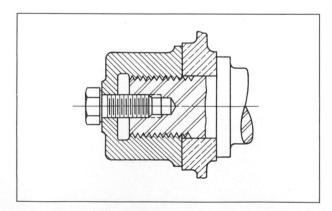

FIG. 14.20   Threads in section.

14.13 American-British Unified Thread

The Unified Thread Standard came into existence after the representatives of the United States, Great Britain, and Canada signed a unification agreement. This accord, which made possible the interchangeability of threads for these countries, created a new thread form (Fig. 14.21) that is a compromise between our own American National Standard design and the British Whitworth. The external thread of the new form has a rounded root and may have either a flat or rounded crest. The Unified thread is a general-purpose thread for screws, bolts, nuts, and other threaded parts (ANSI B1.1). See Sec. 14.2.

14.14 Unified and American Screw-thread Series

The Unified and American screw thread series consists of six series and a selection of special threads that cover special combinations of diameter and pitch. Each series differs from the other by the number of threads per inch for a specific diameter (see Tables 28 and 29 in the Appendix).

The *coarse-thread* series (UNC and NC) is designated UNC for sizes above $\frac{1}{4}$ in. in diameter. This series is recommended for general industrial use.

The *fine-thread* series (UNF and NF), designated UNF for sizes above $\frac{1}{4}$ in., was prepared for use when a fine thread is required and for general use in the automotive and aircraft fields.

The *extra-fine-thread* series (UNEF and NEF) is used for automotive and aircraft work when a maximum number of threads is required for a given length.

The *8-thread* series (8N) is a uniform pitch series for large diameters. It is sometimes used in place of the coarse-thread series for diameters greater than 1 in. This series was originally intended for high-pressure joints.

The *12-thread* series (12UN or 12N) is a uniform pitch series intended for use with large diameters requiring threads of medium-fine pitch. This series is used as a continuation of the fine-thread series for diameters greater than $1\frac{1}{2}$ in.

The *16-thread* series (16UN or 16N) is a uniform pitch series for large diameters requiring a fine-pitch thread. This series is used as a continuation of the extra-fine-thread series for diameters greater than 2 in.

14.15 Unified and American Screw-thread Classes

Classes of thread are determined by the amounts of tolerance and allowance specified. Under the new unified system, classes 1A, 2A, and 3A apply only to external threads; classes 1B, 2B, and 3B apply to internal threads.

Class 2 and Class 3 fits are defined as

Class 2 fit. Represents a high quality of commercial thread product and is recommended for the great bulk of interchangeable screw-thread work.

Class 3 fit. Represents an exceptionally high quality of commercially threaded product and is recommended only in cases where the high cost of precision tools and continual checking is warranted.

14.16 Identification Symbols for Unified Screw Threads

Threads are specified under the unified system by giving the diameter, number of threads per inch, initial letters (UNC, UNF, etc.) and class of thread (1A, 2A, and 3A; or 1B, 2B, and 3B) (see Fig. 14.22).

Unified and American National threads are specified on drawings, in specifications, and in stock lists by thread information given as shown in Fig. 14.23. A multiple-start thread is designated by specifying in sequence the nominal size, pitch, and lead.

14.17 Thread Dimensioning

In general, the thread length dimension shown on a drawing should be the length of the complete (full-form) threads. The incomplete threads should be beyond this dimensioned length.

14.18 Square Threads

Square threads can be completely specified by a note. The nominal diameter is given first, followed by the number of threads per inch and the type of thread (see Fig. 14.23).

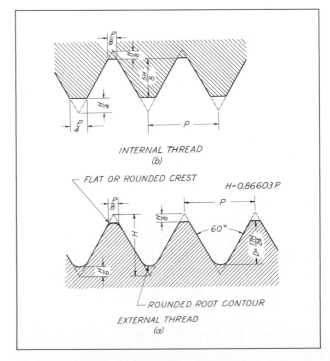

FIG. 14.21 American-British Unified thread.

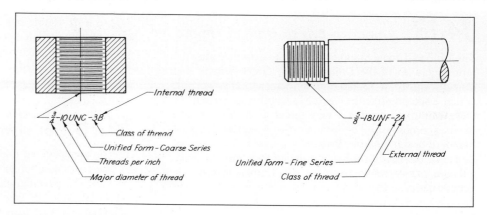

FIG. 14.22 Unified thread identification symbols.

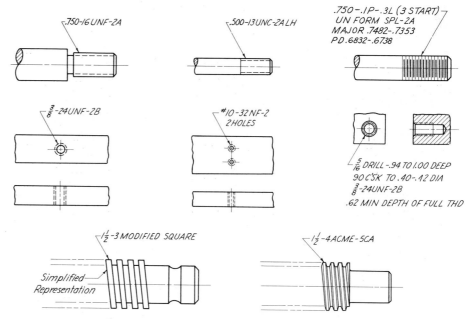

FIG. 14.23 Thread identification symbols.

14.19 Acme and Stubb Acme Threads

Acme threads have been standardized in a preferred (single) series of diameter-pitch combinations by the American National Standards Institute (B1.5–1952). The standard provides two types of Acme threads, the general-purpose and centralizing. The three classes of general-purpose threads (2G, 3G, and 4G) have clearances on all diameters for free movement. The five classes of centralizing threads (2C, 3C, 4C, 5C, and 6C) have a limited clearance at the major diameters of the external and internal threads to ensure alignment of mating parts and to prevent wedging on the flanks of the thread.

Acme threads are specified by giving the diameter, number of threads per inch, type of thread, and class (Fig. 14.23). The letter A or B is added to indicate an external or an internal thread.

EXAMPLES

1⅜–4 ACME–2GA or 1.375–4 ACME–2GA
1¾–4 ACME–2GB–LH
¾–6 ACME–4CB or .750–6 ACME–4CB
2¾–3 ACME–3GA–2-START (2-START indicates a double thread)

14.20 Buttress Threads

The buttress thread is designed for exceptionally high stress along the axis of the thread in one direction only. No pitch-diameter series has been recommended, because of the need for special design of most components. The ANSI B1.9 standard covers three classes of buttress threads. These are 1A, 2A, and 3A for external threads and 1B, 2B, and 3B for internal threads.

Buttress threads are specified by giving, in order, the diameter, threads per inch, type of thread (National Buttress), and class.

EXAMPLES

$\frac{5}{8}$–20 BUTTRESS–2A
10–10 BUTTRESS–3B (2-START) (Optionally, these threads may be designated by giving the pitch followed by the letter P instead of the number of threads per inch.)
.625–.05P BUTTRESS–2A

B / FASTENERS

14.21 Metric Fasteners

Metric fasteners are standard in one metric thread series (see Sec. 14.9). The thread pitches are between those of the coarse-thread and fine-thread series of the present Unified (inch) threads.

Except for metric socket screws, all standard fastener products have one tolerance grade, 6g for externally threaded parts (bolts, machine screws, etc.) and 6H for nuts. The 6H/6g metric threads closely match our present 2A/2B threads for fasteners in that there is an allowance on the external thread and no allowance on the internal thread. See Sec. 14.8. Socket screws have 4g6g threads.

Except for machine and tapping screws, metric screws and bolts have standard thread lengths based on D, the nominal diameter. These thread lengths are

2D + 6 mm for fastener lengths up to 125 mm

2D + 12 mm for lengths over 125 mm up to 200 mm

2D + 25 mm for lengths over 200 mm

If the length of the fastener is shorter than the standard thread length for its diameter, it should normally be threaded for its full length.

14.22 Standard Bolts and Nuts (Fig. 14.24)

Commercial producers of bolts and nuts manufacture their products in accordance with standard specifications given in approved standards.

The present American Standard covers the specifications for three series of bolts and nuts:

1. *Regular series.* The regular series was adopted for general use.

2. *Heavy series.* Heavy bolt heads and nuts are designed to satisfy the special commercial need for greater bearing surface.

3. *Light-series nuts.* Light nuts are used under conditions requiring a substantial savings in weight and material. They are usually supplied with a fine thread.

The amount of machining is the basis for further classification of hexagonal bolts and nuts in both the regular and heavy series as unfinished and semifinished.

Square-head bolts and nuts are standardized as unfinished only.

Unfinished heads and nuts are not washer-faced, nor are they machined on any surface.

Semifinished bolt heads and nuts are machined or treated on the bearing surface to provide a washer face for bolt heads and either a washer face or a circular bearing surface for nuts. Nuts, not washer-faced, have the circular bearing surface formed by chamfering the edges.

Bolts and nuts are *always* drawn across corners in all views. This recognized commercial practice, which violates the principles of true projection, prevents confusion of square and hexagonal forms on drawings.

The chamfer angle on the tops of heads and nuts is 30° on hexagons and 25° on squares, but both are drawn at 30° on bolts greater than 1 in. in diameter.

■ SPECIFICATION OF AMERICAN STANDARD HEX BOLTS

American Standard bolts are specified in parts lists and elsewhere by giving the diameter, number of threads per inch, series, class of thread, length, finish, and type of head.

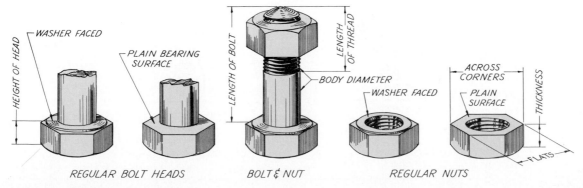

FIG. 14.24 Bolts and nuts.

EXAMPLES

$\frac{1}{2}$–13 UNC–2A $\times$ 1$\frac{3}{4}$
SEMIFIN HEX HD BOLT

Frequently it is advantageous and practical to abbreviate the specification as

$\frac{1}{2}$ $\times$ 1$\frac{3}{4}$ UNC SEMIFIN HEX HD BOLT

Bolt lengths can be considered as $\frac{1}{8}$ in. for bolts $\frac{1}{4}$–3 in. in length, $\frac{1}{4}$ in. for bolts $\frac{3}{4}$–3 in. in length, and $\frac{1}{2}$ in. for bolts 3–6 in. in length. Length increments for square-head bolts are $\frac{1}{8}$ in. for bolts $\frac{1}{4}$–$\frac{3}{4}$ in. in length, and $\frac{1}{4}$ in. for bolts $\frac{3}{4}$–4$\frac{3}{4}$ in. in length.

The minimum thread length for bolts up to and including 6 in. in length will be twice the diameter plus $\frac{1}{4}$ in. For lengths over 6 in. the minimum thread length will be twice the diameter of the bolt plus $\frac{1}{2}$ in. (ANSI B18.2).

■ SPECIFICATION OF METRIC HEX BOLTS

Hex bolts are specified in parts lists and elsewhere by giving in sequence the nominal size, thread pitch, nominal length, material, product name, and, if required, the protective coating.

EXAMPLES

M10 $\times$ 1.5 $\times$ 100 HEX BOLT

M24 $\times$ 3 $\times$ 50 STAINLESS STEEL HEX BOLT

14.23 To Draw Bolt Heads and Nuts

Using the head height dimension taken from the tables, draw the lines representing the top and contact surfaces of the head or nut and the diameter of the bolt. Lay out a hexagon about an inscribed chamfer circle having a diameter equal to the distance across the flats (Fig. 14.25) and project the necessary lines to block in the view. Draw in the arcs after finding the centers, as shown in Fig. 14.25.

A square-head bolt or nut may be drawn by following the steps indicated in Fig. 14.26.

The engineer and experienced drafter wisely resort to some form of template for drawing the views of a bolt head or nut (see Fig. 2.12). To draw the views as shown in Figs. 14.25 and 14.26 consumes valuable time needlessly.

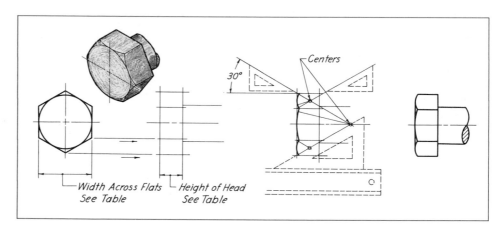

FIG. 14.25 Steps in drawing a hexagonal bolt head.

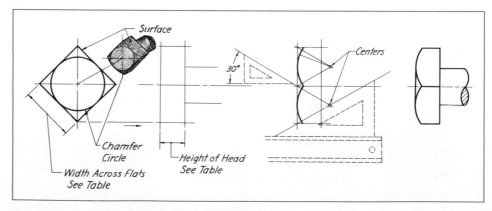

FIG. 14.26 Steps in drawing a square bolt head.

14.24 *Studs*

Studs, or stud bolts, which are threaded on both ends as shown in Fig. 14.27, are used where bolts would be impractical and for parts that must be removed frequently (cylinder heads, pumps, and so on). They are first screwed permanently into the tapped holes in one part before the removable member with its corresponding clearance holes is placed in position. Nuts are used on the projecting ends to hold the parts together.

Since most studs are not standard they must be produced from specifications given on a detail drawing. In dimensioning a stud, the length of thread must be given for both the stud end and nut end along with an overall dimension. The thread information is given by note.

In a bill of material, studs may be specified as follows.

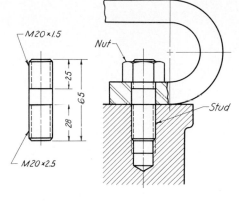

FIG. 14.27 Stud bolt.

■ *UNIFIED INCH STUDS*

EXAMPLES

$\frac{1}{2}$–13 UNC–2A $\times$ $2\frac{3}{4}$ STUD

It is good practice to abbreviate the specification as

$\frac{1}{2}$ $\times$ $2\frac{3}{4}$ STUD

■ *METRIC STUDS*

EXAMPLE

DOUBLE-END STUD, M10 $\times$ 1.5 $\times$ 100, STEEL, CADMIUM PLATED

14.25 *Cap Screws (Fig. 14.28)*

American Standard cap screws are available in four standard heads, usually in finished form. When parts are assembled, the cap screws pass through clear holes in one member and screw into threaded holes in the other (Fig. 14.29). American Standard hexagonal cap screws have a washer face $\frac{1}{64}$ in. thick with a diameter equal to the distance across flats. All cap screws 25 mm (1 in.) or less in length are threaded very nearly to the head.

American Standard cap screws are specified by giving the diameter, number of threads per inch, series, class of thread, length, and type of head.

■ *AMERICAN STANDARD CAP SCREWS*

EXAMPLES

$\frac{5}{8}$–11 UNC–2A $\times$ 2 FIL HD CAP SC

It is good practice to abbreviate the specification as

$\frac{5}{8}$ $\times$ 2 UNC FIL HD CAP SC

■ *METRIC CAP SCREWS*

EXAMPLES

M6 $\times$ 1 $\times$ 40 HEX CAP SCREW

M14 $\times$ 2 $\times$ 80 HEX CAP SC, CADMIUM PLATED

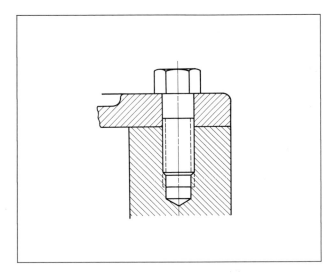

FIG. 14.28 Hexagonal-head cap screw.

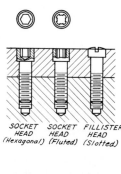

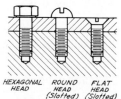

FIG. 14.29 American Standard cap screws.

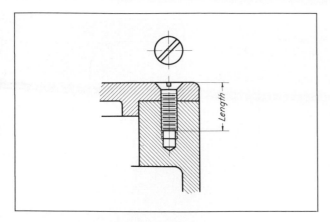

FIG. 14.30 Use of a machine screw.

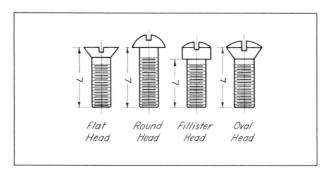

FIG. 14.31 Types of machine screws.

14.26 Machine Screws

American Standard machine screws, which fulfill the same purpose as cap screws, are used chiefly for small work having thin sections (Fig. 14.30). Under the approved American National Standard they range from No. 0 (.060 in. diameter) to $\frac{3}{4}$ in. (.750 in. diameter) and are available in either the American Standard Coarse or Fine-Threaded Series. The four forms of heads shown in Fig. 14.31 have been standardized.

Metric machine screws may have either slotted or cross recess drives with flat countersunk heads, oval countersunk heads, and pan heads. In addition, hex head and hex washer head machine screws are available to the designer. They may be purchased in ten sizes in a range from M2 through M12. Table 19 in the Appendix gives the head dimensions for four types.

Machine screws M3 and smaller have a thread length of 25 mm. Larger diameters have 38 mm of thread length.

■ AMERICAN STANDARD MACHINE SCREWS

To specify American Standard machine screws, give the diameter, threads per inch, thread series, class of thread, length, and type of head.

EXAMPLES

No. 12–24 NC–3 × $\frac{3}{4}$ FIL HD MACH SC

It is good practice to abbreviate by omitting the thread series and class of fit.

No. 12–24 × $\frac{3}{4}$ FIL HD MACH SC

■ METRIC MACHINE SCREWS

Metric machine screws are specified in parts lists and elsewhere by giving in sequence the nominal size, thread pitch, nominal length, product name (including type of head), material, and protective finish, if needed.

EXAMPLES

M10 × 1.5 × 40 SLOTTED FLAT HEAD MACHINE SCREW

M12 × 1.75 × 80 HEX HEAD MACHINE SCREW

14.27 Commercial Lengths: Studs, Cap Screws, Machine Screws

Unless a fastening of any of these types carries a constant and appreciable fatigue stress, the usual practice is to have it enter a distance related to its nominal diameter (Fig. 14.32). If the depth of the hole is not limited, it should be drilled to a depth of 1 diameter beyond the end of the fastener to permit tapping to a distance of $\frac{1}{2}$ diameter below the fastener.

The length of the fastening should be determined to the nearest commercial length that will allow it to fulfill minimum conditions. In the case of a stud, care should be taken that the length allows for a full engagement of the nut. Commercial lengths for fasteners increase by the following increments.

Standard length increments

INCH INCREMENTS

For fastener lengths $\frac{1}{4}$–1 in.: $\frac{1}{8}$ in.

For fastener lengths 1–4 in.: $\frac{1}{4}$ in.

MILLIMETER INCREMENTS

Standard lengths—short fasteners: 8 mm, 12 mm, 14 mm, 16 mm, and 20 mm

For fastener lengths: 20 mm–100 mm = 5 mm increments. 110 mm–200 mm = 10 mm increments.

It is generally recognized that the shortest practical length for any metric fastener should not be less than 1.5 times its diameter. For example, an M8 metric fastener should be 12 mm or longer and an M24 fastener not less than 35 mm in length.

For fastenings and other general-purpose applications, the engagement length should be equal to the nominal diameter (*D*) of the thread when both components are of steel. For steel external threads in cast iron, brass, or bronze, the engagement length should be 1.5*D*. When assembled into aluminum, zinc, or plastic, the engagement should be 2*D*.

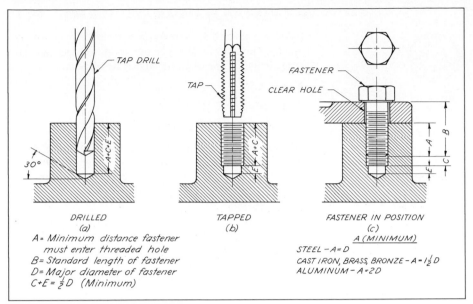

DRILLED
(a)
A = Minimum distance fastener
    must enter threaded hole
B = Standard length of fastener
D = Major diameter of fastener
$C + E = \frac{1}{2}D$ (Minimum)

TAPPED
(b)

FASTENER IN POSITION
(c)
A (MINIMUM)
STEEL — A = D
CAST IRON, BRASS, BRONZE — A = $1\frac{1}{2}$ D
ALUMINUM — A = 2D

FIG. 14.32 Threaded hole and fastener.

14.28 Set Screws

Set screws are used principally to prevent rotary motion between two parts, such as that which tends to occur in the case of a rotating member mounted on a shaft. A set screw is screwed through one part until the point presses firmly against the other part (Fig. 14.33).

The several forms of safety heads shown in Fig. 14.34 are available in combination with any of the points. Headless set screws comply with safety codes and should be used on all revolving parts. The many serious injuries that have been caused by the projecting heads of square-head set screws have led to legislation prohibiting their use in some states [Fig. 14.33(c)]. Tables 20 and 35 in the Appendix give dimensions for metric and American Standard set screws.

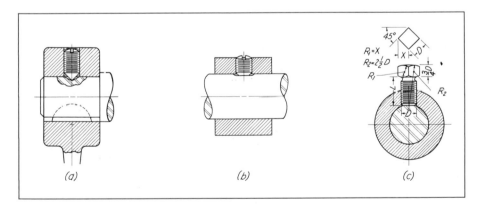

FIG. 14.33 Use of set screws.

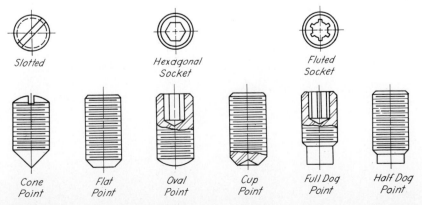

Slotted

Hexagonal Socket

Fluted Socket

Cone Point

Flat Point

Oval Point

Cup Point

Full Dog Point

Half Dog Point

FIG. 14.34 Set screws.

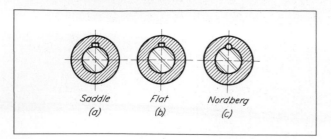

FIG. 14.35   Light-duty keys.

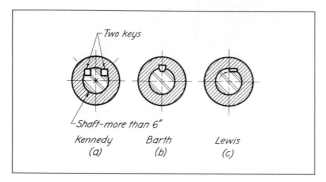

FIG. 14.36   Heavy-duty keys.

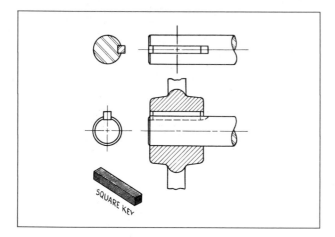

FIG. 14.37   Square key.

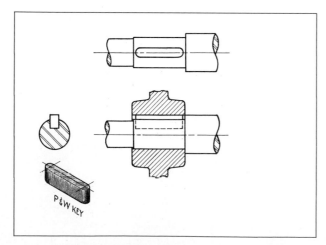

FIG. 14.38   Pratt and Whitney key.

■ *AMERICAN STANDARD SET SCREWS*

American Standard set screws are specified by giving the diameter, number of threads per inch, series, class of thread, length, type of head, and type of point.

EXAMPLES

¼–20 UNC–2A × ½

SLOTTED CONE PT SET SC

The preferred abbreviated form gives the diameter, number of threads per inch, length, type of head, and type of point.

¼–20 × ½ HEX SOCKET CONE PT SET SC

■ *METRIC SET SCREWS*

EXAMPLES

M6 × 1 × 12 SLOTTED CONE PT SET SC

M8 × 1.25 × 14 HEX SOCKET

FLAT PT SET SC

14.29 Keys

Keys are used in the assembling of machine parts to secure them against relative motion, generally rotary, as is the case between shafts, cranks, wheels, and so on. When the relative forces are not great, a round key, saddle key, or flat key is used (Fig. 14.35). For heavier duty, rectangular keys are more suitable (Fig. 14.36).

The square key (Fig. 14.37) and the Pratt and Whitney key (Fig. 14.38) are the two keys most frequently used in machine design. A plain milling cutter is used to cut the keyway for the square key, and an end mill is used for the Pratt and Whitney keyway. Both keys fit tightly in the shaft and in the part mounted on it.

The gib-head key (Fig. 14.39) is designed so that the head remains far enough from the hub to allow a drift pin to be driven to remove the key. The hub side of the key is tapered ⅛ in. per ft to ensure a fit tight enough to prevent both axial and rotary motion. For this type of key, the keyway must be cut to one end of the shaft.

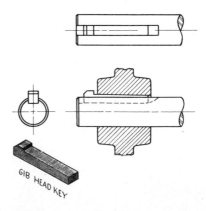

FIG. 14.39   Gib-head key.

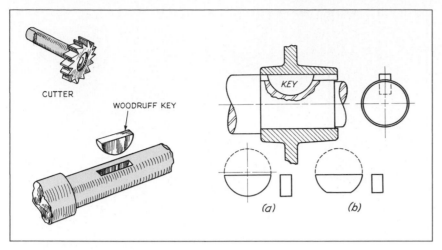

FIG. 14.40 Woodruff key.

14.30 *Woodruff Keys*

A Woodruff key is a flat segmental disc with either a flat or a round bottom (Fig. 14.40). It is always specified by a number, the last two digits of which indicate the nominal diameter in eighths of an inch, while the digits preceding the last two give the nominal width in thirty-seconds of an inch.

A practical rule for selecting a Woodruff key for a given shaft is as follows: Choose a standard key that has a width approximately equal to one-fourth of the diameter of the shaft and a radius nearly equal (plus or minus) to the radius of the shaft. Table 44 in the Appendix gives the dimensions for American Standard Woodruff keys. Table 24 lists dimensions for metric keys.

When Woodruff keys are drawn, it should be remembered that the center of the arc is placed above the top of the key at a distance shown in column E in the table.

14.31 *Taper Pins*

A taper pin is commonly used for fastening collars and pulleys to shafts, as illustrated in Fig. 14.41. The hole for the pin is drilled and reamed with the parts assembled. When a taper pin is to be used, the drawing callout should read as follows:

DRILL AND REAM FOR #4 TAPER PIN WITH PC #6 IN POSITION

Drill sizes and exact dimensions for taper pins are given in Table 45 in the Appendix.

14.32 *Locking Devices*

A few of the many types of locking devices that prevent nuts from becoming loose under vibration are shown in Figs. 14.42 and 14.43.

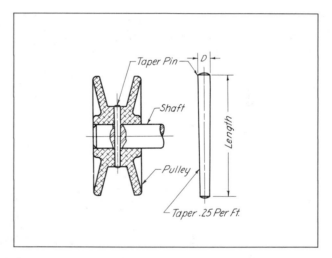

FIG. 14.41 Use of a taper pin.

Figure 14.42 shows six forms of patented spring washers. The ones shown in (D), (E), and (F) have internal and external teeth.

In common use is the castellated nut with a spring cotter pin that passes through the shaft and the slots in the top [Fig. 14.43(a)]. This type is used extensively in automotive and aeronautical work.

Figure 14.43(b) shows a regular nut that is prevented from loosening by an American National Standard jam nut.

In Fig. 14.43(c) the use of two jam nuts is illustrated.

A regular nut with a spring-lock washer is shown in Fig. 14.43(d). The reaction provided by the lock washer tends to prevent the nut from turning.

A regular nut with a spring cotter pin through the shaft, to prevent the nut from backing off, is shown in Fig. 14.43(e).

Special devices for locking nuts are illustrated in Fig. 14.43(f) and (g). A set screw may be held in position with a jam nut, as in (h).

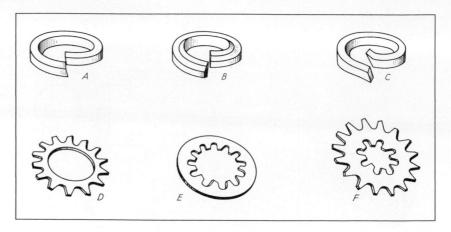

FIG. 14.42   Special lock washers.

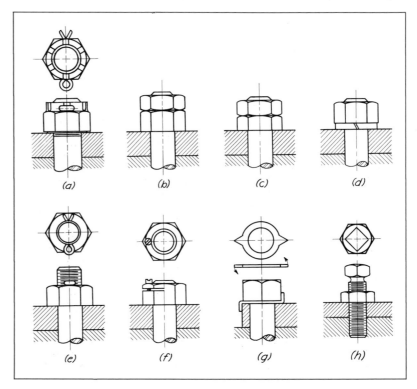

FIG. 14.43   Locking schemes.

14.33   *Areo Thread*

The Areo-thread (helicoil) screw-thread system allows the use of high-strength cap screws and studs in light soft metals, such as aluminum and magnesium, through the use of a phosphor bronze or stainless steel coilspring lining in the tapped hole, as shown in Fig. 14.44. This coil (screw bushing) is formed to fit a modified American Standard thread. Special tools are needed for inserting the coil in the tapped hole.

14.34   *Miscellaneous Bolts, Screws, and Nuts*

Other types of bolts and screws that have been adopted for commercial use are illustrated in Fig. 14.45.

Wood screws have threads proportioned for the holding strength of wood. They are available with different forms of heads (flat, round, and oval).

Some of the fastenings shown in Fig. 14.45 have been standardized by the American National Standards Institute.

FIG. 14.44   Aero thread.

14.35 *Phillips Head*

The Phillips head, shown in Fig. 14.46 for a wood screw, is one of various types of recessed heads. Although special drivers are usually employed for installation, an ordinary screwdriver can be used. Machine screws, cap screws, and many special types of fasteners are available.

C / RIVETING

14.36 *Rivets*

Rivets are permanent fasteners used chiefly for connecting members in such structures as buildings and bridges and for assembling steel sheets and plates for tanks, boilers, and ships. They are cylindrical rods of wrought iron or soft steel, with one head formed when manufactured. A head is formed on the other end after the rivet has been put in place through the drilled or punched holes of the mating parts. A hole for a rivet is generally drilled,

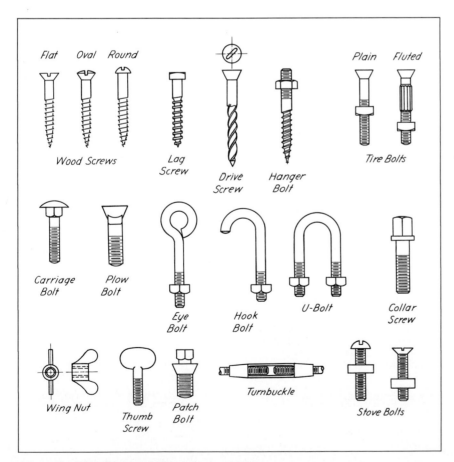

FIG. 14.45   Miscellaneous bolts, screws, and nuts.

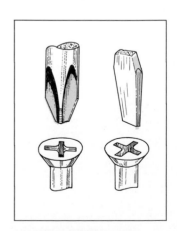

FIG. 14.46   Phillips head screw.

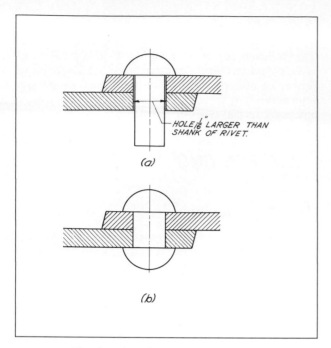

HOLE $\frac{1}{16}$" LARGER THAN
SHANK OF RIVET.

(a)

(b)

FIG. 14.47 **Riveting procedure.**

punched, or punched and reamed $\frac{1}{16}$ in. larger than the diameter of the shank of the rivet [Fig. 14.47(a)]. Figure 14.47(b) illustrates a rivet in position. Small rivets, less than $\frac{1}{2}$ in. in diameter, may be driven cold, but the larger sizes are driven hot. For specialized types of engineering work, rivets are manufactured of chrome-iron, aluminum, brass, copper, and so on. Standard dimensions for small rivets are given in Table 40 in the Appendix.

The type of rivets and their treatment are indicated on drawings by the American National Standard conventional symbols shown in Fig. 25.12.

14.37 Riveted Joints

Joints on boilers, tanks, and so on, are classified as either *lap joints* or *butt joints* (Fig. 14.48). Lap joints are generally used for seams around a circumference. Butt joints are used for longitudinal seams, except on small tanks where the pressure is to be less than 100 lb per sq in.

D/ WELDING

14.38 Welding Processes

For convenience, the various welding processes used in commercial production may be classified into three types: *pressure processes*, *nonpressure processes*, and *casting processes*. The nonpressure processes are arc welding and gas welding. Metallic arc welding is the joining of two pieces of metal through the use of a sustained arc formed between the work and a metal rod held in a holder (Fig. 14.49). The intense heat melts the metal of the work and at the same time heats the end of the electrode, causing small globules to form and cross the arc to the weld. In gas welding, the heat is produced by a burning mixture of two gases, which ordinarily are oxygen and acetylene. The weld is formed by melting a filler rod with the torch flame, along the line of contact, after the metal of the work has been preheated to a molten state. This method is essentially a puddling process, in that the weld is produced by a small moving molten pool that is maintained by the flame constantly directed on it. Resistance welding is a pressure process, the fusion being made through heat and mechanical pressure. The work is heated by a strong electrical current that passes through it until fusion temperature is reached; then pressure is applied to create the weld.

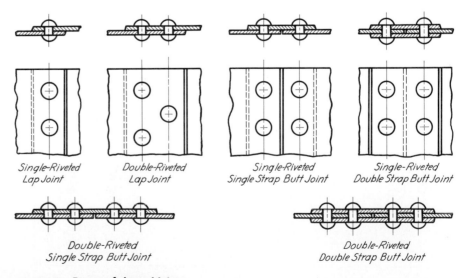

Single-Riveted
Lap Joint

Double-Riveted
Lap Joint

Single-Riveted
Single Strap Butt Joint

Single-Riveted
Double Strap Butt Joint

Double-Riveted
Single Strap Butt Joint

Double-Riveted
Double Strap Butt Joint

FIG. 14.48 **Forms of riveted joints.**

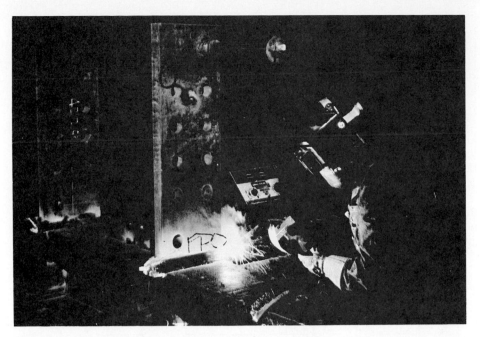

FIG. 14.49 **Arc welding.** (*Courtesy Lincoln Electric Company*)

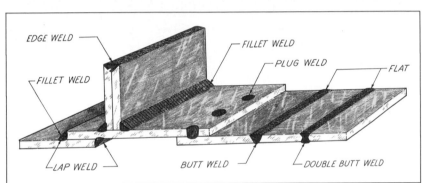

Lighter structure
sheet stronger than cast
(rolled)
TIME SAVE

EDGE WELD

FILLET WELD

FILLET WELD

PLUG WELD

FLAT

LAP WELD

BUTT WELD

DOUBLE BUTT WELD

FIG. 14.50 **Types of welds.**

The forms of resistance welding are projection welding, seam welding, spot welding, and flash welding. In spot welding, the parts are overlapped and welds are made at successive single spots. A seam weld is similar to a spot weld, except that a continuous weld is produced. In projection welding, one part is embossed and welds are made at the successive projections. In making a flash weld, the two pieces to be joined are held end to end in jaws and act as electrodes. At the right instant, after the facing metal has been heated by the arc across the gap, the power is shut off and the two ends are forced together to cool in a fused position.

Thermit welding can be considered a casting process, in that molten iron is run into a mold built around the parts at the point at which they are to be connected. The liquid metal is obtained from a mixture of finely divided iron oxide and aluminum, which is ignited in a crucible. In the chemical reaction that takes place, the oxygen passes from the iron oxide to the aluminum, leaving free molten iron that flows into the mold around the preheated parts forming the joint. The metal of the members being welded fuses with the liquid metal and forms a weld when the joint is cool.

14.39 *Types of Welds*

Figure 14.50 illustrates in pictorial mode some of the various types of welds. Cross-sectional views of the fundamental welds that are commonly encountered are shown in Fig. 14.51.

14.40 *Classification of Weld Joints*

Welded joints are classified in accordance with the method of assembly of the parts at a joint. See Figure 14.52.

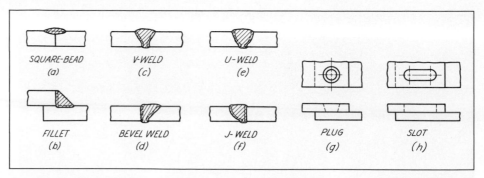

FIG. 14.51 Types of welds.

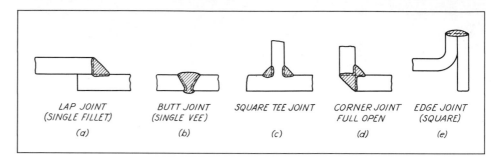

FIG. 14.52 Types of welded joints.

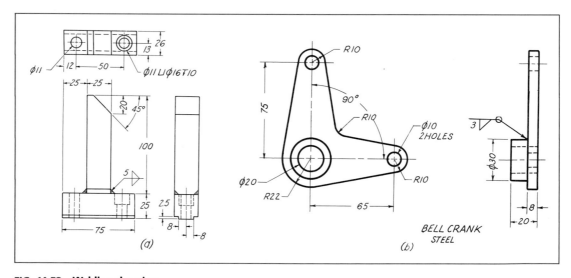

FIG. 14.53 Welding drawing.

14.41 Working Drawings of Welded Parts

Figure 14.53 shows parts that have been constructed by welding rolled plates and bar stock. It should be noted that each joint is completely specified through the use of a welding symbol. A careful study will show that the drawings, except for the absence of fillets and rounds and the fact that properly composed welding symbols are directed to the necessary joints, are very much like casting drawings.

A satisfactory welding design may be produced by a competent designer who possesses a fair amount of ingenuity, and the necessary drawing can be made by any draftsman who has a thorough understanding of the use of the symbols recommended in ANSI Y32.3.

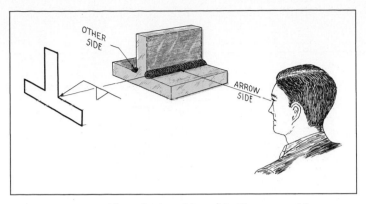

FIG. 14.54 Arrow-side and other-side welds. The arrow-side is considered to be the near side.

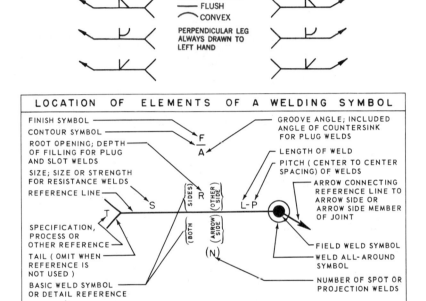

FIG. 14.55 Basic welding symbol.

14.42 Arrow-side and Other-side Welds (Fig. 14.54)

In the present system the joint is the basis of reference. Any joint the welding of which is indicated by a symbol will always have an *arrow side* and an *other side*.

On a drawing, when a joint is represented by a single line (see Fig. 14.54) and the arrow is directed to the line, the arrow-side is considered to be the near side for the reader of the drawing. The side opposite becomes the other side. In applying fillet and groove welding symbols, where the arrow portion connects the welding symbol reference line (shaft) to one side of the joint, this side is accepted as the arrow side of the joint. For plug, slot, and projection welding symbols, where the arrow from the reference line must necessarily be directed to the outer surface of one of the two members at the center line of the required weld,

the member to which the arrow points is recognized as being the arrow-side member. This, of course, would be the near-side to the reader. In general, it can be said that spot, seam, and flash or upset symbols have no arrow-side or other-side significance. However, supplementary symbols, sometimes used with them, can have such significance.

14.43 Welding Symbols

An enlarged drawing of the approved welding symbol is shown in Fig. 14.55, along with explanatory notes that indicate the proper locations of the marks and size dimensions necessary for a complete description of a weld.

The arrow is the basic portion of the symbol, as shown

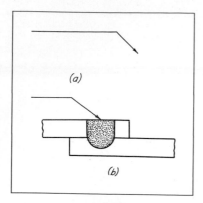

FIG. 14.56   Welding arrow.

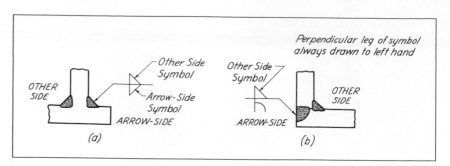

FIG. 14.57   Location of welding symbols.

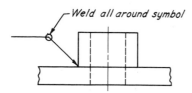

FIG. 14.58   Weld-all-around symbol.

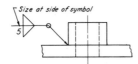

FIG. 14.59   Method of specifying the size of weld (metric).

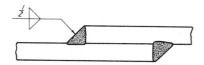

FIG. 14.60   Dimensioning a weld.

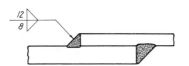

FIG. 14.61   Dimensioning a weld (metric).

in Fig. 14.56(*a*). It points toward the joint where the required weld is to be made, as in Fig. 14.56(*b*).

If the weld is on the arrow-side, the symbol indicating the type of weld is placed below or to the right of the base line, depending on whether that line is horizontal or vertical (Fig. 14.57). If the weld is located on the other side, the symbol should be above or to the left.

To indicate that a weld is to be made all around a connection, as is necessary when a piece of tubing must be welded to a plate, a weld-all-around symbol, a circle is placed as shown in Fig. 14.58.

The size of a weld is given along the base of the arrow, at the side of the symbol, as shown in Fig. 14.59. If the welds on the arrow-side and the other-side of a lap joint are the same size, only one dimension should be given (Fig. 14.60). If they are not the same size, each dimension should be placed beside its associated symbol (Fig. 14.61).

The welding terms associated with the specification of the size of welds are illustrated in Fig. 14.62.

Figure 14.63 shows the common types of single-groove welds and the related symbol for each. The symbols for double-groove welds are illustrated in Fig. 14.64.

The same symbol is used for both plug and slot welds. As illustrated in Fig. 14.65, a hole is formed in one member to receive the weld. When the hole or slot is in the member on the arrow-side, the symbol must be placed below the reference line of the symbol arrow; if in the other-side member, the symbol must be placed above the reference line.

The needed size specifications for plug and slot welds are placed as shown in Fig. 14.66. In the case of plug

FIG. 14.62   Welding terms.

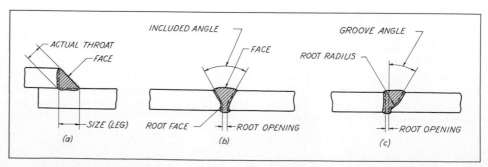

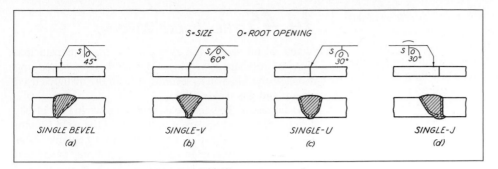

FIG. 14.63 Single-groove welds and symbols.

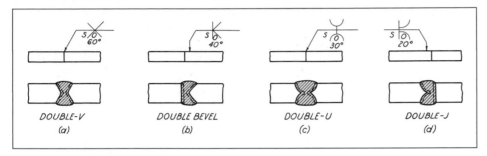

FIG. 14.64 Double-groove welds and symbols.

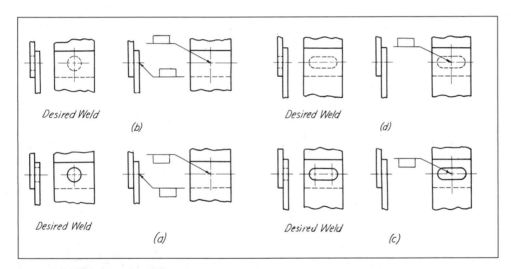

FIG. 14.65 Plug and slot welds.

FIG. 14.66 Size specification—plug and slot welds.

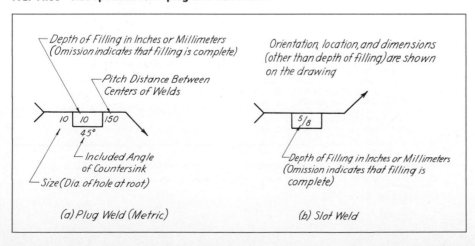

welds, the smallest diameter of the hole, if countersunk, is placed at the left of the symbol. The angle of countersink is added below the symbol, as shown in (a). However, the angle may be omitted if it is user's standard. When the depth of filling of a plug weld is to be less than complete, the depth of filling is entered inside the symbol. If no value is given, it is understood that the depth of filling is to be complete. To complete the size specifications, the pitch (distance between centers) of plug welds is added to the right of the symbol as shown in (a). When plug welds are to be approximately flush without finishing, a flush-contour symbol is added to the weld symbol. If the welds are to be made flush by some mechanical means, a standard user's finish symbol should be given with the flush-contour symbol.

The depth of filling of slot welds is indicated within the symbol in the same manner as for plug welds. Then, as noted in Fig. 14.66(b), all size and location dimensions must be shown on the drawing.

14.44   Gas and Arc Welding Symbols

To satisfy the need for a standard group of symbols that could be understood in all manufacturing plants, conventional symbols are used so that each symbol resembles in a general way, the type of weld it represents. Figure 14.67 shows a condensed table of symbols. With one exception the symbols shown are the same as those first proposed by the American Welding Society.

14.45   Resistance Welding

Figure 14.68 shows the symbols for the four principal types of resistance welding. The method of specifying resistance welds differs somewhat from the methods used for arc and gas welds.

In the case of the spot-weld symbol shown in Fig. 14.68, the size of the weld is given to the left of the weld symbol as a diameter. If desired, the minimum acceptable shear strength in Newtons or in pounds per spot may be entered in the same location. The pitch of welds is specified at the right of the symbol. Should it be desirable to indicate a specific number of welds for a joint, the number is given in parentheses either above or below the weld symbol.

Projection welds are specified in the same manner as spot welds. The size may also be given as a diameter or the minimum acceptable shear strength in Newtons or pounds per weld entered in its place. If it is desired that the exposed surface of one member be flush, the flush symbol may be added.

The size of a seam weld is its width that is given decimally to the left of the weld symbol in hundredths of an inch, as shown in Fig. 14.68. If it is thought to be desirable to specify the minimum acceptable shear strength in pounds or Newtons per linear unit, this may be substituted for the size of the weld at the left of the weld symbol. The weld length is placed at the right of the weld symbol. For intermittent seam welding, both the length of the weld and the pitch is placed in this position as illustrated.

Flash and upset weld symbols have been shown in Fig.

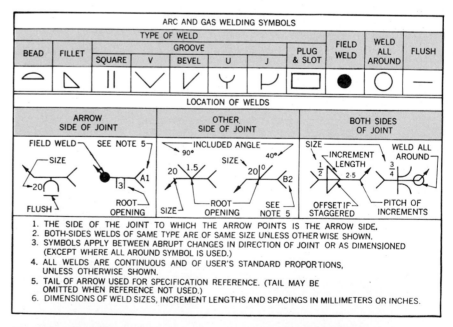

FIG. 14.67 American National Standard arc and gas welding symbols. Examples for arrow-side and other-side of joint show dimensions in millimeters.*

* Based on ANSI Y32.3.

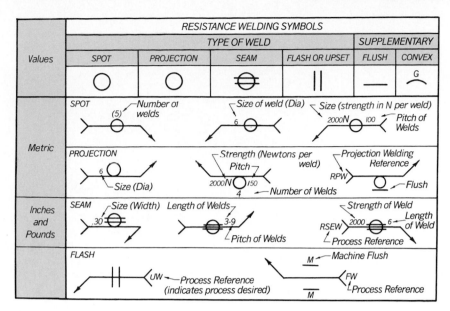

FIG. 14.68   American National Standard resistance welding symbols.*

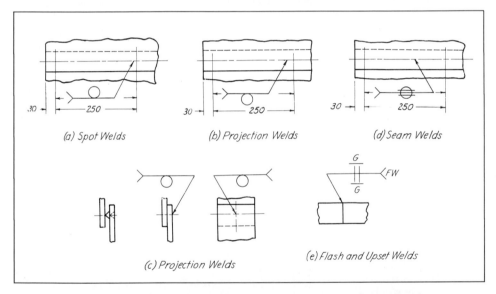

FIG. 14.69   Resistance welding—spot, projection, seam, and flash and upset welds.

14.68. No dimensions are shown. However, process reference (FW or UW) must be given in the tail of the symbol to indicate the process to be used. When flash and upset welds are to be made flush by some mechanical means, the complete symbol would consist of the weld symbol along with contour and machining symbols as shown. The use of symbols for resistance welding in conjunction with welding dimensions has been illustrated in Fig. 14.69.

* Based on ANSI Y32.3.

14.46 *Welded Machine Parts*

Many machine parts often can be constructed of welded rolled shapes at a much lower cost than if they were cast. This is due to the fact that the cost of the preparation of patterns is completely eliminated, less material is required, and labor costs are lower. A welded part is sometimes more desirable for a particular mechanism, because steel is stiffer, stronger in tension, and more resistant to fatigue stresses and sudden impact. Also, aside from the production of new parts, welding can be used to make a

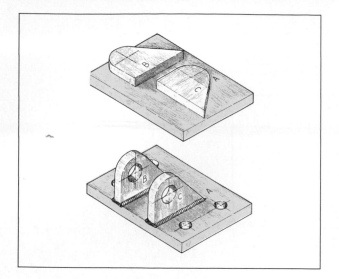

FIG. 14.70   Bracket-welded design.

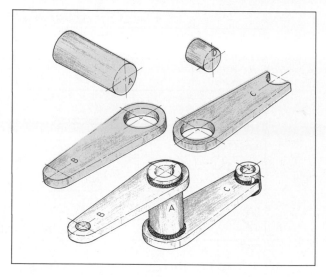

FIG. 14.71   Link-welded design.

machine part to replace a broken cast part when it is necessary to place a machine back in operation in the shortest possible time. Figure 14.70 shows a part that is constructed of plates. Figure 14.71 shows the construction of a link using plates and round bar stock.

As previously stated, a designer is limited only by his own ingenuity. Parts of all shapes and sizes may be produced of readily available rolled forms. Simple bearings, levers, cranks, clevises, gear arms, and even cams can be quickly and easily made.

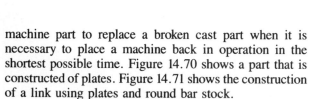 **PIPE THREADS AND FITTINGS**

14.47

Since piping is used in all types of construction of conveying fluids and gases such as oil, water, steam, and chemicals, some knowledge of it is essential not only for the draftsman making drawings but for the engineer who must select and use pipe in the design of machines, power plants, water systems, and so on. There are so many types of fittings and materials used for various purposes that only the most common can be discussed briefly in this chapter. Additional information may be obtained from publications of research associations and from the catalogs of manufacturers.

14.48   *Pipe Materials*

Cast-iron pipe is suitable for underground gas and water mains, plumbing lines, and low-pressure steam systems.

Steel pipe is used chiefly where high temperatures and high pressures are encountered. The addition of such alloys

as nickel, chromium, and the like makes the steel pipe more resistant to corrosion at high temperatures.

Seamless brass pipe is the most satisfactory type for hot-water lines, condenser tubes, and so on, but it is expensive and, therefore, is used only when conditions justify the extra cost.

For lines having turns and bends in inaccessible locations, copper tubing is frequently used. But copper pipe, even though it is flexible and can resist the corrosive action of chemicals, is not always practical. It cannot be used in any system subject to high temperatures and repeated stress.

Lead and lead-lined pipe is widely used for chemical work, particularly where piping is subject to the action of acids.

Galvanized pipe (ordinarily iron pipe that has been dipped in molten zinc to prevent rust) is suitable for lines conveying drinking water.

14.49   *Special Pipe and Tubing*

Special-purpose pipe and tubing, made from a variety of materials such as stainless steel, brass, bronze, aluminum, and plastics, are now being manufactured. Plastic pipe and plastic-lined pipe are being used extensively in the chemical industry in place of metal pipe because plastic does not corrode and has high resistance to a wide range of chemicals. Plastic pipe is easily bent and resists weathering. In the heavy weights it may be threaded. The principal disadvantages are its higher cost and temperature-pressure limitations. However, the higher price will be partially offset by lower installation costs. Where strength is needed, a metal pipe lined with plastic may be used.

Polyvinyl chloride (PVC) is the basic plastic material most widely used for pipe. PVC pipe is very light, has low flow resistance, and can be readily bent and easily assembled by solvent cementing.

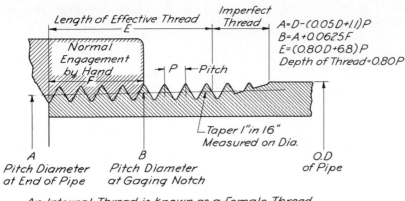

FIG. 14.72   American National taper pipe thread.

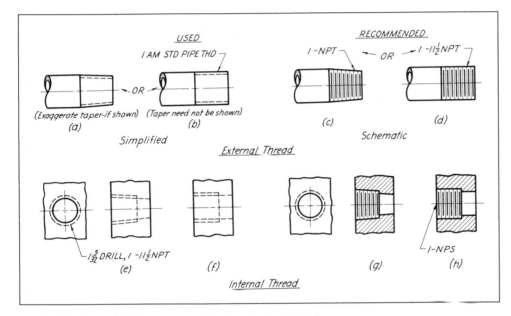

FIG. 14.73   Representation and specification of pipe threads.

14.50   American National Standard Pipe Thread (NPT) for General Use

The American National Standard pipe taper thread, illustrated in Fig. 14.72, is similar to the ordinary American National Standard thread and has the same thread angle; but it is tapered $\frac{1}{16}$ in. per in., to ensure a tight joint at a fitting. The crest is flattened and the root is filled in so that the depth of the thread is .80P. The number of threads per inch for any given nominal diameter can be obtained from Table 46 in the Appendix.

The distance a pipe enters a fitting is fixed for any nominal diameter. Numerical values for this distance may be determined from Table 46.

An American National Standard straight pipe thread, having the same number of threads per inch as the taper thread, is in use for pressure-tight joints for couplings,
for pressure-tight joints for grease and oil fittings, and for hose couplings and nipples. This thread may also be used for free-fitting mechanical joints. Usually a taper external thread is used with a straight internal thread, for pipe material is sufficiently ductile for an adjustment of the threads.

The NPT threads are for general use. When taper pipe threads are used and the joints are made up wrench-tight with a lubricant or sealer, they will be pressure tight.

14.51   Specification of Threads

In specifying pipe threads, the ANSI recommends that the note be formulated using symbolic letters as illustrated in Fig. 14.73. For example, the specification for a 1-in. standard taper pipe thread should read

1–NPT

The letters NPT, following the nominal diameter, indicate that the thread is American National Standard (N), pipe (P), taper (T) thread.

Continuing with the same scheme of using letters, the specification for a 1-in. straight pipe thread would read

1–NPS [American National Standard (N)–pipe (P)–Straight (S)]

The form of note given in Fig. 14.73(*b*), reading 1 AM STD PIPE THD, is quite commonly used in practice.

The letter symbols that have been adopted by the ANSI have the following significance.

N—American National Standard

P—pipe

T—taper

S—straight

C—coupling

H—hose coupling

L—lock nut

M—mechanical joints

R—railing

F—fuel and oil

I—intermediate

SPL—special

See Sec. 14.8 for specification of metric taper threads.

14.52   Threads for Joints and Couplings

Pipe threads are used for several types of joints. The significance of the letters designating the threads for these joints is given in Sec. 14.51. Where the letter S appears in the designation, a straight thread is indicated

NPTR—taper pipe thread for rail fittings. These threads are the same as the NPT threads except the external thread is shortened to permit use of the larger end of the pipe thread.

NPSC—internal straight pipe threads in pipe fittings. These threads are straight threads of the same thread form as the NPT thread shown in Fig. 14.72. NPSC threads form a pressure-tight joint when assembled with an external taper (NPT) pipe thread using a lubricant or sealer.

NPSH—straight pipe thread for loose-fitting mechanical joints for hose couplings. Ordinarily, straight internal and external threads are used for hose-coupling joints. One of several standards for hose threads is based on the American National Standard pipe thread.

NPSL—straight pipe threads for loose-fitting mechanical joints with lock nuts. These threads are straight threads of the same form as the NPT thread. The external NPSL thread has the largest diameter that can be cut on standard pipe. Usually, straight internal threads are used with these straight external threads.

NPSM—straight pipe threads for free-fitting mechanical joints for fixtures.

14.53   Dryseal (American National Standard) Pipe Threads (NPTF, NPSF, and NPSI)

With these threads, truncation at the crest and root is controlled to ensure metal-to-metal contact coincident with, or prior to, flank contact. This contact at root and crest makes a pressure-tight joint to prevent spiral leakage. Although a lubricant or sealer is not required, the use of a lubricant to minimize galling is not objectionable.

NPTF—taper pipe thread. This series of threads covers both external and internal threads. These threads are suitable for pipe joints in practically every type of service.

PTF-SAE SHORT—this series conforms to the NPTF series except that (1) the full thread length of the external thread is shortened by one thread at the small end, and (2) the full thread length of the internal thread is shortened by eliminating one thread at the large end.

NPSF—straight pipe thread (internal only) for fuel and oil. This series of threads is generally used for soft and ductile materials, which will adjust to the taper of external threads in assembly.

NPSI—intermediate straight pipe thread (internal only). This series of threads is suited for hard or brittle materials where the section is heavy and where there will be little expansion with the external taper threads in assembly.

Dryseal threads are specified as follows:

$\frac{1}{4}$—18 DRYSEAL NPTF

$\frac{1}{8}$—27 DRYSEAL NPSI

14.54   Drawing Pipe Threads

The taper on a pipe thread is so slight that it will not attract attention on a drawing unless it is exaggerated. If it is shown at all, it is usually magnified to $\frac{1}{8}$ in. per in.

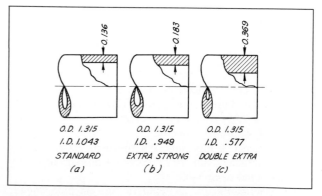

0.136   0.183   0.369

O.D. 1.315	O.D. 1.315	O.D. 1.315
I.D. 1.043	I.D. .949	I.D. .577
STANDARD	EXTRA STRONG	DOUBLE EXTRA
(a)	(b)	(c)

FIG. 14.74   Comparison of different weights of 1 in. wrought-iron pipe.

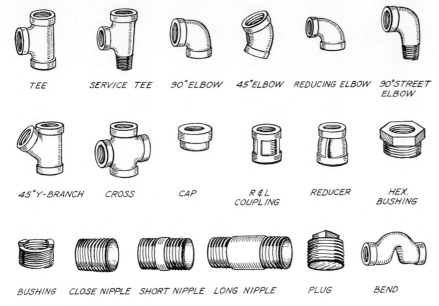

TEE	SERVICE TEE	90° ELBOW	45° ELBOW	REDUCING ELBOW	90° STREET ELBOW
45° Y-BRANCH	CROSS	CAP	R & L COUPLING	REDUCER	HEX. BUSHING
BUSHING	CLOSE NIPPLE	SHORT NIPPLE	LONG NIPPLE	PLUG	BEND

FIG. 14.75 Screwed fittings.

Pipe threads are generally represented by the same conventional symbols used for ordinary American National Standard thread. See Fig. 14.73.

14.55 Specification of Wrought-iron and Steel Pipe

The standardized weights commonly used are the standard, extra-strong, and double-extra-strong. All are specified by the nominal inside diameter.

The nominal inside diameter of standard pipe is less than the actual diameter, because early manufacturers made the wall thickness greater than necessary and, in correcting, took all of the excess from the inside to avoid altering the size of openings in fittings. Metal was added to the inside to increase wall thicknesses for the extra-strong and double-extra-strong. As a result, all three weights of pipe for any given nominal diameter have the same outside diameter and can be used with the same fittings.

Wrought-iron or steel pipe greater than 12 in. in diameter is specified by giving the outside diameter and the thickness of the wall.

Fig. 14.74 illustrates the relative wall thickness of 1-in. standard, extra-strong, double-extra-strong pipe.

14.56 Sizes of Wrought-iron, Steel, and Cast-iron Pipe

The standard-weight pipe is used for normal pressures. It may be purchased in sizes ranging from $\frac{1}{8}$ to 12 in. (nominal diameter). Pipe is received threaded on both ends with a plain coupling attached.

Extra-strong pipe, designed for steam and hydraulic pressures over 125 lb per sq in., is also manufactured in sizes $\frac{1}{8}$–12 in.

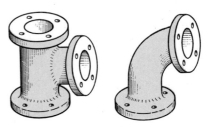

FIG. 14.76 Flanged fittings.

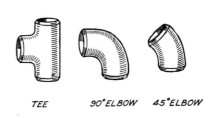

| TEE | 90° ELBOW | 45° ELBOW |

FIG. 14.77 Welded fittings.

Double-extra-strong pipe, designed for extremely high pressures, is furnished in nominal diameters from $\frac{1}{2}$ to 8 in. in the same lengths as the extra-strong.

Cast-iron pipe, in sizes ranging up to 48 in., can be used for pressures up to 350 lb per sq in.

14.57 Pipe Fittings

Fittings are parts, such as elbows, tees, crosses, couplings, nipples, flanges, and so on, that are used to make turns and connections. They fall into three general classes; screwed, welded, and flanged. See Figs. 14.75–14.77.

In small piping systems and for house plumbing, screwed fittings are generally used.

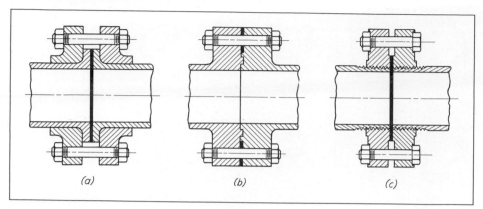

FIG. 14.78   Flanges.

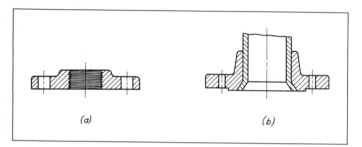

FIG. 14.79   Flanges.

Welded fittings are used where connections are to be permanent. They are manufactured of forged seamless steel having the same thickness as the pipe. In this type of construction, the weld is depended on to seal the joint and to carry the pipeline stresses. Many miles of line having welded fittings are giving satisfactory service to pipeline corporations.

Flanged fittings are used in large piping systems where pressures are high and the connection must be strong enough to carry the weight of large pipes. Table 50 in the Appendix gives the dimensions for American National Standard cast-iron flanged fittings. Several types of flanges and flanged joints are shown in Figs. 14.78 and 14.79.

14.58 Screwed Fittings (Fig. 14.75)

Straight sections of pipe are connected by a short cylindrical fitting (threaded on the inside), which is known as a *coupling*. A *right-and-left coupling*, which can be recognized by the ribs on the outside, is often used to close a system. A *union* is preferable, however, where pipe must be frequently disconnected.

A *cap* is screwed on the end of a pipe to close it.

A *plug* is used to close an opening in a fitting.

A *nipple* is a short piece of pipe that has been threaded on both ends. If it is threaded the entire length, it is called a *close nipple*; if not, it is called a *short* or *long nipple*. Extra-long nipples may be purchased.

A *bushing* is used to reduce the size of an opening in

a fitting when it would be inconvenient to use a reducing fitting.

Tees, *crosses*, and *laterals* form the connections for lines and branches in a piping system.

By standardizing the screwed fittings, the American National Standards Institute has eliminated many difficulties that would arise if each manufacturer produced the varied sizes of elbows, tees, laterals, and so on, according to his own specifications. The adopted dimensions, now recognized by all manufacturers, will be found in Tables 47–49 in the Appendix. See Fig. 14.84.

14.59 Specification of Fittings

A fitting is specified by giving the nominal inside diameter of the pipe for which the openings are threaded, the type of fitting, and the material. If it connects more than one size of pipe, it is called a reducing fitting, and the largest opening of the through run is given first, followed in order by the opposite end and the outlet. Figure 14.80 illustrates the order of specifying reducing fittings. If all of the openings are for the same size of pipe, the fitting is known as a straight tee, cross, and so on. A straight fitting is specified by the size of the openings followed by the name of the fitting (2-in. tee, 4-in. cross, etc.).

14.60 Unions

Screwed or flanged unions connect pipe that must be frequently disconnected for the purpose of making repairs. In many cases, screwed unions are used for making the final closing connection in a line. The union illustrated in Fig. 14.81(*a*) is made up of three separate pieces. The mating parts, *A* and *B*, are screwed on the ends of the two pipes. The third part, the nut, draws them together so that *A* and *B* will be against the gasket, *D*, to ensure a tight joint. In systems having pipes more than 2 in. in diameter, screwed unions are not generally used, because the stronger and more substantial flange unions, such as the one shown at (*c*), become desirable. A screwed union with a ground metal seat is shown at (*b*).

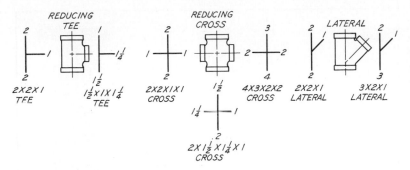

FIG. 14.80 Specification of fittings.

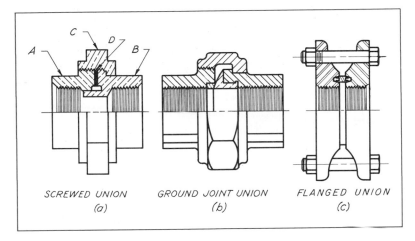

SCREWED UNION
(a)

GROUND JOINT UNION
(b)

FLANGED UNION
(c)

FIG. 14.81 Unions.

14.61 Valves

Valves are used in piping systems to stop or control the movement of fluids and gases. A few of the many forms are illustrated in Fig. 14.82. Of these, the globe valve and gate valve are the two types most frequently used.

Globe valves are used for throttling steam, in both high- and low-pressure steam lines, and to regulate the passage of other fluids. Their design, however, creates a slight retardation to the flow, because the fluid is forced to make a double turn and pass through the opening at 90° to the axis of the pipe. The valve disc is raised or lowered to stop or regulate the flow through a circular opening.

A gate valve allows a straight-line movement of a fluid and offers only slight resistance to the flow. Since the disc moves completely out of the passage and leaves a full opening, this type of valve is particularly suitable for water lines, oil lines, and the like.

A swing-type check valve permits movement in one direction only and prevents any back flow. It will be noted from a study of this valve that the design makes the action automatic. Such valves are used in feedwater lines to boilers. The ball-check valve is preferred for heavy liquids.

The dimensions of the valves given in Fig. 14.82, as well as those for many special types, may be found in the catalogs of manufacturers.

14.62 Piping Drawings

Since standard pipe and fittings can be purchased for almost any purpose, a piping drawing usually shows only the arrangement of a system in some conventional form and gives the size and location of fittings. The drawing may be a freehand sketch, single-line diagram, double-line diagram, or pictorial diagram. Occasionally, when conditions necessitate the design of special valves or the redesign of an existing type, complete working drawings are made.

Single-line drawings or sketches are made in orthographic projection or are drawn as though the entire system were swung into one plane (Fig. 14.83). On these drawings, single lines represent the runs of pipe, regardless of variations in diameters; conventional symbols are used for the fittings. A developed single-line sketch is frequently used for repair work, small jobs, and for making studies and calculations. For more complicated small-scale layouts, a single-line diagram drawn in orthographic projection is more suitable.

Double-line diagrams are drawn when many similar installations are to be made at the plants of various purchasers of pumps, manufacturing equipment, heating equipment, and so on (Fig. 14.84).

A *diagrammatic isometric layout* (Fig. 14.85) showing

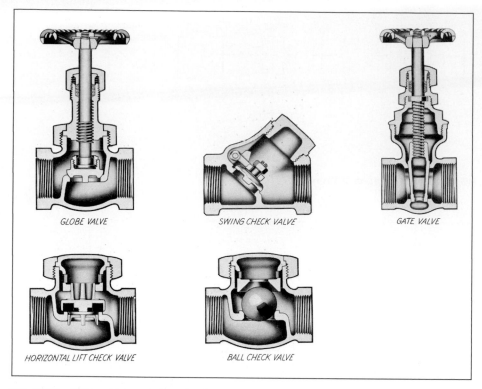

FIG. 14.82   Valves. (*Courtesy Crane Co.*)

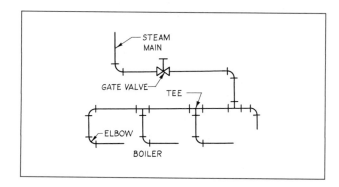

FIG. 14.83   Single-line drawing.

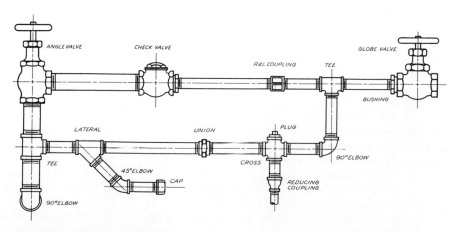

FIG. 14.84   Double-line drawing.

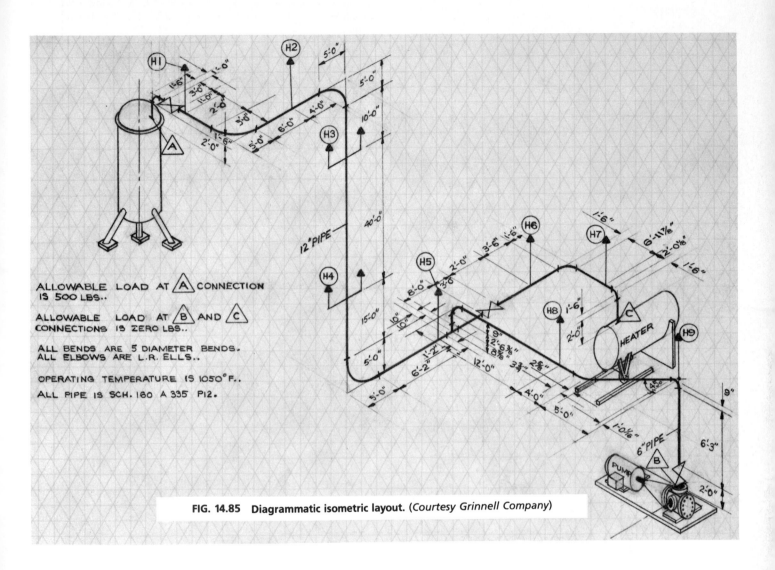

FIG. 14.85 Diagrammatic isometric layout. (*Courtesy Grinnell Company***)**

a piping system in space reveals the changes in direction and the difference in levels more clearly than does any other type of line diagram. Pictorial diagrams are often used for preliminary layouts.

14.63 Dimensions on Piping Drawings

The rules for dimensioning working drawings apply to piping drawings. Fittings and pipes are always located by giving center-to-center distances, because the determination of pipe lengths is generally left to the pipe-fitter. Notes should be used to specify the nominal size and type

of each fitting and the nominal size of the pipe in each run. In addition, it is good practice to indicate, on a flanged valve, the diameter of the handwheel and its distance above the center of the fitting when wide open. It may be necessary to give overall dimensions for other apparatus, if the maximum space to be allowed is important.

14.64 Conventional Symbols

A few of the conventional symbols for fittings that have been approved by the American National Standards Institute can be found in the Appendix.

PROBLEMS

Fasteners

Excellent practice in drawing (or sketching) the representations of threads, threaded fasteners, keys, and rivets is provided by the problems of this chapter.

Use Table 2 in the Appendix if it should be desirable to convert inches and fractions of an inch to millimeters. A metric scale has been given for Problem 1.

1. Draw or sketch the three layouts shown in Fig. 14.86 to full size, using the given scale to determine the measurements. On layout ① complete the drawing to show a suitable fastener on center line *AA*. On layout ② show a M12 ($\frac{1}{2}$ in.) hexagonal-head cap screw on center line *BB*. On layout ③ show a $\frac{3}{8}$-in. button-head rivet on center lines *CC* and a No. 608 Woodruff key (or equivalent metric key) on center line (CL) *DD*. Use the schematic symbol for the representation of threads.

2. (Fig. 14.87). Reproduce the views of the assembly of the alignment bearing. On CLs A show $\frac{1}{4}$-in. button-head rivets (four required). On CL *B* shows a $\frac{5}{16} \times \frac{1}{2}$ American Standard square-head set screw. Do not dimension the views.

3. (Fig. 14.88). Reproduce the views of the assembly of the impeller drive. On CLs A show $\frac{1}{4}$–20 UNC $\times \frac{1}{2}$ round-head machine screws and regular lock washers. On CL *B* show a No. 406 Woodruff key. On CL *C* show a standard No. 2 $\times$ 1$\frac{1}{2}$ taper pin.

4. (Fig. 14.89). Reproduce the views of the assembly of the bearing head. On CLs *A* show M12 $\times$ 1.75 studs with regular lock washers and regular semifinished hexagonal nuts (four required). On CLs *B* show M10 $\times$ 1.5 $\times$ 35 hexagonal-head cap screws (two required). On CL *C* drill through and tap $\frac{1}{8}$-in. pipe thread.

5. (Fig. 14.90). Reproduce the views of the assembly of the air cylinder. On CL *AA* at the left end of the shaft show an M24 $\times$ 3 semifinished hexagonal nut. At the right end show a hole tapped M20 $\times$ 2.5 $\times$ 40 deep. Between the piston and the (right) end plate draw a spring 75 mm OD, five full coils, 6.5 mm wire. On CLs *B* show M10 $\times$ 1.5 $\times$ 30 hexagonal-head cap screws. On CLs *C* draw M6 $\times$ 1 $\times$ 20 flathead machine screws with heads to the left. On CL *D* show a $\frac{1}{4}$-in. standard pipe thread. On CLs *E* show M12 $\times$ 1.75 $\times$ 45 semifinished hexagonal-head bolts. Use semifinished hexagonal nuts. Show visible fasteners on the end view.

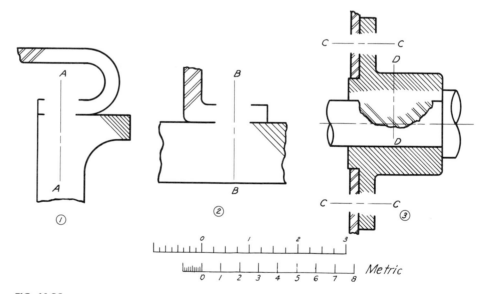

FIG. 14.86

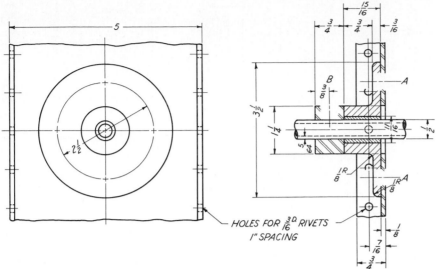

FIG. 14.87 Alignment bearing.

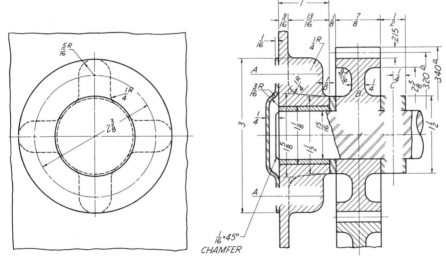

FIG. 14.88 Impeller drive.

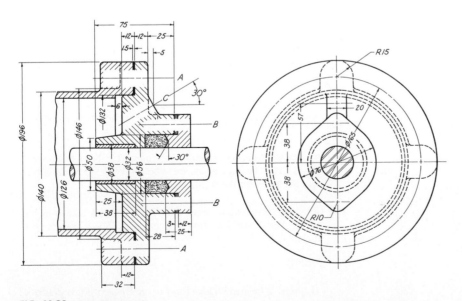

FIG. 14.89 Bearing head.

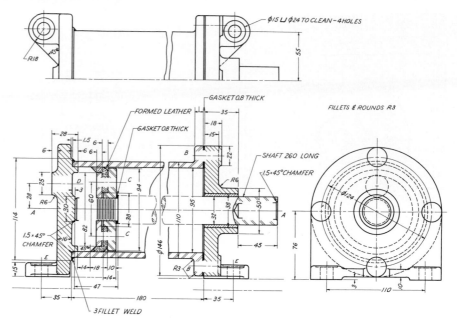

FIG. 14.90   Air cylinder.

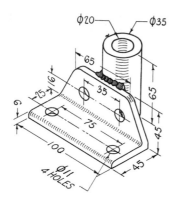

FIG. 14.91   Bracket.

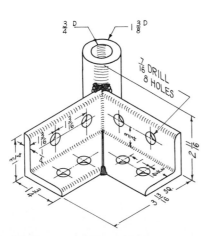

FIG. 14.92   Caster bracket.

Welding

Two problems offering experience in the preparation of welding drawings are given in Figs. 14.91 and 14.92. Others may be had by redesigning many of the cast parts (given at the end of Chapter 5, and in some of the other chapters) in such a way that they may be made of welded steel shapes. The student will find in these problems an opportunity to exercise some of his own ingenuity.

Note: With the approval of his Instructor, the student may convert the inch measurements in Prob. 2 to millimeters by using Appendix Tables 2–8. Since it is common practice under the metric system to use full millimeters for all dimension values that are not critical, the student should do likewise. For example, in Fig. 14.92 the 1 in. value converts to 25.4 millimeters. The student should use 25 millimeters. Twist drill sizes are given in Tables 27 and 52 in the Appendix.

6. Prepare a welding drawing of the bracket shown in Fig. 14.91.

7. Prepare a welding drawing of the caster bracket shown in Fig. 14.92. The length of the tubing is $2\frac{11}{16}$ in.

8. Make a two-view orthographic drawing of the object shown in Fig. 14.50. The dimensions are to be assumed. The plates are $\frac{3}{8}$ in. thick. Show the correct specification for each type of weld.

9. Make a three-view detail drawing of the bracket shown in Fig. 14.70. The dimensions are to be assumed. Show the correct specifications for the welds.

10. Make a two-view detail drawing of the link shown in Fig. 14.71. The dimensions are to be assumed. Show correct specifications for the welds.

Piping

11. Make a freehand sketch (on $\frac{1}{8}$-in. grid paper, if it is available) of a 1-in. nipple connecting, a 1-in. cast-iron elbow, and a

$2 \times 2 \times 1$-in. maleable-iron tee. The distance between centers of fittings is to be 6 in. Enter neatly, in drafting style, the length of the nipple to the nearest $\frac{1}{8}$ in.

12. Make a freehand sketch (on $\frac{1}{8}$-in. grid paper) of a $1 \times 1 \times 1$-in. malleable-iron tee and a 1-in. cast-iron elbow joined by a length of pipe. The distance between centers of fittings is 4 in. Enter the length of the connecting pipe to the nearest $\frac{1}{8}$ in.

13. Make a single-line multiview sketch of the portion of a piping shown in Fig. 14.93.

14. Make a double-line developed drawing of the portion of a piping system shown in Fig. 14.93. Use a 2-in. pipe and screwed fittings. Select a suitable scale. Determine the measurements by transferring distances from the drawing to the open-divided scale in the figure.

15. Dimension the drawing of Problem 14.

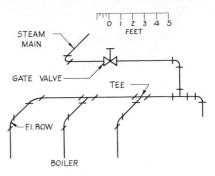

FIG. 14.93 Pictorial line diagram.

16. Make a detail (working) drawing of an assigned part of the air cleaner shown in Fig. 14.94. Determine the dimensions by transferring them from the drawing to the accompanying scale by means of the dividers (study the pictorial representation).

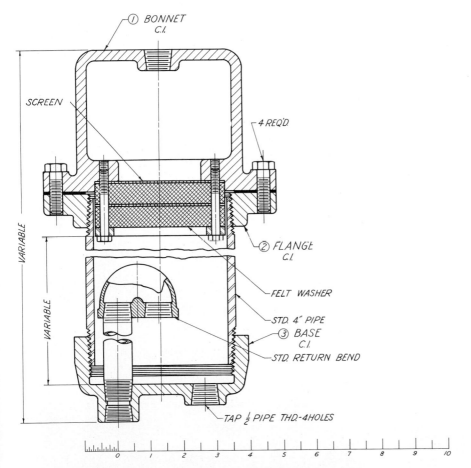

FIG. 14.94 Air cleaner. (*Courtesy A. Schrader's Son Mfg. Co.*)

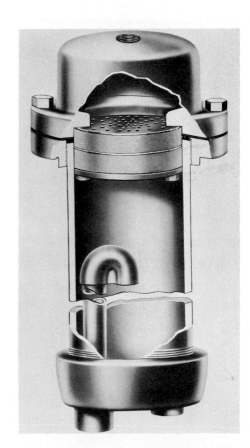

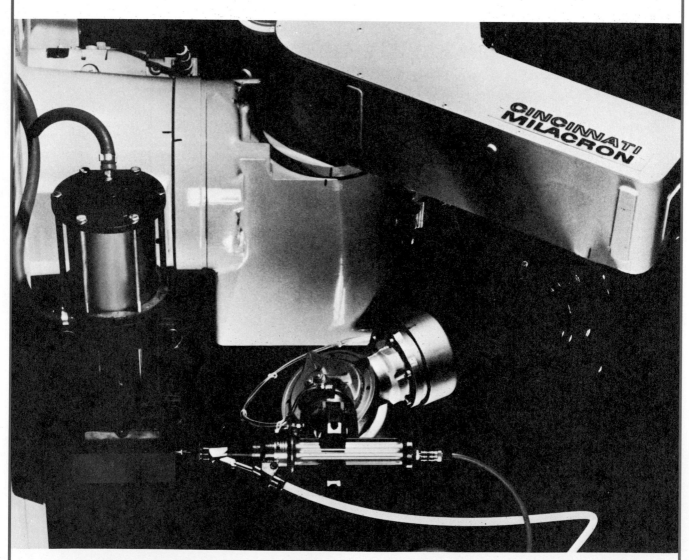

Cincinnati Milacron's Industrial Robot uses an automatic screw driver to insert screws into a workpiece. The ability of the robot to maintain correct tool orientation is essential to this application. (*Courtesy Cincinnati Milacron*)

CHAPTER · 15

Shop Processes and Tool Drawings

15.1 Shop Processes

To be qualified to prepare drawings that will fulfill the requirements of production shops, an engineering drafter must be thoroughly familiar with fundamental shop processes. In preparing working drawings, consider each and every individual process involved in the production of a piece and then specify the processes in terms that the shopman will understand (see the facing page). All too frequently, drawings that specify impractical methods and impossible operations are sent to the shops. Most of these impractical specifications are the result of a drafter's lack of knowledge of what can or cannot be done by skilled technicians using modern machines and tools.

Although an accurate knowledge of shop processes can be acquired only through actual experience in the various shops, it is possible for an apprentice drafter to obtain a working knowledge of the fundamental operations through study and observation. This chapter presents and explains the principal operations in the pattern shop, foundry, forge shop, and machine shop.

15.2 Die Casting

Die casting is an inexpensive method for mass producing certain types of machine parts, particularly those needing no great strength. The castings are made by forcing molten metal or molten alloy into a cavity between metal dies in a die-casting machine. Parts thus produced usually require little or no finishing.

15.3 *Sand Casting*

Castings are formed by pouring molten metal into a mold or cavity formed in sand. The molten metal assumes the shape of the cavity that has been formed in a sand mold by ramming prepared moist sand around a pattern and then removing the pattern. Although a casting shrinks somewhat in cooling, the metal hardens in the exact shape of the pattern used (Fig. 15.1).

A sand mold consists of at least two sections. The upper section, called the *cope*, and the lower section, called the *drag*, together form a box-shaped structure called a *flask*.

When large holes [20 mm (.80 in.) and over] or interior passageways and openings are needed in a casting, dry sand cores are placed in the cavity. Cores exclude the metal from the space they occupy and thus form desired openings. Large holes are cored to avoid an unnecessary boring operation. A dry sand core is formed by ramming a mixture of sand and a binding material into a core box that has been made in the pattern shop. To make a finished core rigid, the core-maker places it in a core oven, where it is baked until it is hard.

The molder when making a mold inserts in the sand a sprue stick that he removes after the cope has been rammed. This resulting hole, known as the *sprue*, conducts the molten metal to the *gate*, which is a passageway cut to the cavity. The adjacent hole, called the *riser*, provides an outlet for excess metal.

15.4 *Pattern Shop*

The pattern shop prepares patterns of all pieces that the foundry is to cast. Although special pattern drawings are frequently submitted, the patternmaker ordinarily uses a drawing of the finished piece that the drafter has prepared for both the pattern shop and the machine shop (see Fig. 15.27). The finish marks on such a drawing are just as important to the patternmaker as to the machinist, for he must allow, on each surface to be finished, extra metal, the amount of which depends on the method of machining and the size of the casting. In general, this amount varies from 1.5 mm (.06 in.), on very small castings, to as much as 20 mm (.80 in.), on large castings.

It is not necessary for the drafter to specify the amount to be allowed for shrinkage, for the patternmaker has available a ''shrink rule,'' which is sufficiently oversize to take care of the shrinkage.

A pattern usually is first constructed of light, strong wood, such as white pine or mahogany, which, if only a few castings are required, may be used in making sand molds. In quantity production, however, where a pattern must be used repeatedly, the wooden one will not hold up, so a metal pattern (aluminum, brass, and so on) is made from it and is used in its place.

Every pattern must be constructed in such a way that it can be withdrawn from each section of the sand mold. If the pattern consists of two halves (split), the plane of separation should be so located that it will coincide with the plane of separation of the cope and the drag (Fig. 15.1). Each portion of the pattern must be slightly tapered, so that it can be withdrawn without leaving a damaged cavity. The line of intersection, where the dividing plane

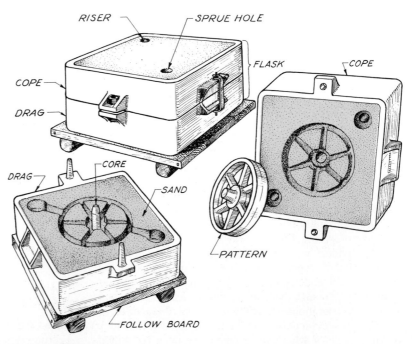

FIG. 15.1 Sand mold.

cuts the pattern, is called the *parting line*. Although this line is rarely shown on a drawing, the draftsman should make certain that the design will allow the patternmaker to establish it. Ordinarily, it is not necessary to specify the slight taper, known as *draft*, on each side of the parting line, for the patternmaker assumes such responsibility when constructing the pattern.

A "filled-in" interior angle on a casting is called a *fillet*, to distinguish it from a rounded exterior angle, which is known as a *round* (Fig. 5.45). Sharp interior angles are avoided for two reasons: They are difficult to cast; and they are likely to be potential points of failure because the crystals of the cooling metal arrange themselves at a sharp corner in a weak pattern. Fillets are formed by nailing quarter rounds of wood or strips of leather into the sharp angles or by filling the angles with wax.

15.5 Foundry

Although a drafter is not directly concerned with the foundry (since the patternmaker takes the drawing and prepares the pattern and core boxes for the molder), it is most important that the drafter be familiar with the operations required in making a sand mold and a casting. Otherwise, it will be difficult to prepare an economical design, since the cost depends on how simple it is to mold and cast.

15.6 Forge Shop

Many machine parts, especially those that must have strength and yet be light, are forged into shape, the heated metal being forced into dies with a drop hammer. Drop forging, since heated metal is made to conform to the shape of a cavity, might be considered a form of casting. However, because dies are difficult to make and are expensive, this method of production is used principally to make parts having an irregular shape that would be costly to machine and could not be made from casting material. Forgings are made of a high-grade steel. Dies are made by experts in their craft who are known simply as die-makers.

Generally, special drawings, giving only the dimensions needed, are made for the forge shop.

15.7 Machine Shop

In general, the drafters are more concerned with machine-shop processes than with the processes in other shops, for all castings and forgings that have been prepared in accordance with their drawings must receive the final machining in the machine shop (see Fig. 15.27). Since all machining operations must be considered in the design and then properly specified, a drafter must be thoroughly familiar with the limitations as well as the possibilities of such common machines as the lathe, drill press, boring

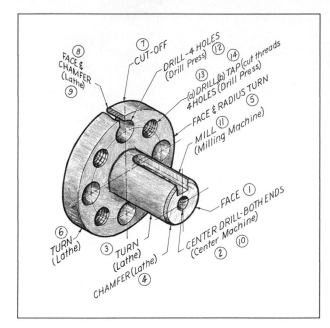

FIG. 15.2 Shop operations required to produce a part from cold-rolled stock. The numbers indicate a possible order in which the operations could be performed in the shops.

machine, shaper, planer, milling machine, and grinder (Fig. 15.2).

15.8 Standard Stock Forms

Many types of metal shapes, along with other materials that are used in the shops for making parts for structures, are purchased from manufacturers in stock sizes. They are made available from the stock department, where rough stock, such as rods, bars, plates, sheet metal, and so on, is cut into sizes desired by the machine shop.

15.9 Lathe

Many common operations, such as turning, facing, boring, reaming, knurling, threading, and so on, may be performed with this widely used machine. In general, however, it is used principally for machining (roughing-out) cylindrical surfaces to be finished on a grinding machine. Removing metal from the exterior surfaces of cylindrical objects is known as *turning* and is accomplished by a sharp cutting tool that removes a thin layer of metal each time it travels the length of a cylindrical surface on the revolving work (Fig. 15.3). The piece, which is supported in the machine between two aligned centers, known as the *dead center* and the *live center*, is caused to rotate about an axis by power transmitted through a lathe dog, chuck, or faceplate. The work revolves against the cutting tool, held in a tool post, as the tool moves parallel to the longitudinal axis

FIG. 15.3 **Lathe operation—turning.**

FIG. 15.4 **Boring on a lathe.**

FIG. 15.5 **Reaming on a lathe.**

FIG. 15.6 **Cutting threads on a lathe.**

of the piece being turned. Cutting an interior surface is known as *boring* (Fig. 15.4). A note is not necessary on a drawing to indicate that a surface is to be turned on a lathe.

When a hole is reamed, it is finished very accurately with a fluted reamer of the exact required diameter. If the operation is performed on a lathe, the work revolves as the nonrotating reamer is fed into the hole by turning the handwheel on the tail stock (see Fig. 15.5).

Screw threads may be cut on a lathe by a cutting tool that has been ground to the shape required for the desired thread. The thread is cut as the tool travels parallel to the axis of the revolving work at a fixed speed (Fig. 15.6).

Knurling is the process of roughening or embossing a cylindrical surface. This is accomplished by means of a knurling tool containing knurl rollers that press into the work as the rollers are fed across the surface (Fig. 15.7).

FIG. 15.7 Knurling.

15.10 *Drill Press*

A drill press is a necessary piece of equipment in any shop because, although it is used principally for drilling, as the name implies (Fig. 15.8), other operations, such as reaming, counterboring, countersinking, and so on, may be performed on it by merely using the proper type of cutting tool. The cutting tool is held in position in a chuck at the end of a vertical spindle that is made to revolve, through power from a motor, at a particular speed suitable for the type of metal being drilled. The most flexible drill press, especially for large work, is the radial type, which is so designed that the spindle is mounted on a movable arm that can be revolved into any desired position for drilling. With this machine, holes may be drilled at various angles and locations without shifting the work, which may be either clamped to the horizontal table or held in a drill vise or drill jig. The ordinary type of drill press without a movable arm is usually found in most shops along with the radial type. A multispindle drill is used for drilling a number of holes at the same time.

Figure 15.9 shows a setup on a drill press for performing the operation of *counterboring*. A counterbore is used to enlarge a hole to a depth that will allow the head of a fastener, such as a fillister-head cap screw, to be brought to the level of the surface of the piece through which it passes. A counterbore has a piloted end having approximately the same diameter as the drilled hole.

Figure 15.10 shows a setup for the operation of *countersinking*. A countersink is used to form a tapering depression that will fit the head of a flathead machine screw or cap screw and allow it to be brought to the level of the surface of the piece through which it passes.

A *plug tap* is used to cut threads in a drilled hole (Fig. 15.11).

A *spotfacer* is used to finish a round spot that will provide a good seat for the head of a screw or bolt on the unfinished surface of a casting. Figure 15.12 shows various cutting tools commonly used for forming holes and cutting threads.

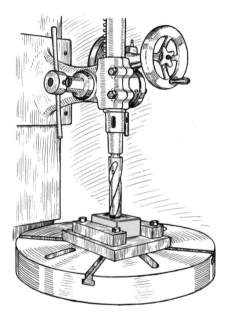

FIG. 15.8 Drill press.

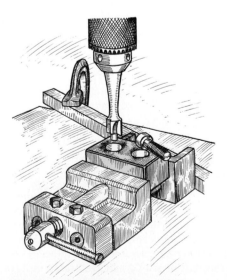

FIG. 15.9 Counterboring on a drill press.

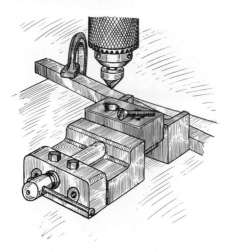

FIG. 15.10 Countersinking on a drill press.

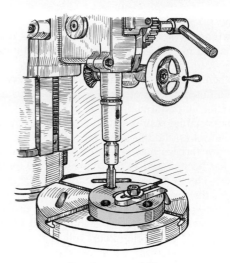

FIG. 15.11 Tapping on a drill press.

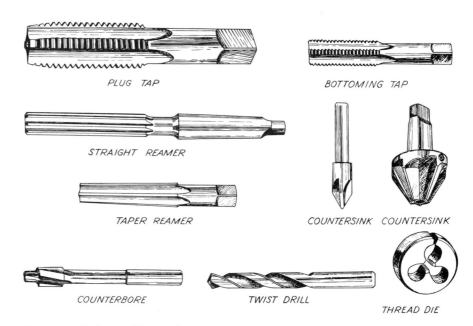

PLUG TAP

BOTTOMING TAP

STRAIGHT REAMER

TAPER REAMER

COUNTERSINK COUNTERSINK

COUNTERBORE

TWIST DRILL

THREAD DIE

FIG. 15.12 Various cutting tools.

15.11 Hand Reaming

A hole may be finished to an accurate size by hand reaming, as shown in Fig. 15.13. The reamer in this illustration is of a special type known as a *line reamer*.

15.12 Boring (Fig. 15.14)

Boring is the operation of enlarging a circular hole for accuracy in roundness or straightness and may be accomplished on a lathe, drill press, milling machine, or boring mill. When the hole is small and of considerable length, the operation may be performed on a lathe. If the hole is large, the work is usually done on a boring mill, of which

there are two types—the vertical and the horizontal. On a vertical boring machine, the work is fastened on a horizontal revolving table, and the cutting tool or tools, which are stationary, advance vertically into it as the table revolves. On a horizontal boring machine, the tool revolves and the work is stationary.

15.13 Milling Machine

A milling machine is used for finishing plane surfaces and for milling gear teeth, slots, keyways, and so on. In finishing a plane surface, a rotating circular cutter removes the metal for a desired cut as the work, fastened to a

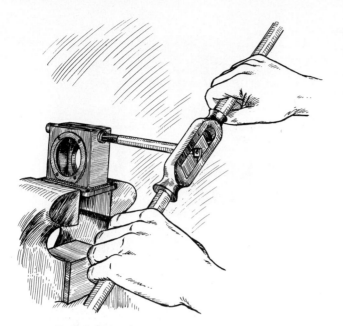

FIG. 15.13   **Hand reaming.**

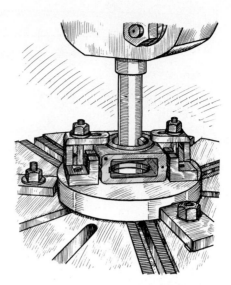

FIG. 15.14   **Boring on a boring mill.**

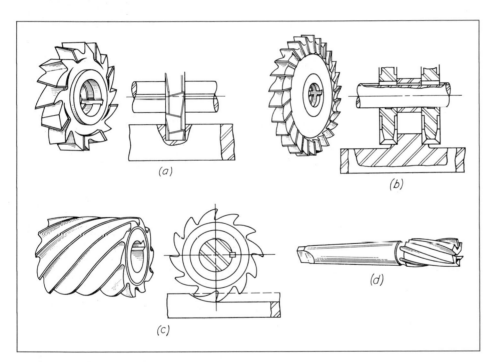

(a)

(b)

(c)

(d)

FIG. 15.15   **Milling cutters and operations.**

moving horizontal bed, is automatically fed against it. Several types of milling cutters are shown in Fig. 15.15.

Figure 15.16 shows a setup for milling gear teeth in a gear blank. Note the form of this particular type of cutter.

15.14   *Shaper (Fig. 15.17)*

A shaper is used for finishing small plane surfaces and for cutting slots and grooves. In action, a fast-moving reciprocating ram carries a tool across the surface of the work, which is fastened to an adjustable horizontal table. The tool cuts only the forward stroke.

15.15   *Planer*

The planer is a machine particularly designed for cutting down and finishing large flat surfaces. The work is fastened to a long horizontal table that moves back and forth under the cutting tool. In action, the tool cuts as the table moves

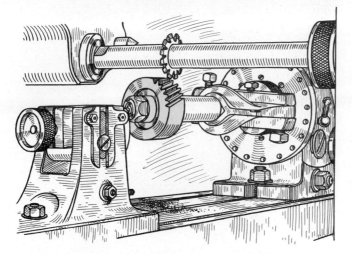

FIG. 15.16 Cutting gear teeth on a milling machine.

FIG. 15.17 Shaper.

the surface against it. Unlike the cutter on the shaper, it is stationary except for a slight movement laterally for successive cuts.

15.16 *Grinding Machine*

A grinding machine has a rotating grinding wheel that, ordinarily, is either an emery wheel (fine or coarse) or some type of high-speed wheel made of carborundum (Fig. 15.18). Grinding consists of bringing the surface to be ground into contact with the wheel. Although grinding machines are often used for "roughing" and for grinding down projections and surfaces on castings, their principal use, as far as a drafter is concerned, is for the final finishing operation that will bring the cylindrical surface of a piece of work down to accurate dimensions. Very fine surface finishes with tolerances as close as 0.025 or 0.050 micrometer (1 or 2 microinches) may be obtained by grinding. Grinding wheels of special form are used for grinding threads when close fits are desired and for forming other shapes.

When a flat surface is to be brought to a super finish, a surface grinder is used. The work, clamped to either a reciprocating or rotating worktable, passes under a rapidly rotating abrasive wheel, as illustrated in Fig. 15.18(*a*).

In grinding the external surface of a shaft, as illustrated in (*d*), the work, mounted between centers, rotates slowly while in contact with the rapidly rotating grinding wheel moving along the work as indicated by the arrows. The grinding wheel is mounted so that it may be moved up to the work or away from it. Normally the depth of cut is about 0.05 mm (.002 in.) per pass. However, finishing cuts may be made as fine as 0.003 mm (.0001 in.) per pass to bring the finished size within the dimensional limits given on a detail drawing.

As might be expected with centerless grinding, the work is not held between centers, as in the case of cylindrical grinding. Rather, it may be said to be fed along against the grinding wheel by a regulating wheel while being supported on a rest blade. See Fig. 15.18(*e*). It should be noted that, with both the grinding wheel and regulating wheel rotating in a clockwise direction, the work is forced to rotate counterclockwise. Pressure from the grinding wheel forces the work against both the work rest and the regulating wheel. A centerless grinder produces work that is round and of constant diameter.

Internal grinding for fine finish and close tolerances is illustrated in (*b*). Form grinding is shown in (*c*).

15.17 *Broaching*

A broach is a tool used to cut keyways and to form square, rectangular, hexagonal, or irregular-shaped holes. It is a hard, tempered cutting tool with serrated cutting edges that enlarge a drilled, punched, or cored hole to a required shape. In operation, each tooth removes a chip of material from the surface except the last few teeth, which are to size (Fig. 15.19). A broach produces a fine finish with the work accurately sized. This is accomplished with a single pass. Broaches, whether used to finish an internal (hole) or an external surface, are either pushed or pulled. A special broaching machine is used for pulling broaches. Some form of press, hydraulic or otherwise, is required for push broaches.

The cutting of a keyway using a slotter (broach) is illustrated in Fig. 15.20. A guide bushing has been inserted in the hole in the piece to hold the tool in position.

Some typical broached contours are shown in Fig. 15.21.

15.18 *Superfinishing: Polishing, Honing, and Lapping*

Polishing consists of bringing a ground surface into contact with a revolving disc of leather or cloth, thus producing a lustrous smoothness that would be impossible to obtain by using even the finest grinding wheel. The operation is specified on a drawing by a note: "polish" or "grind and polish."

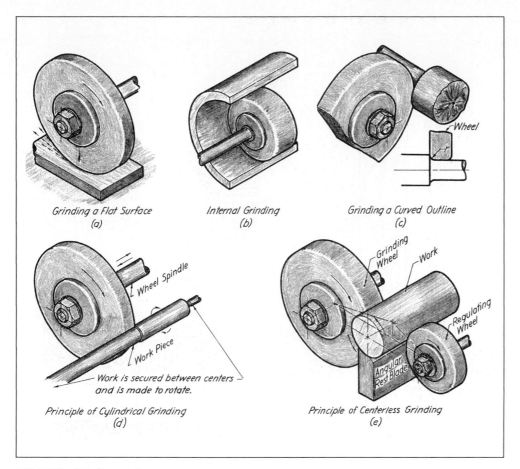

Grinding a Flat Surface
(a)

Internal Grinding
(b)

Grinding a Curved Outline
(c)

Wheel

Wheel Spindle

Work Piece

Work is secured between centers and is made to rotate.

Principle of Cylindrical Grinding
(d)

Grinding Wheel

Work

Regulating Wheel

Angular Rest Blade

Principle of Centerless Grinding
(e)

FIG. 15.18 Grinding surfaces.

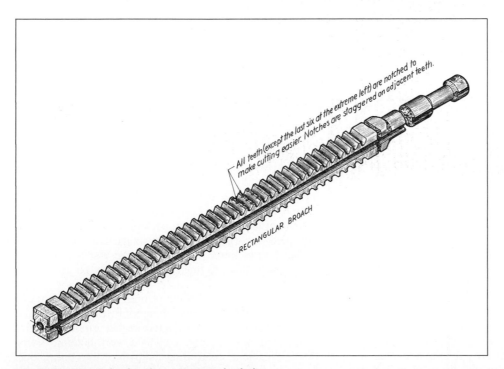

All teeth (except the last six at the extreme left) are notched to make cutting easier. Notches are staggered on adjacent teeth.

RECTANGULAR BROACH

FIG. 15.19 Broach for forming a rectangular hole.

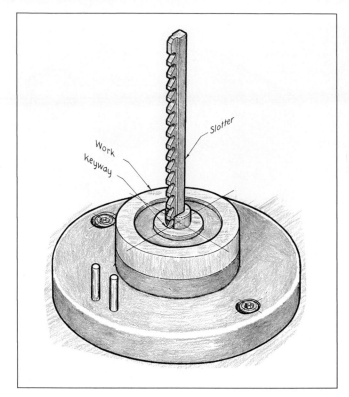

FIG. 15.20   Cutting a keyway.

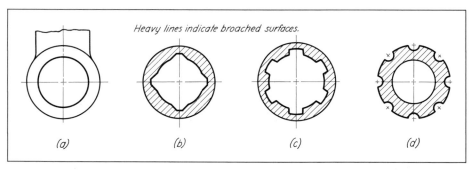

FIG. 15.21   Some typical broached contours.

Honing and *lapping* are methods of producing super-finishes, after grinding, through the use of abrasives. For honing cylindrical bores, fine-grained abrasive sticks which are available in many styles and grades, are used to minimize scratches and to finish the bore to precision limits.

Lapping is a final stock-removing operation that is performed by rubbing the surfaces of parts to be lapped over a lap of soft metal to which has been applied a powdered abrasive mixed with a lubricant.

15.19   *Jigs and Fixtures*

Often, when an operation must be performed many times in making a part in quantities on a general machine, one of two devices, a *jig* or a *fixture*, may be used to facilitate production and ensure accuracy without making repeated measurements. In general, it might be said that jigs and fixtures are devices for holding work while various machining operations are being performed. Such devices play an important role in the present quantity production of interchangeable parts, for their use greatly reduces the amount of labor required in performing accurate machining operations. However, the advantage that may be gained in the form of lower labor cost in the assembly of units having parts that are interchangeable may be just as important.

Although there are some people who use the terms denoting these devices interchangeably, there is a definite difference between a jig and a fixture that is quite generally recognized. Specifically, a jig is a work-holding device that is capable of controlling the path of the cutting tool. Since it is not fastened to the machine, a jig may be moved around so that several holes may be drilled, tapped, or reamed in their proper locations as established by the position of the drill bushings. Although it is usual for the work to be held by the jig, there are cases where the jig may be clamped on the piece. A well-designed jig should permit the work to be quickly inserted and removed.

A fixture is fixed or fastened to the table of the machine to locate and hold the work securely in a definite position. Either the cutting tool is moved into position for the operation or the table is moved under the tool. A fixture does not guide the cutting tool. Fixtures are used for such operations as milling, honing, broaching, grinding, and welding.

The three factors that must be considered and evaluated when determining whether or not the cost of making a jig or a fixture will be justified are (1) the number of parts to be produced, (2) the saving in labor cost, and (3) the possibility that the device may be needed for making replacement parts at a later time. Using simple arithmetic, one can compute and compare the cost of producing one unit in the conventional manner against making the part using a jig or fixture. Knowing the cost of the jig or fixture, the number of units to be produced, and the savings per unit, it will not be difficult to arrive at a decision. Generally, the use of the device should result in a definite saving of time and labor.

More could be given here relating to the economic problems surrounding the use and design of jigs and fixtures but it would be beyond the scope of this chapter.

15.20 Jig Borers

Jig borers are provided with accurate lead screws and special measuring devices that make possible the accurate location and boring of holes. Their use greatly reduces the cost of making jigs since the usual layout work with scriber and punch is not required. It is a common practice in many shops to use a jig borer or its equivalent to machine pieces in small lots rather than go to the expense of making jigs.

15.21 Types of Jigs

There are a number of types of jigs and fixtures that have been developed and are suitable for a wide range of work. The drawing for a simple jig for drilling a jacket tube is shown in Fig. 15.22. The work, shown in phantom line, is usually drawn in red, while the jig itself is drawn in black. In preparing the drawing, the visibility of the lines of the jig itself are as they would be if the work were not

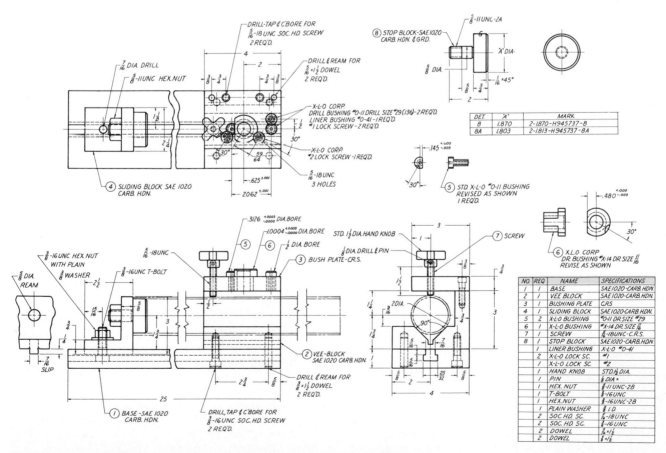

FIG. 15.22 Jig for drilling a jacket tube. (*Courtesy Ross Gear Division of TRW, Inc.*)

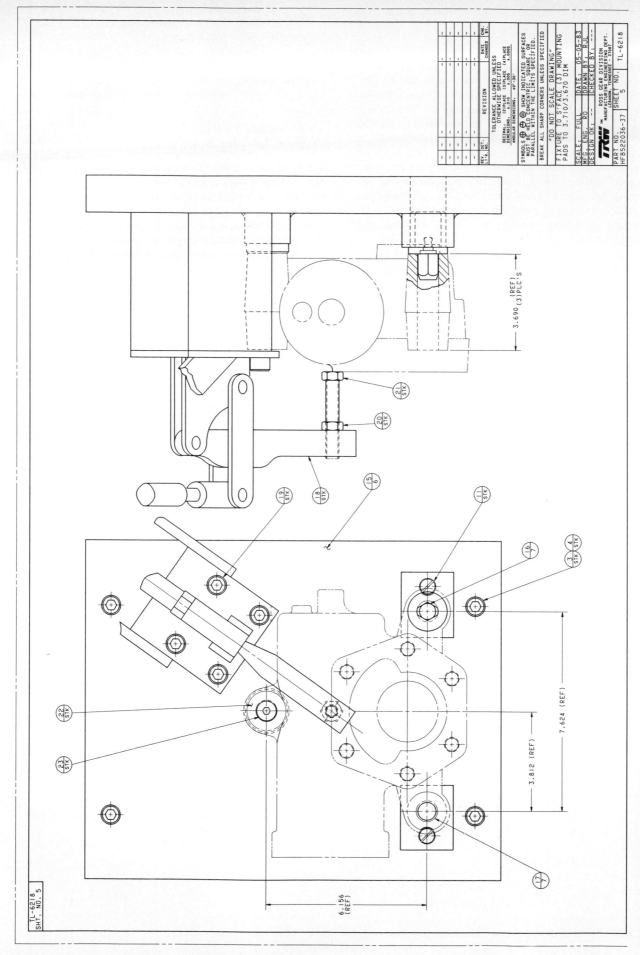

FIG. 15.23 A computer-generated design assembly layout for a fixture to spotface (3) mounting pads. *(Courtesy Ross Gear Division, TRW, Inc.)*

in place. This treatment of visibility along with showing the piece to be drilled in red is universal practice.

15.22 Fixtures

Fixtures are designed for varied uses. They may be classified as vise, boring, milling machine, tapping, lathe, grinding, broaching, honing, welding, or checking fixtures. Most readers should be familiar with the principle of the vise. Checking fixtures, as the name implies, are used to check accuracy of production. Figure 15.23 shows a special fixture that is designed to facilitate the spotfacing of the mounting pads on a part.

15.23 Tool Drawings

When the product drawings for a structure or mechanism have been completed by the engineering department, they are released to the production division, where a *tool routing sheet* and a worksheet are prepared by a process engineer. A routing sheet lists needed manufacturing operations in proper sequence and specifies which tools and machines are to be used. It also gives information for checking production accuracy and may specify the type of inspection instruments. The worksheets contain detailed information concerning the manufacturing processes.

In some cases, clearance specifications between mating parts may be given when critical situations are known to exist. These sheets furnish information needed by the tool designer and tool drafter and ensure that current machine-shop and production practices will be followed.

The drawings for making the holding devices and special tools needed for the various operations are known as *tool drawings*, and they are prepared by the tool engineering branch of the production division. Along with the routing sheet and worksheet, the tool designer and tool draftsman must have catalogs at hand that furnish information about standard parts and devices that may be purchased ready-made in standard arrangements.

In the planning stage of the development of a jig or fixture, freehand sketches may be used by the designer to integrate his thoughts and record information on standard parts. Such sketches assist the designer in preparing the accurate design layouts that are needed, particularly for complicated situations and where different tool drafters will be assigned the task of preparing the working drawings.

The tool drafter starts with an outline of the workpiece traced in red using phantom lines and then draws the outline assembly around the workpiece using black lines (Fig. 15.22). Where a black line of the assembly drawing coincides with one of the red lines of the workpiece, the black line is drawn over the phantom red line. In other words, on tool drawings the black lines are given precedence over the red ones. The practice of showing the workpiece in red and the outline of the jig or fixture in black permits easy reading of the drawing by enabling the reader to distinguish quickly between the workpiece and the jig or fixture holding it. An assembly layout may show locating dimensions and toleranced dimensions for critical conditions.

After the design assembly layout has been completed and approved, prints are made for the use of the tool drafter, who is to make the detail drawings. The drafter, or group of drafters under a squad leader, prepares drawings for all nonstandard and altered standard parts, placing several detail drawings on each detail sheet, as space will permit without overcrowding. Each detail drawing is identified by the same part number shown for it on the assembly layout. In addition to showing the part number in a small circle, it is customary to letter, just below the drawing, the part name, part material, and number required for one unit (Fig. 15.22). In the case of relatively simple jigs and fixtures, the parts are dimensioned, where possible, directly on the layout so that the layout becomes a working drawing. Those parts that cannot be dimensioned conveniently on the layout are detailed in an open space adjacent to the layout (Fig. 15.22). Of course, additional sheets may be used if necessary.

15.24 Dies and Die Drawings

Since standard die-sets may be found in manufacturing catalogs in different styles and sizes, it is the usual practice of tool designers to analyze the workpiece carefully and then select a suitable die-set on which the required punches and dies can be mounted.

A standard die-set has three principal parts, namely, the *punch holder* (upper shoe), *die holder* (lower shoe), and *guideposts*. Other parts of a standard combination are the shank and guidepost bushings. The shank provides for fastening the punch holder to the ram of the press. The guideposts ensure accurate alignment of the punch with the die under all conditions (Fig. 15.24).

Specific information on die design may be found in the several books listed in the bibliography. It is recommended that one consult the handy reference book *Practical Design of Manufacturing Tools, Dies, and Fixtures*, that was prepared under the direction of the Society of Manufacturing Engineers.

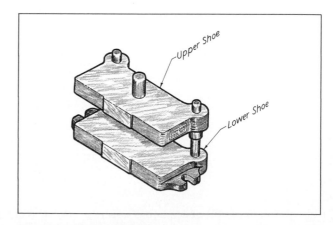

FIG. 15.24 Typical die set.

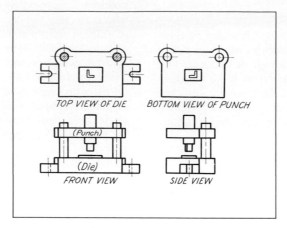

FIG. 15.25 Arrangement of views for a punch and die drawing.

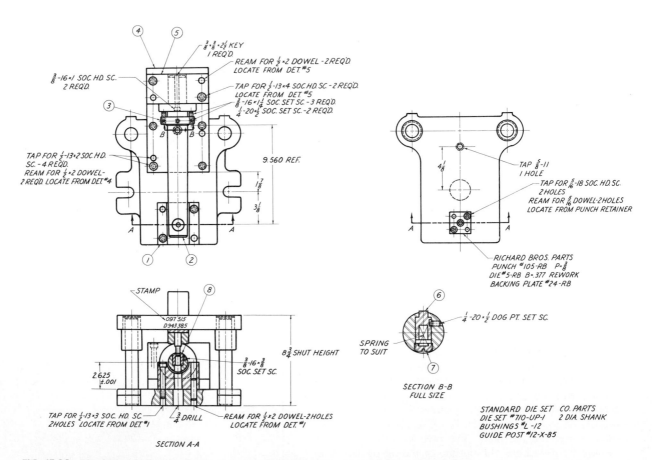

FIG. 15.26 Die design layout assembly drawing. *(Courtesy Ross Gear Division, TRW, Inc.)*

The fundamental press-work operations involving the use of punches and dies are

1. plain blanking,
2. piercing,
3. bending or forming,
4. drawing,
5. coining,
6. assembling.

Die assembly drawings usually show three or four views of the assembled die and the same general procedure is followed in their preparation as for jig and fixture drawings except for a modification of the theoretical arrangement of the views as shown in Fig. 15.25. In this illustration, the front view (front elevation) shows both of the main members (punch and die) in working position. The plan view of the lower shoe (die holder) is in the position of a top view directly above the front view. Aligned horizontally with this latter view is the plan view of the upper shoe (punch holder) shown as it would be seen looking upward from below. This is done so that the diemaker has a plan of the actual layout from which to work and to conserve space. The front view, always in section, shows the die as it would be seen from the front of the press and the views of the upper and lower shoes show all parts attached as they would be when in operation. Small part-section views and auxiliary views are used when needed. If a side view is necessary, it should be drawn in section. The assembly drawing should give general information and show the important dimensions needed by the die-maker, such as the shut height of the die, shrank diameter, the type and size of press, and the press stroke. On the die assembly drawing, each component part is numbered as shown in Fig. 15.26.

When it is necessary to detail a designed die, it is the usual practice to place the details on sheets separate from the assembly drawing.

A material list should be given on the die assembly drawing. This should show the number of each part and specify the rough material size.

15.25 Automatic Machines

In large industrial concerns, most mechanical parts are made on either semiautomatic or fully automatic machines by semiskilled operators. No prospective designer or drafter should ever forgo an opportunity to observe special production machines, for a thorough understanding of all shop machines and methods, is needed if drawings are to be satisfactory for the shops.

Production machines may follow directions given on punched tapes. The use of tapes and computers by industry does not mean that the drafters, technologists, and engineers will have less work to do and that there will be fewer of them. It does mean, however, that those assigned to both areas must keep themselves informed and up-to-date on late developments. The drafter should learn how to dimension the drawing of a part to meet the requirement for programming the machine or machines that will perform the shop operations. Numerically controlled machine tools are discussed in Chapter 18.

Since use of CADD (computer-aided design and drafting) and CAM (computer-aided manufacturing) is increasing, young people now in training should expand their knowledge in both areas. Where CADD/CAM has become a reality for a firm, the departments in both areas will function more closely and will come to react as a single entity. With this happening, engineers, engineering technologists, and drafting technologists may find that they have responsibilities in more than one area.

15.26 Manufacturing Processes and the Detail Drawing

In preparing the detail drawing that is needed for the production of a part, the drafter must give considerable thought to the manufacturing processes that will be required to make the conception become a reality. Machined surfaces must be indicated and the dimensions selected and placed with the manufacturing methods in mind. Shop notes, as they may be needed, must be prepared in accordance with the recommendations given in the appropriate ANSI standards.

The detail working drawing of the inlet cover shown in Fig. 15.27 has been prepared for use by both the pattern shop and the machine shop since these two shops will be involved in the production of the part. It should and does contain all of the dimensions and notes needed, first, to make the pattern for the casting and, second, to machine the cast part to obtain the finished inlet cover ready for use.

The pictorial drawing, not actually part of the detail drawing, gives an interpretation, in shop terms, of the machining operations required by the detail drawing. The sequence of these operations is indicated by the numbers enclosed within circles.

15.27 Measuring Tools

Figures 15.28, 15.29, and 15.30 show a few of the measuring tools commonly available in shops. When great accuracy is not required, calipers are used (Fig. 15.29). The outside calipers are suited for taking external measurements, as, for example, from a shaft. They are adjusted to fit the piece, and then the setting is applied to a rule to make a reading. The inside calipers have outturned toes, which fit them for taking internal measurements, as, for example, in measuring either a cylindrical or a rectangular hole. When extreme accuracy is required, some form of micrometer calipers may be used (Fig. 15.30).

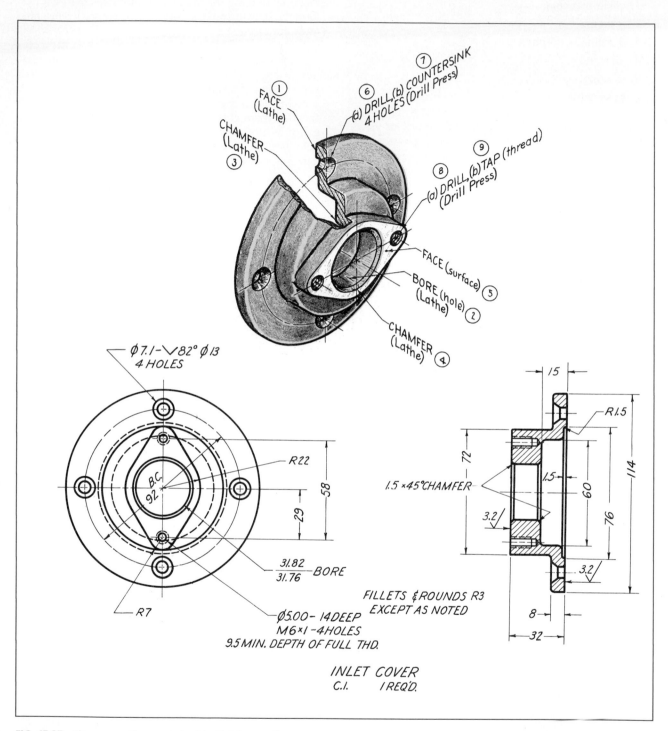

FIG. 15.27 Shop operations required to finish a casting.

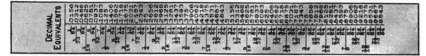

FIG. 15.28   Steel rule.

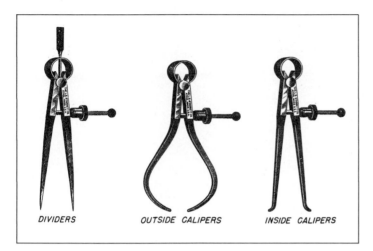

DIVIDERS   OUTSIDE CALIPERS   INSIDE CALIPERS

FIG. 15.29   Calipers.

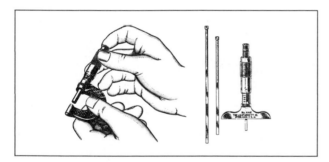

FIG. 15.30   Micrometers.

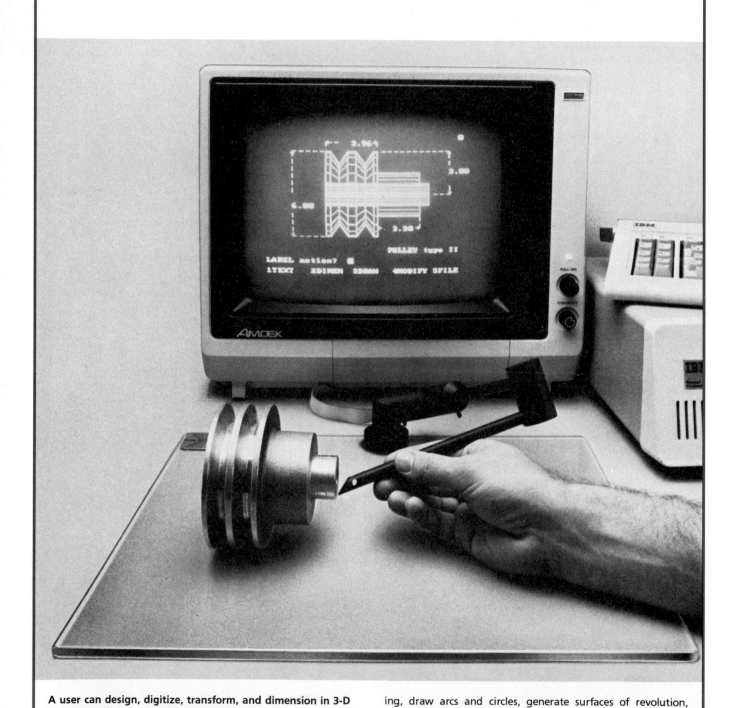

A user can design, digitize, transform, and dimension in 3-D space by utilizing a 3-D digitizer. The space tablet is a digitizer that plots and records the coordinates (x, y, and z) of points in space. All one needs to do is to position the space tablet arm at a desired location and press a button. By joining points together, 3-D wireframe models (real or imaginary) can be created. The system has the capability to scale and rotate a drawing, draw arcs and circles, generate surfaces of revolution, examine a design from orthogonal views, and finally, automatically dimension by giving angles and lengths. With a design-oriented perspective drawing displayed, the push of a single button will instantly and automatically cause the three orthogonal projections of that drawing to appear on the screen. See Chapter 17. (*Courtesy Micro Control Systems, Inc.*)

C H A P T E R · 16

Production Drawings and Process Models

A / PRODUCTION (SHOP) DRAWINGS

16.1 Communication Drawings

These varied types of engineering drawings, ranging from design drawings to exploded pictorial drawings, have one thing in common: they are prepared to convey needed ideas and facts to others. Since all serve the same purpose they may be classed together as "communication drawings," this term being almost all-inclusive.

In this chapter we will be concerned mainly with the types of drawings that are prepared by draftsmen under an engineer's supervision and that are to serve as com-

munications to others beyond the engineering department. The preparation of idea sketches, both multiview and pictorial, has been discussed in detail in Chapters 6 and 12. Charts and graphs, which may also be thought of as communication drawings, are presented in Chapter 19. Charts and graphs are used by engineers to supplement written reports and technical papers.

16.2 Sketches and Design Drawings

The first stage in the development of an idea for a structure or machine is to prepare freehand sketches and to make the calculations required to determine the feasibility of the design. From these sketches the designer prepares a layout, on which an accurate analysis of the design is worked out. It is usually drawn full-size and is executed with instruments in pencil (Fig. 12.25). The layout should be complete enough to allow a survey of the location of parts (to avoid interference), the accessibility for maintenance, the requirements for lubrication, and the method of assembly. See Chapter 12.

Usually, only center distances and certain fixed dimensions are given. The general dimensioning, as well as the determination of material and degree of finish of individual parts, is left for the draftsman, who makes the detail drawings while using the layout drawing as a guide.

Design layouts require both empirical and scientific design. Empirical design involves the use of charts, formulas, tables, and so forth, which have been derived from experimental studies and scientific computations. Scientific design, which requires a broad knowledge of the allied fields, such as mechanics, metallurgy, and mathematics, is used when a new machine is designed to operate under special specified conditions for which data are not available in any handbook.

16.3 Classes of Machine Drawings

There are two recognized classes of machine drawings: detail drawings and assembly drawings.

16.4 Set of Working Drawings

A complete set of working drawings for a machine consists of detail sheets giving all necessary shop information for the production of individual pieces and an assembly drawing showing the location of each piece in the finished machine. In addition, the set may include drawings showing a foundation plan, piping diagram, oiling diagram, and so on.

16.5 Detail Drawing

A *detail drawing should give complete information for the manufacture of a part, describing with adequate dimensions the part's size.* Finished surfaces should be indicated and all necessary shop operations shown. The title should give the material of which the part is to be made and should state the number of the parts that are required for the production of an assembled unit of which the part is a member. Commercial examples of detail drawings are shown in Figs. 16.1–16.4.

Since a machinist will ordinarily make one part at a time, it is advisable to detail each piece, regardless of its size, on a separate individual sheet. In some shops, however, custom dictates that related parts be grouped on the same sheet, particularly when the parts form a unit in themselves. Other concerns sometimes group small parts of the same material together thus: castings on one sheet, forgings on another, special fasteners on still another, and so on.

16.6 Making a Detail Drawing

With a design layout or original sketches as a guide, the procedure for making a detail drawing is as follows:

1. Select the views, remembering that, aside from the view showing the characteristic shape of the object, there should be as many additional views as are necessary to complete the shape description. These may be sectional views that reveal a complicated interior construction, or auxiliary views of surfaces not fully described in any of the principal views.

2. Decide on a scale that will allow, without crowding, a balanced arrangement of all necessary views and the location of dimensions and notes. Although very small parts should be drawn double-size or larger, to show detail and to allow for dimensions, a full-size scale should be used when possible. In general, the same scale should be used for pieces of the same size.

3. Draw the main center lines and block in the general outline of the views with light, sharp 6H pencil lines.

4. Draw main circles and arcs in finished weight.

5. Starting with the characteristic view, work back and forth from view to view until the shape of the object is completed. Lines whose definite location and length are known may be drawn in their finished weight.

6. Put in fillets and rounds.

7. Complete the views by darkening the object lines.

8. Draw extension and dimension lines.

9. Add arrowheads, dimensions, and notes.

10. Complete the title.

11. Check the entire drawing carefully.

16.7 One-view Drawings

Many parts, such as shafts, bolts, studs, and washers, may require only one properly dimensioned view. In the case of each of these parts, a note can imply the complete shape of the piece without sacrificing clearness. Most engineering departments, however, deem it better practice to show two views.

16.8 Detail Titles

Every detail drawing must give information not conveyed by the notes and dimensions, such as the name of the part, part number, material, number required, and so on. The method of recording and the location of this information on the drawing varies somewhat in different drafting rooms. It may be lettered either in the record strip or directly below the views (Figs. 16.1 and 16.2).

If all surfaces on a part are machined, finish marks are omitted and a title note, ''FINISH ALL OVER,'' is added to the detail title.

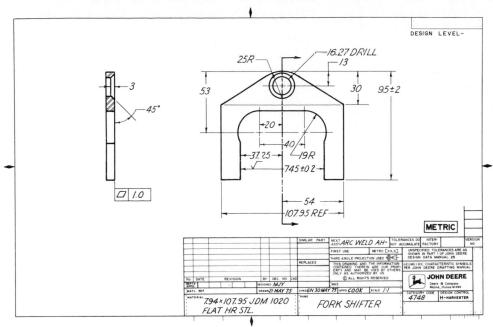

FIG. 16.1 Detail drawing. (*Courtesy John Deere and Company*)

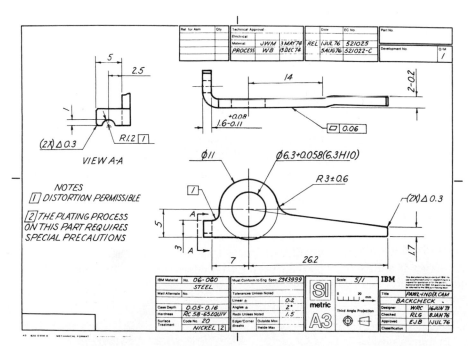

FIG. 16.2 Commercial example of a detail drawing. (*Courtesy International Business Machines Corp.*)

16.9 Title Blocks and Record Strips

The purpose of a title or record strip is to present in an orderly manner the name of the machine, name of the manufacturer, date, scale, drawing number, and other drafting-room information.

Every commercial drafting room has developed its own standard title forms, whose features depend on the processes of manufacture, the peculiarities of the plant organization, and the established customs of particular types of manufacturing. In large organizations, the blank form, along with the borderline, is printed on standard sizes of tracing paper and/or Mylar.

A record strip is a form of title extending almost the entire distance across the bottom of the sheet. In addition to the usual title information, it may contain a section for recording revisions, changes, and so on, with the dates on which they were adopted (Fig. 16.3).

16.10 Contents of the Title (Figs. 16.1 and 16.2)

The title on a machine drawing generally contains the following information.

1. Name of the part.

2. Name of the machined structure. (This is given in the main title and is usually followed by one of two words *details* or *assembly*.) See Fig. 16.7.

3. Name and location of the manufacturing firm.

4. Name and address of the purchasing firm, if the structure has been designed for a particular company.

5. Scale.

6. Date. (Often spaces are provided for the date of completion of each operation in the preparation of the drawing. If only one date is given, it is usually the date of completion of the drawing.)

7. Initials or name of the drafter who made the pencil drawing.

8. Initials of the checker.

9. Initials or signature of the chief drafter, chief engineer, or another in authority who approved the drawing.

10. Initials of the tracer (if drawing has been traced).

11. Drawing number. This generally serves as a filing number and may furnish information in code form. Letters and numbers may be so combined to indicate departments, plants, model, type, order number, filing number, and so on. The drawing number is sometimes repeated in the upper-left-hand corner (in an upside-down position), so that the drawing may be quickly identified if it should become reversed in the file.

Some titles furnish information such as material, part number, pattern number, finish, treatment, estimated weight, superseded drawing number, and so on.

16.11 Corrections and Alterations

Alterations on working drawings are made either by cancellation or by erasure. Cancellations are indicated by parallel inclined lines drawn through the views, lines, notes, or dimensions to be changed.

Superseding dimensions should be placed above or near the original ones. If alterations are made by erasure, the changed dimensions are often underlined.

All changes on a completed or approved drawing should be recorded in a revision record that may be located either adjacent to the title block (Fig. 16.3) or at one corner of the drawing (Fig. 16.4). This note should contain the identification symbol, date, authorization number, character of the revision, and the initials of the drafter and checker who made the change. The identification symbol is a numeral or letter placed in a small circle near the alteration on the body of the drawing.

If the changes are made by complete erasure, record prints should be made for the file before the original is altered. Many companies make record prints whenever changes are extensive.

Since revisions on completed drawings are usually necessitated by unsatisfactory methods of production or by a customer's request, they should never be made by a drafter unless an order has been issued with the approval of the chief engineer's office.

16.12 Pattern-shop Drawings

Sometimes special pattern-shop drawings, giving information needed for making a pattern, are required for large and complicated castings. A patternmaker who receives a drawing that shows finished dimensions must provide for the draft necessary to draw the pattern and for the extra metal for machining, as well as allow for shrinkage by making the pattern oversize. When, however, the draft and allowances for finish are determined by the engineering department, no finish marks appear on the drawing. The allowances are included in the dimensions.

16.13 Forge-shop Drawings

If a forging is to be machined, separate detail drawings usually are made for the forge and machine shops. A forging drawing gives all the nominal dimensions required by the forge shop for a completed rough forging.

16.14 Machine-shop Drawings

Rough castings and forgings are sent to the machine shop to be finished. See Fig. 16.4. Since the machinist is not interested in the dimensions and information for the previous stages, a machine-shop drawing frequently gives only the information necessary for machining.

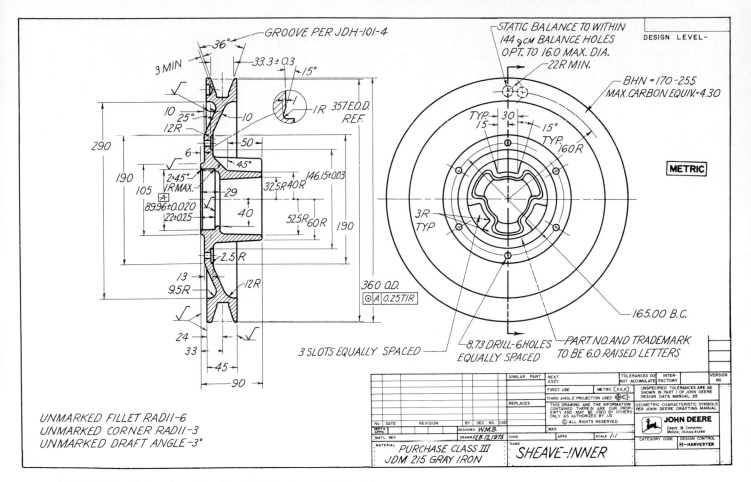

FIG. 16.3 Detail drawing. (*Courtesy John Deere and Company*)

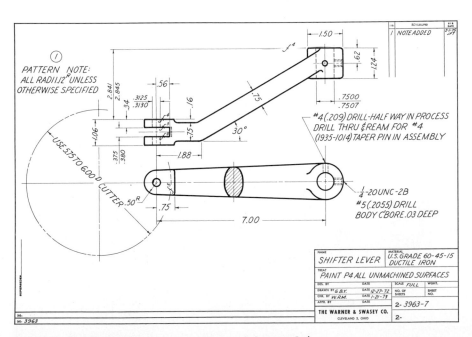

FIG. 16.4 Detail drawing. (*Courtesy Warner & Swasey Co.*)

16.15 Assembly Drawings

A drawing that shows the parts of a machine or machine unit assembled in their relative working positions is an assembly drawing. There are several types of such drawings: design assembly drawings, working assembly drawings, unit assembly drawings, installation diagrams, and so on, each of which will be described separately (Figs. 16.5–16.7 and 16.10–16.12).

16.16 Working Assembly Drawings

A working assembly drawing, showing each piece completely dimensioned, is sometimes made for a simple

mechanism or unit of related parts. No additional detail drawings of parts are required.

16.17 Subassembly (Unit) Drawings

A unit assembly is an assembly drawing of a group of related parts that form a unit in a more complicated machine. Such a drawing would be made for the tail stock of a lathe, the clutch of an automobile, or the carburetor of an airplane. A set of assembly drawings thus takes the place of a complete assembly of a complex machine (Fig. 16.6).

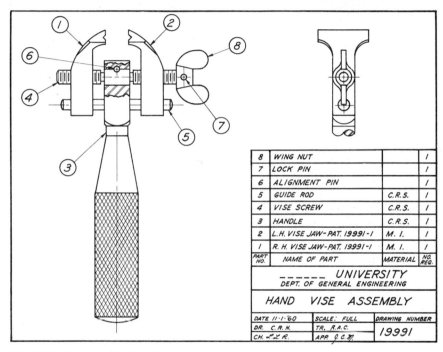

8	WING NUT		1
7	LOCK PIN		1
6	ALIGNMENT PIN		1
5	GUIDE ROD	C.R.S.	1
4	VISE SCREW	C.R.S.	1
3	HANDLE	C.R.S.	1
2	L.H. VISE JAW–PAT. 19991–1	M. I.	1
1	R. H. VISE JAW–PAT. 19991–1	M. I.	1
PART NO.	NAME OF PART	MATERIAL	NO. REQ.

_ _ _ _ _ _ UNIVERSITY
DEPT. OF GENERAL ENGINEERING

HAND   VISE   ASSEMBLY

DATE 11-1-'60	SCALE: FULL	DRAWING NUMBER
DR. C.R.H.	TR. R.A.C.	
CH. E.Z.R.	APP. O.C.M.	19991

FIG. 16.5   An assembly drawing.

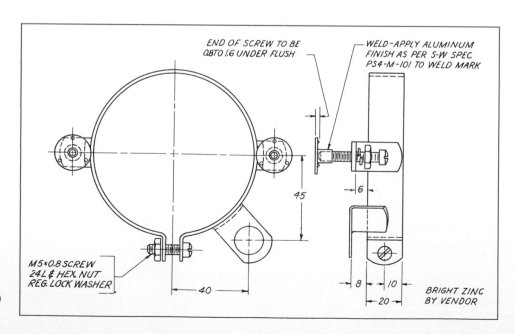

END OF SCREW TO BE 0.8 TO 1.6 UNDER FLUSH

WELD–APPLY ALUMINUM FINISH AS PER S-W SPEC. PS4-M-101 TO WELD MARK

45

6

M5×0.8 SCREW 24 L ℄ HEX. NUT REG. LOCK WASHER

40

8   10
20

BRIGHT ZINC BY VENDOR

FIG. 16.6 Tachometer mounting bracket (unit) assembly.

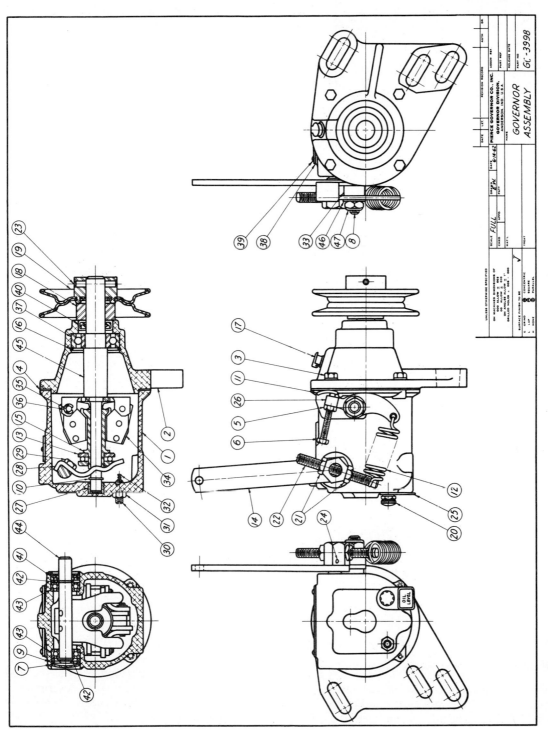

FIG. 16.7 Assembly drawing. *(Courtesy Pierce Governor Co., Inc.)*

PIERCE GOVERNOR ASSEMBLY GC-3998 PARTS LIST			
Key No.	Part Name	Part No.	Quantity
1	Governor Body	G-9042-16	1
2	Governor Flange	G-9138-3	1
3	Sems Fastener	X-1784	4
4	Gasket	X-1425	1
5	Hex. Nut	X-977	1
6	Hex. Head Screw	X-890-4	1
7	Welch Plug	X-2019	1
8	Shoulder Stud	G-9799	1
9	Washer	X-2307	2
10	Snap Ring	X-1923	1
11	Stop Bracket	G-9556	1
12	Governor Spring	SN-1304	1
13	Thrust Bearing	X-1336-A	1
14	Throttle Lever Assembly	A-6325	1
15	Thrust Sleeve	G-10813	1
16	Snap Ring—Internal	X-1921	1
17	Oil Cup	X-2053	1
18	Spacer	G-12614-1	1
19	Governor Pulley	G-10908-1	1
20	Oil Lever Check	X-2054	1
21	Hex. Nut	X-1011-1	2
22	Adj. Screw Eye	G-12306	1
23	Roll Pin	X-2620	1
24	Roll Pin	X-2602	1
25	Oil Lever Tag	X-1945	1
26	Spring Adj. Lever	G-5715	1
27	Bushing	X-2721	1
28	Yoke	G-9838	1
29	Sems Fastener	X-1687	2
30	Bumper Screw	G-5113-1	1
31	Hex. Nut	X-246-4	1
32	Bumper Spring	SN-1481	1
33	Spacer	G-11886-1	1
34	Laminated Weight Assembly	A-2446	4
35	Weight Pin	G-14007	4
36	"E" Retaining Ring	X-2996	4
37	Ball Bearing	X-310	1
38	Name Plate	X-581	1
39	Escutcheon Pin	X-455	2
40	Oil Seal	X-652	1
41	Rocker Shaft Oil Seal	A-6118	1
42	External Snap Ring	X-1904	2
43	Ball Bearing	X-328	2
44	Rocker Shaft	G-11698	1
45	Spider and Shaft Assembly	A-6637	1
46	Washer	X-2026-4	1
47	Elastic Stop Nut	X-1845	1

FIG. 16.8 Parts list—governor assembly (Fig. 16.7).

16.18 Bill of Material or Parts List

A bill of material is a list of parts placed on an assembly drawing just above the title block, or, in the case of quantity production, on a separate sheet. The bill contains the part (item or key) number, descriptive name, material, quantity (number) required, and so on, of each piece. Additional information, such as stock size, pattern number (castings), and so forth, is sometimes listed.

Suggested dimensions for ruling are shown in Fig. 16.9.

For 3.5 mm ($\frac{1}{8}$ in.) letters, the lines should never be spaced closer than 8 mm ($\frac{5}{16}$ in.). Fractions, if used, are made slightly less than full height and are centered between the lines.

When listing standard parts in a bill of material, the general practice is to omit the name of the materials and to use abbreviated descriptive titles. A pattern number may be composed of the commercial job number followed by the assigned number one, two, three, and so on. It is suggested that parts be listed in the following order: (1) castings, (2) forgings, (3) parts made from bar stock, and (4) standard parts.

Sometimes bills of material are first typed on thin paper and then blueprinted. The form may be ruled or printed (Fig. 16.8).

16.19 Title

The title strip on an assembly drawing usually is the same as that used on a detail drawing. It will be noted, when lettering in the block, that the title of the drawing is generally composed of the name of the machine followed by the word *assembly* (Figs. 16.5 and 16.7).

16.20 Making the Assembly Drawing

The final assembly may be traced from the design assembly drawing, but more often it is redrawn to a smaller scale on a separate sheet. Since the redrawing, being done from both the design and detail drawings, furnishes a check that frequently reveals errors, the assembly always should be drawn before the details are accepted as finished and the blueprints are made. The assembly of a simple machine or unit is sometimes shown on the same sheet with the details.

Accepted practices to be observed on assemblies are as follows.

1. *Sectioning.* Parts should be sectioned using the American Standard symbols shown in Fig. 7.31. The practices of sectioning apply to assemblies.

2. *Views.* The main view, which is usually in full section, should show to the best advantage nearly all the individual parts and their locations. Additional views are shown only when they add necessary information that should be conveyed by the drawing.

3. *Hidden lines.* Hidden lines should be omitted from an assembly drawing, for they tend merely to overload it and create confusion. Complete shape description is unnecessary, since parts are either standard or are shown on detail drawings.

4. *Dimensions.* Overall dimensions and center-to-center distances indicating the relationship of parts in the machine as a whole are sometimes given. Detail dimensions are omitted, except on working-assembly drawings.

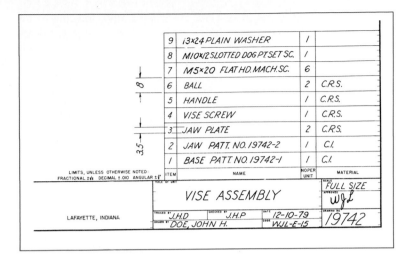

	9	13×24 PLAIN WASHER	1	
	8	M10×12 SLOTTED DOG PT. SET SC.	1	
	7	M5×20 FLAT HD. MACH. SC.	6	
	6	BALL	2	C.R.S.
	5	HANDLE	1	C.R.S.
	4	VISE SCREW	1	C.R.S.
	3	JAW PLATE	2	C.R.S.
	2	JAW PATT. NO. 19742-2	1	C.I.
	1	BASE PATT. NO. 19742-1	1	C.I.

LIMITS, UNLESS OTHERWISE NOTED:
FRACTIONAL ±1/64   DECIMAL ± .010   ANGULAR ±1°

ITEM	NAME	NO. PER UNIT	MATERIAL

TITLE OF UNIT: **VISE ASSEMBLY**   SCALE: *FULL SIZE*   APPROVED: WJL

LAFAYETTE, INDIANA

TRACED BY: J.H.D   CHECKED BY: J.H.P   DATE: 12-10-79
DRAWN BY: DOE, JOHN H.   CODE: WJL-E-15   DRAWING NO: 19742

FIG. 16.9   Bill of material.

5. *Identification of parts.* Parts in a machine or structure are identified on the assembly drawing by numbers that are used on the details and in the bill of material (Fig. 16.5). These should be made at least 5 mm ($\frac{3}{16}$ in.) high and enclosed in a 10 mm ($\frac{3}{8}$ in.) circle. The centers of the circles are located not less than 20 mm ($\frac{3}{4}$ in.) from the nearest line of the drawing. Leaders, terminated by arrowheads touching the parts, are drawn radial with a straightedge. The numbers, in order to be centered in the circles, should be made first and the circles drawn around them. An alternative method used in commercial practice is to letter the name and descriptive information for each part and draw a leader pointing to it in the main view.

16.21 Checking Drawings

Checking, the final assurance that the assembly is correctly designed, should be done by a person (checker or squad foreman) who has not prepared the drawings but who is thoroughly familiar with the principles of the design. The checker must have a broad knowledge of shop practices and assembly methods. In commercial drafting rooms, the most experienced persons are assigned to this type of work. The assembly drawing is checked against the detail drawings and corrections are indicated with either a soft or colored pencil. The checker should

1. Survey the assembly as a whole from the standpoint of operation, ease of assembly, and accessibility for repair work. The type, strength, and suitability of the materials should also be considered.

2. Check each part with the parts adjacent to it, to make certain that proper clearances are maintained. (To determine whether or not all positions are free of interference, it may be necessary to lay out the extreme travel of moving parts to an enlarged scale.)

3. Study all drawings to see that each piece has been illustrated correctly and that all necessary views, types of views, treatments of views, and scales have been shown.

4. Check dimensions by scaling; calculate and check size and location dimensions that affect mating parts; determine the suitability of dimensions from the standpoint of the various departments' needs, such as pattern, forge, machine, assembly shop, and so on; examine views for proper dimensioning; and mark unnecessary, repeated, or omitted dimensions.

5. Check tolerances, making sure the computations are correct and that proper fits have been used, so that there will be no unnecessary production costs.

6. See that finishes and such operations as drilling, reaming, boring, tapping, and grinding are properly specified.

7. Check specifications for material.

8. Examine notes for correctness and location.

9. See that stock sizes have been used for standard parts, such as bolts, screws, keys, and so on. (Stock sizes may be determined from catalogs.)

10. Add any additional explanatory notes that should supply necessary information.

11. Check the bill of material to see that each part is completely and correctly specified.

12. Check items in the title block.

13. Make a final survey of the drawing in its entirety, making certain there is either a check or correction for each dimension, note, and specification.

16.22 Installation Assembly Drawings

An installation drawing gives useful information for putting a machine or structure together. The names of parts, order of assembling parts, location dimensions, and special instructions for operating may also be shown.

16.23   Outline Assembly Drawings

Outline assembly drawings are most frequently made for illustrative purposes in catalogs. Usually they show merely overall and principal dimensions (Fig. 16.10). Their appearance may be improved by the use of line shading.

16.24   Exploded Pictorial Assembly Drawings for Parts Lists and Instruction Manuals

Exploded pictorial assembly drawings are used frequently in the parts lists sections of company catalogs and in instruction manuals. Drawings of this type are easily understood by those with very little experience in reading multiview drawings. Figure 16.11 shows a commercial example of an exploded pictorial assembly drawing.

16.25   Diagram Assembly Drawings

Diagram drawings may be grouped into two general classes: (1) those composed of single lines and conventional symbols, such as piping diagrams, wiring diagrams, and so on (Fig. 16.12); and (2) those drawn in regular projection, such as an erection drawing, which may be shown in either orthographic or pictorial projection.

Piping diagrams give the size of pipe, location of fittings, and so on. To draw an assembly of a piping system in true orthographic projection would add no information and merely entail needless work.

A large portion of electrical drawing is composed of diagrammatic sketches using conventional electrical symbols (Fig. 16.13). Electrical engineers therefore need to know the American National (ANSI) Standard wiring symbols given in the Appendix.

16.26   Chemical Engineering Drawings

In general, the chemical engineer, concerned with plant layouts and equipment design, must be well informed about the types of machinery used in grinding, drying, mixing, evaporation, sedimentation, and distillation, and must be able to design or select conveying machinery. It is obvious that the determining of the sequence of operations, selecting of machinery, arranging of piping, and so on, must be done by a trained chemical engineer who can cooperate with the mechanical, electrical, or civil engineer. To be able to do this, a thorough knowledge of the principles of engineering drawing must be acquired.

Plant layout drawings, the satisfactory development of which requires numerous preliminary sketches (layouts, scale diagrams, flow sheets, and so on), show the location

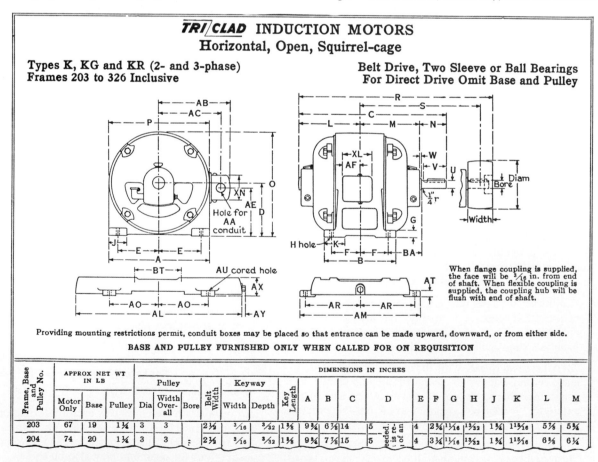

FIG. 16.10   Outline assembly drawing. (*Courtesy General Electric Company*)

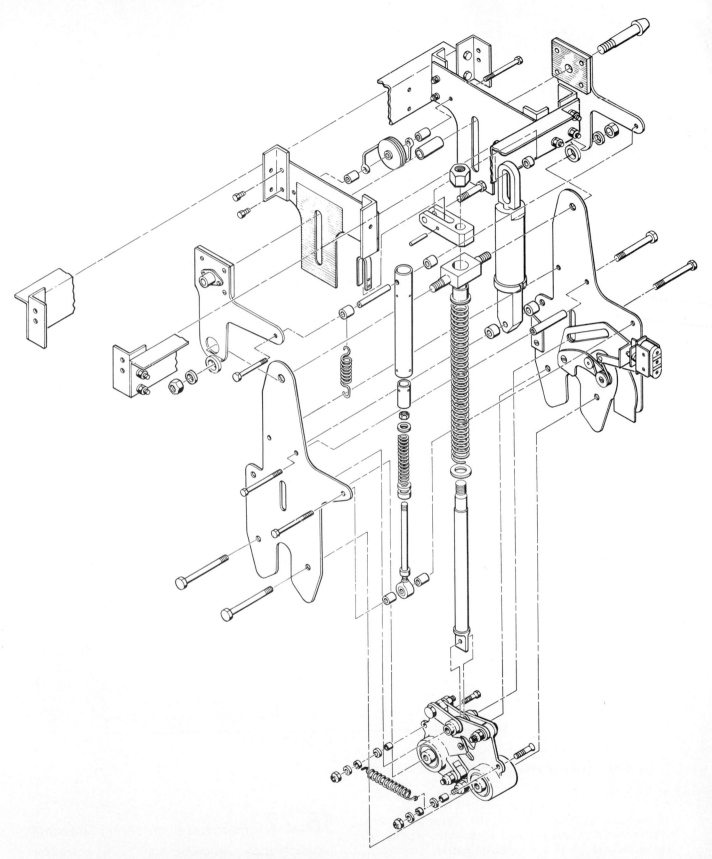

FIG. 16.11   Exploded pictorial assembly drawing. (*Courtesy Lockheed Aircraft Corporation***)**

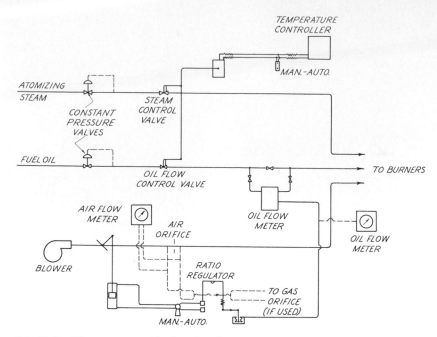

FIG. 16.12 **Diagram assembly drawing.** (*Courtesy Instruments Magazine*)

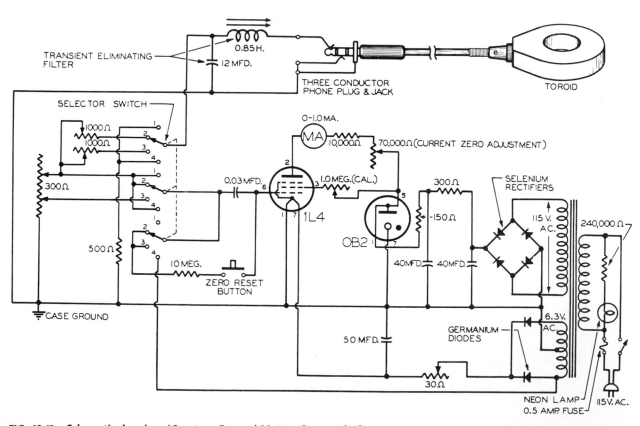

FIG. 16.13 **Schematic drawing.** (*Courtesy General Motors Corporation*)

of machines, equipment, and the like. Often, if the machinery and apparatus are used in the manufacturing of chemicals and are of a specialized nature, a chemical engineer is called on to do the designing. It even may be necessary to build experimental apparatus.

16.27 Electrical Engineering Drawings

Electrical engineering drawings are of two types: machine drawings and diagrammatic assemblies (Fig. 16.13). Working drawings, which are made for electrical machinery, involve all of the principles and conventions of the working drawings of the mechanical engineer. Diagrammatic drawings have been discussed in Sec. 16.25. The practices to be followed in preparing drawings of electrical systems are presented in Chapter 22.

16.28 Civil Engineering Drawings

The civil engineer is concerned with a broad field of construction and with civic planning. The drawings prepared for civil engineers may be in the nature of maps for city, state, and nationwide planning for streets, water systems, sewerage systems, airports, highways, railroads, harbor and waterways; or they may be design, fabrication, and erection drawings for concrete and steel structures, as in the case of buildings and bridges. Information needed for the preparation of drawings of these types may be found in Chapters 23 and 24.

B / PROCESS MODELS

16.29 Piping Models (Fig. 16.14)

Plant models with extensive piping installations usually portray complete process systems. The use of piping models varies widely from company to company. At times, some companies build models only after complete sets of design drawings, including piping assembly drawings, have been prepared. These conventional drawings show the piping along with related mechanical equipment, structural steel components, electrical conduit, and so forth. This usage makes the model an instrument that provides the means for review and complete checking of the entire system before final confirmation of the total design. However, owing to somewhat different design and construction habits, other companies build models without first preparing assembly piping drawings. In this case, the piping system is designed directly on the model with the aid of flow diagrams. When this practice is followed every pipeline and all piping accessories, such as metering instruments, valves, traps, strainers, and so forth, must be placed in proper location. When almost all design work has been done directly on the model and only a few drawings, flow charts, and sketches have been prepared, the model with all related systems shown ultimately becomes the means for communicating design information to the construction crews in the field (Fig. 16.15).

Since in most cases it is difficult to coordinate an entire processing system that involves complicated piping, all phases of the total system should be shown on the model along with the necessary piping. If all of the related mechanical and electrical equipment, along with minor structural components, platforms, ladders, and supports, are properly positioned with related items such as ducts, chutes, instruments, electrical conduits, and lights, and each pipe is traced out and checked to see that requirements are met, it is most likely that all phases of the total system will fit together at the construction site.

When the model is nearing completion, the design should be examined critically for inaccuracies and faulty

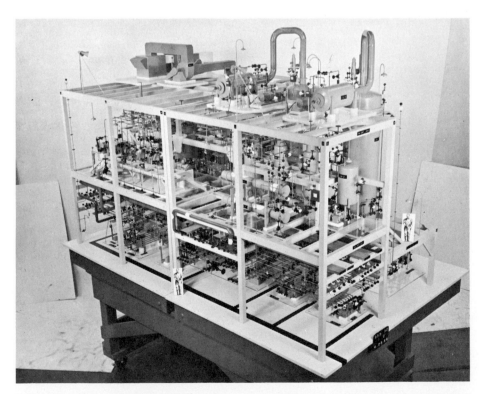

FIG. 16.14 Typical equipment layout model showing processing. (*Courtesy E. I. du Pont de Nemours Company*)

representation. In reviewing the model at this stage those responsible for the project should determine whether the proposal fulfills the requirements of the process and the most economical design has been adopted.

Piping models are built to scales that vary from $\frac{1}{4}$ in. to the foot to 1 in. to the foot. When conventional design and working drawings have been prepared and accuracy of the model is not important, a small-size model prepared to a scale of either $\frac{1}{4}$ or $\frac{3}{8}$ in. to the foot proves to be satisfactory. However, if the design is to be worked out on the model, the scale should be at least $\frac{3}{4}$ in. to the foot.

When the representation of each piece of equipment has been mounted in its proper location on the basic model, consisting essentially of a baseboard and a replica of the structural design, the piping can be installed through and around the formed background (Fig. 16.14). The two techniques that are commonly employed for the piping are known as (1) the *fine-wire* method and (2) the *true-scale* technique. When the fine-wire method is used, a fine wire with fibre discs or rubber sleeves mounted along its length is used to represent the pipe. The wire, normally $\frac{1}{16}$ in. in diameter, represents the center line of the pipe, while the outside diameter of the discs or sleeves indicates the outside diameter of the pipe.

Because fine-wire models sometimes prove unsatisfactory due to built-in inaccuracies and poor representation, many companies are now using the true-scale technique for building true-scale models (Fig. 16.15). In building these models, plastic rod or plastic tubing, having outside diameters equal to the scaled diameters of the various sizes of standard pipe, is used for the piping. A full line of true-scale piping and piping components of plastic may be purchased ready for use. Plastic, as a model material, is easy to work, light in weight, and may be readily joined using solvent cements. It is for these reasons that it is so widely used for models.

After a model has been reviewed and approved it is color-coded, tagged, and made ready for shipment to the plant construction site. On models that have been prepared primarily for confirmation and approval, pipelines are coded to indicate particular systems to which the lines belong. That is, water lines might be shown in blue, while lines carrying a specified chemical might be presented in green. In the case of models for which there are few, if any, conventional drawings, pipelines are coded to indicate their construction specification.

Three-dimensional models enable designers to discover errors, detect interferences, and determine the most economical design more readily than would be possible from drawings alone. In addition, models prove invaluable to those directing construction work in the field. However, drawings will always be needed since not all details can be incorporated and indicated on a model.

C / REPRODUCTION AND DUPLICATION OF ENGINEERING DRAWINGS

16.30   Introduction

Usually it is necessary to duplicate a set of drawings of a machine or structure, one or more copies being made for the office and extra copies for interested persons connected with the home organization or an outside cooperating firm. Sometimes many sets are required. Various

FIG. 16.15   Complete true-scale production engineering model ready for use at the construction site. (*Courtesy Procter and Gamble*)

parts of a machine or structure may be produced in numerous departments and plants that are located miles apart. Each of these departments must have exact copies of the original drawings. To satisfy this demand, several economical processes have been devised.

The various processes may be grouped in accordance with the general similarity in methods. The mechanical processes form a group that includes mimeographing, hectographing, and printing; the photochemical processes, using reflected light, include photography and photocopying; the photochemical processes, requiring transmitted light, include blueprinting, Ozalid printing, Van Dyke printing, and so on. A few processes, such as photolithography, are combinations of methods.

16.31 *Blueprints*

A blueprint may be considered to be a photographic copy of an original drawing in that the process is similar to that of photography. A piece of sensitized paper is exposed to light transmitted through a negative (tracing) and then developed in water to bring out the image. The negative is a translucent sheet of paper or Mylar on which the image of the original drawing has been drawn with opaque lines.

16.32 *Blue-line Prints*

A positive blue-line print, having blue lines on a white background, may be made on regular blueprint paper by exposing with a brown-print negative in the place of the original tracing. Such prints may also be made from an original tracing.

16.33 *Ozalid Prints*

Ozalid prints, depending on the type of Ozalid paper used, show black, blue, or maroon lines on a white background. The prints are exposed in the same manner as for other contact processes. They are developed in controlled dry ammonia vapor and, since no liquid solution is used with this process, a finished print will be an undistorted exact-scale reproduction of the original tracing.

Ozalid prints, such as the one shown in Fig. 16.16, have

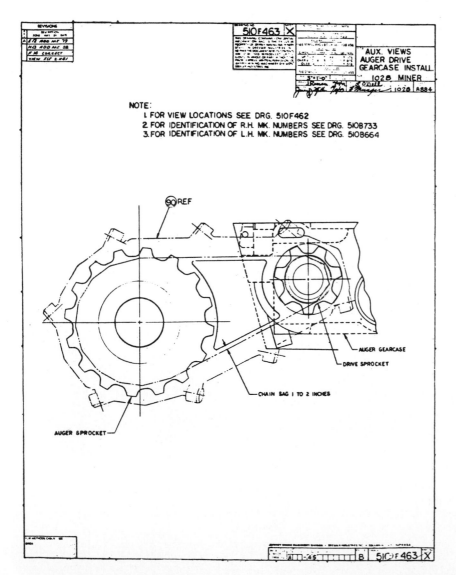

FIG. 16.16 An ozalid print.

largely been replaced by prints using the diazo and electrostatic processes.

16.34 *Black-and-White (Diazo) Prints*

Positive black-line prints may be made directly from an original tracing by using a specially prepared blackprint paper. When developed, the sensitized surface turns black where it has been protected and remains white where exposed. Only two steps are required to obtain a fully developed print, exposure and development. This method of printing is similar to that used for making blueprints, except that the print is developed by a solution in a special developing unit on top of the printer (Fig. 16.17). Black-and-white prints are more desirable than blueprints, because of their greater legibility and the fact that they can be made in only two steps, whereas blueprints require five.

16.35 *Xerox Engineering Print System (Fig. 16.18)*

The Xerox system produces a high-quality reproduction of engineering drawings. It is a complete self-contained system that combines speed, high volume (40 prints per minute) output, push-button control for convenience, on-line folding, and collating. The system uses a dry reproduction process that is safe and odorless since no wet chemicals or gasses are used. Clean, sharp blue-on-white prints are made using ordinary untreated paper.

The printer has two reduction selections in addition to full-size—65% and 50%. Although automatic operation coordinates size of input document with the reduction, the operator by manual control can override the automatic functions for special applications. With this reduction feature, engineering documentation can now be easily and inexpensively reduced for inclusion in design proposals, reports, and project specifications.

FIG. 16.17 Bruning diazo printing and developing machine. (*Courtesy Charles Bruning Company, Inc.*)

16.36 *Photostats*

Photostats are direct photographic reproductions made from original drawings or printed matter by using an automatic machine equipped with a large camera that focuses an image directly on a sensitized paper. After being properly exposed, the print is washed and dried within the machine. Copies may be made to any scale of reduction or enlargement.

This method is used occasionally for preparing drawings for engineering reports and for changing the scales of related drawings so that they may be combined.

16.37 *Electrostatic Prints*

Prints, called photocopies, can be made in considerable number at very little expense. The process employs an electrostatically charged plate.

After the image of the original has been projected onto the plate, a negatively charged powder, which will adhere to the positively charged areas (lines) of the image, is applied over the plate. Then, by means of a positive electrical charge, the powder is transferred to the paper and permanently fixed onto the surface as the last step in the process.

16.38 *Long-distance Xerography (LDX)*

Long-distance xerography networks, utilizing microwave, coaxial cable, or special telephone line, provide a practical means for transmitting and receiving electronically any form of communication, typed or drawn, across a plant area or across the continent. The high-speed, high-capacity facsimile transmission and receiving equipment consists of an LDX scanner and LDX printer. These units

FIG. 16.18 Xerox engineering print system. (*Courtesy Xerox Corporation*)

are shown in Fig. 16.19. The document scanner on the sending end converts images to video transmission signals. At the receiving end the printer electronically restores the transmitted images and produces a black-on-white copy of the original document, sketch, or drawing. The system is easy to use. The original is simply inserted into the feed slot of the scanner. It is returned to the operator in seconds. The scanner senses colors without exposure adjustment. Also see Fig. 12.5.

16.39 *Microfilming Drawings*

Microfilming is an efficient method of storing, reproducing, and transmitting engineering drawings. Some companies have found that a microfilming program will reduce their storage space requirements by as much as 75%.

Reduction in filing space is made possible largely because obsolete drawings may be removed from the files and destroyed after they have been microfilmed.

Of equal importance is the protection offered when microfilm prints can be filed away in fireproof vaults. Since lost original tracings may be quickly replaced by first making enlarged printing intermediates, it can be said that microfilming offers a form of insurance both against careless loss through handling and from fire.

The three sizes of film in general use today are the 16 mm, 35 mm and 70 mm. Although the most common size is the 35 mm, the 16 mm is often used for small drawings. One large corporation uses 70 mm for tracings as large as 95 cm × 130 cm. Tracings, which may be as large as 140 cc × 600 cm, are microfilmed in sections, and related negatives are filed together so that prints can be made at a later time should the need arise.

FIG. 16.19 Transmitting drawings and sketches electronically. (*Courtesy Xerox Corporation*)

Microfilm negatives may be cut and filed individually in envelopes marked with the same number given the tracing. Negatives are made for subsequent revisions, thus permitting the destruction of unnecessary and outdated tracings.

A film may be mounted in an engineering document card, commonly called an *aperture card*. The card shown in Fig. 16.20 is a form of standard data card that has been keypunched for mechanized equipment capable of sorting, filing, and retrieving cards. If necessary, a keypunched and imaged duplicate aperture card can be easily produced for distribution. Groups of microfilmed drawings may be printed together on a larger negative, called a microfiche, for easier review.

An engineer may obtain any needed information con-cerning the progressive development of a part or mechanism by viewing the negative on a microfilm reader or by using a portable projector (Fig. 16.20). When a full-size print, a reduced-scale print, or an intermediate is needed, either can be quickly made by enlargement by using the negative, mounted on an aperture card, in a microfilm printer designed for printing from microfilm.

16.40 Storing Drawings in Digital Form

Drawings prepared on an CADD system are stored digitally on tape or disks and not generally kept as hard-copy originals. See Sec. 17.18.

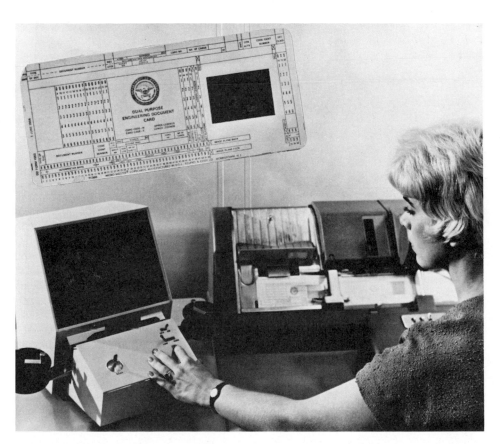

FIG. 16.20 Microfilm reader and aperture card. (*Courtesy Keuffel & Esser Co.*)

PROBLEMS

The four general types of problems presented in this chapter have been designed to furnish practice in the preparation of working drawings or sketches. The first type is composed of dimensioned pictorial drawings of individual pieces taken from a variety of mechanisms. The student should prepare complete working detail drawings of these pieces as they may be assigned by the instructor. It should be recognized that dimensions are not necessarily placed the same on orthographic views as they are on pictorial drawings. To make it possible for the student to apply the principles presented in Chapter 13, no special effort has been made to place dimensions in accordance with the rules of good practice.

The second type of problem is that which shows in pictorial all the parts of a unit mechanism. This gives the student an opportunity to prepare a complete set of working drawings of a simple unit. It is suggested that the detail drawings be prepared before the assembly is drawn.

The third and fourth types provide practice in both reading and preparing drawings, the third requiring the preparation of detail drawings from given assembly drawings, the fourth requiring the making of assembly drawings from the details.

NOTE: The inch values given on the drawings of problems may be converted to millimeters by using Appendix Tables 2 and 3. It is common practice under the metric system to give a dimension that is not critical or a size dimension for a hole or a shaft in full millimeters. For drill sizes see Tables 27 and 52. Metric threads may be found in Tables 9 and 10. See Table 5 for metric tolerance equivalents.

1–13. (Figs. 16.21–16.33). Make a detail drawing of an assigned machine part. Draw all necessary views. Give a detail title with suitable notes concerning material, number required, etc.

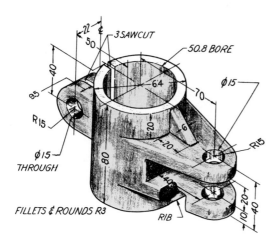

FIG. 16.21   Guide bracket.

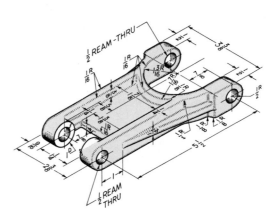

FIG. 16.22   Link.

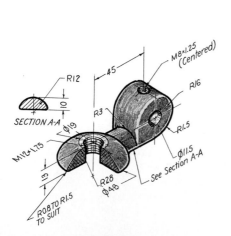

FIG. 16.23   Handle block.

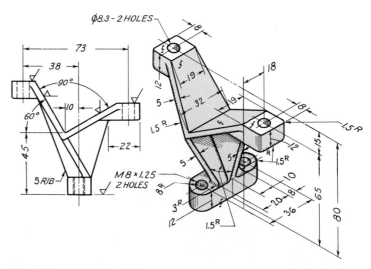

FIG. 16.24   Angle bracket.

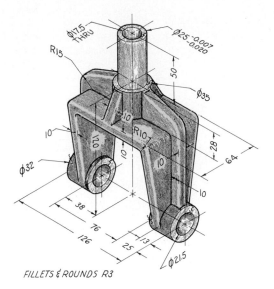

FILLETS & ROUNDS R3

FIG. 16.25   Caster frame.

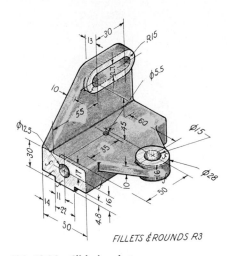

FILLETS & ROUNDS R3

FIG. 16.26   Slide bracket.

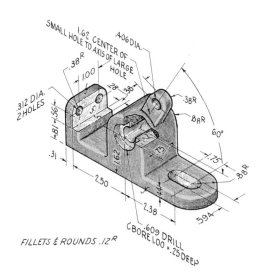

FILLETS & ROUNDS .12ᴿ

FIG. 16.27   Stabilizer bracket.

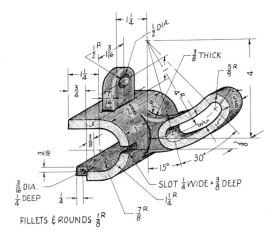

FILLETS & ROUNDS ⅛ᴿ

FIG. 16.28   Gear-shifter link.

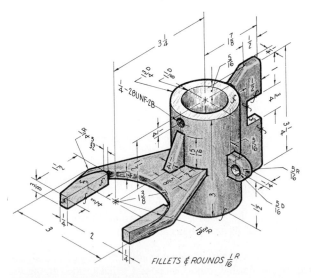

FILLETS & ROUNDS ¹⁄₁₆ᴿ

FIG. 16.29   Gear shifting fork.

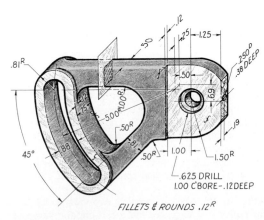

.625 DRILL
1.00 C'BORE-.12DEEP

FILLETS & ROUNDS .12ᴿ

FIG. 16.30   Shifter guide.

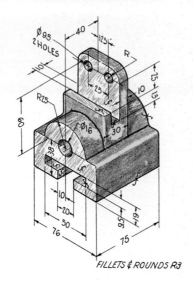

FILLETS & ROUNDS R3

FIG. 16.31 **Slide block.**

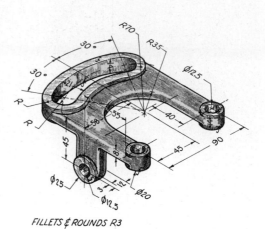

FILLETS & ROUNDS R3

FIG. 16.32 **Shifter arm.** (Dimension values shown in [ ] are in millimeters.)

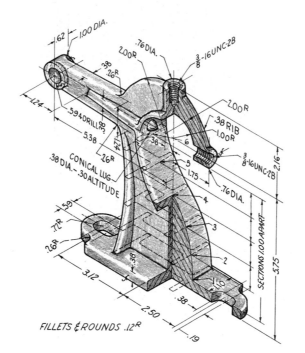

FILLETS & ROUNDS .12R

FIG. 16.33 **Shaft hanger.**

SECTION NUMBER	A	B	C	D	E
1	1.00	2.00	3.00	1.88	1.62
2	.94	1.88	2.82	1.75	1.38
3	.62	1.82	2.44	1.62	1.12
4	0.00	1.88	1.88	1.50	.88
5	.68	2.12	1.44	1.38	.62
6	1.00	—	—	1.25	—

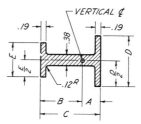

14. (Fig. 16.34). Make a two-view freehand detail sketch of the fan spindle. Determine dimensions by transferring them from the drawing to one of the given scales, by means of the dividers. The material is SAE 1045, CRS. The limits for the spindle may be found in Table 26A, in the Appendix. The last figure in the bearing number is the bearing bore number given in the tables. Use a free running fit between spindle and felt retainer, spindle and felt retaining washer, and spindle and cone-clamp washer.

15. (Fig. 16.34). Make a complete two-view detail drawing of the fan pulley. It is suggested that a half-circular view and a full-sectional view be shown. Determine the dimensions as suggested

in Problem 14. Housing limits for the given bearings will be found in Table 26B in the Appendix.

16. (Fig. 16.35). Make a detail drawing of an assigned part of the pipe stand.

17. (Fig. 16.36). Make a fully dimensioned drawing of an assigned part of the tool holder.

18. (Fig. 16.37). Make a detail working drawing of an assigned part of the flexible joint. Compose a suitable title, giving the name of the part, the material, etc.

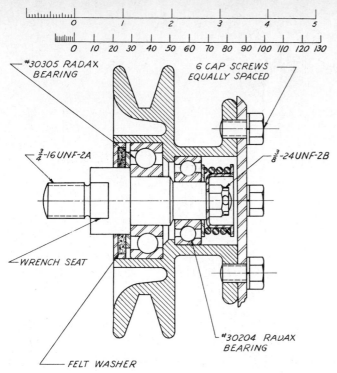

#30305 RADAX
BEARING

6 CAP SCREWS
EQUALLY SPACED

$\frac{3}{4}$-16UNF-2A

$\frac{3}{8}$-24UNF-2B

WRENCH SEAT

#30204 RADAX
BEARING

FELT WASHER

FIG. 16.34   Fan assembly.

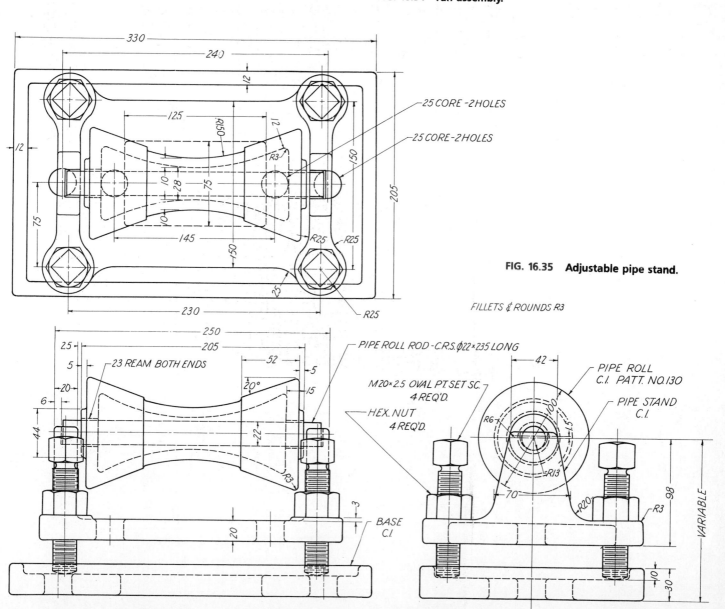

25 CORE -2HOLES

25 CORE -2HOLES

FIG. 16.35   Adjustable pipe stand.

FILLETS & ROUNDS R3

330

240

12

125

R150

R3

R25   R25

R25

230

12

75

10

28

75

150

150

25

145

10

205

PIPE ROLL ROD -C.R.S.Ø22×235 LONG

23 REAM BOTH ENDS

250

205

52

5

5

20

20°

15

6

44

22

R3

3

20

BASE
C.I.

M20×2.5 OVAL PT.SET SC.
4 REQ'D.

HEX. NUT
4 REQ'D.

PIPE ROLL
C.I. PATT. NO.130

PIPE STAND
C.I.

42

100

1.5

R6

R13

70

R20

R3

98

VARIABLE

10

30

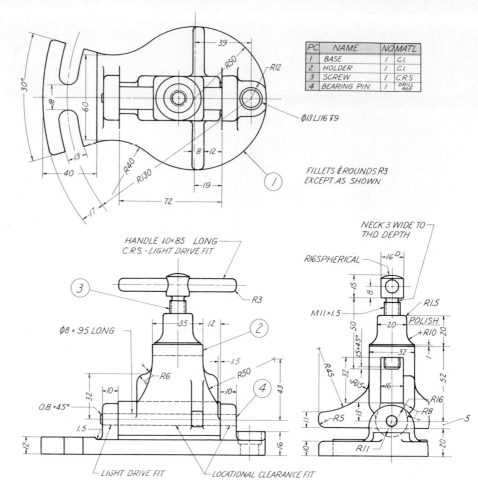

PC	NAME	NO.	MATL
1	BASE	1	C.I.
2	HOLDER	1	C.I.
3	SCREW	1	C.R.S
4	BEARING PIN	1	DRILL ROD

FILLETS & ROUNDS R3
EXCEPT AS SHOWN

FIG. 16.36   Tool holder.

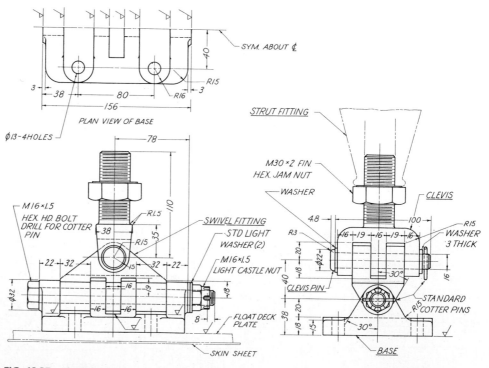

FIG. 16.37   Flexible joint.

19. (Fig. 16.38). Make a detail working drawing of an assigned part of the stud bearing. Compose a suitable title, giving the name of the part, the material, etc.

20. (Fig. 16.39). Make a detail drawing of an assigned part of the conveyer take-up unit.

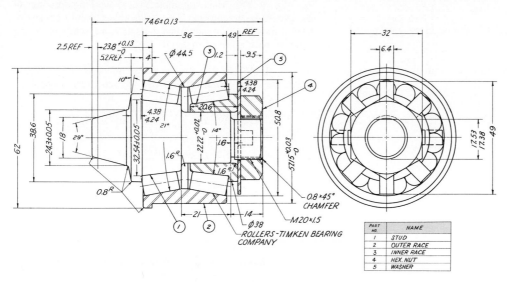

FIG. 16.38 Stud bearing. (*Courtesy Ross Gear Division of TRW, Inc.*)

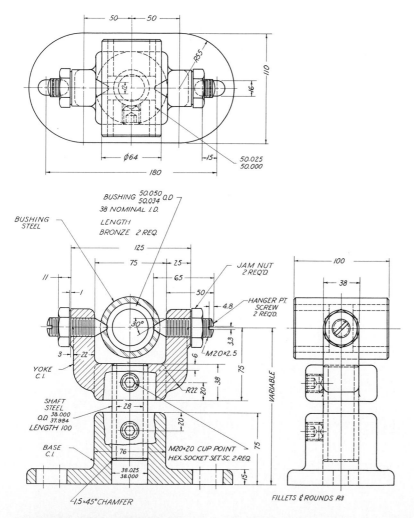

FIG. 16.39 Conveyer take-up unit.

21. (Fig. 16.40). Make a detail drawing of an assigned part of the Simplex ball-bearing screw jack.

22. (Fig. 16.41). Make a detail drawing of an assigned part of the bench arbor press.

23. (Fig. 16.42). Make a detail drawing of an assigned part of the gear pump.

24. (Fig. 16.43). Make a detail drawing of an assigned part of the bench grinder.

25. (Fig. 16.44). Make detail drawings of the tool carrier and the holder. Make an assembly drawing showing all of the parts in their relative positions.

26. (Fig. 16.45). Make a complete set of working drawings of the adjustable attachment. The complete set should consist of detail drawings of the individual parts and an assembly drawing complete with a bill of material.

27. (Fig. 16.46). Make a two-view assembly drawing of the cup center, using the given details. Use the schematic symbol for

PC.NO	NAME	QUAN.	MATERIAL	PC.NO	NAME	QUAN.	MATERIAL
1	STANDARD	1	MALL. IRON	5	GROOVE PIN	3	$\frac{7}{32}D \times \frac{5}{8}$ STEEL ROD
2	SCREW	1	S.A.E. 1120 FORGING	6	LEVER BAR	1	REROLLED RAIL STK.
3	CAP	1	S.A.E. 1045 FORGING	7	$\frac{7}{8}$ DIA. BALL BEARING	1	STD.
4	THRUST WASHER	1	S.A.E. 2315				

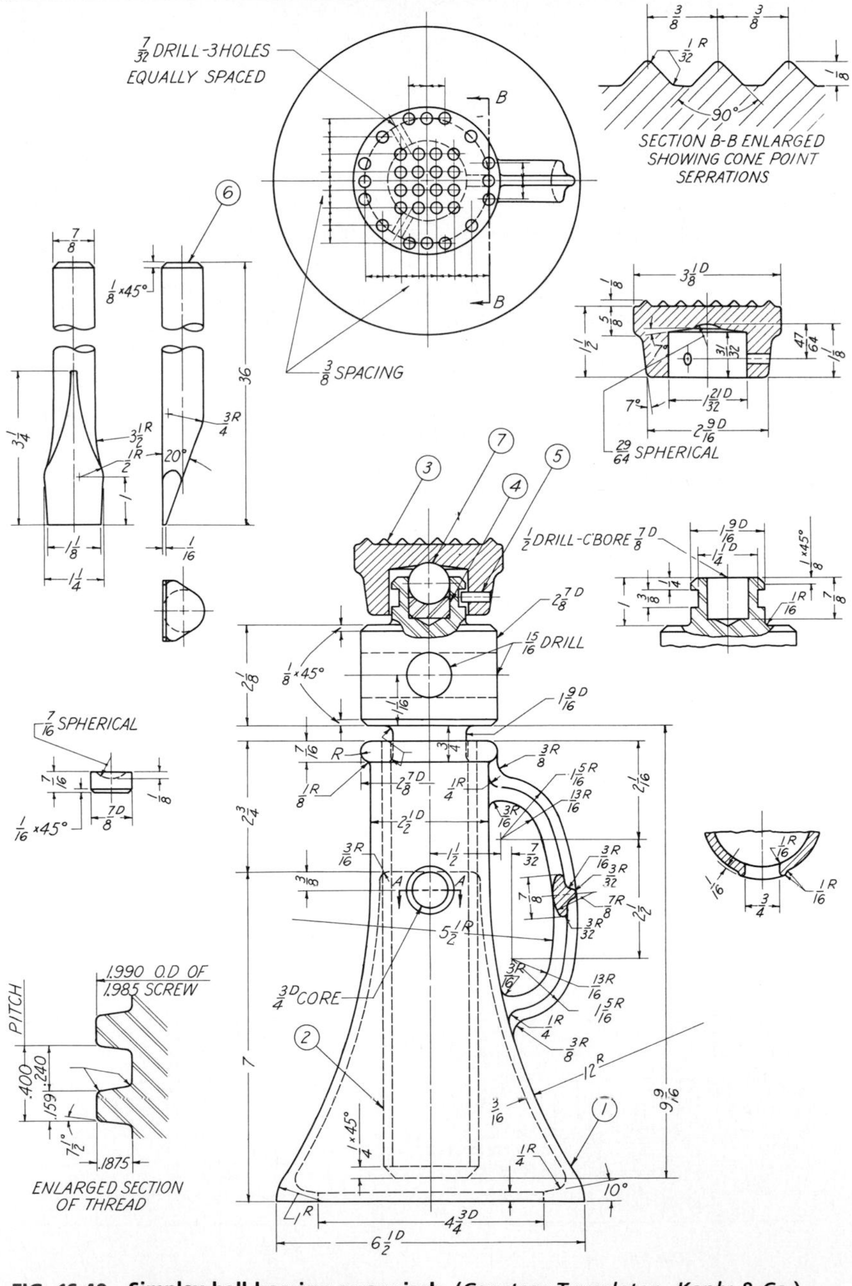

FIG. 16.40   Simplex ball-bearing screw jack. (*Courtesy Templeton, Kenly & Co.*)

screw threads. Study the pictorial drawing carefully before starting the views.

28. (Fig. 16.47). Make an assembly drawing of the radial engine unit, using the given details. It is suggested that one piston be shown in full section so that the relative positions of the parts will be revealed.

29. (Figs. 16.48–16.50). Make an assembly drawing of the hand-clamp vise, using the given details. Use the schematic symbol for screw threads.

30. (Figs. 16.51–16.53). Make an assembly drawing of the blowgun.

31. (Figs. 16.54–16.56). Make an assembly drawing of the hand grinder.

32. (Figs. 16.57–16.59). Make an assembly drawing of the right-angle head, using the given details.

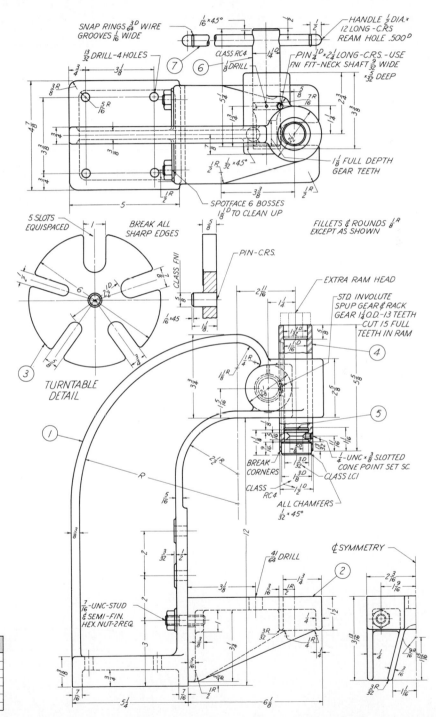

PART NO.	NAME	MATERIAL	NO. REQ.
1	BASE	C.I.	1
2	TABLE	C.I.	1
3	TURNTABLE	C.I.	1
4	RAM	S.A.E. 1045	1
5	RAM HEAD	S.A.E. 1040	1
6	SPINDLE	S.A.E. 1045	1
7	HANDLE	C.R.S.	1

FIG. 16.41   Bench arbor press.

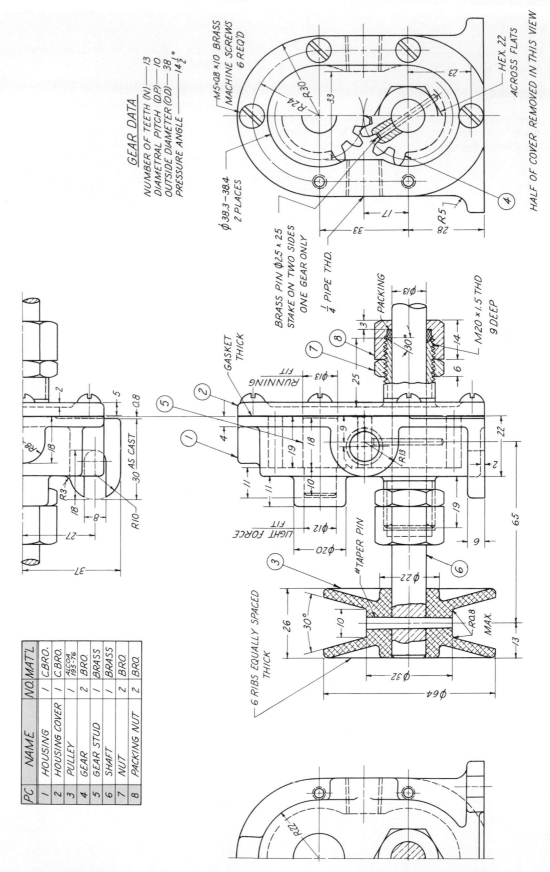

PC	NAME	NO	MAT'L
1	HOUSING	1	C.BRO.
2	HOUSING COVER	1	C.BRO.
3	PULLEY	1	ALCOA 195-T6
4	GEAR	2	BRO.
5	GEAR STUD	1	BRASS
6	SHAFT	1	BRASS
7	NUT	2	BRO.
8	PACKING NUT	2	BRO.

GEAR DATA

NUMBER OF TEETH (N) —— 13
DIAMETRAL PITCH (D.P.) —— 10
OUTSIDE DIAMETER (OD) —— 38.
PRESSURE ANGLE —— 14½°

FIG. 16.42 Gear pump.

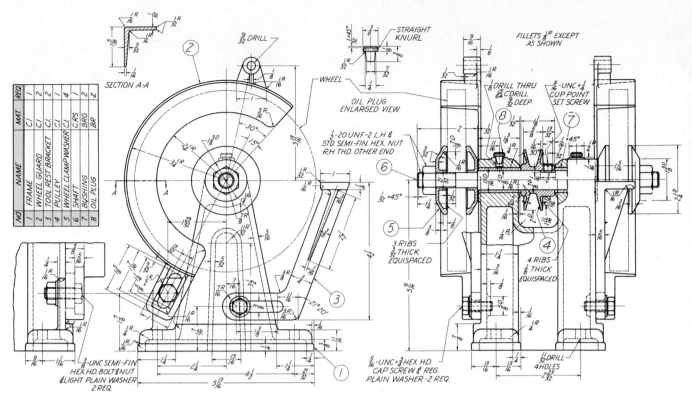

FIG. 16.43 **Bench grinder.**

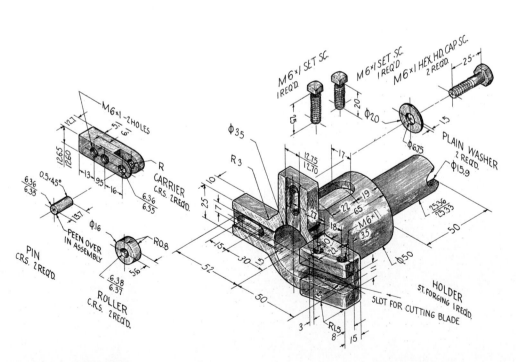

FIG. 16.44 **Tool holder.**

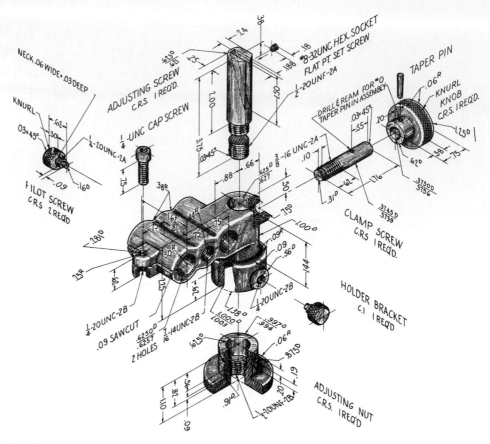

FIG. 16.45 **Adjustable attachment.**

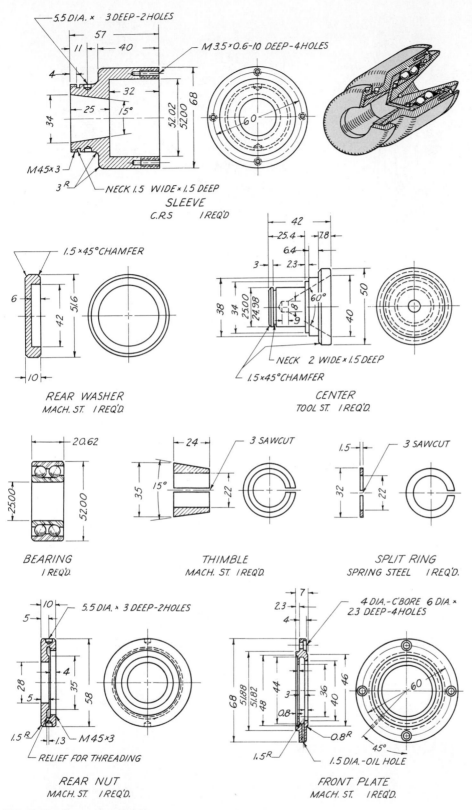

FIG. 16.46 **Cup center details.**

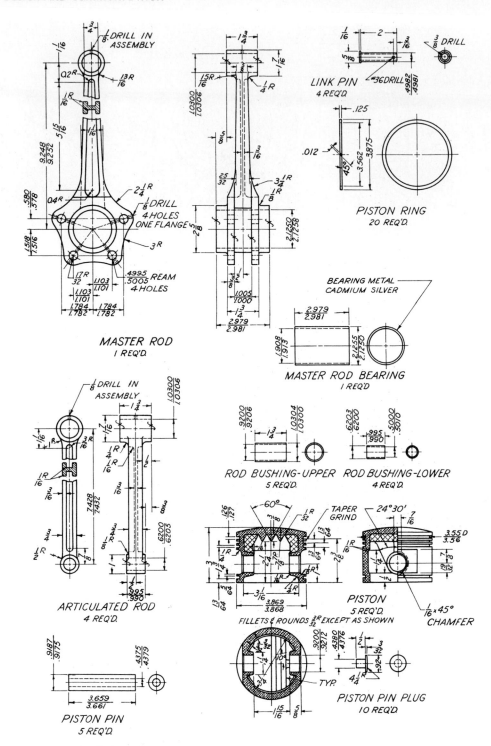

FIG. 16.47 **Radial engine details.**

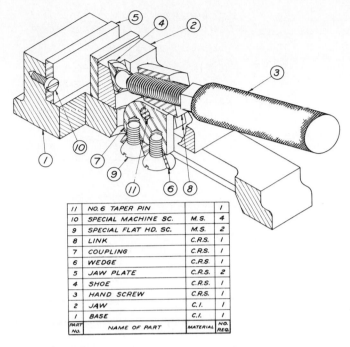

11	NO. 6 TAPER PIN		1
10	SPECIAL MACHINE SC.	M.S.	4
9	SPECIAL FLAT HD. SC.	M.S.	2
8	LINK	C.R.S.	1
7	COUPLING	C.R.S.	1
6	WEDGE	C.R.S.	1
5	JAW PLATE	C.R.S.	2
4	SHOE	C.R.S.	1
3	HAND SCREW	C.R.S.	1
2	JAW	C.I.	1
1	BASE	C.I.	1
PART NO.	NAME OF PART	MATERIAL	NO. REQ.

FIG. 16.48 Hand-clamp vise.

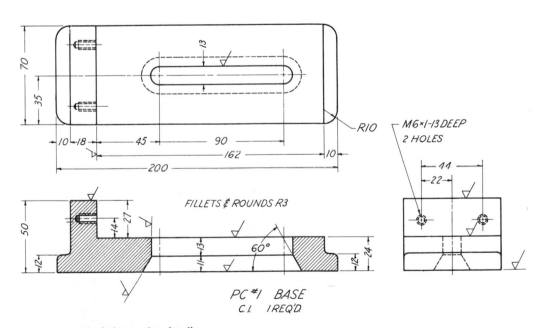

PC #1 BASE
C.I. 1 REQ'D.

FILLETS & ROUNDS R3

FIG. 16.49 Hand-clamp-vise details.

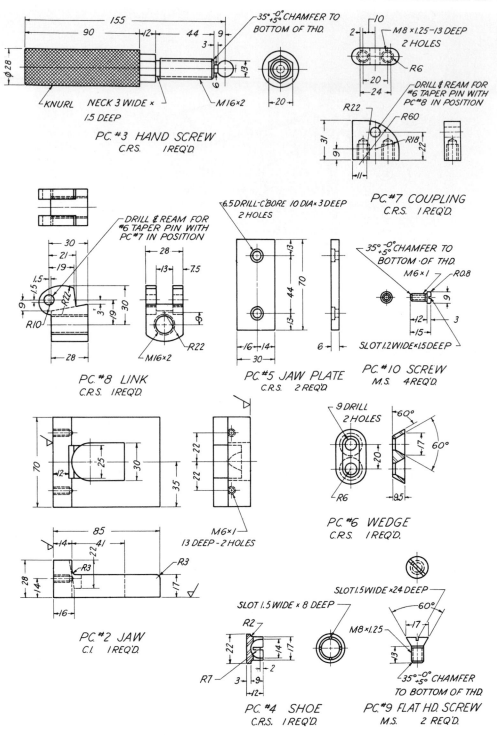

FIG. 16.50 Hand-clamp-vise details.

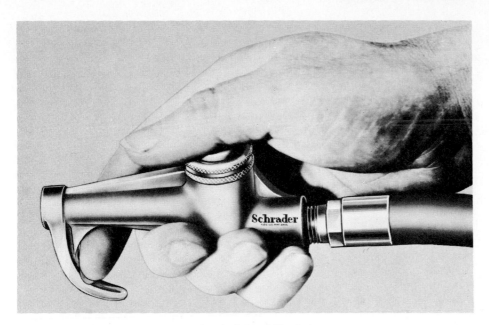

FIG. 16.51 **Blowgun.** (*Courtesy A. Schrader's Son Mfg. Co.*)

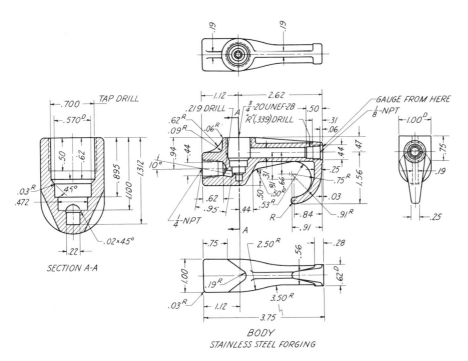

FIG. 16.52 **Blowgun details.**

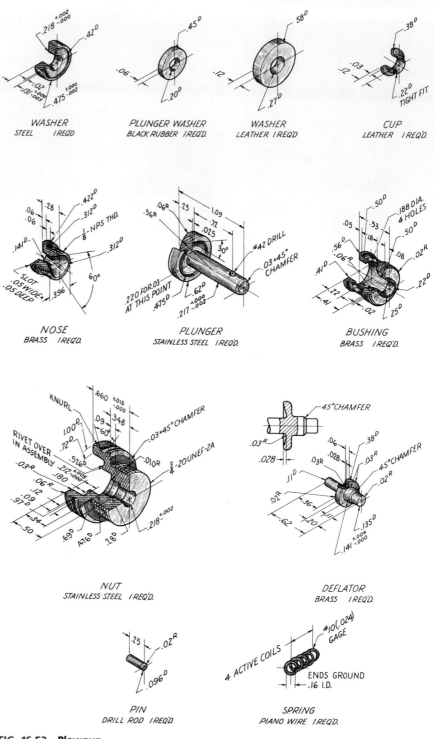

FIG. 16.53 **Blowgun.**

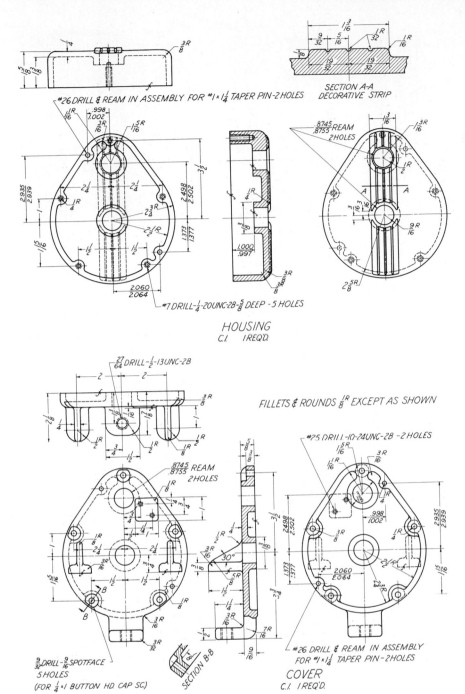

FIG. 16.54 Hand-grinder details. (See Fig. 16.55.)

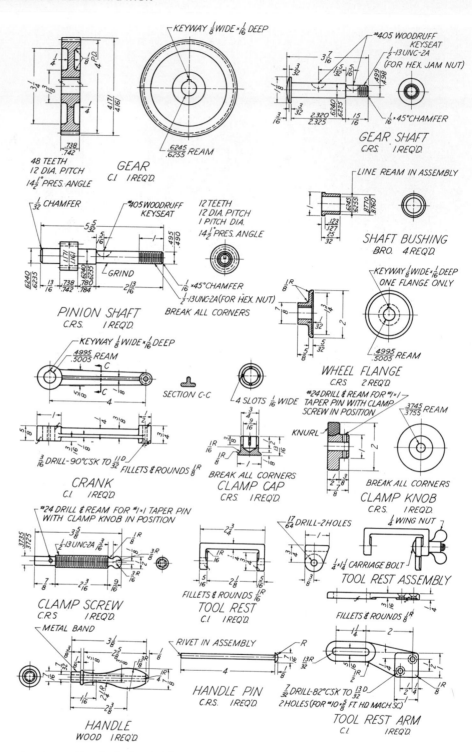

FIG. 16.55   Hand-grinder details.

FIG. 16.56 Hand grinder.

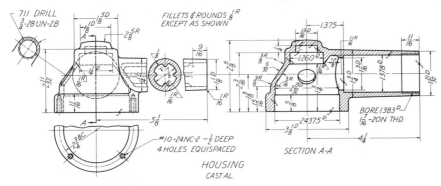

FIG. 16.57 Right-angle-head details. (*Courtesy R. C. Haskins Co.*)

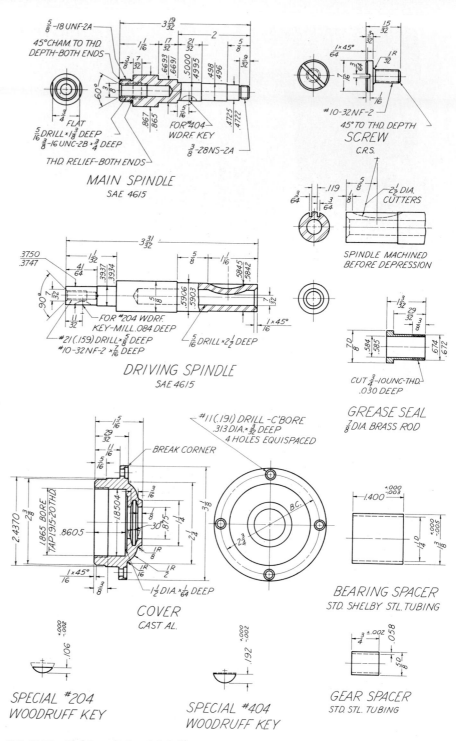

FIG. 16.58   **Right-angle-head details.**

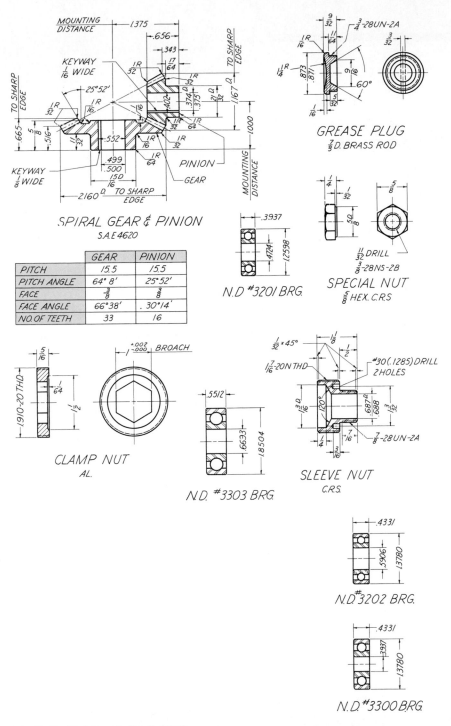

SPIRAL GEAR & PINION
S.A.E 4620

	GEAR	PINION
PITCH	15.5	15.5
PITCH ANGLE	64° 8'	25°52'
FACE	3/8	3/8
FACE ANGLE	66°38'	.30°14'
NO. OF TEETH	33	16

N.D #3201 BRG.

GREASE PLUG
7/8 D. BRASS ROD

SPECIAL NUT
5/8 HEX. C.R.S

CLAMP NUT
AL.

N.D. #3303 BRG.

SLEEVE NUT
C.R.S.

N.D #3202 BRG.

N.D #3300 BRG.

FIG. 16.59 Right-angle-head details.

PART · 5

CAD/CAM: COMPUTER-AIDED DESIGN AND COMPUTER-AIDED MANUFACTURING

Shown are students at work in the Purdue interactive computer-aided drafting system laboratory for advanced study.

(*Courtesy Department of Technical Graphics, School of Technology, Purdue University*)

CHAPTER · 17

Computer-aided Design and Drafting

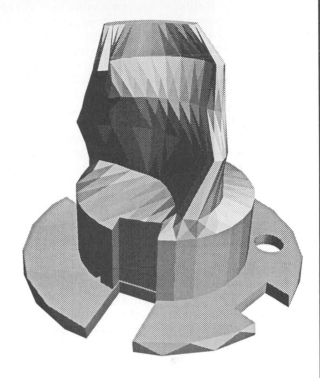

17.1 Introduction

This edition of *Fundamentals of Engineering Drawing* presents the fundamentals of drawing, manual or CADD, that can be applied in any engineering environment. The authors acknowledge that there are differences in CADD systems, especially in terminology. Luckily, these differences are generally superficial. If the general concepts of CADD are understood, they can be applied to any system. For example, one system may *insert* a line while another may *create* a line and a third *model* a line. One system may use a *symbol* while another may use a *template*, *pattern*, *library part*, or *cell*. It is important to understand the function of these concepts in engineering design and

drawing. This chapter gives a general overview, covering why CADD drawings are created, how they are produced, what unique drawing techniques might be encountered in CADD, and how CADD drawings are stored and corrected. At the end of this chapter are presented nine tasks for completing a CADD drawing common to all systems. This is done intentionally without reference to a particular CADD system. Knowledge of a particular CADD system is not necessary to benefit from this chapter. The procedures and terms have been chosen to apply to a broad range of situations. Many of the terms used in this chapter are included in the Glossary of CADD terms in Appendix A. It might be helpful for readers who are totally unfamiliar with the subject to read these terms before continuing with this chapter.

17.2 The Function of a CADD Drawing

In most ways the functions of CADD drawings are much the same as for traditional engineering drawings. All engineering drawings are used to communicate among engineers and technologists, and between technical and non-technical workers. CADD drawings are the way designers communicate with one another and form the basis for the historical documentation of a design. CADD drawings are used extensively in construction, manufacturing, and service industries.

However, CADD drawings differ from traditional manually-produced engineering drawings in a fundamental way. A traditional engineering drawing is the entire description of the design, based on the conception of the design in the engineer's head. Lose the drawing and you may have essentially lost the design, since the engineer responsible for the design may not be close at hand. Since traditional engineering drawings are often directly scaled or measured and the measurements are transferred to a pattern, mold, or machine tool, the engineering drawing *is* the data base.

In contrast, a CADD drawing is only the visual representation of an underlying numeric data base, rather than actually *being* the data base. The CADD drawing is made only to display the data base to humans who have difficulty interpreting a set of numbers. The design exists independently from the drawing. The data base itself may be directly transmitted to machines that make the patterns, molds, or final parts. This is the power of CADD.

17.3 The Components of a CADD Drawing

A CADD engineering drawing consists of two parts: the numeric data base and supporting annotation in the form of dimensions, notes, labels and text. This supporting text is necessary for the drawing to function as a document. As a source of information for machine tools, this textual information is unnecessary because the mathematical description of a part sent directly to a numerical machine tool controlled by a computer provides the machine tool with the complete description of the part, including necessary tolerances. When engineers and drafters look at a drawing of that data base and interpret it, conventional practices must be used to further describe the design in terms understood by a particular audience.

The form of the numeric data base is unique to each CADD system but generally follows a format like that shown in Fig. 17.1. The data base in Fig. 17.1 is organized so that each entry is separated by a space or comma and records an important parameter of the geometry. To get parts designed on different CADD systems to be compatible with one another, data bases may be filtered or translated into an accepted exchange format such as the Drawing Exchange Format (DXF) or the Initial Graphics Exchange Standard (IGES).

The annotation of a drawing is part of the total data base, but not part of the numeric data base, since notes and dimensions are not part of the object's form. CADD drawings should have a border and title block showing the drawing title, drawing number, bill of materials, and history of changes, just as is done on traditional, manually prepared engineering drawings. (See Fig. 17.2.)

A/ FUNDAMENTAL PRACTICES OF CADD DRAWING

17.4 When the Drawing Already Exists

When preparing a CADD engineering drawing some investigation needs to be done. And as happens in manual drawing, a designer using CADD must determine if the object has already been designed by his company or by another company from which it could be purchased as a vendor-supplied item. This requires that vendor catalogs

```
406,1,3HTOP,0,0;                                                           1P      1
410,1,1.000000,0,0,0,0,0,0,0,1,1;                                          3P      2
116,2.000000,1.500000,0.0,0,0;                                             5P      3
100,0.0,2.000000,1.500000,2.200000,1.500000,2.200000,1.500000,             7P      4
0,0;                                                                       7P      5
106,1,3,0.0,1.000000,2.000000,1.000000,2.000000,1.000000,                  9P      6
2.874256,0,0;                                                              9P      7
106,1,3,0.0,3.000000,2.000000,3.000000,2.000000,3.000000,                 11P      8
2.874256,0,0;                                                             11P      9
214,1,0.156000,0.062061,0.0,3.000000,2.724256,2.307130,                   13P     10
2.724256,0,0;                                                             13P     11
214,1,0.156000,0.062061,0.0,1.000000,2.724256,1.568844,                   15P     12
2.724256,0,0;                                                             15P     13
212,1,4,0.582286,0.156000,1,1.570796,0.0,0,0,1.646844,2.646256,           17P     14
0.0,4H2.00,0,0;                                                           17P     15
216,17,15,13,11,9,0,0;                                                    19P     16
116,3.000000,1.000000,0.0,0,0;                                            21P     17
116,3.000000,2.000000,0.0,0,0;                                            23P     18
116,1.000000,2.000000,0.0,0,0;                                           25P     19
116,1.000000,1.000000,0.0,0,0;                                           27P     20
110,3.000000,2.000000,0.0,3.000000,1.000000,0.0,0,0;                     29P     21
110,1.000000,2.000000,0.0,3.000000,2.000000,0.0,0,0;                     31P     22
110,1.000000,1.000000,0.0,1.000000,2.000000,0.0,0,0;                     33P     23
110,1.000000,1.000000,0.0,3.000000,1.000000,0.0,0,0;                     35P     24
406,1,7HDRAWING,0,0;                                                     37P     25
406,2,5.000000,5.000000,0,0;                                            39P     26
404,1,3,0.0,0.0,0,0,2,37,39;                                            41P     27
S      2G      4D      42P     27                                         T       1
```

FIG. 17.1 An example of a geometric data base. This is a portion of a part file in Initial Graphics Exchange Standard format.

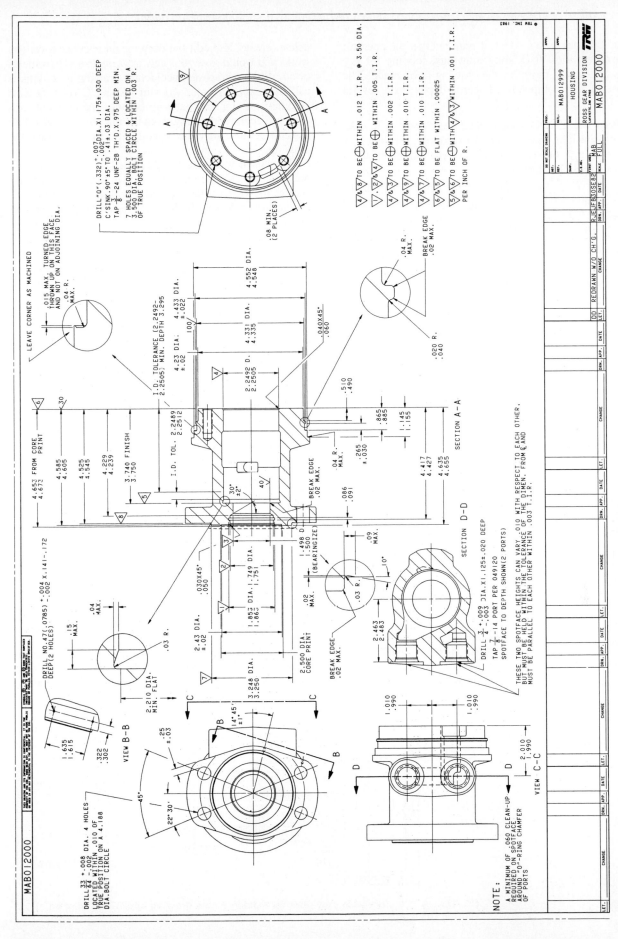

FIG. 17.2 Detail drawing of the housing for the hydraulic motor represented by the data shown in Fig. 17.1. (Courtesy Ross Gear Division, TRW, Inc.)

457

be studied, and with CADD, it requires that the operator search directories of existing drawings, looking for key words that match those for the drawing to be made. The success of this computerized search depends on several factors.

• Does the CADD system enter the drawings into an intelligent data base, where for example, all drawings having a gear of a certain diameter and number of teeth can be quickly identified?

• If not, does the CADD system have a method of naming and numbering drawings in a way that the contents of each drawing is easily identified?

• Are all of the drawings on the system in a central storage location and available to the designer?

If an existing drawing is found, it may not need to be changed beyond adding a new entry in the title block (Fig. 17.3). One drawing can be used for a number of applications by making the appropriate entries in the title block. If, however, a name or number change is advisable, it is important to remember that in CADD one always works on a *copy* of the original drawing. The new drawing can be renamed, renumbered, and saved as a drawing separate from the original.

17.5 Editing an Existing Drawing

If you find a drawing that is very similar to the one you need to make, or a part design exists that can be turned into the drawing you need, you must make several changes to both the numeric data base and the annotations.

A drawing is stored by a name, called its *file name*. The file name should be descriptive of the drawing. The exact procedure for naming files varies from company to company. As an example, the file name 120AXBRKT4HD might be interpreted as a *heavy duty*(HD)*four inch*(4) *axle*(AX)*bracket*(BRKT) for *machine number 120*(120). To turn this into a new drawing, say a three inch heavy duty axle bracket used on machine 110, you would save the old drawing with a new name. Then you would have the start of a new drawing, 110AXBRKT3HD. Using the CADD procedures available on your system, you would

delete, create, and edit the drawing as necessary. As you work, you would save your new drawing every 30 minutes or so, to protect against accidental drawing loss. This is called *backing up* your drawing. Some CADD systems automatically make a back-up copy of the drawing each time you delete something from the data base, while other systems require that the operator periodically save the drawing.

When the new drawing is completed, save it with similar drawings to make finding it easier in the future. For instance, the bracket may be saved with all other drawings for machine 110 or with all other bracket drawings or with all other axle drawings.

17.6 When a New Drawing Must Be Created

A new drawing is made *only* after you are convinced that the object has not already been designed and that no similar design exists on which the new drawing could quickly and efficiently be based.

If no drawing exists that can be used as the basis for the new drawing, the part must be defined starting with the very first line. This is an expensive and time-consuming task and only marginally more cost effective than using manual methods. But remember, when you are finished you will have more than just a drawing. You will have the data base of the design from which any number of subsequent drawings can be made and by which instructions to machine tools can be given.

17.7 Input Methods

Several methods are commonly used to input data and instructions into a CADD system, the most effective of which is the digitizing process. For example, a drafter might digitize a sketch provided by a designer. To do this, the CADD operator would tape the sketch to a digitizing tablet and trace it using either a stylus or a puck. (See Fig. 17.4.) In this way lines, arcs, circles, and other geometric shapes are converted into digital data. If desired, these data may be called up and displayed on the terminal

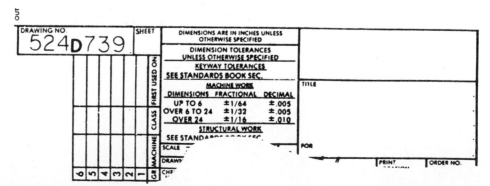

FIG. 17.3 A title block with provisions for making a new drawing application from an existing drawing.

FIG. 17.4 A puck is used as the input device for identifying geometry on an engineering drawing. The display in the background records the progress of the digitizing process. (*Courtesy Computervision Corp.*)

where any necessary adjustments can be made. CADD operators are able to mirror views, rotate the geometry into pictorial representations for visual checking, and "zoom in" to see the smallest detail. An automated method of digitizing is called scanning (Fig. 17.5) and is appropriate for digitizing existing engineering drawings. The scanning process uses an optical device which is programmed to recognize geometric primitives and text characters.

When digitizing, a cursor symbol (a point, intersecting lines, blinking arrow, etc.) corresponding to the digitizing position appears on the screen. The stylus need not actually touch the surface of the tablet to produce this symbol, but it is by this process of digitizing that drawing coordinates are entered into the CADD computer (Fig. 17.6).

Some CADD systems have unique menus for the input of operator commands as shown in Fig. 17.7. Commands

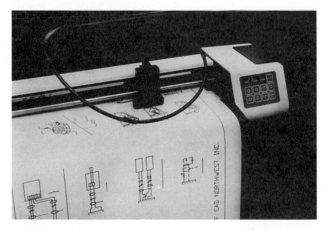

FIG. 17.5 A digital plotter can be converted into an automatic digitizer. (*Courtesy Houston Instruments*)

FIG. 17.6 Performing drawing manipulations using a menu-driven program. At the work station are Professor Jerry Smith and Systems Manager Dennis Short. (*Courtesy Department of Technical Graphics, School of Technology, Purdue University*)

459

are entered either by making choices from menus or by typing the commands from the keyboard. Use of keyboard commands requires knowing the appropriate rules of syntax, and such commands are susceptible to typing mistakes. A system using menu-driven commands is generally easier to learn and more efficient to use than one requiring the entering of commands.

17.8 CADD Menus

An *application program* accomplishes a specific task and each application program has its own menu. A mechanical CADD program would have a mechanical menu while an architectural CADD program would utilize an architectural menu. Commercial applications use grid menus to convey instructions to the computer (Fig. 17.7). Commands are activated by contacting the touch-sensitive tablet like that shown in Fig. 17.8. Note that the processes of entering

data and entering commands from a tablet are essentially the same, but commands have reserved locations on the digitizing tablet. Examples of special purpose menus include

* Mechanical design and drafting,
* Printed circuit board design,
* Electrical schematics,
* Hydraulic schematics,
* Plant and piping design,
* Structural design,
* Architectural design.

Two methods exist for organizing CADD menus: *flat* and *hierarchial*. A flat menu requires the operator to reenter an almost identical string of commands as the first part of each command. A hierarchial menu works within a

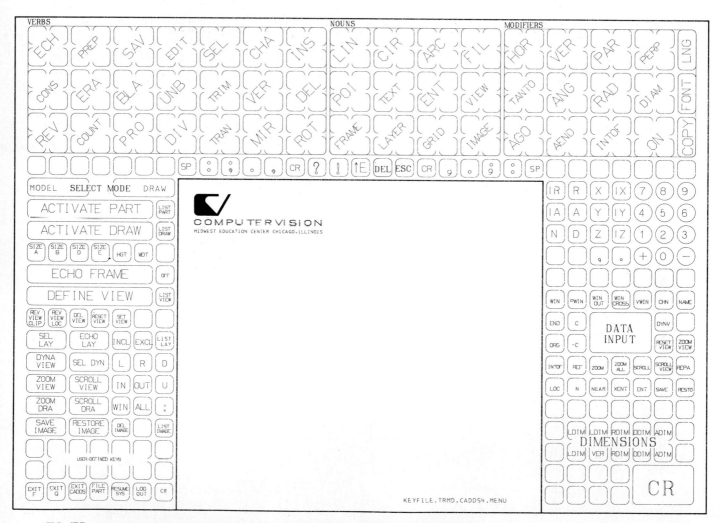

FIG. 17.7 Menu form used in conjunction with a digitizing board as an input device. The boxes around the outside activate specific commands or functions while the open area in the middle is used for inputting geometric position. (*Courtesy Computervision Corporation*)

level of the command (creating a line parallel to another line for example) without the need to reenter that part of the command. Consider these two commands to be made one after the other in a flat menu system.

```
CREATE LINE PARALLEL LENGTH 4.345
CREATE LINE PARALLEL LENGTH 3.879.
```

The second command would have to be entered from the beginning even though only the length of the parallel line changed. Using a hierarchial menu the commands would appear

```
CREATE LINE PARALLEL LENGTH 4.345
3.879.
```

Note that the CADD system assumes that the operator wishes to continue making parallel lines until instructed otherwise. It is easy to see that when making complex CADD drawings, the use of hierarchial menus can shorten the drawing time considerably.

17.9 Using a Tablet

A tablet used with a stylus or puck controls the cursor on the computer terminal so that the operator can point to and create elements like lines, circles, arcs, and other geometric forms in combinations as needed. The menu part of the tablet is divided into a matrix of squares with each square assigned a specific function (see Fig. 17.8).

Thus, when a selected square is touched, a programmed message is sent to the computer. The tablet and stylus or puck provide the instructions necessary for all drawing tasks and may be thought of as the paper and pencil of the CADD operator.

All CADD systems use instructions to tell the computer what to do. These instructions comprise the CADD system's *command language*. All CADD systems have subtle differences among the commands; however in general they take the form verb, noun, modifier, mask.

An example of this would be INSERT LINE HORIZONTAL END OF ENTITY. INSERT is the verb, LINE is the noun, HORIZONTAL is the modifier, and END OF ENTITY is the mask that allows the CADD operator to start the horizontal line at the exact end of another line.

17.10 Selection of a Grid or Scale

One may think of the grid used in creating a CADD drawing as being equivalent to a piece of grid paper on a drawing board since it is used in essentially the same manner. The grid paper in traditional drafting is slipped under the drawing paper so that when finished, the grid does not appear on the drawing. A CADD grid can be turned on or off and will not show up on prints or plots of the final drawing.

On the grid, one unit of spacing may represent a value as small as 0.005 mm or as large as 500 m (.0001 in. to 1000 ft). When necessary, the scale ratio of the X-axis to Y- or Z-axes may be different. The CADD operator may

FIG. 17.8 Entering computer commands. (*Courtesy Computervision Corporation*)

change the grid scale (the world increment assigned to each grid unit) and the computer will automatically scale the graphic data to fit the display size. Therefore, the operator can think in terms of full-size or world dimensions. However, even though the dimensions are full-size, the computer is scaling the data to fit a specific drawing area and working in device units (see Sec. 4.11). Note the use of a reduced scale in the designing of a highway route in Fig. 17.9.

In summary, when graphical data is input in world units, the computer makes the necessary calculations to scale the data to fit a specific drawing area. The output will be accurately scaled and if dimensions are added, they will be in world units.

17.11 Drawing Activities Unique to CADD

Because the computer is such a powerful tool in engineering design and drawing, many methods used to draw manually have been replaced by new methods unique to CADD. The following sections introduce CADD operations called functions, which are used extensively in all CADD systems. The CADD functions in the following sections are universal and not representative of a particular piece of software. In examples that contain commands, the actual commands appear in capital letters, while other information supplied by the operator is enclosed in [brackets].

17.12 Layers

CADD systems have overlaying capabilities similar to the use of translucent paper overlays. Overlays on different layers are generally used when making CADD drawings but not when building the part description. Layers are used to group related graphic elements or text and to separate groups to make work on both detail and assembly drawings easier. A complete assembly drawing may be prepared first and details drawn on successive overlays; or, individual details can be built up, each on a different layer, until an assembly results from the combined overlays. Layers are particularly important in facility engineering or architectural design. By using layers, a building designer can place the foundation, floor plan, heating plan, and electrical plans on separate layers where they can be selectively displayed, aligned, and edited. An overlay is the drawing made on a separate layer. See Fig. 17.10 where the drawing of a subdivision was done on one layer and the extensive dimensions and notes on another.

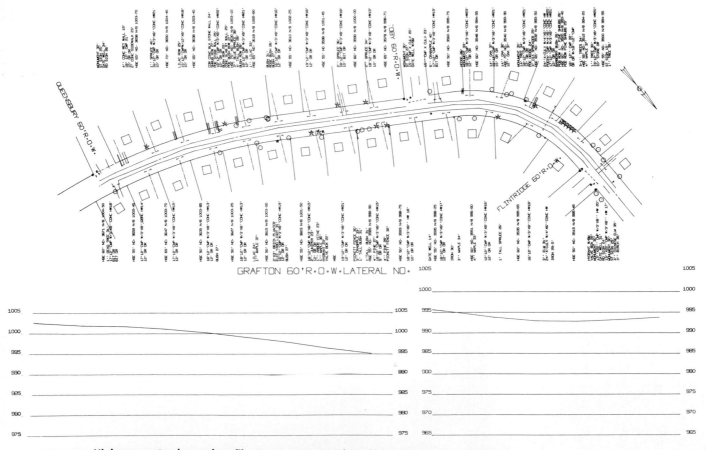

FIG. 17.9 Highway route plan and profile. Program systems have been developed to produce civil engineering plans for highway, sewer, and drainage construction. (*Courtesy Johnson & Anderson, Inc., Consulting Engineers, Pontiac, Michigan*)

A layer is the logical grouping of geometry and notation that is referenced by an index number, called the layer number. All geometry belonging to a layer can be independently displayed, deleted, plotted, scaled, rotated, or otherwise manipulated. If we assign the border and title block to layer 1, four additional layers might suffice for the detail drawing of a single part that was not too complex. The second and third overlays, arbitrarily assigned to layers 21 and 22, might show front and side views of the part, while all dimensions might be assigned to layer 121. When layers 1, 21, 22, and 121 are displayed together, a complete drawing results. Most companies have strict guidelines for what information is assigned to which layer.

To prepare a set of drawings for a product such as an air cylinder, the CADD operator would first draw each part on a separate layer. Then the overlays needed for the assembly drawing would be combined. This method of constructing an assembly drawing requires that each view be on a separate layer so that selected views may be used in preparing the assembly. For the entire assembly to appear on the screen, the operator would enter a command like DISPLAY LAYERS followed by a list of the necessary layer numbers. The final assembly drawing would include the desired layers plus a layer for the border and title block.

17.13 View Manipulation

The drawing manipulation functions of a CADD system include zoom, rotate, pan, and scroll. A view manipulation changes the manner in which the operator sees the data base. It does not alter the data base itself nor does it change the relationship of the geometry to the world axis system. Each function will be discussed briefly.

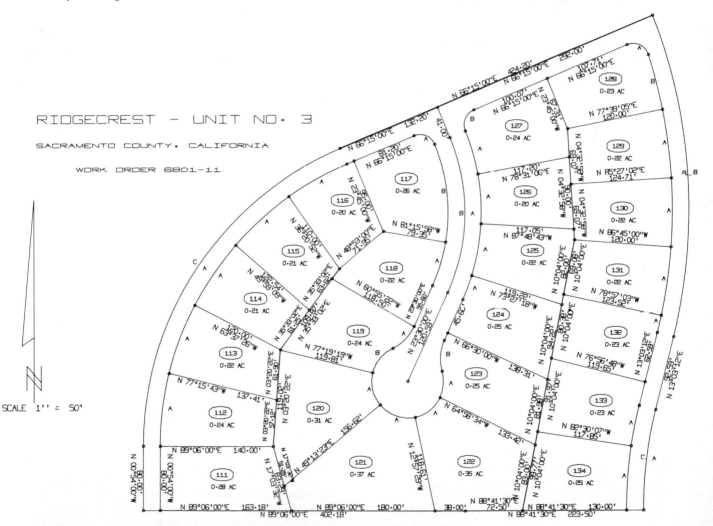

FIG. 17.10 Plot of a subdivision. The computer program designated SAMPS (Subdivision and Mapping System) plots a complete subdivision map with bearings, distances, and other information ordinarily given on a map of this type. As developed by PMT Associates of Sacramento, California, the program can also be used for plotting control networks, for surveying jobs, and for primary control of aerial photographs. The program provides for choice of plotting scale, rotation of plotting axes, plotting of lines and points, plotting of a north arrow, and annotation of distances and bearings. (*Courtesy PMT Associates—Engineers, Land Surveyors, Planners*)

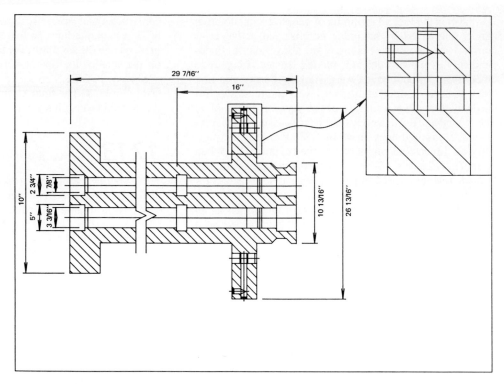

FIG. 17.11 The zoom-in function. The zoom-in view appears at the upper-right. (*Courtesy Cadapple, T & W Systems, Inc.*)

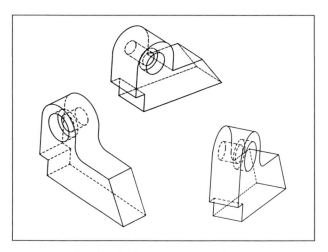

FIG. 17.12 View Manipulation-Rotation. (*This drawing was prepared by Dennis Short, Interactive Computer Graphics Systems, Technical Graphics Department, Purdue University*)

Zoom is the capability of a CADD system to either enlarge or reduce the drawing or a portion of the drawing. Zoom is required because a drawing may be so large that small detail cannot be distinguished. By creating a window, a rectangular area on the screen enclosing specific entities needing attention and using the zoom-in capabilities of the system, the operator can obtain an enlarged view of a specific portion of the drawing for close-up work (Fig. 17.11). This is done with the command ZOOM WINDOW [first corner] [opposite diagonal corner]. Zoom out is the opposite of zoom in. ZOOM OUT 2 displays twice as much of the object as can currently be seen. ZOOM ALL displays the entire object.

Rotate command enables a CADD operator to turn the world axis system and all geometry in the workplace through a selected angle about one or more device axes (Fig. 17.12). The geometry itself does not move in relation to the world axes. This rotation is accomplished by identifying the axis of rotation and by entering the required angle. If desired, a printer or plotter can be activated to draw several three-dimensional views of the object showing it at different angles. This function is activated by a command like ROTATE [axis] [angle].

Commands to *scroll* and *pan* allow the operator to move smoothly from one place on the drawing to another. Scroll usually is limited to vertical movement while pan is used to describe a left-right change in viewing position, although pan may also be used to designate diagonal change. A command to move a point on the drawing from one spot to another might be PAN [point] [new position of point]. Figure 17.13 shows the image of the computer terminal superimposed on on a drawing and the directions of movement which would constitute scroll and pan.

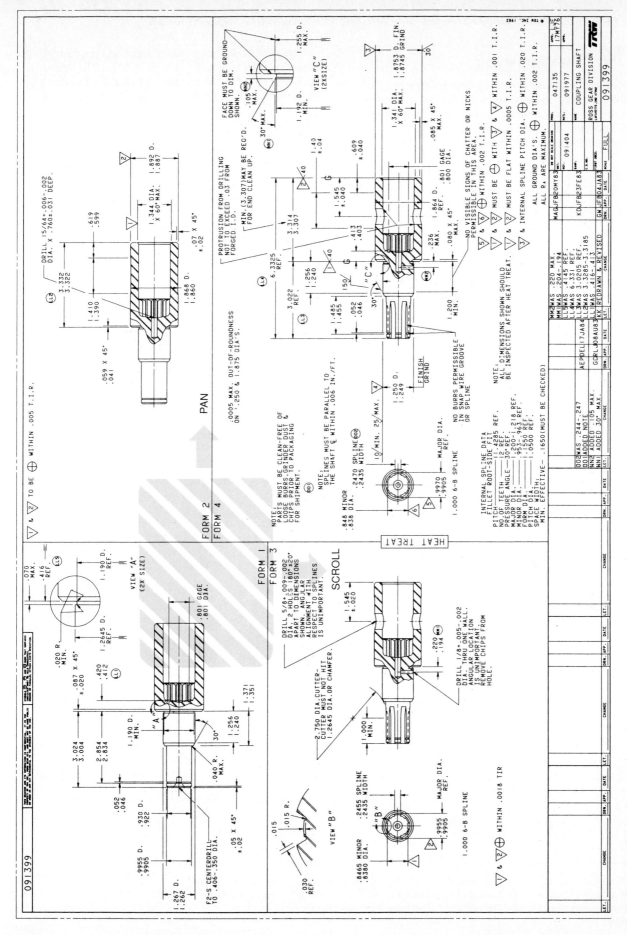

FIG. 17.13 Detail drawing of the coupling shaft for the hydraulic motor shown in Fig. 17.15.
(Courtesy Ross Gear Division, TRW, Inc.)

17.14 Transformation

A transformation, unlike a viewing manipulation, actually alters the data base in some way, usually by an addition; and during a transformation the operator's viewing position remains constant. The transformation functions to be discussed are translate, mirror, copy, rotate, and rotate copy.

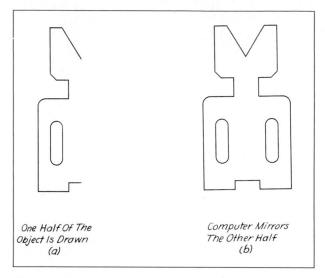

One Half Of The
Object Is Drawn
(a)

Computer Mirrors
The Other Half
(b)

FIG. 17.14   A mirrored and copied view. (*This drawing was prepared by Dennis Short, Interactive Computer Graphics Systems, Technical Graphics Department, Purdue University*)

A *translate* command causes identified geometry to be moved relative to the X-, Y-, or Z-axis. The spatial relationship of the geometry does not change, but the entire group is moved to another location. Translation may be initiated by a command like TRANSLATE [geometry] [reference point] [new position of reference point].

Objects that are symmetrical about an axis may be *mirrored* (Fig. 17.14); that is, the operator may create one half of an object and then automatically produce the second half by mirroring. If a modifier like COPY is used, the two halves of the object will be completed; otherwise the first half will be moved to the second half. This is accomplished by specifying what forms are to be mirrored and about which axis the mirroring is to take place. This function is activated by the command like MIRROR COPY [geometry] [axis].

A *rotate transformation* command moves geometry in a circular path about a stable axis system and differs from the rotate-viewing manipulation discussed in the previous section. Remember that a viewing manipulation does not change the geometry-to-axis relationship. A rotation transformation is accomplished by a command like ROTATE [geometry] [axis] [angle].

A *rotate copy* or *translate copy* command allows a CADD operator to efficiently display chosen forms around a circular arc or along a line. These commands might take the form ROTATE COPY [geometry] [axis] [angle] [number of copies] or TRANSLATE COPY [geometry] [X-offset] [Y-offset] [Z-offset] [number of copies]. An example of rotating and copying circles is shown in Fig. 17.15.

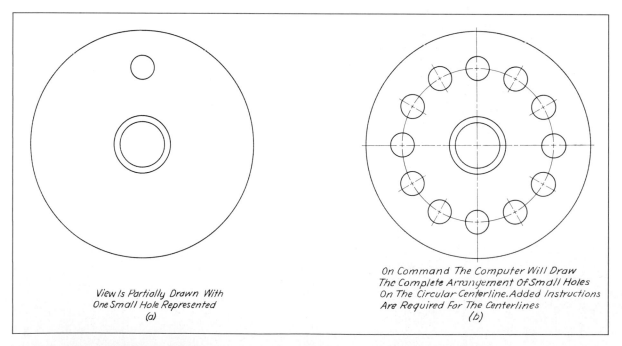

View Is Partially Drawn With
One Small Hole Represented
(a)

On Command The Computer Will Draw
The Complete Arrangement Of Small Holes
On The Circular Centerline. Added Instructions
Are Required For The Centerlines
(b)

FIG. 17.15   Move and duplicate function. The first hole was positioned at the radius at 90°. The remaining eleven holes were produced in one command, ROTATE COPY [30 degrees] [11 copies]. (*This drawing was prepared by Dennis Short, Technical Graphics Department, Purdue University*)

17.15 *Filleting*

A fillet is a tangent arc between two intersecting lines (see Sec. 3.22) or, in a three-dimensional sense, a partial cylinder, sphere, or torus between two surfaces. CADD systems often provide a function to create a fillet semi-automatically, allowing the operator either to trim or not to trim the lines (Fig. 17.16). This is accomplished by a command like CREATE FILLET [radius] [line 1] [line 2] or CREATE FILLET [radius] [surface 1] [surface 2]. A three-dimensional computer model is shown in Fig. 17.17 containing several filleted surfaces.

17.16 *Crosshatching*

In producing sectional views, areas cut by a cutting plane are shown crosshatched. CADD systems may vary slightly in how they handle this. Some systems require that the area be completely bounded by closed lines. Other systems will automatically close an area if an opening is left. The general form of the command is CROSSHATCH [pattern number] [line 1] [line 2] [line 3] . . . [line n]. An example of an assembly drawing including areas of cross-hatching is shown in Fig. 17.18.

17.17 *Automatic and Semiautomatic Dimensioning*

Dimensioning involves the selection and placement of dimensional values for size and location. Fully automatic dimensioning requires minimal input from the operator and will create dimensions on a drawing based on an internal set of rules conforming to ANSI Y14.5 standards. (Chapter 13 contains a complete discussion of dimensioning.) The operator may decide afterward to remove one or more of these automatic dimensions or to add a dimension or note that the system did not include. Semiautomatic dimensioning requires that the operator identify

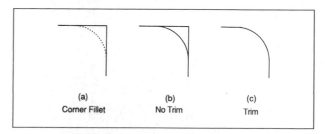

FIG. 17.16 The fillet function.

FIG. 17.17 A computer-generated pictorial illustration of the drum shown at the start of Chapter 7. Note the fillet inside and relief slots on the top surface. (*Courtesy Computervision Corporation*)

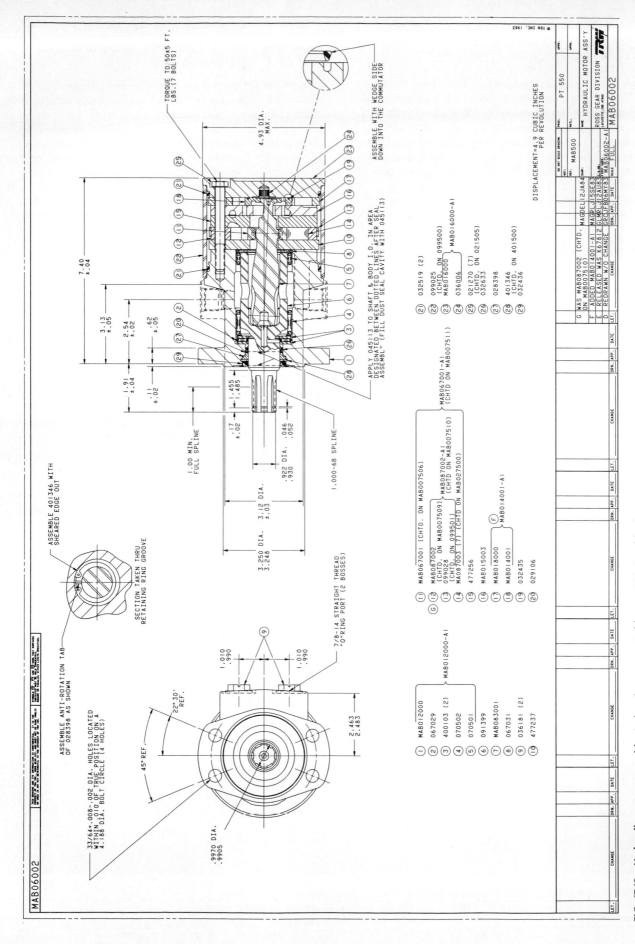

FIG. 17.18 Hydraulic motor assembly. A drawing generated by an IBM Graphics System.
(Courtesy Ross Gear Division, TRW, Inc.)

the type of dimension desired (horizontal, vertical, inclined, radial, or diametrial), the feature to be dimensioned, and the position of the dimension value on the drawing. The CADD system would then supply the dimension number, witness lines, dimension lines, and line terminators (arrow heads) based on the operator-specified position. A typical dimensioning command might be CREATE DIMENSION HORIZONTAL [point 1] [point 2] [text position]. In Fig. 17.19 horizontal, vertical, angular, in-

clined, radial, and diameter dimensions are shown along with several notes.

To supplement dimensions, a CADD operator can create a notation, note, or number value needed at a particular location on a drawing. The operator will have a choice of fonts and can select text height, aspect ratio, justification, angle, and slant. These parameters, once set, affect all subsequent text in dimensions and notes and do not have to be respecified each time a dimension or note is created.

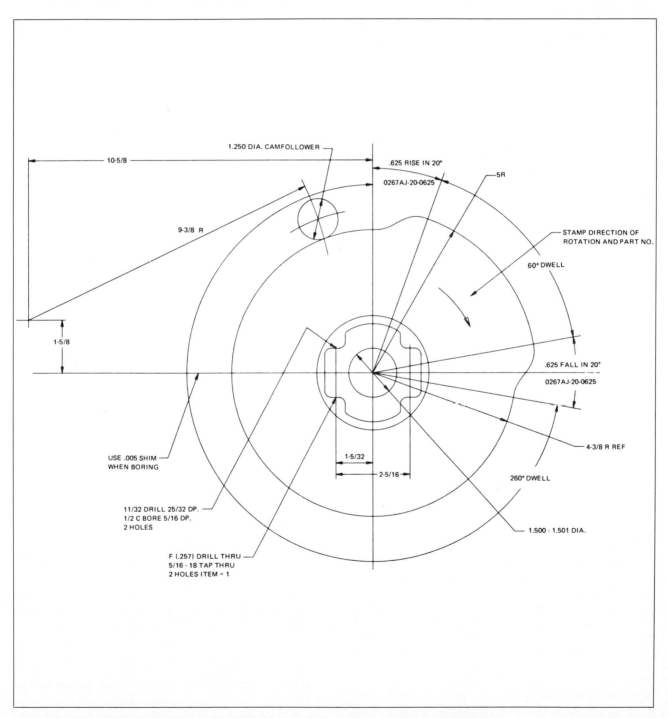

FIG. 17.19   Dimensioning techniques with CADD. (*Courtesy Westinghouse Electric Corporation*)

The placement of text is accomplished by digitizing the proper location. As an example, to place a note stating "REMOVE ALL BURRS AND POLISH" would require the command CREATE NOTE/REMOVE ALL BURRS AND POLISH/[text position]. The slashes (/) that set off the text string are called *delimiters*. Delimiters define for the computer the text that is to be put in the note.

17.18 Storing Drawings in Digital Form

As a CADD operator works on a drawing, the drawing is held in the temporary memory of the computer. When instructed, the computer will record the drawing on a permanent storage medium, such as disk or tape. Unlike manually prepared designs, drawings made on a CADD system and stored on tape or disks are not generally stored on paper as well. Prints of CADD drawings are made so that they may be checked or distributed. The original CADD drawing is stored in electronic form at the manufacturing site where it can easily be recalled for use. Additional electronic copies of the CADD drawings, called back-up copies, can be stored in remote locations—underground in fireproof bank vaults for instance—for safekeeping.

Both the filing and retrieving of CADD drawings are quick and easy operations. To file a drawing, a simple command to the computer is required. A typical filing command might be FILE [drawing name] [file type], where the file type could be PART, SYMBOL, or DRAWING.

Retrieving a drawing from storage is called *loading a drawing*. The CADD computer checks to see if the operator has made any additions to the drawing currently on the screen. If there have been, the operator is given the opportunity to save this altered drawing before returning it to storage and loading another. A typical command for retrieving a drawing would be: LOAD DRAWING [drawing name] [file type].

17.19 Checking and Correcting CADD Drawings

A full discussion of checking and correcting drawings can be found in Sec. 16.21. The items mentioned there apply to both traditional and CADD drawings.

When the initial work has been completed on a set of working drawings for a project, the CADD operator may wish to submit the drawings for checking. This can be done by producing hard copies on paper or by placing copies of the drawings in a location in the computer where they can be reviewed on the display terminal. As noted before, the checker should be a person familiar with the principles of the design who has not been involved in the preparation of the drawings to be checked and corrected. A checker should have production experience coupled with a broad knowledge of drawing standards, manufacturing practices, and assembly methods. The checker should

study the assembly together with the detail drawings and mark all necessary changes or corrections in colored pencil. The checker should also verify all dimensions, notes, and specifications for correctness and be certain that they are in a form that can be used for manufacturing. Any omissions should be indicated. If the checking is done on the computer, changes and corrections should be placed on a layer separate from the original.

When the marked copy has been returned for changes, the CADD operator can call up the original drawing and make the needed alterations with little difficulty.

B / GENERAL STEPS—NEW CADD DRAWING

17.20 Making a CADD Drawing

There are nine general steps that must be performed with any CADD system to complete a new drawing. During the process, part, symbol, and drawing files will be created. The difference between these files are that part files contain only geometry, symbol files contain symbol geometry that can be combined with symbols and parts, and drawing files contain part geometry, symbols, and dimensions and notes. The precise commands would of course vary from one system to the next but the tasks would remain the same. When commands are given, they appear in a general form.

STEP 1. Analyze the geometry of the design. Break the object into component parts. Can the object be defined with common geometric shapes? Are there features that repeat and can be created once and copied at the appropriate position, scale, and rotation? Is the object symmetrical? Can the shape be created and then mirrored to completion? Plan an approach. For example, the gasket in Fig. 17.20 is comprised of a quadrant that can be mirrored and copied to complete one half of the design which can then be mirror copied to form the entire design.

STEP 2. Locate the origin. Choose an appropriate location in the design to coincide with X0, Y0, Z0, the world axis system origin. This accomplishes several ends. First, it reduces the magnitude of absolute coordinate values. Second, it allows you to zoom in on the form and keep the origin on the screen. Lastly, the eventual combining of the geometry with a border and title block is facilitated if the origin of the border and the part are both at their geometric centers.

STEP 3. Select a scale that is appropriate for the size of the object. If a grid is desired, specify that the grid be defined using the smallest practicable unit, such as .125 in. or 1 mm.

STEP 4. Create the object full size in world units. As you work, save the geometry every few minutes to prevent loss of more than a small portion of work should there be a power failure or other unforseen accident. To save the part as you work:

SAVE PART (part name)

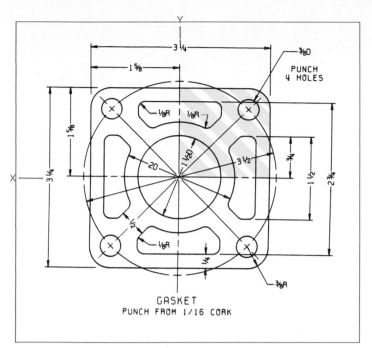

FIG. 17.20 Gasket drawing prepared using a CADD drafting system and a CalComp electrostatic printer/plotter. The upper left-hand quadrant can be mirror copied about the Y-axis, and the top half can then be mirror copied about the X-axis.

STEP 5. Save the completed part as a figure or symbol. Symbols can generally be merged with parts; however, parts cannot usually be merged with other parts. For example,

```
SAVE FIGURE (figure name)
```

STEP 6. Turn part geometry into an engineering drawing. Locate and display a standard drawing border and title block on its own layer. For example,

```
ACTIVE LAYER (layer number)
LOAD FIGURE (border name)
```

STEP 7. Merge the part figure with the border. Merge the symbol on a layer separate from the border. Choose a scale that makes the part geometry readable. Very large objects will be reduced. Very small objects will be enlarged. Many objects will be full-size. When the part and border are combined, save as a drawing. For example,

```
MERGE (symbol name) (scale) (rotation)
(position)
SAVE DRAWING (drawing name)
```

STEP 8. Dimension the drawing. On a dimensioning layer, create dimensions, notes, and title block information. For example,

```
ACTIVE LAYER (layer number)
CREATE RADIAL DIMENSION (arc) (leader
side)
CREATE VERTICAL DIMENSION (point 1)
(point 2) (text position)
CREATE NOTE/PURCHASE FROM ALLIED
FASTENER COMPANY/(text position)
```

STEP 9. Save the completed drawing.

```
SAVE DRAWING (drawing name) (file type)
```

These steps provide for the efficient creation of a CADD drawing. The steps were organized to accomplish the following ends:

- The efficient use of terminal time by adequate pre-planning,

- Using the power of CADD (rotate, copy, mirror, merge) rather than using CADD simply as an electronic drafting board,

- Saving the geometry without border, dimensions, or notes to create a part file that is more efficient for manufacturing,

- Merging the finished part file into the border for an efficient way to make a drawing.

17.21 *Examples of CADD Drawings*

Figures 17.21–17.24 give the reader a sense of the diversity of drawings that can be produced by CADD applications. Figure 17.21 is a three-dimensional model of a process plant. It functions identically with a traditional engineering model like that shown in Chapter 16. The entire plant can be designed, all clearances checked, and even the operation of the process simulated before construction drawings are created. Figure 17.22 is typical of CADD use in the automotive industry and Fig. 17.23 demonstrates CADD's applicability in structural design. Figure 17.24 shows a design drawing for a fixture to be used in the shop to press a bearing into a side cover of an air press.

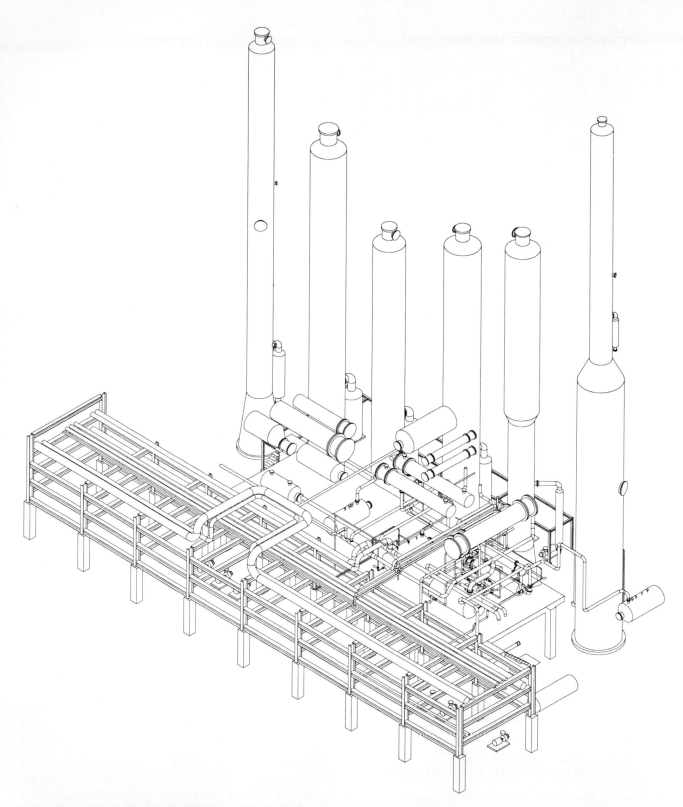

FIG. 17.21 Computer generated piping installation drawing. The computer may check the design for inaccuracies, conflicts, or faulty representations. (*Courtesy Computervision Corporation*)

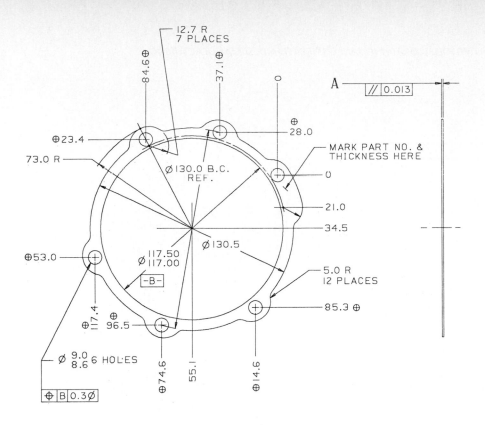

SYMBOL CODE	
PARALLELISM	//
TRUE POSITION	⊕
DENOTES BASIC DIMENSIONS	⊕

CHANGE IN DESIGN, COMPOSITION, PROCESSING OR
IN-PROCESS GAGING FROM PART PREVIOUSLY APPROVED
FOR PART PRODUCTION REQUIRES PRIOR PRODUCT
ENGINEERING APPROVAL.

ENGINEERING APPROVAL & TEST OF SAMPLES FROM
EACH SUPPLIER IS REQUIRED PRIOR TO FIRST
PRODUCTION SHIPMENT OF PARTS.

FOR ENGINEERING APPROVED SOURCE SEE
ENGINEERING NOTICE.

FIG. 17.22 Conventional drawing prepared by a plotter. (*Courtesy Ford Motor Company*)

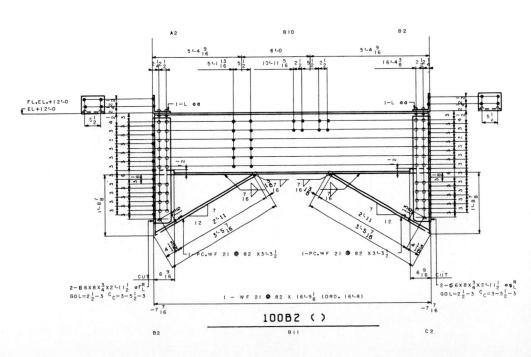

FIG. 17.23 Structural drawing prepared on a plotter. Made using the CONSTRUCTS System of automated drafting. (*Courtesy Control Data Corporation*)

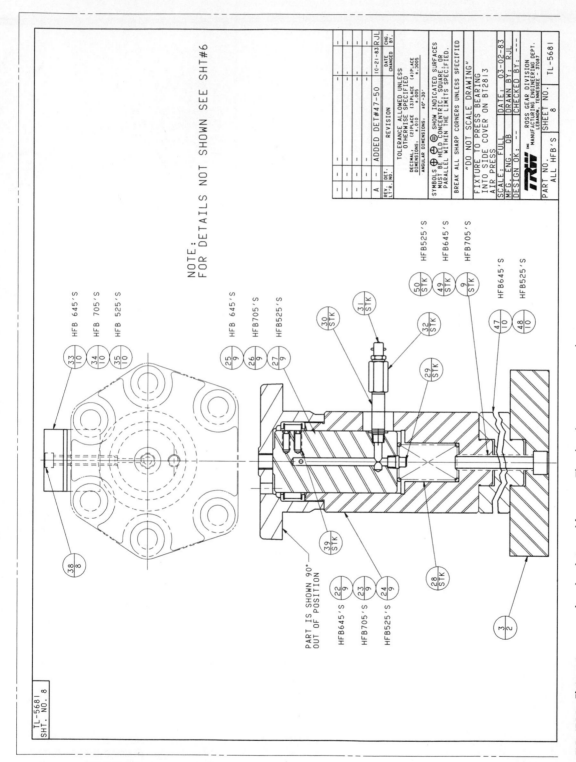

FIG. 17.24 Fixture to press a bearing into a side cover. The drawing was prepared on an interactive CADD workstation and produced on an electrostatic printer. (*Courtesy Ross Gear Division, TRW, Inc.*)

PROBLEMS

The following problems are suitable for students at those institutions where computer-aided drafting programs are available. These problems have been selected to offer the opportunity to develop the basic skills needed for the use of the computer and related peripheral equipment associated with present-day interactive computer-aided graphics systems. These problems may be prepared on most CADD systems without added special hardware.

Other exercises and problems at the basic level may be found at the end of Chapters 5–11, 13 and 16. Simple exercises may be selected from the first of the problems for Chapter 5. It is suggested that instructors prepare lesson assignments of their own to develop their students as they see fit.

1. (Fig. 17.25). Make a complete three-view orthographic drawing of the rear support bracket. Also prepare a pictorial representation.

2. (Fig. 17.26). Make a detail drawing of the motor base. The ribs are 10 mm thick. At points *A*, four holes are to be drilled for 12 mm bolts that are to be 64 mm center to center in one direction and 80 mm in the other. At points *B* four holes are to be drilled for 12 mm bolts that fasten the motor base to a steel column. Fillets and rounds are *R*3.

3. (Fig. 17.27). Make a detail drawing of the tool rest and/or the tool rest bracket. The rectangular top surface of the tool rest is to be 30 mm above the center line of the hole for the 12 mm bolt. The overall dimensions of the top are 30 × 65 mm. It is to be 6 mm thick. The overall dimensions of the rectangular pad of the bracket are 32 × 48 mm. The center line of the adjustment slot is 15 mm above the center line of the top holes in the rectangular pad and the distance from center line to center line of the slot is 35 mm. The bracket is to be fastened to a housing with 6 mm pan head machine screws.

4. (Fig. 17.28). Make a complete orthographic drawing of the tube holder.

5. (Fig. 17.29). Make a detail drawing of an assigned part of the air cylinder.

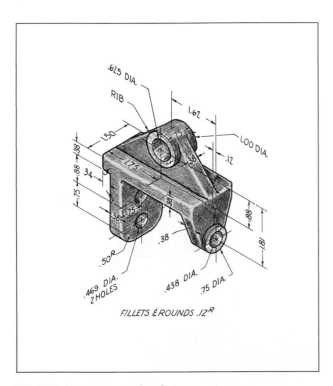

FIG. 17.25 Rear support bracket.

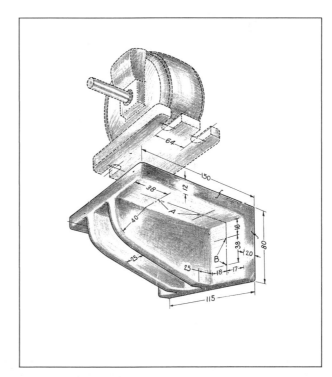

FIG. 17.26 Motor base.

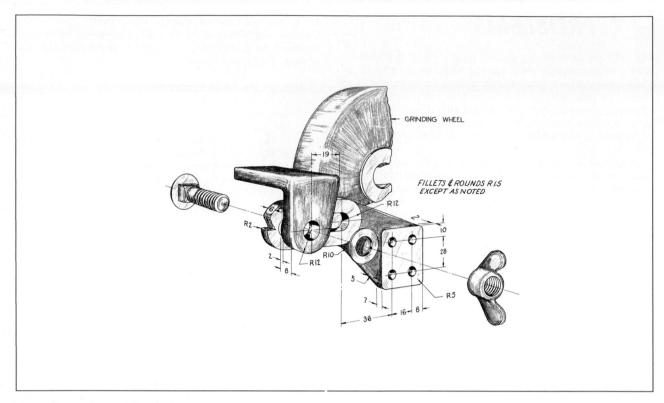

FIG. 17.27 Tool rest and tool rest bracket.

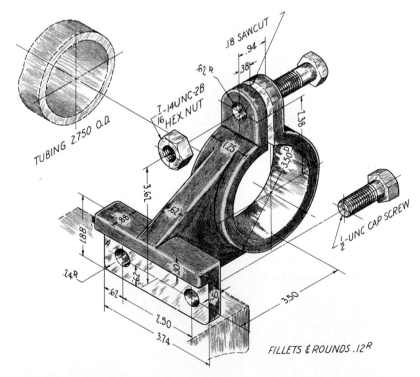

FIG. 17.28 Tube holder.

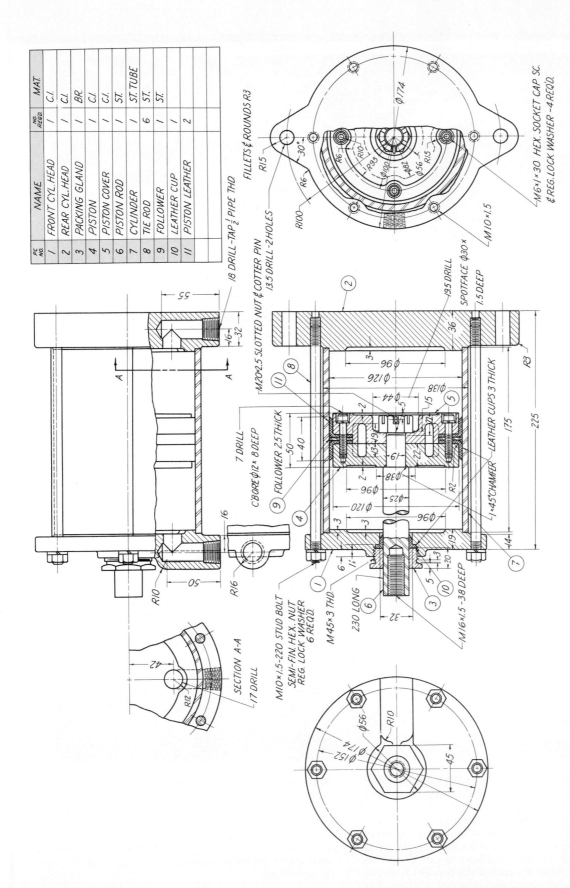

FIG. 17.29 Air cylinder.

PC NO.	NAME	NO. REQD.	MAT.
1	FRONT CYL. HEAD	1	C.I.
2	REAR CYL. HEAD	1	C.I.
3	PACKING GLAND	1	BR.
4	PISTON	1	C.I.
5	PISTON COVER	1	C.I.
6	PISTON ROD	1	ST.
7	CYLINDER	1	ST. TUBE
8	TIE ROD	6	ST.
9	FOLLOWER	1	ST.
10	LEATHER CUP	1	
11	PISTON LEATHER	2	

A Cincinnati Milacron computer-controlled T3 Industrial Robot serves two (NC) Milacron Cinturn Series C turning centers. Parts enter the cell as rough castings and leave it completely machined and inspected. The sequence is as follows. (1) the T3 Robot takes a part from the pallet on the conveyor to the first turning center (not shown). Entering from the rear of the machine, the robot removes the part just machined and loads the part from the pallet. (2) The robot takes the part it just removed from the first machine to an automatic gauging station. If the part is found to be within tolerances at this point, the robot takes it to the second turning center (as shown above). (3) The robot removes a finished part from the second turning center and loads the part just gauged for a second operation. (4) The robot takes the finished part to the gauging station. If the part is good, it is returned to the pallet, and the T3 Robot picks up another part and the sequence is repeated. (*Courtesy Cincinnati Milacron, Inc.*)

Numerically Controlled Machine Tools and Robots

18.1 Introduction

In use at the present time are production machines that follow directions given on punched tapes and gauges that measure electrically to 0.025 or 0.050 micrometer (1 or 2 microinches). The use of tapes and computers by industry does not mean that the drafters and engineers will have less work to do and that there will be fewer of them. It does mean, however, that those assigned to both areas must keep themselves informed and up-to-date on late developments. The drafter should learn how to dimension the drawing of a part to meet the requirements for programming the machine or machines that will perform the shop operations (Fig. 18.8). One must have a more thor-

ough understanding of basic fundamentals than in the past and at the same time be willing to accept change.

18.2 Numerically Controlled Equipment

The term *numerical control* (NC), as applied to automated production machines, denotes a method of electronically controlling the operation and motions of a machine tool. Early machine tools with built-in intelligence were numerically controlled milling and profiling machines that provided faster and less expensive means for producing aircraft parts. Now, numerical control has been applied to other machine tools, such as lathes, drilling and boring machines, welding and flame cutting machines, punching machines, and inspection devices.

There are two basic types of numerical control.

1. contouring or continuous path
2. positioning or point-to-point.

In the case of the contouring system, generally applied to milling machines, the path is continuously controlled. Normally, the path consists of very short straight-line segments, which approximate specified circular or curved lines. This system is used for machining items such as dies, gears, cams, and pieces having contoured surfaces. For most types of contour machining, the computer is used for instruction calculations.

The cutting tool is controlled with respect to its final position for the point-to-point system as it is primarily applied to drilling and boring operations (Fig. 18.8). The path that a tool may take when directed from one point to another when drilling is relatively unimportant since the tool is not in contact with the workpiece during the movement. When the operations are not complex, the instruction tape may be prepared without the aid of a computer.

A numerically controlled machine tool must have complete instructions to perform an operation at a specified location. This is accomplished by the machine control unit that contains the data processing equipment needed to translate programmed information into operating commands (electrical pulses) for the machine-tool system with its servomechanism drives and related feedback and measuring devices. The control unit consists of an input section, an interpolating or command-generating section, and a servoloop (drive) section. The input section, with tape reader and temporary storage capabilities, decodes and processes the programmed information for eventual transfer to the interpolating section (Fig. 18.1). The instructions, presented as a series of holes punched in tape (either paper or aluminum-Mylar), are read by a tape reader that may be either photoelectric or mechanical. Provided with both temporary and active storage areas, the complete control unit can receive new information from tape for temporary storage while using information in active storage. The information transferred into the interpolating section (active storage) is processed by logic circuitry that periodically sends signals (pulses) to the servoloop section. The digital or pulse information received by the servoloop section is, in turn, converted to an analog form and directed to the servomechanisms of the machine tool. Each command pulse (converted to an analog voltage) will create a definite movement of a slide, the actual movement itself being accomplished by some form of auxiliary equipment, such as a hydraulic motor, a hydraulic cylinder, or an electric motor (a.c. or d.c.) attached to the machine tool itself.

As stated before, the information on the control tape must provide complete instructions in a logical sequence. With a print or a sketch at hand, the parts programmer prepares a process (program) sheet (Fig. 18.4). On this sheet he lists the movements and operations necessary to machine the piece to the dimensions specified. This in-

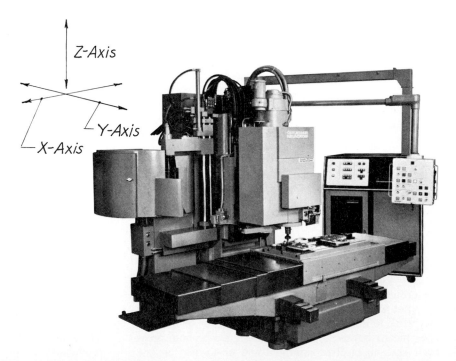

FIG. 18.1   Cintimatic Vertical Machining Center automatically drills, taps, mills, and bores in one setting. Work table positioning accuracy is ±0.025 mm (±.001 in.) in 610 mm (24 in.) at 10 m/min. (400 ipm.) and its repeatability is ±0.013 mm (±.0005 in.). (*Courtesy Cincinnati Milacron, Inc.*)

FIG. 18.2 **Tape (NC)-controlled turning machine that automatically performs all turning operations, including facing, grooving, contouring, threading, boring, drilling, and reaming.** The Acramatic control for the lathe shown above features (1) complete integrated circuit electronics; (2) four-digit, feed-rate coding in either millimeters or inches per minute; (3) automatic acceleration and deceleration without special coding; (4) incremental programming; (5) 300-row-per-second tape reader; (6) buffer storage; (7) manual data input; (8) linear interpolation; (9) circular interpolation; and (10) threading. The last two features are optional. (*Courtesy Cincinnati Milacron, Inc.*)

formation, in order, is then punched on the control tape by a machine (tape punch) that resembles an electric typewriter. As the operator types the digits, symbols, or letters, the device translates the information into a binary-coded-decimal form and at the same time punches the tape (see Fig. 17.4). Manually programming a machine tool to machine a part is not an easy task. It may be both frustrating and time consuming, for records show that it may require as many as 250–400 hrs to prepare a program for a particularly complicated piece.

Numerical control may be the most important metalworking development of the twentieth century. It will grow rapidly in use, and as metalworking assumes a new aspect, new job patterns will result that will require higher skills and considerably more technical training for the individual. Much NC machining is performed by general-purpose NC machining centers (Fig. 18.2).

18.3 Use of Cartesian Coordinates (Fig. 18.3)

The standard Cartesian coordinate system is used to designate machine-tool axes and is the basis for all programming for numerical control. Most numerical-control systems use only the first quadrant for dimensioning purposes in accordance with the recommendation given in the EIA

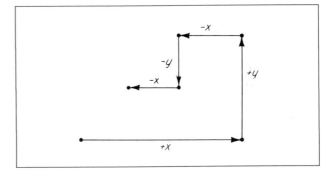

FIG. 18.3 **Cartesian coordinate system.**

(Electronics Industry Association) standards. This practice of considering the workpiece to be positioned in a right-handed Cartesian system provides an accepted frame of reference that enables a programmer to relate a workpiece to almost any numerically controlled machine tool in general use. The geometry of the workpiece is described within the framework of X, Y, and Z axes by relating all critical points of the part to the origin. The position of a point is established by giving its distances from the origin, measured parallel to the mutually perpendicular axes.

Although the angular orientation and the location of the origin is not the same for all machines (see Figs. 18.1 and

18.2), the *X* and *Y* axes usually lie in a horizontal plane. In the plane of the *X* and *Y* axes, meaurements taken along the *X* axis to the right are plus (+), while those taken to the left in the opposite direction are considered to be minus (−). The *Z* axis, perpendicular to both the *X* and *Y* axes, is always parallel to the axis of the cutting tool. Finally, the longest motion of which a machine is capable is usually in the direction of the *X* axis. See Fig. 18.1. These axes are referred to as *device axes*; they describe how the machine tool operates on the workpiece.

In preparing programs one must always remember that the machine axis designation is always based on a coordinate system associated with that particular machine. See also Sec. 4.11 where the relationship of device and world axes systems is discussed.

18.4 Programming Process and Manual Tape Preparation (Fig. 18.4)

It is essential for those who are concerned with numerical control, as it applies to design, drafting and production, to have a thorough understanding of the programming of numerically controlled equipment. The programming process is known as *part programming*. Part programming, however, may involve the use of million-dollar computers and computer programs in the process of producing needed numerical control tapes.

The usual steps in the programming process are as follows.

1. Make an analysis of the part and consider its relation to the machine tool.

2. Consider the tooling required to produce the workpiece. That is, determine the type of holding device, the style and size of cutters, and so forth.

3. Determine the operations required and the sequence in which they must be performed.

Some parts may be programmed exceedingly fast because of their simplicity. This may be done with the aid of a desk calculator and trigonometry tables. In the case of the simplest NC machine tools, which have only the *X* and *Y* distances controlled by perforated tape, parts of the total operation may be performed by the machinist manually in accordance with instructions that are included in the program manuscript. After deciding on the machining sequence, the programmer must list on the program sheet (see Fig. 18.4) all tool movements, operations, tool selections, speeds, feeds, coolant control, and tool changes.

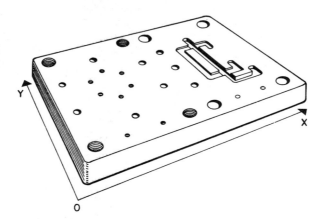

SEQ. NO.	PREP. FUNCT.	X POSITION	Y POSITION	MISC. FUNCT.	POS. NO.	FEED MM/REV.	SPEED RPM	DEPTH OF CUT	TOOL REMARKS
H046	G81	X42000	Y15000	M57	30	.13	1060	35	17 MM DRILL
N047			Y35500		31				
N048		X28000			32				
N049		X22500			33				
N050		X9000			34				
N051			Y15000		35				
N052		X22500			36				
N053		X28000			37				
N054		X16000	Y25000	M06	38				
N055	G84	X16000	Y25000	M58	20	1.19	490	20	M6 X 1 TAP
N056		X28000	Y15000		21				

FIG. 18.4   Program sheet showing a portion of the programming for the workpiece shown (metric).
(Courtesy Cincinnati Milacron, Inc.)

A portion of the programming for the workpiece shown in Fig. 18.4 is given for sequences numbered H046–N056 (see column 1). The sequence numbers show the order in which the machining operations are to be performed. Column 2 (from the left), preparatory function, is for programming the operating mode, such as drill, tap, mill, or bore. The function code G81 calls for the machine to drill.

The X and Y columns are for listing the hole locations. These are programmed in positive numbers in the first quadrant. The 0 position for the X and Y axes corresponds to the left-front corner of the table. It should be noted that, aside from calling for a drilling cycle, the sequence number H046 also calls for the table to move to the hole position 420.00 mm along the X axis and 150.00 mm along the Y axis.

Entries in the column headed "Miscellaneous Function" are for such purposes as specifying what depth of collar to use. The code M57 calls for the use of depth stop No. 7.

The sequence numbers that follow through N054 call for the table to move to the hole positions called for in the drilling cycle. At sequence number N054, the code M06 (see column 5) stops the automatic drilling cycle. At this point the tool, feed, or speeds are changed as called for in the manuscript instructions. In looking ahead to sequence number N055, we see that preparatory function G84 has been specified. This command cycle code calls for a tapping sequence that instructs the machine to move rapidly to gage height, feed to depth, reverse spindle, feed out, and then move to upper limit. No other instructions are needed to perform this operation.

Sequence number N056 calls for a move to a new hole position in the same tapping sequence.

By following the solid black arrows in Fig. 18.5, it can be noted that the next step is punching the information given on the program form (process sheet) into the tape. Before this is done, however, it is wise to have the completed draft checked by a second programmer. If no errors are found, the manuscript is turned over to a typist, who enters the programmed numerical commands on a Friden Flexowriter or similar machine having a keyboard much like that of an electric typewriter. As the operator types, two outputs result. One is in the form of a master document that is called a print-out. This is an exact copy of the manuscript that was typed. The other output, made simultaneously, is a control tape with a series of perforations. See Fig. 18.9. A second tape may then be made using the data tape verification unit and the first tape. The automatic retyping of the process sheet produces a second tape. Mechanical proofreading is thus provided since the second tape will not punchout if it differs in any way from the first tape.

The tape used is standard 1-in. (25.4 mm)-wide paper or Mylar-coated tape. Using a word address format, this eight-track tape is perforated in accordance with EIA Standard RS–273 (Fig. 18.6). Tab and delete codes are permitted. The code perforations are punched across the 1-in. (25.4-mm) width of the tape. The eight positions that are available for code holes are called *channels* (Fig. 18.6).

Tape feed holes, located between the third and fourth channels, ensure the proper positioning of the tape in the tape punch as well as in the tape reader. The upper portion of the tape shown in Fig. 18.7 shows a block containing X and Y commands.

18.5 *Computer-assisted Programming*

Intelligent machining programs offer the most reasonable approach to contour machining (Fig. 18.11). The work of the part programmer in this case is considerably simplified since the computer assumes the tedious task of making the necessary computations as well as the logical decisions. It is not even necessary to have intimate knowledge of the specific codes required for a particular control unit. With accurate postprocessing, all control units can be programmed by using the same problem-oriented language (APT–ADAPT–REMAPT) for the part program input to a computer. However, since both APT and ADAPT are basically computer programs, the part programmer and computer programmer will find it necessary to work closely together. This is true because the part programmer, using the standard Cartesian system, defines the part geometrically as being made up of planes, circles, conicals, and so forth.

18.6 *Time-shared Computer Services*

Even in very small industrial plants a parts programmer can now take advantage of a million-dollar computer to prepare control tapes for NC machine tools in a few minutes. This has been made possible through a time-sharing concept where a computer terminal can be installed on the programmer's desk that gives access to a remote computer. The computer is linked to the terminal by public telephone lines. In some cities, it is possible for users to dial local telephone numbers to gain access to remote computers. This can be done without incurring long-distance charges. The programmer, at the keyboard of a teletypewriter, can use library programs stored in the computer to describe and analyze part geometry, calculate X and Y coordinates, generate data for machine-tool controllers, and receive as computer-output punched EIA coded tape at the terminal. It is common to prepare NC tape for two- and three-axis machine tools. The programs available can also be used to prepare control tapes for contouring operations.

18.7 *Preparation of NC Tapes*

The part programs needed to produce the workpieces shown in Figs. 18.8 and 18.10 were prepared utilizing a computer and APT post-processing.

The first twenty lines in Fig. 18.9 are computer input statements for drilling and tapping the upper plate shown in Fig. 18.8. The remainder of the print-out (lines 01–N38) gives X and Y positions, tool sizes, specific instruc-

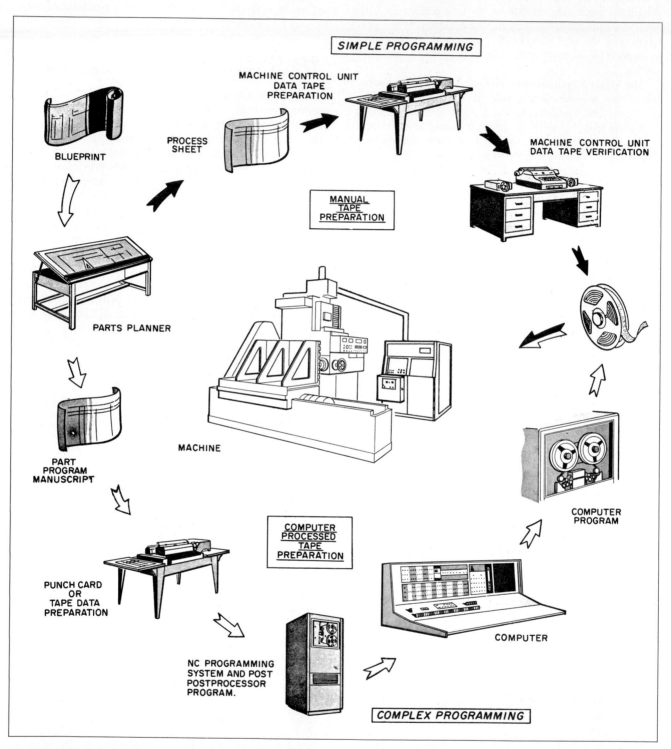

FIG. 18.5   Simple and complex programming. (*Courtesy Bendix Corporation*)

I INCH WIDE, EIGHT TRACK TAPE

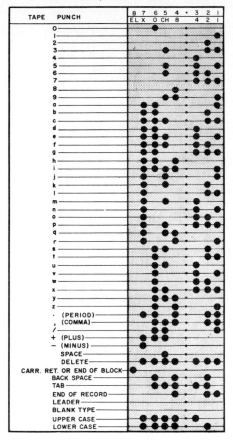

FIG. 18.6 EIA standard coding.

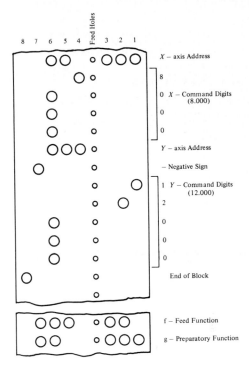

FIG. 18.7 Two portions of a tape illustrating format. Upper portion shows block containing *X* and *Y* commands.

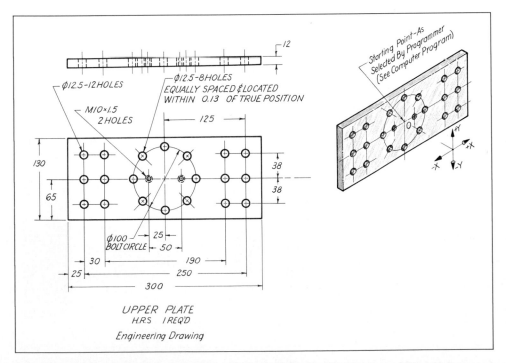

FIG. 18.8 Positional dimensioning using *X* and *Y* coordinates. This plate was drilled and tapped on the NC machine shown in Fig. 18.1.

```
DATE=

CARDS PART=JR  000028      83
PARTNO FIG 19.8
PPRINT MACHN= 10VC-1000 PROG'D BY J.M. RITENOUR
PPRINT LOAD PART IN VISE - CENTER OF PART TO BE ON X 0 AND Y 0.
CALL/STP,XO=0,YO=0,MACHN=38,CLS=2.54,XS=254,YS=254,MET=1,TLIST=1
CALL/MTLF
DATA=DXX(1)/-125,-95,-95,-125,-125,-95,-35,35,-50,-35,35,0,35,35,50,$
35,35,0,95,125,125,95,95,125,25,-25/
DATA=DYY(1)/38,38,0,0,-38,-38,-35,35,0,35,35,50,35,35,0,-35,35,-50,$
-38,-38,0,0,38,38,0,0/
CALL/RDPAT,NP=20,K=1,N=1          $$(20)12.7 HOLES
CALL/RDPAT,NP=2,K=2,N=21          $$(2) M10 X 1.5 THRDS
PPRINT ALL TOOLS ARE METRIC.
DATA=D(1)/12.5,8,8,10/
DATA=TAN(1)/38020125,38020088,38080010/
CALL/TLLD2,TN=1,TP=2,DTH=12.7,FP=1
CALL/TLLD2,TP=3,DTH=12.7,FP=2
CALL/TLLD8,TP=3,P=1.5,DTH=12.7,FP=2,ORD=1
CALL/FIN
CALL/TOOLS,GLIST=1,ETNO=3
```

```
CCCC MMMM    MMMM           C I N C I N N A T I    A P T    P O S T P R O C E S S O R
CCCCC MMMMM  MMMMM
CCC    MMM MMM                          C I N A C 6   52
CCC    MMM MMM   C I N C I N N A T I
CCCCC MMMMM  MMMMM      CINCINNATI ACRAMATIC CNC MACHINING CENTER
CCCC MMMM    MMMM M I L A C R O N
```

	ABSX	ABSY	ABSZ	MMPM	RPM	TIME	CLRC

```
DATE 08/14/79    TIME 1255    SNUMB 7370T
FIG 19.8
MACHN= 10VC-1000 PROG'D BY J.M. RITENOUR
LOAD PART IN VISE - CENTER OF PART TO BE ON X 0 AND Y 0.
ALL TOOLS ARE METRIC.
* * * * * * * * * * * * * * * * * * * * * * * * * * * *
TOOL   +01.=     TWIST DRILL, DIAM=      +012.5000
* * * * * * * * * * * * * * * * * * * * * * * * * * * *

LEADER/    1.00
```

		ABSX	ABSY	ABSZ	MMPM	RPM	TIME	CLRC
!$ 01 G00 F13 T1 M06$								
02 G81 X-125000 Y38000 Z-15388 R0 F208 S698 T2 M13$		-125.000	38.000	-15.388	207	698	.359	287
N3 X-95000$		-125.000	38.000	-15.388	207	698	.093	358
N4 Y0$		-95.000	38.000	-15.388	207	698	.102	359
N5 X-125000$		-95.000	.000	-15.388	207	698	.103	360
N6 Y-38000$		-125.000	.000	-15.388	207	698	.102	361
N7 X-95000$		-125.000	-38.000	-15.388	207	698	.103	362
N8 X-35350 Y-35350$		-95.000	-38.000	-15.388	207	698	.102	363
N9 X-50000 Y0$		-35.350	-35.350	-15.388	207	698	.107	364
N10 X-35350 Y35350$		-50.000	.000	-15.388	207	698	.103	365
N11 X0 Y50000$		-35.350	35.350	-15.388	207	698	.103	366
N12 X35350 Y35350$		.000	50.000	-15.388	207	698	.103	367
N13 X50000 Y0$		35.350	35.350	-15.388	207	698	.103	368
N14 X35350 Y-35350$		50.000	.000	-15.388	207	698	.103	369
N15 X0 Y-50000$		35.350	-35.350	-15.388	207	698	.103	370
N16 X95000 Y-38000$		.000	-50.000	-15.388	207	698	.103	371
N17 X125000$		95.000	-38.000	-15.388	207	698	.112	372
N18 Y0$		125.000	-38.000	-15.388	207	698	.102	373
N19 X95000$		125.000	.000	-15.388	207	698	.103	374
N20 Y38000$		95.000	.000	-15.388	207	698	.102	375
N21 X125000$		95.000	38.000	-15.388	207	698	.103	376
N22 G00 Z3000$		125.000	38.000	-15.388	207	698	.102	377
023 G00 X125000 Y38000 Z2540 F208 S698 M13$		125.000	38.000	3.000	207	698	.000	446
		125.000	38.000	2.540	207	698	.000	450

```
* * * * * * * * * * * * * * * * * * * * * * * * * * * *
TOOL   +02.=     TWIST DRILL, DIAM=      +008.8000
* * * * * * * * * * * * * * * * * * * *
```

		ABSX	ABSY	ABSZ	MMPM	RPM	TIME	CLRC
024 G00 F208 T2 M06$							.239	465
025 G81 X25000 Y0 Z-14622 R0 F201 S992 T3 M13$		25.000	.000	-14.622	200	992	.113	528
N26 X-25000$		-25.000	.000	-14.622	200	992	.104	529
N27 G80 R2540$		-25.000	.000	-14.622	200	992	.001	602
N28 G00 Z5540$		-25.000	.000	5.540	200	992	.000	604
029 G00 X254000 Y254000 Z5080 F201 S992 M13$		254.000	254.000	5.080	200	992	.069	608
N30 G01 M00$								611

```
TAPE LENGTH =  1.20 METERS         MACHINING TIME =  2.952

OPERATOR - PLEASE CLEAN AND PREPARE ALL HOLES TO BE TAPPED
```

		ABSX	ABSY	ABSZ	MMPM	RPM	TIME	CLRC
031 G00 X254000 Y254000 Z2540 F201 S992 M13$		254.000	254.000	2.540	200	992	.001	624

```
* * * * * * * * * * * * * * * * * * * * * * * * * * * *
TOOL   +03.=     STD TAP, SIZE=      +010.0000-    +001.50
* * * * * * * * * * * * * * * * * * * * * * * * * * * *
```

		ABSX	ABSY	ABSZ	MMPM	RPM	TIME	CLRC
032 G00 F201 T3 M06$							.239	643
033 G84 X-25000 Y0 Z-20200 R0 F380 S253 M13$		-25.000	.000	-20.200	379	253	.193	693
N34 X25000$		25.000	.000	-20.200	379	253	.133	694
N35 G00 Z3000$		25.000	.000	3.000	379	253	.000	716
036 G00 X25000 Y0 Z2540 F380 S253 M13$		25.000	.000	2.540	379	253	.000	718
N37 X254000 Y254000 Z5080$		254.000	254.000	5.080	379	253	.067	722
N38 G01 M02$								767

```
LEADER/    1.00

TAPE LENGTH =  1.65 METERS         MACHINING TIME =  3.587
```

FIG. 18.9 Computer print-out of input data used to prepare the tape needed to drill and tap the upper plate shown in Fig. 18.8 (*Courtesy Cincinnati Milacron, Inc.*)

tions, and so forth. Omitted from the complete print-out are the POSTPROCESSOR SUMMARY (slide motion recap and machining recap), the CINDIV GENERATOR ASSEMBLY NUMBER LISTINGS, and the LIST OF REQUIRED GAUGES. Each statement of the print-out in Fig. 18.9 contains a unit of information that is complete in itself as an instruction related to a specific machining operation.

Contouring control, used for milling, turning, and similar operations, is much more complex than point-to-point control. This is so because the workpiece must be completely described and the path of the cutting tool must be rigidly controlled along a tight course.

The adjustment cone shown in Fig. 18.10 was produced from a solid 76-mm bar on the turning machine shown in Fig. 18.2. The input statements that ultimately led to the preparation of the needed control tape are given in Fig. 18.11. Not shown are the 300 or more lines of computer output that followed the no errors detected-processing continuing statement. The APT POSTPROCESSOR portion alone consisted of 72 lines.

18.8 CAD/CAM and Robots in Industry

As we approach the beginning of the twenty-first century, we are faced with a second industrial revolution. This revolution in production processes is now well underway as our national and international tool-makers, aided by new CAD/CAM technology, are blending the capabilities of computer-directed NC machines and robots to satisfy fabrication needs. CAD/CAM can be thought of as being divided into three general areas:

1. design drawing and drafting,
2. planning and scheduling,
3. fabrication and machining.

Once the data base has been established in the design and drafting phases, CAM, encompassing all of the manufacturing processes, takes over and the data base is used for overall machine control, tooling, quality control, and material movement. Even selective assembly, where parts

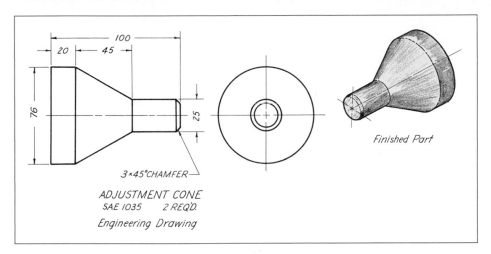

FIG. 18.10 Engineering drawing of adjustment cone.

```
CCCCC
C
C                          C I N T U R N   G E N E R A T O R   L E V E L A   (A002)   PAGE  1
C
C
CCCCCCINCINNATI                              (C)C O P Y R I G H T   1974
       M     M
      MM    MM
      M M M M
      M  M  M              C I N C I N N A T I   M I L A C R O N   I N C.
      M     M
      M     MILACRON
   2    LEADER/1
   3    PARTNO ADJUSTMENT CONE
   4    DATA=Z(1)/100,65,0/
   5    DATA=D(1)/25,76,77/
   6    CALL/ST,MACH=C12D65,METRIC=4,MAT=MTJ,CH=.4
  43    C(1)=3
  44    C(2)=45
  45    A(2)=29.8833
  46    FD(1)=.26
  47    FD(2)=.26
  48    FC(1)=.26
  49    FC(2)=.26
  50    CALL/FC,DD=2.5,NR1=1.19,FPR1=.38,DTH1=6.3,DWL=0
  51    CALL/RT,DTH1=6.35,DTH2=6.35,DTH3=6.35,NR1=1.2,FPR1=.5
  52    CALL/FT,NR1=1.19,RET=1.27,DWL=0
  53    FINI

*** NO ERRORS DETECTED - PROCESSING CONTINUING ***
```

FIG. 18.11 Print-out of statements used to prepare the control tape for production of the adjustment cone. Contour control has been utilized here. (*Courtesy Cincinnati Milacron, Inc.*)

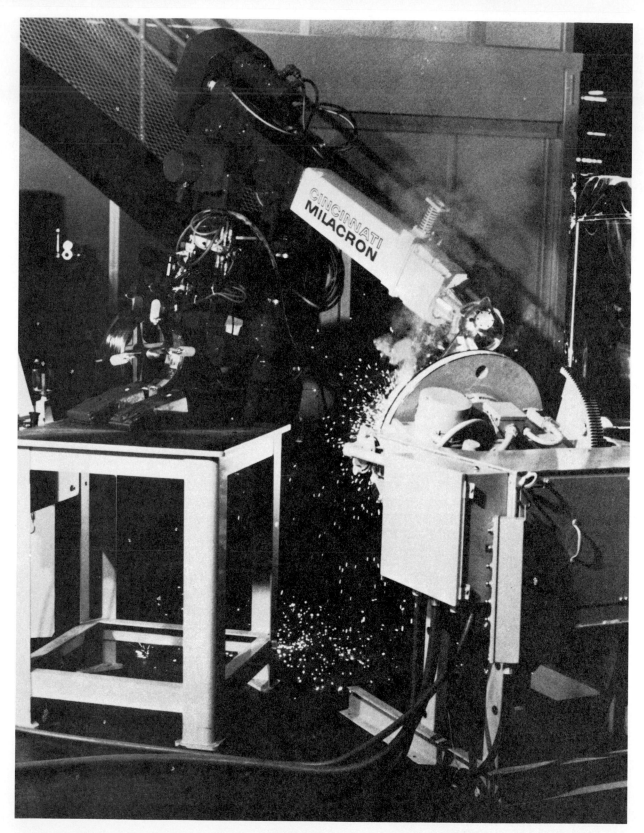

FIG. 18.12   Shown is a T3-746 Robot performing an arc seam welding operation. Included with the T3 Robot are welding packages that can be tailored to various applications. Included in a typical package are the robot, the welding power supply, gun, wire reels, feeders, and part-positioning tables. The necessary interfaces are provided between the various elements of the package in order to provide optimum production capabilities. (*Courtesy Cincinnati Milacron*)

must be fitted together with close tolerances, can be accomplished through CAM.

At this time, plants are being built that are equipped with robots having camera-aided artificial vision and the intelligence to check the work done by other robots stationed along an assembly line. NC machines and robots are now doing routine tasks like setting fasteners and making welds. (See Fig. 18.12.) Automotive painting is almost entirely handled by robots. Shown on the left-hand facing page at the start of this chapter is a computer controlled robot capable of serving two NC turning centers.

In looking toward the future, one can visualize computers controlling machines and processes at the fabrication level. These computers in turn will be under the control of intermediate computers at the department level. Above these intermediate departmental computers will be the master computers controlling the activities taking place between the several departments involved with the needed fabrication and assembly processes.

PART · 6

GRAPHIC METHODS FOR ENGINEERING COMMUNICATION, DESIGN, AND COMPUTATION

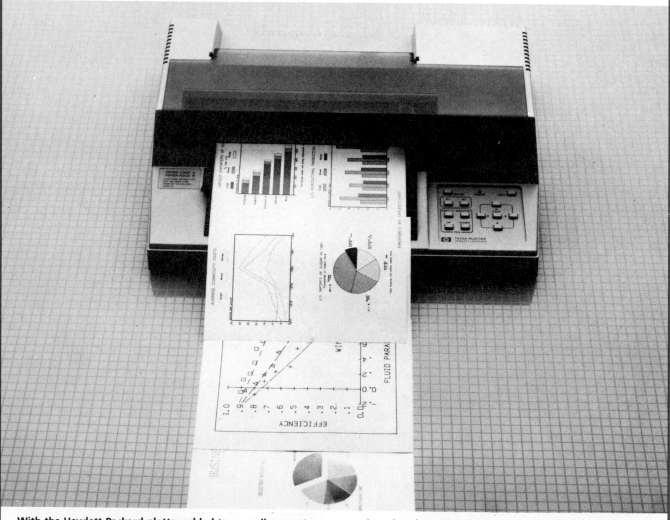

With the Hewlett-Packard plotter added to a small computer, a versatile computer graphics system can be created. A system that can plot data in any form that a user may choose, such as line graphs, bar charts, and even pie charts. The information gathered and presented may be retained either on paper or overhead transparency film. (*Courtesy Hewlett-Packard Corporation*)

Graphic Methods for Engineering Communication and Computation

 A/ GRAPHS AND CHARTS

19.1 Introduction

A properly designed graphical representation will convey data and facts more rapidly and effectively than a verbal, written, or tabulated description. A visual impression is easily comprehended and requires less mental effort than would be necessary to ascertain the facts from complex tables and reports (Figs. 19.1 and 19.22). It is because of this that many types of graphs and charts have been developed to present scientific, statistical, and technical information. Note how quickly the relationship presented by the line graph in Fig. 19.2 can be interpreted.

Engineers, even though they are concerned mainly with technical graphs, should be familiar also with the popular forms, for every industrial concern frequently must prepare popular types of graphs to strengthen their relationship with the public.

It is impossible to treat exhaustively the subject of graphical representation in a single chapter. Only a few of the most common forms used to analyze economic, scientific, and technical data can be discussed in detail. Many of the principles followed in the construction of engineering graphs, however, apply to the other types.

Those who are interested in further study of graphical methods for problem solving in the area of graphical mathematics (graphical algebra and graphical calculus) will find full coverage in Chapter 20.

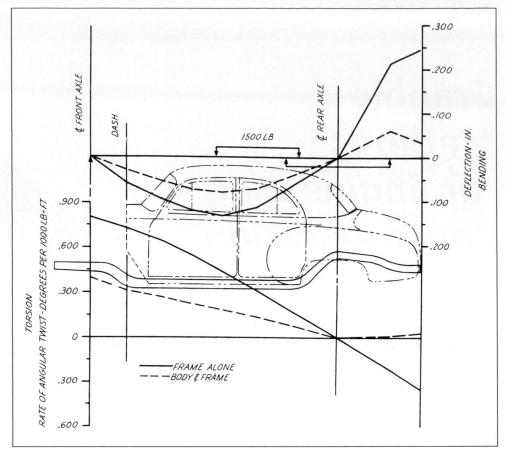

FIG. 19.1 **Engineering graph.** (*Courtesy General Motors Corporation. Reprinted from the General Motors Engineering Journal with permission.*)

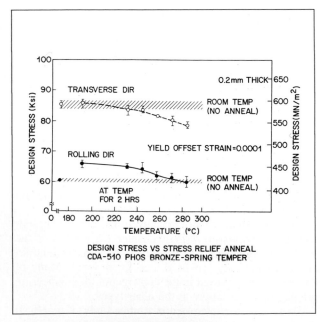

FIG. 19.2 **Engineering graph prepared for publication.** (*Courtesy Bell Laboratories*)

The same drafting skill is required in the execution of a graph as in making any other type of technical drawing. Good appearance is important and can be achieved only with the help of good lettering and smooth, uniform, and properly contrasted lines.

19.2 Classification of Charts, Graphs, and Diagrams

Graphs, charts, and diagrams may be divided into two classes in accordance with their use and then further subdivided according to type.

Use

1. Scientific
2. Popular

Type

1. Rectilinear charts
2. Semilogarithmic charts
3. Logarithmic charts

4. Barographs and area and volume charts
5. Percentage charts
6. Polar charts
7. Trilinear charts
8. Alignment charts (nomographs)
9. Pictorial charts

19.3 Quantitative and Qualitative Charts and Graphs

In general, charts and diagrams are used for one of two purposes, either to read values or to present a comparative picture relationship between variables. If a chart or graph is prepared for reading values, it is called a *quantitative graph*; if prepared for presenting a comparative relationship, it is called a *qualitative graph*. Obviously, some charts serve both purposes and cannot be classified strictly as either type. One of these purposes, however, must be predominant. Since a number of features in the preparation depend on the predominant purpose, such purpose must be determined before attempting to construct a graph.

19.4 Ordinary Rectangular Coordinate Graphs

Most engineering graphs prepared for laboratory and office use are drawn on ruled, rectangular graph paper and are plotted in the first quadrant (upper-right-hand), with the intersection of the X (horizontal)-axis and Y (vertical)-axis at the lower left used as the zero point or origin of coordinates. The paper is ruled with equispaced horizontal and vertical lines, forming small rectangles. The type most commonly used for chart work in experimental engineering is $8\frac{1}{2} \times 11$ in. and is ruled to form $\frac{1}{20}$-in. squares [Fig. 19.3(a)], every fifth line being heavy. Another type of paper frequently used, which is suitable for most laboratory reports in technical schools, has rulings that form 1-mm and 1-cm squares [Fig. 19.3(b)]. Other rulings run $\frac{1}{10}$, $\frac{1}{8}$, or $\frac{1}{4}$ in. apart. Ordinarily the ruled lines are spaced well apart on charts prepared for reproduction in popular and technical literature (Figs. 19.6 and 19.8). The principal advantage of having greater spacing between the lines is that large squares or rectangles tend to make the graph easier to read. Ready printed graph papers are available with various rulings in several colors.

Ordinary coordinate line graphs are used extensively because they are easily constructed and easily read. The known relationship between the variables is expressed by one or more continuous lines, which may be straight, broken, or curved.

The graph in Fig. 19.4 shows the approximate barometric pressure at different heights above sea level.

A graphical representation may be drawn easily and correctly if, after the required data have been assembled, careful consideration is given to the principles of curve drawing discussed in the following sections.

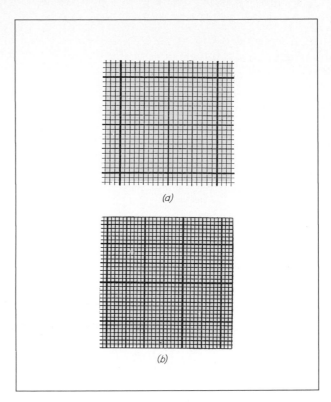

FIG. 19.3 **Types of graph paper.**

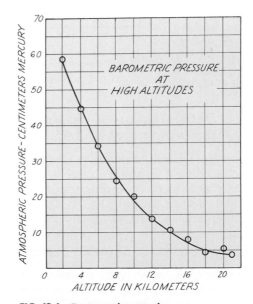

FIG. 19.4 **Rectangular graph.**

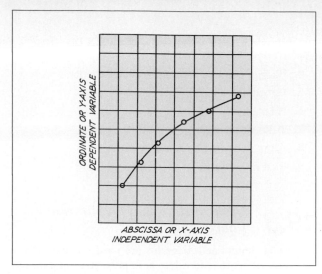

FIG. 19.5 **Independent and dependent variables.**

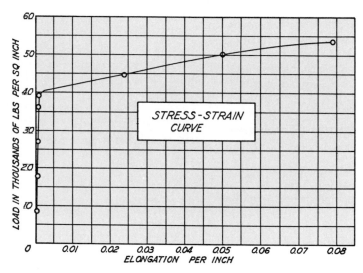

FIG. 19.6 **Stress-strain diagram.**

19.5 Determination of the Variables for Ordinate and Abscissa

The independent variable, the quantity arbitrarily varied during the experiment, usually is chosen for the abscissa (Fig. 19.5). Certain kinds of experimental data, however, such as a stress-strain diagram (Fig. 19.6), are plotted with the independent variable along the ordinate.

19.6 Selection of Suitable Scales

Since the slope of the curve, in its entirety as well as at intermediate points, provides a visual impression of the degree of change in the dependent variable for a given increment of the independent variable, care must be exercised in selecting the scales. The choice of scales is the controlling factor in creating the correct impression of the relationship to be shown by the graph.

The range of the scales should be such as to ensure full and effective use of the coordinate area in attaining the objective of the graph. Should a visual comparison of

plotted magnitudes be desired, the zero line should be included. That is, should the chart or graph be qualitative, both the ordinate and abscissa generally should have zero value at the intersection of the axes, as in Figs. 19.2 and 19.6. However, if the chart is quantitative, the intersection of the axes need not be at the origin.

In the case of arithmetic scales, the scale numbers on the graph and the space between coordinate rulings should generally correspond to 1, 2, or 5 units of measurement that may be multiplied (or divided) by 1, 10, 100, 1000, and so forth. Other units could be used except for the fact that they create situations wherein it becomes difficult to interpolate values. For example, one square should equal one of the following:

0.01	0.1	1	10	100	etc.
0.02	0.2	2	20	200	etc.
0.04	0.4	4	40	400	etc.
0.05	0.5	5	50	500	etc.
etc.	etc.	etc.	etc.	etc.	etc.

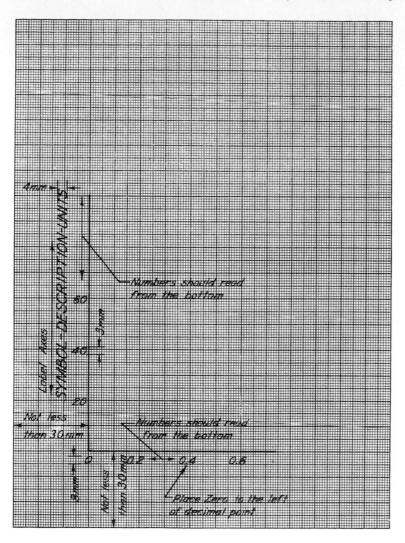

FIG. 19.7 Graph construction.

Along the horizontal (independent variable) scale, the values increase from the left to the right, while those along the vertical (dependent variable) scale should increase from bottom to top.

19.7 Locating the Axes and Marking the Values of the Variables

On graphs prepared for laboratory reports and not for publication, the axes should be located 30 mm (1 in.) or more inside the border of the coordinate ruling (Fig. 19.7). When selecting the scale units and locating the axes, it should be remembered that the abscissa may be taken either the long way or short way of the coordinate paper, depending on the range of the scales.

The use of many digits in the scale numbers is to be avoided. This can be easily accomplished by using a suitable designation in the scale caption like—IN THOUSANDS.

EXAMPLES

FORCE-kN
VELOCITY-km/h
PRESSURE, MM OF HG $\times$ 10^{-5};
RESISTANCE, THOUSANDS OF OHMS.

The numbers should read from the bottom when possible (Fig. 19.7). For the sake of good appearance, they never should be crowded. Always place a zero to the left of the decimal point when the quantity is less than unity.

Usually, only the heavy coordinate lines are marked to indicate their values or distance from the origin, and, even then, the values may be shown only at a regular selected interval (Fig. 19.7). The numbers should be placed to the left of the Y-axis and just below the X-axis.

When several curves representing different variables are to appear on the same graph, a separate axis generally is required for each variable (Fig. 19.8). In this case, a corresponding description should be given along each axis. The axes should be grouped at the left or at the bottom of the graph, unless it is desirable to place some at the right or along the top.

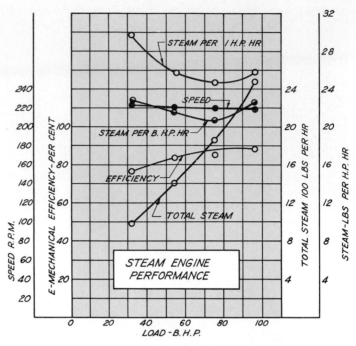

FIG. 19.8 Representation of several curves on a graph.

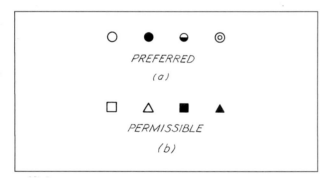

FIG. 19.9 Identification symbols.

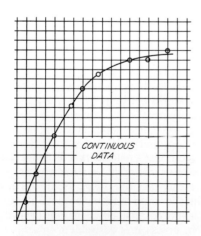

FIG. 19.10 Continuous curve.

19.8 Indicating Plotted Points Representing the Data

If the data represent a set of experimental observations, the plotted points of a single-curve graph should be marked by small circles approximately 3 mm (.1 in.) in diameter (Fig. 19.9). The following practice is recommended: open circles, filled-in circles, and partially filled-in circles (○ ● ◗) rather than crosses, squares, and triangles should be used to differentiate observed points of several curves on a graph. Filled-in symbols may be made smaller than those not filled in.

Mathematical curves are frequently drawn without distinguishing marks at computed positions.

19.9 Drawing a Curve

Since most physical phenomena are continuous, curves on engineering graphs usually represent an average of plotted points (Fig. 19.10). Discontinuous data should be plotted with a broken line, as shown in Fig. 19.11.

It is preferable to represent curves by solid lines. If more than one curve appears on a graph, differentiation may be secured by varied types of lines; but the most important curve should be represented by a solid one. A very fine line should be used for a quantitative curve if values are to be read accurately. A heavy line 0.7 mm (.03 in.) in width is recommended for a qualitative curve. It should be observed in Figs. 19.10 and 19.11 that the curve line does not pass through open circles.

When several curves are to appear on a graph, it is recommended that a solid line be used for the most important curve. Differentiation between the curves may be secured by using different types of lines for the other curves (dashed, etc.). In addition, the curves should bear suitable designations and, if practicable, be identified by brief labels placed horizontally or along the curves rather than by letters and numbers that require the reader to refer to a key (Fig. 19.8).

19.10 Labeling the Scales

Each scale caption should give a description of the variable represented and the unit of measurement. The captions on engineering graphs frequently contain an added identifying symbol, such as

N-EFFICIENCY-PERCENT
P-OUTPUT-HP

All lettering should be readable from the bottom and right side of the graph (not the left side). When space is limited, standard abbreviations should be used, particularly for designating the unit of measurement. To avoid confusing the reader, the draftsman should use only recognized word contractions.

19.11 Titles, Legends, Notes, and So On

The title of a graph should be clear, concise, complete, and symmetrical. It should give the name of the curve, the source of the data, the date, and other important information (Fig. 19.12). It should be so placed that it gives a balanced effect to the completed drawing (Fig. 19.8). In addition to the title, a wiring diagram, pictorial diagram, formula, or explanatory note is often necessary to give a clear picture of the nature of the experiment. For example, if there is any great irregularity in the plotted points or a condition that may have affected the values as shown by the data, a note of explanation should be given. A legend or key is sometimes included to explain a set of curves in greater detail.

19.12 Procedure for Making a Graphical Representation in Ink

1. Select the type of coordinate paper.

2. Determine the variables for ordinate and abscissa.

3. Determine the scale units.

4. Locate the axes and mark the scale values in pencil.

5. Plot the points representing the data. [Many drafters ink the symbol (○ ◑), indicating the points at this stage.]

6. Draw the curve. If the curve is to strike an average among the plotted points a trial curve should be drawn in pencil. If the curve consists of a broken line, as is the case with discontinuous data, the curve need not be drawn until the graph is traced in ink.

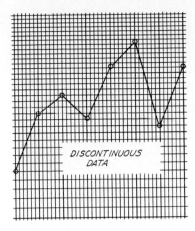

FIG. 19.11 **Discontinuous data.**

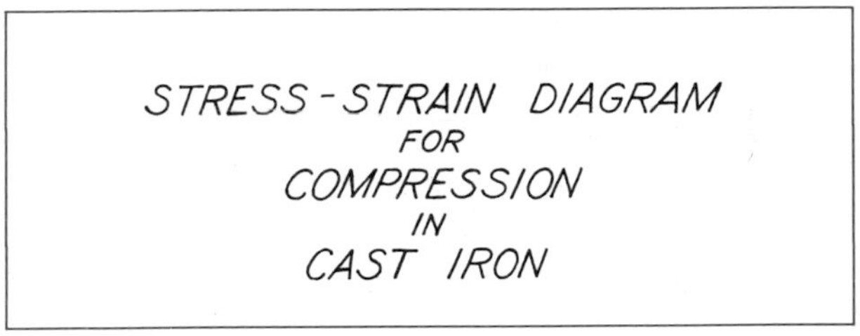

DISCONTINUOUS DATA

STRESS-STRAIN DIAGRAM
FOR
COMPRESSION
IN
CAST IRON

FIG. 19.12 **Title.**

7. Label the axes directly in ink.

8. Letter the title, notes, and so on. The title should be lettered on a trial sheet that can be used as a guide for lettering directly in ink on the graph.

9. Check the work and complete the diagram by tracing the curve in ink.

19.13 Logarithmic Graphs

Logarithmic coordinate graphs are constructed on prepared paper on which the parallel horizontal and parallel vertical rulings are spaced proportional to the logarithms of numbers (Fig. 19.13). This type of graph has two principal advantages over the ordinary coordinate type. First, the error in plotting or reading values is a constant percentage, and, second, an algebraic equation of the form $y = ax^b$ appears as a straight line if x has a value other than zero. The exponent b may be either plus or minus.

The equation for a falling body, $D = \frac{1}{2}gt^2$, is represented in Figs. 19.13 and 19.14. Observe that the plotted points form a parabolic curve on ordinary coordinate graph paper (Fig. 19.14) and a straight line on logarithmic paper (Fig. 19.13). To draw the line on the graph in Fig. 19.13, it is necessary to calculate and locate only two points, while in Fig. 19.14, several points must be plotted to establish the location of the corresponding curved-line representation. The line on Fig. 19.13 has a slope of 2:1, because the exponent of t is 2. Therefore, the line could be drawn

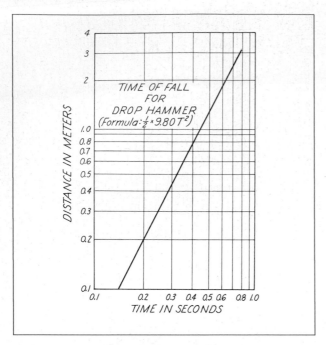

FIG. 19.13 **Logarithmic graph.**

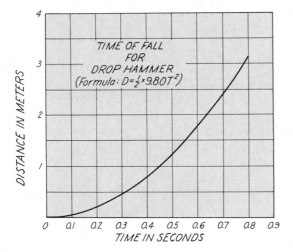

FIG. 19.14 **Coordinate graph.**

using one point and the slope, instead of plotting two points and joining them with a straight line.

Log paper is available with rulings in one or more cycles for any range of values to be plotted. Part-cycle and split-cycle papers may also be purchased.

19.14 Semilogarithmic Graphs

Semilogarithmic paper has ruled lines that are spaced to a uniform scale in one direction and to a logarithmic scale in the other direction (Fig. 19.15). Charts drawn on this form of paper are used extensively in scientific studies, because functions having values in the form of geometric progressions are represented by straight lines. In any case,

the main reason for the use of semilogarithmic paper is that the slope of the resulting curve indicates rate of change rather than amount of change, the opposite being true in the case of curves on ordinary coordinate graph paper. Persons who are interested may determine the rate of increase or decrease at any point by measuring the slope. A straight line indicates a constant rate of change. In commercial work this form of paper is generally called *ratio paper*, and the charts are known as *rate-of-change charts*.

As previously stated, the choice of a type of graph paper depends on the information to be revealed. Curves drawn on uniform coordinate graph paper to illustrate the percentage of expansion or contraction of sales, and so on, present a misleading picture. The same data plotted on semilogarithmic paper would reveal the true rate of change to the business management. For this reason, semilogarithmic paper should be used whenever percentage of change rather than quantity of change is to be shown. In scientific work, when the value of one variable increases in a geometric progression and the other in an arithmetic progression, this form is valuable.

19.15 Bar Charts

Bar charts are used principally in popular literature covering economic and industrial surveys. They are a simple diagrammatic form giving a pictorial summary of statistical data and can be easily understood by the average person. Logarithmic and uniform coordinate graphs are less suited for this purpose, because few people know the procedure for reading curves or understand their picture qualities.

Whenever values or quantities are illustrated, as in Fig. 19.16, by consecutive heavy bars whose lengths are proportional to the amounts they represent, the resulting representation is called a *bar chart*.

The bars on this type of diagram may be drawn either horizontally or vertically, but all should start at the same zero line. Their lengths should be to some fixed scale, the division values of which may be given in the margin along the bottom or left side of the graph. When it is necessary to give the exact values represented, the figures should be placed along each bar in a direction parallel to it. To place the values at the end gives the illusion of increasing the length of the bars. Usually, the names of the items are lettered to the left of the vertical starting line on a horizontal chart and below the starting line on a vertical chart.

19.16 Area (Percentage) Charts

An area diagram can be used profitably when it is desirable to present pictorially a comparison of related quantities in percentage. This form of representation illustrates the relative magnitudes of the component divisions of a total of the distribution of income, the composition of the population, and so on. Two common types of the various forms of area diagrams used in informative literature are

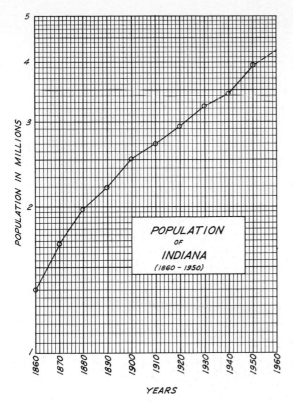

FIG. 19.15 Semilogarithmic graph.

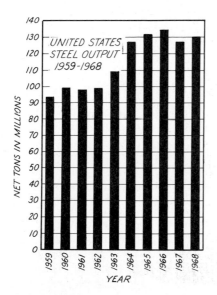

FIG. 19.16 Bar chart.

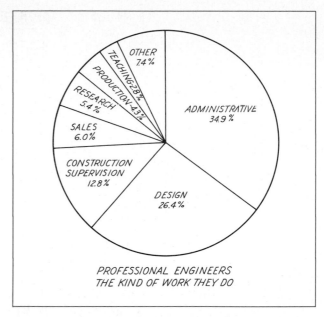

FIG. 19.17 Pie chart.

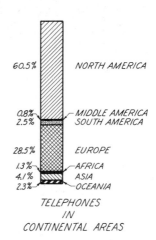

FIG. 19.18 Percentage bar chart.

illustrated in Figs. 19.17 and 19.18. Percentages, when represented by sectors of a circle or subdivisions of a bar, are easy to interpolate.

The pie chart (Fig. 19.17) is the most popular form of area diagram, as well as the easiest to construct. The area of the circle represents 100% and the sectors represent percentages of the total. To make the chart effective, a description of each quantity and its corresponding percentage should be lettered in its individual sector. All

lettering should be completed before the areas are crosshatched or colored if this is to be done. The percentage bar chart shown in Fig. 19.18 fulfills the same purpose as the pie chart. The overall area of the bar represents 100%. Note that each percentage division is crosshatched in a different direction. The descriptions may be placed on either side of the bar; the percentages should be on the bar or at the side.

19.17 Polar Charts

Certain types of technical data can be more easily plotted and better represented on polar coordinate paper. Polar charts may be drawn by self-recording instruments. Polar diagrams and plotted polar curves representing various kinds of scientific data are very common. Polar curves are

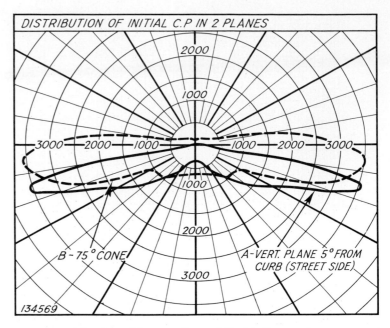

FIG. 19.19   Polar chart. (*Courtesy General Electric Company*)

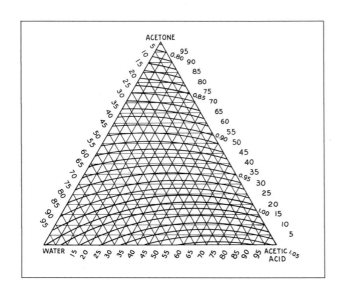

FIG. 19.20   Trilinear chart. (*Courtesy American Chemical Society*)

used to represent the intensity of diffused light, intensity of heat, and so on. The polar chart in Fig. 19.19 gives, in terms of candlepower, the intensity of light in two planes.

19.18 *Trilinear Charts*

Trilinear charts are used principally in the study of the properties of chemical compounds, mixtures, solutions, and alloys (Fig. 19.20). Basically this is a 100% chart the use of which, owing to its geometric form, is limited to the investigation of that which is composed of three constituents or variables. Its use depends on the geometric principle that the sum of the three perpendiculars from

any point is equal to the altitude. If the altitude represents 100%, the perpendiculars will represent the percentages of the three variables composing the whole.

The ruling can be accomplished conveniently by dividing any two sides of the triangle into the number of equal-percentage divisions desired and drawing through these points lines parallel to the sides of the triangle.

19.19 *Chemical Engineering Charts*

Figure 19.21 shows a type of flow chart that must be prepared frequently by chemical engineers in industrial practice.

19.20 *Pictorial Charts*

Pictorial charts are quite generally used to present data in reports prepared for nontechnical readers. Usually, such charts present comparisons of populations (Fig. 19.22), expenditures, costs, and so forth. Stacks of silver dollars may represent expenditures; sizes of animals can represent livestock production; and human figures can present employment data.

B/ **EMPIRICAL EQUATIONS**

19.21 *Empirical Equations*

In all phases of engineering work considerable experimentation is done with physical quantities, and the engineer (or engineering technologist) is usually the person most concerned with the behavior of quantities in relation

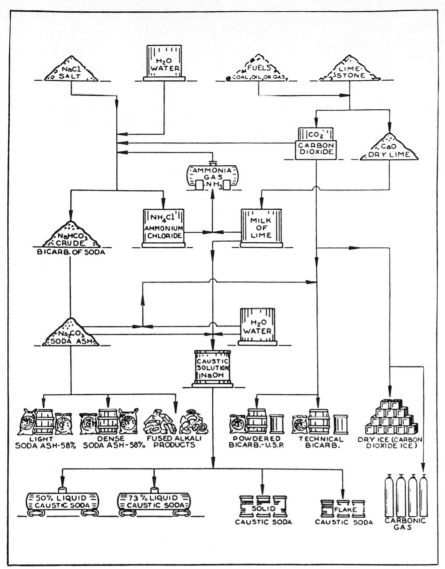

FIG. 19.21 **Flow chart of ammonia-soda operations.** (*Courtesy Chemical Industries Magazine*)

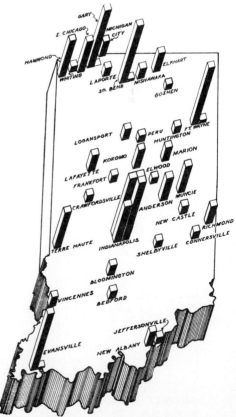

FIG. 19.22 **Population chart.** (*Courtesy Indiana State Planning Board*)

to one another. Often it is known that the subject of the experiment obeys some physical law which can be expressed by a mathematical equation, but the exact equation is unknown. Then the person performing the experiment is faced with the task of finding an equation to fit the data that has been obtained. The three articles that follow discuss means of arriving at an equation from a graphical study of the data. An equation determined in this manner is an *empirical equation*. Since the unknown law may be quite complex, a single empirical equation to fit the whole range of data may not exist. However, in the majority of such cases a series of the various empirical equations with limited coverage (parameters) can be found.

19.22 Equations of the Form $y = a + bx$

If the plotted points, representing the data, lie in what appears to be a straight line when plotted on rectangular coordinate paper, the equation of the data is a linear or first-degree equation of the form $y = a + bx$, where b is the slope of the line and a is the y-intercept ($x = 0$).

After it has been decided that the relationship between quantities is linear, the next step is to draw the best average straight line that will be representative of the data (see Fig. 19.23). This line, extended if necessary to the y-axis, establishes the y-intercept ($x = 0$) which will be the value of a in the equation. The slope can be determined from any two points along the line. However, for accuracy the points should be selected as far apart as possible. The value of b is expressed as

$$b = \frac{y_2 - y_1}{x_2 - x_1}$$

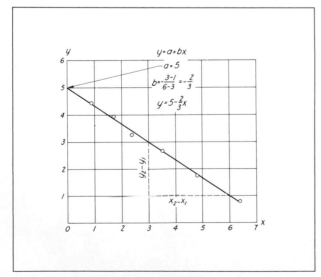

FIG. 19.23 Determination of empirical equations of the form $y = a + bx$ (first degree).

Thus the slope b in Fig. 19.23 is calculated, once two points have been selected, as

$$b = -\frac{3 - 1}{6 - 3} = -\frac{2}{3} \qquad \text{(slope)}$$

With the slope now known and the value of a having been read directly on the y-axis as 5, the equation of the line can be written as

$$y = 5 - \frac{2}{3}x$$

If it is not reasonable to include $x = 0$ in the plot of the data, then a pair of simultaneous equations are set up using two points on the line.

$$y_1 = a + bx_1$$
$$y_2 = a + bx_2$$

This system can then be solved for a and b.

19.23 Equations of the Form $y = ax^b$

If the data plots very nearly in a straight line on logarithmic graph paper, the equation of the data is a power equation of the form $y = ax^b$. Power equations of this type, where one quantity varies directly as some power of another, appear as either parabolic or hyperbolic curves when plotted on rectangular coordinate paper, depending upon whether the exponent is positive or negative. For positive values of b (except for unity) the curves are parabolic; negative values produce hyperbolic curves.

When the equation is placed in logarithmic form and rewritten as $\log y = \log a + b \log x$, it is now in the same form as the equation of a straight line. In this case, when $x = 1$, then $y = a$ because $\log 1 = 0$.

Hence, after the data have been plotted (y vs. x) on logarithmic paper as in Fig. 19.24, a representative (average) straight line should be drawn. The line extended to the y-axis establishes a at the y intercept ($x = 1$). The value of a can be read directly, since the spacing is logarithmic. As before, two points along the line are used in the equation that follows to determine b.

$$b = \frac{\log y_2 - \log y_1}{\log x_2 - \log x_1}$$

For the two points selected along the line in Fig. 19.24 we calculate

$$b = \frac{1.477 - 0.699}{1.772 - 0.398} = .57$$

Thus, the equation is $y = 3x^{.57}$.

If $x = 1$ is not included in the plot of the data, then one may resort to a pair of simultaneous equations.

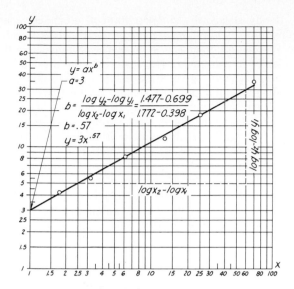

FIG. 19.24 Determination of empirical equations of the form $y = ax^b$ (power).

$$\log y_1 = \log a + b \log x_1$$

$$\log y_2 = \log a + b \log x_2$$

In solving for a (also b) the solution gives the value for $\log a$. The real value of a (and b) must be found in log tables.

19.24 Equations of the Form $y = ab^x$

If the data plots as a nearly straight line on semilogarithmic graph paper (Fig. 19.25), the equation of the data is an exponential equation of the form $y = ab^x$. This equation may be rewritten as $\log y = \log a + x \log b$, which once again is in the same form as the equation of a straight line. In this case, when $x = 0$, then $y = a$.

With the data for the exponential equation plotted directly on semilogarithmic graph paper, one must again draw the most representative line along the path of the plotted points. This line extended to the y-axis ($x = 0$) determines the value of a. Again, two points must be selected along the line and their coordinates substituted in the equation

$$\log b = \frac{\log y_2 - \log y_1}{x_2 - x_1}$$

Then, with the logarithm of the value of b known, the real value of b as needed for the equation can be determined from log tables.

For the two points selected for the line in Fig. 19.25,

$$\log b = \frac{1.778 - 1.176}{5 - 2}$$

$$= \frac{0.602}{3} = .201$$

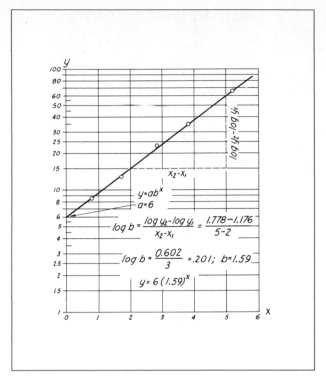

FIG. 19.25 Equations of the form $y = ab^x$.

Hence,

$$b = 1.59$$

With the value of a read on the y-axis as 6, the formula for the data may be written as $y = 6 \, (1.59)^x$.

If $x = 0$ is not included in the plot of the data, then one must solve the simultaneous equations

$$\log y_1 = \log a + x_1 \log b$$

$$\log y_2 = \log a + x_2 \log b$$

Then, since the solution gives both a and b in terms of logarithms, log tables must be used to determine the real values of a and b.

When natural logarithms are available, the equation $y = ab^x$ may be changed to $y = Ae^{mx}$. This latter equation, in turn, may be written as $\ln y = \ln A + mx$, which is still the equation of a straight line. The value of A can be determined as before at $x = 0$, and

$$m = \frac{\ln y_2 - \ln y_1}{x_2 - x_1}$$

where the value of m is found directly.

When $x = 0$ is not included, the equations to be used are

$$\ln y_1 = \ln A + mx_1$$

$$\ln y_2 = \ln A + mx_2$$

C / ALIGNMENT CHARTS

19.25 Alignment Charts (Nomographs)

The purpose of alignment charts is to eliminate many of the laborious calculations necessary to solve formulas containing three or more variables. Such a chart is often complicated and difficult to construct, but if it can be used repeatedly, the labor involved in making it will be justified. In the commercial field, these charts appear in varied forms, which may be very simple or very complicated (Fig. 19.26).

Briefly stated, the simplest form of alignment chart consists of a set of three or more inclined or vertical scales so spaced and graduated as to represent graphically the variables in a formula. The scales may be divided into logarithmic units or some other types of functions, depending upon the form of equation. As illustrated in Fig. 19.27, the unknown value may be found by aligning a straight-edge to the points representing known values on two of the scales. With a scale or triangle so placed, the numerical value representing the solution of the equation can be read on the third scale at the point of intersection.

Since alignment charts in varied forms are frequently prepared by engineers and technologists, it is desirable that students studying in the fields dealing with the sciences have some knowledge of the fundamental principles un-

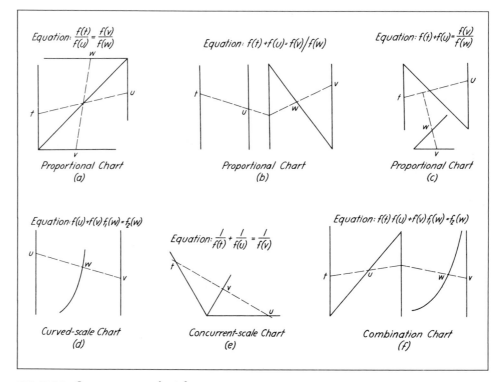

FIG. 19.26   Some common chart forms.

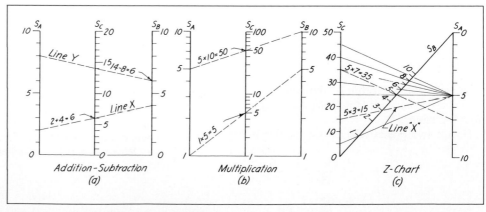

FIG. 19.27   Parallel-scale chart and Z-chart.

derlying their construction. However, in any brief treatment, directed toward a beginner, it is impossible to explain fully the mathematics involved in the construction of the many and varied types (Fig. 19.26). Therefore, our attention here must be directed toward an understanding of a few of the less complicated straight-line forms with the hope that the student will gather sufficient knowledge to construct simple charts for familiar equations (Fig. 19.34).

19.26 Forms of Alignment Charts

Examples of some of the forms that alignment charts may have are shown in outline in Fig. 19.26. Examples of forms of proportional charts are illustrated in (a), (b), and (c). Miscellaneous forms are shown in (d), (e), and (f). One may obtain information needed for the construction of proportional-type charts, concurrent scale charts, four-variable N-charts, and charts having a curved scale from any of the several books on nomography that are listed in the Bibliography of this text.

19.27 Construction of Alignment Charts

In this limited study, the explanation for the constructions will be based on the principles of plane geometry. The two forms to be considered for formulas are the parallel-scale chart and the Z-chart, also called an N-chart (Fig. 19.27).

Without giving thought at this time to the geometry underlying the construction of alignment charts and to the selection of scales, the methods that might be used to construct simple charts for graphical addition and subtraction and graphical multiplication and division might well be considered [see Fig. 19.27(a) and (b)].

A parallel-scale alignment chart of the type shown in (a), prepared for the purpose of making additions and subtractions, could be constructed as follows.

STEP 1. Draw three vertical straight lines spaced an equal distance apart.

STEP 2. Draw a horizontal base line. This line will align and establish the origins (0) of the three scales.

STEP 3. Using an engineer's decimal or metric scale, mark off a series of equal lengths on scales S_A and S_B. Start at the base line in each case. Mark the values of the graduations upward on both scales starting with 0 at the base line.

STEP 4. Mark off on the S_C scale a series of lengths that are half as long as those on scales S_A and S_B. Number the graduation marks starting at the base line.

In using this chart to add two numbers, say 2 and 4, one may align the ruling edge of a triangle through 2 on the S_A scale and 4 on the S_B scale, then, read their sum at the point where the edge of the triangle crosses the S_C scale (see line X). To subtract one number from another, say 8 from 14, the edge of the triangle should be placed so as to pass through 8 on scale S_A and 14 on S_C. The

difference, read on the S_B scale, will be 6, as shown by line Y.

If logarithmic scales are used for this form of chart as in (b) instead of natural scales as in (a), a chart for multiplication and division results, for the log of the product of two numbers is equal to the sum of the logs of the factors. Thus, by a method of addition a product can be obtained.

Necessary information for the construction of logarithmic scales is given in Sec. 19.29.

A Z-chart (also called an N-chart), which will give the product of two numbers, is shown in (c). For example, if line X is assumed to represent the edge of a triangle so placed as to pass through 5 on the S_A scale and 3 on the S_B scale, it can be seen that $3 \times 5 = 15$.

A simple Z-chart that has been prepared solely for straight multiplication will have outside vertical scales of uniform spacing. The length of the scales and the spacing of the graduation marks can be arbitrarily determined as long as the chart will fit on the paper that is available and provided that the graduations cover the desired range in each case. The two vertical scales begin with 0 (zero) value at opposite ends of the diagonal scale so that the values of the graduations read in increasing magnitude upward on one scale and downward on the other.

19.28 Definitions

Before starting a discussion on the construction of scales, it is necessary that the student have an understanding of the meaning of the following terms and expressions that are commonly used when constructing alignment charts for solving equations:

Constant. A quantity whose value remains unchanged in an equation.

Variable. A quantity capable of taking values in an equation. A variable is designated by some letter, usually one of the latter letters of the alphabet.

Function of a variable. A mathematical expression for a combination of terms containing a variable, usually expressed in abbreviated form as $f(x)$ which is understood to mean "function of x." An equation usually contains several functions of different variables, such as $f(r) + f(s) = f(t)$; or $f(u) \cdot f(v) = f(w)$.

Functional modulus. A proportionality multiplier that is used to bring a range of values of a particular function within a selected length for a scale. For instance, with the upper and lower limits of a function known and a definite length L chosen for the scale, the value of the functional modulus (m) can be found by dividing L by the amount of the difference between the upper and lower limits of the function. The scale equation for determining m may be written

$$m = \frac{L}{f(u_2) - f(u_1)}$$

where $f(u_2)$ and $f(u_1)$ are the upper and lower limits.

u	0	1	2	3	4	5	6	7	8	9	10
u^2	0	1	4	9	16	25	36	49	64	81	100
$u^2/2$	0	0.5	2.0	4.5	8.0	12.5	18.0	24.5	32.0	40.5	50.0
$m(u^2/2)$	0	1.25	5	11.25	20	31.25	45	61.25	80	101.25	125

FIG. 19.28   Table.

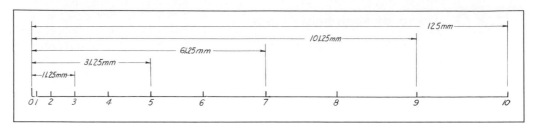

FIG. 19.29   Functional scale for $u^2/2$ (distances in millimeters).

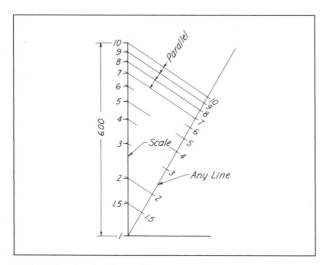

FIG. 19.30   Graduating a log scale.

Scale. A graduated line that may be either straight or curved. When the graduation marks are equally spaced, that is, when the distance between marks is the same for equal increments of the variable as the variable increases in magnitude, the scale is known as a uniform scale. When the lengths to the graduation marks are laid off to correspond to scale values of the function of a variable, the scale is called a *functional scale*.

19.29   Construction of a Functional Scale

Let it be supposed that it is necessary to construct a functional scale, 125 mm in length, for $f(u) = u^2/2$ with u to range from 0 to 10. It will be found desirable to make the necessary computations by steps and to record the scale data in tabular form (Fig. 19.28).

STEP 1. Record the values of u in the table.

STEP 2. Compute the values of the function.

STEP 3. Determine the functional modulus m.

STEP 4. Multiply the recorded values of the function by the functional modulus.

In this case and in many other cases, the functional modulus may be chosen by inspection. For this problem, the overall length (L) of the scale will be 125 mm when $m = 2.5$. If the scale equation is used to determine the functional modulus, then

$$m = \frac{L}{f(u_2) - f(u_1)} = \frac{125}{(10^2/2) - 0}$$

$$= \frac{125}{50} = 2.5$$

All that now remains to be done to construct the scale is to lay off the computed distances along the line for the scale and mark the values at the corresponding interval points (Fig. 19.29).

Although a logarithmic scale might be constructed in this same manner, much time can be saved by using the graphic method shown in Fig. 19.30 for subdividing the scale between its end points. To apply this method to a scale that has already been laid off to a predetermined length with the end points of the range (say 1 to 10) marked, the steps of the construction are as follows:

STEP 1. Draw a light construction line through point 1 making any convenient angle with the scale line.

STEP 2. Using a printed log scale, mark off points on the auxiliary line.

STEP 3. Draw a line through the 10 point on the construction line and the 10 point on the scale; then, through the remaining points draw lines parallel to this line through the 10s. These will divide the scale in proportion to the logarithms of numbers from 1 to 10.

19.30 Parallel-scale Charts for Equations of the Form $f(t) + f(u) = f(v)$

An alignment chart that is designed for solving an equation that can be set up to take this form will have three parallel functional scales that may be either uniform or logarithmic depending upon the equation. More information will be presented later concerning parallel-scale charts with logarithmic scales. In this section, attention will be directed to charts having scales with uniform spacing.

Before one can start constructing the chart, one must determine by calculation certain necessary information. First, determine how the scales are to be graduated, and then calculate the ratio for the scale spacing.

To be competent to design parallel-scale charts, one must have a full understanding of the geometric basis for their construction. The explanation to follow is associated with the line layout in Fig. 19.31. Three parallel scales S_A, S_B, and S_C are shown with the origins t_0, u_0, and v_0 on line AB. Line (1) and line (2) are drawn parallel to AB through points v and u, respectively, on the isopleth. By similar triangles (shown shaded), we calculate

$$\frac{L_t - L_v}{a} = \frac{L_v - L_u}{b}$$

Now if

$$m_t = \frac{L_t}{f(t) - f(t_0)}$$

then

$$L_t = m_t[f(t) - f(t_0)]$$

When the function of t_0 is zero, the equation becomes

$$L_t = m_t f(t)$$

Similarly, when v_0 is zero,

$$L_v = m_v f(v)$$

and, when u_0 is zero,

$$L_u = m_u f(u)$$

Substituting these values,

$$\frac{m_t f(t) - m_v f(v)}{a} = \frac{m_v f(v) - m_u f(u)}{b}$$

Collecting terms, we find

$$m_t f(t) + \left(\frac{a}{b}\right) m_u f(u) = \left(\frac{a}{b}\right) m_v f(v) + m_v f(v)$$

$$= m_v \left(1 + \frac{a}{b}\right) f(v)$$

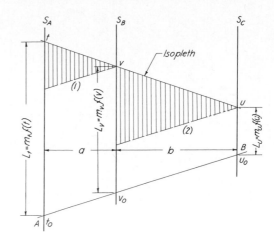

FIG. 19.31 Geometric basis for construction of parallel-scale alignment charts.

But $f(t) + f(u) = f(v)$ only if the coefficients of the three terms are equal; therefore,

$$m_t = \left(\frac{a}{b}\right) m_u = \left(1 + \frac{a}{b}\right) m_v$$

and

$$\frac{m_t}{m_u} = \frac{a}{b} \tag{1}$$

Now, since

$$m_t = \left(1 + \frac{a}{b}\right) m_v$$

and

$$\frac{a}{b} = \frac{m_t}{m_u}$$

then

$$m_t = \left(1 + \frac{m_t}{m_u}\right) m_v$$

$$m_v = \frac{m_t}{\left(1 + \dfrac{m_t}{m_u}\right)}$$

and, finally,

$$m_v = \frac{m_t m_u}{m_t + m_u} \tag{2}$$

Now suppose that one wants to construct a chart having the form of $t + u = v$ and that t is to have a range from

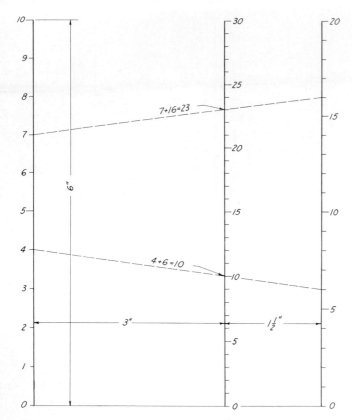

FIG. 19.32 Alignment chart for equation of the form $f(t) + f(u) = f(v)$.

0 to 10 and u from 0 to 20 (Fig. 19.32). It has been determined that the scale lengths should be 6 in. Then

$$m_t = \frac{6}{10.0} = \frac{6}{10} = .6$$

$$m_u = \frac{6}{20.0} = \frac{6}{20} = .3$$

and

$$m_v = \frac{m_t m_u}{m_t + m_u} = \frac{.6 \times .3}{.6 + .3}$$

$$= \frac{.18}{.90} = .20 \tag{2}$$

To determine the ratio of the scale spacing,

$$\frac{m_t}{m_u} = \frac{a}{b} = \frac{.6}{.3} = \frac{2}{1} \tag{1}$$

For convenience, distance a between scales can be made 3 in. Distance b must then be $1\frac{1}{2}$ in. to satisfy the proportion of 2 to 1. The total width of the chart will be $4\frac{1}{2}$ in.

Since in this particular case the modulus for each of the scales is in full tenths and the scales are to be uniformly divided, an engineer's scale may be used to mark off the scales of the chart. With other conditions, it would be necessary to prepare either a table such as the one shown in Fig. 19.28 or to divide the line between its end-points using a geometric method.

19.31 Parallel-scale Charts for Equations of the Form $f(t) \cdot f(u) = f(v)$

Equations of this form may be rewritten so as to take the form $f(t) + f(u) = f(v)$ by using logarithms for both sides of the equation. For example, suppose that it is desirable to prepare a chart for $M = Wl$, a formula commonly used by designers and engineers. In this equation, M is the maximum bending moment at the point of support of a cantilever beam, W is the concentrated load, and l is the distance from the point of support to the load.

GIVEN:

$$M = Wl$$

REWRITTEN:

$$\log M = \log W + \log l$$

Another example would be the formula for determining the discharge of trapezoidal weirs.

GIVEN:

$$Q = 3.367 \, Lh^{3/2}$$

REWRITTEN:

$$\log Q = \log 3.367 + \log L + 1.5 \log h$$

In this last equation, Q is discharge in cubic feet per second, L is the length of the crest in feet (width of weir), and h is the observed head (depth of water).

For the purpose of our discussion, the formula $P = I^2R$ will be used where P is power in watts, I is current in amperes, and R is resistance in ohms (Fig. 19.33). It has been determined that I must vary from 1 to 10 amps and R from 1 to 10 ohms. A chart of this type might be used to determine the power loss in inductive windings. The length of the scales is to be 5 in.

The steps for the construction are as follows.

STEP 1. Write the equation in standard form.

$$\log P = 2 \log I + \log R$$

STEP 2. Determine the moduli m_I and m_R for the outside scales.

$$m_I = \frac{5}{2 \log 10 - 2 \log 1} = \frac{5}{2} = 2.5$$

$$L_I = 2.5(2 \log I) = 5 \log I$$

$$m_R = \frac{5}{\log 10 - \log 1} = \frac{5}{1} = 5$$

$$L_R = 5 \log R$$

STEP 3. Determine m_P and L_P for the P-scale.

$$m_P = \frac{2.5 \times 5}{2.5 + 5} = \frac{12.5}{7.5} = \frac{5}{3}$$

$$L_P = \frac{5}{3} \log P$$

STEP 4. Determine the ratio for the spacing of the scales.

$$\frac{m_I}{M_R} = \frac{2.5}{5} = \frac{1}{2} \qquad \text{(ratio)}$$

STEP 5. Draw three vertical lines 1 in. and 2 in. apart and add a horizontal base line. By using these selected values, the ratio for spacing will be maintained and the chart will have good proportion.

STEP 6. Graduate the scales for I and R using the method shown in Fig. 19.30. In this particular case, both logarithmic scales will be alike and will range from 1 on the base line upward to 10.

STEP 7. Graduate the P-scale. By substituting values in the equation, it will be found that P will range from 1 W to 1000 W; therefore, the scale will be a three-cycle logarithmic scale, and the graphic method may be used for locating the graduation marks.

Figure 19.34 shows another parallel-scale alignment chart. This particular one could be used to determine the volume of a cylindrical tank when the diameter and height are known.

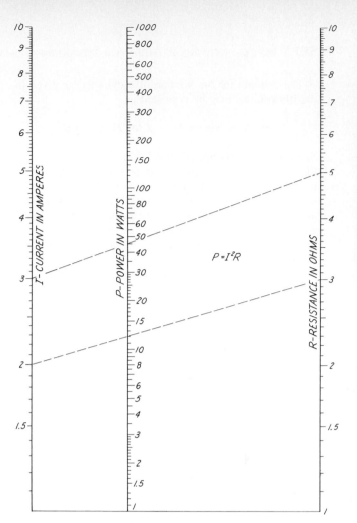

FIG. 19.33 Alignment chart for the equation $P = I^2R$.

FIG. 19.34 Alignment chart for $V = (\pi/4)D^2H$ (volume of a cylinder).

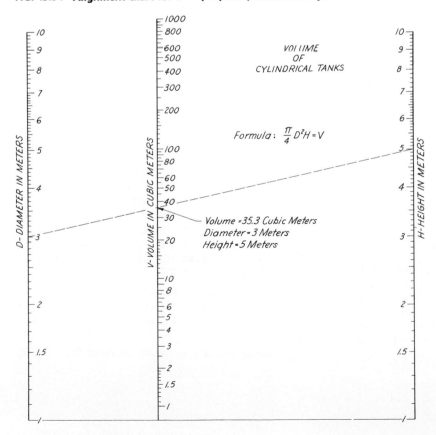

The formula for the volume of a cylinder, as given in the illustration, may be rewritten as

$$\log V = \log (\pi/4) + 2 \log D + \log H$$

Suppose that it has been decided, as in this case, that both the diameter (D) and the height (H) are to vary from 1 meter to 10 meters, and that the length of the two outside scales is to be 15 cm. For the moment, the constant term $\log (\pi/4)$ can be ignored, for it can be accounted for at a later time by shifting the V-scale for the volume upward until it is in the position for furnishing a correct reading for a particular value as computed using the formula.

The moduli can be determined as previously explained.

$$m_D = \frac{15}{2 \log 10 - 2 \log 1} = 7.5$$

and

$$L_D = 7.5(2 \log D) = 15 \log D$$

$$m_H = \frac{15}{\log 10 - \log 1} = 15$$

and

$$L_H = 15 \log H$$

$$m_V = \frac{m_D \times m_H}{m_D + m_H} = \frac{7.5 \times 15}{7.5 + 15} = \frac{112.5}{22.5} = 5$$

and

$$L_V = 5(\log V)$$

Scale spacing ratio is calculated as

$$\frac{m_D}{m_H} = \frac{7.5}{15} = \frac{1}{2}$$

For convenience, the scales can be spaced at 5 cm and 10 cm, giving a chart that is a square in form.

The scales for D and H may be graduated by using the graphical method illustrated in Fig. 19.30. The scale for V will be a three-cycle scale with a 15 cm range of length between the 1 and 1000 values. Since the value of V at the base line must result from the substitution in the equation of the values of 1 and 1 for the other two scales, the resulting value of .7854 cu m is the volume at the base line. The most convenient procedure to follow in graduating the V-scale is to start with the .7854 value at the base line. By so doing, the constant term, $\log (\pi/4)$, is taken into account. Should one desire to determine the distance to be laid off along the V-scale from the base line to the 1, it will be found to be equal to 5 times the difference between the logs of the numbers in this particular case, for the value of m_v is 5.

19.32 Three-scale Alignment Chart— Simplified (Graphical) Construction

After you have acquired a thorough understanding of the theory of alignment charts, which includes a full knowledge of the geometric basis underlying their construction, you will be in a position to simplify your construction work and make it purely graphical through the use of tie lines. Of course, this can only be done after the chart form has been identified, the equation converted, and the ranges for the variables (as needed) have been decided upon. The use of tie lines to locate check points on the third scale of a three-scale alignment chart is shown in Fig. 19.35.

GIVEN:

$$M = \frac{wl^2}{12}$$

the equation for the bending moment (maximum) of a beam, having a uniformly distributed load and fixed at both ends. In this equation, w is the load in pounds per foot of length and l is the span in feet.

It was decided that the alignment chart would fulfill most needs if w had limits from 10 to 200 lb per ft and l had limits from 5 to 25 ft.

EQUATION REWRITTEN:

$$\log w + 2 \log l = \log M + \log 12$$

The steps for the construction of the chart, with only minor calculations that may be performed mentally, are as follows:

1. Draw two vertical lines (say 5 in. long) at any convenient distance apart.

2. Graduate the w-scale (left-hand line) from the lower point 10 to the uppermost point 200. (In this case, use was made of the method shown in Fig. 19.30 to locate the scale values between 10 and 200.)

3. Graduate the l-scale, with the lower point as 5 and the point at the top of the scale as 25. (Scale values located as in step 2.)

4. Using a value of 10 for l and 500 for M yields $w = 60$ for the tie-line from the 60 value on the w-scale to the 10 value on the l-scale. Again, with $l = 20$ and $M = 500$, w is found to be 15 for the second tie-line through $M = 500$. The intersection of these two tie-lines is a check point on the third vertical scale, a point that establishes not only the position of the scale line but the 500 value as well.

5. Locate a second check point on the M-scale by drawing a third tie line. In this case it was found to be convenient to use $w = 200$ and $l = 12$ to yield $M = 2400$.

6. Graduate the M-scale as shown (read Sec. 19.29).

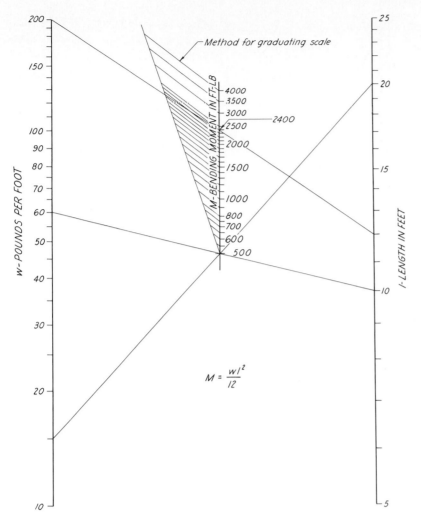

FIG. 19.35 **Alignment chart-simplified construction.**

19.33 *Graphical Construction of a Z-chart*

A simplified graphical construction may be employed to construct a Z-chart (N-chart) such as the one shown in Fig. 19.36, for the equation $V = 2.467\,Dd^2$, which gives the volume of a torus, a solid taking the form illustrated in Fig. 3.71. The given equation was considered to be so arranged that the vertical V- and D-scales would be uniformly graduated. The d-scale on the diagonal representing a variable to a power, is nonuniformly graduated. For this form of chart, one should recall that the vertical scales begin at zero and run in opposite directions (see Sec. 19.27). The diagonal connecting these two zero points completes the usual N-shape.

The steps in making the construction are as follows.

STEP 1. Draw the scale lines for the variables V and D at any convenient distance apart, in this case 9 cm. Then, lay off the D-scale to some selected length, say 15 cm, and locate the zero point of V scale opposite the 10 of the D-scale.

STEP 2. Graduate the V-scale uniformly, being certain that the range is sufficient to make possible a full reading when using the maximum values of the other scales.

STEP 3. Graduate the D-scale for the selected range of values, in this case from 0 to 10 cm.

STEP 4. Graduate the d-scale by selecting one convenient value of D and substituting it repeatedly in the original equation, each time with a different value of d which is to be marked on the diagonal scale. In so doing, values of V are found that establish the ends of tie-lines from the pole point (10) selected on the D scale. For example, when the values 10 and 1 are substituted for D and d, respectively, in the given equation, V is found to be 24.67 cu cm. With this known, a tie-line drawn from the pole to 24.67 on the V-scale, will locate $d = 1$ on the diagonal scale. The volumes for a range of d from 1 to 5 cm are shown in the table in Fig. 19.36. These values were used to draw the tie-lines shown.

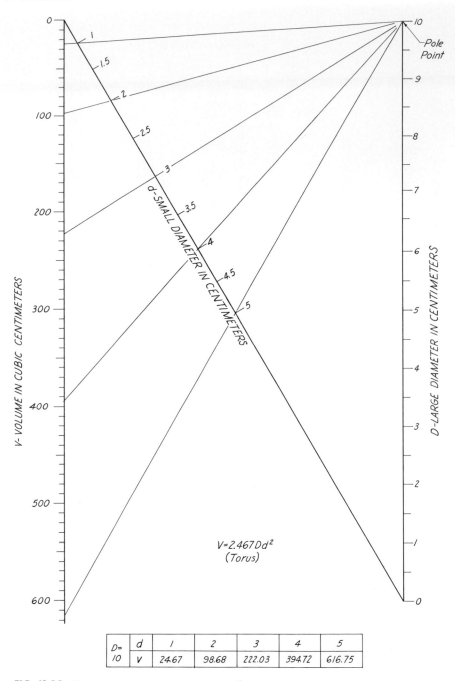

FIG. 19.36 **Chart for equation** ***V* = 2.467 *Dd²*.**

19.34 *Four-variable Relationship—Parallel-scale Alignment Chart*

A four-variable parallel-scale alignment chart may be constructed for an equation of the form $f(t) + f(u) + f(v) = f(w)$ when the equation has been rewritten as follows.

$$f(t) + f(u) = f(k)$$
$$f(k) + f(v) = f(w)$$

When rewritten in this form, the four-variable chart can then be constructed as two three-scale nomographs with the k-scale common to both. The k-scale, which serves as a pivot line in using the chart, need not be graduated, but it must be remembered that it does have a modulus and does represent a function in both equations.

In Fig. 19.37, a four-variable alignment chart is shown for the addition of numbers. For this particular construction, it was arbitrarily decided that t should have a range from 0 to 5, u from 0 to 10, and v from 0 to 20. The procedure for the construction of this chart is as follows.

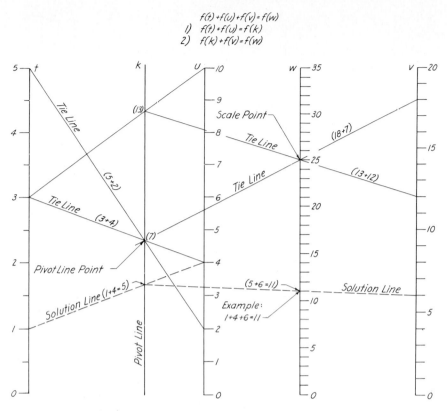

FIG. 19.37 Alignment chart for the equation f(t) + f(u) + f(v) = f(w).

STEP 1. Draw the vertical lines (stems) for the *t*- and *u*-scales at any convenient distance apart and, by means of the engineer's scale or graphically, graduate *t* from 0 to 5 and *u* from 0 to 10.

STEP 2. Draw the two intersecting tie-lines between the *t*- and *u*-scales that determine the position of the pivot line *k* at their point of intersection (7). Intersecting tie-lines could have been drawn from other graduation marks along the *t*- and *u*-scales as long as the sum of the values at the ends of one of the selected tie-lines equals the sum of the values at the ends of the other tie-line intersecting it.

STEP 3. Draw a third tie-line to locate another check point along the pivot line *k*. In this case, the tie-line was drawn from 3 on the *t*-scale to 10 on the *u*-scale to locate the mark for 13 on the *k*-scale.

STEP 4. Decide upon the best location for the *v*-scale, then draw the stem line and graduate it uniformly from 0 to 20.

STEP 5. Using the 7 and 13 value marks that have now been established on the *k*-scale, draw the two tie lines (18 + 7 = 25 and 13 + 12 = 25) that will determine the location of the *w*-scale at their point of intersection as well as the position of the 25 graduation mark.

STEP 6. Draw the *w*-scale and graduate it from 0 to 35, the 35 value representing the largest sum obtainable from the use of the *t*-, *u*-, and *v*-scales as they have been graduated for this particular chart.

The procedure for finding the sum of the numbers 1, 4, and 6 is illustrated by the broken lines extending between the scales. First, a straightedge can be laid across the

t- and *u*-scales through the 1 and 4 values, as shown, to determine the position on the pivot line (*k*) that the graduation mark would have that represents their sum, in this case 5. Then the straightedge should be shifted so that its edge will pass through the 5 position on the pivot line and the graduation mark for the 6 value on the *v*-scale. The sum of the three given numbers can then be read as 11 on the *w*-scale.

The four-scale alignment chart, shown in Fig. 19.38, for the computation of simple interest, may be easily constructed once the equation $I = PRT$ has first been written in the logarithmic form of

$$\log P + \log R + \log T = \log I,$$

and then rewritten as

$$\log P + \log R = \log k \qquad (1)$$

and

$$\log k + \log T = \log I. \qquad (2)$$

The equation as first rewritten is in the form $f(t) + f(u) + f(v) = f(w)$ and as arranged in (1) and (2) in the form of

$$f(t) + f(u) = f(k)$$
$$f(k) + f(v) = f(w)$$

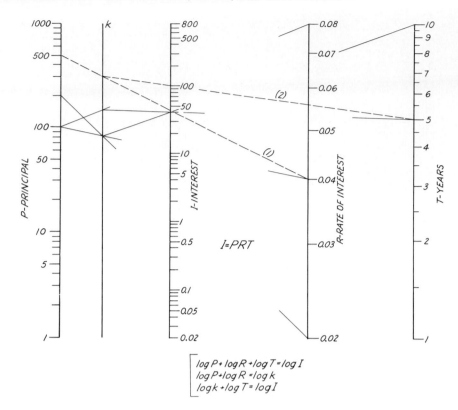

FIG. 19.38 Alignment chart for the equation *I* = *PRT*.

Now, since equations (1) and (2) have the same form as those for the previous problem, the alignment chart for the equation *I = PRT* is constructed similarly except that logarithmic scales must be used in this case in place of natural (uniformly divided) scales. As before, tie-lines were used to position the *k*- and *I*-scales and to locate needed check points. The procedure for reading the chart is illustrated by the broken lines, numbered (1) and (2).

The reader is urged to follow through the construction of the alignment chart for the equation *I = PRT* mentally, step by step. At the start, however, you must recognize that the *k*-scale belongs first to the *PRk* chart and then to the *kTI* three-scale chart that together make the complete chart for the four variables *P, R, T,* and *I*.

 ## PROBLEMS

Technical Graphs and Charts

The following problems have been designed to emphasize the fundamental principles underlying the preparation and use of technical graphs and charts. If desired, the tables given in the Appendix may be used to prepare a graph or chart in metric.

1. Determine the values for the following equations, as assigned, and plot the curve in each case for quantitative purposes.

Parabola	$Y = 4x^2$, x from 0 to 5
Ellipse	$Y^2 = 100 - 2x^2$
Sines	$Y = \sin x$, x from 0° to 360°
Cosines	$Y = \cos x$, x from 0° to 360°
Logarithms	$Y = \log x$, x from 1 to 10
Reciprocals	$Y = 1/x$, x from 1 to 10

2. The freezing temperatures for two common antifreeze solutions for various compositions are given below.

Prestone			Denatured Alcohol		
% by Vol.	Temp. (°F)	Temp. (°C)	% by Vol.	Temp. (°F)	Temp. (°C)
10	25	−4	10	25	−4
20	16.5	−8.5	20	17.5	−8
25	11	−11.5	25	14	−10
30	5	−15	30	5	−15
35	−3	−19.5	35	−1	−18
40	−12	−24.5	40	−11	−24
45	−25	−31.5	45	−18	−28
50	−38	−39	50	−25	−31.5
55	−47	−44	55	−32	−35.5
			65	−45	−43

Prepare a chart for these data, mainly quantitative in character, from which the required per cent volume can be read for any desired freezing temperature. Use the type of paper shown in Fig. 19.3(*b*).

3. Approximate barometric pressures at different heights above sea level are given below. Prepare a qualitative chart for the given data on rectangular coordinate paper.

Note that the curve would be straight on semilogarithmic paper.

H, Altitude (miles)	B, Barometric Pressure (inches Hg)
0	29.92
1	24.5
2	20.0
3	16.2
4	13.45
5	11.0
6	8.9
7	7.28
8	5.95
9	4.87
10	4.0

4. Data on rate of growth are frequently plotted on semilogarithmic paper, because the slope of the curve then represents the rate of growth. On semilogarithmic paper, plot the data for the enrollment in a university.

Place the vertical axis 1 unit from the left edge and the horizontal axis at the extreme bottom of the page ruling. Letter in an appropriate title.

Also draw the curve of the general trend and enter on the chart, just below the title, what the enrollment would be in 1975 if the same rate of growth is maintained.

Year	Enrollment
1956	5,745
1957	5,831
1958	5,906
1959	6,238
1960	6,984
1961	7,313
1962	6,011
1963	5,112
1964	4,838
1965	4,621
1966	10,221
1967	11,320
1968	10,210
1969	10,101
1970	9,871
1971	9,211
1972	9,071
1973	9,647
1974	10,101
1975	10,456
1976	11,132
1977	11,656
1978	12,231
1979	12,910
1980	13,621
1981	14,222
1982	15,011
1983	15,976
1984	16,211
1985	18,013

5. In a hydraulics laboratory, the construction of a quantitative curve that would give the weight of water contained in tubes of various diameters and lengths was desired. This was accomplished by filling tubes of known diameters with water to a depth of 1 ft and observing the weight of water thus added. The water was kept at a temperature for maximum density and the following data were obtained:

D = Diameter of Tube (inches)	W = Weight of 1-foot Column of Water
2	1.362
2½	2.128
3	3.064
3½	4.171
4	5.448
4½	6.895
5	8.512
5½	10.299
6	12.257
6½	14.385
7	16.683
7½	19.152
8	21.790

On a sheet of graph paper [Fig. 19.3(b)], plot the above data. Place the axes 3 cm in from the edges. Letter the title in any convenient open space.

6. Owing to uncontrollable factors, such as lack of absolute uniformity of material or test procedure, repeated tests of samples of material do not give identical results. Also, it has been observed in many practical situations that

a. Large departures from the average seldom occur.
b. Small variations from average occur quite often.
c. The variations are equally likely to be above average and below average.

The foregoing statements are borne out by the accompanying data showing the results of 4000 measurements of tensile strength of malleable iron.

On a sheet of coordinate graph paper [Fig. 19.3(a) or (b)] prepare a graph showing frequency of occurrence of various strength values as ordinates and tensile strength as abscissa. Draw a smooth symmetrical curve approximating the given data.

Range of Tensile Strength Values (pounds per square inch)	Number of Observations
Under 45,000	0
45,000–45,999	1
46,000–46,999	2
47,000–47,999	3
48,000–48,999	6
49,000–49,999	20
50,000–50,999	232
51,000–51,999	376
52,000–52,999	590
53,000–53,999	740
54,000–54,999	771
55,000–55,999	604
56,000–56,999	383
57,000–57,999	184
58,000–58,999	60
59,000–59,999	20
Over 60,000	8
	4000

7. On a sheet of paper of the type shown in Fig. 19.3(b), using India ink, plot a curve to represent the data given above. (*Note:* For stress-strain diagrams, although the load is the independent variable, it is plotted as ordinate, contrary to the general rule as given in Sec. 19.5. Figure 19.6 shows a similar chart. In performing tests of this nature, some load is imposed before any readings of elongation are taken.

It is suggested that the label along the abscissa be marked "Strain, 0.00001 in. per in." then fewer figures will be required along the axis.

Stress (pounds per square inch)	Strain (inches per inch)
3,000	.0001
5,000	.0002
10,000	.00035
15,000	.00054
20,000	.00070
25,000	.00090
30,000	.00106
32,000	.00112
33,000	.00130
34,000	.00140

8. Make a vertical multiple bar chart showing the enrollment at _____ University from 1961 to 1970. Obtain data from Problem 4.

9. Make a semilogarithmic graph showing the enrollment of your school for the last 20 years.

Alignments Charts

The following problems have been designed to emphasize the fundamental principles underlying the preparation and use of alignment charts.

10–21. Prepare an alignment chart for the given equation. Chart scales should have a sufficient number of division marks to enable the user to obtain some reasonably accurate results (readings). In each problem the range for two of the variables has been given. The range of the third variable must make possible the use of the full range of each of the other two variables.

10. Construct an alignment chart for the multiplication of numbers from 1 to 100 (read Sec. 19.27 and study Fig. 19.27).

11. Construct an alignment chart of the form $t + u = v$. Let t vary from 0 to 10 and u from 0 to 15 (read Sec. 19.30 and study Fig. 19.32).

12. Construct an alignment chart for determining the area of a triangle. The student is to determine for himself the range for each of the scales.

13. Construct an alignment chart for determining the volume of a cylinder. The diameter is to range from 1 to 5 m and the height from 1 to 10 m. The volume is to be in cubic meters (read Sec. 19.31).

14. Make an alignment chart for determining the volume of a paraboloid, $V = 1/8\pi ab^2$, where a is the length (measured along the axis) and b is the diameter of the base. Let a vary from 1 to 20 in. and b from 1 to 10 in.

15. Construct an alignment chart for the maximum bending moment at the point of support of a cantilever beam, formula $M = Wl$. W is the concentrated load at the end of the beam and l is the distance from the point of support to the load (read Sec. 19.31). Let W vary from 100 to 1000 lb and l from 10 to 20 ft. M will be in ft-lb.

16. Make an alignment chart for the discharge of trapezoid weirs, formula $Q = 3.367 Lh^{3/2}$. Q is discharge in cubic feet per second; L is length of crest in feet (width of weir); and h is the head (depth of water). Let L vary from 1 to 10 ft and h from .5 to 2 ft (read Secs. 19.31 and 19.32).

17. Make an alignment chart for the formula $P = I^2R$ as explained in Sec. 19.31. Let I vary from 1 to 20 amp and R from 1 to 10 ohms.

18. Make an alignment chart for the formula $R = E/I$, where E is the electromotive force in volts; I is current in amperes; and R is resistance in ohms. Let R vary from 1 to 20 ohms and I from 1 to 100 amp (read Secs. 19.31 and 19.32).

19. Make an alignment chart for the formula $I = bd^3/36$, where I is the moment of inertia of a triangular section; b is the length of the base in inches; and d is the depth of the section (altitude of triangle). Let b vary from 1 to 10 in. and d from 1 to 10 in. (read Secs. 19.31 and 19.32).

20. Make an alignment chart for the formula $M = wl^2/8$, where M is the bending moment in foot-pounds, w is the load in pounds per foot, and l the length of span in feet. Let w vary from 10 to 200 lb per ft and l from 5 to 25 ft (see Fig. 19.35).

21. Make an alignment chart (Z-chart) for the equation for the volume of a right circular cylinder, $V = \pi r^2 h/144$, where V is the volume, r is the radius of the base in millimeters, and h is the height in millimeters. Let r vary from 0 to 50 mm and h from 0 to 500 mm.

$$y=Ae^{mx}$$

$$y=ab^x$$

$$y=ax^b$$

$$y=a+bx$$

$$y=3x^{.57}$$

This manned reusable space vehicle will be used for a wide variety of missions. The Shuttle will consist of a reusable orbiter, mounted "piggyback" at launch on a large expendable propellant tank and two recoverable and reusable solid-propellant rockets. The various stages of launch have been illustrated. Graphics played a very large role in the development of the Shuttle. (*Courtesy National Aeronautics and Space Administration*)

CHAPTER · 20

Graphical Mathematics

A / GRAPHICAL ALGEBRA

20.1 Introduction

Persons who have a thorough understanding of the many graphical methods available frequently find it more desirable to use a graphical procedure for making a calculation than to resort to an algebraic approach, particularly when repeated calculations of the same nature are required. These people, no matter in what phase of engineering endeavor they may be found (research, design, development or production), have discovered that a graphical method can be used to present a needed visual relationship of the variables, save valuable time, minimize the possibility of error, and provide a sure way to explain the given calculation.

The student must realize that graphical methods, while theoretically precise, give only approximate solutions. The accuracy is limited by the size of the paper, the tools available, and the skill of the "graphician." However, in many problems the initial data is also approximate; it is dependent upon the visual reading of instruments provided with graphical scales.

As stated before, when the solution has been accomplished by graphical methods, that solution may often be used for subsequent problems at a great saving in time. Furthermore, in most cases the solution by graphical methods can be arrived at in a shorter time than through the use of pure mathematical methods.

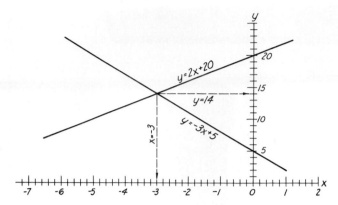

FIG. 20.1 Solution of linear simultaneous equations.

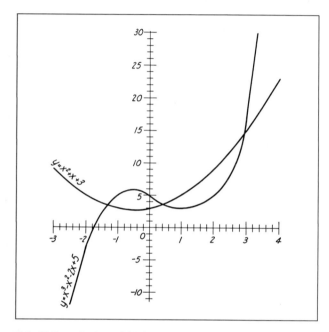

FIG. 20.2 Solution of higher-order equations.

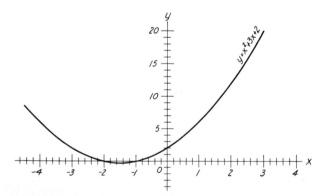

FIG. 20.3 Graphical determination of roots of the equation $y = x^2 + 3x + 2$.

It is desirable for an engineer to be well versed in both the graphical and mathematical sciences and be aware of the limitations of both so that sound judgment can be exercised concerning which method to use for any given situation.

20.2 Simultaneous Equations

Simultaneous equations are frequently encountered in engineering work. There are times when the algebraic solution of a pair of equations is complicated and difficult, if not almost impossible, making it necessary for an engineer to know how to find the solution by graphical means. Situations that make a graphical approach desirable arise when the data have been obtained as the result of an experiment and when the equations, although known, are so complicated that the algebraic solution would be too time consuming. In some cases, the graphical solution may, by giving the roots to the desired accuracy or by showing where the roots are located, permit a saving of time in the mathematical solution. Figure 20.1 shows the solution of two linear simultaneous equations. Figure 20.2 shows the solution of two higher-order equations. The equations are

$$y = x^2 + x + 3$$
$$y = x^3 - x^2 - 2x + 5$$

The mathematical solution of this pair of equations would be very time consuming. The graphical solution is largely a matter of plotting. It should be noted that there are three roots that satisfy both equations.

20.3 Determination of the Roots of Quadratic Equations

There are three methods of solution that can be used for the determination of the roots (points where $y = 0$) of quadratic equations, each method having its own advantages. First, the equation can be plotted directly as illustrated in Fig. 20.3. The roots are the points where the curve crosses the X-axis. If the curve crosses the X-axis, there are two real and different roots; if the curve is tangent to the X-axis, there are two real but equal roots; and, finally, if the curve does not cross the X-axis, there are two imaginary roots. Imaginary roots are always in pairs, equal but opposite in sign. The advantage of this method is that an immediate picture of the conditions of the equation is obtained. This method is applicable for any order equation. However, a new curve must be drawn for each equation.

Second, the equation can be rewritten as two simultaneous equations, one being the equation of a straight line. For example, the equation

$$y = x^2 + Ax - C$$

can be rewritten as

$$y = x^2$$
$$y = -Ax + C$$

This pair of equations can be solved as explained in Sec. 20.2 and the x-values of their solution will be the roots of the original quadratic equation. Figure 20.4 shows this method applied to the same quadratic equation as represented in Fig. 20.3. If the straight line cuts the curve, there are two real and different roots; if the straight line is tangent to the curve, there are two real but equal roots; and finally, if the straight line does not cut the curve, there are two imaginary roots. The advantage of this method is that the curve representing $y = x^2$ need be drawn only once. Each subsequent quadratic is solved by drawing a different straight line.

For the third method, based on geometric construction, the quadratic must be written in the form

$$y = x^2 + Ax + C$$

The steps to be followed in making the construction shown in Fig. 20.5 are (1) Draw a set of X- and Y-axes and lay off A along the X-axis, with opposite sign, to locate point a. (2) At a erect a perpendicular and lay off C with the same sign to locate c. Finally, (3) Connect c to a point at plus one unit on the Y-axis with a straight line and draw the circle that has this line as a diameter. The points where this circle cuts the X-axis are the roots that are sought. Again there is the possibility of real and different; real but equal; and imaginary roots. The existing conditon would be revealed by the position of the circle. The advantage of this method is that all construction is done with a straightedge, scale, and compass. No irregular curves need be drawn with the likelihood of errors in construction. However, a completely new construction is required for each equation.

B / GRAPHICAL CALCULUS

20.4 Graphical Calculus

In solving engineering problems, it is frequently desirable and often necessary to present a graphical analysis of empirical data. Even though it is often possible to make an evaluation through the use of analytical calculus, a graphical representation is more meaningful because it is pictorial in character. Graphical integration and differentiation are particularly desirable for problems for which only a set of values are known, or for curves that have been produced mechanically, as in the case of indicator diagrams, or if the results cannot be determined by the analytical methods of calculus.

The following sections are devoted to the graphical rules and methods for determining derived curves. Discussion of the interpretation of results has been intentionally omitted since interpretation is not usually graphical and, therefore, not within the scope of this text.

20.5 Graphical Integration

In deriving curves of a higher order, the principle is applied that the area bounded by two successive ordinates, the curve, and the axis is equal to the difference in magnitude of the corresponding ordinates of the integral curve. Figure 20.6 illustrates this principle of graphical integration. In

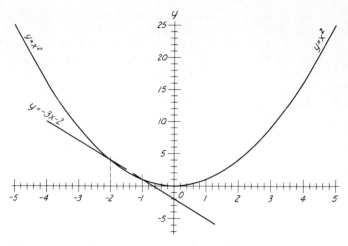

FIG. 20.4 Graphical determination of roots of the equation $y = x^2 + 3x + 2$.

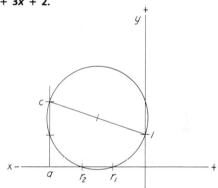

FIG. 20.5 Graphical determination of roots of the equation $y = x^2 + 3x + 2$.

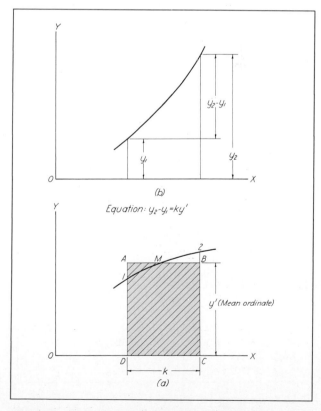

FIG. 20.6 Illustration of the principle of integration.

(*a*) an increment of a curve is shown enlarged. The area under the curve will be approximately equal to the area of the shaded rectangle *ABCD* when the line *AB* is drawn so that the area *AM*1 above the curve is approximately equal to the area *MB*2 below the curve. With a little practice one will find it easy to establish a line such as *AB* quite accurately by eye if a strip of acetate or a triangle is used through which the curve can be seen.

By applying the principle of graphical integration to a series of increments, an integral curve may be drawn as shown in Fig. 20.7. At this point, it should be recognized that since the difference between successive ordinates represents increase in area, the difference between the final ordinate and the initial ordinate represents the total area between these ordinates that is bounded by the curve and the *X*-axis.

The scale selected for the *Y*-axis of the integral curve need not be the same as the scale for the given curve.

Portions of a lower-order curve that are above the *X*-axis are considered to be positive whereas areas below with negative ordinates are recognized as negative [see Fig. 20.8(*a*)]. Since the negative area between any two ordinates on the lower-order curve represents only the

difference in the length of the corresponding ordinates on the integral curve, the length of y_7 is less than the length of y_6 by an amount equal to the negative area. Also, because areas represent only differences in length of successive ordinates of the integral curve, the initial point on the integral curve might have any value and still fulfill its purpose. For example, either integral curve shown in Fig. 20.8(*b*) is a satisfactory solution for the curve in (*a*).

Figure 20.9 shows the derived curves for a falling drop hammer. It is common practice, when drawing related curves, to place them in descending order as shown; that is, the lower-order curve is placed below.

In (*a*) the straight line represents a uniform acceleration of 9.81 m per sec per sec, which is the acceleration for a freely falling body. The initial velocity is 0. The units along the *X*-axis represent time in seconds, and the units along the *Y*-axis represent acceleration in meters per second per second.

Since the acceleration is uniform, the velocity will be a straight line of constant slope, (*b*). The length of the last ordinate is equal to the total area under the acceleration curve, 7.85 m per sec.

The distance curve, which is obtained by integrating

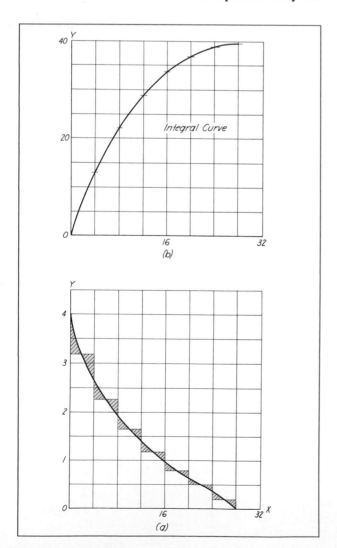

FIG. 20.7   Integration of a curve.

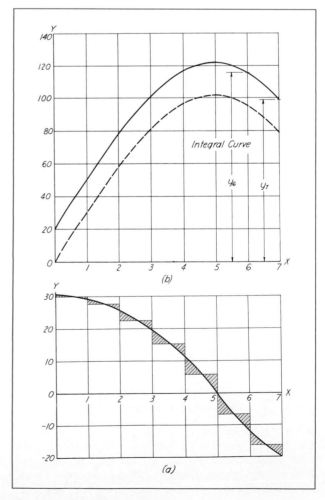

FIG. 20.8   Integration of a curve.

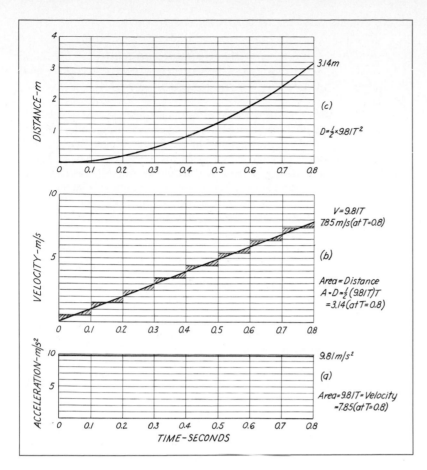

FIG. 20.9 **Derived curves for a falling drop hammer.**

the velocity time curve, is shown in (c). The length of the ordinate at any interval point is equal to the total area below the velocity curve between the origin and the point $[D = A = \frac{1}{2}(9.81T)T]$. See Figs. 20.6, 20.7, and 20.8.

20.6 To Integrate a Curve by the Ray Polygon Method

An integral curve may be drawn by a purely graphical process known as the *ray polgon method*. This method of integrating the area under a curve is illustrated in Fig. 20.10. Divide the X-axis into intervals and draw ordinates at the division points. Then, select the pole point P at

some convenient location that will make the distance d equal to any number of the full units assigned to the X-axis. The selection of the number of units for the distance d determines the length of the scale along a Y-axis for the integral curve. To establish relationship between y_1 and y_0 the following equation based on similar right triangles can be written

$$y_1 : k = y_0 : d$$
$$y_1 \cdot d = k \cdot y_0$$
$$y_1 = \frac{k \cdot y_0}{d} = \frac{k}{d} \cdot y_0$$

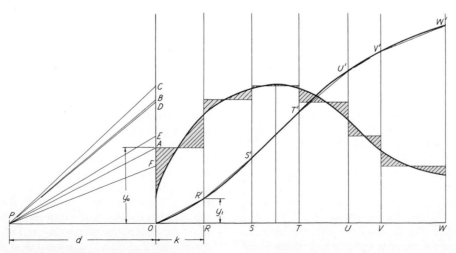

FIG. 20.10 **Use of the ray polygon.**

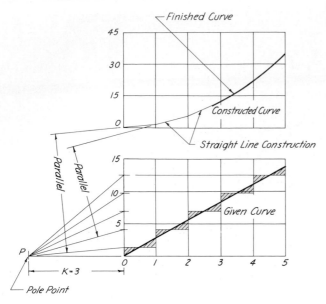

FIG. 20.11   Pole method.

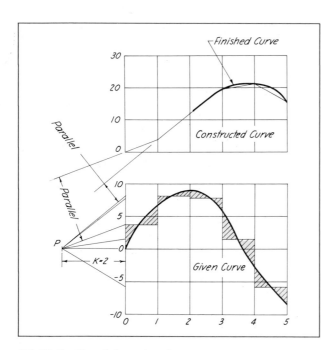

FIG. 20.12   Use of the pole point.

20.7 Pole Method Applied to the Construction of an Integral Curve

The first of the several examples that have been selected to illustrate the use of the *pole method* for the construction of an integral graph, has a sloping straight line, plotted on rectangular coordinates, as the given curve of lower order (see Fig. 20.11). It should be noted in this case that the integral graph has been placed directly above the lower graph and in projection with it so that the same intervals may be used for both. As shown, ordinates were first drawn to divide the chart area into vertical strips. With this done the horizontal lines were drawn next, one for each of the strips at mean-ordinate height, to form the five rectangular areas that are assumed to approximate the area under the curve (sloping straight line). Then, these mean-ordinate heights must be projected horizontally to the Y-axis. With the pole distance k determined and the pole point P located along the X-axis extended, ray lines are drawn from P to the points on the Y-axis representing the heights of the mean ordinates. The manner in which the distance k affects the steepness of the integral curve, as illustrated in Fig. 20.13, will be discussed later. The first point on the integral curve was found by drawing a line through the origin parallel to the ray for the corresponding interval below, in this case the lowest ray. The second segment is then drawn parallel to the ray for the second interval through the end of the first segment. Other intervals are treated similarly and then the smooth curve, that is, the integral curve, is drawn through the points obtained. Since the finished curve has in reality been drawn through the end points of chords by this method, most persons identify this construction procedure by calling it the *chordal method*.

In Fig. 20.12, the chordal method using a pole has been applied to the integration of an irregular curve. The intervals between the ordinates need not be equal as shown. In fact, the number and spacing of the ordinates will usually be determined by the shape of the curve and one soon discovers that it is wise to space ordinates closer together where there is a sharp change or a reversal of curvature and farther apart where the curve is flat. For this illustration, an example was purposely selected that would have a portion of the curve below the X-axis. This area below represents negative area that must be subtracted from the positive area above the X-axis. When a pole is used, this subtraction is taken care of graphically by the negative slope of the rays for the intervals at the location where the curve lies below the X-axis.

Ordinarily, the modulus of the ordinate (Y-axis) scale of the integral curve must be smaller than the modulus of the given (lower-order) curve if the graph for the integral curve is to be of reasonable size. When the height of the derived curve is somewhat near the same as the height of the given curve, the general overall appearance of the graphs will be satisfactory and the scales will ordinarily be large enough to be read with reasonable accuracy. At the very beginning, when it becomes necessary to determine the maximum value to be read on the ordinate scale for the integral curve, one must first remember that the value will be equal numerically to the value for the total

Determine the mean ordinate for each strip and transfer its height to the Y-axis as length OA, OB, OC, and so forth. Draw rays from P to points A, B, C, D, E, and F.

To construct the integral curve, start at O and draw a line parallel to PA cutting the first vertical through R at R'. Through R' draw a line parallel to PB until it cuts the second vertical through S at S'. Repeat this procedure to obtain points T', U', and so on. Points R', S', T', U', V', and W' are points on the required integral curve. In Fig. 20.10 the integral curve is constructed on the same coordinate axes as the lower-order curve.

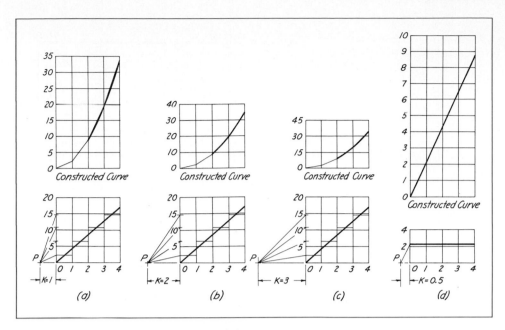

FIG. 20.13 **Pole distance and steepness of the curve.**

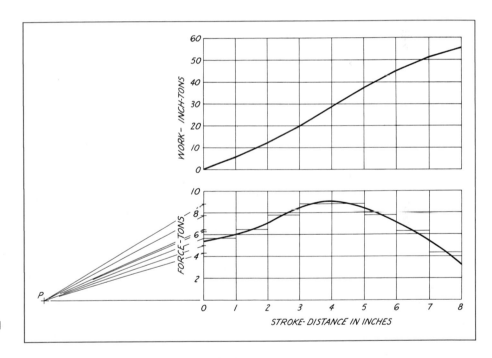

FIG. 20.14 **Example of graphical integration.**

area under the given curve. This value may be found either by adding the areas of the individual rectangles or by calculating the area of a single large rectangle that is estimated to be equal to the total area under the curve. Once this maximum value is known it is relatively easy to select a value for k that will give good proportion to the integral curve and at the same time permit one to make accurate readings.

When the pole distance used is equal to the length of one abscissa unit as shown in Fig. 20.13(*a*), the same distance represents the same number of units on both scales. In (*b*), where a pole distance of 2 has been used,

ten units are represented on the scale of the constructed curve by the same distance that represents 5 units on the scale of the curve below. As can be easily seen, a three-unit pole distance reduces the height of the constructed curve still more (*c*). When the pole distance is one-half unit, the constructed graph becomes very high, too high to be appropriate in the case shown in (*d*). However, there may be conditions encountered now and then when the use of a pole distance of one-half unit is desirable. An area graph is shown in Fig. 20.14, where a five-unit pole distance was used to construct the work (integral) curve.

An interesting application of graphical integration is

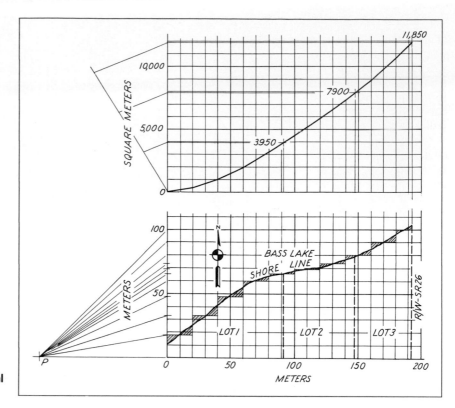

FIG. 20.15 Example of graphical integration.

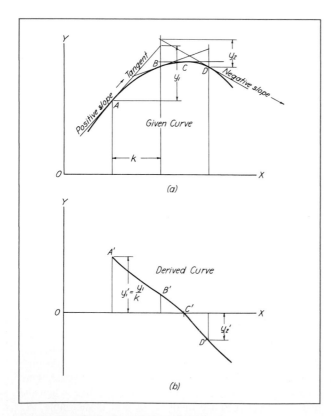

FIG. 20.16 Illustration of the principle of differentiation.

shown in Fig. 20.15. Suppose that an irregularly shaped plot of land, lying along a lake shore and a road, is to be divided into three equal lots. The area to be divided was first plotted (meters against meters) on the lower graph. Its boundary lines in addition to the shore line at the north, are the X and Y axes, representing property lines, and the right-of-way line of S.R. 26. It was decided that the division lines were to run due north and south, perpendicular to the south property line. With the lower curve drawn, the intergral curve was constructed by the pole-and-ray method as shown and the total area was found to be 11,850 square meters, an area that, in this particular case, can be divided evenly into three lots, each having 3,950 square meters. Horizontal lines, drawn through points representing 3,950 and 7,900 values on the scale of the area graph to the area curve and thence downward to the shore line, located the end points of the division lines for the lots. A graphical method was used to divide the area as shown to the left of the area scale.

20.8 Graphical Differentiation

Curves of a lower order are derived through the application of the principle that *the ordinate at any point on the derived curve is equal to the slope of a tangent line at the corresponding point on the given curve.* The slope of a curve at a point is the tangent of the angle with the X-axis formed by the tangent to the curve at the point. For all practical purposes, when constructing a derivative curve, the slope may be taken as the rise of the tangent line parallel to the Y-axis in one unit of distance along the X-axis, or the slope of the tangent equals y_1/k as shown in Fig. 20.16.

Figure 20.16 illustrates the application of this principle of graphical differentiation. The length of the ordinate y_1' at point A' on the derived curve is equal to the slope y_1/k at point A on the given curve as shown in (a). When the slope is zero as at point C, the length of the ordinate is zero and point C' lies on the X-axis for the derived curve. When the slope is negative, as shown at D, the ordinate is negative and lies below the X-axis.

The graph shown in Fig. 20.17 is composed of segments of straight lines. Since the slope is constant for the interval 0–1, the derivative curve in the interval is a horizontal line. In the interval 1–2, the slope is also constant but of a lesser magnitude. Thus, the derivative curve is composed of straight line segments as shown in (b).

At this point in the discussion of graphical calculus it becomes possible to determine the relationship between the principles of integration and differentiation and to show that one is derived from the other.

From inspection of the graphs shown in Fig. 20.17, equations may be formulated as follows:

By the principle of differentiation,

$$y_2' = \frac{y_2 - y_1}{k_2}$$

in the interval 1–2, where

$$\frac{y_2 - y_1}{k_2}$$

represents the slope of AB. The area under the curve in (b) in the interval 1–2 is equal to

$$y_2' \cdot k_2 = \frac{y_2 - y_1}{k_2} \cdot k_2$$

$$= y_2 - y_1 \qquad \text{(integral curve)}$$

and

$$y_2' = \frac{y_2 - y_1}{k_2} \qquad \text{(differential curve)}$$

In constructing a derivative curve, the determination of the tangent lines is often difficult because the direction of a tangent at a particular point is usually not well defined by the curvature of the graph. Two related schemes that may be used for constructing tangents are shown in Fig. 20.18(a) and (b). In (a) the tangent is drawn parallel to a chord of the curve, the arc of which is assumed to approximate the arc of a parabola. A sufficiently accurate location for the point of tangency T_1 may be determined by drawing a line from the midpoint of the chord to the arc parallel to an assumed direction for the axis of the parabola. When working with small segments of the curve, one may assume the axis to be either horizontal or vertical.

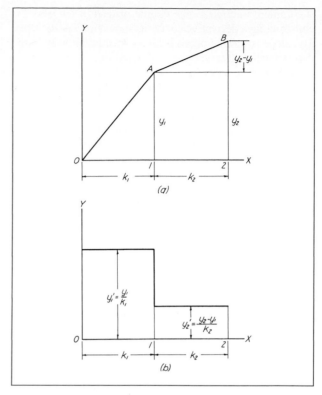

FIG. 20.17 Graphical differentiation.

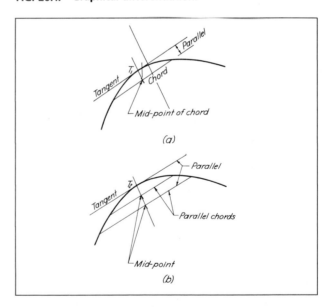

FIG. 20.18 Construction of a tangent line.

A more accurate construction is shown in (b), where the tangent is drawn parallel to two parallel chords. The point of tangency T_2 is determined by connecting the midpoints of the chords and extending this line to the curve. This line determines the direction of the axis.

The construction in (b) using two chords to establish a tangent is applicable to any curve that may be approximated by a portion of a circle, ellipse, parabola, or hyperbola.

Since a tangent is assumed to be parallel to a chord, it

is common practice to use chords instead of tangents for constructing a derivative curve, as shown in Fig. 20.19(a). The slope is plotted on an ordinate located midway in the corresponding interval of the derived curve.

A derivative curve can also be drawn using the ray polygon method as explained in Sec. 20.6 in reverse (see Fig. 20.20). The lines *PA*, *PB*, and *PC* of the ray polygon are drawn parallel to the tangents at points T_1, T_2, and T_3. Point *Q* is found by drawing a line horizontally from point *A* to the ordinate through the point of contact of the tangent parallel to *PA*. Points *R* and *S* are found similarly.

20.9 Drawing a Derivative Curve by the Pole Method

Before attempting to construct a derivative curve by the pole method, the student must have a clear understanding of the "slope law" as it applies to the differentiation, and must be able to think of the derived curve being sought

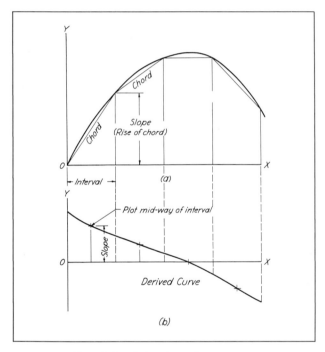

FIG. 20.19   Use of chords in place of tangents.

as the curve of slopes (slope locus) for the given curve. In making a start, the given curve should first be analyzed and certain facts noted concerning the slope at specific points along the curve. For example, in Fig. 20.21, it should be recognized that the slope for the first and second intervals is constant and positive, a fact indicated by the single straight line sloping upward across both strips. In the third interval (2 to 3), the curve has a constant zero slope; while again, for the fourth and fifth intervals, the slope is constant and positive. Finally, the curve for the last three intervals has a constant negative slope, because the straight line, this time downward, indicates decreasing values.

A pole distance must now be selected, remembering that the distance in abscissa units determines the ordinate scale ratio for both curves. Then, with the location of the pole point *P* established, rays are drawn, parallel to the straight lines of the given curve until they intersect the *Y*-axis for the derived curve. These points of intersection, projected horizontally, determine the derived curve that is composed of straight lines in this particular case.

In Fig. 20.22, the derived curve (slope locus) was obtained from the given (upper) curve by using tangents. Through the pole point *P*, a ray was drawn parallel to each of the several tangent lines. Then, the intersection point of each ray with the *Y*-axis was projected horizontally across to the corresponding ordinate as shown. The drawing of a smooth curve through these points completed the graph.

Chords were used instead of tangent lines in Fig. 20.23. The procedure that is followed here may be thought of as being the reverse of the chordal method illustrated in Fig. 20.11. As the initial step for the construction, chords of the given curve were drawn as shown. In selecting these chords, an effort was made in every case to see that the curve would be nearly symmetrical about a perpendicular bisector of the subtended chord. Also, as is appropriate and desirable, shorter chords were used where the curvature is sharper. After the pole point *P* had been located and the axes established for the derivative curve, rays were drawn through *P*, parallel to the chords, to an intersection with the *Y*-axis. Next, the tangent point (such as T_1) was located for each of the subtended arcs of the given curve, using the method shown below in (*b*). In doing this it was

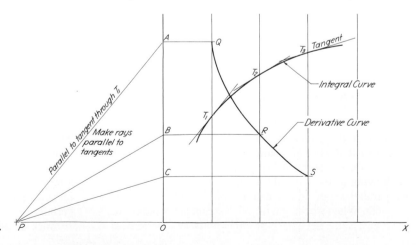

FIG. 20.20   Ray polygon method.

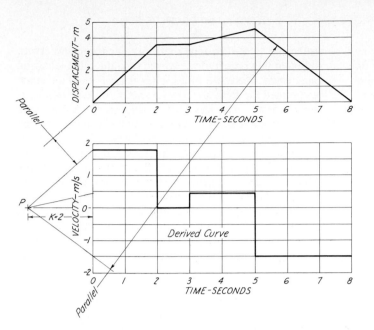

FIG. 20.21   Derivative curve constructed by the pole method.

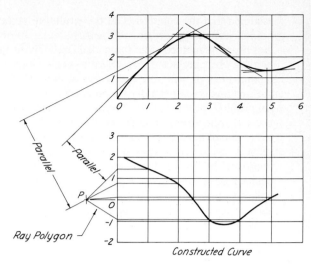

FIG. 20.22   Use of tangents.

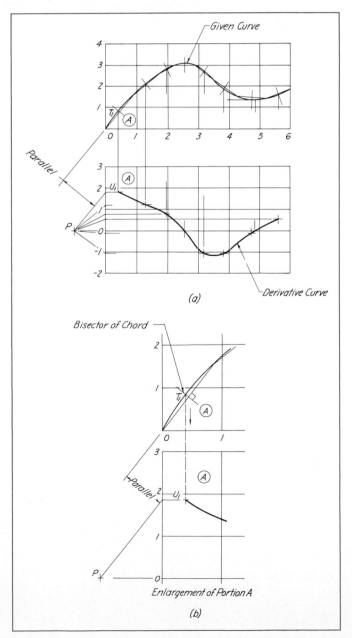

(a)

Given Curve

Derivative Curve

Parallel

Bisector of Chord

Parallel

Enlargement of Portion A

(b)

FIG. 20.23   Chordal method.

assumed that the chords would be parallel to tangents at points, such as T_1, where the perpendicular bisector of the chord would cut the subtended arc. These tangent points, when projected downward to corresponding horizontal lines drawn from the points along the Y-axis, at the rays, determine the derivative curve. For example, in (*b*) point U_1 in interval A is found by projecting T_1 downward to a horizontal line drawn through the point of intersection of the Y-axis and the related ray through P. The related ray in this case is the one that is parallel to the chord of the arc of the given curve in interval A. The desired derivative curve was completed by drawing a smooth curve through U_1 and the several other points shown, all of which were located in the same manner.

When using the chordal method to determine a derivative curve, some people prefer to use equally spaced ordinates and then confine each chord to a single interval. Although this procedure proves to be quite satisfactory in many cases, there are times when desired results cannot be obtained for there are either no points or too few points given to draw a smooth curve at critical locations. In such cases, some extra construction will be needed. However, the easiest and best solution would have resulted from a careful choice of chords of the given curve at the start.

Figure 20.24 shows the differentiation of a curve using chords instead of tangents. The given distance-time curve was plotted from data obtained for a passenger train leaving a small station near a large city. The velocity and acceleration curves reveal that the train moves with a constant acceleration for approximately 100 sec until it reaches a velocity of 55 miles per hr. From this point it travels with a constant velocity toward its destination.

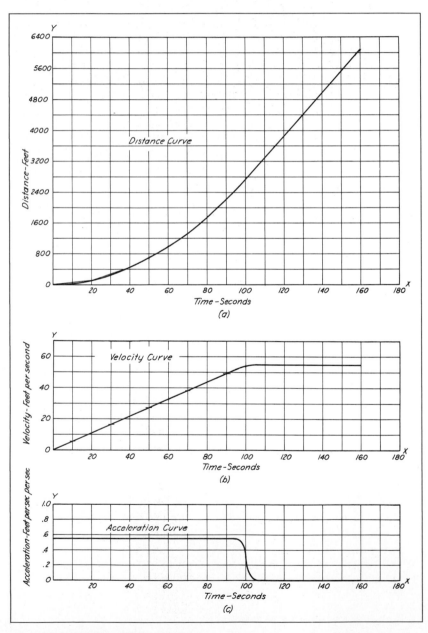

FIG. 20.24 Graphical differentiation.

PROBLEMS

Graphical Algebra

1. Find the roots of the simultaneous equations

$$y = 2x^2 - x + 4$$

and

$$y = x^2 + 3x + 9$$

2. Find the roots of the following equation by the method shown in Fig. 20.4.

$$y = x^2 + 2x - 3$$

3. Find the roots of the following equation by the method shown in Fig. 20.5.

$$y = 3x^2 + 6x + 3$$

4. Find the roots of the following equation by the method shown in Fig. 20.4.

$$y = 3x^2 + 12x + 12$$

Graphical Calculus

The following problems have been designed to emphasize the fundamental principles underlying graphical integration and differentiation and to offer the student an opportunity to acquire a working knowledge of the methods that are commonly used by professional engineers.

5. A beam 3 m long is uniformly loaded at 30 kg per meter as shown in Fig. 20.25. Plot distance along the X-axis. Draw (1) the integral curve to show the shearing force, and (2) the second integral curve to show the bending moment. Note: Under the SI system a force is given in Newtons. Read Sec. 9.23.

6. A beam 15 ft long and supported at both ends is loaded uniformly at 18 lb per foot as shown in Fig. 20.26. Plot distance in feet along the X-axis and load along the Y-axis. Draw (1) the integral curve to show the shearing force, and (2) the second integral curve to show the bending moment.

7. Plot the points given in the table and draw a smooth curve. Construct the derivative curve. Write the equation of the derivative curve. Equation of given curve $y = \frac{1}{2}x^2 + 10$.

x	0	1	2	3	4	5	6	7	8	9	10
y	10	10.5	12	14.5	18	22.5	28	34.5	42	50.5	60

8. Construct the distance–time, velocity–time, and acceleration–time curves for an automobile moving as follows: time = 10 sec, acceleration = 1.5 m per sec per sec throughout the interval, initial velocity = 0.

9. The passenger train for which derived curves are shown in Fig. 20.24 is brought to a stop with a constant negative acceleration of 1.0 ft per sec per sec. Before applying the brakes the train was traveling with a constant velocity of 55 ft per sec. Construct the curves showing acceleration–time, velocity–time, and distance–time relationships.

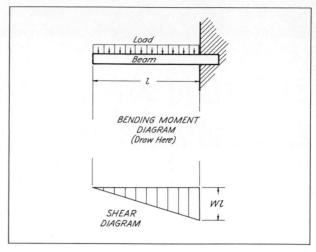

FIG. 20.25

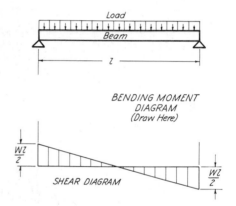

FIG. 20.26

DESIGN AND COMMUNICATION DRAWING IN SPECIALIZED FIELDS

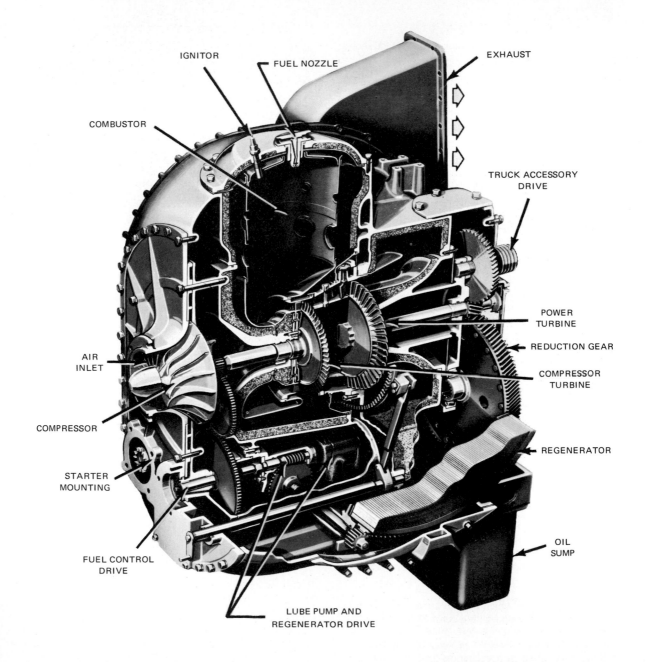

IGNITOR

FUEL NOZZLE

EXHAUST

COMBUSTOR

TRUCK ACCESSORY DRIVE

POWER TURBINE

REDUCTION GEAR

AIR INLET

COMPRESSOR TURBINE

COMPRESSOR

REGENERATOR

STARTER MOUNTING

OIL SUMP

FUEL CONTROL DRIVE

LUBE PUMP AND REGENERATOR DRIVE

New research prototype gas turbine engine, a low-pressure, regenerative power plant. The compressor in the 375-horsepower engine turns at 37,500 revolutions per minute, the output shaft at only 3000 revolutions per minute. It weighs 1700 pounds installed, with a length of 40 inches and a height of 39 inches. The reader should note the machine elements that can be readily seen. A few of these elements will be discussed in this chapter. (*Courtesy Ford Motor Company*)

Design and Selection of Machine Elements: Gears, Cams, Linkages, Springs, and Bearings

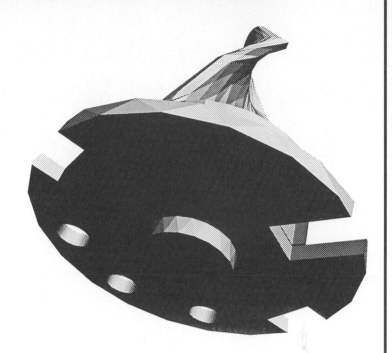

A / GEARS

21.1 Gears

It is important for the designer of mechanical systems to know the general proportions and nomenclature pertaining to gearing. In Fig. 21.1 the nomenclature for bevel gears is shown. It will be noted that, in general, the definitions pertaining to gear-tooth parts can be represented in a right section of the gear.

The theory of gears is a part of the study of mechanisms. In working drawings of gears and toothed wheels it is necessary to draw at least one tooth of each gear. Some of the terms used in defining gear teeth are shown in Fig. 21.2.

Two systems of generating tooth curves are in general use, the *involute system* and the *cycloidal system*. The curve most commonly used for gear-tooth profiles is the involute of a circle.

An *involute* is the curve generated by a point on a straightedge as the straightedge is rolled on a cylinder. It also may be defined as the curve generated by a point on a taut string as the string is unwrapped from a cylinder. The circle from which the involute is developed is called the *base circle*.

A method of constructing an involute curve is shown

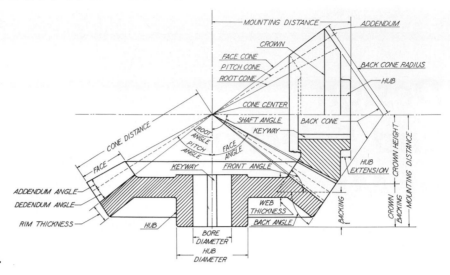

FIG. 21.1 Bevel-gear nomenclature.

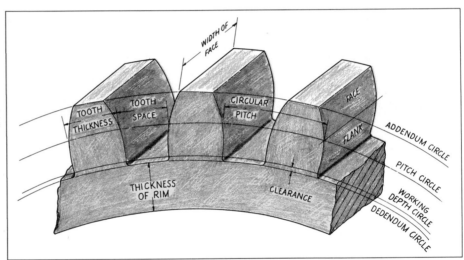

FIG. 21.2 Spur-gear nomenclature.

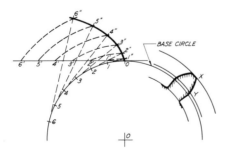

FIG. 21.3 Involute tooth.

in Fig. 21.3. Starting with point *O* on the base circle, divide the base circle into a convenient number of equal arcs of length 0–1, 1–2, 2–3, and so forth. (Where the lengths of the divisions on the base circle are not too great, the chord can be taken as the length of the arc.) Draw a tangent to the base circle at point *O*, and divide this line to the left of *O* into equal parts of the same lengths as the arcs. Next, draw tangents to the circle from points 1, 2, 3, and so on. With the center of the base circle *O* as a

pivot, draw concentric arcs from 1′, 2′, 3′, and so forth, until they intersect the tangent lines drawn from 1, 2, 3, and so forth. The intersection of the arcs and the tangents are points on the required involute curve, such as 1″, 2″, 3″, and so forth. The illustration in Fig. 21.3 shows the portion *XY* of the tooth outline as part of the involute curve.

The *cycloidal system*, as the name implies, has tooth curves of cycloidal form. A *cycloid* is the curve generated by a point on the circumference of a circle as the circle rolls on a straight line. If the circle rolls on the outside of another circle, the curve generated is called an *epicycloid*; if it rolls on the inside of another circle, the curve generated is called a *hypocycloid*. In Fig. 21.4, let *R* be the radius of the fixed circle and *r* be the radius of the rolling circle. Draw through *O* a circle arc, *AB*, concentric with the fixed circle. Lay off on the rolling circle a convenient number of divisions, such as 0–1, 1–2, 2–3, and so forth; then divide the fixed-circle circumference into divisions, of the same length, such as 0–1′, 1′–2′, 2′–3′, and so on. Through these points on the fixed circle,

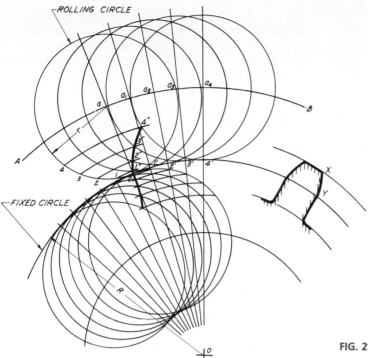

FIG. 21.4 Tooth curves—cycloidal form.

draw radii and extend them to intersect the arc *AB*, thus producing points a_1, a_2, a_3, and so on. These points will be the centers of the successive positions of the rolling circle. Draw the positions of the rolling circle, using the centers a_1, a_2, a_3, and so forth. Next draw, on the rolling circle with the center *O* of the fixed circle as the pivot point, concentric arcs through points 1, 2, 3, and so forth. The intersection of these arcs with the rolling circles about a_1, a_2, a_3, and so forth, determines points, such as 1″, 2″, 3″, and so forth, on the epicyclic curve. The illustration in Fig. 21.4 shows *XY* of the tooth outline as part of the epicyclic curve.

The hypocyclic-curve construction is the same as that for the epicyclic curve. In the construction of the hypocyclic curve, if the rolling circle has a diameter equal to one-half of the diameter of the fixed circle, the hypocyclic curve thus generated will be a radial line of the fixed circle.

GEAR TERMS

1. The *addendum circle* is drawn with its center at the center of the gear and bounds the ends of the teeth. See Fig. 21.2.

2. The *dedendum circle*, or *root circle*, is drawn with its center at the center of the gear and bounds the bottoms of the teeth. See Fig. 21.2.

3. The *pitch circle* is a right section of the equivalent cylinder the toothed gear may be considered to replace.

4. *Pitch diameter* is the diameter of the pitch circle.

5. The *addendum* is the radial distance from the pitch circle to the outer end of the tooth.

6. The *dedendum* is the radial distance from the pitch circle to the bottom of the tooth.

7. The *clearance* is the difference between the dedendum of one gear and the addendum of the mating gear.

8. The *face of a tooth* is that portion of the tooth surface lying outside the pitch circle.

9. The *flank of a tooth* is that portion of the tooth surface lying inside the pitch circle.

10. The *thickness* of a tooth is measured on the arc of the pitch circle. It is the length of an arc and not the length of a straight line.

11. The *tooth space* is the space between the teeth measured on the pitch circle.

12. *Backlash* is the difference between the tooth thickness of one gear and the tooth space on the mating gear, measured on the pitch circles.

13. The *circular pitch* of a gear is the distance between a point on one tooth and the corresponding point on the adjacent tooth, measured along the arc of the pitch circle. The circular pitches of two gears in mesh are equal.

14. The *diametral pitch* is the number of teeth per inch of pitch diameter. It is obtained by dividing the number of teeth by the pitch diameter.

15. The *face of a gear* is the width of its rim measured parallel to the axis. It should not be confused with the face of a tooth, for the two are entirely different.

16. The *pitch point* is on the line joining the centers of the two gears where the pitch circles touch.

17. The *common tangent* is the line tangent to the pitch circles at the pitch point.

18. The *pressure angle* is the angle between the line of action and the common tangent.

19. The *line of action* is a line drawn through the pitch point at an angle (equal to the pressure angle) to the common tangent.

20. The *base circle* is used in involute gearing to generate the involutes that form the tooth outlines. It is drawn from the center of each pair of mating gears tangent to the line of action.

21. When two gears mesh with each other, the larger is called the *gear* and the smaller the *pinion*.

It should be noted that *circular pitch* is a linear dimension expressed in inches, whereas *diametral pitch* is a ratio. There must be a whole number of teeth on the circumference of a gear. Thus it is necessary that the circumference of the pitch circle, divided by the circular pitch, be a whole number.

For circular pitch, let P' be the circular pitch in inches, D the pitch diameter, and T the number of teeth. Then

$$TP' = \pi D, \qquad T = \frac{\pi D}{P'},$$

$$P' = \frac{\pi D}{T}, \quad \text{and} \quad D = \frac{TP'}{\pi}$$

For diametral pitch, let P be the diametral pitch, D the pitch diameter, and T the number of teeth. Then

$$T = PD, \qquad D = \frac{T}{P},$$

$$\text{and} \qquad P = \frac{T}{D}$$

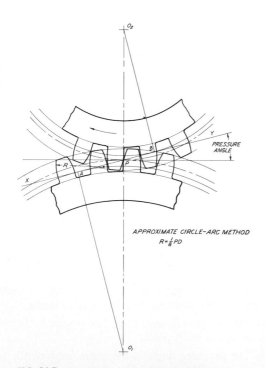

APPROXIMATE CIRCLE-ARC METHOD
$R = \frac{1}{8} PD$

FIG. 21.5   To draw a pair of spur gears.

The Brown and Sharpe $14\frac{1}{2}°$ involute system has been adopted as one of the American standards and is commonly known as the $14\frac{1}{2}°$ *composite system*. The tooth proportions of this system are given in terms of the diametral pitch P and circular pitch P'.

Pressure angle $= 14\frac{1}{2}°$

Addendum (inches) $= 1/\text{diametral pitch} = 1/P$

Dedendum (inches) $= \text{addendum plus clearance} = (1/P) + .05P'$

Clearance $= .05 \times \text{circular pitch} = .05P'$

Whole depth of tooth $= 2 \times \text{addendum} + \text{clearance} = 2 \times (1/P) + .05P'$

Working depth of tooth $= 2 \times \text{addendum} = 2 \times (1/P)$

Thickness of tooth $= \text{circular pitch}/2 = P'/2$

Width of tooth space $= \text{circular pitch}/2 = P'/2$

Minimum radius of fillet $= \text{clearance} = .05P'$

In the above calculations the backlash is zero. Actually, however, it is common practice to provide backlash, and this is accomplished by using standard cutters and cutting the teeth slightly deeper than for standard teeth.

21.2   To Lay Out a Pair of Standard Involute Spur Gears

The following facts are known regarding the laying out of a pair of standard spur gears: (1) number of teeth on each gear—large gear 24, small gear 16; (2) diametral pitch $= 2$; (3) pressure angle $= 14\frac{1}{2}°$.

To draw a pair of spur gears, determine the pitch diameters as follows.

$$D = \frac{T}{P} = \frac{24}{2} = 12 \text{ in.} \quad \text{(for large gear)}$$

$$D = \frac{T}{P} = \frac{16}{2} = 8 \text{ in.} \quad \text{(for small gear)}$$

In Fig. 21.5, with radii O_1P and O_2P equal to 6 and 4 in., respectively, draw the pitch circles and, through P, draw the common tangent. Draw the line of action XY at an angle of $14\frac{1}{2}°$ to the common tangent. Drop perpendiculars from the centers O_1 and O_2, cutting the line of action at A and B, respectively. O_1A and O_2B are the radii of the base circles that can now be drawn.

From Sec. 21.1, determine the addendum and dedendum of the teeth, and draw in the respective addendum and dedendum circles.

Divide the pitch circle of the smaller gear into 16 equal parts and the pitch circle of the larger gear into 24 equal parts, which will give the circular pitch. Assuming that no allowance is made for backlash, bisect the circular pitch on each of the gears, which will give 32 equal divisions on the small gear and 48 equal divisions on the large gear.

At any point on the base circle of each gear, develop an involute (see Fig. 21.3) and draw in the curves between the base and addendum circles through alternate points on the pitch circles. This produces one side of all the teeth in each gear. The curve for the other side of the tooth is the reverse of the side just drawn. The part of the tooth between the base and dedendum circles is part of a radial line drawn from the base circles to the centers of the gears. The tooth is finished by putting in a small fillet between the working depth and dedendum circles. Since the detailing of gear teeth is normally done only for display purposes, most drafters and designers find that a gear template is satisfactory for showing the gear tooth profile rather than to lay out the gear tooth shape meticulously by hand.

21.3 Metric Spur Gears*

Involute spur gears are also designed and manufactured to metric specifications. As with metric thread specifications, there is a similarity of design criteria to those used when manufacturing to inch measurements, but there are a number of definitions and terms which are different.

In metric gear specifications, diametral pitch is replaced by the term *module*, which is the reciprocal of diametral pitch. The module for a metric gear is an actual distance in millimeters. Tooth form for metric gears is full depth. The generating curve is the involute, and the pressure angle is 20°.

The following definitions are used for metric gear design.

* Prepared by Larry D. Goss, University of Southern Indiana.

Module = pitch circle diameter/number of teeth

Addendum = module
Dedendum = 1.25 × module

Circular pitch = module × π

Center distance between meshed gears = module
$$\times \frac{\text{total teeth of the two gears}}{2}$$

Overall diameter of gear = pitch circle diameter + (2 × module)

or

= (number of teeth + 2)
× module

Other definitions may be derived from these.
Metric gears are available commercially in the following modules: 0.5 to 4.0 by 0.25 increments, 4.0 to 7.0 by 0.5 increments, and 7.0 to 10.0 by 1.0 increments.

Inch-sized gears must have identical pitch to mesh properly; similarly, metric gears must have identical modules for proper engagement.

21.4 Dimensioning Gears

On a detail drawing, a gear may be represented by a one-view section, except for the larger sizes, where it is necessary to show spokes, and for the small sizes, when a full description of some feature would not be given. A second view was drawn in Fig. 21.6 to reveal the shape of the keyway. In dimensioning, it is recommended that the dimensions be given on the view or views, as shown

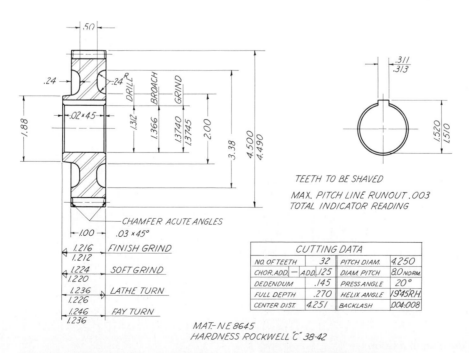

TEETH TO BE SHAVED

MAX. PITCH LINE RUNOUT .003
TOTAL INDICATOR READING

CUTTING DATA			
NO. OF TEETH	32	PITCH DIAM.	4.250
CHOR. ADD.	—ADD..125	DIAM. PITCH	8.0 NORM.
DEDENDUM	.145	PRESS ANGLE	20°
FULL DEPTH	.270	HELIX ANGLE	19°45' RH.
CENTER DIST.	4.251	BACKLASH	.004-.008

MAT- N.E 8645
HARDNESS ROCKWELL "C" 38-42

FIG. 21.6 Detail drawing of a helical gear. (Courtesy Fairfield Mfg. Co.)

in Figs. 21.6 and 21.7, and that the cutting data be incorporated in an accompanying table.

21.5 Working Drawings of Gears

Figure 21.6 shows a working drawing of a helical gear. In practice it is customary to show one view in section and just enough of the circular view to supply needed information for the shop. Individual teeth are never shown on the circular view unless the drawing is to be used for display purposes. In some drafting rooms where practice requires a full circular view, the addendum and root circles are drawn using a phantom line and the pitch circle is given as a center line.

It should be noted that the dimensions for the gear blank are given on the sectioned view, while the necessary data for cutting the teeth are given in a table. This is in accordance with the practice recommended in Sec. 21.4.

A working drawing of a spiral bevel gear is shown in Fig. 21.7. The same practices apply as for spur gears. Before starting to prepare a working drawing of a bevel gear one should study Fig. 21.1, which illustrates bevel-gear nomenclature.

Figure 21.8 shows a working drawing of a worm.

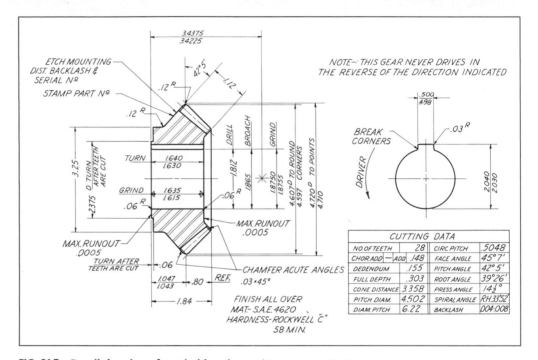

CUTTING DATA			
NO. OF TEETH	28	CIRC. PITCH	.5048
CHOR. ADD	ADD .148	FACE ANGLE	45°7'
DEDENDUM	.155	PITCH ANGLE	42°5'
FULL DEPTH	.303	ROOT ANGLE	39°26'
CONE DISTANCE	3.358	PRESS ANGLE	14½°
PITCH DIAM.	4.502	SPIRAL ANGLE	RH.33°52
DIAM. PITCH	6.22	BACKLASH	004:008

FIG. 21.7   Detail drawing of a spiral bevel gear. (Courtesy Fairfield Mfg. Co.)

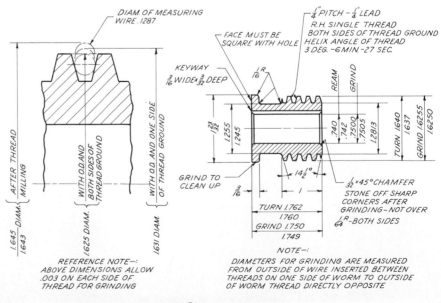

FIG. 21.8   Detail drawing of a worm.
(Courtesy Warner & Swasey Co.)

B / CAMS

21.6 Cams

A *cam* is a plate, cylinder, or any solid having a curved outline or curved groove that, by its oscillating or rotating motion, gives a predetermined motion to another piece, called the *follower*, in contact with it. The cam plays a very important part in the operation of many classes of machines. Cam mechanisms are commonly used to operate valves in automobiles and stationary and marine internal combustion engines. They also are used in automatic screw machines, clocks, locks, printing machinery, and in nearly all kinds of machinery that we generally regard as "automatic machines." The applications of cams are practically unlimited, and their shapes or outlines are found in wide variety.

All cam mechanisms consist of at least three parts, (1) the cam, which has a contact surface either curved or straight; (2) the follower, whose motion is produced by contact with the cam surface; and (3) the frame, which supports the cam and guides the follower.

The most common type of cam is the disc or plate cam. Here the cam takes the form of a revolving disc or plate, the circumference of the disc or plate forming the profile with which the follower makes contact. In Figs. 21.9 and 21.10, two simple examples of a disc cam and follower are shown. In Fig. 21.9, the cam is given a motion of rotation, thus causing the follower to rise and then return again to its initial position. In cams of this type it is necessary to use some external force, such as the spring, to keep the follower in contact with the cam at all times. Contact between the follower and the cam is made through a roller, which serves to reduce friction. It is sometimes necessary to use a flat-faced follower, instead of the roller type, an example of which is shown in Fig. 21.10. The follower face that comes in contact with the cam is usually provided with a hardened surface, to prevent excessive wear.

Another type of cam is one in which the follower is constrained to move in a definite path without the application of external forces. See Fig. 21.11. In this type, two contact surfaces of the follower bear on the cam at the same time, thus controlling the motion of the follower in two directions.

21.7 Design of a Cam

The design of a cam profile is governed by the requirements with respect to the motion of the follower. In the layout of a cam, the initial position, displacement, and character

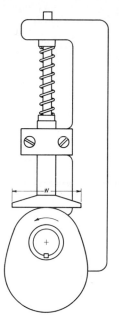

FIG. 21.10 Disc cam and follower.

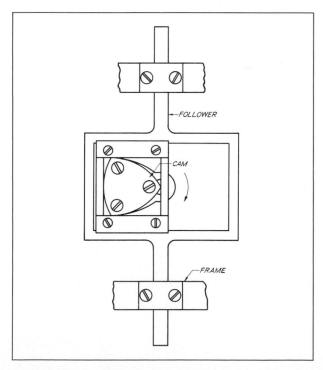

FIG. 21.11 Cam and follower.

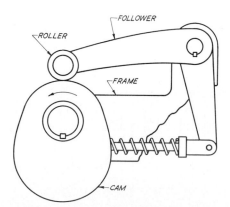

FIG. 21.9 Disc cam and follower.

of the motion of the follower are generally known. It is convenient to make first a graphical representation of the follower movement, a procedure that is called *making a displacement diagram*. This is a linear curve in which the length of the diagram represents the time for one revolution of the cam. The height of the diagram represents the total displacement of the follower; the length is made to any convenient length and is divided into equal time intervals, the total representing one rotation of the cam.

In Fig. 21.12 is shown a displacement diagram in which the follower rises 2 in. during 180° of rotation of the cam, then rests for 30° and returns to its initial position for the remainder of the cam revolution. Cam outlines should be designed to avoid sudden changes of motion at the beginning and end of the follower stroke. This can be accomplished by having a uniformly accelerated and decelerated motion at the beginning and end of the constant-velocity curve. The construction for uniformly accelerated motion is shown in Fig. 21.12. On a line, *OX*, making any convenient angle with *OA*, mark off any unit of length in this figure equal to *Oa*. The next point, *b*, is found by marking off, from *O*, 4 units of length. Point *c* is found by marking off 9 units of length. Next, project the intersection (point *s*) of time unit 3 and the constant-velocity line over to line *OA*, thus locating point *t*. Connect points *c* and *t* with a straight line and draw parallel lines from *a* and *b* intersecting line *OA*. From these intersections draw lines parallel to *ts*, intersecting the time-unit lines 1 and 2, respectively. These intersections are points on the displacement curve. With uniformly decelerated motion, the series of points are laid off in the reverse order, such as 9–4–1. It will be noted that the units are laid off according to the square of the time unit. Thus, if there were 4 time units, the acceleration curve would be laid off according to the ratio of 1, 4, 9, 16, and the deceleration, 16, 9, 4, 1.

The construction for the displacement diagram for simple harmonic motion is shown in the same figure. A semicircle is drawn as shown, the follower displacement being used as a diameter, and is then divided into a convenient number of parts equal to the number of cam displacement units. Horizontal projection lines are drawn from the semicircle, and the intersections of these lines with the cam displacement lines are points on the displacement curve. Thus, the projection of point 15 on the semicircle to time-unit line 15 locates one point on the displacement curve for simple harmonic motion.

The next step is that of finding the cam profile necessary to produce these movements. The construction is shown in Fig. 21.13. Select a base circle of convenient size, and on it lay off radial lines according to the number of time units of cam displacement. Draw line *OB* extended to *W*, and on it lay off the distances y_1, y_2, y_3, and so forth, obtained from the displacement diagram, from the center of the roller shown in the starting position, thus locating points B_1, B_2, B_3, and so forth. With *O* as a center, draw arcs B_1–B_1', B_2–B_2', B_3–B_3', and so forth, and at B_1', B_2', B_3', and so forth, draw in the circles representing the diameter of the roller. To complete the cam outline, draw a smooth curve tangent to the positions of the roller.

C / LINKAGES

21.8 Linkages*

The link is the most common machine element. A link is a rigid bar that transmits force and velocity. An assemblage of links that produces a prescribed motion is called a *linkage*. Since linkages appear in nearly all phases of mechanical design, a knowledge of linkage design and analysis is necessary for all those designers who may be involved in the creation of useful mechanisms. See the page facing the beginning of this chapter.

21.9 Position and Clearance Analysis

In the preliminary design stage, a designer is frequently called upon to find the position of each link as the machine operates. When the layout has been completed, it must be checked to be sure that the linkage does not interfere with any other moving machine elements. Sufficient clearance must be left between rapidly moving machine parts to ensure that any deflection in the parts will still allow adequate clearance.

Mathematics, primarily geometry and trigonometry, can be used to calculate accurately both the position and clearances. On the other hand, graphical solutions provide sufficient accuracy with a minimum amount of effort and time. A graphical solution also allows the designer to visualize the problem easily and rapidly. Figure 21.14

---

* Secs. 21.8–21.12 prepared by Wesley L. Baldwin, Mechanical Engineering Technology, Purdue University.

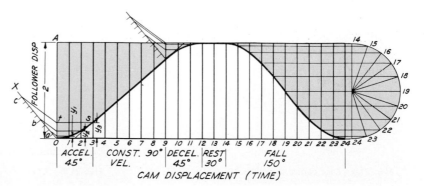

FIG. 21.12 Displacement diagram.

CAM DISPLACEMENT (TIME)

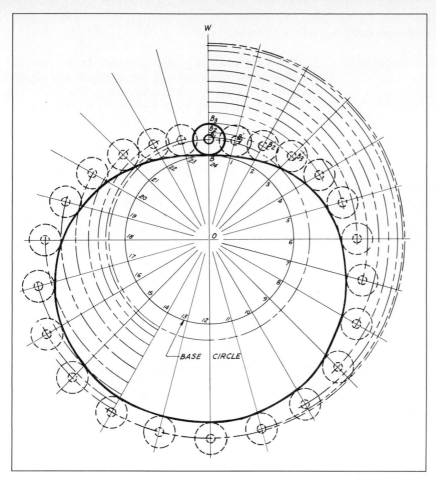

FIG. 21.13 Construction for cam profile.

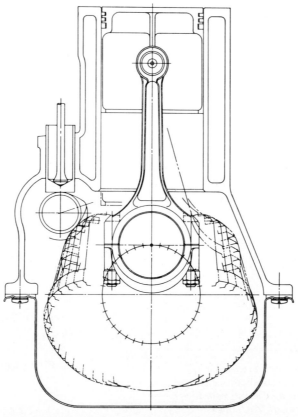

FIG. 21.14 Graphical solution for the clearances between moving machine elements. (*Courtesy General Motors Corporation*)

541

shows how the layout man can analyze clearances between moving machine parts.

21.10   *Four-bar Linkage*

The most common type of linkage found in machines is called 4-bar linkage. This linkage is made up of four pin-connected links. Figure 21.15 shows a typical 4-bar linkage. Notice that link 1 is the distance between the two pin connections attached to the ground. The 4-bar linkage has one degree of freedom. This means that once the position of a link is defined with respect to another link, the remaining links are also in a fixed position. For example, if the angle θ is defined in Fig. 21.15, then the angles β and ψ can have only one value.

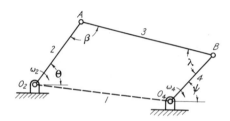

FIG. 21.15   Typical four-bar linkage.

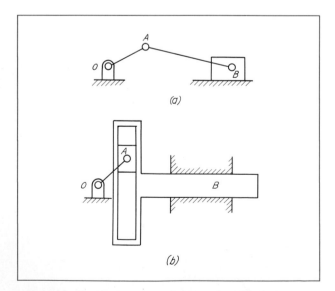

FIG. 21.16   Examples of mechanisms that are similar to four-bar linkages: (a) scotch yoke and (b) slider crank. (*Extracted from* Kinematics and Linkage Design *with the permission of Allen S. Hall, Professor of Mechanical Engineering, Purdue University*)

The 4-bar linkage is also important to the designer because many other linkages are analyzed in a similar manner. The slider crank and the scotch yoke, shown in Fig. 21.16, are examples of this type of mechanism.

21.11   *Velocity Analysis of a Link*

A link does not have to be straight. It may be curved so that it does not interfere with another moving element of a mechanism. For analysis, however, we can replace a curved link by a straight link connecting the two ends. Figure 21.17 shows a curved link represented by a straight link. When the link is in motion, the velocity of points A and B will be different. The velocities of these two ends will be designated V_A and B_B, respectively. Because velocity is a vector quantity, we may resolve each vector, V_A and V_B, into two components. Consider the components of V_A and V_B parallel to and perpendicular to the straight line connecting points A and B. In Fig. 21.17 the components parallel to the straight line connecting A and B are designated V_A^t, and V_B^t. The perpendicular components are designated V_A^n, and V_B^n.

Because the link is solid, V_A^t, and V_B^t must be equal. If V_A^t, and V_B^t were not equal, then the line joining A and B would have a changing length. The difference in velocity between points A and B must then be due entirely to the difference between V_A^n, and V_B^n. This difference between V_A^n, and V_B^n is called the relative velocity between points A and B. The relative velocity is designated $V_{B/A}$, when V_A^n, is subtracted from V_B^n.

Since V_A^t, and V_B^t are equal, $V_{B/A}$ is also equal to the difference between V_A and V_B. The difference between V_A and V_B, called the relative velocity $V_{B/A}$, lies in a direction

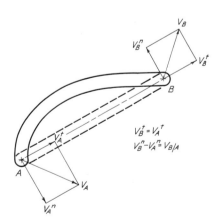

FIG. 21.17   Link in motion.

perpendicular to the straight line joining A and B. The magnitude of $V_{B/A}$ is the same as the difference between $V_A{}^n$, and $V_B{}^n$. Figure 21.17 shows these relationships.

21.12   *Velocity Analysis of a 4-bar Linkage*

A typical 4-bar linkage has link 2 driven at a constant angular velocity ω_2. To calculate the angular velocity, ω_4, of link 4

1. Draw the linkage to scale in the desired position.
2. Calculate the velocity of point A.

$$A_A = \omega_2(O_2A)$$

V_A is the velocity of point A and is perpendicular to (O_2A), ω^2 is the angular velocity of link 2 measured in radians/second, and (O_2A) is the length of link 2.

3. Construct vector V_A to scale.
4. From the tip of V_A construct a line perpendicular to link 3. This line lies in the direction of $V_{B/A}$.

5. From the tail of vector V_A construct a line perpendicular to (O_4B) to the line constructed in step 3. This line lies in the direction of V_B.
6. The line drawn in step 5 is the scaled velocity of point B.
7. Calculate ω_4.

$$\omega_4 = \frac{V_B}{O_4\,B}$$

D/   SPRINGS

21.13   *Springs*

In production work, a spring is largely a matter of mathematical calculation rather than drawing, and it is usually purchased from a spring manufacturer, with the understanding that it will fulfill specified conditions. A spring follows the helical path of the screw thread. For this reason the steps in the layout of the representation for a spring are similar to the screw thread. Pitch distances are marked off, and the coils are given a slope of one-half of the pitch. Figure 21.18(*a*) shows a partial layout of a tension spring. Other types of ends are shown in (*b*). A compression spring layout, with various types of ends, is illustrated in Fig.

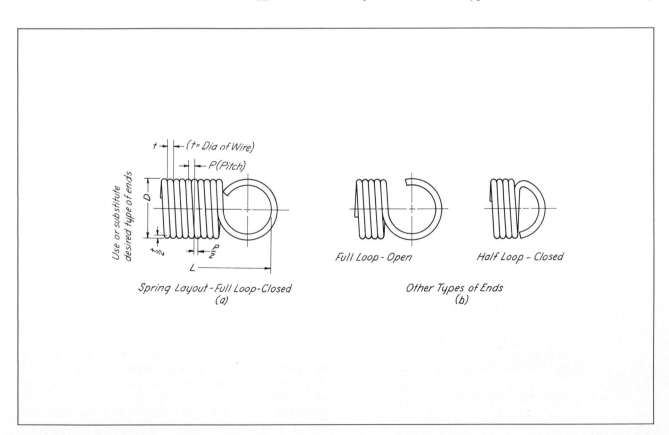

Spring Layout-Full Loop-Closed
(a)

Full Loop - Open   *Half Loop - Closed*

Other Types of Ends
(b)

FIG. 21.18   Tension springs.

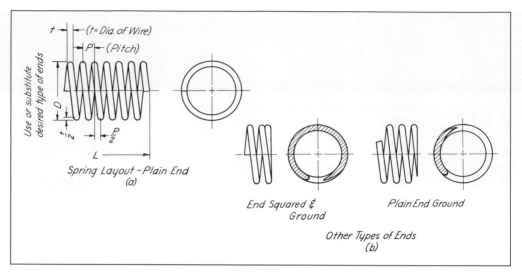

FIG. 21.19 Compression springs.

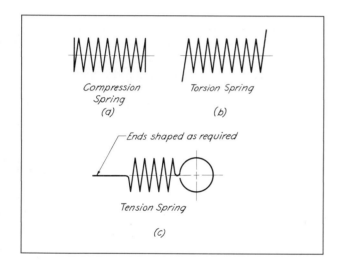

FIG. 21.20 Single-line representation of springs.*

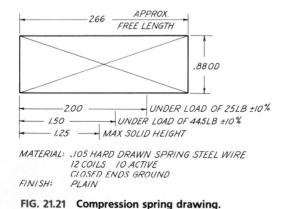

FIG. 21.21 Compression spring drawing.

* ANSI Z14.1–1946.

21.19. Single-line symbols for the representation of springs are shown in Fig. 21.20.

When making a detail working drawing of a spring, it should be shown to its free length. On either an assembly or detail drawing, a fairly accurate representation, neatly drawn, will satisfy all requirements.

It is common practice in industry to rely on a printed spring drawing, accompanied by a filled-in printed form, to convey the necessary information for the production of a needed spring by a reliable manufacturer. The best procedure is to give the spring characteristics along with a list of necessary dimensions and then depend on an experienced spring designer, at the plant where the spring will be produced, to finalize the design.

The method or representing and dimensioning a compression spring is shown in Fig. 21.21. The spring is represented by a rectangle with diagonals (printed or drawn). Pertinent information is added as dimensions and notes. Either an ID (inside diameter) or an OD (outside diameter) dimension is given, depending on how the spring is to be used.

A method of representing and dimensioning an extension spring is shown in Fig. 21.22. Although drawings of this type may be printed forms, it is often necessary to prepare a drawing showing the ends and a few coils, since extension springs may have any one of a wide variety of types of ends. Needed information is presented in tabular form, as shown, with or without a complete spring design. When no printed form is available, a drawing similar to the one shown in Fig. 21.22 must be prepared.

A *torsion spring* offers resistance to a torque load. The extended ends form the torque arms, which are rotated about the axis of the spring. One method of representing and dimensioning torsion springs is shown in Fig. 21.23. A printed form may be used when there is sufficient uniformity in product requirements to warrant the preparation of a printed form or several printed forms. When a printed

(Give either I D or O D - not both)

Specify { Wire size - material (kind and grade)
Number of coils (right hand - left hand)
Type of ends
Load (__ at __ inches inside hooks)
Load rate (__ per I inch deflection)
Maximum extended length
Finish, etc.

FIG. 21.22   Extension spring drawing.

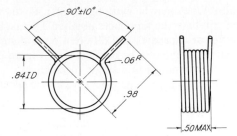

MATERIAL: .059 MUSIC WIRE
6¾ COILS - RIGHT HAND - NO INITIAL TENSION
TORQUE : 2.5 INCH LB AT 155° DEFLECTION SPRING MUST
DEFLECT 180° WITHOUT PERMANENT SET AND
MUST OPERATE FREELY ON .750 DIA SHAFT
FINISH: CADMIUM PLATE

FIG. 21.23   Torsion spring representation and dimensioning.

form is not available, a drawing similar to the one shown must be prepared.

The term *flat spring* includes all springs made of a strip material. One method of representing and dimensioning flat springs is shown in Fig. 21.24.

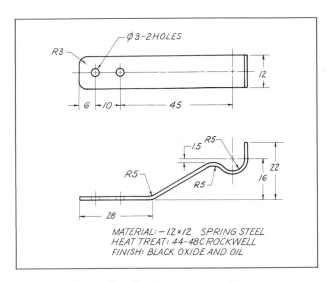

MATERIAL: – 1.2 × 12   SPRING STEEL
HEAT TREAT: 44 - 48 C ROCKWELL
FINISH: BLACK OXIDE AND OIL

FIG. 21.24   Flat spring drawing.

E   BEARINGS

21.14   Bearings

A draftsman ordinarily is never called on to make a detail drawing of a ball or roller bearing, because bearings of these two types are precision-made units that are purchased from bearing manufacturers. All drafters working on machine drawings, however, should be familiar with the various types commonly used and should be able to represent them correctly on an assembly drawing. An engineer will find it necessary to determine shaft-mounting fits and housing-mounting fits from a manufacturer's handbook, in order to place the correct limits on shafts and housings. Figure 21.25 shows two types of ball bearings. A roller bearing is shown in Fig. 21.26.

Ball bearings may be designed for loads either perpendicular or parallel to the shaft. In the former, they are known as *radial bearings* and in the latter, as *thrust bearings*. Other types, designated by various names, are made to take both radial and thrust loads, either light or heavy. In most designs, bearings are forced to take both radial and thrust loads. Ball bearings are designated by a letter and code number, the last number of which represents the

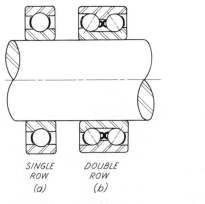

SINGLE   DOUBLE
ROW   ROW
(a)   (b)

FIG. 21.25   Ball bearings.

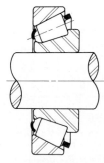

FIG. 21.26   Roller bearing.

bearing bore. They may be extra light, light, medium, or heavy and still have the same bore number. That is, bearings of different capacities and different outer diameters can fit shafts having the same nominal size. Figure 21.27 shows typical mountings in a single mechanism.

Tables 25 and 26 in the Appendix give all the necessary information on the bearings used in the problems in this text.

Roller bearings are designed for both radial and thrust loads. The bearing consists of tapered rollers that roll between an inner and an outer race. The rollers are enclosed in a retainer (cage) that keeps them properly spaced.

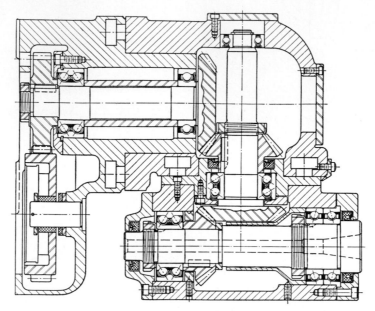

FIG. 21.27 Ball bearings. (*Courtesy New Departure Bearing Co.*)

 # PROBLEMS

Note: With the approval of his instructor, the student may convert the inch measurements in the problems that follow to millimeters by using Tables 2–8 in the Appendix. Since it is common practice under the metric system to use full millimeters for all dimension values that are not critical, the student should do likewise. For example, 3 in. converts to 76.2 millimeters. The student should use 76 millimeters.

Prob. No.	Gear				Pinion				Pressure Angle
	Circular Pitch (in.)	Diametral Pitch	Pitch Diameter	No. of Teeth	Circular Pitch (in.)	Diametral Pitch	Pitch Diameter	No. of Teeth	
1	1.31			24	1.31			16	$14\frac{1}{2}°$
2		2		20		2		14	$14\frac{1}{2}°$
3		2.5	10			2.5	8		$14\frac{1}{2}°$
4	1.0			30	1.0			20	$14\frac{1}{2}°$
5	2.0			18	2.0			12	$14\frac{1}{2}°$
6		3	8			3	6		$14\frac{1}{2}°$

Gears and Cams

Gear Problems

1–6. Following the method shown in Sec. 21.2, lay out a pair of standard involute spur gears as assigned from the given table. The pinion is the driver

Cam Problems

Cam Data

Diameter of cam shaft	$1\frac{1}{4}$ in.
Diameter of cam hub	$2\frac{1}{4}$ in.
Diameter of roller	1 in.

Keyway	$\frac{1}{4} \times \frac{1}{8}$ in.
Diameter of base circle	$2\frac{3}{4}$ in.
Follower displacement	2 in.
Scale: full-size	
Cam rotation: as noted	

Determine points on the cam profiles at intervals of 15°.

7. Using the data above, design a plate cam to satisfy the following conditions: (a) a rise of 2 in. in 180°, with constant velocity, except for uniform acceleration for the first 30° and uniform deceleration for the last 45°; (b) rest 30°; and (c) return with simple harmonic motion. Use clockwise cam rotation.

8. Same as Problem 7, excpet that the follower is of the flat-face type and is $2\frac{1}{2}$ in. wide.

9. Using the data for Problem 7, design a plate cam to satisfy the following conditions: (a) rise of 2 in. during 180°, the first 45° of which is uniformly accelerated motion, the next 60° being constant velocity, and the last 75° of rise being uniformly decelerated motion; (b) rest 15°; and (c) return to starting position with simple harmonic motion. Use counterclockwise cam rotation.

10. Same as Problem 9, except that the follower is of the flat-face type and is $2\frac{1}{2}$ in. wide.

11. Using the data for Problem 7, design a plate cam to satisfy the following conditions: (a) rise to 2 in. during 150°, by simple harmonic motion; (b) rest 30°; and (c) return to starting position during remainder of the revolution, with uniformly accelerated and decelerated motion, the value of the deceleration being twice that of the acceleration. Use clockwise cam rotation.

12. Using the data for Problem 7, except that the follower is to be of the flat-face type, $2\frac{1}{2}$ in. wide, design a plate cam to satisfy the following conditions: (a) rise of 2 in. with simple harmonic motion, in 120°; (b) rest 30°; (c) return in 150°, with constant velocity, except for uniform acceleration for the first 45° and uniform deceleration for the last 30° of fall; and (d) rest the balance of the revolution. Use counterclockwise cam rotation.

Linkage Problems*

13. A 4-bar linkage has link 1 = 6 in., link 2 = 2 in., link 3 = 8 in., link 4 = 5 in. Draw the linkage in all positions for $\theta = 0°$ to 360° in 30° increments. Plot a graph of θ vs. ψ.

14. If link 2 in Problem 13 has a constant angular velocity of 10 radians per sec, find the angular velocity of link 4. Use the 12 positions indicated in Problem 13. Plot θ vs. ω_4.

* Prepared by Wesley L. Baldwin, Mechanical Engineering Technology, Purdue University.

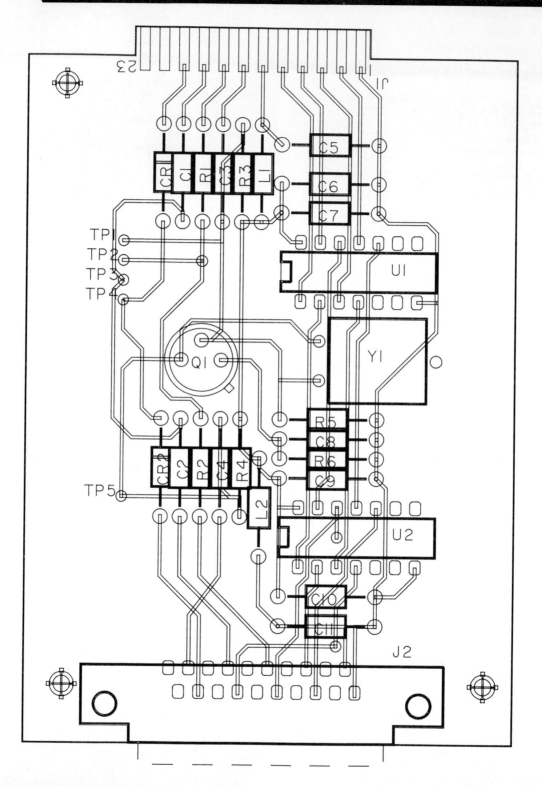

Electronic drafting involves the preparation of schematics, connection diagrams, wiring diagrams, block diagrams, and printed circuit board layouts. These require the use of symbols. Constructing electronic drawings manually is time consuming even if templates are used. On the other hand, computer-generated electronic drawings, such as the one shown, are done with a considerable saving of time because an electronic symbol can be produced almost instantly. (*The drawing was prepared by Dennis Short, Systems Manager, Department of Technical Graphics, School of Technology, Purdue University.*)

Electronic Drawings*

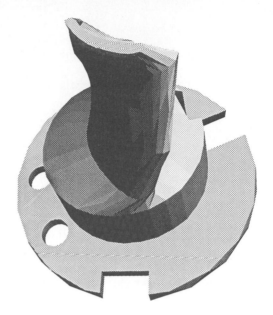

22.1 Introduction

A number of drawings are needed in the design, manufacture, and use of electronic equipment (Fig. 22.1). The preparation of these drawings, particularly schematic and connection diagrams, requires a working knowledge of specialized symbolic forms and conventions. These conventions vary somewhat according to the product and the field of application, which, broadly speaking, may be classified as communications and commercial equipment, military, scientific research, or industrial control.

Engineering sketches are first prepared to show the layout of a proposed system or circuit or the features of existing equipment. Finished drawings are then made by a drafter who follows the instructions given in the sketch.

Usually, many details are left to the drafter's discretion; however, it is the engineer's responsibility to make sure that the final work is correct and complete. In all instances, electronic drawings should be prepared in accordance with the relevant technical standards and specifications so that they will be acceptable for the particular applications.

22.2 Electronic Drawings and Diagrams

The most specialized work in electronic drafting concerns the preparation of symbolic diagrams. In contrast to mechanical drawings, which represent objects, symbolic diagrams impart technical information in abstract form.

* Prepared by George Shiers, Santa Barbara City College.

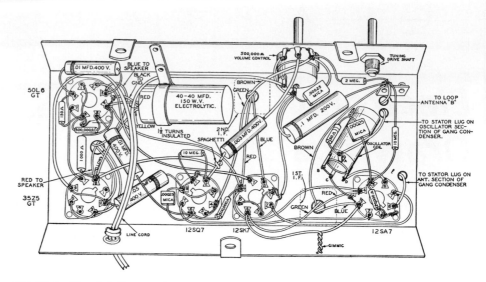

FIG. 22.1 Pictorial diagram. (*Courtesy Allied Radio Corp.*)

Since these diagrams are intended to show the function of a system or a circuit, they lack intrinsic dimensions and, in general, do not show physical details of the parts.

Important examples of these specialized drawings are *block diagrams* which show the overall organization of a system; *schematic diagrams*, which show the component parts and electrical details of a circuit; and *connection diagrams*, which show the wiring and connections between the component parts of an assembly. Other drawings and artwork are needed in the design and processing of printed circuits.

All these drawings are interrelated with the physical assembly and with each other. In many instances further material, such as lists, tables, and charts, may be needed to supplement assembly drawings and electronic diagrams, particularly for production breakdowns, test procedures, and service manuals.

22.3 Graphical Standards

Technical publications dealing with standardized engineering practices are available to guide the engineer and draftsman in the preparation of electronic diagrams. These standards have been established by technical committees representing professional societies, trade associations, government agencies, and various manufacturers and users. The most useful standards for electronic purposes are those that concern graphical symbols, letter symbols, reference designations, abbreviations, color codes, and electrical diagrams, as given in the following list.

ANSI—American National Standards Institute, Inc.
  Y14.15—Electrical Diagrams
  Y32.2 —Graphical Symbols for Electrical Diagrams
  Z10.1 —Abbreviations for Scientific and Engineering Terms
  Z10.5 —Letter Symbols for Electrical Quantities
  Z32.13—Abbreviations for Use on Drawings

EIA—Electronic Industries Association
  GEN–101—Color Coding for Numerical Values
  GEN–102—Preferred Values
  REC–108—Color Coding

IEEE—The Institute of Electrical and Electronics Engineers, Inc.
  (These publications were formerly issued by The Institute of Radio Engineers, Inc.)
  51 IRE 21.S1—Standards on Abbreviations of Radio-Electronic Terms
  54 IRE 21.S1—Standards on Graphical Symbols for Electrical Diagrams
  57 IRE 21.S1—Standards on Letter Symbols and Mathematical Signs
  57 IRE 21.S2—Standards on Reference Designations for Electrical and Electronic Equipment
  57 IRE 21.S3—Standards on Graphical Symbols for Semiconductor Devices

MIL—U.S. Government Printing Office
  MIL–STD–12 —Abbreviations for Use on Drawings
      –15 —Electrical and Electronic Symbols
      –16 —Electrical and Electronic Reference Designations
      –122—Color Code for Chassis Wiring for Electronic Equipment
      –283—Letter Symbols for Electrical and Electronic Quantities

NEMA—National Electrical Manufacturers Association
  IC–1959— Industrial Control

NMTBA—National Machine Tool Builders' Association
  Electrical Standards for Industrial Equipment
  Machine Tool Electrical Standards

These and related standards have been developed in the interest of clear documentation and accurate terminology. Their use promotes efficient engineering methods leading

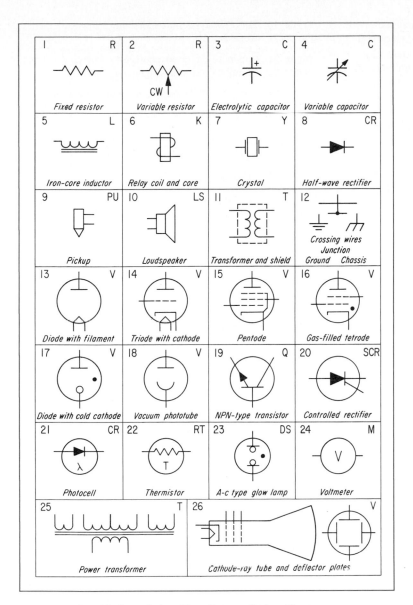

FIG. 22.2 **Graphical symbols with reference designations.**

to economies in time, materials, and labor. An acquaintance with these standards, therefore, forms an essential part of the background of engineers, draftsmen, technicians, and supervisors engaged in the electronics industry.

22.4 *Graphical Symbols*

A graphical symbol is a geometrical design that represents an electronic device or component part in a circuit. Most symbols are composed of two or more basic elements, each of which represents a functional part of the device. Some commonly used electronic symbols are shown approximately full-size, with their letter designations, in Fig. 22.2.

Graphical symbols should be drawn in proportion to one another and with clear details so that the significance will be unmistakable. Symbol templates are widely used for drawing schematic diagrams. Printed designs on adhesive sheets (appliqués) are also employed, particularly for ink work and drawings prepared for publication. These drafting aids save time, avoid tedious construction of individual symbols, and promote uniformity. CADD systems make use of extensive symbology in the form of patterns. These patterns are created with a central origin and with gravity points to show where wires may be connected to components.

22.5 *Reference Designations*

Component designations, such as R1, C3, etc., are added to each symbol to indicate the class of component and its position in a circuit. These designations are assigned during the development stages of a product and usually appear as component markings on the actual equipment.

Other designations or symbol references may be required in a drawing to show the part value, type, electrical rating, or specific characteristics of a component. Examples are 15K, for a 15,000-Ω (ohms) resistor, where K = 1000; 6.8MH (or mH.), for a coil with an inductance of 6.8 millihenrys; 2/400, as applied to a 2-μF (microfarad—MF) capacitor with a maximum working rating of 400 V (volts) d.c.; and 5–25, as applied to the range in micromicrofarads ($\mu\mu$F, or MMF) of an adjustable capacitor.

The following examples are with reference to Fig. 22.2.

ITEM 2. The letters CW indicate the direction of the slider when the shaft is rotated clockwise. An arrowhead may be used instead.

ITEM 3. The plus sign indicates a polarized capacitor, such as an electrolytic type.

ITEM 4. The arrowhead signifies variability.

ITEM 5. Parallel lines indicate a magnetic core.

ITEM 11. Broken lines signify a shield or screen, either electrical or magnetic, or a mechanical enclosure. A broken line that connects two or more symbols also denotes a mechanical linkage

ITEM 12. Crossing lines that are not connected are drawn straight; a loop or saddle at the crossover point is not necessary. The junction dot is optional and may be eliminated provided the junction is obvious. Connections to a ground point or to a chassis or frame are distinguished by different symbols.

ITEM 15. Grids in electron tubes are numbered starting with the one nearest the cathode. A suppressor grid (No. 3) may be connected to the cathode internally, as shown, or to a separate pin in the tube base.

ITEM 16. Gaseous tubes are distinguished from high-vacuum devices by a solid dot conveniently located in the symbol, as shown.

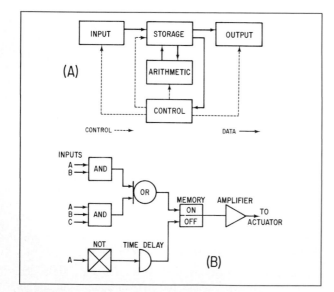

FIG. 22.3 Typical block diagrams: (A) computer system; (B) control system with special symbols. [*From G. Shiers, Electronic Drafting (Englewood Cliffs, N.J.: Prentice-Hall, Inc.)*]

ITEM 19. For *PNP*-type transistors the arrowhead on the emitter is reversed.

ITEM 21. Conductive type of photocell employing a semiconductor. "Lambda" signifies light sensitivity.

ITEM 22. The letter T denotes temperature sensitivity.

ITEM 24. Meter function is shown by an appropriate letter designation.

22.6 Data Sheets

Manufacturers' data sheets, technical bulletins, and product manuals contain full particulars of each class of device. These publications, such as tube, diode, transistor, and rectifier handbooks, should be consulted for basing diagrams, lead arrangements, socket and terminal connections, polarities, part codings, and the like.

22.7 Block Diagrams

Rectangular and other symbols in block diagrams may represent integral parts of a system, self-contained units, complete circuits, or single functional stages, according to the purpose of the drawing. The blocks are customarily located and joined with flow lines so that the overall progression is from left to right. Auxiliary items, such as power sources, are placed below the major blocks. In general, all blocks should be drawn the same size, with other symbols in proportion, and arranged in an orderly manner with equal spacings between adjacent blocks and flow lines.

The lines to each block also follow a similar sequence: input on the left, output on the right, with auxiliary lines usually at the bottom. In more complex drawings, where the flow paths may be somewhat circulatory, the symbols should be located to preserve the most logical sequence in the clearest possible manner with a minimum of crossovers.

Two representative block diagrams are shown in Fig. 22.3. In the computer system diagram (A) the blocks are drawn large enough to accommodate the lettering. The paths for control signals and data are distinguished by different line symbols, and a legend is furnished accordingly. The flow directions are indicated by arrowheads, which may be placed either on or beside the lines.

The control system diagram (B) includes some special block-type symbols, also some lettering adjacent to the symbols. Graphical symbols may be used in block diagrams, particularly for auxiliary items, such as antennas, grounds, switches, and various input and output devices.

22.8 Schematic Diagrams

A schematic diagram shows the electrical functions of a circuit without regard to the physical layout of the actual parts. Circuit components, wiring devices, and the interconnections are represented in a schematic by graphical symbols. Suitable part designations and other lettering

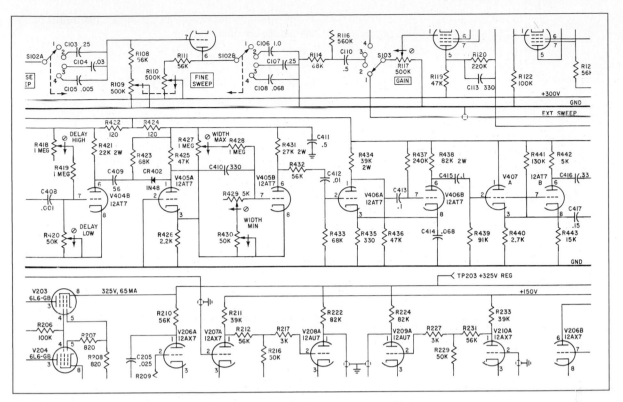

FIG. 22.4 Part of a complex drawing showing a typical schematic pattern. [*From G. Shiers, Electronic Drafting (Englewood Cliffs, N.J.: Prentice-Hall, Inc.)*]

must be furnished for reference. These diagrams are used in designing and developing circuits, in manufacturing and testing a product, and for installing and servicing complete equipment.

Part of a complex schematic, shown in Fig. 22.4, illustrates how a large amount of circuit data can be presented in an orderly manner. In laying out this kind of diagram, care is required during the planning stages to ensure proper distribution of the parts and to avoid undue crowding, wasted space, or excessive crossovers. The finished diagram should show the circuit in a logical sequence proceeding from left to right, as for block diagrams.

22.9 *Schematic Projection*

Most electronic circuits consist of several functional blocks, or stages. Each stage normally contains an active device, such as a tube or transistor, and the associated circuit components. Usually, stages are connected in cascade, whereby the output from one stage is fed to the input of the succeeding stage. In general, the symbols representing a single stage are grouped together. Each stage requires the four following basic connections; signal input and output, power supply line, and common line.

Using horizontal projection, stages are located between the power bus line and the common line, as shown in Figs. 22.4 and 22.6. When the vertical method of projection is used, the power bus is omitted and the circuit layout is arranged somewhat differently with the power supply lines located vertically between the stages, as shown in Fig. 22.5. In both methods the diagram may be

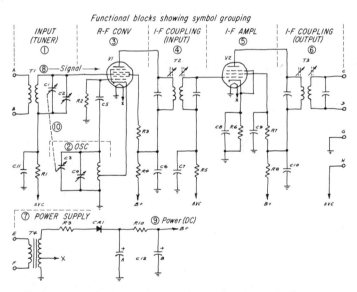

FIG. 22.5 Typical engineering sketch, marked up to show symbol grouping. [*From G. Shiers, Electronic Drafting (Englewood Cliffs, N.J.: Prentice-Hall, Inc.)*]

complete, showing all bus lines and common circuit paths (Fig. 22.6); or these lines may be isolated, with terminating arrowheads and designations to show the respective connections (Fig. 22.5). As a rule, horizontal projection is commonly used for military and some industrial applications, whereas the vertical projection method is favored for most commercial and radio communication equipment.

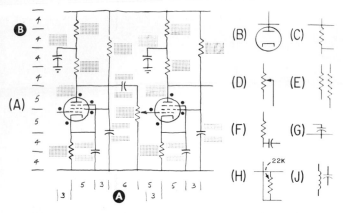

FIG. 22.6 Trial layout: (A) preliminary layout for a two-stage amplifier showing typical spacings; (B) through (J), layout errors to be avoided. [*From G. Shiers*, Electronic Drafting (*Englewood Cliffs, N.J.: Prentice-Hall, Inc.*)]

22.10 Circuit Analysis

In the early stages of electronic design suitable components are selected and connected to achieve a certain result. An engineering sketch, such as Fig. 22.5, is drawn to show the proposed circuit. Component types and values may be given in this sketch or in a separate parts list. With the parts mounted on a temporary base, or "breadboard," the circuit is then tested and modified as may be required. As soon as the circuit features have been established a finished schematic is drawn from the engineering data.

The first step in laying out a schematic diagram is to analyze the circuit sequence and electrical details so that the technical requirements and data are clearly understood. Major elements common to all schematic diagrams that should be recognized are (1) graphical symbols; (2) symbol grouping; (3) flow paths for signals and power; (4) operating functions; and (5) reference designations, electrical values, and similar identities.

It may be helpful to mark up a sketch temporarily in order to show the preferred symbol grouping, as indicated by the broken lines in Fig. 22.5. In this diagram the functional blocks are marked ① – ⑦ . Flow paths for the signal ⑧ and the operating power ⑨ are shown, also mechanical linkage ⑩ between the variable capacitors, which actually consist of two units in a common assembly.

Symbols may be divided where they refer to a multifunction device, such as a dualtriode, multicircuit switch, or a multiple-contact relay. Symbols for each part of a device, with suitable references, are placed in the respective circuit paths. Clarity is improved, and drafting time is saved by the elimination of long lines and the reduction of crossovers.

Figure 22.4 shows the use of split symbols where each half of tubes V405, 406, and 407 are drawn apart to allow space for the interstage circuits. These sections may be located anywhere in a diagram, as for V206 in the bottom circuit. This diagram also shows the omission of tube heaters, again in the interest of simplicity and clarity. Conventionally, tube heaters and associated items, such

as low-voltage pilot lamps, are drawn as subcircuits below the main part of a diagram.

22.11 Modular Layouts

Because of the abstract nature of the data and the "dimensionless" form, schematic diagrams are best laid out according to a modular plan. Cross-section or quadrille paper is suitable for developing most diagrams. Symbol sizes, symbol locations, and the needed drawing areas can be estimated or specified easily and accurately in terms of grid units with this method. If necessary, a large diagram may be broken down into sections containing several stages. Each section may then be developed separately in trial layouts, revised, and fitted together ready for the final drawing.

22.12 Trial Layouts

Figure 22.6(A) is an example of a trial layout, with typical spacings in grid units marked on each axis. The shaded areas represent spaces for required lettering. The solid dots denote tube pin numbers. Symbols should be aligned on common grid lines but staggered if required to permit adequate lettering. Common layout faults to be avoided are also shown. These consist of lines too close to symbols (B), (C), (D), and (G); symbols too close (E), (F), and (J); inconsistent symbols (E); insufficient lettering space (H); and symbols off-center (J).

The grid units shown in Fig. 22.6 represent typical dimensions suitable for most schematic layouts. Thus, lengthwise: four-six units for a capacitor, eight units for a resistor or inductor, and nine or ten units between the plate and cathode connections of a tube, or between the emitter and collector connections of a transistor. Horizontal spacings will depend on the specific circuits, symbol arrangements, amount of lettering, and the method of projection.

Sufficient space should be allotted for auxiliary circuits, such as filters, and for switches, terminal blocks, connectors, and the related lettering. If the lettering size corresponds with the grid scale it will be in proportion to the symbols. Complex diagrams containing several hundred symbols, or other diagrams designed for a compact space, as in equipment manuals, can be drawn to a smaller scale, such as $\frac{1}{10}$ in., with lettering to correspond. Using a modular plan and trial layout as described, a finished diagram may be drawn to an enlarged or reduced scale by selecting a suitable grid background.

22.13 Placement of Designations

The parts in a schematic diagram have to be identified for descriptive purposes, cross references, circuit testing, and parts replacement. Numbers should be assigned in sequence from left to right and top to bottom. Lettering should be placed close to and, where possible, centered on the respective symbol.

Typical forms of reference designations and their locations are shown in Fig. 22.7. Several conventions are (A) part designation, R41; ohms value, 1800; wattage rating, 10; type, wire-wound (WW); (B) terminal numbers; (C) lead colors, T3; (E) terminal markings, C29; test point voltage, TP3; relay K12 coil resistance, 500 Ω.

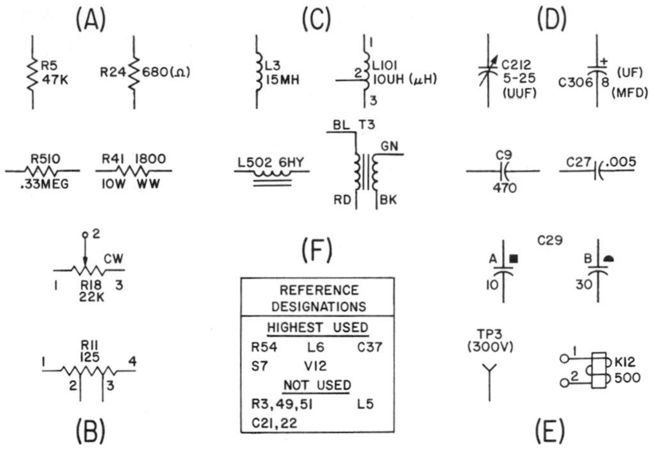

22.14 *Drawing Notes*

Repetitious lettering in schematic diagrams may be avoided if items with common characteristics are explained in the general notes. Typical examples are

1. All resistors are half-watt except where specified.

2. Capacitor values given decimally are in microfarads; all others are in micromicrofarads.

3. All tolerances are ± 10%, except where specified.

4. Components marked with an asterisk (*) apply only to Model No. *XYZ*.

Other significant data, such as control settings, meter types, test conditions, and calibration procedures, are generally included in these notes, along with explanation of special symbols or critical values. Test points, test voltages, and waveform pictures may be given as supplementary references in the body of the diagram. A reference table [see Fig. 22.7(F)], is generally furnished for the more complex diagrams. Such a table shows the extent of part numbering and any numbers in a sequence that are not used or that have been deleted through revisions. Since lettering is an important part of symbolic diagrams, all notations should be carefully compiled and checked to ensure clarity and completeness and to avoid ambiguity.

22.15 *Industrial Schematics*

Because most industrial control circuits contain electrical devices that frequently serve specialized functions, these diagrams may have to be laid out somewhat differently from the more usual radio, television, or instrument schematics. Typical examples of the different symbols required for industrial diagrams are those that denote circuit breakers, solenoids and actuators, timers, rotary machines, power tubes, relays, contactors, and special-purpose switches. The approved graphical symbols for these devices (some of which are alternative forms) are given in the respective standards. A few of these symbols are shown full-size in the numbered blocks in Fig. 22.8.

Several variables, particularly time, motion, and sequence, are involved in most industrial processes. Other factors, such as heat, light, color, weight, size, thickness, and alignment, may enter into the control or operation of equipment and the treatment of materials. Consequently, industrial schematics, or elementary diagrams, as they are also called, should be laid out in a logical sequence according to the functions of the individual circuits.

FIG. 22.7 Reference designations. [*From G. Shiers, Electronic Drafting (Englewood Cliffs, N.J.: Prentice-Hall, Inc.)*]

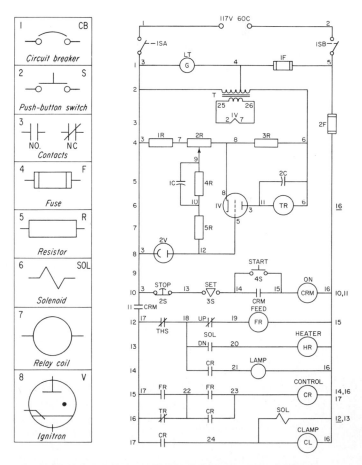

FIG. 22.8 Industrial-type graphical symbols and a typical elementary diagram of an industrial control circuit.

22.16 Control Circuit Layouts

The type of circuit format commonly used for industrial diagrams is shown in Fig. 22.8. Because of the layout, with individual circuits drawn between vertical supply lines, this is sometimes referred to as a *ladder diagram*. The power source is placed at the top with the circuits that follow located more or less in the order in which the operating sequence progresses.

As will be seen, contactor and relay coils and the associated contacts are separated, and the symbols for them are drawn in the circuits where they functionally belong. Circuit lines are evenly spaced in the interest of neatness and clarity. Similarly, symbols are evenly located and staggered to permit lettering.

Usually, all control devices are connected to the left-hand bus, while operating coils, lamps, solenoids, etc., are connected to the right-hand common line. In some layouts, the control circuits and the power circuits may be separated and drawn as two diagrams on the same sheet. Industrial diagram layouts may be prepared according to the modular principles previously described in Sec. 22.11.

22.17 Industrial Designations

Although basically serving the same purpose, the method of designating components in industrial diagrams differs somewhat from communications practice. Device designations are comprised of a standard letter or letter sequence that denotes the part and it particular function. Thus, with reference to Fig. 22.8: TR—timing relay; CRM—control relay master; FR—feed relay; CR—control relay; and SOL—solenoid. Other variations to be found include ET—electron tube; P—potentiometer; RH—rheostat; and SS—selector switch. Two or more similar devices are differentiated by prefix numbers, for example, 1V, 2V, 1R, and 2R. Component types and values may be given within a symbol, for example, 1200 for a resistor value in ohms; or G for a lamp with a green jewel.

The positions of component parts in a diagram are given with reference to the circuit level, which is identified by a number on the left-hand side. Thus, the solenoid on level 16 in Fig. 22.8 has associated contacts, which are shown on levels 12 and 13. The UP contact is normally closed, as denoted by the slant line in the symbol, and by the underlined reference number. The other numbers adjacent to the wire lines identify individual wires. Corresponding numbers are marked on the wires in the equipment, as discussed in Sec. 22.20.

22.18 Wiring Diagrams

A drawing that shows the wiring between components is known as a connection, or wiring diagram. A similar drawing showing the external connections between self-contained units is referred to as an interconnection diagram. These drawings are used for original wiring and assembly, circuit tracing, equipment installation, testing, and maintenance.

The best method of presenting wiring and connection data depends on the wiring method, type of equipment, and the purpose of the drawing. The most common types of connection diagrams are presented in Fig. 22.9. Separate line diagrams are not needed for all equipment; sometimes a wiring list provides the necessary information. Again, the wiring may be superimposed on a plan view of the assembly, or it may be shown in a pictorial view (Fig. 22.1).

22.19 Wires and Cables

Particulars concerning wire types and sizes are frequently included with connection data. Typical examples of wires and cables with the respective graphical symbols are shown in Fig. 22.10.

Stranded conductors are used for most chassis wiring. The wire size may be specified by the number of strands and the strand size or by the equivalent American Wire Gauge. For example, 7/30 (22) signifies a conductor with seven strands of No. 30 gauge equivalent in cross-sectional area to a solid wire, size AWG 22. These conductors are usually covered with a thermoplastic insulating material, such as polyvinyl chloride (PVC), in single colors or with a colored stripe, as shown in Fig. 22.10(A).

A solid conductor of bare tinned copper (BTC) wire is generally used for rigid bus bars or for short connections (jumpers) between adjacent terminals. The other types shown in Fig. 22.10 are a twisted pair (B), shielded cable (C), multiconductor cable (D), and a coaxial cable (E). The outer metallic braids serve as shields to minimize radiation or interference.

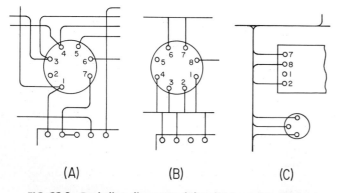

(A)   (B)   (C)

FIG. 22.9 Basic line diagrams: (A) point to point; (B) base line; (C) trunk-line or highway. [*From G. Shiers*, Electronic Drafting (*Englewood Cliffs, N.J.: Prentice-Hall, Inc.*)]

22.20 *Wire Coding*

Wire markings are necessary for original wiring inspection, testing, and troubleshooting. These markings may consist of insulation colors or number-letter sequences stamped on or attached to the wires. Such identities are given by code in line diagrams. Wire grades may be designated by a code letter that denotes type of insulation, voltage rating, temperature limits, etc.

The standard color code universally employed for electronic applications consists of color markings and their numerical equivalents. This code, with the two-letter abbreviations, is as follows:

0, black (BK);

1, brown (BN);

2, red (RD);

3, orange (OR);

4, yellow (YL);

5, green (GN);

6, blue (BL);

7, violet (VI);

8, gray (GY);

9, white (WH).

A single-letter code is sometimes used, as shown in Fig. 22.13(B).

Wire-coding methods are based on the need to identify specific wires; circuit function; wire size, type, and number; wire destination; and terminal connection. In the case of multicolored wires, for example, a code such as 14–22–A would signify a brown wire (1) with a yellow tracer (4), size 22, segment A. A two-section code showing the wire destination and the wire details, such as R7/3–C20/6, would signify a blue wire (6), size 20, type C, connected to terminal 3 on resistor R7.

22.21 *Wiring Methods*

Basic methods for interconnecting components within a selfcontained assembly are as follows: direct or point-to-point wiring; grouped or cableform wiring; and printed circuits. These methods may be combined in a single assembly.

An example of point-to-point wiring will be seen in Fig. 22.1, where small parts are suspended by their leads and other wires take the shortest route between terminals. This type of wiring, which can be visually traced quite easily, is suitable for simple assemblies and special circuits that operate at high frequencies.

In other applications, particularly military, industrial control, scientific research, and high-grade commercial equipment, all parts must be individually mounted with the wiring properly secured. Usually, small parts are grouped together on component boards, with the wires tied in bundles or cableforms. These wiring assemblies, also called *cable harnesses*, are usually prefabricated on a wiring jig. Wire markings are essential in order to distinguish the wires and trace the connections.

22.22 *Connection Diagram Layouts*

The component parts in a connection diagram are represented in simplified outline by blocks or pictorial symbols. Sufficient detail should be included to enable the reader to identify the parts and the terminations. Graphical symbols may also be used. Although these drawings need not be done to scale, the layout should be in approximate agreement with the assembly, with similar placement of terminals, connections, and identifying features. These diagrams should be developed using a modular layout, as described in Sec. 22.11.

To reveal every component, perpendicular surfaces, such as the sides of a box-type chassis, are turned 90° and developed in the major plane. Connection diagrams therefore present a plan view of the wiring in which the turned surfaces are normally placed adjacent to the respective edges. Connections to certain components, such as multigang switches, may require a detailed view or a supplementary view of the part. If necessary, a symbol may be divided for clarity. In such instances, the sections should be suitably aligned and properly labeled, with specific references given to clarify the views.

22.23 *Point-to-point Diagrams*

In a point-to-point diagram, each wire is shown by a separate line. These diagrams show the wiring side of an assembly, such as for the back view of a control panel

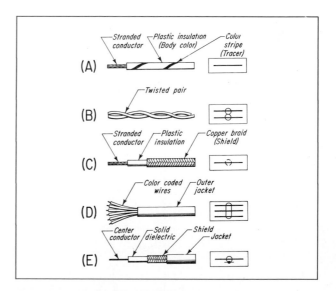

FIG. 22.10 Typical wires and cables with symbols. [*From G. Shiers*, Electronic Drafting (*Englewood Cliffs, N.J.: Prentice-Hall, Inc.*)]

presented in Fig. 22.11. The wire lines are spaced apart for clarity and to permit lettering and therefore do not necessarily represent the actual wire routing. Wires, connections, and component parts must be suitably identified. Other references should correspond with panel or chassis markings, terminal numbers or letters, and schematic designations.

22.24 Base-line Diagrams

A wiring diagram may be simplified if the wire lines from the components are terminated into common base lines. These lines, drawn horizontally and vertically between

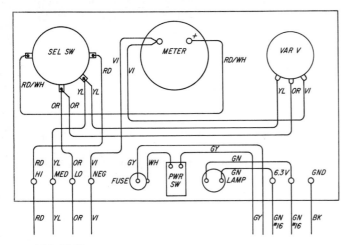

FIG. 22.11 Point-to-point wiring diagram of a panel layout. [*From G. Shiers*, Electronic Drafting (*Englewood Cliffs, N.J.: Prentice-Hall, Inc.*)]

component symbols, may or may not represent the routing of the actual wires.

Figure 22.12(A) is a base-line diagram, which partially shows the wiring between a potentiometer (R5), a tube socket (V3), a component board, a test point, and related items. The wire identities showing wire numbers and segments, for example, 5A, 5B, and so on, are inserted in the wire lines. The actual connections would be given in a wiring list, which would also specify other wire details, such as wire type, size, length, and type of terminations.

The diagrams shown in Fig. 22.12(B) refers to the terminal wiring on chassis designated "C." In this example, the wire coding includes wire size, color, and destination or address, for example, a red (R) wire, size 16, on terminal 20, is connected to terminal 9 on chassis "B." The other end of this wire would be designated C20 (R16) in the diagram and marked accordingly.

22.25 Highway Diagrams

In another type of connection diagram the feeder lines from the component terminals are merged into a single trunk line, or highway. The routing for each lead should be indicated by a slant line, as shown in Fig. 22.13(A), or by radial turnoffs, as shown in (B).

Since the wire identities are lost by grouping, a code number is required to show the destination. With reference to Fig. 22.13(A), for example, the lead connected to terminal 2 on item 2 is designated 6/E(7), which signifies a violet wire (7), which is connected at the other end to terminal E on item 6. Such items, for example, component boards, tube sockets, switches, connectors, and so on, may be arbitrarily numbered, as shown, or they may be

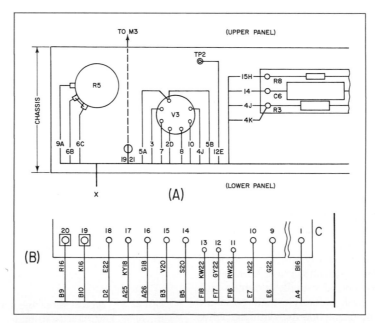

FIG. 22.12 Typical base-line diagrams: (A) partial chassis wiring; (B) terminal-strip interconnections. [*From G. Shiers*, Electronic Drafting (*Englewood Cliffs, N.J.: Prentice-Hall, Inc.*)]

designated according to the class of item using standard letter symbols (such as V5, TB2, etc.) which agree with the schematic references.

Figure 22.13(B) shows separate highways for associated wires, a method that may be preferred because it facilitates circuit tracing. Thus, highway A carries wires from left to right to terminals 30, 24, 12, 8, and 6; also wires to the right from terminals 28, 16, and 6. These (and perhaps other wires) are carried upward to other parts of the drawing, a fact that is not specifically shown in (A).

22.26 Printed Circuits

Savings in space and weight and other advantages may be realized if conventional wiring is replaced by a printed circuit. Such a circuit may be produced by depositing a conductive pattern on a dielectric base, by stamping or etching copper foil, or by special processes which may include the formation of circuit components with the conductive pattern.

A suitable grade of insulating material, such as phenolic stock with copper foil bonded to the sheet, is used for general applications. Glass, ceramic, and synthetic dielectrics are used for special purposes. Typically, the circuit pattern is reproduced on the foil by photographic means, followed by an etching process. A large-scale black-and-white drawing showing the pattern in silhouette is prepared as a master layout. A photographic negative is made from this artwork.

In the usual construction, components are mounted on the plain side of the board, opposite the etched pattern. Component leads are inserted through holes or eyelets in the conductor pattern, the connections being made either by hand soldering or by dipping the wiring side of the assembled board in a bath of molten solder.

22.27 Conductor Patterns

Conductor patterns must be compatible with various mechanical and electrical requirements, notably, size, shape, and layout of the board; circuit connections; and electrical values. Adequate contact areas at the terminal points and interconnections must be incorporated in the design. Several basic conductor shapes are shown in Fig. 22.14. Conductor patterns should be drawn solid black with clean edges, free-flowing lines, and smooth contours. Fillets should be provided at junctions and land areas.

A conductor width of .031 in. is commonly used for general-purpose boards where the current is not more than a few hundred milliamperes. The spacing between conductors depends on the working voltage; .062 in. is usual where the peak voltage difference does not exceed 300 V. Conductor locations and spacings are also governed by electrical factors, such as capacitance, intercircuit coupling, shielding, and other requirements, which may be critical in certain designs.

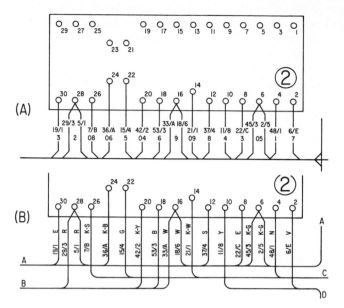

(A)

(B)

FIG. 22.13 Typical trunk-line or highway diagrams showing a component board and associated wiring: (A) single trunk; (B) separate highways. [*From G. Shiers, Electronic Drafting (Englewood Cliffs, N.J.: Prentice-Hall, Inc.)*]

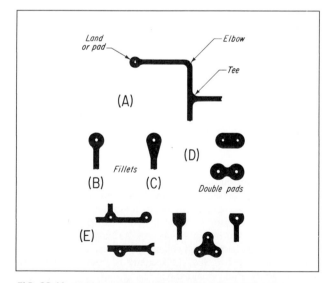

FIG. 22.14 Basic conductor shapes for printed circuits. [*From G. Shiers, Electronic Drafting (Englewood Cliffs, N.J.: Prentice-Hall, Inc.)*]

22.28 *Printed-circuit Layouts*

Quadrille or cross-section paper is ideal for sketching printed-circuit layouts. The pattern can be drawn freehand in pencil, using the grid lines, with the divisions representing a suitable scale. Terminal pads should preferably be located on standard .1-in. intersections, with increments of .025 in., according to the space requirements and the component dimensions.

At this stage the objective is to develop connecting paths as short and direct as possible without any crossovers. Component parts may have to be reoriented or interchanged to provide access for every conductor. When an obstruction is unavoidable, however, a jumper or insulated wire is used as a bridge at the crossover point. Several sketches are generally needed before a feasible pattern can be established.

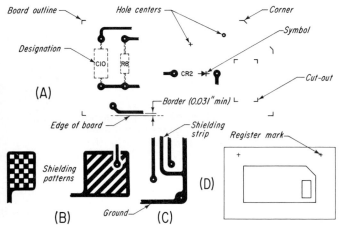

FIG. 22.15   Printed circuit artwork: (A) master layout details; (B) shielding patterns; (C) shielding strips and ground; (D) overlay and register marks. [From G. Shiers, Electronic Drafting (Englewood Cliffs, N.J.: Prentice-Hall, Inc.)]

22.29 *Printed-circuit Artwork*

Master layouts for printed-circuit work are made to an enlarged scale, which may range from 2:1 to 20:1, depending on the size of the part and the intricacy of detail. This enlargement facilitates drawing and improves the accuracy of the final full-size copy. Accurate locations of the terminals and other center points are essential because land areas, pads, edge-connector strips, and so on, must match the holes and cutouts in the board. One major dimension, placed outside the board area, should be given for reference.

Some important details of master layouts are shown in Fig. 22.15. If component designations, part symbols, and other lettering are given on the master drawing they will appear in copper on the board. Such markings, sometimes alternatively printed or silkscreened on the board, are valuable service aids. Other markings, for example, hole centers and cutouts, may be needed for experimental models, hand assemblies, and small-quantity productions.

Polyester drafting film is preferred for master layouts. The drawing may be done in black ink or with adhesive drafting aids. These aids, of appliqués, especially developed for printed-circuit work, consist of black tapes and precut matching shapes, such as elbows, tees, junctions, and pads, as shown in Fig. 22.14. In another method, the surface of a specially prepared film may be scribed and stripped to produce either a positive or negative master.

Separate drawings are required for multiple-sided boards, silkscreen lettering stencils, and for making press tools and jigs used in quantity productions. In such cases, each drawing is aligned by means of register marks, as shown in Fig. 22.15(D), to ensure satisfactory matching of the important features. Printed-circuit artwork must be accurately and skillfully prepared because, unlike other technical drawings, the qualities of the drawing will be fully reproduced in the finished product.

/ PROBLEMS

Suitable layouts for the following problems should be developed freehand on opaque cross-section paper. Finished drawings should be made on plain vellum with a cross-section paper beneath it to serve as a grid background. Templates with appropriate cutouts should be used for drawing symbols. Some of the problems should be finished with adhesive drafting aids and in India ink in order to provide acquaintance with these techniques. The publications listed in Sec. 22.3 should be consulted when selecting symbols and designations.

1. Draw a block diagram representing the circuit of the chassis shown in Fig. 22.1. Letter the blocks as follows: (1) pentagrid converter; (2) IF (intermediate-frequency) amplifier; (3) detector, AVC, audio amplifier; (4) power amplifier; and (5) half-wave rectifier. Include graphical symbols for the antenna (to block 1) and the loudspeaker (from block 4). Show block 5 as an auxiliary circuit, and include the power line.

2. Draw a schematic diagram of Fig. 22.5. Use horizontal projection and a $\frac{1}{8}$-in. grid scale.

3. Make a schematic diagram of the control panel wiring shown in Fig. 22.11.

4. Draw a schematic diagram to correspond with the parts and circuit of the chassis shown in Fig. 22.1. Use a symbol similar to item 11 of Fig. 22.2 with separate variable tuning cores for the IF transformers. The connections are as follows: blue-plate, red-B plus, green-grid, white-return, or common-line. The antenna and oscillator sections of the tuning capacitor (not shown) are ganged. The line switch and the volume control also are ganged. Use vertical projection, as shown in Fig. 22.5, and a $\frac{1}{8}$-in. grid scale.

5. Make a schematic diagram of the 400 series circuit shown in the middle level of Fig. 22.4. Show the circuit from V404B up to the grid of V407A. Renumber all symbols starting from

1 for each class. Include all designations and appropriate notes. Use horizontal projection and a $\frac{1}{10}$-in. grid scale.

6. Redraw the timing circuit (1V, 2V, from the input to level 8) shown in Fig. 22.8, using horizontal projection and a suitable scale. Show a double-wound transformer, similar to item 25 in Fig. 22.2 with two separate secondary windings, instead of T. Show a linecord terminated by a 3-pin plug, with the third pin grounded. Use zigzag resistor symbols. Capacitor designated 2C is an electrolytic type. Relay TR has two sets of contacts: single-pole, normally open; single-pole, double-throw. Include a terminal block with five terminals suitably numbered and identified for these connections.

7. The underside view of a small chassis (7 × 5¼ × 2 in.) is shown in Fig. 22.16. This unit includes a power supply consisting of power transformer T; full-wave vacuum rectifier V1, type 5U4–GB; filter choke L; triple-unit electrolytic filter capacitor C1; 5-W resistor R1; and part of the circuit shown in Fig. 22.4. This circuit contains tube V406 (V2) and the associated components from C412 to pin 2, V407. With reference to Fig. 22.4, the respective component designations are C2–C412 through C5–C415 and R2–R433 through R8–R439. Chassis connections are made via a 10-wire terminal strip (TB).

   Make a connection diagram to correspond with the circuit described, and the following connections. Use either the point-to-point or the base-line method. [*Note:* Component board connections are denoted thus: (T)–top, (B)–bottom.]

   TB1–T9–V2/4–V2/5, TB2–T10–V2/9. TB3–C2(B). TB4–T7–C4(B)–chassis lug (ground). TB5–R1B–LB–C1B. TB6–R7(T)–R1A–C1C. TB7–R8(T). TB8 (spare). TB9–T1. TB10–T2. T3–V1/8. T4–LA–C1A. T5–V1/2. T6–V1/4. T8–V1/6.

8. Make a working sketch of a schematic to correspond with the circuit given in Problem 7. Use quadrille paper with eight divisions per inch. Include all designations and part values (in addition to those shown in Fig. 22.4) as follows: Power transformer primary, 1, 2; 5-V heater winding, 3, 4 (center tap), 5; 325–0–325-V secondary, 6, 7 (center tap), 8; 6.3-V heater winding, 9, 10. L, 20H; C1A, C1B, 20 mF; C1C, 10 mF; R1, 1800 Ω, 5-W, wire-sound. Show a divided terminal block with the power line input on the left and the other connections on the right.

9. A printed-circuit board with the required land areas and strip connections is shown in Fig. 22.17. Make a master layout drawing of this board twice full size to suit a conductor width of .031 in. A minimum clearance of .050 in. should be maintained between conductors and between a conductor and the edge of the board. Land dimensions in inches are ID, .040; OD, .110. The strip connector widths of .10 in. should be maintained for .20 in. All tolerances on the conductor pattern are ±0.015 in. The connections are as follows: A–1–5–9, B–2–4–12–14, C–8–

10–19, D–17, E–13–20, 3–15–16–18, and 6–7–11. The grid pattern and land numbers are for reference only.

10. Make a working sketch of a schematic to correspond with the printed circuit given in Problem 9. The components are as follows (terminal connections are with reference to Fig. 22.17):

   R1, 9–10; R2, 5–6; R3, 16–17; R4, 14–15; R5, 1–2. C1, 7–8; C2, 3–4. Transistors are *PNP*-type; Q1, base 18, emitter 20, collector 19; Q2, base 11, emitter 13, collector 12. Label the strip connections as follows: *A*, negative 15 *V*; *B*, output; *C*, trigger; *D*, positive; *E*, ground.

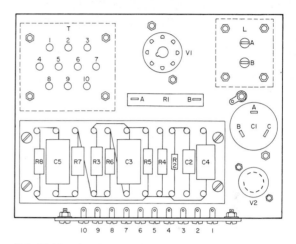

FIG. 22.16 Chassis assembly.

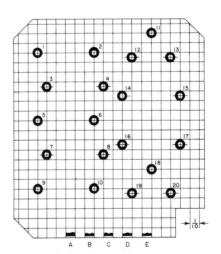

FIG. 22.17 Printed circuit board.

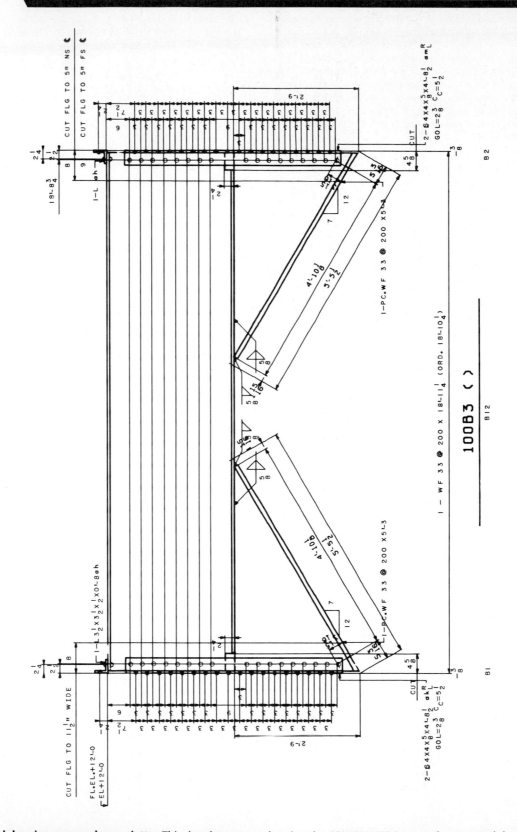

Structural drawing prepared on a plotter. This drawing was made using the CONSTRUCTS System of automated drafting. See Section 25.15. (*Courtesy Control Data Corporation*)

Structural
Drawings*

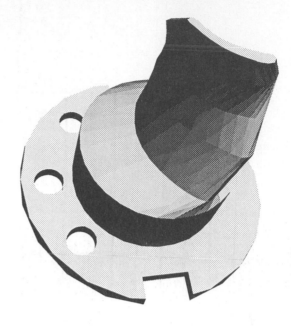

23.1

Although structural drawings are prepared in accordance with the general principles of projection, they differ somewhat from machine drawings in certain practices. These differences, which have gradually developed due to the type of raw materials used and methods of fabrication, have become established drawing-room customs that are recognized universally throughout industry and must be understood and adhered to by every prospective structural engineer.

Steel structures vary widely and include almost everything fabricated from rolled shapes and plates.

Fabrication consists of shearing, flame cutting, punching, bending, forging, and machining, then fitting and aligning the parts, and finally permanently fastening the assembly by bolting, riveting, or welding. Although small roof trusses and girder bridges may be assembled as complete units in the shop, the size of most structures makes necessary the fabrication of subassemblies and shipment in knock-down form.

Sections of the principal shapes (angles, I-beams, channels, wide-flange sections, and plates) are shown in Fig. 23.1.

The dimensions of the various standard shapes and other available information required by a structural detailer are given in structural steel handbooks published by different

---

* Prepared in collaboration with R. S. Green, Ohio State University, and M. T. Ward, American Bridge Company.

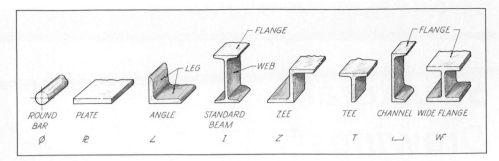

FIG. 23.1 Structural shapes.

manufacturers and by the American Institute of Steel Construction.

23.2 Equipment of a Structural Drafter

The equipment needed by a structural drafter is the same as for any other line of industrial drafting, with a few additions. Tables of trigonometric functions and logarithms and tables of slopes and rises are a necessity. A copy of the structural handbook published by the American Institute of Steel Construction must be readily available.

Some companies furnish each drafter with a book that gives drawing-room standards. Included are typical drawings illustrating the arrangement of views, approved dimensioning practices, and notes for various types of structures. Some information also may be given about plant equipment.

It is suggested that the structural draftsman in training obtain a copy of the latest edition of the AISC text, *Structural Steel Detailing*. Mastery of the instructions in this excellent text should qualify one who is starting in this field to do acceptable work in structural detailing with any structural steel fabricator.*

A CLASSES OF STRUCTURAL DRAWINGS

23.3 Classes of Structural Drawings

Most of the large steel fabricators maintain a design office and a detailing office. The former prepares design drawings and estimates costs in the preparation of bids and frequently serves in a consulting capacity on designs furnished by a customer. The detailing office, which is usually located at the fabricating plant, orders material and prepares shop and erection plans from the design sheets.

A set of specifications covering special conditions, unit stresses, materials to be used, and so forth, is considered as part of the design information.

---

* *Structural Steel Detailing*, New York: American Institute of Steel Construction, Inc.

Shop detail drawings show all of the information necessary for shop fabrication.

Erection plans, which are prepared primarily for use in the field, consist of line diagrams giving dimensions, shipping marks, and notes in sufficient detail to guide the erector in assembling the parts to complete the finished structure.

23.4 Design Drawings

Design drawings usually are line diagrams showing the shape of a structure, the principal dimensions, structural sections, and in some cases the stresses to be used in detailing the connections. Figure 23.2 is an example of a design drawing.

For the use of the layout people a set of design drawings may contain elaborate design details, showing the type of connections, thickness of gusset plates, and the number of rivets. Figure 23.3 is the design detail for joint L_0 of the truss shown in Fig. 23.2. Design drawings of this type, however, are furnished only for important or complicated structures. Quite often the design of connections becomes a detailer's task.

A design drawing of a light industrial building is shown in Fig. 23.4. On this drawing the designer has given the detailer the information necessary to prepare the shop drawings for the structural frame. The composite plan view shows the top and bottom chord bracing and the braced bays that require sway frames. Although the size of the eave struts has been indicated on the plan, their location is given on the typical wall section. The size and location of the purlins and girts have been shown. It should be noted that sag rods have been used to align the purlins and girts.

The cross section, shown at the far right, has been taken in the direction of the 60-ft width of the structure. This view shows the sizes of the columns and truss components. It may be noted that the designer has given the axial tension and compression stresses in the truss and knee braces. This is information needed by the detailer for developing adequate connections. Two sets of stresses have been indicated: those produced by gravity (vertical) loads and those that might result from wind. Wind stresses have been preceded by a ± to indicate either tension or compression, since the wind may blow from either direction against the building. The listing of stresses, as was done

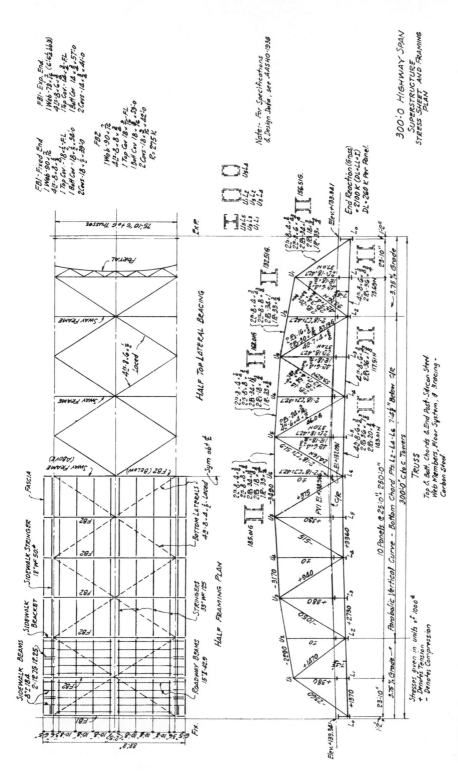

FIG. 23.2 Stress sheet.

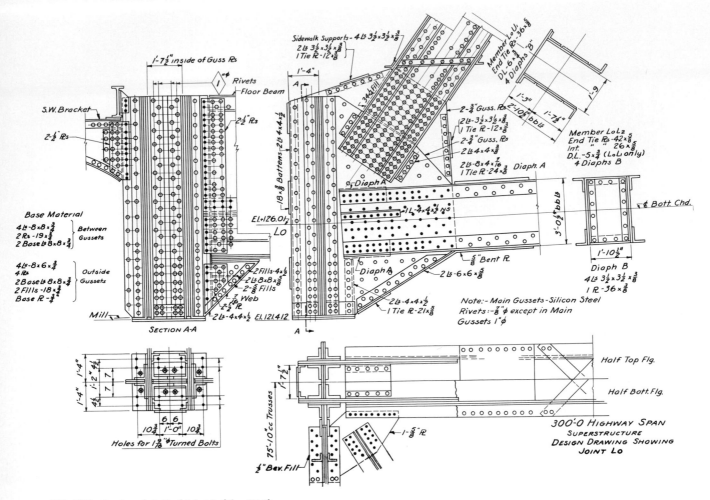

FIG. 23.3 **Design detail of joint *L*₀ (Fig. 23.2).**

in Fig. 23.4, is important because the detailer must know if any of the double-angle truss members will be subject to both tension and compression.

Design drawings of trusses should show those dimensions that are needed to establish the necessary working points. However, the exact position of working points for diagonal bracing is sometimes left for the detailer to determine.

23.5 Layouts

The first step in the development of structural steel detail drawings is the drawing of the layout sheets. These are intermediate drawings that are used only in the drafting room of the fabricating shop. Layouts are used for ordering material, obtaining early approval of details, and coordinating the work of the several draftsmen who may be employed on the project. As a general rule, layouts are made only when the complexity of the work demands a carefully scaled picture. For example, layouts would be necessary for all types of skewed work and for truss joints. Layouts may be drawn to any appropriate scale, but usually the detail drawing and layout scales are the same. Layouts

are not completely dimensioned, but the layout person may indicate any dimensions to be used.

Figure 23.5 is a layout of the cross frame of a deck-girder bridge. The detailer takes layouts of this kind, in addition to the design drawings for the project, and completely details the structure.

23.6 Detail Drawings

The making of the detail drawing is the final step in the process of creating structural steel working drawings (Fig. 23.6). These drawings must be clear and concise, to enable persons in the shop to do their portion of the work efficiently. To ensure accuracy, a thorough check of all arrangements and dimensions is made in the drafting room by a checker.

Parts to be riveted together in the shop are detailed in their assembled positions in the structure, instead of being detailed individually, as is the practice for machine work. Figure 23.7 is a detail drawing of a cross frame. It describes each plate and main member and shows the relations of the various elements of the structure to one another. When the structure is too large to be completely

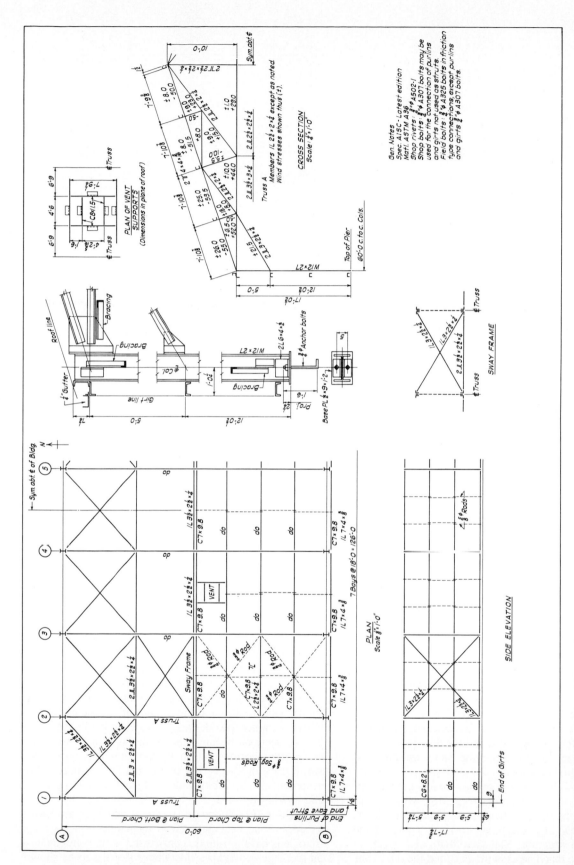

FIG. 23.4 Design drawing of an industrial building. (Courtesy American Institute of Steel Construction, Inc.)

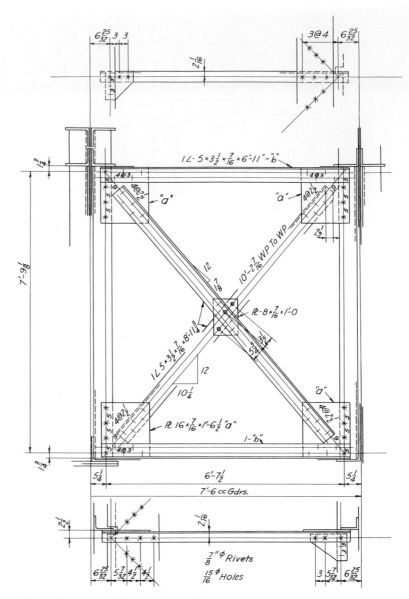

FIG. 23.5 Layout for internal cross frame.

assembled in the shop and shipped in one piece, an assembly or erection diagram becomes necessary.

The scales in general use are not large enough to permit direct scaling of dimensions. One of the scales most commonly used in structural work is $\frac{3}{4}$ in. = 1 ft. Often structural members are too long to be drawn to scale and yet be contained on the sheet of drawing paper. In this event, the transverse dimensions and the details are drawn to one scale and the longitudinal dimensions are shortened or drawn to a smaller scale.

Structural steel drawings incorporate a few practices of projection that differ from other types of work. In all structural work the view of the structure that corresponds to the front view in a machine drawing is termed the *elevation* of the structure. The view that corresponds to the top view in a machine drawing is called the *plan view* of the structure. The view below the cross frame in Fig.

23.7 is a plan view of the lower-chord member and is not a bottom view. Likewise, the two views at the bottom of the sheet in Fig. 23.3 are not bottom views of the joint but are sectional views looking downward. The left view shows the location of the anchor bolts and the right view shows both the top and the bottom of the member L_0–L_1 where it frames into the joint. Shopmen prefer the use of bottom sections because the elevation of a piece, while being fitted, corresponds to the elevation on the drawings. Any fitting or inspection of the bottom is more easily accomplished by looking down than by crawling under and looking upward.

Angular dimensions on structural drawings are shown as slopes that are expressed in inches per foot. A slope triangle is a right triangle constructed with its hypotenuse on the gage or working line whose slope is to be shown. The longer of the two legs is always given as 12 in., and

FIG. 23.6 **Detailer at work.** (*Courtesy American Institute of Steel Construction, Inc.*)

the length of the shorter leg determines the slope. These slope triangles are not drawn to scale but are constructed in any convenient size (Fig. 23.7).

Structural drawings are currently made in pencil on translucent bond paper. Pencil work requires particular care on the part of the drafter to make the lines black enough to blueprint clearly. The outlines of the structural members should be wider than the gage lines and dimension lines because the contrast makes the blueprints easier to read.

Pencil tracings on special cloth or ink tracings on cloth are required by certain customers.

Even when roof trusses are to be transported to the site in more than one piece, the several component sections are usually detailed in their assembled position. Although the truss shown in Fig. 23.8 might not be too large to be shipped as a single piece, the details, as prepared, show the usual separation that would be used for the fabrication and shipment of a large truss.

23.7 Procedure for Making a Layout of a Gusset Plate

The general procedure for making the layout of a gusset plate that connects two members of a roof truss to the bottom chord member is given in Fig. 23.9. Each member is composed of two angles.

STEP 1. Calculate the slopes of the diagonal members and draw in the working lines (Fig. 23.9). Use a scale of $1\frac{1}{2}$ in. = 1 ft or 3 in. = 1 ft. (Ordinarily the working lines will be gage lines.)

STEP 2. Determine the correct dimensions and gages (Fig. 23.10) and draw in the lines representing the outstanding and perpendicular legs of the angles on the proper sides of the working lines. Draw in the clearance line at a preferred distance above the bottom chord angle and cut the diagonal angles perpendicular to their axes so that the corners fall on the clearance line.

STEP 3. Locate an initial rivet in each diagonal at the desired edge distance from the sheared edge of the member. Since it is customary in structural drawing to give the distance from the working point to the first rivet to the nearest $\frac{1}{4}$ in., it usually will be found necessary to change either the edge distance or the clearance to meet this requirement. Locate the remaining rivets in each diagonal, using minimum spacing (3 × diameter of rivet), so that the plate will not be larger than is absolutely necessary.

STEP 4. Draw in the edges of the gusset plate, after giving some consideration to the factors involved in an economical treatment of the design. The points to be observed are as follows: (1) Allow not less than minimum edge distance *e* from the center of each rivet to the nearest edge of the plate (Fig. 23.11). (2) Allow no corners of the plate to project beyond the angle. (3) Design the plate so that there will be a minimum number of cuts, for each cut increases the labor cost. (4) Make at least two edges parallel at a distance apart equal to a standard

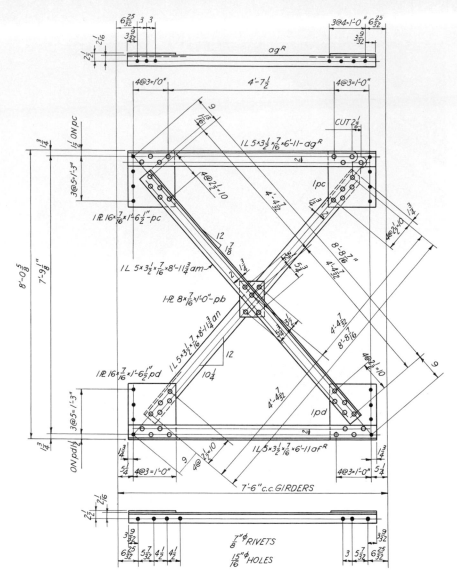

FIG. 23.7 Detail drawing of a cross frame.

plate width, so that unnecessary cuts and material waste may be avoided.

The shorter dimension usually is considered to be the width and the longer dimension, the length. If the longer dimension is across the plate between parallel sides, however, it should be given as the width, because the plate may be cut from a long plate of that width. The length dimension usually is given to the nearest $\frac{1}{4}$ in. The dimensions of a plate are always those of a rectangle from which the plate may be cut.

DIMENSIONING PRACTICES*

23.8 Structural Notations

    ′ = foot or feet
    ″ = inch or inches
    φ = diameter

    # = pound or pounds
    ∠ = angle
    I = I-beam
    ⊔ = channel
    W̄ = wide-flange section

23.9 Sizes of Standard Members

The following structural specifications and abbreviations are those adopted by the American Institute of Steel Construction.

    Plates. Width (in inches) × thickness × length. (*Pl* $15 \times \frac{3}{8} \times 1'$–10.) If it is a connection plate on a truss, cross frame, and so on, which is fabricated in the shop, the specification will be followed by the letters *pa, pb, pc,* or *pd,* and so on, which indicate the location of the plate. (*Pl* $15 \times \frac{3}{8} \times 1'$–10 *pa.*)

* See Appendix A Glossary of Structural Drafting Terms.

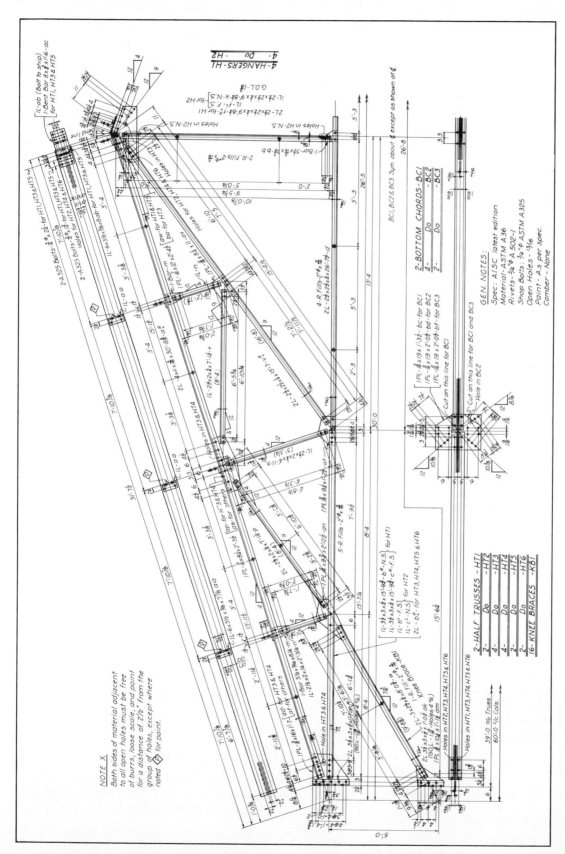

FIG. 23.8 Riveted truss. (Courtesy American Institute of Steel Construction, Inc.)

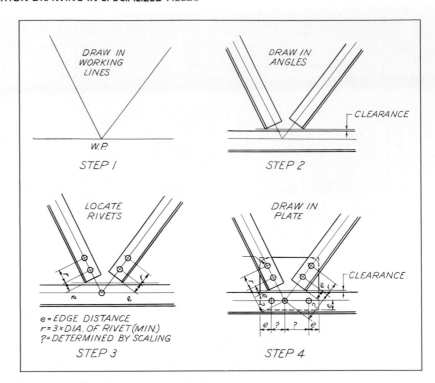

FIG. 23.9 Steps for making layout of gusset.

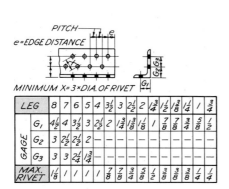

FIG. 23.10 Gage distances.

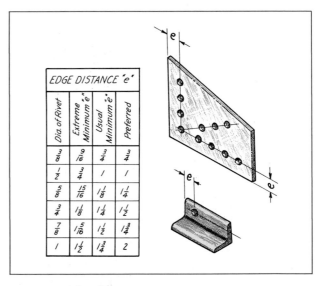

FIG. 23.11 Edge distance e.

Angles—equal legs. Size of leg × size of leg × thickness × length. (∠ $3\frac{1}{2}$ × $3\frac{1}{2}$ × $\frac{3}{8}$ × 7′–6.)

Angles—unequal legs. Size of leg shown × size of outstanding leg × thickness × length. (∠ 4 × 3 × $\frac{5}{16}$ × 24′–7.)

I-Beams. Depth of I-weight per foot × length. (12 I 31.8 × 15′–4.)

Channels. Depth of ⊔-weight per foot × length. 10 ⊔ 15.3 × 15′–10.)

Wide-flange sections. Depth of WF weight per foot × length. (24 WF 74 × 12′–6.)

23.10 Detailing Information

The type of rivets and their treatment are indicated on structural drawings by the American National Standard conventional symbols shown in Fig. 23.12.

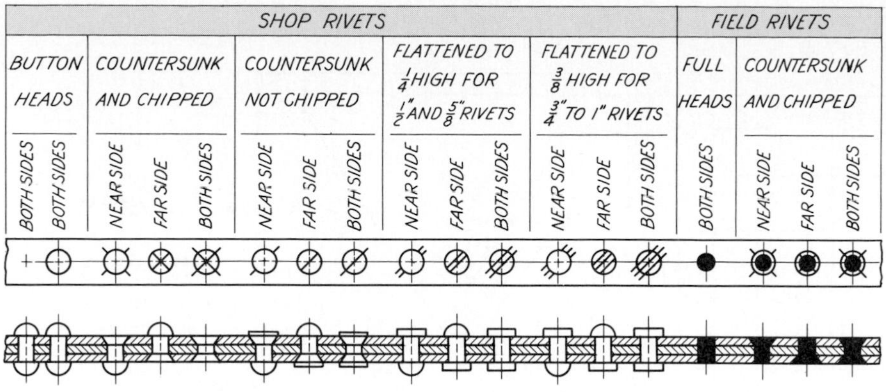

SHOP RIVETS						FIELD RIVETS	
BUTTON HEADS	COUNTERSUNK AND CHIPPED	COUNTERSUNK NOT CHIPPED	FLATTENED TO $\frac{1}{4}$ HIGH FOR $\frac{1}{2}$ AND $\frac{5}{8}$ RIVETS	FLATTENED TO $\frac{3}{8}$ HIGH FOR $\frac{3}{4}$ TO 1" RIVETS		FULL HEADS	COUNTERSUNK AND CHIPPED
BOTH SIDES | BOTH SIDES	NEAR SIDE | FAR SIDE | BOTH SIDES	NEAR SIDE | FAR SIDE | BOTH SIDES	NEAR SIDE | FAR SIDE | BOTH SIDES	NEAR SIDE | FAR SIDE | BOTH SIDES		BOTH SIDES	NEAR SIDE | FAR SIDE | BOTH SIDES

FIG. 23.12 Conventional symbols for rivets.

The holes for field rivets are indicated in solid black on a drawing, while shop rivets are shown by open circles having the same diameter as the rivet head (Fig. 23.13). Rivets should be drawn with either a drop pen or a bow pencil. In practice, the circles representing rivets are often drawn free-hand on pencil drawings.

23.11 Location of Dimension Lines

Since shop personnel are never permitted to scale a drawing, all dimensions must be placed in such a manner that they will be easily understood. Principal dimensions are generally obtained from the design sheets; other dimensions necessary for detailing are found in tables or are determined by the detailer.

Dimensions for rivet spacing, minor location and size dimensions, and so on, are placed close to the view, while the longer dimensions, such as overall lengths, are placed farther away so that extension lines will not cross dimension lines (Figs. 23.14 and 23.15).

Dimension figures are generally placed above continuous (unbroken) dimension lines, which are made narrow and black. These lines usually should be placed off the view, but oftentimes added clarity may be obtained by putting a few dimensions in an open area on the view itself. Dimension lines ordinarily should not be placed less than $\frac{3}{8}$ in. apart or closer to the view than $\frac{1}{2}$ in. All of the above rules for the location of dimension lines may be modified to suit the available space (Fig. 23.16).

23.12 Dimensions and Notes in Structural Detailing

1. Figures may be compressed without reducing their height, in order to place them in a limited space between arrowheads.

2. Figures can be placed to one side, with a leader to the dimension line, if the available space is very small.

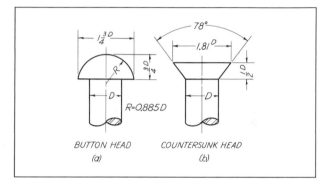

FIG. 23.13 Dimensions of rivet heads.

3. Figures and notes must read from the bottom and the right side of the sheet because shopmen are accustomed to reading from these positions (Fig. 23.14).

4. For dimensions less than 1 ft, the inch marks (") may be omitted (Fig. 23.15).

5. With the exception of widths of plates and depths of sections, all dimensions of 1 ft or more are expressed in feet and inches (Fig. 23.14).

Correct	Incorrect
$\frac{1}{4}$	$0\frac{1}{4}$
9	$0'-9''$
10	$10''$
$1'-0$	$12''$
$2'-3\frac{1}{4}$	$2'-03\frac{1}{4}''$
$4'-0\frac{1}{4}$	$4'\frac{1}{4}''$

6. Usually, dimensions for rivet spacing are given in multiples of $\frac{1}{8}$ in. or, preferably, $\frac{1}{4}$ in. It is not desirable to use multiples of $\frac{1}{16}$ or $\frac{1}{32}$ in., except in rare cases.

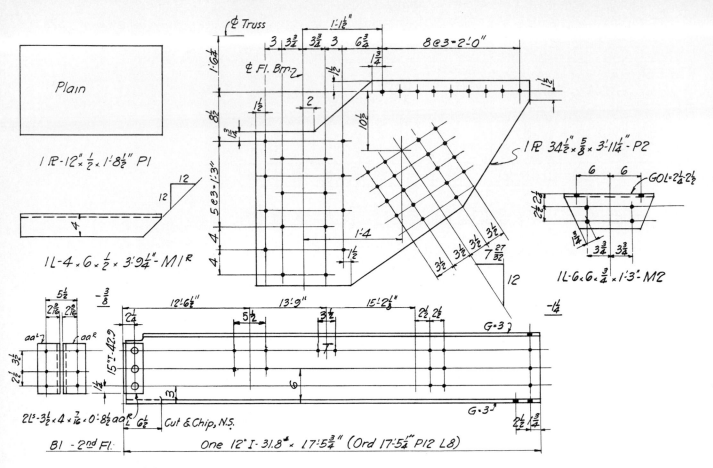

Plain

$1 \, \text{R} - 12" \times \frac{1}{2} \times 1' \cdot 8\frac{1}{2}"$ P1

$1 L \cdot 4 \times 6 \times \frac{1}{2} \times 3' \cdot 9\frac{1}{4}" - M1^R$

$1 \, \text{R} \, 34\frac{1}{2}" \times \frac{5}{8} \times 3' \cdot 11\frac{1}{4}" - P2$

$1 L \cdot 6 \times 6 \times \frac{3}{4} \times 1' \cdot 3" - M2$

$2 L^s \cdot 3\frac{1}{2} \times 4 \times \frac{7}{16} \times 0' \cdot 8\frac{1}{2} \, aa^R$

B1 - 2nd Fl.

One 12" I - 31.8# × 17'-5¾" (Ord 17'-5¼" P12 L8)

FIG. 23.14 Structural dimensioning.

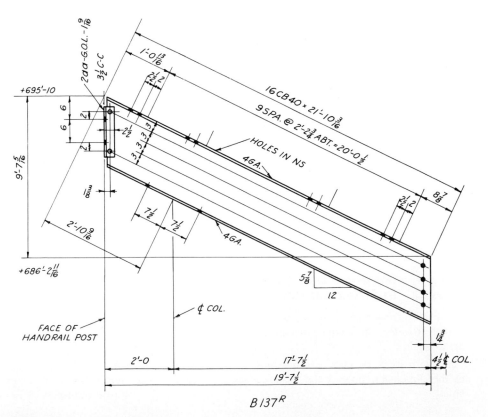

B 137^R

FIG. 23.15 Structural dimensioning. (*Courtesy American Bridge Co.*)

574

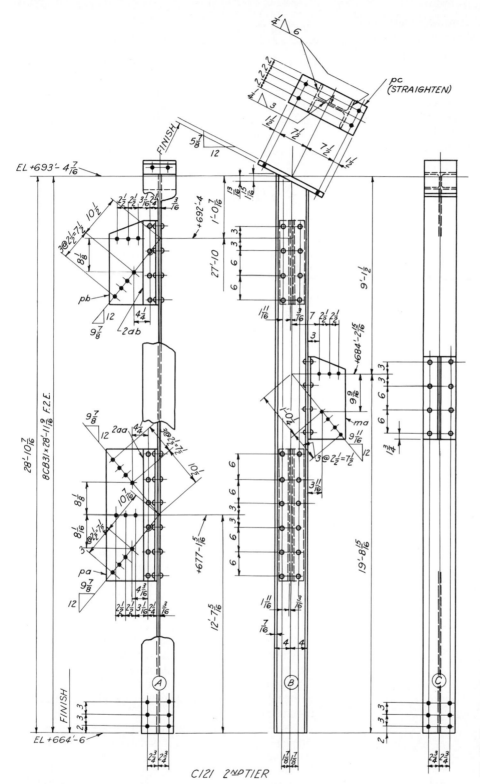

FIG. 23.16   Column details. *(Courtesy American Bridge Co.)*

C121 2ND TIER

7. Decimals found in tables should be converted to fractions to the nearest $\frac{1}{16}$ in. (except for machine-shop drawings for gears, shafts, and so forth).

8. To avoid complications that arise when corrections are made, dimensions shown on one view should not be repeated on another.

9. Rivets and holes are located by dimensions from center to center (Fig. 23.14).

10. Edge distances are frequently omitted, unless they are necessary to ensure clearances with connecting parts. (Shopworkers understand that the distances on opposite edges are to be made equal.)

11. Dimensions *always* should be given to the center lines of I-beams and to the backs of angles and channels. (See Appendix, gage definition.)

12. When three or more rivet spaces for a line of rivets are equal, they should be dimensioned as a group (4 @ 3 = 1'–0). Staggered rivets are dimensioned as though they were on one gage line (Fig. 23.14).

13. Since a workman must use a rule or tape to lay off angles, a slope triangle should be shown to give the inclination of a working line (Fig. 23.15).

14. A person in the shop should never be compelled to add or subtract to obtain a necessary dimension.

15. A general note is usually placed on a detail drawing giving painting instructions, size of rivets, size of open holes, reaming instructions, and so on.

16. Members that are shipped separately for field erection are given a shipping mark of a letter and number that appears on the detail drawing and on the erection plan (Fig. 23.14).

17. The size of a member is indicated by a specification (in the form of a note) parallel to it (Fig. 23.14).

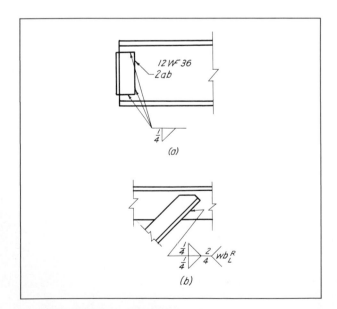

FIG. 23.17 Welded connections.

18. The width of plate is always given in inches (Fig. 23.14).

23.13 Machine-shop Work

The structural detailer will occasionally be called on to prepare detail drawings for castings for bridge shoes and roadway expansion joints and at times drawings for complicated gearing and shafting for movable bridges. Regular machine drawing practices apply, the principal difference being that bridge machinery in general is ponderous with single castings often weighing 10 or 15 tons.

23.14 Structural Welding

Since a large portion of structural work is either partially or completely welded, a working knowledge of the use of welding symbols (Chapter 14) is a requirement for a structural draftsman (Fig. 23.17).

Sizes and location of welds are usually given on the design drawing, but very often connections and minor details are left to the detailer.

An example of a welded truss of light construction is shown in Fig. 23.18.

23.15 Computer-prepared Structural Drawings

CADD systems have been programmed to produce complete shop (detail) drawings for the structural steel fabricating industry by automated drafting methods (see Chapter 17). At present, these systems make shop drawings of beams, columns and bracing for most types of steel building frames. The Control Data Corporation (CDC) CONSTRUCTS System contains all of the specifications of the American Institute of Steel Construction and recognizes the standard practices for fitting and connecting structural steel shapes to carry the design loads indicated on the engineer's design drawings. The drawing shown on the left-hand facing page at the start of this chapter was made using the Control Data Structural System.

In the application of programming systems that have been developed to prepare shop drawings to fabricate different types of structural members, a technician fills out an input form utilizing information furnished on the structural design drawing. This information, in turn, is fed into a computer for computation of the dimensions necessary for the fabrication of each structural member in a building framing plan.

The output from the initial run of the program is a complete set of tabulated computations and information. This print-out should be inspected for obvious errors. At this time, if there is a special reason, the person making the check may introduce nonstandard treatment of connections of members.

A second run through the computer produces a magnetic tape with the drafting instructions needed to direct the offline plotter that will prepare the drawings.

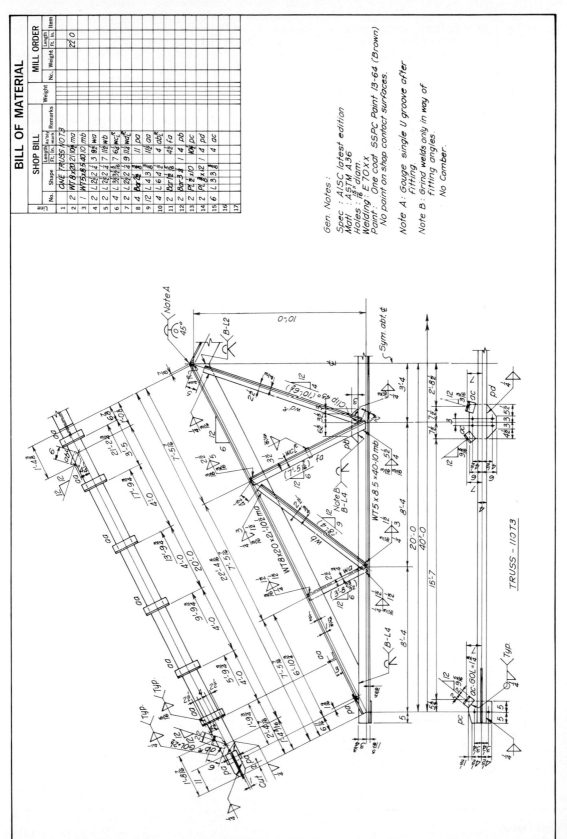

FIG. 23.18 Welded roof truss. (Courtesy American Institute of Steel Construction, Inc.)

PROBLEMS

The following problems are intended to furnish experience in the preparation of layouts and to emphasize the principles of structural drawing.

1. Make a pencil layout, plan, and elevation of joint *A*, Fig. 23.19. Follow carefully the steps outlined in Sec. 23.7 for making a layout of a gusset plate. The following requirements must be observed.

Use $\frac{3}{4}$-in. ϕ rivets.

Make the minimum allowable clearance between members $= \frac{1}{4}$ in.

Use minimum rivet spacing $= 3 \times$ diameter of rivets.

Use the preferred edge distance given in the table in Fig. 23.11.

Use standard gage distances, as given in the table in Fig. 23.10.

Make the width of the plate equal to the width of a standard plate. The variation of plate widths is by inches.

The bearing plate is a $9 \times \frac{3}{4} \times 1'{-}0$ steel plate. The rivet

pitch in the plan view should be equal to the pitch of the rivets in the vertical leg, and they must be so located that the distances *c*, *f*, and *k* are equal to or exceed the minimum values for these distances for a $\frac{3}{4}$-in. rivet as given in structural handbooks. The minimum value of *f* for a $\frac{3}{4}$-in. rivet is $1\frac{1}{4}$ in.; the minimum value of *k* is $1\frac{3}{4}$ in.; and the minimum value of *c* is $1\frac{1}{4}$ in. These three values represent minimum driving clearance for a $\frac{3}{4}$-in. rivet.

The holes for the anchor bolts are $\frac{13}{16}$ in. in diameter. Letter the correct specifications for the gusset plate.

2. Make a pencil drawing of the cross frame shown in Fig. 23.20 on the next page. The following requirements must be observed.

Use $\frac{7}{8}$-in. rivets.

Use a minimum rivet pitch of $2\frac{1}{2}$ in.

The elevation of the cross frame is the only view that is required.

Use standard gage and preferred edge distances, as shown in Figs. 23.10 and 23.11.

Use Fig. 23.7 as a model for the placing of complete dimensions.

The open holes are spaced at $3\frac{1}{2}$-in. pitch.

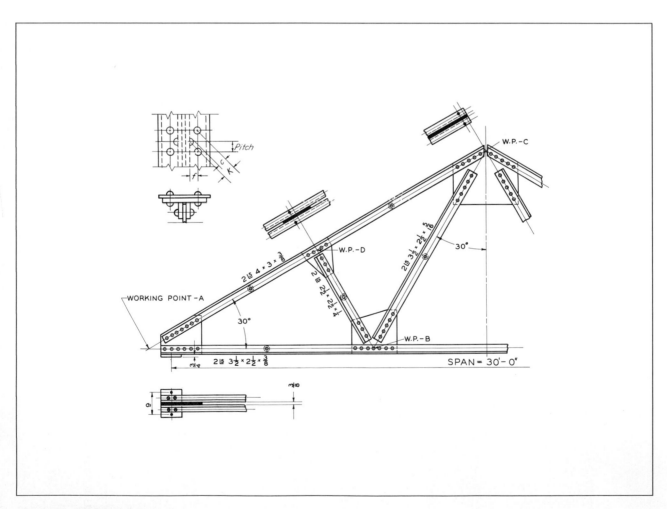

FIG. 23.19 Roof truss layout.

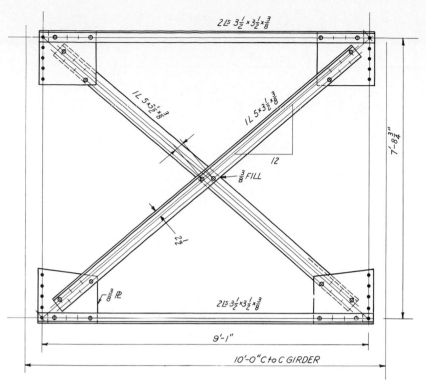

FIG. 23.20   Layout of interior cross frame for 100-ft railroad deck girder span.

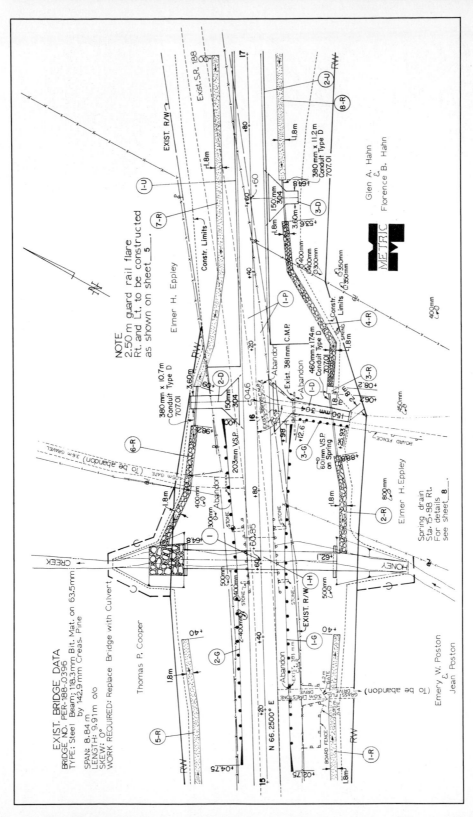

Plan view for a portion of a highway project. Distances and size values are in meters and millimeters. (*Courtesy of Ohio Department of Transportation*)

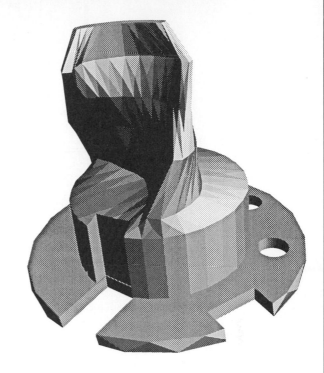

CHAPTER · 24

Topographic and Engineering Map Drawings

24.1 Map Drawing*

A map is a drawing that represents a portion of the earth's surface area. Since it usually represents a relatively small part and the third dimension (the height) is not shown except in some cases by contour lines, a map may be thought of as a one-view orthographic projection (Fig. 24.1). Various forms of maps have been devised to satisfy different requirements. Land maps reveal only the natural and man-made features along with imaginary division lines and geometrical measurements in only two dimensions. Others, such as topographical maps, show three dimensions, by representing height by means of contours.

* See Figs. 17.9 and 17.10 for engineering maps prepared on a plotter.

24.2 Engineering Maps and Plats—Metric

Ideally, the scales used for preparing topographic maps, engineering maps, and plats should be from the 1, 2, and 5 series, though the 2.5 scale is used to some extent. The United States Geological Survey (U.S.G.S.) recommends that the inch-foot scales be replaced by the 1, 2, 5 series as follows:

Inch-foot	SI
1″ = 80′	1 : 1000
1″ = 100′	1 : 1000
1″ = 200′	1 : 2000
1″ = 400′	1 : 5000
1″ = 500′	1 : 5000

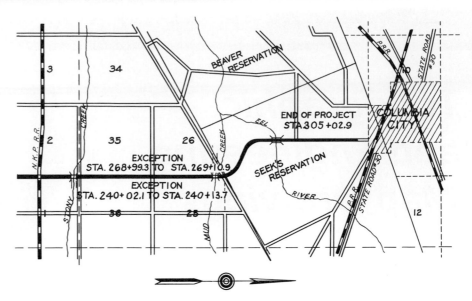

FIG. 24.1 **Section from an engineering map (metric).**

U.S.G.S. maps are prepared to scales of 1:25,000, 1:50,000, and 1:100,000.

When a land area is small, say city lot size, the area should be given in square meters. Large areas may be given in hectares or in cases of very large areas, such as state and national parks, in square kilometers.

For new residential subdivisions, where lengths, widths, and angles are unrestricted, roads should be made 10 m, 12 m, or 15 m wide. Arterial thoroughfares and streets should be laid out to be a width of 20 m, 22 m, or 25 m as may be required for parking and expected traffic. Utility easements should be either 2 m, 3 m, or 5 m in width.

For metric maps of site plans, contour intervals can be either 0.1 m, 0.2 m, or 0.5 m. For topographic maps of rough terrain the U.S.G.S. commonly uses intervals of 1 m, 1.5 m, 2 m, 50 m, or 100 m.

24.3 Classification of Maps

Maps of interest to the engineer may be grouped for study in four general classes, in accordance with their purpose and method of preparation. The recognized classes are

1. topographical,
2. cadastral,
3. engineering,
4. aerial photographic.

24.4 Topographic Maps

Topographic maps, although they are drawn to a relatively small scale, contain much detail. All natural features, such as lakes, streams, forests, fields, mines, and so on, and important permanent man-made creations, such as

buildings, bridges, and houses, may be represented if necessary to fulfill the purpose of a map. Topographic maps, prepared by the United States Geological Survey to a scale of 1:24000, naturally do not contain fine detail (Fig. 24.2). The form of the surface of the ground is represented by contour lines. Any one contour line passes through points of the same elevation and closes on itself either on the map or beyond its limits. Closed contour lines represent either a hill or a depression. Figure 24.6 shows a topographic map.

24.5 Cadastral Maps

This group consists of city plats, city development maps, town maps, county maps, and maps prepared to show ownership (particularly for the purposes of governmental control and taxation). These maps, although they show practically no detail, must be accurate and for this reason are drawn to a large scale. Property lines, political boundaries, and a few important features, such as streams, roads, and towns, may be given on township and county maps, to enable a reader to identify particular locations.

24.6 Engineering Maps (see Chapter Opening Facing Page)

Working maps prepared for engineering projects are known as engineering maps. They may be drawn for either reconnaissance or construction purposes. They usually are made to a large scale and accurately show the location of all property lines and important features. On maps of a topographic nature, practically all natural and man-made features along a right-of-way or on a site are shown, and the form of the surface of the ground is indicated by means of contours.

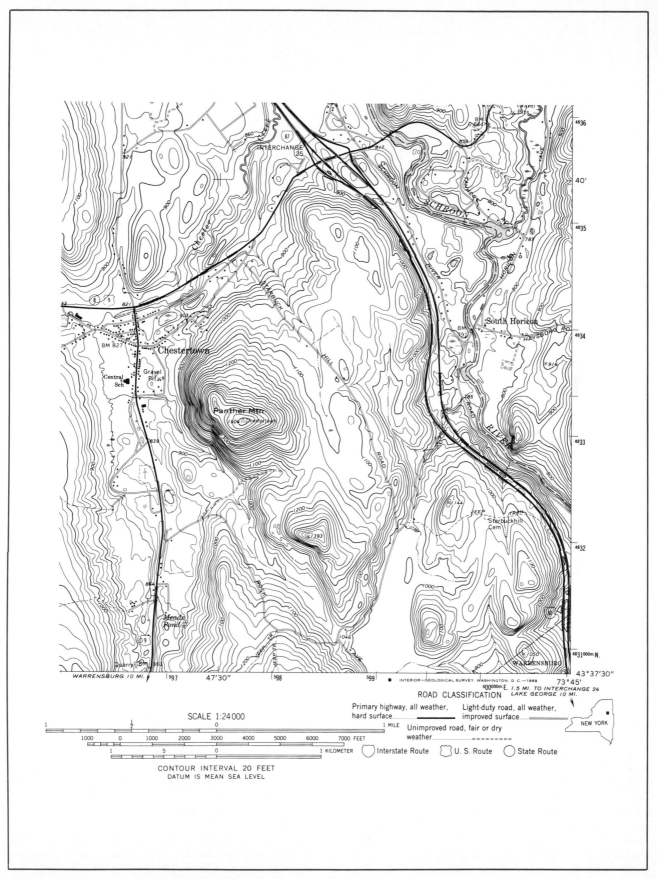

FIG. 24.2 Portion of a typical map prepared by the U.S. Geological Survey.

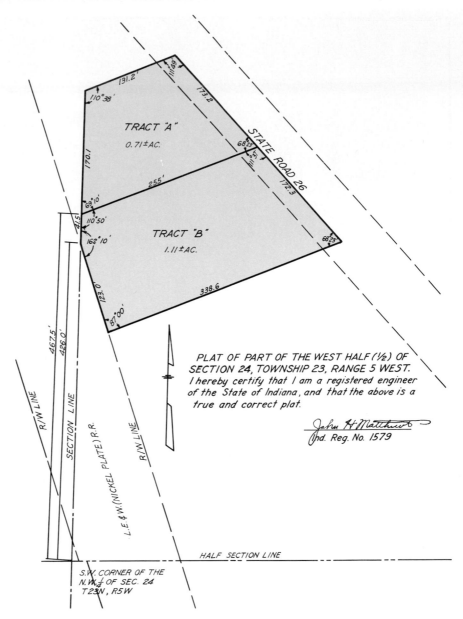

FIG. 24.3 Plat of a land survey.

24.7 *Plats of Land Surveys*

A plat of a tract of land should contain a complete description of the land surveyed. It should show the lengths and bearings (or included angles) of the bounding sides and division lines, the included acreages, the locations of the monuments, and the names of the owners of the adjoining properties. Figure 24.3 shows a plat of a typical land survey. Note that a clear, concise title is lettered in the large open area. A certification of the survey is generally required by law. In most states, a plat must bear the seal of a licensed surveyor.

24.8 *Subdivision Plats*

A plat of a real-estate development should show the measurements and angles of the survey of the whole tract of land, the sizes of the included lots, the widths of the streets and drives, and the locations of all monuments. Plats of subdivisions must be complete and accurate, since they are filed as public records in the county recorder's office. Sufficient information must be given to enable a surveyor to locate the corners of any lot with precision when making a resurvey at a later date. Figure 24.4 is a plat of a city subdivision.

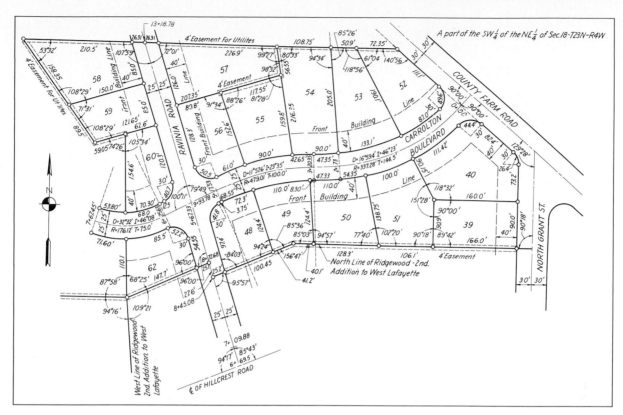

FIG. 24.4 Plot of a city subdivision (also see Fig. 17.10).

24.9 Plats and Partial City Maps

Maps made from subdivision plats or city maps are prepared by the engineering departments of cities and public utilities. The purpose of these partial maps is to record special information concerning such things as proposed improvement projects, the location of lines of transportation, and the location of existing and proposed water mains, sewers, and so on. It is not necessary for such a map to contain all the information given on the subdivision plat from which it is made. The locations of monuments and angles generally are not shown. The widths of streets and the sizes of lots may or may not be shown, depending on the usefulness of such information. A few important buildings may be indicated for the sake of aiding the reader in orientation. Figure 24.5 shows the proposed locations of water mains and fire hydrants in a portion of a city.

24.10 Topographic Drawing

As previously stated, a topographic map is a reproduction, to scale, of a large area. On a complete topographic map, the natural and artificial features are represented by recognized conventional symbols and the form of the ground is shown by contours. Excessive detail should be avoided, and only necessary surface features should be shown. All

names and required notes should be lettered in a position where they can be easily read; a complete title should be lettered, in the lower-right-hand corner. Ordinarily, single-stroke lettering is preferable on topographic maps prepared solely for construction projects, while vertical lettering is more desirable on finished maps where effect and pleasing appearance are important. Figure 24.6 shows a topographic map.

The scale to be used for a topographic map depends on the size of the area and the amount of the detail that must be shown. Scales range from 1 in. = 100 ft to 1 in. = 4 miles. Maps prepared by the United States Geological Survey have in the past usually been drawn to 1:62,500, which is almost 1 in. to a mile.

For drawings prepared in metric, the U.S.G.S. uses 1:1000 in place of 1″ = 100′. Other commonly used metric scales are: 1 = 25,000, 1:50,000, and 1:100,000.

24.11 Use of Standard Topographic Symbols (U.S.G.S.)

Recognized signs and symbols are used to represent the natural and artificial features on a topographic map. Many of these symbols have been designed to bear some pictorial resemblance to the feature or object represented and, for convenience, may be grouped as follows: (1) physical features (buildings, highways, railroads, and so on); (2) ground formations (elevations and depressions);

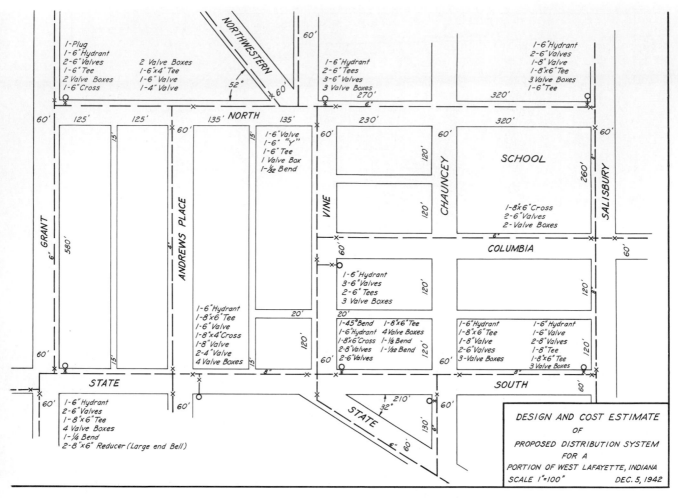

FIG. 24.5 Plat or partial city map showing the proposed location of water mains and fire hydrants.

(3) water-surface features (rivers, lake, and streams); and (4) vegetation growths (grass, trees, and cultivated crops).

Conventional symbols used by the United States Geological Survey for representing human works and structures are shown in Fig. 24.7. Symbols for natural land formations, water features, and vegetation growths, both natural and cultivated, are shown in Figs. 24.8–24.10.

24.12 Topographic Symbols

Topographic symbols are drawn either freehand or mechanically, depending on the character of the features to be represented. For example, the symbols representing natural features are drawn freehand, while those representing artificial works are drawn mechanically. See Figs. 24.7–24.10. On topographic maps prepared for engineering projects, the symbols are drawn in ink. When colors are used, as in finished maps, the artificial features (buildings, bridges, railroads, and so on) are drawn in black,

the contours in brown, the water features in blue, and vegetation growths in green.

Although the size of symbols may vary somewhat with the size of the map, *they are never shown to scale but are always exaggerations*. The usual mistake of the beginner is to draw symbols too large or too close together. Either fault produces a disagreeable appearance and tends to attract the reader's attention away from more important features. The symbols representing prominent features are made to stand out from those of lesser importance by being drawn slightly larger and with heavier lines.

The beginner should study carefully the symbols as given in the various illustrations, so that essential points in this construction will not be missed. For instance, it should be noted that the symbol for a tree (Fig. 24.10) is composed of separate lines, irregularly located, and not of one closed line drawn without lifting the pen. The symbols for grass, corn, and other vegetation should be placed with the bases of the tufts and stocks parallel to the lower borderline.

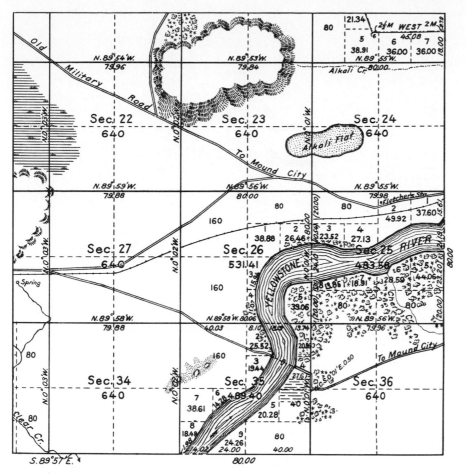

FIG. 24.6 Topographic map. (A portion of a map taken from a manual prepared by the U.S. Department of the Interior.)

24.13 Drawing Water Lines

Water lining, used to indicate water surfaces, is done entirely freehand. The starting line (shoreline) should be fairly heavy, and each successive line should decrease in width until the center of the body of water is reached. See Fig. 24.9. The line next to the shoreline should be drawn parallel to it throughout its entire length, and the space between should be equal to the width of the shoreline. The spacing between succeeding lines should increase gradually to the center, but the change should be so slight that no marked increase will be noticeable. Each added line should show fewer of the small irregularities of the shoreline, the last few following only the prominent ones.

If several bodies of water are to be indicated on the same map, a good way to obtain uniformity is to draw all the shorelines first, then all the lines next to the shorelines, and so on, working back and forth from one body of water to another until the representations are completed. Excessive waviness gives these lines an unnatural appearance and should be avoided.

24.14 Contour Lines (Fig. 24.2)

A contour line is a line through points of the same elevation on the surface of the ground. Theoretically, the contour lines on a map may be thought of as the lines of intersection of a series of horizontal planes and the ground surface. In practice, the imaginary planes are equally spaced vertically so that the contour intervals will be equal and the horizontal distances between contours on a map will indicate the steepness of the rise or descent of the surface. The closer together they are, the greater the slope and, conversely, the farther apart they are, the less the slope. An arrangement of contour lines that close indicates either a hill or a depression (Fig. 24.8). The case, whatever it is, usually can be determined by reading the values of the elevations of the contours. Usually each fifth contour is drawn heavier than the others and has a break in it where its elevation above a datum plane is recorded. If a U.S.G.S. bench mark is used, the datum plane will be at mean sea level.

The selection of the contour interval (vertical distance between contour planes) for a topographic survey is de-

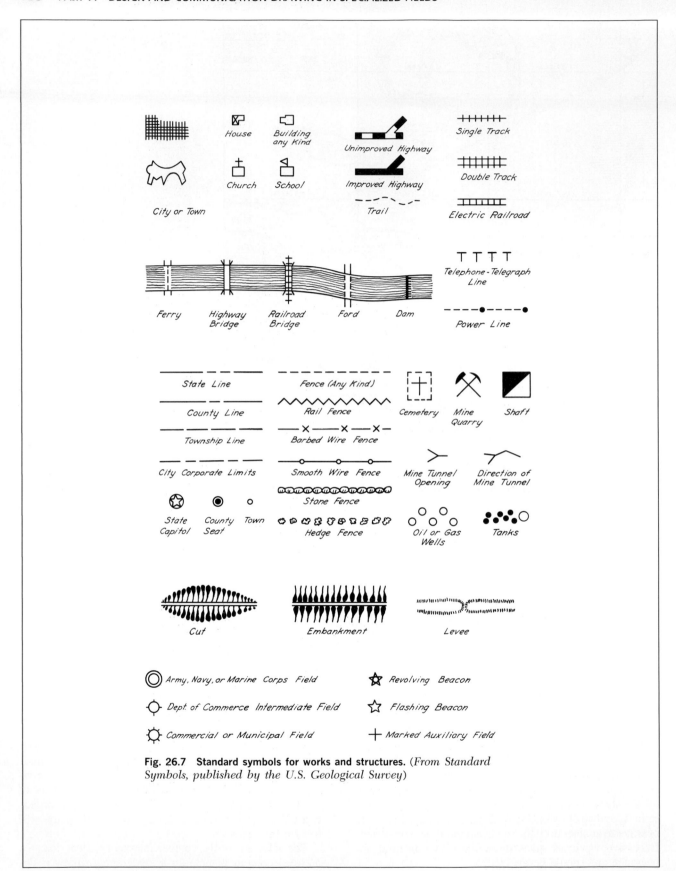

Fig. 26.7 Standard symbols for works and structures. (*From Standard Symbols, published by the U.S. Geological Survey*)

FIG. 24.7 **Standard symbols for works and structures.**
(*From Standard Symbols, published by the U.S. Geological Survey*)

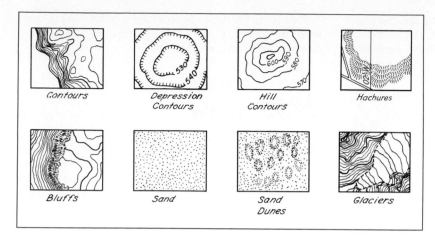

FIG. 24.8 **Relief (natural land formations).**

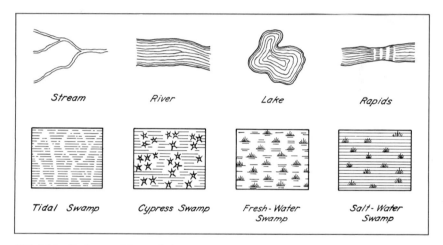

FIG. 24.9 **Hydrographic symbols.**

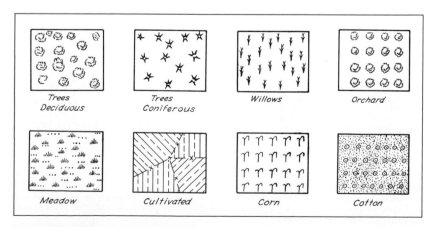

FIG. 24.10 **Vegetation symbols.**

termined by the nature of the ground forms and the purpose for which the map will be prepared. For instance, if the area is relatively level, a 1- or 2-ft interval probably would be desirable, while, if the area is rugged, an interval of 50 or even 100 ft might be used. For a metric drawing, if the area is rather level, the contour interval may be 0.1 m, 0.2 m, or 0.5 m. For more rugged areas, the interval, could be 1 m, 2 m, 50 m, or in some cases 100 m in mountainous country.

Contour lines are plotted from survey notes made in the field. In the case of small areas, the usual method for locating contours is to divide the area into squares and take level readings at every intersection and at intermediate points where a pronounced change in slope takes place (Fig. 24.11). On the assumption that the slope of the ground is uniform between two points, the contours are sketched in by interpolating between the readings to establish the points at which the contours cross the survey lines. The interpolation may be done by eye or by calculation.

Frequently, contours are determined from level readings taken along known lines or from stadia survey notes.

When extreme accuracy is necessary and the land is fairly level, contours are established by finding and locating points directly on each contour.

24.15 Profiles

The profile of a line is prepared to show the relative elevation of every point on the line. Theoretically, it may be thought of as a view showing the line of intersection of a vertical plane through the line and the ground surface. Figure 24.12 shows a plan view and a profile of a section of a proposed highway project. When the plan and profile are on the same sheet, as in the horizontal scale for the profile should be the same as the scale for the plan view.

Profiles are drawn by plotting elevations computed from level readings taken at regular intervals on the ground and at points where the ground changes in slope. Since the slope between adjacent points is assumed to be uniform, a profile consists of a series of straight lines joining successive points.

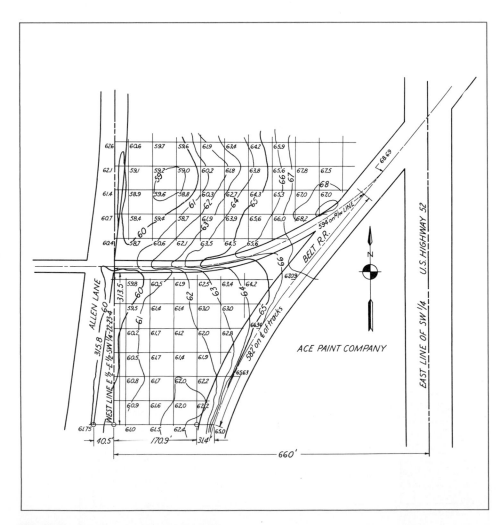

FIG. 24.11 Contour map of a small area.

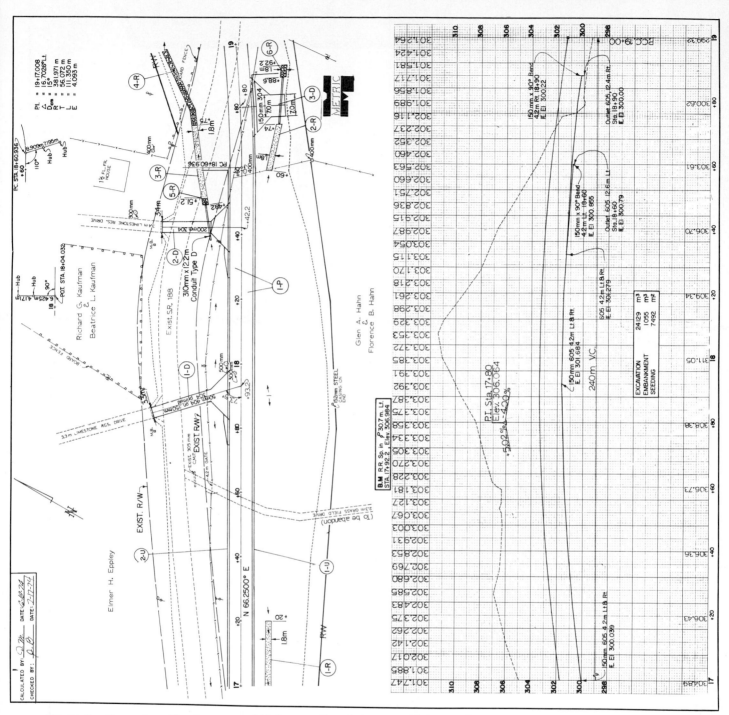

FIG. 24.12 **Plan and profile of a proposed highway project.** (Distances and sizes are in meters and millimeters). This drawing was prepared manually. It is suggested that this drawing be compared with the one shown in Fig. 17.9. (*Courtesy of Ohio Department of Transportation*)

Glossary

◢ Shop Terms

◢ Computer-aided Drafting Terms

◢ Structural Drafting Terms

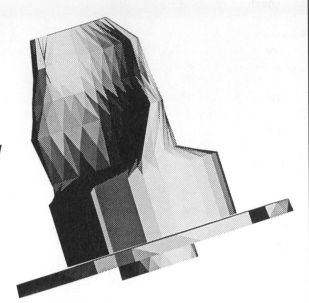

GLOSSARY OF COMMON SHOP TERMS

Anneal (*v*) To heat a piece of metal to a particular temperature and then allow it to cool slowly for the purpose of removing internal stresses.

Bore (*v*) To enlarge a hole using a boring bar in order to make it smooth, round, and coaxial. Boring is usually done on a lathe or boring mill.

Boss (*n*) A circular projection, which is raised above a principal surface of a casting or forging.

Braze (*v*) To join two pieces of metal by the use of hard solder. The solder is usually a copper–zinc alloy.

Broach (*v*) To machine a hole to a desired shape, usually other than round. The cutting tool, known as a broach, is pushed or pulled through the rough-finished hole. It has transverse cutting edges.

Burnish (*v*) To smooth or apply a brilliant finish.

Bushing (*n*) A removable cylindrical sleeve, which is used to provide a bearing surface.

Carburize (*v*) To harden the surface of a piece of low-grade steel by heating in a carbonizing material to increase the carbon content and then quenching.

Case-harden (*v*) To harden a surface as described above or through the use of potassium cyanide.

Chamfer (*v*) To bevel an external edge or corner.

Chase (*v*) To cut screw threads on a lathe using a chaser, a tool shaped to the profile of a thread.

Chill (*v*) To cool the surface of a casting suddenly so that the surface will be white and hard.

Chip (*v*) To cut away or remove surface defects with a chisel.

Collar (*n*) A cylindrical part fitted on a shaft to prevent a sliding movement.

Color-harden (*v*) A piece is color-hardened mainly for the sake of appearance. (*See* Case-harden).

Core (*v*) To form a hole or hollow cavity in a casting through the use of a core.

Counterbore (*v*) To enlarge the end of a cylindrical hole to a certain depth, as is often done to accommodate the head of a fillister-head screw. (*n*) The name of the tool used to produce the enlargement.

Countersink (*v*) To form a conical enlargement at the end of a cylindrical hole to accommodate the head of a screw or rivet. (*n*) The name of the tool used to form a conical shaped enlargement.

Crown (*n*) The angular or curved contour of the outer surface of a part, such as on a pulley.

Die (*n*) A metal block used for forming or stamping operations. A thread-cutting tool for producing external threads.

Die casting (*n*) A casting that has been produced by forcing a molten alloy having an aluminum, copper, zinc, tin, or lead base into a metal mold composed of two halves.

Die stamping (*n*) A piece that has been cut or formed from sheet metal through the use of a die.

Draw (*v*) To form metal, which may be either cold or hot, by a distorting or stretching process. To temper steel by gradual or intermittent quenching.

Drill (*v*) To form a cylindrical hole in metal. (*n*) A revolving cutting tool designed for cutting at the point.

Drop forging (*n*) A piece formed while hot between two dies under a drop hammer.

Face (*v*) To machine on a lathe a flat face, which is perpendicular to the axis of rotation of the piece.

Feather (*n*) A rectangular sliding key, which permits a pulley to move along the shaft parallel to its axis.

File (*v*) To shape, finish, or trim with a fine-toothed metal cutting tool, which is used with the hands.

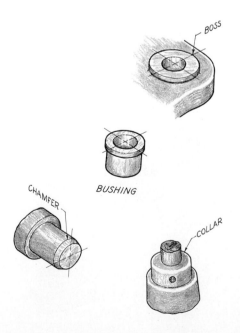

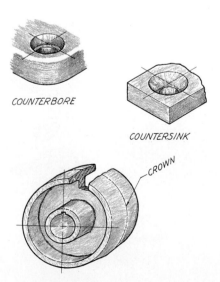

Fillet (*n*) A rounded filling, which increases the strength at the junction of two surfaces that form an internal angle.

Fit (*n*) The tightness of adjustment between the contacting surfaces of mating parts.

Flange (*n*) The top and bottom member of a beam. A projecting rim added on the end of a pipe or fitting for making a connection.

Forge (*v*) To shape hot metals by hammering, using a hand-hammer or machine.

Galvanize (*v*) To coat steel or iron by immersion in a bath of zinc.

Graduate (*v*) To mark off or divide a scale into intervals.

Grind (*v*) To finish a surface through the action of a revolving abrasive wheel.

Kerf (*n*) A groove or channel cut by a saw or some other tool.

Key (*n*) A piece used between a shaft and a hub to prevent the movement of one relative to the other.

Keyway or Keyseat (*n*) A longitudinal groove cut in a shaft or a hub to receive a key. A key rests in a keyseat and slides in a keyway.

Knurl (*v*) To roughen a cylindrical surface to produce a better grip for the fingers.

Lap (*v*) To finish or polish with a piece of soft metal, wood, or leather impregnated with an abrasive.

Lug (*n*) A projection or ear, which has been cast or forged as a portion of a piece to provide a support or to allow the attachment of another part.

Malleable casting (*n*) A casting that has been annealed to toughen it.

Mill (*v*) To machine a piece on a milling machine by means of a rotating toothed cutter.

Neck (*v*) To cut a circumferential groove around a shaft.

Pack-harden (*v*) To case-carburize and harden.

Pad (*n*) A low, projecting surface, usually rectangular.

Peen (*v*) To stretch or bend over metal using the peen end (ball end) of a hammer.

Pickel (*v*) To remove scale and rust from a casting or forging by immersing it in an acid bath.

Plane (*v*) To machine a flat surface on a planer, a machine having a fixed tool and a reciprocating bed.

Polish (*v*) To make a surface smooth and lustrous through the use of a fine abrasive.

Punch (*v*) To perforate a thin piece of metal by shearing out a circular wad with a nonrotating tool under pressure.

Ream (*v*) To finish a hole to an exact size using a rotating fluted cutting tool known as a reamer.

Rib (*n*) A thin component of a part that acts as a brace or support.

Rivet (*n*) A headed shank, which more or less permanently unites two pieces. (*v*) To fasten steel plates with rivets.

Round (*n*) A rounded external corner on a casting.

Sandblast (*v*) To clean the surface of castings or forgings by means of sand forced from a nozzle at a high velocity.

Shear (*v*) To cut off sheet or bar metal through the shearing action of two blades.

Shim (*n*) A thin metal plate, which is inserted between two surfaces for the purpose of adjustment.

Spline (*n*) A keyway, usually for a feather key. (*See* Feather.)

Spotface (*v*) To finish a round spot on the rough surface of a casting at a drilled hole for the purpose of providing a smooth seat for a bolt or screw head.

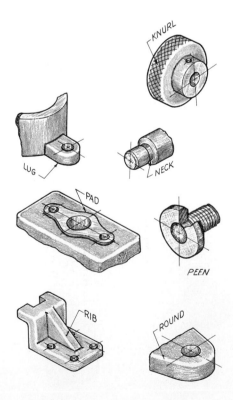

Spot weld (*v*) To weld two overlapping metal sheets in spots by means of the heat of resistance to an electric current between a pair of electrodes.

Steel casting (*n*) A casting made of cast iron to which scrap steel has been added.

Swage (*v*) To form metal with a swag block, a tool so constructed that through hammering or pressure the work may be made to take a desired shape.

Sweat (*v*) To solder together by clamping the pieces in contact with soft solder between and then heating.

Tack weld (*n*) A weld of short intermittent sections.

Tap (*v*) To cut an internal thread, by hand or with power, by screwing into the hole a fluted tapered tool having thread-cutting edges.

Taper (*v*) To make gradually smaller toward one end. (*n*) Gradual diminution of diameter or thickness of an elongated object.

Taper pin (*n*) A tapered pin used for fastening hubs or collars to shafts.

Temper (*v*) To reduce the hardness of a piece of hardened steel through reheating and sudden quenching.

Template (*n*) A pattern cut to a desired shape, which is used in layout work to establish shearing lines, to locate holes, etc.

Tumble (*v*) To clean and smooth castings and forgings through contact in a revolving barrel. To further the results, small pieces of scrap are added.

Turn (*v*) To turn-down or machine a cylindrical surface on a lathe.

Undercut (*n*) A recessed cut.

Upset (*v*) To increase the diameter or form a shoulder on a piece during forging.

Weld (*v*) To join two pieces of metal by pressure or hammering after heating to the fusion point.

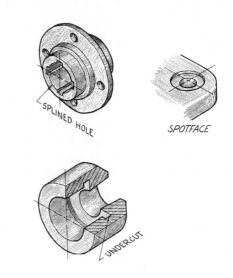

SPLINED HOLE

SPOTFACE

UNDERCUT

GLOSSARY OF COMPUTER-AIDED DRAFTING TERMS

The following terms are those most commonly encountered when operating a CADD computer. The list is by no means exhaustive. Rather, these terms will provide the student with a working vocabulary to better understand CADD applications of engineering drawing.

Attribute (*n*) The characteristic of a single drawing component such as text *height* or line *thickness*; a specific example of a drawing parameter.

Axis (*n*) A direction in space; one of the principal axes of Y (height), X (width), and Z (depth).

Back up (*v*) To make a copy of a file separate from the original.

Bit map graphics (*n*) A method of organizing graphics into a grid of dots or bits and characterized by a lack of changable entity attributes such as line width or circle diameter; raster graphics.

CAD (*n*) Computer Aided Design; the use of computers to model and test the performance of a design while providing visual feedback to the designer.

CADD (*n*) Computer Aided Design and Drafting; the use of computers to assist in the description of engineering designs in terms of engineering drawings.

CAM (*n*) Computer Aided Manufacturing; the use of computers to assist in the planning, execution, monitoring, and necessary adjustment of the manufacturing process.

CIM (*n*) Computer Integrated Manufacturing; the overall term used for the systems integration of all manufacturing activities.

Cartesian coordinates (*n*) A position in space relative to the intersection of the X, Y, and Z axes (the origin); absolute coordinates relative to XO, YO, ZO; relative or delta coordinates from a temporary origin.

Character string (*n*) A pattern of alphabetic, numeric, or punctuation characters used to annotate the geometric data base.

Coordinate (*n*) A location in space specified by X, Y, and Z positions.

Coordinate system (*n*) A method of specifying exact locations in space such as the cartesian or polar coordinate systems.

Command (*n*) An instruction from the operator to the computer.

Computer graphics (*n*) The use of computing machinery to assist in the creation, storage, and display of visual imagery.

Construction plane (*n*) A 2-D plane in 3-D space on which geometry is constructed; the method by which a wire frame model is constructed.

Crosshatch or Hatch (*n*) The filling of an area with a pattern of parallel lines or similar symbols.

Cursor (*n*) The graphic symbol on a display screen corresponding to the current position in either world or device coordinates.

Data base (*n*) The organized description of data; in *geometric data base*, the organized description of a part's geometry.

Default (*n*) A pre-defined setting within a CADD program established when the program was created; a setting that an operator may change during a work session that reverts to the predefined setting when the session is ended; a setting that an operator may permanently change.

Delete (*v*) To remove information from the data base.

Device coordinate (*n*) The coordinate location on the screen of a point in space.

Digitizer (*n*) An input device which translates an X-Y or X-Y-Z position into digital information understood by a CADD program; a tablet.

Digitizing (*n*) The identification of coordinate points by using a digitizer; the assigning of coordinate values by keyboard entry.

Display (*n*) An output device which accurately presents the image of a geometric data base; a display terminal such as a video terminal (VT), a cathode ray tube (CRT), or a direct view storage tube (DVST).

Edit (*v*) To change the parameters or attributes of drawing components; to change a text string.

Engineering workstation (*n*) An easily networked computer able to fit on an engineer's desk capable of rapid, accurate creation and display of engineering drawings and designs.

Entity (*n*) A drawing component that cannot be readily broken apart; a basic drawing shape or object resident with the CADD program such as a line, arc, circle, or point; a symbol which is stored as a single grouped graphic.

File (*n*) The CADD drawing as stored in the computer; any information stored as a unit in the computer such as a font file, a pattern file, or plot file; when used as a verb, the act of saving a file.

Fill (*n*) Crosshatching of an area with a fill or crosshatch pattern.

Firmware (*n*) Program instructions that are permanently built into electronic circuitry.

Font (*n*) In *text font*, all of the characters of a particular type style and size; as in *line font*, all available line styles (solid, dashed, center, phantom, user defined).

Geometric data base (*n*) The organization of information in the computer that describes the physical nature of an object.

Graphic application (*n*) A program intended to perform a specific task such as a piping application, an electronic circuit board application, or a bridge beam analysis application.

Grid (*n*) A matrix or pattern of dots displayed on the terminal used to facilitate drawing and design; the spacing of the pattern of dots set to coincide with major features of the design.

Hard copy (*n*) A print or plot of a drawing on paper or film.

Hardware (*n*) The physical equipment necessary for computer drawing; the computer, terminal, keyboard, tablet, plotter and storage disks or tapes.

Hidden line editing (*n*) The automatic changing of line type parameter from solid to hidden to correctly show visibility; the manual changing of such lines.

Input device (*n*) A device which facilitates entering data or instructions into the computer; the keyboard, mouse, joy stick, or tablet with puck or stylus.

Interactive computer graphics (*n*) Computer drawing that shows the operator's actions and the response of the computer to such actions in graphical form.

Layer (*n*) The organization of associated geometry, dimensions, or notes by an index number known as the layer number; (*v*) to organize a drawing into groups that can be displayed, edited, or plotted individually.

Load (*v*) To bring a file from storage to current memory to be worked on.

Mask (*n*) An instruction that limits the possible responses to an operator command such as a position mask.

Menu (*n*) A list of possible commands from which the operator selects.

Mirror (*v*) To revolve selected geometry 180° about the normal view of an axis as differentiated from revolution which is done about the point view of an axis.

Modifier (*n*) An instruction that limits the possible responses to an operator's command.

Mouse (*n*) A hand-held input device used for entering X-Y data or instructions.

Network (*n*) A group of computers linked together for the purpose of sharing information (data bases or drawings) or resources (printers or plotters).

Object graphics (*n*) The method of organizing graphics into objects or entities.

Output device (*n*) A device like a terminal, plotter, or printer used for viewing a data base.

Origin (*n*) The position XO, YO, ZO in Cartesian space; the position radius O, Angle O in polar space; the insertion point of a symbol.

Pan (*v*) To move the drawing left, right, up or down without altering the object's relationship to the world coordinate system (no translation or rotation) or the object-viewer distance (no zoom).

Parameter (*n*) A setting used in making a drawing such as a scale, line type, or text font that is applied to the drawing globally.

Part (*n*) The completed description of an object.

Part file (*n*) The saved description of an object.

Pattern (*n*) A grouping of graphics used as a symbol; the design of lines used to fill an area.

Plotter (*n*) An output device that represents the data base as drawn lines. In a drum plotter, paper is fed around a cylindrical drum and both paper and pens move. In a flat bed plotter, sheet paper is held stationary on a large table and only the pens move. In a belt bed plotter, the paper is held against a moving flat belt with moving pens. A photo plotter uses a light beam to draw on light sensitive film.

Polar coordinates (*n*) Designate a position in space at a specific angle and at a specific radius from an origin.

Primitives (*n*) The lowest level graphic shapes available to construct more complex shapes; entities that cannot be ungrouped into more simplistic parts.

Printer (*n*) An output device that changes object graphic files into raster images. Examples are the dot matrix printer, an electrostatic printer, and a laser printer.

Prompt (*n*) An instruction from the computer to the operator.

Puck (*n*) A hand-held input device used in conjunction with a tablet for recording X and Y coordinates.

Raster graphics (*n*) Drawings comprised of a grid of dots; a bit map; the opposite of object graphics.

Rotate (*v*) To move circularly in a plane perpendicular to an axis; positive rotation is counterclockwise (see mirror).

Snap (*n*) An invisible grid that acts like a magnet to attract the cursor to grid intersection points.

Software (*n*) Instructions written in a language understood by computing machinery.

Solid model (*n*) The complete description of an object in the computer consistent with its 3-D properties (material, mass, weight, surface finish, etc.).

Storage media (*n*) The disks or tapes used for the permanent storage of data and drawings.

Surface model (*n*) A hollow 3-D model comprised only of surfaces with no solid material inside.

Surface shading (*n*) The assigning of values to the surfaces of a 3-D computer model consistent with light source position and material characteristics.

Stylus (*n*) An electronic input device held like a pencil.

Symbol (*n*) A graphic created and stored for subsequent use; a library part or figure such as a cell, a pattern, or a template.

Tablet (*n*) A device used to record X and Y positions. Tablets are used either to digitize geometry or input operator's commands from a menu on the tablet.

Translate (*v*) To move in a linear fashion in relation to the X, Y, and Z axes.

Workspace (*n*) The world coordinate space in which a CADD operator works.

World coordinate (*n*) The position of a point given in the same units that is used to create the actual object; the design in full scale.

Window (*n*) A section of the computer terminal set aside for displaying information independently of information shown in other windows.

Wire frame model (*n*) A transparent 3-D model comprised of vertices, connecting lines, or surface elements but having no solid planes or mass.

Zoom (*v*) To change the apparent size by getting closer to or farther from the object; *zoom in* increases the apparent size and *zoom out* reduces the apparent size.

2-D (*a*) Having dimensional characteristics in X and Y directions.

2½-D (*a*) Having the appearance of three-dimensionality but with only X and Y information in the geometric data base; a 2-D pictorial representation.

3-D (*a*) Having dimensional characteristics in X, Y, and Z directions.

GLOSSARY OF STRUCTURAL DRAFTING TERMS

Anchor bolt (*n*) A bolt used to fasten steel girders, columns, trusses, and so on, to masonry.

Beam (*n*) A horizontal structural member (usually an I-beam or WF beam)

Bearing plate (*n*) A plate used under the end of a truss, beam or girder, to increase the bearing area.

Chip (*v*) To chip off projecting parts with a chisel (pneumatic chisel).

Clearance (*n*) A general term applying to an opening or space between two adjacent pieces, without which interference would result (Fig. 23.9)

Column (*n*) A general term for a vertical member supporting beams or trusses.

Countersink (*v*) The operation of chamfering the edges of a hole to receive the conical head of a bolt, rivet, or screw.

Detail (*v, n*) To prepare a working structural drawing. (Working drawings are called details.) (Fig. 23.7).

Driving clearance (*n*) The distance from the center of a rivet to the nearest obstruction that would interfere with the driving of the rivet.

Edge distance (*n*) The distance from the center of a rivet or hole to the edge of the member (Fig. 23.9).

Erection diagram (*n*) An assembly diagram drawn to aid the erector in placing members of the structure in their proper positions.

Fabrication (*n*) The shop work of converting rolled shapes to complete structural members.

Field clearance (*n*) Minimum distance between unfinished edges that abut when erected at the site. Usually $\frac{1}{2}$–$\frac{3}{4}$ in.

Field rivet (*n*) A rivet driven into the structure at the site of construction.

Filler (*n*) A plate or washer used to fill up space between two surfaces.

Flame cut (*v*) To cut by hand or by a machine-guided oxyacetylene torch.

Flange (*n*) A general term for the outstanding part of a member (Fig. 23.1).

Gage (*n*) The distance from the back of an angle or channel to the center line of a row of rivets, or the distance between the center lines of two rows of rivets (Fig. 23.10).

Gage line (*n*) The center line of a row of rivets (Fig. 23.10).

Girder (*n*) A horizontal member built up of plates and angles.

Grip (*n*) The combined thickness of members connected by a rivet.

Gusset plate (*n*) A connection plate used to connect several members of a truss (Fig. 23.9).

Layout (*n*) A preliminary scale drawing made in the detailing department prior to detailing, for the purpose of ordering material and coordinating the work of the several detailers (Fig. 23.5).

Leg (*n*) The name for either of the two flanges of an angle (Fig. 23.1).

Line diagram (*n*) A drawing in which cach member is represented by a single line (Fig. 23.2).

Pitch (*n*) The center-to-center longitudinal distance between adjacent rivets (Fig. 23.10).

Plate (*n*) A flat piece of structural steel having a rectangular cross section (Fig. 23.1).

Punch (*v*) To make a hole by forcing a nonrotating tool through the material.

Ream (*v*) To enlarge and finish a punched hole, using a rotating fluted cutter.

Rivet (*n*) A cylindrical rod of steel that is used to fasten together members of a steel structure. It has one head formed when manufactured; the other is formed after the rivet is in position.

Shape (*n*) The structural term for rolled steel having any cross section (except a steel plate) (Fig. 23.1).

Shop clearance (*n*) Minimum distance between unfinished abutting edges of members assembled in the shop. Usually $\frac{1}{4}$ in.

Shop drawing (*n*) A working drawing made for the shop. Commonly called a *detail drawing* (Fig. 23.7).

Slope (*n*) The inclination of a line designated by a slope triangle expressed in inches of rise to a base of 12 in. (Fig. 23.7).

Span (*n*) The center-to-center distance between the supports of a beam, girder, or truss.

Staggered rivets (*n*) Rivets spaced alternately in parallel rows (Fig. 23.10).

Stitch rivets (*n*) Rivets spaced at intervals along a built-up member to cause the component parts to act as a unit.

Stress sheet (*n*) A drawing having a line diagram on which are recorded the stresses in the main members of a structure (Fig. 23.2).

Truss (*n*) A rigid framed structure, in the form of a series of triangles, which acts as a beam (Fig. 23.8).

Web (*n*) The thin portion between the flanges of a member (Fig. 23.1).

Weldment (*n*) Any welded assembly or subassembly (Fig. 23.18)

Working point (*n*) The point of intersection of working lines (usually gage lines) (Fig. 23.8).

ANSI Abbreviations and Symbols for Electrical Diagrams and Pipe Fittings

ANSI ABBREVIATIONS AND SYMBOLS*

alternating current	AC	left hand	LH
aluminum	AL	linear	LIN
American Standard	AMER STD	long	LG
approved	APPD	longitude	LONG.
average	AVG	machine	MACH
Babbitt	BAB	malleable iron	MI
ball bearing	BB	material	MATL
brass	BRS	maximum	MAX
Brinell hardness number	BHN	meter	M
bronze	BRZ	miles	MI
Brown & Sharpe	B & S	miles per hour	MPH
cast iron	CI	millimeter	MM
center line	CL or ℄	minimum	MIN
center to center	C to C	minute (angular measure)	' or MIN
centimeter	CM	minute (time)	MIN
chemical	CHEM	outside diameter	OD
circular	CIR	pattern	PATT
circular pitch	CP	phosphor bronze	PH BRZ
cold-rolled steel	CRS	piece	PC
copper	COP	pitch	P
counterbore	CBORE	pitch diameter	PD
countersink	CSK	plate	PL
cubic	CU	pound	# or LB
cubic foot	CU FT	pounds per square foot	PSF
cubic inch	CU IN	pounds per square inch	PSI
cubic yard	CU YD	Pratt & Whitney	P & W
cylinder	CYL	quantity	QTY
degree	DEG or °	radius	R or RAD
detail drawing	DET DWG	required	REQD
diagonal	DIAG	revolution per minute	RPM
diameter	DIA	right hand	RH
diametral pitch	DP	round	RD
direct current	DC	round bar	φ
drawing	DWG	screw	SCR
drawn	DR	second (angular measure)	''
effective	EFF	second (time)	SEC
electric	ELEC	section	SECT
engineer	ENGR	Society of Automotive Engineers	SAE
external	EXT	square	SQ
fabricate	FAB	square foot	SQ FT
fillister	FIL	square inch	SQ IN
finish	FIN.	standard	STD
foot	' or FT	steel	STL
gallon	GAL	steel casting	STL CSTG
galvanized iron	GI	thousand	M
grind	GRD	thread	THD
harden	HDN	ton	TON
hexagon	HEX	traced	TR
horsepower	HP	volt	V
hour	HR	watt	W
impregnate	IMPREG	weight	WT
inch	'' or IN.	Woodruff	WDF
inside diameter	ID	wrought iron	WI
internal	INT	yard	YD
lateral	LAT	year	YR

* Only those abbreviations that are commonly used on engineering drawings have been given in the list above. These have been selected from the long and comprehensive list given in ANSI Z32.13. The professional draftsman should have this standard close at hand for use when needed.

Graphical symbols for electrical diagrams*

AMPLIFIER General	**DEVICE, VISUAL SIGNALING** Annunciator	**PATH, TRANSMISSION** General
ANTENNA General	**ELEMENT, THERMAL** Thermal Cutout, Flasher	Crossing, Not Connected
	Thermal Relay	Junction
		Junction, Connected Paths
ARRESTER General Multigap	or	Pair
BATTERY Multicell		Assembled Conductors; Cable Coaxial
BREAKER, CIRCUIT General	**FUSE** General	2-Conductor Cable
	Fusible Element	Grouping Leads
CAPACITOR General	**GROUND**	or
COIL Blowout (Broken line not part of symbol) Operating	**HANDSET** General	**RECEIVER** General Headset
CONNECTOR Female Contact Male Contact	**INDUCTOR WINDING** General or	**RECTIFIER** General
CONTACT, ELECTRIC Fixed ∘ or → Locking Nonlocking Rotating Closed Open ⟂ or	**LAMP** Ballast Lamp Incandescent	**REPEATER** 1-Way
		RESISTOR General
	MACHINE, ROTATING Basic Generator Motor Wound Rotor Armature	**SWITCH** General Single Throw Double Throw Knife
CORE Magnetic (General) Magnet or Relay	Winding Symbols 1-Phase 2-Phase	**THERMOCOUPLE** Temperature-Measuring
COUPLER, DIRECTIONAL General	3-Phase Wye 3-Phase Delta	**TRANSFORMER** Magnetic Core Shielded

*ANSI Y32.2.

Conventional symbols for fittings*

GRAPHICAL SYMBOLS FOR PIPE FITTINGS					
	FLANGED	SCREWED	BELL & SPIGOT	WELDED	SOLDERED
1) BUSHING		⊳	⊏	✕	⊕
2) CAP		⊐	⊃		
3) CROSS-STRAIGHT SIZE					
4) ELBOW — 45 DEGREE					
— 90 DEGREE					
— TURNED DOWN					
— TURNED UP					
5) LATERAL					
6) PLUG		◁	⊏		
7) TEE — STRAIGHT SIZE					
— OUTLET UP					
— OUTLET DOWN					
8) UNION					
9) REDUCER — CONCENTRIC					
— ECCENTRIC					
10) CHECK VALVE — STRAIGHT WAY					
11) GATE VALVE					
12) GLOBE VALVE					
13) ANGLE VALVE GLOBE — ELEVATION					
GLOBE — PLAN					

* Based on ANSI Z32.23.

Trigonometric Functions

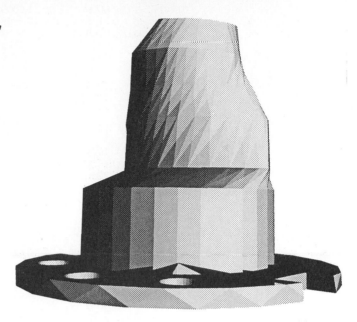

TABLE 1 *Trigonometric Functions*

Angle	Sine	Cosine	Tan	Cotan	Angle
0°	0.0000	1.0000	0.0000	∞	90°
1°	0.0175	0.9998	0.0175	57.290	89°
2°	.0349	.9994	.0349	28.636	88°
3°	.0523	.9986	.0524	19.081	87°
4°	.0698	.9976	.0699	14.301	86°
5°	.0872	.9962	.0875	11.430	85°
6°	.1045	.9945	.1051	9.5144	84°
7°	.1219	.9925	.1228	8.1443	83°
8°	.1392	.9903	.1405	7.1154	82°
9°	.1564	.9877	.1584	6.3138	81°
10°	.1736	.9848	.1763	5.6713	80°
11°	.1908	.9816	.1944	5.1446	79°
12°	.2079	.9781	.2126	4.7046	78°
13°	.2250	.9744	.2309	4.3315	77°
14°	.2419	.9703	.2493	4.0108	76°
15°	.2588	.9659	.2679	3.7321	75°
16°	.2756	.9613	.2867	3.4874	74°
17°	.2924	.9563	.3057	3.2709	73°
18°	.3090	.9511	.3249	3.0777	72°
19°	.3256	.9455	.3443	2.9042	71°
20°	.3420	.9397	.3640	2.7475	70°
21°	.3584	.9336	.3839	2.6051	69°
22°	.3746	.9272	.4040	2.4751	68°
23°	.3907	.9205	.4245	2.3559	67°
24°	.4067	.9135	.4452	2.2460	66°
25°	.4226	.9063	.4663	2.1445	65°
26°	.4384	.8988	.4877	2.0503	64°
27°	.4540	.8910	.5095	1.9626	63°
28°	.4695	.8829	.5317	1.8807	62°
29°	.4848	.8746	.5543	1.8040	61°
30°	.5000	.8660	.5774	1.7321	60°
31°	.5150	.8572	.6009	1.6643	59°
32°	.5299	.8480	.6249	1.6003	58°
33°	.5446	.8387	.6494	1.5399	57°
34°	.5592	.8290	.6745	1.4826	56°
35°	.5736	.8192	.7002	1.4281	55°
36°	.5878	.8090	.7265	1.3764	54°
37°	.6018	.7986	.7536	1.3270	53°
38°	.6157	.7880	.7813	1.2799	52°
39°	.6293	.7771	.8098	1.2349	51°
40°	.6428	.7660	.8391	1.1918	50°
41°	.6561	.7547	.8693	1.1504	49°
42°	.6691	.7431	.9004	1.1106	48°
43°	.6820	.7314	.9325	1.0724	47°
44°	.6947	.7193	.9657	1.0355	46°
45°	.7071	.7071	1.0000	1.0000	45°
Angle	Cosine	Sine	Cotan	Tan	Angle

Metric Tables*

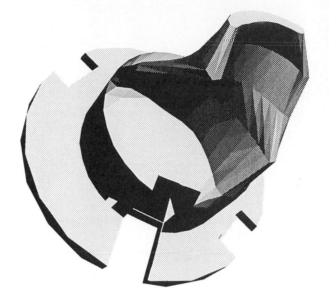

TABLE

* Refer to Appendix F for a listing of applicable standards.

TABLE 2 Inch—Millimeter Equivalents

4ths	8ths	16ths	32nds	64ths	To 4 places	To 3 places	To 2 places	To 4 places
				1/64	.0156	.016	.02	0.3969
			1/32		.0312	.031	.03	0.7938
				3/64	.0469	.047	.05	1.1906
		1/16			.0625	.062	.06	1.5875
				5/64	.0781	.078	.08	1.9844
			3/32		.0938	.094	.09	2.3813
				7/64	.1094	.109	.11	2.7781
	1/8				.1250	.125	.12	3.1750
				9/64	.1406	.141	.14	3.5719
			5/32		.1562	.156	.16	3.9688
				11/64	.1719	.172	.17	4.3656
		3/16			.1875	.188	.19	4.7625
				13/64	.2031	.203	.20	5.1594
			7/32		.2188	.219	.22	5.5563
				15/64	.2344	.234	.23	5.9531
1/4					.2500	.250	.25	6.3500
				17/64	.2656	.266	.27	6.7469
			9/32		.2812	.281	.28	7.1438
				19/64	.2969	.297	.30	7.5406
		5/16			.3125	.312	.31	7.9375
				21/64	.3281	.328	.33	8.3344
			11/32		.3438	.344	.34	8.7313
				23/64	.3594	.359	.36	9.1281
	3/8				.3750	.375	.38	9.5250
				25/64	.3906	.391	.39	9.9219
			13/32		.4062	.406	.41	10.3188
				27/64	.4219	.422	.42	10.7156
		7/16			.4375	.438	.44	11.1125
				29/64	.4531	.453	.45	11.5094
			15/32		.4688	.469	.47	11.9063
				31/64	.4844	.484	.48	12.3031
					.5000	.500	.50	12.7000
				33/64	.5156	.516	.52	13.0969
			17/32		.5312	.531	.53	13.4938
				35/64	.5469	.547	.55	13.8906
		9/16			.5625	.562	.56	14.2875
				37/64	.5781	.578	.58	14.6844
			19/32		.5938	.594	.59	15.0813
				39/64	.6094	.609	.61	15.4781
	5/8				.6250	.625	.62	15.8750
				41/64	.6406	.641	.64	16.2719
			21/32		.6562	.656	.66	16.6688
				43/64	.6719	.672	.67	17.0656
		11/16			.6875	.688	.69	17.4625
				45/64	.7031	.703	.70	17.8594
			23/32		.7188	.719	.72	18.2563
				47/64	.7344	.734	.73	18.6531
3/4					.7500	.750	.75	19.0500
				49/64	.7656	.766	.77	19.4469
			25/32		.7812	.781	.78	19.8438
				51/64	.7969	.797	.80	20.2406
		13/16			.8125	.812	.81	20.6375
				53/64	.8281	.828	.83	21.0344
			27/32		.8438	.844	.84	21.4313
				55/64	.8594	.859	.86	21.8281
	7/8				.8750	.875	.88	22.2250
				57/64	.8906	.891	.89	22.6219
			29/32		.9062	.906	.91	23.0188
				59/64	.9219	.922	.92	23.4156
		15/16			.9375	.938	.94	23.8125
				61/64	.9531	.953	.95	24.2094
			31/32		.9688	.969	.97	24.6063
				63/64	.9844	.984	.98	25.0031
					1.0000	1.000	1.00	25.4000

TABLE *3* Inches to Millimeters*

| Inches | Inches | | | | | | | | | |
	0	1	2	3	4	5	6	7	8	9
0–9	0	25.4	50.8	76.2	101.6	127.0	152.4	177.8	203.2	228.6
10–19	254.0	279.4	304.8	330.2	355.6	381.0	406.4	431.8	457.2	482.6
20–29	508.0	533.4	558.8	584.2	609.6	635.0	660.4	685.8	711.2	736.6
30–39	762.0	787.4	812.8	838.2	863.6	889.0	914.4	939.8	965.2	990.6
40–49	1016.0	1041.4	1066.8	1092.2	1117.6	1143.0	1168.4	1193.8	1219.2	1244.6
50–59	1270.0	1295.4	1320.8	1346.2	1371.6	1397.0	1422.4	1447.8	1473.2	1498.6
60–69	1524.0	1549.4	1574.8	1600.2	1625.6	1651.0	1676.4	1701.8	1727.2	1752.6
70–79	1778.0	1803.4	1828.8	1854.2	1879.6	1905.0	1930.4	1955.8	1981.2	2006.6
80–89	2032.0	2057.4	2082.8	2108.2	2133.6	2159.0	2184.4	2209.8	2235.2	2260.6
90–99	2286.0	2311.4	2336.8	2362.2	2387.6	2413.0	2438.4	2463.8	2489.2	2514.6

*Based on 1 in. = 25.4 mm.
Examples: To obtain the millimeter equivalent of 14 in., read the value below 4 in. in the horizontal line of values for 10–19 in. (355.6). The equivalent value of 52.4 in. is: 1320.8 *plus* 10.16 = 1330.96, the 10.16 value being obtained by moving the decimal point left in the equivalent value for 4 in.

TABLE *4* Metric Equivalents—Millimeters to Decimal Inches*

mm = in.	mm = in.	mm = in.	mm = in.	mm = in.
1 = 0.0394	21 = 0.8268	41 = 1.6142	61 = 2.4016	81 = 3.1890
2 = 0.0787	22 = 0.8661	42 = 1.6535	62 = 2.4409	82 = 3.2283
3 = 0.1181	23 = 0.9055	43 = 1.6929	63 = 2.4803	83 = 3.2677
4 = 0.1575	24 = 0.9449	44 = 1.7323	64 = 2.5197	84 = 3.3071
5 = 0.1969	25 = 0.9843	45 = 1.7717	65 = 2.5591	85 = 3.3465
6 = 0.2362	26 = 1.0236	46 = 1.8110	66 = 2.5984	86 = 3.3858
7 = 0.2756	27 = 1.0630	47 = 1.8504	67 = 2.6378	87 = 3.4252
8 = 0.3150	28 = 1.1024	48 = 1.8898	68 = 2.6772	88 = 3.4646
9 = 0.3543	29 = 1.1417	49 = 1.9291	69 = 2.7165	89 = 3.5039
10 = 0.3937	30 = 1.1811	50 = 1.9685	70 = 2.7559	90 = 3.5433
11 = 0.4331	31 = 1.2205	51 = 2.0079	71 = 2.7953	91 = 3.5827
12 = 0.4724	32 = 1.2598	52 = 2.0472	72 = 2.8346	92 = 3.6220
13 = 0.5118	33 = 1.2992	53 = 2.0866	73 = 2.8740	93 = 3.6614
14 = 0.5512	34 = 1.3386	54 = 2.1260	74 = 2.9134	94 = 3.7008
15 = 0.5906	35 = 1.3780	55 = 2.1654	75 = 2.9528	95 = 3.7402
16 = 0.6299	36 = 1.4173	56 = 2.2047	76 = 2.9921	96 = 3.7795
17 = 0.6693	37 = 1.4567	57 = 2.2441	77 = 3.0315	97 = 3.8189
18 = 0.7087	38 = 1.4961	58 = 2.2835	78 = 3.0709	98 = 3.8583
19 = 0.7480	39 = 1.5354	59 = 2.3228	79 = 3.1102	99 = 3.8976
20 = 0.7874	40 = 1.5748	60 = 2.3622	80 = 3.1496	100 = 3.9370

*To nearest fourth decimal place.

TABLE 5 Metric Tolerance Equivalents: Conversion of Tolerances—Decimal-inch to Millimeters*

Design practice Decimal, in.	Metric, mm	Calculated equivalents of decimal-inch tolerances†	Design practice Decimal, in.	Metric, mm	Calculated equivalents of decimal-inch tolerances°
0.0001	0.003	0.00254	0.004	0.1	0.1016
0.0002	0.005	0.00508	0.005	0.13	0.1270
0.0003	0.008	0.00762	0.006	0.15	0.1524
0.0004	0.01	0.01016	0.007	0.18	0.1778
0.0005	0.013	0.01270	0.008	0.2	0.2032
0.0006	0.015	0.01524	0.009	0.23	0.2286
0.0007	0.018	0.01778	0.010	0.25	0.254
0.0008	0.02	0.02032	0.015	0.4	0.381
0.0009	0.023	0.02286	0.02	0.5	0.508
0.001	0.025	0.0254	0.03	0.8	0.762
0.0015	0.04	0.0381	0.04	1	1.016
0.002	0.05	0.0508	0.06	1.5	1.524
0.0025	0.06	0.0635	0.08	2	2.032
0.003	0.08	0.0762	0.10	2.5	2.540

*To be used for tolerance conversions only. Table gives commonly used inch tolerances.
†Calculated on the basis of 1 in. = 25.4 mm.
Additional information relating to the conversion of tolerances may be obtained from the SAE Standard-Dual Dimensioning-SAE J390.

TABLE 6 English—Metric Equivalents

Measures of length

1 Meter = 39.37 Inches = 3.281 Feet = 1.094 Yards
1 Centimeter = 0.3937 Inch
1 Millimeter = $\frac{1}{25}$ Inch (approximately)
1 Kilometer = 0.621 Mile
1 Inch = 2.540 Centimeters = 25.400 Millimeters
1 Foot = 0.305 Meter
1 Mile = 1.609 Kilometers

Measures of surface

1 Square Meter = 10.764 Square Feet = 1.196 Square Yards
1 Square Centimeter = 0.155 Square Inch
1 Square Millimeter = 0.00155 Square Inch
1 Square Yard = 0.836 Square Meter
1 Square Foot = 0.0929 Square Meter
1 Square Inch = 6.452 Square Centimeters = 645.2 Square Millimeters

Measures of volume

1 Cubic Meter = 35.314 Cubic Feet = 1.308 Cubic Yards
1 Cubic Decimeter = 61.023 Cubic Inches = 0.0353 Cubic Foot
1 Cubic Centimeter = 0.061 Cubic Inch
1 Liter = 61.223 Cubic Inches = 0.0353 Cubic Foot = 0.2642 (U.S.) Gallon
1 Cubic Foot = 28.317 Cubic Decimeters = 28.317 Liters
1 Cubic Inch = 16.387 Cubic Centimeters

Measures of weight

1 Kilogram = 2.2046 Pounds
1 Metric Ton = 2204.6 Pounds = .9842 Ton (2240 pounds)
1 Ounce (avoirdupois) = 28.35 Grams
1 Pound = 0.4536 Kilogram

Speed measurements

1 Kilometer per Hour = 0.621 Mile per Hour

TABLE 7 Metric Units—Symbols and Formulas

Quantity	Unit	SI Symbol	Formula
General			
Length	millimeter	mm	—
	meter	m	—
	kilometer	km	—
Area	square meter	—	m^2
	hectare (10,000 square meters)	ha	—
Volume	cubic centimeter	—	cm^3
	cubic meter	—	m^3
	milliliter	ml	—
	liter	l	—
Mass	gram	g	—
	kilogram	kg	—
	ton (1000 kilograms)	t	—
Speed	meter per second	—	m/s
	kilometer per hour	—	km/h
Time	second	s	—
	minute, hour, day, month, etc.	—	—
Temperature	degree Celsius	°C	—
Thermodynamic temperature	kelvin	K	—
*Power	watt	W	J/s
	kilowatt	kW	—
*Energy, work	joule	J	N·m
	kilowatt hour	—	kW·h
Electric current	ampere	A	—
Electric potential difference	volt	V	W/A
Electric resistance	ohm	Ω	V/A
Space and Time			
Velocity	meter per second	—	m/s
Acceleration	meter per second squared	—	m/s^2
Angular velocity	radian per second	—	rad/s
Angular acceleration	radian per second squared	—	rad/s^2
*Frequency	hertz	Hz	cycle/s
Mechanics			
Density	kilogram per cubic meter	—	kg/m^3
Momentum	kilogram-meter per second	—	kg·m/s
Moment of inertia	kilogram-meter squared	—	$kg·m^2$
*Force	newton	N	$kg·m/s^2$
Torque or moment of force	newton meter	—	N·m
*Pressure and stress	pascal	Pa	N/m^2
Light			
*Luminous flux	lumen	lm	—
*Illuminance	lux	lx	lm/m^2

* Derived units with special names.

TABLE 8 English—Metric Conversion Multipliers*

To convert from	To	Multiplier
Centigrade (°C)	Fahrenheit (°F)	$\frac{9}{5}$(°C) + 32
Centimeters (cm)	feet	0.0328
Centimeters	inches	0.3937
Centimeters	meters	0.01
Cubic centimeters (cm³)	cubic feet	0.00003531
Cubic centimeters	cubic inches	0.06102
Cubic centimeters	liters	0.0010
Cubic centimeters	cubic meter	0.0000010
Cubic inches (c in.³)	cubic centimeters	16.3872
Cubic inches	cubic meters	0.000016
Cubic feet	cubic centimeters	28,317.08
Cubic feet	cubic meters	0.0283
Cubic meters	cubic feet	35.3133
Cubic millimeters (mm³)	cubic centimeters	0.001
Cubic yards	cubic meters	0.7646
Degrees (arc)	radians	0.01745
Fahrenheit (°F)	centigrade (°C)	$\frac{5}{9}$(°F-32)
Feet (ft)	centimeters	30.4801
Feet	meters	0.3048
Foot-pound-force	joule (J)	1.3558
Foot²	meter²	0.0929
Foot³	meter³	0.0283
Foot/second	meter/second	0.3048
Gallons (U.S.)	cubic meters	0.0038
Gallons (U.S.)	liters	3.7878
Grams (g)	kilograms	0.0010
Grams	milligrams	1000
Grams/cubic centimeter	kilograms/cubic meter	1000
Hectometers (hm)	meters	100
Horsepower (hp)	kilogram-meters/second	76.042
Horsepower	metric horsepower	1.0139
Horsepower, metric	kilogram-meters/second	75.0
Inches (in.)	centimeters	2.54
Inches	meters	0.0254
Inches²	meters²	0.0006452
Inches³	meters³	0.00001639
Inches of mercury	grams/square centimeter	34.542
Kilogram-meters/second	horsepower	0.01305
Kilogram-meters/second	horsepower, metric	0.01333
Kilograms (kg)	grams	1000
Kilograms	tons	0.0011
Kilograms	tons, metric	0.001
Kilogram-force (kgf)	newton (N)	9.8066
Kiloliters (kl)	liters	1000
Kilometers (km)	meters	1000
Kilometers	statute miles	0.6214
Liters (l)	cubic centimeters	1000
Liters	cubic feet	0.035313
Liters	cubic inches	61.02398
Liters	U.S. gallons	0.2641
Megameters	meters	100,000
Meters (m)	U.S. miles	0.000622
Microns	meters	0.000001
Miles, statute (mi)	kilometers	1.6093
Miles, statute	meters	1,609.34
Miles/hour (mph)	kilometers/hour	1.6093
Miles/hour	meters/second	0,4470
Milligrams	grams	0.001
Millimeters (mm)	inches	0.03937
Millimeters	meters	0.001
Millimeters	microns	1000
Ounces, avoirdupois	grams	28.3495
Ounces, fluid	milliliters	29.57
Ounces, U.S. fluid	liters	0.0296

TABLE 8 (cont.) English—Metric Conversion Multipliers

To convert from	To	Multiplier
Pints, liquid	liters	0.4732
Pounds, avoirdupois	grams	453.5924
Poundal	newton (N)	0.1383
Poundal/foot²	pascal (Pa)	1.4882
Pound-force (avoirdupois)	newton (N)	4.4482
Pound-force-foot	newton-meter (Nm)	1.3558
Pound-force/foot²	pascal (Pa)	47.88
Pound-force/inch² (psi)	pascal (Pa)	6894.8
Pound-mass (avoirdupois)	kilogram	0.4536
Pound-mass/foot²	kilogram/meter²	4.8824
Pound-mass/foot³	kilogram/meter³	16.0185
Pound-mass/inch³	kilogram/meter³	2767.99
Quarts, liquid	liters	0.9464
Quintals	grams	100,000
Radians	degrees, arc	57.2958
Square centimeters	square feet	0.001076
Square centimeters	square inches	0.1550
Square centimeters	square millimeters	100
Square decameters	square meters	100
Square decimeters	square meters	0.01
Square feet	square centimeters	929.0341
Square hectometers	square meters	10,000
Square inches	square centimeters	6.4516
Square inches	square millimeters	645.1625
Square kilometers	hectares	100
Square kilometers	square meters	1,000,000
Square kilometers	square miles	0.3861
Square meters	square feet	10.7639
Square meters	square yards	1.1960
Square miles	square kilometers	2.590
Square millimeters (mm²)	square inches	0.00155
Square millimeters	square meters	0.000001
Square yards	square meters	0.8361
Tons, long	kilograms	1016.0470
Tons, metric	kilograms	1000
Tons, short	kilograms	907.18
Yards	meters	0.9144

*A complete listing of conversion factors prepared for use with the computer may be found in the ASTM–Metric Practice Guide (E380–72). Definitions of derived units of the International System (SI), such as the joule, the newton, and the pascal, will be found in this publication.

Energy—the joule is the meter-kilogram-second unit of work that is equal to the work done by a force of one newton when the point of application is displaced one meter in the direction of the force (Nm).

Force—the newton is the standard meter-kilogram-second unit of force. It is equal to the force that produces an acceleration of one meter per second per second when applied to a body having a mass of one kilogram ($kg \cdot m/s^2$).

Stress or pressure—the pascal is the SI unit of pressure or stress. The pascal (Pa) is the pressure or stress of one newton per square meter (N/m^2).

TABLE **9**   **ISO Metric Screw Threads—Fine and Coarse Series**

Nominal size (mm)	Series with graded pitches*				Thread-diameter preference†		
	Coarse	Tap drill‡	Fine	Tap drill‡	1	2	3¶
1.6	**0.35**	1.25	—	—	1.6	—	—
1.8	**0.35**	1.45	—	—	—	1.8	—
2	**0.4**	1.60	—	—	2	—	—
2.2	**0.45**	1.75	—	—	—	2.2	—
2.5	**0.45**	2.05	—	—	2.5	—	—
3	**0.5**	2.50	—	—	3	—	—
3.5	**0.6**	2.90	—	—	—	3.5	—
4	**0.7**	3.30	—	—	4	—	—
4.5	**0.75**	3.75	—	—	—	4.5	—
5	**0.8**	4.20	—	—	5	—	—
5.5	—	—	—	—	—	—	5.5
6	**1**	5.00	—	—	6	—	—
7	**1**	6.00	—	—	—	—	7
8	**1.25**	6.75	**1**	7.00	8	—	—
9	**1.25**	7.75	—	—	—	—	9
10	**1.5**	8.50	**1.25**	8.75	10	—	—
11	**1.5**	9.50	—	—	—	—	11
12	**1.75**	10.00	**1.25**	10.50	12	—	—
14	**2**	12.00	**1.5**	12.50	14	—	—
15	—	—	—	—	—	—	15
16	**2**	14.00	**1.5**	14.50	16	—	—
17	—	—	—	—	—	—	17
18	**2.5**	15.50	**1.5**	16.50	—	18	—
20	**2.5**	17.50	**1.5**	18.50	20	—	—
22	**2.5**	19.50	**1.5**	20.50	—	22	—
24	**3**	21.00	**2**	22.00	24	—	—
25	—	—	—	—	—	—	25
26	—	—	—	—	—	—	26
27	**3**	24.00	**2**	25.00	—	27	—
28	—	—	—	—	—	—	28
30	**3.5**	26.50	**2**	28.00	30	—	—
32	—	—	—	—	—	—	32
33	**3.5**	29.50	**2**	31.00	—	33	—
35	—	—	—	—	—	—	35
36	**4**	32.00	**3**	33.00	36	—	—
38	—	—	—	--	—	—	38
39	**4**	35.00	**3**	36.00	—	39	—
40	—	—	—	—	—	—	40
42	**4.5**	37.50	**3**	39.00	42	—	—
45	**4.5**	40.50	**3**	42.00	—	45	—
48	**5**	43.00	**3**	45.00	48	—	—
50	—	—	—	—	—	—	50
52	**5**	47.00	**3**	49.00	—	52	—
55	—	—	—	—	—	—	55
56	**5.5**	50.50	**4**	52.00	56	—	—
58	—	—	—	—	—	—	58
60	**5.5**	54.50	**4**	56.00	—	60	—
62	—	—	—	—	—	—	62
64	**6**	58.00	**4**	60.00	64	—	—
65	—	—	—	—	—	—	65
68	**6**	62.00	**4**	64.00	—	68	—

From ANSI B1.13M, Metric Screw Threads—M Profile.
*The pitches shown in bold type are those which have been established by ISO as a selected coarse- and fine-thread series for commercial threads and fasteners.
†Select thread diameter from columns 1, 2, or 3, with preference for selection being in that order.
‡For an approximate 75% thread, use the formula: nominal OD minus .97 × pitch.
¶For diameters listed in this column, see Table 11—Series With Constant Pitches.

TABLE 10 ISO Metric Screw Threads—Series with Constant Pitches

Nominal size (mm)	Pitches (mm)											
	6	4	3	2	1.5	1.25	1	0.75	0.5	0.35	0.25	0.2
1.6	—	—	—	—	—	—	—	—	—	—	—	0.2
1.8	—	—	—	—	—	—	—	—	—	—	—	0.2
2	—	—	—	—	—	—	—	—	—	—	0.25	—
2.2	—	—	—	—	—	—	—	—	—	—	0.25	—
2.5	—	—	—	—	—	—	—	—	—	0.35	—	—
3	—	—	—	—	—	—	—	—	—	0.35	—	—
3.5	—	—	—	—	—	—	—	—	—	0.35	—	—
4	—	—	—	—	—	—	—	—	0.5	—	—	—
4.5	—	—	—	—	—	—	—	—	0.5	—	—	—
5	—	—	—	—	—	—	—	—	0.5	—	—	—
5.5	—	—	—	—	—	—	—	—	0.5	—	—	—
6	—	—	—	—	—	—	—	0.75	—	—	—	—
7	—	—	—	—	—	—	—	0.75	—	—	—	—
8	—	—	—	—	—	—	1	0.75	—	—	—	—
9	—	—	—	—	—	—	1	0.75	—	—	—	—
10	—	—	—	—	—	1.25	1	0.75	—	—	—	—
11	—	—	—	—	—	—	1	0.75	—	—	—	—
12	—	—	—	—	1.5	1.25	1	—	—	—	—	—
14	—	—	—	—	1.5	1.25	1	—	—	—	—	—
15	—	—	—	—	1.5	—	1	—	—	—	—	—
16	—	—	—	—	1.5	—	1	—	—	—	—	—
17	—	—	—	—	1.5	—	1	—	—	—	—	—
18	—	—	—	2	1.5	—	1	—	—	—	—	—
20	—	—	—	2	1.5	—	1	—	—	—	—	—
22	—	—	—	2	1.5	—	1	—	—	—	—	—
24	—	—	—	2	1.5	—	1	—	—	—	—	—
25	—	—	—	2	1.5	—	1	—	—	—	—	—
26	—	—	—	—	1.5	—	1	—	—	—	—	—
27	—	—	—	2	1.5	—	1	—	—	—	—	—
28	—	—	—	2	1.5	—	1	—	—	—	—	—
30	—	—	3	2	1.5	—	1	—	—	—	—	—
32	—	—	—	2	1.5	—	—	—	—	—	—	—
33	—	—	3	2	1.5	—	—	—	—	—	—	—
35	—	—	—	—	1.5	—	—	—	—	—	—	—
36	—	—	—	2	1.5	—	—	—	—	—	—	—
38	—	—	—	—	1.5	—	—	—	—	—	—	—
39	—	—	—	2	1.5	—	—	—	—	—	—	—
40	—	—	3	2	1.5	—	—	—	—	—	—	—
42	—	4	3	2	1.5	—	—	—	—	—	—	—
45	—	4	3	2	1.5	—	—	—	—	—	—	—
48	—	4	3	2	1.5	—	—	—	—	—	—	—
50	—	—	3	2	1.5	—	—	—	—	—	—	—
52	—	4	3	2	1.5	—	—	—	—	—	—	—
55	—	4	3	2	1.5	—	—	—	—	—	—	—
56	—	4	3	2	1.5	—	—	—	—	—	—	—
58	—	4	3	2	1.5	—	—	—	—	—	—	—
60	—	4	3	2	1.5	—	—	—	—	—	—	—
62	—	4	3	2	1.5	—	—	—	—	—	—	—
64	—	4	3	2	1.5	—	—	—	—	—	—	—
65	—	4	3	2	1.5	—	—	—	—	—	—	—
68	—	4	3	2	1.5	—	—	—	—	—	—	—
70	6	4	3	2	1.5	—	—	—	—	—	—	—
72	6	4	3	2	1.5	—	—	—	—	—	—	—
75	—	4	3	2	1.5	—	—	—	—	—	—	—
76	6	4	3	2	1.5	—	—	—	—	—	—	—

TABLE **11** *Metric Hex Bolts and Hex Cap Screws**

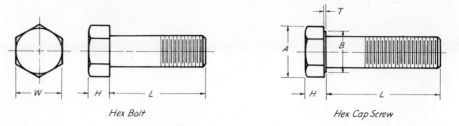

Hex Bolt   Hex Cap Screw

Nominal size and thread pitch	Hex bolts and hex cap screws		Bolts	Hex cap screws		
	W	A	H	H	B	T
	Width across flats max	Width across corners max	Head height max	Head height max	Washer face diameter min	Washer face thickness max
M5 × 0.8	8.00	9.24	3.58	3.65	7.0	0.5
M6 × 1	10.00	11.55	4.38	4.15	8.9	0.5
M8 × 1.25	13.00	15.01	5.68	5.50	11.6	0.6
M10 × 1.5	16.00	18.48	6.85	6.63	14.6	0.6
M12 × 1.75	18.00	20.78	7.95	7.76	16.6	0.6
M14 × 2	21.00	24.25	9.25	9.09	19.6	0.6
M16 × 2	24.00	27.71	10.75	10.32	22.5	0.8
M20 × 2.5	30.00	34.64	13.40	12.88	27.7	0.8
M24 × 3	36.00	41.57	15.90	15.44	33.2	0.8
M30 × 3.5	46.00	53.12	19.75	19.48	42.7	0.8
M36 × 4	55.00	63.51	23.55	23.38	51.1	0.8
M42 × 4.5	65.00	75.06	27.05	26.97	59.8	1.0
M48 × 5	75.00	86.60	31.07	31.07	69.0	1.0
M56 × 5.5	85.00	98.15	36.20	36.20	78.1	1.0
M64 × 6	95.00	109.70	41.32	41.32	87.2	1.0
M72 × 6	105.00	121.24	46.45	46.45	96.3	1.2
M80 × 6	115.00	132.72	51.58	51.58	105.4	1.2
M90 × 6	130.00	150.11	57.74	57.74	119.2	1.2
M100 × 6	145.00	167.43	63.90	63.90	133.0	1.2

*ANSI 18.2.3.1M–1979, Metric Hex Cap Screws; ANSI 18.2.3.5M–1979, Metric Hex Bolts.
All dimensions are in millimeters.
Dimension values are for drawing purposes only.

TABLE *12* **Metric Heavy Hex Bolts and Heavy Hex Screws***

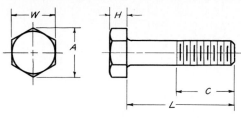

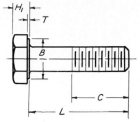

Metric Heavy Hex Bolt   Metric Heavy Hex Screw

Heavy hex bolts and heavy hex screws			Hex bolts†	Heavy hex screws*			Thread length (basic)— bolts and screws		
	W	A	H	H_1	B	T	C(Ref)		
Normal size and thread pitch	Width across flats max	Width across corners max	Head height max	Head height max	Washer face diameter min	Washer face thickness max	Lengths ⩽125	Lengths >125 and ⩽200	Lengths >200
M12 × 1.75	21.00	24.25	7.95	7.76	19.6	0.6	30	36	49
M14 × 2	24.00	27.71	9.25	9.09	22.5	0.6	34	40	53
M16 × 2	27.00	31.18	10.75	10.32	25.3	0.8	38	44	57
M20 × 2.5	34.00	39.26	13.40	12.88	31.4	0.8	46	52	65
M24 × 3	41.00	47.34	15.90	15.44	38.0	0.8	54	60	73
M30 × 3.5	50.00	57.74	19.75	19.48	46.6	0.8	66	72	85
M36 × 4	60.00	69.28	23.55	23.38	55.9	0.8	78	84	97

* ANSI B18.2.3.3M Metric Heavy Hex Screws.
† ANSI B18.2.3.6M Metric Heavy Hex Bolts.
Dimension values are for drawing purposes only.

TABLE **13** **Metric Hex Socket Cap Screws***

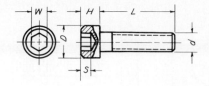

| d | Head | | Socket | |
| Nominal screw diameter and thread pitch | D | H | W | S |
	Head diameter max	Head height max	Hex socket size nom	Hex socket depth min
M1.6 × 0.35	3.0	1.6	1.5	0.80
M2 × 0.4	3.8	2.0	1.5	1.00
M2.5 × 0.45	4.5	2.5	2.0	1.25
M3 × 0.5	5.5	3.0	2.5	1.50
M4 × 0.7	7.0	4.0	3.0	2.00
M5 × 0.8	8.5	5.0	4.0	2.50
M6 × 1	10.0	6.0	5.0	3.00
M8 × 1.25	13.0	8.0	6.0	4.00
M10 × 1.5	16.0	10.0	8.0	5.00
M12 × 1.75	18.0	12.0	10.0	6.00
M14 × 2	21.0	14.0	12.0	7.00
M16 × 2	24.0	16.0	14.0	8.00
M20 × 2.5	30.0	20.0	17.0	10.00
M24 × 3	36.0	24.0	19.0	12.00
M30 × 3.5	45.0	30.0	22.0	15.00
M36 × 4	54.0	36.0	27.0	18.00
M42 × 4.5	63.0	42.0	32.0	21.00
M48 × 5	72.0	48.0	36.0	24.00

*ANSI B18.3.1M–1982, Socket Head Cap Screws.
Dimension values for drawing purposes only.

TABLE *14* Hexagon Socket Head Shoulder Screws*

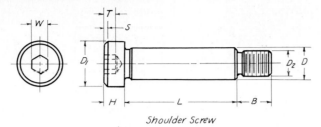

Shoulder Screw

	D	D_1	H	S	W	T	—	—	D_2	B	—	—
Basic shoulder diameter (mm)	Shoulder diameter max	Head diameter max	Head height max	Cham-fer or radius max	Hexa-gon socket size nom	Key engage-ment socket depth	Shoulder neck diameter min	Shoulder neck width max	Nominal screw diameter and thread pitch	Thread length max	Thread neck diameter max	Thread neck width max
6.5	6.487	10.00	4.50	0.6	3	2.4	5.92	2.5	M5 × 0.8	9.75	3.86	2.4
8.0	7.987	13.00	5.50	0.8	4	3.3	7.42	2.5	M6 × 1	11.25	4.58	2.6
10.0	9.987	16.00	7.00	1.0	5	4.2	9.42	2.5	M8 × 1.25	13.25	6.25	2.8
13.0	12.984	18.00	9.00	1.2	6	4.9	12.42	2.5	M10 × 1.5	16.40	7.91	3.0
16.0	15.984	24.00	11.00	1.6	8	6.6	15.42	2.5	M12 × 1.75	18.40	9.57	4.0
20.0	19.980	30.00	14.00	2.0	10	8.8	19.42	2.5	M16 × 2	22.40	13.23	4.8
25.0	24.980	36.00	16.00	2.4	12	10.0	22.42	3.0	M20 × 2.5	27.40	16.57	5.6

*ANSI B18.3.3M–1979, Hexagon Socket Head Shoulder Screws Metric Series.
Dimension values are for drawing purposes only.

TABLE *15* Hexagon Socket Button Head Cap Screws*

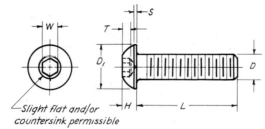

—Slight flat and/or countersink permissible

Hexagon Socket Button Head Cap Screw

D	D_1	H	S	W	T	L
Nominal size and thread pitch	Head diameter max	Head height max	Head side height ref	Hex socket size nom	Key engage-ment min	Maximum standard length nom
M3 × 0.5	5.70	1.65	0.38	2	1.04	12
M4 × 0.7	7.60	2.20	0.38	2.5	1.30	20
M5 × 0.8	9.50	2.75	0.50	3	1.56	30
M6 × 1	10.50	3.30	0.80	4	2.08	30
M8 × 1.25	14.00	4.40	0.80	5	2.60	40
M10 × 1.5	17.50	5.50	0.80	6	3.12	40
M12 × 1.75	21.00	6.60	0.80	8	4.16	60
M16 × 2	28.00	8.80	1.50	10	5.20	60

*ANSI B18.3.4M–1979, Hexagon Socket Button Head Cap Screws—Metric Series.
Dimension values are for drawing purposes only.

TABLE **16** *Metric Hex Nuts (Style 1 and Style 2) and Hex Jam Nuts**

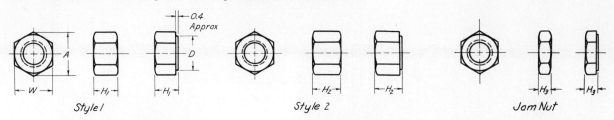

Style 1   Style 2   Jam Nut

	W	A	D	H₁	H₂	H₃	
Nominal size and thread pitch	*Width across flats max*	*Width across corners max*	*Bearing face diameter min*	*Nut thickness*			*Washer face thickness max*
				Style 1 max	*Style 2 max*	*Jam max*	
M1.6 × 0.35	3.20	3.70	2.4	1.3	—	—	—
M2 × 0.4	4.00	4.62	3.1	1.6	—	—	—
M2.5 × 0.45	5.00	5.77	4.1	2.0	—	—	—
M3 × 0.5	5.50	6.35	4.6	2.4	2.9	—	—
M3.5 × 0.6	6.00	6.93	5.1	2.8	3.3	—	—
M4 × 0.7	7.00	8.08	5.9	3.2	3.8	—	—
M5 × 0.8	8.00	9.24	6.9	4.7	5.1	2.7	—
M6 × 1	10.00	11.55	8.9	5.2	5.7	3.2	—
M8 × 1.25	13.00	15.01	11.6	6.8	7.5	4.0	—
M10 × 1.5	16.00	18.48	14.6	8.4	9.3	5.0	—
M12 × 1.75	18.00	20.78	16.6	10.8	12.0	6.0	—
M14 × 2	21.00	24.25	19.6	12.8	14.1	7.0	—
M16 × 2	24.00	27.71	22.5	14.8	16.4	8.0	—
M20 × 2.5	30.00	34.64	27.7	18.0	20.3	10.0	0.8
M24 × 3	36.00	41.57	33.2	21.5	23.9	12.0	0.8
M30 × 3.5	46.00	53.12	42.7	25.6	28.6	15.0	0.8
M36 × 4	55.00	63.51	51.1	31.0	34.7	18.0	0.8

Style 1: M1.6–M36
Style 2: M3–M36
Jam M5–M36

* ANSI B18.2.4.1M–1979, Metric Hex Nuts, Style 1; ANSI B18.2.4.2M–1979, Metric Hex Nuts, Style 2; ANSI B18.2.4.5M–1979, Metric Hex Jam Nuts.
All dimensions are in millimeters.
Values are for drawing purposes only.

TABLE *17* **Metric Hex Slotted Nuts***

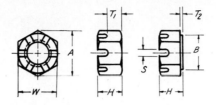

Nominal size and thread pitch	W Width across flats max	A Width across corners max	B Bearing face diameter min	H Nut thickness max	T_1 Unslotted thickness max	S Width of slot max	Depth of slot max	T_2 Washer face thickness max
M5 × 0.8	8.00	9.24	6.9	5.10	3.2	2.0	1.9	—
M6 × 1	10.00	11.55	8.9	5.70	3.5	2.4	2.2	—
M8 × 1.25	13.00	15.01	11.6	7.50	4.4	2.9	3.1	—
M10 × 1.5	16.00	18.48	14.6	9.30	5.2	3.4	4.1	—
M12 × 1.75	18.00	20.78	16.6	12.00	7.3	4.0	4.7	—
M14 × 2	21.00	24.25	19.6	14.10	8.6	4.3	5.5	—
M16 × 2	24.00	27.71	22.5	16.40	9.9	5.3	6.5	—
M20 × 2.5	30.00	34.64	27.7	20.30	13.3	5.7	7.0	0.8
M24 × 3	36.00	41.57	33.2	23.90	15.4	6.7	8.5	0.8
M30 × 3.5	46.00	53.12	42.7	28.60	18.1	8.5	10.5	0.8
M36 × 4	55.00	63.51	51.1	34.70	23.7	8.5	11.0	0.8

* ANSI B18.2.4.3M–1979, Metric Slotted Hex Nuts.
Dimension values are for drawing purposes only.
Threads are general-purpose threads.

TABLE **18** **Metric Slotted, Recessed, and Hex Head Machine Screws**

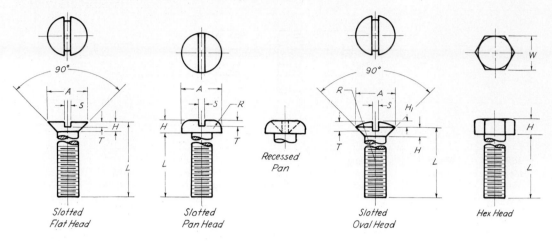

Slotted Flat Head   Slotted Pan Head   Recessed Pan   Slotted Oval Head   Hex Head

Nominal size and thread pitch	Flat head slotted			Pan head slotted				Oval head slotted					Slot width	Hex head	
	A Actual	H Max	T Max	A Max	H Max	R Max	T Min	A Actual	H Max	H_1 Max	R Approx	T Max	S Max	W Max	H Max
M2 × 0.4	3.60	1.20	0.6	3.90	1.35	0.8	0.55	3.60	1.20	0.50	3.8	1.0	0.7	3.20	1.27
M2.5 × 0.45	4.60	1.50	0.7	4.90	1.65	1.0	0.73	4.60	1.50	0.60	5.0	1.2	0.8	4.00	1.40
M3 × 0.5	5.50	1.80	0.9	5.80	1.90	1.2	0.80	5.50	1.80	0.75	5.7	1.5	1.0	5.00	1.52
M3.5 × 0.6	6.44	2.10	1.0	6.80	2.25	1.4	0.95	6.44	2.10	0.90	6.5	1.7	1.2	5.50	2.36
M4 × 0.7	7.44	2.32	1.1	7.80	2.55	1.6	1.15	7.44	2.32	1.00	7.8	1.9	1.4	7.00	2.79
M5 × 0.8	9.44	2.85	1.4	9.80	3.10	2.0	1.35	9.44	2.85	1.25	9.9	2.3	1.5	8.00	3.05
M6 × 1	11.87	3.60	1.8	12.00	3.90	2.5	1.70	11.87	3.60	1.60	12.2	3.0	1.9	10.00	4.83
M8 × 1.25	15.17	4.40	2.1	15.60	5.00	3.2	2.20	15.17	4.40	2.00	15.8	3.7	2.3	13.00	5.84
M10 × 1.5	18.98	5.35	2.6	19.50	6.20	4.0	2.70	18.98	5.35	2.50	19.8	4.5	2.8	15.00	7.49
M12 × 1.75	22.88	6.35	3.1	23.40	7.50	4.8	3.20	22.88	6.35	3.00	23.8	5.3	2.8	18.00	9.50

Dimension values are for drawing purposes only.

TABLE **19** **Metric Round Head Square Neck Bolts**

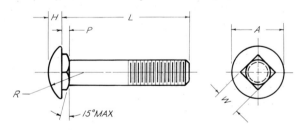

Nominal bolt size and thread pitch	A Head diameter max	H Head height max	R Head radius approx	W Width of square max	P Depth of square max
M5 × 0.8	11.3	3.0	7.2	5.48	3.0
M6 × 1	13.8	3.6	8.9	6.78	3.0
M8 × 1.25	17.0	4.5	11.2	8.58	3.0
M10 × 1.5	20.6	5.5	13.7	10.58	4.0
M12 × 1.75	24.7	6.5	16.5	12.70	4.0
M14 × 2	28.3	7.5	19.1	14.70	4.0
M16 × 2	33.1	8.8	22.6	20.84	5.0
M20 × 2.5	39.8	10.8	27.3	24.84	5.0

Dimension values are for drawing purposes only.

TABLE 20 Metric Set Screws‡

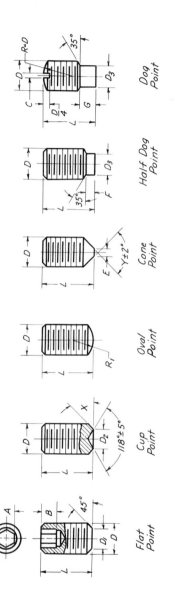

Flat Point   Cup Point   Oval Point   Cone Point   Half Dog Point   Dog Point

Nominal size of screw D‡ (mm)	Thread pitch	Nom screw lengths (mm)	Hexagon socket size nom A nom	B†	Slotted* headless C	R	Flat‡ D₁ max	Cup‡ D₂ max	X	Oval‡ R₁ max	Cone‡ Y	E max	Half‡ dog D₃ max	F max	Full* dog G
1.6	0.35	1.5, 2, 2.5, 3	0.7		—		0.80	0.80		1.60		0.16	0.8	0.53	—
2	0.4	1.5, 2, 2.5, 3, 4	0.9		—		1.00	1.00		1.90		0.2	1.00	0.64	—
2.5	0.45	2, 2.5, 3, 4	1.3		—		1.50	1.20		2.28		0.25	1.50	0.78	—
3	0.5	2, 2.5, 3, 4, 5	1.5	On a drawing the socket depth may be made equal the distance A	0.4	Radius = D	2.00	1.40	Angle X shall be 45° for screws of length equal to nominal dia. and longer	2.65	Cone point angle 90° for these lengths and over	0.3	2.00	0.92	2.40
4	0.7	2.5, 3, 4, 5, 6	2		0.6		2.50	2.00		3.80		0.4	2.50	1.20	2.80
5	0.8	3, 4, 5, 6, 8	2.5		0.8		3.50	2.50		4.55		0.5	3.50	1.37	2.80
6	1	4, 5, 6, 8	3		1		4.00	3.00		5.30		1.5	4.00	1.74	3.35
8	1.25	5, 6, 8, 10	4		1.2		5.50	5.00		6.80		2.0	5.50	2.28	4.70
10	1.5	6, 8, 10, 12	5		1.6		7.00	6.00		8.30		2.5	7.00	2.82	4.90
12	1.75	8, 10, 12, 16	6		—		8.50	8.00		9.80		3.0	8.50	3.35	6.40
16	2	10, 12, 16, 20	8		—		12.00	10.00		12.80		4.0	12.00	4.40	8.15
20	2.5	12, 16, 20, 25	10		—		15.00	14.00		15.80		5.0	15.00	5.45	8.85
24	3	16, 20, 25, 30	12		—		18.00	16.00		18.80		6.0	18.00	6.49	9.65

* From ISO Tables. Not in ANSI B18.3.6M-1979.
† Depth varies for different types of points and lengths. See ANSI B18.3.6M-1979.
‡ Compiled from ANSI B18.3.6M-1979, Metric Series—Hexagon Socket Set Screws.
Dimension values are for drawing purposes only.

TABLE 21 Metric Plain and Lock Washers

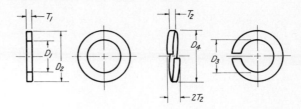

Nominal size			M4	M5	M6	M8	M10	M12	M16	M20	M24
Diameters	Plain	D_1	4.3	5.3	6.4	8.4	10.5	13	17	21	25
		D_2	9	10	12.5	17	21	24	30	37	44
	Lock	D_3	4.1	5.1	6.1	8.1	10.2	12.2	16.2	20.2	24.5
		D_4	7.6	9.2	11.8	14.8	18.1	21.1	27.4	33.6	40
Thickness		T_1	0.8	1	1.6	1.6	2	2.5	3	3	4
		T_2	0.9	1.2	1.6	2	2.2	2.5	3.5	4	5

Dimension values are for drawing purposes only.

TABLE 22 Preferred ISO (Metric) Fits-hole Basis. In Agreement with ANSI B4.2

Basic size Over	To	Loose running Hole H11	Loose running Shaft c11	Free running Hole H9	Free running Shaft d9	Close running Hole H8	Close running Shaft f7	Sliding Hole H7	Sliding Shaft g6	Loc. clearance Hole H7	Loc. clearance Shaft h6	Loc. trans. Hole H7	Loc. trans. Shaft k6	Loc. trans. Hole H7	Loc. trans. Shaft n6	Loc. interf. Hole H7	Loc. interf. Shaft p6	Medium drive Hole H7	Medium drive Shaft s6	Force Hole H7	Force Shaft u6
1	3	+60/0	-60/-120	+25/0	-20/-45	+14/0	-6/-16	+10/0	-2/-8	+10/0	-6/0	+10/0	+6/0	+10/0	+10/+4	+10/0	+12/+6	+10/0	+20/+14	+10/0	+24/+18
3	6	+75/0	-70/-145	+30/0	-30/-60	+18/0	-10/-22	+12/0	-4/-12	+12/0	-8/0	+12/0	+9/+1	+12/0	+16/+8	+12/0	+20/+12	+12/0	+27/+19	+12/0	+31/+23
6	10	+90/0	-80/-170	+36/0	-40/-76	+22/0	-13/-28	+15/0	-5/-14	+15/0	-9/0	+15/0	+10/+1	+15/0	+19/+10	+15/0	+24/+15	+15/0	+32/+23	+15/0	+37/+28
10	18	+110/0	-95/-205	+43/0	-50/-93	+27/0	-16/-34	+18/0	-6/-17	+18/0	-11/0	+18/0	+12/+1	+18/0	+23/+12	+18/0	+29/+18	+18/0	+39/+28	+18/0	+44/+33
18	24	+130/0	-110/-240	+52/0	-65/-117	+33/0	-20/-41	+21/0	-7/-20	+21/0	-13/0	+21/0	+15/+2	+21/0	+28/+15	+21/0	+35/+22	+21/0	+48/+35	+21/0	+54/+41
24	30	+130/0	-110/-240	+52/0	-65/-117	+33/0	-20/-41	+21/0	-7/-20	+21/0	-13/0	+21/0	+15/+2	+21/0	+28/+15	+21/0	+35/+22	+21/0	+48/+35	0	+61/+48
30	40	+160/0	-120/-280	+62/0	-80/-142	+39/0	-25/-50	+25/0	-9/-25	+25/0	-16/0	+25/0	+18/+2	+25/0	+33/+17	+25/0	+42/+26	+25/0	+59/+43	+25/0	+76/+60
40	50	+160/0	-130/-290	+62/0	-80/-142	+39/0	-25/-50	+25/0	-9/-25	+25/0	-16/0	+25/0	+18/+2	+25/0	+33/+17	+25/0	+42/+26	+25/0	+59/+43	0	+86/+70
50	65	+190/0	-140/-330	+74/0	-100/-174	+46/0	-30/-60	+30/0	-10/-29	+30/0	-19/0	+30/0	+21/+2	+30/0	+39/+20	+30/0	+51/+32	+30/0	+72/+53	+30/0	+106/+87
65	80	+190/0	-150/-340	+74/0	-100/-174	+46/0	-30/-60	+30/0	-10/-29	+30/0	-19/0	+30/0	+21/+2	+30/0	+39/+20	+30/0	+51/+32	0	+78/+59	0	+121/+102
80	100	+220/0	-170/-390	+87/0	-120/-207	+54/0	-36/-71	+35/0	-12/-34	+35/0	-22/0	+35/0	+25/+3	+35/0	+45/+23	+35/0	+59/+37	+35/0	+93/+71	+35/0	+146/+124
100	120	+220/0	-180/-400	+87/0	-120/-207	+54/0	-36/-71	+35/0	-12/-34	+35/0	-22/0	+35/0	+25/+3	+35/0	+45/+23	+35/0	+59/+37	0	+101/+79	0	+166/+144
120	140	+250/0	-200/-450	+100/0	-145/-245	+63/0	-43/-83	+40/0	-14/-39	+40/0	-25/0	+40/0	+28/+3	+40/0	+52/+27	+40/0	+68/+43	+40/0	+117/+92	+40/0	+195/+170
140	160	+250/0	-210/-460	+100/0	-145/-245	+63/0	-43/-83	+40/0	-14/-39	+40/0	-25/0	+40/0	+28/+3	+40/0	+52/+27	+40/0	+68/+43	+40/0	+125/+100	+40/0	+215/+190
160	180	+250/0	-230/-480	+100/0	-145/-245	+63/0	-43/-83	+40/0	-14/-39	+40/0	-25/0	+40/0	+28/+3	+40/0	+52/+27	+40/0	+68/+43	0	+133/+108	0	+235/+210
180	200	+290/0	-240/-530	+115/0	-170/-285	+72/0	-50/-96	+46/0	-15/-44	+46/0	-29/0	+46/0	+33/+4	+46/0	+60/+31	+46/0	+79/+60	+46/0	+151/+122	+46/0	+265/+236
200	225	+290/0	-260/-550	+115/0	-170/-285	+72/0	-50/-96	+46/0	-15/-44	+46/0	-29/0	+46/0	+33/+4	+46/0	+60/+31	+46/0	+79/+60	+46/0	+159/+130	+46/0	+287/+258
225	250	+290/0	-280/-570	+115/0	-170/-285	+72/0	-50/-96	+46/0	-15/-44	+46/0	-29/0	+46/0	+33/+4	+46/0	+60/+31	+46/0	+79/+60	0	+169/+140	0	+313/+284

Values in micrometers (0.001 mm)

Example: Basic Hole ⌀50.000; Sliding Fit H7g6

Tolerances: Hole $^{+25}_{0}$; Shaft $^{-9}_{-25}$. Limit Dimensions: Hole ⌀50$^{+0.025}_{0}$ or $^{50.025}_{50.000}$; Shaft ⌀50$^{-0.009}_{-0.025}$ or $^{49.991}_{49.975}$

TABLE 23A Preferred Tolerance Zones for External (Shaft) Dimensions (Metric)

Nominal size (mm) Over	To (incl.)	c11		d9		f7		g6		h6	h7	h9	h11	k6		n6		p6		s6		u6*	
		−	−	−	−	−	−	−	−	−	−	−	−	+	+	+	+	+	+	+	+	+	+
1	3	60	120	20	45	6	16	2	8	0 6	0 10	0 25	0 60	6	0	10	4	12	6	20	14	24	18
3	6	70	145	30	60	10	22	4	12	0 8	0 12	0 30	0 75	9	1	16	8	20	12	27	19	31	23
6	10	80	170	40	76	13	28	5	14	0 9	0 15	0 36	0 90	10	1	19	10	24	15	32	23	37	28
10	18	95	205	50	93	16	34	6	17	0 11	0 18	0 43	0 110	12	1	23	12	29	18	39	28	44	33
18	24	110	240	65	117	20	41	7	20	0 13	0 21	0 52	0 130	15	2	28	15	35	22	48	35	54	41
24	30																					61	48
30	40	120	280	80	142	25	50	9	25	0 16	0 25	0 62	0 160	18	2	33	17	42	26	59	43	76	60
40	50	130	290																			86	70
50	65	140	330	100	174	30	60	10	29	0 19	0 30	0 74	0 190	21	2	39	20	51	32	72	53	106	87
65	80	150	340																	78	59	121	102
80	100	170	390	120	207	36	71	12	34	0 22	0 35	0 87	0 220	25	3	45	23	59	37	93	71	146	124
100	120	180	400																	101	79	166	144
120	140	200	450	145	245	43	83	14	39	0 25	0 40	0 100	0 250	28	3	52	27	68	43	117	92	195	170
140	160	210	460																	125	100	215	190
160	180	230	480																	133	108	235	210
180	200	240	530	170	285	50	96	15	44	0 29	0 46	0 115	0 290	33	4	60	31	79	50	151	122	265	236
200	225	260	550																	159	130	287	258
225	250	280	570																	169	140	313	284

Compiled from ANSI B4.2 Values in micrometers (0.001 mm)
* Numerical values for tolerance zones in this area not defined.
Example: ø $50^{-0.009}_{-0.025}$ or 49.991–49.975(g6)

TABLE 23B Preferred Tolerance Zones for Internal (Hole) Dimensions

Nominal size (mm) Over	To (incl.)	H7	H8	H9	H11	C11		D9		F8		G7		K7*		N7		P7		S7		U7*	
		+	+	+	+	+	+	+	+	+	+	+	+	+	−	−	−	−	−	−	−	−	−
1	3	10 0	14 0	25 0	60 0	120	60	45	20	20	6	12	2	0	10	4	14	6	16	14	24	18	28
3	6	12 0	18 0	30 0	75 0	145	70	60	30	28	10	16	4	3	9	4	16	8	20	15	27	19	31
6	10	15 0	22 0	36 0	90 0	170	80	76	40	35	13	20	5	5	10	4	19	9	24	17	32	22	37
10	18	18 0	27 0	43 0	110 0	205	95	93	50	43	16	24	6	6	12	5	23	11	29	21	39	26	44
18	24	21 0	33 0	52 0	130 0	240	110	117	65	53	20	28	7	6	15	7	28	14	35	27	48	33	54
24	30																					40	61
30	40	25 0	39 0	62 0	160 0	280	120	142	80	64	25	34	9	7	18	8	33	17	42	34	59	51	76
40	50					290	130															61	86
50	65	30 0	46 0	74 0	190 0	330	140	174	100	76	30	40	10	9	21	9	39	21	51	42	72	76	106
65	80					340	150													48	78	91	121
80	100	35 0	54 0	87 0	220 0	390	170	207	120	90	36	47	12	10	25	10	45	24	59	58	93	111	146
100	120					400	180													66	101	131	166
120	140	40 0	63 0	100 0	250 0	450	200	245	145	106	43	54	14	12	28	12	52	28	68	77	117	155	195
140	160					460	210													85	125	175	215
160	180					480	230													93	133	195	235
180	200	46 0	72 0	115 0	290 0	530	240	285	170	122	50	61	15	13	33	14	60	33	79	105	151	219	265
200	225					550	260													113	159	241	287
225	250					570	280													123	169	267	313

Completed from ANSI B4.2 Values in micrometers (0.001 mm)
* Numerical values for tolerance zones in this area not defined.
Example: ø $50^{+0.025}_{0}$ or $^{50.025}_{50.000}$ (H7).

TABLE *24* **Metric Woodruff Keys Selected Sizes**

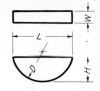 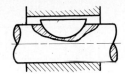

| Nominal size | | Shaft dia mm | Key data | | Depth of keyseat in shaft |
| W | H | | L mm | D Dia | |
mm					
1	1.4	3 to 4	3.82	4	1
1.5	2.6	4 to 6	6.76	7	2
2	2.6	6 to 8	6.76	7	1.8
2	3.7		9.66	10	2.9
2.5	3.7	8 to 10	9.66	10	2.9
3	3.7		9.66	10	2.5
3	5		12.65	13	3.8
3	6.5		15.72	16	5.3
4	5	10 to 12	12.65	13	3.5
4	6.5		15.72	16	5
4	7.5		18.57	19	6
5	6.5	12 to 17	15.72	16	4.5
5	7.5		18.57	19	5.5
5	9		21.63	22	7
6	7.5	17 to 22	18.57	19	5.1
6	9		21.63	22	6.6
6	11		27.35	28	8.6
8	9	22 to 30	21.63	22	6.2
8	11		27.35	28	8.2
8	13		31.43	32	10.2
10	11	30 to 38	27.35	28	7.8
10	13		31.43	32	9.8
10	16		43.08	45	12.8

Not American Standard.

TABLE *25*A *Single-row Radial Bearings—Type 3000**

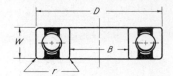

See table below for explanation of letters

Bearing no.	Bore B		Diameter D		Width W		Balls		Radius r†
	mm	in.	mm	in.	mm	in.	Diam	No.	
3200	10	.3937	30	1.1811	9	.3543	$7/32$	7	.025
3300	10	.3937	35	1.3780	11	.4331	$1/4$	7	.025
3201	12	.4724	32	1.2598	10	.3937	.210	8	.025
3301	12	.4724	37	1.4567	12	.4724	$9/32$	7	.04
3202	15	.5906	35	1.3780	11	.4331	.210	9	.025
3302	15	.5906	42	1.6535	13	.5118	$5/16$	7	.04
3203	17	.6693	40	1.5748	12	.4724	$9/32$	8	.04
3303	17	.6693	47	1.8504	14	.5512	$11/32$	7	.04
3204	20	.7874	47	1.8504	14	.5512	$5/16$	8	.04
3304	20	.7874	52	2.0472	15	.5906	$13/32$	7	.04
3205	25	.9843	52	2.0472	15	.5906	$5/16$	9	.04
3305	25	.9843	62	2.4409	17	.6693	$13/32$	8	.04
3206	30	1.1811	62	2.4409	16	.6299	$11/32$	9	.04
3306	30	1.1811	72	2.8346	19	.7480	$15/32$	8	.04
3207	35	1.3780	72	2.8346	17	.6693	$7/16$	9	.04
3307	35	1.3780	80	3.1496	21	.8268	$17/32$	8	.06
3208	40	1.5748	80	3.1496	18	.7087	$15/32$	9	.04
3308	40	1.5748	90	3.5433	23	.9055	$19/32$	8	.06
3209	45	1.7717	85	3.3465	19	.7480	$15/32$	10	.04
3309	45	1.7717	100	3.9370	25	.9843	$21/32$	8	.06
3210	50	1.9685	90	3.5433	20	.7874	$15/32$	11	.04
3310	50	1.9685	110	4.3307	27	1.0630	$23/32$	8	.08
3211	55	2.1654	100	3.9370	21	.8268	$1/2$	11	.06
3311	55	2.1654	120	4.7244	29	1.1417	$25/32$	8	.08

* For radial or combined loads from either direction where thrust is to be resisted by a single bearing and is not great enough to require use of angular contact type.
† Radius *r* indicates maximum fillet radius in housing or on shaft, which bearing radius will clear.
Pages 140–143 from Vol. I, *New Departure Handbook.*

TABLE *25B* *Radax Bearings—Type 30,000**

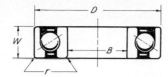

See table below for explanation of letters

Bearing no.	Bore B		Diameter D		Width W		Balls		Radius r†
	mm	in.	mm	in.	mm	in.	Diam	No.	
30204			47	1.8504	14	.5512	$\frac{11}{32}$	10	
30304	20	.7874	52	2.0472	15	.5906	$\frac{3}{8}$	10	.04
30404			72	2.8346	19	.7480	$\frac{9}{16}$	8	
30205			52	2.0472	15	.5906	$\frac{11}{32}$	11	.04
30305	25	.9843	62	2.4409	17	.6693	$\frac{7}{16}$	10	.04
30405			80	3.1496	21	.8268	$\frac{5}{8}$	9	.06
30206			62	2.4409	16	.6299	$\frac{3}{8}$	12	.04
30306	30	1.1811	72	2.8346	19	.7480	$\frac{1}{2}$	10	.04
30406			90	3.5433	23	.9055	$\frac{11}{16}$	9	.06
30207			72	2.8346	17	.6693	$\frac{7}{16}$	12	.04
30307	35	1.3780	80	3.1496	21	.8268	$\frac{9}{16}$	11	.06
30407			100	3.9370	25	.9843	$\frac{3}{4}$	9	.06
30208			80	3.1496	18	.7087	$\frac{1}{2}$	12	.04
30308	40	1.5748	90	3.5433	23	.9055	$\frac{5}{8}$	11	.06
30408			110	4.3307	27	1.0630	$\frac{13}{16}$	10	.08
30209			85	3.3465	19	.7480	$\frac{1}{2}$	13	.04
30309	45	1.7717	100	3.9370	25	.9843	$\frac{11}{16}$	11	.06
30409			120	4.7244	29	1.1417	$\frac{7}{8}$	10	.08
30210			90	3.5433	20	.7874	$\frac{1}{2}$	14	.04
30310	50	1.9685	110	4.3307	27	1.0630	$\frac{3}{4}$	11	.08
30410			130	5.1181	31	1.2205	$\frac{15}{16}$	10	.08
30211			100	3.9370	21	.8268	$\frac{9}{16}$	14	.06
30311	55	2.1654	120	4.7244	29	1.1417	$\frac{13}{16}$	12	.08
30411			140	5.5118	33	1.2992	1	10	.08
30212			110	4.3307	22	.8661	$\frac{5}{8}$	14	.06
30312	60	2.3622	130	5.1181	31	1.2205	$\frac{7}{8}$	12	.08
30412			150	5.9055	35	1.3780	$1\frac{1}{16}$	10	.08

* Single-row angular contact; provide maximum capacity for one-direction thrust loads. Mounted two bearings opposed for combined loads or thrust from either direction.
† Radius r indicates maximum fillet radius in housing or on shaft, which bearing radius will clear.
Page 143 from Vol. I, *New Departure Handbook.*

TABLE 26A Shaft Mounting Fits

Bearing and bore numbers	Bearing bore				Shaft revolving				Shaft stationary			
	Diameters inches°		Diameters millimeters		Diameters inches		Diameters millimeters		Diameters inches		Diameters millimeters	
	max	min	max	min	max	min	max	min	max	min	max	min
0	.3937	.3934	10.000	9.992	.3939	.3936	10.005	9.997	.3935	.3932	9.995	9.987
1	.4724	.4721	12.000	11.992	.4726	.4723	12.005	11.997	.4722	.4719	11.995	11.987
2	.5906	.5903	15.000	14.992	.5908	.5905	15.005	14.997	.5904	.5901	14.995	14.987
3	.6693	.6690	17.000	16.992	.6695	.6692	17.005	16.997	.6691	.6688	16.995	16.987
4	.7874	.7870	20.000	19.990	.7877	.7873	20.008	19.998	.7871	.7867	19.993	19.983
5	.9843	.9839	25.000	24.990	.9846	.9842	25.008	24.998	.9840	.9836	24.993	24.983
6	1.1811	1.1807	30.000	29.990	1.1814	1.1810	30.008	29.998	1.1808	1.1804	29.993	29.983
7	1.3780	1.3775	35.000	34.987	1.3784	1.3779	35.010	34.997	1.3776	1.3771	34.990	34.977
8	1.5748	1.5743	40.000	39.987	1.5752	1.5747	40.010	39.997	1.5744	1.5739	39.990	39.977
9	1.7717	1.7712	45.000	44.987	1.7721	1.7716	45.010	44.997	1.7713	1.7708	44.990	44.977
10	1.9685	1.9680	50.000	49.987	1.9689	1.9684	50.010	49.997	1.9681	1.9676	49.990	43.977
11	2.1654	2.1648	55.000	54.985	2.1659	2.1653	55.013	54.998	2.1649	2.1643	54.988	54.973
12	2.3622	2.3616	60.000	59.985	2.3627	2.3621	60.013	59.998	2.3617	2.3611	59.988	59.973
13	2.5591	2.5585	65.000	64.985	2.5596	2.5590	65.013	64.998	2.5586	2.5580	64.988	64.973
14	2.7559	2.7553	70.000	69.985	2.7564	2.7558	70.013	69.998	2.7554	2.7548	69.988	69.973

TABLE 26B Housing Mounting Fits

Bearing and bore numbers Series lgt		med	hvy	Bearing outer diam.				Housing stationary				Housing revolving			
				Diameters inches°		Diameters millimeters		Diameters inches		Diameters millimeters		Diameters inches		Diameters millimeters	
lgt	med	hvy		max	min	max	min	max	min	max	min	max	min	max	min
0				1.1811	1.1807	30.000	29.990	1.1815	1.1810	30.010	29.997	1.1810	1.1805	29.997	29.984
1				1.2598	1.2593	32.000	31.987	1.2603	1.2597	32.013	31.998	1.2597	1.2591	31.997	31.982
2	0			1.3780	1.3775	35.000	34.987	1.3785	1.3779	35.013	34.998	1.3779	1.3773	34.997	34.982
	1			1.4567	1.4562	37.000	36.987	1.4572	1.4566	37.013	36.998	1.4566	1.4560	36.997	36.982
3				1.5748	1.5743	40.000	39.987	1.5753	1.5747	40.013	39.998	1.5747	1.5741	39.997	39.982
	2			1.6535	1.6530	42.000	41.987	1.6540	1.6534	42.013	41.998	1.6534	1.6528	41.997	41.982
4	3			1.8504	1.8499	47.000	46.987	1.8509	1.8503	47.013	46.998	1.8503	1.8497	46.997	46.982
5	4			2.0472	2.0466	52.000	51.985	2.0479	2.0471	52.018	51.998	2.0472	2.0464	52.000	51.980
6	5			2.4409	2.4403	62.000	61.985	2.4416	2.4408	62.018	61.998	2.4409	2.4401	62.000	61.980
7	6	4		2.8346	2.8340	72.000	71.985	2.8353	2.8345	72.018	71.998	2.8346	2.8338	72.000	71.980
8	7	5		3.1496	3.1490	80.000	79.985	3.1503	3.1495	80.018	79.998	3.1496	3.1488	80.000	79.980
9				3.3465	3.3457	85.000	84.980	3.3473	3.3463	85.020	84.995	3.3466	3.3456	85.003	84.978
10	8	6		3.5433	3.5425	90.000	89.980	3.5441	3.5431	90.020	89.995	3.5434	3.5424	90.003	89.978
11	9	7		3.9370	3.9362	100.000	99.980	3.9378	3.9368	100.020	99.995	3.9371	3.9361	100.003	99.978
12	10	8		4.3307	4.3299	110.000	109.980	4.3315	4.3305	110.020	109.995	4.3308	4.3298	110.003	109.978
13	11	9		4.7244	4.7236	120.000	119.980	4.7252	4.7242	120.020	119.995	4.7245	4.7235	120.003	119.978
14				4.9213	4.9203	125.000	124.975	4.9223	4.9211	125.025	124.995	4.9214	4.9202	125.003	124.973
15	12	10		5.1181	5.1171	130.000	129.975	5.1191	5.1179	130.025	129.995	5.1182	5.1170	130.002	129.972
16	13	11		5.5118	5.5108	140.000	139.975	5.5128	5.5116	140.025	139.995	5.5119	5.5107	140.002	139.972
17	14	12		5.9055	5.9045	150.000	149.975	5.9065	5.9053	150.025	149.995	5.9056	5.9044	150.002	149.972
18	15	13		6.2992	6.2982	160.000	159.975	6.3002	6.2990	160.025	159.995	6.2993	6.2981	160.002	159.972

*Inch values compiled from *New Departure Handbook*.

TABLE 27 Small Drills—Metric*

Metric drill diameter	Diameter (in.)	Metric drill diameter	Diameter (in.)	Metric drill diameter	Diameter (in.)	Metric drill diameter	Diameter (in.)
0.200	.0078	2.300	.0905	5.70	.2244	9.20	.3622
0.250	.0098	2.35	.0925	5.75	.2264	9.25	.3642
0.300	.0118	2.400	.0945	5.80	.2283	9.30	.3661
0.350	.0138	2.45	.0964	5.90	.2323	9.40	.3701
0.400	.0157	2.50	.0984	6.00	.2362	9.50	.3740
0.450	.0177	2.60	.1024	6.10	.2401	9.60	.3779
0.500	.0197	2.70	.1063	6.20	.2441	9.70	.3819
0.550	.0216	2.75	.1083	6.25	.2461	9.75	.3838
0.600	.0236	2.80	.1102	6.30	.2480	9.80	.3858
0.650	.0256	2.90	.1142	6.40	.2520	9.90	.3898
0.700	.0275	3.00	.1181	6.50	.2559	10.00	.3937
0.750	.0295	3.10	.1220	6.60	.2598	10.50	.4134
0.800	.0315	3.20	.1260	6.70	.2638	11.00	.4331
0.850	.0335	3.25	.1279	6.75	.2657	11.50	.4527
0.900	.0354	3.30	.1299	6.80	.2677	12.00	.4724
0.950	.0374	3.40	.1338	6.90	.2716	12.50	.4921
1.000	.0394	3.50	.1378	7.00	.2756	13.00	.5118
1.050	.0413	3.60	.1417	7.10	.2795	13.50	.5315
1.100	.0433	3.70	.1457	7.20	.2835	14.00	.5512
1.150	.0453	3.75	.1476	7.25	.2854	14.50	.5709
1.200	.0472	3.80	.1496	7.30	.2874	15.00	.5905
1.250	.0492	3.90	.1535	7.40	.2913	15.50	.6102
1.300	.0512	4.00	.1575	7.50	.2953	16.00	.6299
1.350	.0531	4.10	.1614	7.60	.2992	16.50	.6496
1.400	.0551	4.20	.1653	7.70	.3031	17.00	.6693
1.450	.0571	4.25	.1673	7.75	.3051	17.50	.6890
1.500	.0590	4.30	.1693	7.80	.3071	18.00	.7087
1.550	.0610	4.40	.1732	7.90	.3110	18.50	.7283
1.600	.0630	4.50	.1772	8.00	.3150	19.00	.7480
1.650	.0650	4.60	.1811	8.10	.3189	19.50	.7677
1.700	.0669	4.70	.1850	8.20	.3228	20.00	.7874
1.750	.0689	4.75	.1870	8.25	.3248	20.50	.8071
1.800	.0709	4.80	.1890	8.30	.3268	21.00	.8268
1.850	.0728	4.90	.1929	8.40	.3307	21.50	.8464
1.900	.0748	5.00	.1968	8.50	.3346	22.00	.8661
1.950	.0768	5.10	.2008	8.60	.3386	22.50	.8858
2.000	.0787	5.20	.2047	8.70	.3425	23.00	.9055
2.050	.0807	5.25	.2067	8.75	.3445	23.50	.9252
2.100	.0827	5.30	.2087	8.80	.3464	24.00	.9449
2.150	.0846	5.40	.2126	8.90	.3504	24.50	.9646
2.200	.0866	5.50	.2165	9.00	.3543	25.00	.9842
2.250	.0886	5.60	.2205	9.10	.3583	25.50	1.0039

* Drills beyond the range of this table increase in diameter by increments of 0.50 mm—26.00, 26.50, 27.00, etc.

Inch Tables*

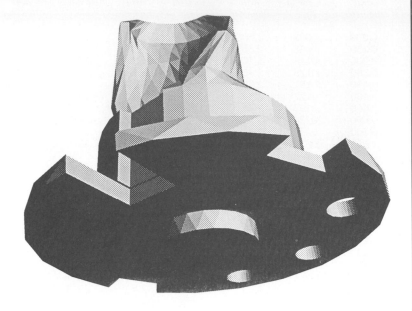

* Refer to Appendix F for a listing of applicable standards.

TABLE 28 *Standard Unified Thread Series**

Present Unified thread Nominal size—diameter			Coarse (NC) (UNC)		Fine (NF) (UNF)		Extra-fine (NEF) (UNEF)	
Inch		Metric Equiv.‡	Threads per inch	Tap drill†	Threads per inch	Tap drill†	Threads per inch	Tap drill†
.060	0	1.52	—	—	80	3/64	—	—
.073	1	1.85	64	No. 53	72	No. 53	—	—
.086	2	2.18	56	No. 50	64	No. 50	—	—
.099	3	2.51	48	No. 47	56	No. 45	—	—
.112	4	2.84	**40**	No. 43	48	No. 42	—	—
.125	5	3.17	40	No. 38	44	No. 37	—	—
.138	6	3.50	**32**	No. 36	40	No. 33	—	—
.164	8	4.16	**32**	No. 29	36	No. 29	—	—
.190	10	4.83	**24**	No. 25	**32**	No. 21	—	—
.216	12	5.49	24	No. 16	28	No. 14	32	No. 13
.250	1/4	6.35	**20**	No. 7	**28**	No. 3	32	No. 2
.3125	5/16	7.94	**18**	F	**24**	I	32	K
.375	3/8	9.52	**16**	5/16	**24**	Q	32	S
.4375	7/16	11.11	**14**	U	**20**	25/64	**28**	Y
.500	1/2	12.70	**13**	27/64	**20**	29/64	**28**	15/32
.5625	9/16	14.29	**12**	31/64	**18**	33/64	24	17/32
.625	5/8	15.87	**11**	17/32	**18**	37/64	24	19/32
.6875	11/16	17.46	—	—	—	—	24	41/64
.750	3/4	19.05	**10**	21/32	**16**	11/16	**20**	45/64
.8125	13/16	20.64	—	—	—	—	**20**	49/64
.875	7/8	22.22	**9**	49/64	**14**	13/16	**20**	53/64
.9375	15/16	23.81	—	—	—	—	**20**	57/64
1.000	1	25.40	**8**	7/8	**12**	59/64	**20**	61/64
1.0625	1 1/16	26.99	—	—	—	—	**18**	1
1.125	1 1/8	28.57	**7**	63/64	**12**	1 3/64	18	1 5/64
1.1875	1 3/16	30.16	—	—	—	—	18	1 9/64
1.250	1 1/4	31.75	**7**	1 7/64	**12**	1 11/64	18	1 13/64
1.3125	1 5/16	33.34	—	—	—	—	18	1 17/64
1.375	1 3/8	34.92	**6**	1 13/64	**12**	1 19/64	18	1 5/16
1.4375	1 7/16	36.51	—	—	—	—	18	1 3/8
1.500	1 1/2	38.10	**6**	1 21/64	**12**	1 27/64	18	1 29/64
1.5625	1 9/16	39.69	—	—	—	—	18	1 1/2
1.625	1 5/8	41.27	—	—	—	—	18	1 9/16
1.6875	1 11/16	42.86	—	—	—	—	18	1 5/8
1.750	1 3/4	44.45	**5**	1 35/64	—	—	**16**	1 11/16
2.000	2	50.80	4½	1 25/32	—	—	**16**	1 15/16
2.250	2 1/4	57.15	4½	2 1/32	—	—	—	—
2.500	2 1/2	63.50	4	2 1/4	—	—	—	—
2.750	2 3/4	69.85	4	2 1/2	—	—	—	—
3.000	3	76.20	4	2 3/4	—	—	—	—
3.250	3 1/4	82.55	4	3	—	—	—	—
3.500	3 1/2	88.90	4	3 1/4	—	—	—	—
3.750	3 3/4	95.25	4	3 1/2	—	—	—	—
4.000	4	101.60	4	3 3/4	—	—	—	—

* Adapted from ANSI B1.1.
Bold type indicates Unified threads. To be designated UNC or UNF.
Unified Standard—Classes 1A, 2A, 3A, 1B, 2B, and 3B.
For recommended hole-size limits before threading, see Tables 38 and 39, ANSI B1.1.
† Tap drill for a 75% thread (not Unified—American Standard).
Bold-type sizes smaller than ¼ in. are accepted for limited applications by the British, but the symbols NC or NF, as applicable, are retained.
‡ The values listed as metric equivalents of decimal inch values have been given to assist user in selecting the closest metric size to be found in Table 10. Adherence to diameter preference is recommended, if feasible. For a metric thread use the tap drill size recommended in Table 10.

TABLE **29** *Standard Unified Special Threads* (8-Pitch, 12-Pitch, and 16-Pitch Series)*

Nominal size—diameter		Threads per inch			Nominal size—diameter		Threads per inch		
mm†	inch				mm	inch			
12.70	.500 $\frac{1}{2}$	—	**12**	—	44.45	1.750 $1\frac{3}{4}$	**8**	**12**	**16**
14.29	.5625 $\frac{9}{16}$	—	**12**	—	46.04	1.8125 $1\frac{13}{16}$	—	—	**16**
15.88	.625 $\frac{5}{8}$	—	**12**	—	47.62	1.875 $1\frac{7}{8}$	8	12	**16**
17.46	.6875 $\frac{11}{16}$	—	**12**	—	49.21	1.9375 $1\frac{15}{16}$	—	—	**16**
19.05	.750 $\frac{3}{4}$	—	**12**	**16**	50.80	2.000 2	**8**	**12**	**16**
20.64	.8125 $\frac{13}{16}$	—	**12**	**16**	52.39	2.0625 $2\frac{1}{16}$	—	—	**16**
22.22	.875 $\frac{7}{8}$	—	**12**	**16**	53.98	2.125 $2\frac{1}{8}$	8	12	**16**
23.81	.9375 $\frac{15}{16}$	—	**12**	**16**	55.56	2.1875 $2\frac{3}{16}$	—	—	**16**
25.40	1.000 1	8	**12**	**16**	57.15	2.250 $2\frac{1}{4}$	**8**	**12**	**16**
26.99	1.0625 $1\frac{1}{16}$	—	**12**	**16**	58.74	2.3125 $2\frac{5}{16}$	—	—	**16**
28.58	1.125 $1\frac{1}{8}$	8	**12**	**16**	60.32	2.375 $2\frac{3}{8}$	—	12	**16**
30.16	1.1875 $1\frac{3}{16}$	—	**12**	**16**	61.91	2.4375 $2\frac{7}{16}$	—	—	**16**
31.75	1.250 $1\frac{1}{4}$	8	**12**	**16**	63.50	2.500 $2\frac{1}{2}$	**8**	**12**	**16**
33.34	1.3125 $1\frac{5}{16}$	—	**12**	**16**	66.68	2.625 $2\frac{5}{8}$	—	12	**16**
34.92	1.375 $1\frac{3}{8}$	8	**12**	**16**	69.85	2.750 $2\frac{3}{4}$	**8**	**12**	**16**
36.51	1.4375 $1\frac{7}{16}$	—	**12**	**16**	73.02	2.875 $2\frac{7}{8}$	—	12	**16**
38.10	1.500 $1\frac{1}{2}$	8	**12**	**16**	76.20	3.000 3	**8**	**12**	**16**
39.69	1.5625 $1\frac{9}{16}$	—	—	**16**	79.38	3.125 $3\frac{1}{8}$	—	12	**16**
41.28	1.625 $1\frac{5}{8}$	8	12	**16**	82.55	3.250 $3\frac{1}{4}$	**8**	**12**	**16**
42.86	1.6875 $1\frac{11}{16}$	—	—	**16**	85.72	3.375 $3\frac{3}{8}$	—	12	**16**

* Adapted from ANSI B1.1. Bold type indicates Unified threads (UN).
† The values listed as metric equivalents of decimal inch values have been given to assist user in selecting the closest metric size to be found in Tables 8 and 9. Adherence to diameter preference is recommended, if feasible.

TABLE **30** *Standard Acme and Stub Acme Threads**

Nominal size—diameter		Threads per inch	Nominal size—diameter		Threads per inch
mm	inch		mm	inch	
6.35	.250 $\frac{1}{4}$	16	31.75	1.250 $1\frac{1}{4}$	5
7.94	.3125 $\frac{5}{16}$	14	34.92	1.375 $1\frac{3}{8}$	4
9.52	.375 $\frac{3}{8}$	12	38.10	1.500 $1\frac{1}{2}$	4
11.11	.4375 $\frac{7}{16}$	12	44.45	1.750 $1\frac{3}{4}$	4
12.70	.500 $\frac{1}{2}$	10	50.80	2.000 2	4
15.88	.625 $\frac{5}{8}$	8	57.15	2.250 $2\frac{1}{4}$	3
19.05	.750 $\frac{3}{4}$	6	63.50	2.500 $2\frac{1}{2}$	3
22.22	.875 $\frac{7}{8}$	6	69.85	2.750 $2\frac{3}{4}$	3
25.40	1.000 1	5	76.20	3.000 3	2
28.58	1.125 $1\frac{1}{8}$	5			

* In agreement with ANSI B1.5 and B1.8.

TABLE **31** *Buttress Threads* (Suggested Combinations of Diameters and Pitches)*

Nominal size—diameter								Associated pitches (tpi)
mm	inch	mm	inch	mm	inch	mm	inch	
12.70	.500	14.29	.5625	15.88	.625	17.46	.6875	20 16 12
19.05	.750	22.22	.875	25.40	1.000	—	—	— 16 12 10
28.58	1.125	31.75	1.250	34.92	1.375	38.10	1.500	— 16 12 10 8 6
44.45	1.750	76.20	3.000	57.15	2.250	63.50	2.500	— 16 12 10 8 6 5 4
69.85	2.750	76.20	3.000	88.90	3.500	101.60	4.000	— 16 12 10 8 6 5 4

* In agreement with ANSI B1.9.

TABLE 32 American National Standard Wrench-head Bolts and Nuts—Regular Series*

Bolt diameter, nominal size		Bolt heads				Nuts				
		Width across flats	Height of head			Width across flats	Thickness			
							Regular		Jam	
		Unfinished and semifinished square and hexagon†	Unfinished square	Unfinished hexagon	Semifinished hexagon	Unfinished hexagon, semifinished hexagon and hexagon jam	Unfinished square and hexagon	Semifinished hexagon and hexagon slotted	Unfinished hexagon	Semifinished hexagon
.250	1/4	[3/8 sq. / 7/16 hex.]	11/64	11/64	5/32	7/16	7/32	7/32	5/32	5/32
.3125	5/16	1/2	13/64	7/32	13/64	1/2	17/64	17/64	3/16	3/16
.375	3/8	9/16	1/4	1/4	15/64	9/16	21/64	21/64	7/32	7/32
.4375	7/16	5/8	19/64	19/64	9/32	11/16	3/8	3/8	1/4	1/4
.500	1/2	3/4	21/64	11/32	5/16	3/4	7/16	7/16	5/16	5/16
.5625	9/16	13/16	—	—	23/64	7/8	31/64	31/64	5/16	5/16
.625	5/8	15/16	27/64	27/64	25/64	15/16	35/64	35/64	3/8	3/8
.750	3/4	1 1/8	1/2	1/2	15/32	1 1/8	21/32	41/64	7/16	27/64
.875	7/8	1 5/16	19/32	37/64	35/64	1 5/16	49/64	3/4	1/2	31/64
1.000	1	1 1/2	21/32	43/64	39/64	1 1/2	7/8	55/64	9/16	35/64
1.125	1 1/8	1 11/16	3/4	3/4	11/16	1 11/16	1	31/32	5/8	39/64
1.250	1 1/4	1 7/8	27/32	27/32	25/32	1 7/8	1 3/32	1 1/16	3/4	23/32
1.375	1 3/8	2 1/16	29/32	29/32	27/32	2 1/16	1 13/64	1 11/64	13/16	25/32
1.500	1 1/2	2 1/4	1	1	15/16	2 1/4	1 5/16	1 9/32	7/8	27/32
1.625	1 5/8	2 7/16	1 3/32	—	—	2 7/16	—	1 25/64	—	29/32
1.750	1 3/4	2 5/8	—	1 5/32	1 3/32	2 5/8	—	1 1/2	—	31/32
1.875	1 7/8	2 13/16	—	—	—	2 13/16	—	1 39/64	—	1 1/32
2.000	2	3	—	1 11/32	1 7/32	3	—	1 23/32	—	1 3/32
2.250	2 1/4	3 3/8	—	1 1/2	1 3/8	3 3/8	—	1 59/64	—	1 13/64
2.500	2 1/2	3 3/4	—	1 21/32	1 17/32	3 3/4	—	2 9/64	—	1 23/64
2.750	2 3/4	4 1/8	—	1 13/16	1 11/16	4 1/8	—	2 23/64	—	1 37/64
3.000	3	4 1/2	—	2	1 7/8	4 1/2	—	2 37/64	—	1 45/64

* ANSI B18.2.1; ANSI B18.2.2.
Thread-bolts: coarse-thread series, class 2A.
Thread-nuts: unfinished; coarse series, class 2B: semifinished; coarse, fine, or 8-pitch series.
† Square bolts in ¼–1⅝ in. sizes (nominal) only.

TABLE **33** *American National Standard Finished Hexagon Castle Nuts**

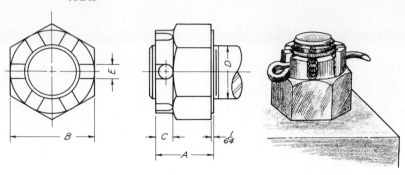

Nominal size— diameter D		Threads per inch		Thickness A	Width across flats, B	Slot		Diameter of cylindrical part min
		UNC	UNF			Depth, C	Width, E	
.250	$\frac{1}{4}$	20	28	$\frac{9}{32}$	$\frac{7}{16}$	.094	.078	.371
.3125	$\frac{5}{16}$	18	24	$\frac{21}{64}$	$\frac{1}{2}$	.094	.094	.425
.375	$\frac{3}{8}$	16	24	$\frac{13}{32}$	$\frac{9}{16}$	.125	.125	.478
.4375	$\frac{7}{16}$	14	20	$\frac{29}{64}$	$\frac{11}{16}$	.156	.125	.582
.500	$\frac{1}{2}$	13	20	$\frac{9}{16}$	$\frac{3}{4}$	.156	.156	.637
.5625	$\frac{9}{16}$	12	18	$\frac{39}{64}$	$\frac{7}{8}$	.188	.156	.744
.625	$\frac{5}{8}$	11	18	$\frac{23}{32}$	$\frac{15}{16}$	.219	.188	.797
.750	$\frac{3}{4}$	10	16	$\frac{13}{16}$	$1\frac{1}{8}$	.250	.188	.941
.875	$\frac{7}{8}$	9	14	$\frac{29}{32}$	$1\frac{5}{16}$	.250	.188	1.097
1.000	1	8	12	1	$1\frac{1}{2}$	.281	.250	1.254
1.125	$1\frac{1}{8}$	7	12	$1\frac{5}{32}$	$1\frac{11}{16}$	.344	.250	1.411

* ANSI B18.2.2.
Thread may be coarse or fine-thread series, class 2B tolerance; unless otherwise specified, fine-thread series will be furnished.

TABLE **34** *American National Standard Machine Screw and Stove Bolt Nuts**

Diameter		Nominal size													
		0	1	2	3	4	5	6	8	10	12	$\frac{1}{4}$	$\frac{5}{16}$	$\frac{3}{8}$
Across flats (nominal)		$\frac{5}{32}$	$\frac{5}{32}$	$\frac{3}{16}$	$\frac{3}{16}$	$\frac{1}{4}$	$\frac{5}{16}$	$\frac{5}{16}$	$\frac{11}{32}$	$\frac{3}{8}$	$\frac{7}{16}$	$\frac{7}{16}$	$\frac{9}{16}$	$\frac{5}{8}$
Across corners (min) {Hexagonal		.171	.171	.205	.205	.275	.344	.344	.378	.413	.482	.482	.621	.692
{Square		.206	.206	.247	.247	.331	.415	.415	.456	.497	.581	.581	.748	.833
Thickness (nom)		$\frac{3}{64}$	$\frac{3}{64}$	$\frac{1}{16}$	$\frac{1}{16}$	$\frac{3}{32}$	$\frac{7}{64}$	$\frac{7}{64}$	$\frac{1}{8}$	$\frac{1}{8}$	$\frac{5}{32}$	$\frac{3}{16}$	$\frac{7}{32}$	$\frac{1}{4}$

* ANSI B18.6.3.
Dimensions in inches.
Square nuts—threads are UNC, Class 2B.
Hexagonal nuts—threads are UNC or UNF, Class 2B.

TABLE 35 American National Standard Set Screws*

Point types illustrated: FLAT POINT, CUP POINT, CONE POINT, OVAL POINT, DOG POINT, HALF DOG POINT.

Nominal size—diameter, D	D	Slotted headless A	B	R₁	Hexagonal socket (min) C	D†	Square-head F_{max} F	H	R₂	Cup and flat K_{max} K	Cone Y	Oval R₃	Full- and half-dog N_{max} L	M	N
5	.125	.023	.031	.125	1/16	.050	—	—	—	.067		.094	.060	.030	.083
6	.138	.025	.035	.138	1/16	.050	—	—	—	.074		.109	.070	.035	.092
8	.164	.029	.041	.164	5/64	.062	—	—	—	.087		.125	.080	.040	.109
10	.190	.032	.048	.190	3/32	.075	.1875	9/64	15/64	.102		.141	.090	.045	.127
12	.216	.036	.054	.216	3/32	.075	.216	5/32	35/64	.115		.156	.110	.055	.144
1/4	.250	.045	.063	.250	1/8	.100	.250	3/16	5/8	.132		.188	.125	.063	.156
5/16	.3125	.051	.078	.313	5/32	.125	.3125	15/64	25/32	.172		.234	.156	.078	.203
3/8	.375	.064	.094	.375	3/16	.150	.375	9/32	15/16	.212		.281	.188	.094	.250
7/16	.4375	.072	.109	.438	7/32	.175	.4375	21/64	1 3/32	.252		.328	.219	.109	.297
1/2	.500	.081	.125	.500	1/4	.200	.500	3/8	1 1/4	.291		.375	.250	.125	.344
9/16	.5625	.091	.141	.563	1/4	.200	.5625	27/64	1 13/32	.332		.422	.281	.140	.391
5/8	.625	.102	.156	.625	5/16	.250	.625	15/32	1 9/16	.371		.469	.313	.156	.469
3/4	.750	.129	.188	.750	3/8	.300	.750	9/16	1 7/8	.450		.563	.375	.188	.563
7/8	.875	—	—	—	1/2	.400	.875	21/32	2 3/16	—		—	—	—	—
1	1.000	—	—	—	9/16	.450	1.000	3/4	2 1/2	—		—	—	—	—
1 1/8	1.125	—	—	—	9/16	.450	1.125	27/32	2 13/16	—		—	—	—	—
1 1/4	1.250	—	—	—	5/8	.500	1.250	15/16	3 1/16	—		—	—	—	—
1 3/8	1.375	—	—	—	5/8	.500	1.375	1 1/16	3 3/8	—		—	—	—	—
1 1/2	1.500	—	—	—	3/4	.600	1.500	1 1/8	3 3/4	—		—	—	—	—

Points — angle notes (constant for all sizes):

- Cup and flat: $W = 80°\text{–}90°$ (Draw as 90°.); $X = 118° \pm 5°$
- Cone: Y — When L equals nominal diameter or less, $Y = 118° \pm 2°$. When L exceeds nominal diameter, $Y = 90° \pm 2°$.
- Full- and half-dog: $Z = 100°\text{–}110°$

* ANSI B18.6.2; ANSI B18.3.
† Dimensions apply to cup and flat-point screws 1 diam in length or longer. For screws shorter than 1 diam in length, and for other types of points, socket to be as deep as practicable.

TABLE 36 American National Standard Machine Screws*

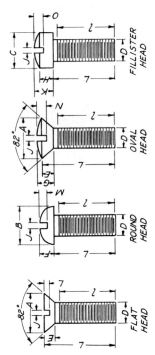

FLAT HEAD · ROUND HEAD · OVAL HEAD · FILLISTER HEAD

| Size (number) and threads per inch | | | Standard dimensions (max) | | | | | | | | | | | | |
Nominal size Diameter, D	Coarse	Fine	Head diameter A	B	C	Height dimensions E	F	G	H	K	Slot width J	Slot depth L	M	N	O
.060	—	80	.119	.113	.096	.035	.053	.056	.045	.059	.023	.015	.039	.030	.025
.073	64	72	.146	.138	.118	.043	.061	.068	.053	.071	.026	.019	.044	.038	.031
.086	56	64	.172	.162	.140	.051	.069	.080	.062	.083	.031	.023	.048	.045	.037
.099	48	56	.199	.187	.161	.059	.078	.092	.070	.095	.035	.027	.053	.052	.043
.112	40	48	.225	.211	.183	.067	.086	.104	.079	.107	.039	.030	.058	.059	.048
.125	40	44	.252	.236	.205	.075	.095	.116	.088	.120	.043	.034	.063	.067	.054
.138	32	40	.279	.260	.226	.083	.103	.128	.096	.132	.048	.038	.068	.074	.060
.164	32	36	.332	.309	.270	.100	.120	.152	.113	.156	.054	.045	.077	.088	.071
.190	24	32	.385	.359	.313	.116	.137	.176	.130	.180	.060	.053	.087	.103	.083
.216	24	28	.438	.408	.357	.132	.153	.200	.148	.205	.067	.060	.096	.117	.094
.250 1/4	20	28	.507	.472	.414	.153	.175	.232	.170	.237	.075	.070	.109	.136	.109
.3125 5/16	18	24	.635	.590	.518	.191	.216	.290	.211	.295	.084	.088	.132	.171	.137
.375 3/8	16	24	.762	.708	.622	.230	.256	.347	.253	.355	.094	.106	.155	.206	.164
.4375 7/16	14	20	.812	.750	.625	.223	.328	.345	.265	.368	.094	.103	.196	.210	.170
.500 1/2	13	20	.875	.813	.750	.223	.355	.354	.297	.412	.106	.103	.211	.216	.190
.5625 9/16	12	18	1.000	.938	.812	.260	.410	.410	.336	.466	.118	.120	.242	.250	.214
.625 5/8	11	18	1.125	1.000	.875	.298	.438	.467	.375	.521	.133	.137	.258	.285	.240
.750 3/4	10	16	1.375	1.250	1.000	.372	.547	.578	.441	.612	.149	.171	.320	.353	.281

* ANSI B18.6.3.
Thread length—screws 2 in. in length or less are threaded as close to the head as practicable. Screws longer than 2 in. should have a minimum thread length of 1¾ in.

TABLE **37** *American National Standard Cap Screws**

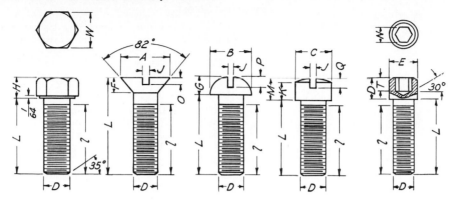

Nominal size—diameter		Head diameter					Height dimensions					Slot width	Slot depth			Socket dimensions	
		A	B	C	E	W	F_{av}	G	H_{nom}	K	M	J	O	P	Q	N_{min}	T_{min}
.250	1/4	.500	.437	.375	3/8	7/16	.140	.191	5/32	.172	.216	.075	.068	.117	.097	3/16	.120
.3125	5/16	.625	.562	.437	7/16	1/2	.177	.245	13/64	.203	.253	.084	.086	.151	.115	7/32	.151
.375	3/8	.750	.625	.562	9/16	9/16	.210	.273	15/64	.250	.314	.094	.103	.168	.142	5/16	.182
.4375	7/16	.8125	.750	.625	5/8	5/8	.210	.328	9/32	.297	.368	.094	.103	.202	.168	5/16	.213
.500	1/2	.875	.812	.750	3/4	3/4	.210	.354	5/16	.328	.413	.106	.103	.218	.193	3/8	.245
.5625	9/16	1.000	.937	.812	13/16	13/16	.244	.409	23/64	.375	.467	.118	.120	.252	.213	3/8	.276
.625	5/8	1.125	1.000	.875	7/8	15/16	.281	.437	25/64	.422	.521	.133	.137	.270	.239	1/2	.307
.750	3/4	1.375	1.250	1.000	1	1 1/8	.352	.546	15/32	.500	.612	.149	.171	.338	.283	9/16	.370
.875	7/8	1.625	—	1.125	1 1/8	1 5/16	.423	—	35/64	.594	.720	.167	.206	—	.334	9/16	.432
1.000	1	1.875	—	1.312	1 5/16	1 1/2	.494	—	39/64	.656	.803	.188	.240	—	.371	5/8	.495
1.125	1 1/8	2.062	—	—	1 1/2	1 11/16	.529	—	11/16	—	—	.196	.257	—	—	3/4	.557
1.250	1 1/4	2.312	—	—	1 3/4	1 7/8	.600	—	25/32	—	—	.211	.291	—	—	3/4	.620
1.375	1 3/8	2.562	—	—	1 7/8	2 1/16	.665	—	27/32	—	—	.226	.326	—	—	3/4	.682
1.500	1 1/2	2.812	—	—	2	2 1/4	.742	—	15/16	—	—	.258	.360	—	—	1	.745

* ANSI B18.3. ANSI B18.6.2.
Basically, threads may be coarse, fine, or 8-thread series; class 2A for plain (unplated) cap screws.
Minimum thread length will be 2D + 1/4 in. for lengths up to and including 6 in.
Socket-head cap screws—thread coarse or fine, class 3A. Thread length: coarse, 2D + 1/2 in; fine, 1 1/2 D + 1/2 in.
Standard dimensions are maximum except as noted.

TABLE **38A** *American National Standard Plain Washers (Type A)**

Washer size—nominal		Light-SAE (N = narrow)			Standard-plate (W = wide)		
		ID, A	OD, B	Thickness, H	ID, A	OD, B	Thickness, H
.250	1/4	.281	.625	.065	.312	.734	.065
.312	5/16	.344	.688	.065	.375	.875	.083
.375	3/8	.406	.812	.065	.438	1.000	.083
.438	7/16	.469	.922	.065	.500	1.250	.083
.500	1/2	.531	1.062	.095	.562	1.375	.109
.562	9/16	.594	1.156	.095	.625	1.469	.109
.625	5/8	.656	1.312	.095	.688	1.750	.134
.750	3/4	.812	1.469	.134	.812	2.000	.148
.875	7/8	.938	1.750	.134	.938	2.250	.165
1.000	1	1.062	2.000	.134	1.062	2.500	.165
1.125	1 1/8	1.250	2.250	.134	1.250	2.750	.165
1.250	1 1/4	1.375	2.500	.165	1.375	3.000	.165
1.375	1 3/8	1.500	2.750	.165	1.500	3.250	.180
1.500	1 1/2	1.625	3.000	.165	1.625	3.500	.180

* ANSI B27.2.
Plain washers are specified as follows: ID × OD × thickness − .375 × .875 × .083 PLAIN WASHER.

TABLE *38*B American National Standard Lock Washers (Selected Sizes)*

Washer size— nominal		Regular (light)			Extra-duty (heavy)		
		ID (min), C	OD (max), D	Thickness (min), T	ID (min), C	OD (max), D	Thickness (min), T
.250	1/4	.255	.489	.062	Same as	.535	.084
.312	5/16	.318	.586	.078	for regular	.622	.108
.375	3/8	.382	.683	.094	lock washers	.741	.123
.438	7/16	.446	.779	.109		.839	.143
.500	1/2	.509	.873	.125		.939	.162
.562	9/16	.572	.971	.141		1.041	.182
.625	5/8	.636	1.079	.156		1.157	.202
.750	3/4	.763	1.271	.188		1.361	.241
.875	7/8	.890	1.464	.219		1.576	.285
1.000	1	1.017	1.661	.250		1.799	.330
1.125	1 1/8	1.144	1.853	.281		2.019	.375
1.250	1 1/4	1.271	2.045	.312		2.231	.417
1.375	1 3/8	1.398	2.239	.344		2.439	.458
1.500	1 1/2	1.525	2.430	.375		2.638	.496

* ANSI B27.1.
Lock washers are specified by giving nominal size and series (⅜ regular lock washer).

TABLE *39* American National Standard Cotter Pins*

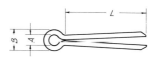

Pin diameter			Eye diameter		Drill size recommended (hole/diameter)	Clevis pin or shaft diameter
Nominal	Max	Min	Inside, A	Outside, B		
.031 1/32	.032	.028	1/32	1/16	3/64—.0469	1/8
.047 3/64	.048	.044	3/64	3/32	1/16—.0625	3/16
.062 1/16	.060	.056	1/16	1/8	5/64—.0781	1/4
.078 5/64	.076	.072	5/64	5/32	3/32—.0938	5/16
.094 3/32	.090	.086	3/32	3/16	7/64—.1094	3/8
.125 1/8	.120	.116	1/8	1/4	9/64—.1406	1/2
.156 5/32	.150	.146	5/32	5/16	11/64—.1719	5/8
.188 3/16	.176	.172	3/16	3/8	13/64—.2031	—
.219 7/32	.207	.202	7/32	7/16	15/64—.2344	—
.250 1/4	.225	.220	1/4	1/2	17/64—.2656	—

* ANSI B5.20.
For shafts up to ⅜ in. diam, select a cotter pin that is approximately equal to one-fourth of the shaft diameter.
For larger sizes, use a cotter pin that is from one-fourth to one-sixth of the shaft diameter.

TABLE 40 American National Standard Small Rivets*

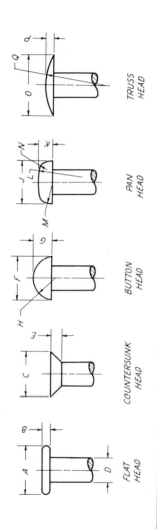

FLAT HEAD COUNTERSUNK HEAD BUTTON HEAD PAN HEAD TRUSS HEAD

Rivet diameter, D	Flat		CSK.		Button			Pan					Truss		
	Diameter, A_{max}	Height, B_{max}	Diameter, C_{max}	Height, E	Diameter, F_{max}	Height, G_{max}	Radius, H	Diameter, J_{max}	Height, K_{max}	Radius, L	Radius, M	Radius, N	Diameter, O_{max}	Height, P_{max}	Radius, Q
.062 1/16	.140	.027	.118	.027	.122	.052	.055	.118	.040	.217	.052	.019	—	—	—
.094 3/32	.200	.038	.176	.040	.182	.077	.084	.173	.060	.326	.080	.030	.226	.038	.239
.125 1/8	.260	.048	.235	.053	.235	.100	.111	.225	.078	.429	.106	.039	.297	.048	.314
.156 5/32	.323	.059	.293	.066	.290	.124	.138	.279	.096	.535	.133	.049	.368	.059	.392
.188 3/16	.387	.069	.351	.079	.348	.147	.166	.334	.114	.641	.159	.059	.442	.069	.470
.219 7/32	.453	.080	.413	.094	.405	.172	.195	.391	.133	.754	.186	.069	.515	.080	.555
.250 1/4	.515	.091	.469	.106	.460	.196	.221	.444	.151	.858	.213	.079	.590	.091	.628
.281 9/32	.579	.103	.528	.119	.518	.220	.249	.499	.170	.963	.239	.088	.661	.103	.706
.313 5/16	.641	.113	.588	.133	.572	.243	.276	.552	.187	1.070	.266	.098	.732	.113	.784
.344 11/32	.705	.124	.646	.146	.630	.267	.304	.608	.206	1.176	.292	.108	.806	.124	.862
.375 3/8	.769	.135	.704	.159	.684	.291	.332	.663	.225	1.286	.319	.118	.878	.135	.942
.406 13/32	.834	.146	.763	.172	.743	.316	.358	.719	.243	1.392	.345	.127	.949	.145	1.028
.438 7/16	.896	.157	.823	.186	.798	.339	.387	.772	.261	1.500	.372	.137	1.020	.157	1.098

* ANSI B18.1.
The length of a rivet is measured from the underside (bearing surface) of the head to the end of the shank except in the case of a rivet with a countersunk head.
The length of a countersunk-head rivet is measured from the top of the head to the end of the shank.

TABLE 41 American National Standard Square and Flat Keys*

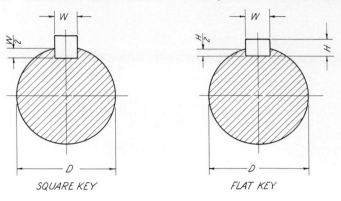

SQUARE KEY   FLAT KEY

Shaft diameter—nominal		Square stock, W	Rectangular stock, W × H	Shaft diameter—nominal		Square stock, W	Rectangular stock, W × H
Over	To (inclusive)			Over	To (inclusive)		
5/16	7/16	3/32	—	1 3/4	2 1/4	1/2	1/2 × 3/8
7/16	9/16	1/8	1/8 × 3/32	2 1/4	2 3/4	5/8	5/8 × 7/16
9/16	7/8	3/16	3/16 × 1/8	2 3/4	3 1/4	3/4	3/4 × 1/2
7/8	1 1/4	1/4	1/4 × 3/16	3 1/4	3 3/4	7/8	7/8 × 5/8
1 1/4	1 3/8	5/16	5/16 × 1/4	3 3/4	4 1/2	1	1 × 3/4
1 3/8	1 3/4	3/8	3/8 × 1/4	4 1/2	5 1/2	1 1/4	1 1/4 × 7/8

* ANSI B17.1.
All dimensions in inches.

TABLE 42 American National Standard Plain Taper and Gib-head Keys†

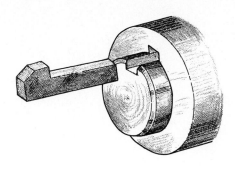

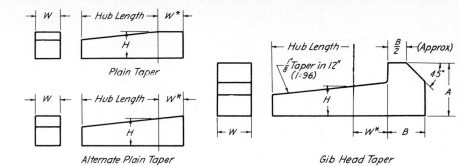

Plain Taper

Alternate Plain Taper

Gib Head Taper

Plain and Gib Head Taper Keys Have a ⅛" Taper in 12"

Plain taper and gib-head keys square and rectangular				Gib-head			
Shaft diameter— nominal		Square type, $W = H$	Rectangular type, $W \times H$	Square		Rectangular	
Over	To (inclusive)			Head height, A	Length, B	Head height, A	Length, B
5/16	7/16	—	—	—	—	—	—
7/16	9/16	1/8	1/8 × 3/32	1/4	7/32	3/16	1/8
9/16	7/8	3/16	3/16 × 1/8	5/16	9/32	1/4	3/16
7/8	1 1/4	1/4	1/4 × 3/16	7/16	11/32	5/16	1/4
1 1/4	1 3/8	5/16	5/16 × 1/4	9/16	13/32	3/8	5/16
1 3/8	1 3/4	3/8	3/8 × 1/4	11/16	15/32	7/16	3/8
1 3/4	2 1/4	1/2	1/2 × 3/8	7/8	19/32	5/8	1/2
2 1/4	2 3/4	5/8	5/8 × 7/16	1 1/16	23/32	3/4	5/8
2 3/4	3 1/4	3/4	3/4 × 1/2	1 1/4	7/8	7/8	3/4
3 1/4	3 3/4	7/8	7/8 × 5/8	1 1/2	1	1 1/16	7/8

† ANSI B17.1.
* For locating position of dimension H.
For longer sizes, see standard.
All dimensions in inches.

TABLE **43** *American National Standard Woodruff Keys**

Key no.	Nominal size, $A \times B$	Height of key		Distance below center, E	Depth of keyseat in shaft $+.005$ $-.000$
		C_{max}	D_{max}		
202	$\frac{1}{16} \times \frac{1}{4}$	.109	.109	$\frac{1}{64}$	.0728
202.5	$\frac{1}{16} \times \frac{5}{16}$	.140	.140	$\frac{1}{64}$	.1038
203	$\frac{1}{16} \times \frac{3}{8}$	.172	.172	$\frac{1}{64}$	.1358
204	$\frac{1}{16} \times \frac{1}{2}$	.203	.194	$\frac{3}{64}$	.1668
302.5	$\frac{3}{32} \times \frac{5}{16}$	.140	.140	$\frac{1}{64}$	.0882
303	$\frac{3}{32} \times \frac{3}{8}$	.172	.172	$\frac{1}{64}$	.1202
304	$\frac{3}{32} \times \frac{1}{2}$	.203	.194	$\frac{3}{64}$	.1511
305	$\frac{3}{32} \times \frac{5}{8}$	.250	.240	$\frac{1}{16}$	.1981
403	$\frac{1}{8} \times \frac{3}{8}$	.172	.172	$\frac{1}{64}$	.1045
404	$\frac{1}{8} \times \frac{1}{2}$	.203	.194	$\frac{3}{64}$	.1355
405	$\frac{1}{8} \times \frac{5}{8}$	.250	.240	$\frac{1}{16}$	.1825
406	$\frac{1}{8} \times \frac{3}{4}$	.313	.303	$\frac{1}{16}$	.2455
505	$\frac{5}{32} \times \frac{5}{8}$	.250	.240	$\frac{1}{16}$	.1669
506	$\frac{5}{32} \times \frac{3}{4}$	.313	.303	$\frac{1}{16}$	.2299
507	$\frac{5}{32} \times \frac{7}{8}$	.375	.365	$\frac{1}{16}$	.2919
605	$\frac{3}{16} \times \frac{5}{8}$	.250	.240	$\frac{1}{16}$	.1513
606	$\frac{3}{16} \times \frac{3}{4}$	.313	.303	$\frac{1}{16}$	.2143
607	$\frac{3}{16} \times \frac{7}{8}$	.375	.365	$\frac{1}{16}$	.2763
608	$\frac{3}{16} \times 1$	.438	.428	$\frac{1}{16}$	.3393
609	$\frac{3}{16} \times 1\frac{1}{8}$	.484	.475	$\frac{5}{64}$	.3853
610	$\frac{3}{16} \times 1\frac{1}{4}$	.547	.537	$\frac{5}{64}$	.4483
707	$\frac{7}{32} \times \frac{7}{8}$	.375	.365	$\frac{1}{16}$	.2607
708	$\frac{7}{32} \times 1$	.438	.428	$\frac{1}{16}$	.3237
709	$\frac{7}{32} \times 1\frac{1}{8}$	.484	.475	$\frac{5}{64}$	.3697
710	$\frac{7}{32} \times 1\frac{1}{4}$	.547	.537	$\frac{5}{64}$	.4327
806	$\frac{1}{4} \times \frac{3}{4}$	.313	.303	$\frac{1}{16}$	.1830
807	$\frac{1}{4} \times \frac{7}{8}$	.375	.365	$\frac{1}{16}$	.2450
808	$\frac{1}{4} \times 1$	.438	.428	$\frac{1}{16}$	.3080
809	$\frac{1}{4} \times 1\frac{1}{8}$	.484	.475	$\frac{5}{64}$	.3540
810	$\frac{1}{4} \times 1\frac{1}{4}$	.547	.537	$\frac{5}{64}$	.4170
811	$\frac{1}{4} \times 1\frac{3}{8}$	.594	.584	$\frac{3}{32}$	.4640
812	$\frac{1}{4} \times 1\frac{1}{2}$	.641	.631	$\frac{7}{64}$	.5110
1008	$\frac{5}{16} \times 1$	.438	.428	$\frac{1}{16}$	.2768
1009	$\frac{5}{16} \times 1\frac{1}{8}$	.484	.475	$\frac{5}{64}$	.3228
1010	$\frac{5}{16} \times 1\frac{1}{4}$	.547	.537	$\frac{5}{64}$	.3858
1011	$\frac{5}{16} \times 1\frac{3}{8}$	.594	.584	$\frac{3}{32}$	.4328
1012	$\frac{5}{16} \times 1\frac{1}{2}$	.641	.631	$\frac{7}{64}$	.4798
1208	$\frac{3}{8} \times 1$	.438	.428	$\frac{1}{16}$	.2455
1210	$\frac{3}{8} \times 1\frac{1}{4}$	.547	.537	$\frac{5}{64}$	.3545
1211	$\frac{3}{8} \times 1\frac{3}{8}$	.594	.584	$\frac{3}{32}$	.4015
1212	$\frac{3}{8} \times 1\frac{1}{2}$	.641	.631	$\frac{7}{64}$	.4485

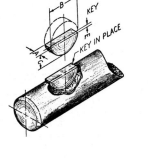

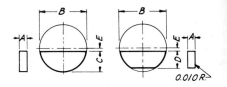

* ANSI B17.2.
All dimensions in inches. Key numbers indicate the nominal key dimensions. The last two digits give the nominal diameter in eighths of an inch and the digits preceding the last two give the nominal width in thirty-seconds of an inch.
Examples: No. 204 indicates a key $\frac{2}{32} \times \frac{4}{8}$ or $\frac{1}{16} \times \frac{1}{2}$.
No. 808 indicates a key $\frac{8}{32} \times \frac{8}{8}$ or $\frac{1}{4} \times 1$.

TABLE 44 *Pratt & Whitney Keys*

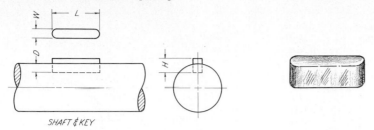

SHAFT & KEY

Key no.	L*	W	H	D	Key no.	L*	W	H	D
1	1/2	1/16	3/32	1/16	22	1 3/8	1/4	3/8	1/4
2	1/2	3/32	9/64	3/32	23	1 3/8	5/16	15/32	5/16
3	1/2	1/8	3/16	1/8	F	1 3/8	3/8	9/16	3/8
4	5/8	3/32	9/64	3/32	24	1 1/2	1/4	3/8	1/4
5	5/8	1/8	3/16	1/8	25	1 1/2	5/16	15/32	5/16
6	5/8	5/32	15/64	5/32	G	1 1/2	3/8	9/16	3/8
7	3/4	1/8	3/16	1/8	51	1 3/4	1/4	3/8	1/4
8	3/4	5/32	15/64	5/32	52	1 3/4	5/16	15/32	5/16
9	3/4	3/16	9/32	3/16	53	1 3/4	3/8	9/16	3/8
10	7/8	5/32	15/64	5/32	26	2	3/16	9/32	3/16
11	7/8	3/16	9/32	3/16	27	2	1/4	3/8	1/4
12	7/8	7/32	21/64	7/32	28	2	5/16	15/32	5/16
A	7/8	1/4	3/8	1/4	29	2	3/8	9/16	3/8
13	1	3/16	9/32	3/16	54	2 1/4	1/4	3/8	1/4
14	1	7/32	21/64	7/32	55	2 1/4	5/16	15/32	5/16
15	1	1/4	3/8	1/4	56	2 1/4	3/8	9/16	3/8
B	1	5/16	15/32	5/16	57	2 1/4	7/16	21/32	7/16
16	1 1/8	3/16	9/32	3/16	58	2 1/2	5/16	15/32	5/16
17	1 1/8	7/32	21/64	7/32	59	2 1/2	3/8	9/16	3/8
18	1 1/8	1/4	3/8	1/4	60	2 1/2	7/16	21/32	7/16
C	1 1/8	5/16	15/32	5/16	61	2 1/2	1/2	3/4	1/2
19	1 1/4	3/16	9/32	3/16	30	3	3/8	9/16	3/8
20	1 1/4	7/32	21/64	7/32	31	3	7/16	21/32	7/16
21	1 1/4	1/4	3/8	1/4	32	3	1/2	3/4	1/2
D	1 1/4	5/16	15/32	5/16	33	3	9/16	27/32	9/16
E	1 1/4	3/8	9/16	3/8	34	3	5/8	15/16	5/8

* The length L may vary but should always be at least $2W$.

TABLE 45 American National Standard Taper Pins*

Number of pin†	Diameter at large end		Maximum length, L	Approx. shaft diameter	Drill size‡
00000	.094	3/32	1	.250	No. 47 (.0785)
0000	.109	7/64	1	.312	No. 42 (.0935)
000	.125	1/8	1	.375	No. 37 (.1040)
00	.141	9/64	1¼	.438	No. 31 (.1200)
0	.156	5/32	1½	.500	No. 28 (.1405)
1	.172	11/64	2	.562	No. 25 (.1495)
2	.193	—	2½	.625	No. 19 (.1660)
3	.219	7/32	3	.750	No. 12 (.1890)
4	.250	1/4	3	.812	No. 3 (.2130)
5	.289	—	3	.875	1/4 (.2500)
6	.341	—	4	1.000	9/32 (.2812)
7	.409	—	4	1.250	11/32 (.3438)
8	.492	—	4	1.500	13/32 (.4062)

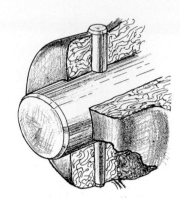

* ANSI B5.20.
† For 7/0, 6/0, 9, and 10 see the standard. Pins 11–14 are special sizes.
‡ Suggested sizes; not American National Standard.
Drill size is for reamer. The small diameter of a pin is equal to the large diameter minus .02083 × **L**, where **L** is the length.

TABLE 46 American National Standard Wrought-iron and Steel Pipe*

					Standard weight		Heavy			
					Nominal wall thickness		Extra heavy		Double extra heavy	
Nominal size	Outside diameter (all weights)	Threads per inch	Tap drill sizes†	Distance pipe enters fitting	Wrought iron	Steel	Wrought iron	Steel	Wrought iron	Steel
1/8	.405	27	11/32	5/16	.069	.068	.099	.095	—	—
1/4	.540	18	7/16	7/16	.090	.088	.122	.119	—	—
3/8	.675	18	19/32	7/16	.093	.091	.129	.126	—	—
1/2	.840	14	23/32	9/16	.111	.109	.151	.147	.307	.294
3/4	1.050	14	15/16	9/16	.115	.113	.157	.154	.318	.308
1	1.315	11½	1 5/32	11/16	.136	.133	.183	.179	.369	.358
1¼	1.660	11½	1½	11/16	.143	.140	.195	.191	.393	.382
1½	1.900	11½	1 23/32	11/16	.148	.145	.204	.200	.411	.400
2	2.375	11½	2 3/16	3/4	.158	.154	.223	.218	.447	.436
2½	2.875	8	2 5/8	1 1/16	.208	.203	.282	.276	.567	.552
3	3.500	8	3¼	1 1/8	.221	.216	.306	.300	.615	.600
3½	4.000	8	3¾	1 3/16	.231	.226	.325	.318	—	—
4	4.500	8	4¼	1 3/16	.242	.237	.344	.337	.690	.674
5	5.563	8	5 5/16	1 5/16	.263	.258	.383	.375	.768	.750
6	6.625	8	6 3/8	1 3/8	.286	.280	.441	.432	.884	.864
8	8.625	8	—	—	.329	.322	.510	.500	.895	.875

* ANSI B36.10, ANSI B2.1.
All dimensions in inches.
† Not American Standard. See ANSI B36.10 for sizes larger than 8 in.

TABLE 47 American National Standard Malleable-iron Screwed Fittings*

Diagrams: ELBOW · TEE · CROSS · 45° ELBOW · 45° Y-BRANCH · STREET ELBOW · 45° STREET ELBOW · STREET TEE · STRAIGHT COUPLINGS · STRAIGHT REDUCING COUPLINGS

Nominal pipe size	A Center to end, elbows, tees, and crosses	B_{min} Length of thread	C Center to end, 45° elbows	E_{min} Width of band	F_{min} Inside diameter of fitting	F_{max}	G Metal thickness	H Outside diameter of band	J Center to male end, elbows and tees	K Center to male end, 45° elbows	L_{min} Length of external thread	M Length of reducing couplings	N_{max} Port diameter, male end	T Center to end, inlet	U Center to end, outlet	V End to end	W Length of straight couplings	Thickness of ribs of couplings
1/8	.69	.25	—	.200	.405	.435	.090	.693	1.00	—		—	.20	—	—	—	.96	.090
1/4	.81	.32	.73	.215	.540	.584	.095	.844	1.19	.94	.2638	1.00	.26	—	—	—	1.06	.095
3/8	.95	.36	.80	.230	.675	.719	.100	1.015	1.44	1.03	.4018	1.13	.37	.50	1.43	1.93	1.16	.100
1/2	1.12	.43	.88	.249	.840	.897	.105	1.197	1.63	1.15	.4078	1.25	.51	.61	1.71	2.32	1.34	.105
3/4	1.31	.50	.98	.273	1.050	1.107	.120	1.458	1.89	1.29	.5337	1.44	.69	.72	2.05	2.77	1.52	.120
1	1.50	.58	1.12	.302	1.315	1.385	.134	1.771	2.14	1.47	.5457	1.69	.91	.85	2.43	3.28	1.67	.134
1¼	1.75	.67	1.29	.341	1.660	1.730	.145	2.153	2.45	1.71	.6828	2.06	1.19	1.02	2.92	3.94	1.93	.145
1½	1.94	.70	1.43	.368	1.900	1.970	.155	2.427	2.69	1.88	.7068	2.31	1.39	1.10	3.28	4.38	2.15	.155
2	2.25	.75	1.68	.422	2.375	2.445	.173	2.963	3.26	2.22	.7235	2.81	1.79	1.24	3.93	5.17	2.53	.173
2½	2.70	.92	1.95	.478	2.875	2.975	.210	3.589	3.86	2.57	.7565	3.25	2.20	1.52	4.73	6.25	2.88	.210
3	3.08	.98	2.17	.548	3.500	3.600	.231	4.285	4.51	3.00	1.1375	3.69	2.78	1.71	5.55	7.26	3.18	.231
3½	3.42	1.03	2.39	.604	4.000	4.100	.248	4.843	—	—	1.2000	—	—	—	—	—	—	—
4	3.79	1.08	2.61	.661	4.500	4.600	.265	5.401	5.69	3.70	1.3000	4.38	3.70	2.01	6.97	8.98	3.69	.265
5	4.50	1.18	3.05	.780	5.563	5.663	.300	6.583	6.86	—	1.4063	—	4.69	—	—	—	—	—
6	5.13	1.28	3.46	.900	6.625	6.725	.336	7.767	8.03	—	1.5125	—	5.67	—	—	—	—	—

* ANSI B16.3.
For use under maximum working steam pressures of 150 lb per sq in.

645

TABLE **48** *125-lb American National Standard Cast-iron Screwed Fittings**

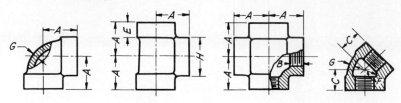

Nominal pipe size	Center to end, elbows, tees, and crosses, A	Length of thread, B_{min}	Center to end, 45° elbows, C	Width of band, E_{min}	Inside diameter of fitting		Metal thickness, G_{min}	Outside diameter of band, H_{min}
					F_{min}	F_{max}		
¼	.81	.32	.73	.38	.540	.584	.110	.93
⅜	.95	.36	.80	.44	.675	.719	.120	1.12
½	1.12	.43	.88	.50	.840	.897	.130	1.34
¾	1.31	.50	.98	.56	1.050	1.107	.155	1.63
1	1.50	.58	1.12	.62	1.315	1.385	.170	1.95
1¼	1.75	.67	1.29	.69	1.660	1.730	.185	2.39
1½	1.94	.70	1.43	.75	1.900	1.970	.200	2.68
2	2.25	.75	1.68	.84	2.375	2.445	.220	3.28
2½	2.70	.92	1.95	.94	2.875	2.975	.240	3.86
3	3.08	.98	2.17	1.00	3.500	3.600	.260	4.62
3½	3.42	1.03	2.39	1.06	4.000	4.100	.280	5.20
4	3.79	1.08	2.61	1.12	4.500	4.600	.310	5.79
5	4.50	1.18	3.05	1.18	5.563	5.663	.380	7.05
6	5.13	1.28	3.46	1.28	6.625	6.725	.430	8.28
8	6.56	1.47	4.28	1.47	8.625	8.725	.550	10.63
10	8.08	1.68	5.16	1.68	10.750	10.850	.690	13.12
12	9.50	1.88	5.97	1.88	12.750	12.850	.800	15.47

* ANSI B16.4.
All dimensions in inches.

TABLE 49 **American National Standard Pipe Plugs and Caps***

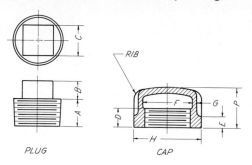

PLUG   CAP

Nominal pipe size	Plug			Cap†						
	Length of thread	Height of square	Width across flats	Length of thread	Width of band	Inside diameter of fitting	Metal thickness	Outside diameter of band	Height	Thickness of ribs
	A_{min}	B_{min}	C_{nom}	D_{min}	E_{min}	F_{max}	G	H_{min}	P_{min}	
1/8	.37	.24	9/32	.25	.200	.435	.090	.693	.53	—
1/4	.44	.28	3/8	.32	.215	.584	.095	.844	.63	—
3/8	.48	.31	7/16	.36	.230	.719	.100	1.015	.74	—
1/2	.56	.38	9/16	.43	.249	.897	.105	1.197	.87	.105
3/4	.63	.44	5/8	.50	.273	1.107	.120	1.458	.97	.120
1	.75	.50	13/16	.58	.302	1.385	.134	1.771	1.16	.134
1¼	.80	.56	15/16	.67	.341	1.730	.145	2.153	1.28	.145
1½	.83	.62	1 1/8	.70	.368	1.970	.155	2.427	1.33	.155
2	.88	.68	1 5/16	.75	.422	2.445	.173	2.963	1.45	.173
2½	1.07	.74	1½	.92	.478	2.975	.210	3.589	1.70	.210
3	1.13	.80	1 11/16	.98	.548	3.600	.231	4.285	1.80	.231
3½	1.18	.86	1 7/8	1.03	.604	4.100	.248	4.843	1.90	.248
4	—	—	—	1.08	.661	4.600	.265	5.401	2.08	.265
5	—	—	—	1.18	.780	5.663	.300	6.583	2.32	.300
6	—	—	—	1.28	.900	6.725	.336	7.767	2.55	.336

* ANSI B16.3.
† The outside radius of top of cap is equal to 3 × F.

TABLE 50 American National Standard Cast-iron Pipe Flanges and Flanged Fittings*

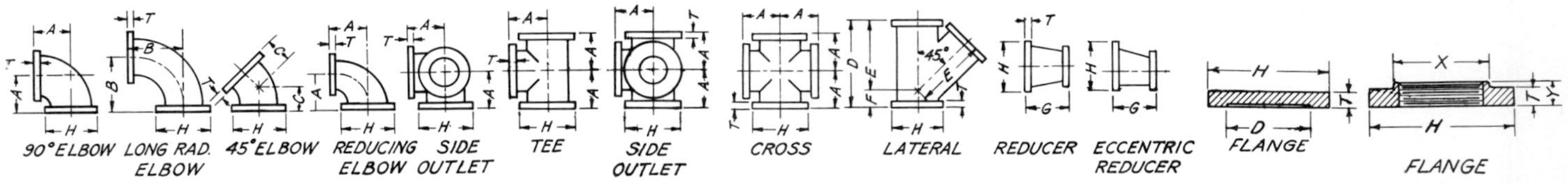

Nominal pipe size D	Center to face, elbow, tees, etc. A	Center to face, long radius elbow B	Center to face, 45° elbow C	Face to face, lateral D	Center to face, lateral E	Center to face, "Y" and lateral F	Face to face, reducer G	Diameter of flange H	Thickness of flange T	Diameter of hub X	Length of hub Y	Diameter of holes in flanges	Number of bolts for flanges	Diameter of bolts for flanges	Length of bolts for flanges	Diameter of bolt circle
1	3 1/2	5	1 3/4	7 1/2	5 3/4	1 3/4	—	4 1/4	7/16	1 15/16	11/16	5/8	4	1/2	1 3/4	3 1/8
1 1/4	3 3/4	5 1/2	2	8	6 1/4	1 3/4	—	4 5/8	1/2	2 5/16	13/16	5/8	4	1/2	2	3 1/2
1 1/2	4	6	2 1/4	9	7	2	—	5	9/16	2 9/16	7/8	5/8	4	1/2	2	3 7/8
2	4 1/2	6 1/2	2 1/2	10 1/2	8	2 1/2	5	6	5/8	3 1/16	1	3/4	4	5/8	2 1/4	4 3/4
2 1/2	5	7	3	12	9 1/2	2 1/2	5 1/2	7	11/16	3 9/16	1 1/8	3/4	4	5/8	2 1/2	5 1/2
3	5 1/2	7 3/4	3	13	10	3	6	7 1/2	3/4	4 1/4	1 3/16	3/4	4	5/8	2 1/2	6
3 1/2	6	8 1/2	3 1/2	14 1/2	11 1/2	3	6 1/2	8 1/2	13/16	4 13/16	1 1/4	3/4	8	5/8	2 3/4	7
4	6 1/2	9	4	15	12	3	7	9	15/16	5 5/16	1 5/16	3/4	8	5/8	3	7 1/2
5	7 1/2	10 1/4	4 1/2	17	13 1/2	3 1/2	8	10	15/16	6 7/16	1 7/16	7/8	8	3/4	3	8 1/2
6	8	11 1/2	5	18	14 1/2	3 1/2	9	11	1	7 9/16	1 9/16	7/8	8	3/4	3 1/4	9 1/2
8	9	14	5 1/2	22	17 1/2	4 1/2	11	13 1/2	1 1/8	9 11/16	1 3/4	7/8	8	3/4	3 1/2	11 3/4
10	11	16 1/2	6 1/2	25 1/2	20 1/2	5	12	16	1 3/16	11 15/16	1 15/16	1	12	7/8	3 3/4	14 1/4
12	12	19	7 1/2	30	24 1/2	5 1/2	14	19	1 1/4	14 1/16	2 3/16	1	12	7/8	3 3/4	17

* ANSI B16.1.
For use under maximum working pressures of 125 lb per sq in.

TABLE 51A-1 American National Standard Running and Sliding Fits

Nominal size range (in.)† Over To	Class RC 1 Limits of clearance	Class RC 1 Hole H5	Class RC 1 Shaft g4	Class RC 2 Limits of clearance	Class RC 2 Hole H6	Class RC 2 Shaft g5	Class RC 3 Limits of clearance	Class RC 3 Hole H7	Class RC 3 Shaft f6	Class RC 4 Limits of clearance	Class RC 4 Hole H8	Class RC 4 Shaft f7
0–12	.1 / .45	+.2 / 0	−.1 / −.25	.1 / .55	+.25 / 0	−.1 / −.3	.3 / .95	+.4 / 0	−.3 / −.55	.3 / 1.3	+.6 / 0	−.3 / −.7
.12–24	.15 / .5	+.2 / 0	−.15 / −.3	.15 / .65	+.3 / 0	−.15 / −.35	.4 / 1.12	+.5 / 0	−.4 / −.7	.4 / 1.6	+.7 / 0	−.4 / −.9
.24–40	.2 / .6	.25 / 0	−.2 / −.35	.2 / .85	+.4 / 0	−.2 / −.45	.5 / 1.5	+.6 / 0	−.5 / −.9	.5 / 2.0	+.9 / 0	−.5 / −1.1
.40–71	.25 / .75	+.3 / 0	−.25 / −.45	.25 / .95	+.4 / 0	−.25 / −.55	.6 / 1.7	+.7 / 0	−.6 / −1.0	.6 / 2.3	+1.0 / 0	−.6 / −1.3
.71–1.19	.3 / .95	+.4 / 0	−.3 / −.55	.3 / 1.2	+.5 / 0	−.3 / −.7	.8 / 2.1	+.8 / 0	−.8 / −1.3	.8 / 2.8	+1.2 / 0	−.8 / −1.6
1.19–1.97	.4 / 1.1	+.4 / 0	−.4 / −.7	.4 / 1.4	+.6 / 0	−.4 / −.8	1.0 / 2.6	+1.0 / 0	−1.0 / −1.6	1.0 / 3.6	+1.6 / 0	−1.0 / −2.0
1.97–3.15	.4 / 1.2	+.5 / 0	−.4 / −.7	.4 / 1.6	+.7 / 0	−.4 / −.9	1.2 / 3.1	+1.2 / 0	−1.2 / −1.9	1.2 / 4.2	+1.8 / 0	−1.2 / −2.4
3.15–4.73	.5 / 1.5	+.6 / 0	−.5 / −.9	.5 / 2.0	+.9 / 0	−.5 / −1.1	1.4 / 3.7	+1.4 / 0	−1.4 / −2.3	1.4 / 5.0	+2.2 / 0	−1.4 / −2.8
4.73–7.09	.6 / 1.8	+.7 / 0	−.6 / −1.1	.6 / 2.3	+1.0 / 0	−.6 / −1.3	1.6 / 4.2	+1.6 / 0	−1.6 / −2.6	1.6 / 5.7	+2.5 / 0	−1.6 / −3.2

Limits are in thousandths of an inch. Limits for hole and shaft are applied algebraically to the basic size to obtain the limits of size for the parts. Symbols H5, g5, etc., are hole and shaft designations used in ABC system.

* Tables 52A-1 through 52E are adapted from ANSI B4.1.

† For diameters greater than those listed in Tables 52A-1 through 52E, see Standard.

TABLE 51 A-2 American National Standard Running and Sliding Fits

Nominal size range (in.) Over To	Class RC 5			Class RC 6			Class RC 7			Class RC 8			Class RC 9		
	Limits of clearance	Standard limits		Limits of clearance	Standard limits		Limits of clearance	Standard limits		Limits of clearance	Standard limits		Limits of clearance	Standard limits	
		Hole H8	Shaft e7		Hole H9	Shaft e8		Hole H9	Shaft d8		Hole H10	Shaft c9		Hole H11	Shaft
0–.12	.6 / 1.6	+.6 / −0	−.6 / −1.0	.6 / 2.2	+1.0 / −0	−.6 / −1.2	1.0 / 2.6	+1.0 / 0	−1.0 / −1.6	2.5 / 5.1	+1.6 / 0	−2.5 / −3.5	4.0 / 8.1	+2.5 / 0	−4.0 / −5.6
.12–.24	.8 / 2.0	+.7 / −0	−.8 / −1.3	.8 / 2.7	+1.2 / −0	−.8 / −1.5	1.2 / 3.1	+1.2 / 0	−1.2 / −1.9	2.8 / 5.8	+1.8 / 0	−2.8 / −4.0	4.5 / 9.0	+3.0 / 0	−4.5 / −6.0
.24–.40	1.0 / 2.5	+.9 / −0	−1.0 / −1.6	1.0 / 3.3	+1.4 / −0	−1.0 / −1.9	1.6 / 3.9	+1.4 / 0	−1.6 / −2.5	3.0 / 6.6	+2.2 / 0	−3.0 / −4.4	5.0 / 10.7	+3.5 / 0	−5.0 / −7.2
.40–.71	1.2 / 2.9	+1.0 / −0	−1.2 / −1.9	1.2 / 3.8	+1.6 / −0	−1.2 / −2.2	2.0 / 4.6	+1.6 / 0	−2.0 / −3.0	3.5 / 7.9	+2.8 / 0	−3.5 / −5.1	6.0 / 12.8	+4.0 / 0	−6.0 / −8.8
.71–1.19	1.6 / 3.6	+1.2 / −0	−1.6 / −2.4	1.6 / 4.8	+2.0 / −0	−1.6 / −2.8	2.5 / 5.7	+2.0 / 0	−2.5 / −3.7	4.5 / 10.0	+3.5 / 0	−4.5 / −6.5	7.0 / 15.5	+5.0 / 0	−7.0 / −10.5
1.19–1.97	2.0 / 4.6	+1.6 / −0	−2.0 / −3.0	2.0 / 6.1	+2.5 / −0	−2.0 / −3.6	3.0 / 7.1	+2.5 / 0	−3.0 / −4.6	5.0 / 11.5	+4.0 / 0	−5.0 / −7.5	8.0 / 18.0	+6.0 / 0	−8.0 / −12.0
1.97–3.15	2.5 / 5.5	+1.8 / −0	−2.5 / −3.7	2.5 / 7.3	+3.0 / −0	−2.5 / −4.3	4.0 / 8.8	+3.0 / 0	−4.0 / −5.8	6.0 / 13.5	+4.5 / 0	−6.0 / −9.0	9.0 / 20.5	+7.0 / 0	−9.0 / −13.5
3.15–4.73	3.0 / 6.6	+2.2 / −0	−3.0 / −4.4	3.0 / 8.7	+3.5 / −0	−3.0 / −5.2	5.0 / 10.7	+3.5 / 0	−5.0 / −7.2	7.0 / 15.5	+5.0 / 0	−7.0 / −10.5	10.0 / 24.0	+9.0 / 0	−10.0 / −15.0
4.73–7.09	3.5 / 7.6	+2.5 / −0	−3.5 / −5.1	3.5 / 10.0	+4.0 / −0	−3.5 / −6.0	6.0 / 12.5	+4.0 / 0	−6.0 / −8.5	8.0 / 18.0	+6.0 / 0	−8.0 / −12.0	12.0 / 28.0	+10.0 / 0	−12.0 / −18.0

TABLE 51 B-1   American National Standard Locational Clearance Fits

Nominal size range (in.) Over To	Class LC 1 Limits of clearance	Class LC 1 Standard limits Hole H6	Class LC 1 Standard limits Shaft h5	Class LC 2 Limits of clearance	Class LC 2 Standard limits Hole H7	Class LC 2 Standard limits Shaft h6	Class LC 3 Limits of clearance	Class LC 3 Standard limits Hole H8	Class LC 3 Standard limits Shaft h7	Class LC 4 Limits of clearance	Class LC 4 Standard limits Hole H10	Class LC 4 Standard limits Shaft h9	Class LC 5 Limits of clearance	Class LC 5 Standard limits Hole H7	Class LC 5 Standard limits Shaft g6
0–.12	0 / .45	+.25 / –0	+0 / –.2	0 / .65	+.4 / –0	+0 / –.25	0 / 1	+.6 / –0	+0 / –.4	0 / 2.6	+1.6 / –0	+0 / –1.0	.1 / .75	+.4 / –0	–.1 / –.35
.12–.24	0 / .5	+.3 / –0	+0 / –.2	0 / .8	+.5 / –0	+0 / –.3	0 / 1.2	+.7 / –0	+0 / –.5	0 / 3.0	+1.8 / –0	+0 / –1.2	.15 / .95	+.5 / –0	–.15 / –.45
.24–.40	0 / .65	+.4 / –0	+0 / –.25	0 / 1.0	+.6 / –0	+0 / –.4	0 / 1.5	+.9 / –0	+0 / –.6	0 / 3.6	+2.2 / –0	+0 / –1.4	.2 / 1.2	+.6 / –0	–.2 / –.6
.40–.71	0 / .7	+.4 / –0	+0 / –.3	0 / 1.1	+.7 / –0	+0 / –.4	0 / 1.7	+1.0 / –0	+0 / –.7	0 / 4.4	+2.8 / –0	+0 / –1.6	.25 / 1.35	+.7 / –0	–.25 / –.65
.71–1.19	0 / .9	+.5 / –0	+0 / –.4	0 / 1.3	+.8 / –0	+0 / –.5	0 / 2	+1.2 / –0	+0 / –.8	0 / 5.5	+3.5 / –0	+0 / –2.0	.3 / 1.6	+.8 / –0	–.3 / –.8
1.19–1.97	0 / 1.0	+.6 / –0	+0 / –.4	0 / 1.6	+1.0 / –0	+0 / –.6	0 / 2.6	+1.6 / –0	+0 / –1	0 / 6.5	+4.0 / –0	+0 / –2.5	.4 / 2.0	+1.0 / –0	–.4 / –1.0
1.97–3.15	0 / 1.2	+.7 / –0	+0 / –.5	0 / 1.9	+1.2 / –0	+0 / –.7	0 / 3	+1.8 / –0	+0 / –1.2	0 / 7.5	+4.5 / –0	+0 / –3	.4 / 2.3	+1.2 / –0	–.4 / –1.1
3.15–4.73	0 / 1.5	+.9 / –0	+0 / –.6	0 / 2.3	+1.4 / –0	+0 / –.9	0 / 3.6	+2.2 / –0	+0 / –1.4	0 / 8.5	+5.0 / –0	+0 / –3.5	.5 / 2.8	+1.4 / –0	–.5 / –1.4
4.73–7.09	0 / 1.7	+1.0 / –0	+0 / –.7	0 / 2.6	+1.6 / –0	+0 / –1.0	0 / 4.1	+2.5 / –0	+0 / –1.6	0 / 10	+6.0 / –0	+0 / –4	.6 / 3.2	+1.6 / –0	–.6 / –1.6

TABLE 51 B-2 American National Standard Locational Clearance Fits

Nominal size range (in.) Over To	Class LC 6 Limits of clearance	Class LC 6 Hole H9	Class LC 6 Shaft f8	Class LC 7 Limits of clearance	Class LC 7 Hole H10	Class LC 7 Shaft e9	Class LC 8 Limits of clearance	Class LC 8 Hole H10	Class LC 8 Shaft d9	Class LC 9 Limits of clearance	Class LC 9 Hole H11	Class LC 9 Shaft c10	Class LC 10 Limits of clearance	Class LC 10 Hole H12	Class LC 10 Shaft	Class LC 11 Limits of clearance	Class LC 11 Hole H13	Class LC 11 Shaft
0–12	.3	+1.0	−.3	.6	+1.6	−.6	1.0	+1.6	−1.0	2.5	+2.5	−2.5	4	+4	−4	5	+6	−5
	1.9	0	−.9	3.2	0	−1.6	3.6	−0	−2.0	6.6	−0	−4.1	12	−0	−8	17	−0	−11
.12–24	.4	+1.2	−.4	.8	+1.8	−.8	1.2	+1.8	−1.2	2.8	+3.0	−2.8	4.5	+5	−4.5	6	+7	−6
	2.3	0	−1.1	3.8	0	−2.0	4.2	−0	−2.4	7.6	−0	−4.6	14.5	−0	−9.5	20	−0	−13
.24–40	.5	+1.4	−.5	1.0	+2.2	−1.0	1.6	+2.2	−1.6	3.0	+3.5	−3.0	5	+6	−5	7	+9	−7
	2.8	0	−1.4	4.6	0	−2.4	5.2	−0	−3.0	8.7	−0	−5.2	17	−0	−11	25	−0	−16
.40–71	.6	+1.6	−.6	1.2	+2.8	−1.2	2.0	+2.8	−2.0	3.5	+4.0	−3.5	6	+7	−6	8	+10	−8
	3.2	0	−1.6	5.6	0	−2.8	6.4	−0	−3.6	10.3	−0	−6.3	20	−0	−13	28	−0	−18
.71–1.19	.8	+2.0	−.8	1.6	+3.5	−1.6	2.5	+3.5	−2.5	4.5	+5.0	−4.5	7	+8	−7	10	+12	−10
	4.0	0	−2.0	7.1	0	−3.6	8.0	−0	−4.5	13.0	−0	−8.0	23	−0	−15	34	−0	−22
1.19–1.97	1.0	+2.5	−1.0	2.0	+4.0	−2.0	3.0	+4.0	−3.0	5	+6	−5	8	+10	−8	12	+16	−12
	5.1	0	−2.6	8.5	0	−4.5	9.5	−0	−5.5	15	−0	−9	28	−0	−18	44	−0	−28
1.97–3.15	1.2	+3.0	−1.2	2.5	+4.5	−2.5	4.0	+4.5	−4.0	6	+7	−6	10	+12	−10	14	+18	−14
	6.0	0	−3.0	10.0	0	−5.5	11.5	−0	−7.0	17.5	−0	−10.5	34	−0	−22	50	−0	−32
3.15–4.73	1.4	+3.5	−1.4	3.0	+5.0	−3.0	5.0	+5.0	−5.0	7	+9	−7	11	+14	−11	16	+22	−16
	7.1	0	−3.6	11.5	0	−6.5	13.5	−0	−8.5	21	−0	−12	39	−0	−25	60	−0	−38
4.73–7.09	1.6	+4.0	−1.6	3.5	+6.0	−3.5	6	+6	−6	8	+10	−8	12	+16	−12	18	+25	−18
	8.1	0	−4.1	13.5	0	−7.5	16	−0	−10	24	−0	−14	44	−0	−28	68	−0	−43

Limits are in thousandths of an inch. Limits for hole and shaft are applied algebraically to the basic size to obtain the limits of size for the parts. Symbols H9, f8, etc., are hole and shaft designations used in ABC system.

TABLE 51c American National Standard Locational Transition Fits

Nominal size range (in.) Over To	Class LT 1 Fit	Class LT 1 Standard limits Hole H7	Shaft js6	Class LT 2 Fit	Class LT 2 Standard limits Hole H8	Shaft js7	Class LT 3 Fit	Class LT 3 Standard limits Hole H7	Shaft k6	Class LT 4 Fit	Class LT 4 Standard limits Hole H8	Shaft k7	Class LT 5 Fit	Class LT 5 Standard limits Hole H7	Shaft n6	Class LT 6 Fit	Class LT 6 Standard limits Hole H7	Shaft n7
0–.12	−.10 / +.50	+.4 / −0	+.10 / −.10	−.2 / +.8	+.6 / −0	+.2 / −.2							−.5 / +.15	+.4 / −0	+.5 / +.25	−.65 / +.15	+.4 / −0	+.65 / +.25
.12–.24	−.15 / +.65	+.5 / −0	+.15 / −.15	−.25 / +.95	+.7 / −0	+.25 / −.25							−.6 / +.2	+.5 / −0	+.6 / +.3	−.8 / +.2	+.5 / −0	+.8 / +.3
.24–.40	−.2 / +.8	+.6 / −0	+.2 / −.2	−.3 / +1.2	+.9 / −0	+.3 / −.3	−.5 / +.5	+.6 / −0	+.5 / +.1	−.7 / +.8	+.9 / −0	+.5 / +.1	−.8 / +.2	+.6 / −0	+.8 / +.4	−1.0 / +.2	+.6 / −0	+1.0 / +.4
.40–.71	−.2 / +.9	+.7 / −0	+.2 / −.2	−.35 / +1.35	+1.0 / −0	+.35 / −.35	−.5 / +.6	+.7 / −0	+.5 / +.1	−.8 / +.9	+1.0 / −0	+.6 / +.1	−.9 / +.2	+.7 / −0	+.9 / +.5	−1.2 / +.2	+.7 / −0	+1.2 / +.5
.71–1.19	−.25 / +1.05	+.8 / −0	+.25 / −.25	−.4 / +1.6	+1.2 / −0	+.4 / −.4	−.6 / +.7	+.8 / −0	+.6 / +.1	−.9 / +1.1	+1.2 / −0	+.7 / +.1	−1.1 / +.2	+.8 / −0	+1.1 / +.6	−1.4 / +.2	+.8 / −0	+1.4 / +.6
1.19–1.97	−.3 / +1.3	+1.0 / −0	+.3 / −.3	−.5 / +2.1	+1.6 / −0	+.5 / −.5	−.7 / +.9	+1.0 / −0	+.7 / +.1	−1.1 / +1.5	+1.6 / −0	+1.1 / +.1	−1.3 / +.3	+1.0 / −0	+1.3 / +.7	−1.7 / +.3	+1.0 / −0	+1.7 / +.7
1.97–3.15	−.3 / +1.5	+1.2 / −0	+.3 / −.3	−.6 / +2.4	+1.8 / −0	+.6 / −.6	−.8 / +1.1	+1.2 / −0	+.8 / +.1	−1.3 / +1.7	+1.8 / −0	+1.3 / +.1	−1.5 / +.4	+1.2 / −0	+1.5 / +.8	−2.0 / +.4	+1.2 / −0	+2.0 / +.8
3.15–4.73	−.4 / +1.8	+1.4 / −0	+.4 / −.4	−.7 / +2.9	+2.2 / −0	+.7 / −.7	−1.0 / +1.3	+1.4 / −0	+1.0 / +.1	−1.5 / +2.1	+2.2 / −0	+1.5 / +.1	−1.9 / +.4	+1.4 / −0	+1.9 / +1.0	−2.4 / +.4	+1.4 / −0	+2.4 / +1.0
4.73–7.09	−.5 / +2.1	+1.6 / −0	+.5 / −.5	−.8 / +3.3	+2.5 / −0	+.8 / −.8	−1.1 / +1.5	+1.6 / −0	+1.1 / +.1	−1.7 / +2.4	+2.5 / −0	+1.7 / +.1	−2.2 / +.4	+1.6 / −0	+2.2 / +1.2	−2.8 / +.4	+1.6 / −0	+2.8 / +1.2

Limits are in thousandths of an inch. Limits for hole and shaft are applied algebraically to the basic size to obtain the limits of size for the mating parts. "Fit" represents the maximum interference (minus values) and the maximum clearance (plus values). Symbols H7, js6, etc., are hole and shaft designations used in ABC system.

TABLE **51**D **American National Standard Locational Interference Fits**

Nominal size range (in.) Over To	Class LN 1			Class LN 2			Class LN 3		
	Limits of interference	Standard limits		Limits of interference	Standard limits		Limits of interference	Standard limits	
		Hole H6	Shaft n5		Hole H7	Shaft p6		Hole H7	Shaft r6
0–.12	0 .45	+.25 −0	+.45 +.25	0 .65	+.4 −0	+.65 +.4	.1 .75	+.4 −0	+.75 +.5
.12–.24	0 .5	+.3 −0	+.5 +.3	0 .8	+.5 −0	+.8 +.5	.1 .9	+.5 −0	+.9 +.6
.24–.40	0 .65	+.4 −0	+.65 +.4	0 1.0	+.6 −0	+1.0 +.6	.2 1.2	+.6 −0	+1.2 +.8
.40–.71	0 .8	+.4 −0	+.8 +.4	0 1.1	+.7 −0	+1.1 +.7	.3 1.4	+.7 −0	+1.4 +1.0
.71–1.19	0 1.0	+.5 −0	+1.0 +.5	0 1.3	+.8 −0	+1.3 +.8	.4 1.7	+.8 −0	+1.7 +1.2
1.19–1.97	0 1.1	+.6 −0	+1.1 +.6	0 1.6	+1.0 −0	+1.6 +1.0	.4 2.0	+1.0 −0	+2.0 +1.4
1.97–3.15	.1 1.3	+.7 −0	+1.3 +.7	.2 2.1	+1.2 −0	+2.1 +1.4	.4 2.3	+1.2 −0	+2.3 +1.6
3.15–4.73	.1 1.6	+.9 −0	+1.6 +1.0	.2 2.5	+1.4 −0	+2.5 +1.6	.6 2.9	+1.4 −0	+2.9 +2.0
4.73–7.09	.2 1.9	+1.0 −0	+1.9 +1.2	.2 2.8	+1.6 −0	+2.8 +1.8	.9 3.5	+1.6 −0	+3.5 +2.5

Limits are in thousandths of an inch. Limits for hole and shaft are applied algebraically to the basic size to obtain the limits of size for the parts. Symbols H7, p6, etc., are hole and shaft designations used in ABC system.

TABLE 51E American National Standard Force and Shrink Fits

Nominal size range (in.) Over To	Class FN 1 Limits of interference	FN1 Hole H6	FN1 Shaft	Class FN 2 Limits of interference	FN2 Hole H7	FN2 Shaft s6	Class FN 3 Limits of interference	FN3 Hole H7	FN3 Shaft t6	Class FN 4 Limits of interference	FN4 Hole H7	FN4 Shaft u6	Class FN 5 Limits of interference	FN5 Hole H8	FN5 Shaft x7
0–.12	.05 / .5	+.25 / −0	+.5 / +.3	.2 / .85	+.4 / −0	+.85 / +.6				.3 / .95	+.4 / −0	+.95 / +.7	.3 / 1.3	+.6 / −0	+1.3 / +.9
.12–.24	.1 / .6	+.3 / −0	+.6 / +.4	.2 / 1.0	+.5 / −0	+1.0 / +.7				.4 / 1.2	+.5 / −0	+1.2 / +.9	.5 / 1.7	+.7 / −0	+1.7 / +1.2
.24–.40	.1 / .75	+.4 / −0	+.75 / +.5	.4 / 1.4	+.6 / −0	+1.4 / +1.0				.6 / 1.6	+.6 / −0	+1.6 / +1.2	.5 / 2.0	+.9 / −0	+2.0 / +1.4
.40–.56	.1 / .8	+.4 / −0	+.8 / +.5	.5 / 1.6	+.7 / −0	+1.6 / +1.2				.7 / 1.8	+.7 / −0	+1.8 / +1.4	.6 / 2.3	+1.0 / −0	+2.3 / +1.6
.56–.71	.2 / .9	+.4 / −0	+.9 / +.6	.5 / 1.6	+.7 / −0	+1.6 / +1.2				.7 / 1.8	+.7 / −0	+1.8 / +1.4	.8 / 2.5	+1.0 / −0	+2.5 / +1.8
.71–.95	.2 / 1.1	+.5 / −0	+1.1 / +.7	.6 / 1.9	+.8 / −0	+1.9 / +1.4				.8 / 2.1	+.8 / −0	+2.1 / +1.6	1.0 / 3.0	+1.2 / −0	+3.0 / +2.2
.95–1.19	.3 / 1.2	+.5 / −0	+1.2 / +.8	.6 / 1.9	+.8 / −0	+1.9 / +1.4	.8 / 2.1	+.8 / −0	+2.1 / +1.6	1.0 / 2.3	+.8 / −0	+2.3 / +1.8	1.3 / 3.3	+1.2 / −0	+3.3 / +2.5
1.19–1.58	.3 / 1.3	+.6 / −0	+1.3 / +.9	.8 / 2.4	+1.0 / −0	+2.4 / +1.8	1.0 / 2.6	+1.0 / −0	+2.6 / +2.0	1.5 / 3.1	+1.0 / −0	+3.1 / +2.5	1.4 / 4.0	+1.6 / −0	+4.0 / +3.0
1.58–1.97	.4 / 1.4	+.6 / −0	+1.4 / +1.0	.8 / 2.4	+1.0 / −0	+2.4 / +1.8	1.2 / 2.8	+1.0 / −0	+2.8 / +2.2	1.8 / 3.4	+1.0 / −0	+3.4 / +2.8	2.4 / 5.0	+1.6 / −0	+5.0 / +4.0
1.97–2.56	.6 / 1.8	+.7 / −0	+1.8 / +1.3	.8 / 2.7	+1.2 / −0	+2.7 / +2.0	1.3 / 3.2	+1.2 / −0	+3.2 / +2.5	2.3 / 4.2	+1.2 / −0	+4.2 / +3.5	3.2 / 6.2	+1.8 / −0	+6.2 / +5.0
2.56–3.15	.7 / 1.9	+.7 / −0	+1.9 / +1.4	1.0 / 2.9	+1.2 / −0	+2.9 / +2.2	1.8 / 3.7	+1.2 / −0	+3.7 / +3.0	2.8 / 4.7	+1.2 / −0	+4.7 / +4.0	4.2 / 7.2	+1.8 / −0	+7.2 / +6.0
3.15–3.94	.9 / 2.4	+.9 / −0	+2.4 / +1.8	1.4 / 3.7	+1.4 / −0	+3.7 / +2.8	2.1 / 4.4	+1.4 / 0	+4.4 / +3.5	3.6 / 5.9	+1.4 / −0	+5.9 / +5.0	4.8 / 8.4	+2.2 / −0	+8.4 / +7.0
3.94–4.73	1.1 / 2.6	+.9 / −0	+2.6 / +2.0	1.6 / 3.9	+1.4 / −0	+3.9 / +3.0	2.6 / 4.9	+1.4 / −0	+4.9 / +4.0	4.6 / 6.9	+1.4 / −0	+6.9 / +6.0	5.8 / 9.4	+2.2 / −0	+9.4 / +8.0

Limits are in thousandths of an inch. Limits for hole and shaft are applied algebraically to the basic size to obtain the limits of size for the parts. Symbols H7, s6, etc., are hole and shaft designations used in ABC system.

TABLE 52 Twist Drill Sizes*

	Number sizes								Letter sizes			
No. size	Decimal equivalent	Metric equivalent	Closest metric drill (mm)	No. size	Decimal equivalent	Metric equivalent	Closest metric drill (mm)		Size letter	Decimal equivalent	Metric equivalent	Closest metric drill (mm)
1	.2280	5.791	5.80	41	.0960	2.438	2.45		A	.234	5.944	5.90
2	.2210	5.613	5.60	42	.0935	2.362	2.35		B	.238	6.045	6.00
3	.2130	5.410	5.40	43	.0890	2.261	2.25		C	.242	6.147	6.10
4	.2090	5.309	5.30	44	.0860	2.184	2.20		D	.246	6.248	6.25
5	.2055	5.220	5.20	45	.0820	2.083	2.10		E	.250	6.350	6.40
6	.2040	5.182	5.20	46	.0810	2.057	2.05		F	.257	6.528	6.50
7	.2010	5.105	5.10	47	.0785	1.994	2.00		G	.261	6.629	6.60
8	.1990	5.055	5.10	48	.0760	1.930	1.95		H	.266	6.756	6.75
9	.1960	4.978	5.00	49	.0730	1.854	1.85		I	.272	6.909	6.90
10	.1935	4.915	4.90	50	.0700	1.778	1.80		J	.277	7.036	7.00
11	.1910	4.851	4.90	51	.0670	1.702	1.70		K	.281	7.137	7.10
12	.1890	4.801	4.80	52	.0635	1.613	1.60		L	.290	7.366	7.40
13	.1850	4.699	4.70	53	.0595	1.511	1.50		M	.295	7.493	7.50
14	.1820	4.623	4.60	54	.0550	1.397	1.40		N	.302	7.671	7.70
15	.1800	4.572	4.60	55	.0520	1.321	1.30		O	.316	8.026	8.00
16	.1770	4.496	4.50	56	.0465	1.181	1.20		P	.323	8.204	8.20
17	.1730	4.394	4.40	57	.0430	1.092	1.10		Q	.332	8.433	8.40
18	.1695	4.305	4.30	58	.0420	1.067	1.05		R	.339	8.611	8.60
19	.1660	4.216	4.20	59	.0410	1.041	1.05		S	.348	8.839	8.80
19	.1610	4.089	4.10	60	.0400	1.016	1.00		T	.358	9.093	9.10
21	.1590	4.039	4.00	61	.0390	0.991	1.00		U	.368	9.347	9.30
22	.1570	3.988	4.00	62	.0380	0.965	0.95		V	.377	9.576	9.60
23	.1540	3.912	3.90	63	.0370	0.940	0.95		W	.386	9.804	9.80
24	.1520	3.861	3.90	64	.0360	0.914	0.90		X	.397	10.084	10.00
25	.1495	3.797	3.80	65	.0350	0.889	0.90		Y	.404	10.262	10.50
26	.1470	3.734	3.75	66	.0330	0.838	0.85		Z	.413	10.491	10.50
27	.1440	3.658	3.70	67	.0320	0.813	0.80					
28	.1405	3.569	3.60	68	.0310	0.787	0.80					
29	.1360	3.454	3.50	69	.0292	0.742	0.75					
30	.1285	3.264	3.25	70	.0280	0.711	0.70					
31	.1200	3.048	3.00	71	.0260	0.660	0.65					
32	.1160	2.946	2.90	72	.0250	0.635	0.65					
33	.1130	2.870	2.90	73	.0240	0.610	0.60					
34	.1110	2.819	2.80	74	.0225	0.572	0.55					
35	.1100	2.794	2.80	75	.0210	0.533	0.55					
36	.1065	2.705	2.70	76	.0200	0.508	0.50					
37	.1040	2.642	2.60	77	.0180	0.457	0.45					
38	.1015	2.578	2.60	78	.0160	0.406	0.40					
39	.0995	2.527	2.50	79	.0145	0.368	0.35					
40	.0980	2.489	2.50	80	.0135	0.343	0.35					

*Fraction-size drills range in size from one-sixteenth—4 in. and over in diameter—by sixty-fourths.

TABLE 53   *Standard Wire and Sheet-metal Gages**

Gage number	(A) Brown & Sharpe or American	(B) American Steel & Wire Co.	(C) Piano wire	(E) U.S. St'd.	Gage number
0000000	.6513	.4900	—	.5000	0000000
000000	.5800	.4615	.004	.4688	000000
00000	.5165	.4305	.005	.4375	00000
0000	.4600	.3938	.006	.4063	0000
000	.4096	.3625	.007	.3750	000
00	.3648	.3310	.008	.3438	00
0	.3249	.3065	.009	.3125	0
1	.2893	.2830	.010	.2813	1
2	.2576	.2625	.011	.2656	2
3	.2294	.2437	.012	.2500	3
4	.2043	.2253	.013	.2344	4
5	.1819	.2070	.014	.2188	5
6	.1620	.1920	.016	.2031	6
7	.1443	.1770	.018	.1875	7
8	.1285	.1620	.020	.1719	8
9	.1144	.1483	.022	.1563	9
10	.1019	.1350	.024	.1406	10
11	.0907	.1205	.026	.1250	11
12	.0808	.1055	.029	.1094	12
13	.0720	.0915	.031	.0938	13
14	.0641	.0800	.033	.0781	14
15	.0571	.0720	.035	.0703	15
16	.0508	.0625	.037	.0625	16
17	.0453	.0540	.039	.0563	17
18	.0403	.0475	.041	.0500	18
19	.0359	.0410	.043	.0438	19
20	.0320	.0348	.045	.0375	20
21	.0285	.0317	.047	.0344	21
22	.0253	.0286	.049	.0313	22
23	.0226	.0258	.051	.0281	23
24	.0201	.0230	.055	.0250	24
25	.0179	.0204	.059	.0219	25
26	.0159	.0181	.063	.0188	26
27	.0142	.0173	.067	.0172	27
28	.0126	.0162	.071	.0156	28
29	.0113	.0150	.075	.0141	29
30	.0100	.0140	.080	.0125	30
31	.0089	.0132	.085	.0109	31
32	.0080	.0128	.090	.0102	32
33	.0071	.0118	.095	.0094	33
34	.0063	.0104	.100	.0086	34
35	.0056	.0095	.106	.0078	35
36	.0050	.0090	.112	.0070	36
37	.0045	.0085	.118	.0066	37
38	.0040	.0080	.124	.0063	38
39	.0035	.0075	.130	—	39
40	.0031	.0070	.138	—	40

*Dimensions in decimal parts of an inch.

(A) Standard in United States for sheet metal and wire (except steel and iron).

(B) Standard for iron and steel wire (U.S. Steel Wire Gage).

(C) American Steel and Wire Company's music (or piano) wire gage sizes. Recognized by U.S. Bureau of Standards.

(E) U.S. Standard for iron and steel plate. However, plate is now generally specified by its thickness in decimals of an inch.

American National Standards and ISO Standards*

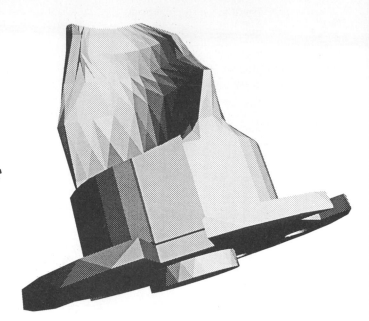

AMERICAN NATIONAL STANDARDS

A few of the more than 500 standards approved by the American National Standards Institute are listed. Copies may be obtained from the American Society of Mechanical Engineers, 345 East 47 Street, New York, N.Y. 10017.

■ INCH—THREADS, FASTENERS, PIPE, AND DIMENSIONING

B1.1 Unified and American Screw Threads for Screws, Bolts, Nuts, and Other Threaded Parts

B1.5 Acme Screw Threads

B1.8 Stub Acme Screw Threads

B1.9 Buttress Screw Threads

B2.1 Pipe Threads

B2.4 Hose Coupling Screw Threads

B4.1 Preferred Limits and Fits for Cylindrical Parts

B4.2 Preferred Metric Limits and Fits

B5.10 Machine Tapers, Self-holding and Steep Taper Series

B5.15 Involute Splines, Side Bearing

B5.20 Machine Pins

B16.1 Cast-iron Pipe Flanges and Flanged Fittings, Class 25, 125, 250, and 800 lb

* International Organization for Standardization

B16.3 Malleable-iron Screwed Fittings, 150 lb

B16.4 Cast-iron Screwed Fittings for Maximum WSP of 125 and 250 lb

B16.5 Steel Pipe Flanges and Flanged Fittings

B16.9 Steel Butt-welding Fittings

B17.1 Shafting and Stock Keys

B17.2 Woodruff Keys and Keyseats

B18.1 Small Solid Rivets

B18.2.1 Square and Hex Bolts and Screws

B18.2.2 Square and Hex Nuts

B18.3 Socket Set Screws and Socket-head Cap Screws

B18.1.2 Large Rivets

B18.5 Round-head Bolts

B18.6.2 Hexagon-head Cap Screws, Slotted-head Cap Screws, Square-head Set Screws, and Slotted Headless Set Screws

B36.1 Welded and Seamless Pipe (ASTM A53–44)

B36.2 Welded Wrought-iron Pipe

B36.10 Wrought-iron and Wrought-steel Pipe

B48.1 Inch-Millimeter Conversion for Industrial Use

B94.6 Knurling

▪ METRIC THREADS AND FASTENERS

ANSI B1.13M Metric Screw Threads—M Profile

ANSI B18.2.3.1M Metric Hex Cap Screws

ANSI B18.2.3.2M Metric Formed Hex Screws

ANSI B18.2.3.3M Metric Heavy Hex Screws

ANSI B18.2.3.4M Metric Hex Flange Screws

ANSI B18.2.3.5M Metric Hex Bolts

ANSI B18.2.3.6M Metric Heavy Hex Bolts

ANSI B18.2.4.1M Metric Hex Nuts, Style 1

ANSI B18.2.4.2M Metric Hex Nuts, Style 2

ANSI B18.2.4.3M Metric Slotted Hex Nuts

ANSI B18.2.4.5M Metric Hex Jam Nuts

ANSI/ASME B18.3.1M Socket Head Cap Screws (Metric Series)

ANSI B18.3.3M Hexagon Socket Head Shoulder Screws Metric Series

ANSI B18.3.4M Hexagon Socket Button Head Cap Screws Metric Series

ANSI B18.3.6M Hexagon Socket Set Screws Metric Series

ANSI B1.13M Metric Screw Threads—M Profile

ANSI B1.20.4 Dryseal Pipe Threads (Metric Translation of B1.20.3) R1982

▪ DRAWING STANDARDS

Y1.1 Abbreviations

Y14.1 Size and Format

Y14.2M Line Conventions and Lettering

Y14.3 Multi and Sectional View Drawing

Y14.4 Pictorial Drawing

Y14.5 Dimensioning and Tolerancing

Y14.5M Dimensioning and Tolerancing

Y14.6 Screw Threads

Y14.6M Screw Threads (Metric Supplement)

Y14.7.1 Spur, Helical and Racks

Y14.7.2 Bevel and Hypoid

Y14.9 Forgings

Y14.10 Metal Stampings

Y14.11 Plastics

Y14.12 Die Castings

Y14.13 Springs, Helical and Flat

Y14.14 Mechanical Assemblies

Y14.15 Electrical and Electronic Diagrams

Y14.15a Interconnection Diagrams

Y14.17 Fluid Power Diagrams

Y14.26.3 Computer-aided Preparation of Product Definition Data Dictionary of Terms

Y14.32.1 Chassis Frames

Y14.36 Surface Texture Symbols

▪ GRAPHICAL AND LETTER SYMBOLS*

Y1.1 Abbreviations for Use on Drawings and in Text (Graphical)*

Y10.2 Hydraulics (Letter)

Y10.3 Mechanics for Solid Bodies (Letter)

Y10.4 Heat and Thermodynamics (Letter)

Y10.7 Aeronautical Sciences (Letter)

Y10.8 Structural Analysis (Letter)

Y10.9 Radio (Letter)

Y10.11 Acoustics (Letter)

Y10.12 Chemical Engineering (Letter)

Y10.14 Rocket Propulsion (Letter)

Y32.2 Electrical and Electronic Diagrams (Graphical)

Y32.3 Welding (Graphical)

Y32.4 Plumbing (Graphical)

Y32.7 Use on Railroad Maps and Profiles (Graphical)

Y32.10 Fluid Power Diagrams (Graphical)

Z32.2.3 Pipe Fittings, Valves and Piping (Graphical)

Z32.2.4 Heating, Ventilating and Air Conditioning (Graphical)

Z32.2.6 Heat-power Apparatus (Graphical)

* Y is the new letter assigned to standards for abbreviations, charts and graphs, drawings, graphical symbols, and letter symbols. The Z will be changed to Y as the standards are revised and reaffirmed.

INTERNATIONAL ORGANIZATION FOR STANDARDIZATION (ISO)

ISO/R 128 Engineering Drawing—Principles of Presentation

ISO/R 129 Engineering Drawing—Dimensioning

ISO/R 406 Inscription of Linear and Angular Tolerances

ISO/R 1101/I Tolerances of Form and of Position—Part 1

ISO 1101/II Tolerances of Form and of Position—Part 2

ISO 1302 Method of Indicating Surface Texture

ISO/R 1660 Tolerances of Form and of Position—Part 3 Dimensioning and Tolerancing of Profiles

ISO/R 1661 Tolerancing of Form and of Position—Part 4

ISO 2162 Representation of Springs

ISO 2203 Conventional Representation of Gears

ISO 3040 Dimensioning and Tolerancing Cones

ISO 3098 Lettering—Part I

R225 Bolts, Screws, and Studs—Dimensioning

R278 Hexagon Bolts and Nuts

R228/1 Slotted and Castle Nuts with Metric Thread

R724 ISO General-Purpose Metric Screw Threads

R887 Washers for Hexagon Bolts and Nuts—Metric Series

R1207 Slotted Cheese (Fillister) Head Screws

R1234 Split Pins—Metric Series

R1478 Tapping Screw Thread

R1479 Hexagon Head Tapping Screws—Metric Series

R1481 Slotted Pan Head Tapping Screws

R1501 ISO Miniature Screw Threads

R1580 Slotted Pan Head Screws

ISO 2009 Slotted Countersunk (Flat) Head Screws—Metric Series

ISO 2010 Slotted Raised Countersunk (Oval) Head Screws

ISO 2306 Drills for Use Prior to Tapping Screw Threads

ISO 2339 Taper Pins, Unhardened—Metric Series

ISO 2340 Clevis Pins—Metric Series

ISO 2342 Slotted Headless Screws—Metric Series

ISO 2343 Hexagon Socket Set Screws

Bibliography of Engineering Drawing and Allied Subjects

Aeronautical Drafting and Engineering

Anderson, N. H., *Aircraft Layout and Detail Design*, 2nd ed., New York: McGraw-Hill.

Blueprint Reading

Hornung, W. J., *Blueprint Reading*, Englewood Cliffs, N.J.: Prentice-Hall.

Norcross, C., *Aircraft Blueprints and How to Read Them*, New York: McGraw-Hill.

CAD/CAM

Bertoline, A. R., *Fundamentals of CAD*, Albany, N.Y.: Delmar.

Besant, C. C., *Computer-aided Design and Manufacturing*, New York: Wiley.

Computer-aided Design Drawing and Drafting

Chasen, S. H., *Geometric Principles and Procedures for Computer Graphic Applications*, Englewood Cliffs, N.J.: Prentice-Hall.

Foley, J. D., and A. Van Dorn, *Fundamentals of Computer Graphics*, Reading, Mass.: Addison-Wesley.

Giloi, W. K., *Interactive Computer Graphics*, Englewood Cliffs, N.J.: Prentice-Hall.

Goetsch, D. L., *Introduction to Computer-aided Drafting*, Englewood Cliffs, N.J.: Prentice-Hall.

Lamit, G., and V. Paige, *Computer-aided Design and Drafting*, Columbus, Ohio: Merrill.

Lange, J. C., and D. Shanahan, *Interactive Computer Graphics*, New York: Wiley.

Newman, W. M., and R. F. Sproull, *Interactive Computer Graphics*, New York: McGraw-Hill.

Thornhill, R. B., *Engineering Graphics and Numerical Control*, New York: McGraw-Hill.

Descriptive Geometry

Paré, E. G., R. O. Loving, and I. L. Hill, *Descriptive Geometry*, 2nd ed., New York: Macmillan.

Slaby, S. M., *Three-dimensional Descriptive Geometry*, New York: Harcourt.

Electrical Drawing

Bishop, C. C., C. T. Gilliam, and others, *Electrical Drafting and Design*, 3rd ed., New York: McGraw-Hill.

Shiers, G., *Electronic Drafting*, Englewood Cliffs, N.J.: Prentice-Hall.

———, *Electronic Drafting Techniques and Exercises*, Englewood Cliffs, N.J.: Prentice-Hall.

Engineering Design

Alger, J. R., and C. V. Hayes, *Creative Synthesis in Design*, Englewood Cliffs, N.J.: Prentice-Hall.

Asimow, M., *Introduction to Design*, Englewood Cliffs, N.J., Prentice-Hall.

Dreyfuss, H., *The Measure of Man*, New York: Whitney Library of Design.

Edel, D. H., *Introduction to Creative Design*, Englewood Cliffs, N.J.: Prentice-Hall.

Starr, M. K., *Product Design and Decision Theory*, Englewood Cliffs, N.J.: Prentice-Hall.

U.S. Patent Office, *Patent Laws*, Washington, D.C.: Superintendent of Documents.

———, *Patents and Inventions: An Information Aid for Inventors*, Washington, D.C.: Superintendent of Documents.

White, W. J., and S. Schneyer, *Pocket Data for Human Factor Engineering*, Buffalo, N.Y.: Cornell University.

Engineering Graphics

Earle, J. H., *Engineering Design Graphics,* Reading, Mass.: Addison-Wesley.

French, T. E., and C. J. Vierck, *Engineering Drawing*, 12th ed., New York: McGraw-Hill.

Giesecke, F. E., and others, *Technical Drawing*, 7th ed., New York: Macmillan.

Engineering Graphics Problems

Earle, J. H., *Engineering Graphics and Design Problems* (Series 1, 2, and 3), Reading, Mass.: Addison-Wesley.

Giesecke, F. E., and others, *Technical Drawing Problems*, 3rd ed., New York: Macmillan.

Goss, L. D. and W. J. Luzadder, *Problems in Engineering Drawing* (Series 2), Englewood Cliffs, N.J.: Prentice-Hall.

Levens, A. S., and A. E. Edstrom, *Problems in Mechanical Drawing*, New York: McGraw-Hill.

Luzadder, W. J., and K. E. Botkin, *Problems in Engineering Drawing*, 9th ed., Englewood Cliffs, N.J.: Prentice-Hall.

Vierck, C. J., and R. I. Hang, *Engineering Drawing Problems*, New York: McGraw-Hill.

Graphical Representation and Computation

Davis, D. S., *Empirical Equations and Nomography*, New York: McGraw-Hill.

Douglass, R. D., and D. P. Adams, *Elements of Nomography*, New York: McGraw-Hill.

Handbooks

American Institute of Steel Construction, *Steel Construction*, New York.

Kent, W., *Mechanical Engineer's Handbook*, New York: Wiley.

Knowlton, A. E., *Standard Handbook for Electrical Engineers*, New York: McGraw-Hill.

Le Grand, R., ed., *New American Machinist's Handbook*, New York: McGraw-Hill.

Kinematics—Machine Design

Hinkle, R. T., *Kinematics of Machines*, Englewood Cliffs, N.J.: Prentice-Hall.

Maleev, V. L., *Machine Design*, 3rd ed., Scranton, Pa.: International Textbook.

Spotts, M. F., *Design of Machine Elements*, 5th ed., Englewood Cliffs, N.J.: Prentice-Hall.

Metric System (Système Internationale d'Unités—SI)

American Society for Testing Materials, *Metric Practice Guide*, New York.

Beloit Tool Co., *Discover Why Metrics*, Roscoe, Ill.: Swani.

———, *U.S.A. Goes Metric*, Roscoe, Ill.: Swani.

Foster, L. W., *Geo-metrics*, Reading, Mass.: Addison-Wesley.

Numerical Control

International Business Machines Corp., *Automatic Programming of Machine Tools*, Poughkeepsie, N.Y.

Shop Practice

Begeman, M. L., *Manufacturing Processes*, 4th ed., New York: Wiley.

Doyle, L. E., *Metal Machining*, Englewood Cliffs, N.J.: Prentice-Hall.

———, J. L. Leach, J. L. Morris, and G. F. Schrader, *Manufacturing Processes and Materials for Engineers*, Englewood Cliffs, N.J.: Prentice-Hall.

Lascoe, O. D., and others, *Machine Shop*, Chicago, American Technical Society.

Structural Drafting

American Institute of Steel Construction, *Structural Steel Detailing*, New York.

Bishop, C. T., *Structural Drafting*, New York: Wiley.

Lothers, S. E., *Design in Structural Steel*, Englewood Cliffs, N.J.: Prentice-Hall.

Welding

Lincoln Electric Co., *Simple Blueprint Reading*, Cleveland.

———, *Procedure Handbook of Arc Welding Design and Practice*, Cleveland.

Index